Monographs on oceanographic methodology 10

In this series:

1. *Determination of photosynthetic pigments in seawater*
2. *Zooplankton sampling*
3. *A guide to the measurement of marine primary production under some special conditions*
4. *Zooplankton fixation and preservation*
5. *Coral reefs: research methods*
6. *Phytoplankton manual*
7. *Mathematical models in biological oceanography*
8. *Mangrove research methodology*
9. *Seagrass research methods*
10. *Phytoplankton pigments in oceanography: guidelines to modern methods (this volume)*

Phytoplankton pigments in oceanography: guidelines to modern methods

Edited by S.W. Jeffrey, R.F.C. Mantoura
and S.W. Wright

Sponsored by SCOR and UNESCO

UNESCO Publishing

The designations employed and the presentation of material throughout this publication do not imply the expression of any opinion whatsoever on the part of UNESCO concerning the legal status of any country, territory, city or area or of its authorities, or concerning the delimitation of its frontiers or boundaries.

Published in 1997 by the United Nations Educational, Scientific and Cultural Organization
7 place de Fontenoy, 75700 Paris

Typeset by Mouflon, Paris
Printed by Imprimerie Jouve, Mayenne
Dépôt légal : décembre 1996

ISBN 92-3-103275-5

This Monograph is dedicated to
Dr G.F. Humphrey, Professor T.R. Parsons
and the late Dr C.J. Lorenzen,
whose pioneering work and vision laid the foundations
for accurate chlorophyll measurements in the oceans.

Preface

Publication by UNESCO of the series of 'Monographs on Oceanographic Methodology' has followed a recommendation adopted by the Scientific Committee on Oceanic Research (SCOR) almost 30 years ago at its first meeting in Halifax in 1963. Since then, nine monographs on methodologies used by the international oceanographic and marine communities have been published. These have usually resulted from the activities of an international SCOR Working Group created to stimulate or focus interest in a particular field of research. Not all working group data result in the publication of monographs; many are more suited to reports, technical papers in marine science or the handbook series (see SCOR Handbook, 1991).

The present publication results from the activities of SCOR Working Group 78, *Determination of Photosynthetic Pigments in Seawater*, established in 1985 under the chairmanship of Dr R.F.C. Mantoura. The first meeting, held in Plymouth in 1986, considered intercalibration of pigment methods, and recommended 'hands-on' workshops for measuring chlorophylls and derivatives, carotenoids and derivatives, and field applications (comparing phytoplankton samples from the open ocean, inshore and estuarine habitats). The objectives and achievements, as well as SCOR's support for marine pigment technology since 1966, are outlined in Chapter 1. The intercalibration workshops were held in 1988 and 1989, but the key scientists did not have dedicated time for writing the monograph until mid-1992.

In May 1992 Drs Jeffrey, Mantoura and Wright met at the Marine Laboratories of the Commonwealth Scientific and Industrial Research Organisation (CSIRO) in Hobart, Australia, to assess the intercalibration data and plan publication of the results. They decided to prepare one comprehensive monograph, 'Phytoplankton Pigments in Oceanography: Guidelines to Modern Methods', which would give all the results of the workshops (comparing classic with modern pigment techniques), and would include a section of data for the identification of all phytoplankton pigments of relevance to oceanography. Detailed protocols for preparing pigment standards, extracting and storing pigments, and spectrophotometric, fluorometric and chromatographic techniques would be included. In order to base the experimental sections in the literature, five review chapters were planned. Thus all data and protocols for the accurate analysis of plant pigments, by either classic or advanced technologies, would be available in a single volume.

In June 1993, Drs Jeffrey and Wright met with Dr Mantoura at the Plymouth Marine Laboratory (PML), UK. It was agreed that Dr Jeffrey should assume responsibility for the book production from that time.

The scientific opinions expressed in this work are those of the authors and should not be interpreted as the views of SCOR or UNESCO. The equipment and materials noted in the text are those that gave the best results at the time of the experimental phase of Working Group 78; they will surely be superseded. The data and recommendations were 'state-of-the-art' when this monograph was compiled, but these, too, will be improved, as scientific knowledge of the oceans advances.

The Intergovernmental Oceanographic Commission (IOC) of UNESCO lent support to this monograph in light of its relevance to many international programmes. The publication has also been supported by CSIRO, PML and the Australian Antarctic Division.

Contents

Contributors

Dr R. G. Barlow. Plymouth Marine Laboratory, Prospect Place, West Hoe, Plymouth PL1 3DH, United Kingdom.

Dr T. Bjørnland Department of Biotechnology, Section Chemistry, PO Box 40, N–1432, Ås–NL4, Norway.

Dr H. Claustre. Laboratoire de Physique et Chimie Marines, CNRS, UA 353, Observatoire Océanologique de Villefranche, BP08, F-06230, Villefranche-sur-mer, France.

Ms D. G Cummings. Plymouth Marine Laboratory, Prospect Place, West Hoe, Plymouth PL1 3DH, United Kingdom.

Dr N. Engel. Institute for Organic Chemistry, University of Fribourg, Fribourg, Switzerland.

Dr E. J. H. Head. Biological Oceanography Division, Bedford Institute of Oceanography, PO Box 1006, Dartmouth, Nova Scotia B2Y 4A2, Canada.

Dr S. W. Jeffrey. CSIRO Division of Fisheries, Marine Laboratories, GPO Box 1538, Hobart, Tasmania 7001, Australia.

Ms J.-M. LeRoi. CSIRO Division of Fisheries, Marine Laboratories, GPO Box 1538, Hobart, Tasmania 7001, Australia.

Ms C. A. Llewellyn. Plymouth Marine Laboratory, Prospect Place, West Hoe, Plymouth PL1 3DH, United Kingdom.

Dr R. F. C. Mantoura. Plymouth Marine Laboratory, Prospect Place, West Hoe, Plymouth PL1 3DH, United Kingdom.

Dr C. E. Morales. Instituto de Fomento Pesquero, Depto. Oceanografia Pesquera, Huito 374, Casilla 8-V, Valparaiso, Chile.

Dr E. E. Pfündel. Institut für Pflanzengenetik, Correnstr. 3, D–06466 Gatersleben, Germany.

Dr R. J. Porra. CSIRO Plant Industry, GPO Box 1600, Canberra ACT 2601, Australia.

Dr D. J. Repeta. Department of Chemistry, Woods Hole Oceanographic Institution, Woods Hole MA 02543, USA.

Dr M. Vesk. Electron Microscope Unit, University of Sydney, Sydney, NSW 2006, Australia.

Dr N. A. Welschmeyer. Moss Landing Marine Laboratory, PO Box 450, Moss Landing 95039, USA.

Dr S. W. Wright. Australian Antarctic Division, Channel Highway, Kingston, Tasmania 7050, Australia.

Acknowledgements

We warmly thank the many people who supported our work on this monograph and contributed to its successful completion:

- SCOR for the opportunity to review, update, intercalibrate and develop pigment methods used in oceanography, for generous financial support of three workshops, for travel funds, and for continued encouragement to complete the present monograph
- The heads of our institutions who approved and supported our commitment to this project
  Dr P. C. Young, Chief, CSIRO Division of Fisheries, Hobart, Australia
  Dr B. L. Bayne and Dr M. Whitfield, Director and Acting Director, respectively, Plymouth Marine Laboratories (PML), Plymouth, United Kingdom
  Dr P. Quilty, Assistant Director, Australian Antarctic Division, Hobart, Australia
- Dr P. C. Young, Chief, CSIRO Division of Fisheries, Hobart, for special financial support to Dr Jeffrey for word processing, graphics and editing services, so that all manuscripts could be brought to a uniform standard for UNESCO.
- The Director, PML, for a grant towards publication costs.
- Our many reviewers, especially Dr D. J. Mackey, CSIRO Division of Oceanography, Hobart, whose thoughtful criticisms contributed significantly to all sections of the monograph. (Each chapter was reviewed by at least two experts.)
  Dr J. M. Anderson
  Dr R. Barlow
  Mr J. Barrett
  Dr T. Bjørnland
  Dr S. Blackburn
  Ms D. Cummings
  Dr P. Donkin
  Dr W. Gieskes
  Dr G. M. Hallegraeff
  Prof. P. Hynninen
  Dr S. W. Jeffrey
  Ms C. A. Llewellyn
  Dr D. J. Mackey
  Dr R. F. C. Mantoura
  Dr P. Nichols
  Prof. T. R. Parsons
  Dr R. Porra
  Dr D. Repeta
  Prof. H. Scheer
  Dr J. Volkman
  Dr S. W. Wright
- Our CSIRO editor, Dr Vivienne Mawson, who edited all sections of the monograph to a high standard before they were sent to UNESCO for publishing.
- Mr G. Wright and Dr H. Moore of UNESCO, Paris, who gave continual encouragement and guided the production of the monograph.
- Ms Toni Cracknell, CSIRO, Hobart, for excellent wordprocessing support to Dr Jeffrey throughout the preparation and editing of the manuscripts, and Ms Gina Hadolt, also of Hobart, who formatted the Part IV Data sheets.

- Mr David Taylor for graphics support to Dr Jeffrey at CSIRO, Hobart, particularly the Part IV graphics sheets.
- Ms Victoria Henderson of Hobart, who designed the cover, and AutoPlot Visual Systems, Hobart who produced the artwork.
- Mrs Clodagh Jones of Hobart, who assisted with the preparation of the index.
- Our colleagues and families, who encouraged and supported us during the long months of effort required to complete this project for the oceanographic community.

S. W. Jeffrey
R. F. C. Mantoura
S. W. Wright
July 1995

Part I

Literature reviews: background to modern pigment oceanography

1

Development of pigment methods for oceanography: SCOR-supported Working Groups and objectives

S. W. Jeffrey and R. F. C. Mantoura

1.1 THE CONTEXT OF THIS MONOGRAPH

1.1.1 *The importance of chlorophyll measurements in the sea*

Chlorophyll is the principal photosynthetic pigment of the plant kingdom both in terrestrial and marine environments. Its value as a biomass indicator of oceanic microscopic marine plants has been recognized for over 40 years.

The floating microscopic marine plant populations, known as phytoplankton (*phyton* = plant; *planktos* = wandering), have a truly global distribution (Fig. 1.1) and they contribute over 25% of the total vegetation of the planet (Jeffrey and Hallegraeff, 1990). The tens of thousands of species (see Chapter 2) present a variety of form and metabolism equalling, and even surpassing, those of terrestrial plants.

Chlorophyll, and its function of converting light energy to chemical energy *via* the process of photosynthesis, possibly began evolving in the ocean about 2,000 million years ago (Callot, 1991; Scheer, 1991). Photosynthesis had a dramatic impact on the biogeochemistry of the earth, changing its environment from oxygen-poor, UV-rich and chemically-reducing to the oxygen-rich, geochemically more corrosive environment of the present day (Lovelock, 1979; Walker *et al.* 1983). More than 1,000 million tons of chlorophyll may be produced and destroyed each year both on land and in the sea, with chlorophyll turning over 40 times a year in the ocean because of the short generation times of phytoplankton (Hendry *et al.*, 1987). Indeed, the chlorophyll that sediments to the ocean floor in senescent or dead phytoplankton accumulates as a complex family of "geoporphyrins" in oil and shales (Treibs, 1936), enabling geochemists to link the organic geochemical record to the past biological productivity of the ocean (Baker and Palmer, 1978; Eckard *et al.* 1991).

Classically, phytoplankton are recognized as the basis of all animal production in the open sea, supporting food webs upon which the world's fisheries are based. It is important to be able to monitor these phytoplankton populations since failures in the abundance and timing of algal blooms can lead to the collapse of whole year classes of fisheries (Lasker, 1975).

Nutrient enrichments due to naturally occurring oceanographic processes or man-made enrichments (e.g., pollution of various kinds) can result in damaging phytoplankton blooms, which if toxic, can decimate not only wild fisheries, but also farmed fish and shellfish in inshore waters as well (Okaichi *et al.* 1989). Although the animals may not be affected by algal toxins, they can accumulate them in their tissues and cause the illness or death of people who eat them. Monitoring of phytoplankton populations is, of necessity, becoming standard practice in many countries with aquaculture ventures (Underdal *et al.* 1989).

The present biological productivity of the ocean is of central concern to a world concerned about global climate change. Indeed, phytoplankton can themselves influence global chemical budgets and climate change by a number of mechanisms. Three areas of importance are: the utilization of carbon dioxide through photosynthesis thus affecting the global carbon dioxide budget (Williamson and Gribbin, 1991; see Fig. 1.1); a contribution to seasonal warming of the surface layers of the ocean (Sathyendranath *et al.* 1991*a*) by absorbing and scattering light; and the production of quantities of volatile compounds (e.g., dimethyl sulphide) which escape into the atmosphere and act as cloud-seeding nuclei, particularly in the north Atlantic (Malin *et al.* 1992).

Because chlorophyll's photosynthetic function makes it a unique indicator of oceanic plant biomass and productivity, it is probably the most frequently measured biochemical parameter in oceanography. Measurements of chlorophyll distribution in the oceans have revealed areas of contrasting fertility, from the oligotrophic ocean gyres with their low concentrations of chlorophyll in surface waters ($<0.05\ \mu g\ l^{-1}$) and their characteristic maxima at depth ($0.1–0.5\ \mu g\ l^{-1}$; 100–150 m), to the chlorophyll-rich waters found in upwelling areas, and along continental shelf fronts and coastal seas and estuaries (chlorophyll = $1–10\ \mu g\ l^{-1}$). Chlorophyll measurements have been used to follow diurnal, seasonal and long-term changes in biological productivity in contrasting oceanic regimes. Further, the biological involvement of atmospheric CO_2 in the micro- and meso-scale ocean has recently been demonstrated from the tight coupling between pCO_2 and surface chlorophyll in the north-east Atlantic (Watson *et al.* 1991).

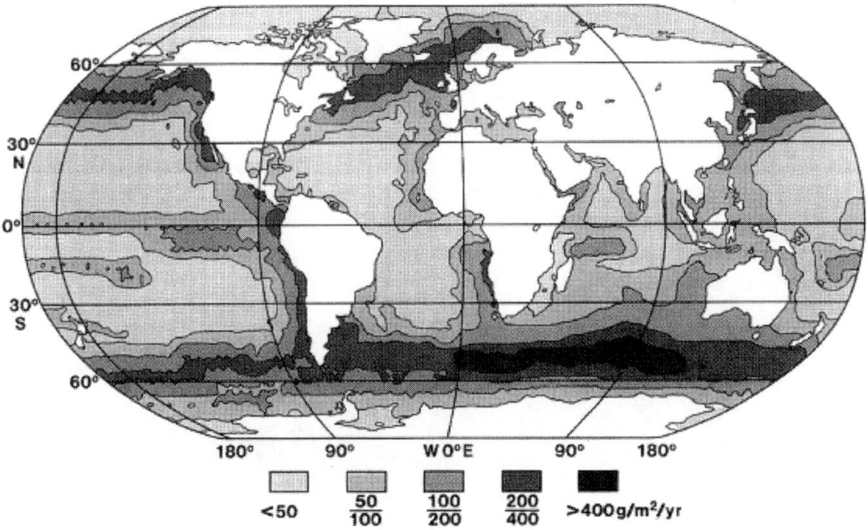

Figure 1.1
Distribution of phytoplankton abundance and production in the oceans — ranges mapped in g m^{-2} year^{-1}, dry matter. (After Bunt, 1975).

Chlorophylls and associated carotenoid pigments are now being used to map the chemotaxonomic composition of phytoplankton in the oceans (e.g., Pacific: Jeffrey and Hallegraeff, 1987; Ondrusek *et al.* 1991; Atlantic: Gieskes and Kraay, 1986*b*; Antarctic: Buma *et al.* 1986; Wright 1987), and also to track the flux, transformation and fate of phytogenic carbon (Repeta and Gagosian, 1987; Barlow *et al.* 1993). Disappearance of native pigments and formation of degradation products have also been used to quantitate grazing by micro- and macro-zooplankton (Burkill *et al.* 1987; Conover *et al.* 1986; Kleppel *et al.* 1988).

Until recently, chlorophyll was measured from ships by taking discrete water samples and analysing them. The species composition was evaluated by sampling with plankton nets or harvesting from water samples in a plankton centrifuge, and counting and identifying the species microscopically. Today, with the advent of remote sensing, ocean temperature and colour can now be monitored on a global scale from space (Aiken *et al.* 1992; see Fig. 1.2). The former indicates hydrographic discontinuities and the latter phytoplankton abundance, provided the ground truthing is accurate. During the limited life-time of the Nimbus 7 Coastal Zone Colour Scanner satellite (1978–1986), literally thousands of exquisite, though non-quantitative, uncalibrated maps of chlorophyll distribution were produced for whole ocean basins (Hovis *et al.* 1980). It is essential that the full potential of this technology is realized with future satellites (e.g., SeaWIFS) by accurate ground-truthing of chlorophyll and accessory pigments. We predict that, since chlorophyll is the only biological parameter measurable from space, pigments will serve as basic or proxy parameters for global mapping of components of the oceanic carbon cycle, including total, regenerated and 'new' production (Sathyendranath *et al.* 1991*b*), seasonal CO_2 'drawdown'

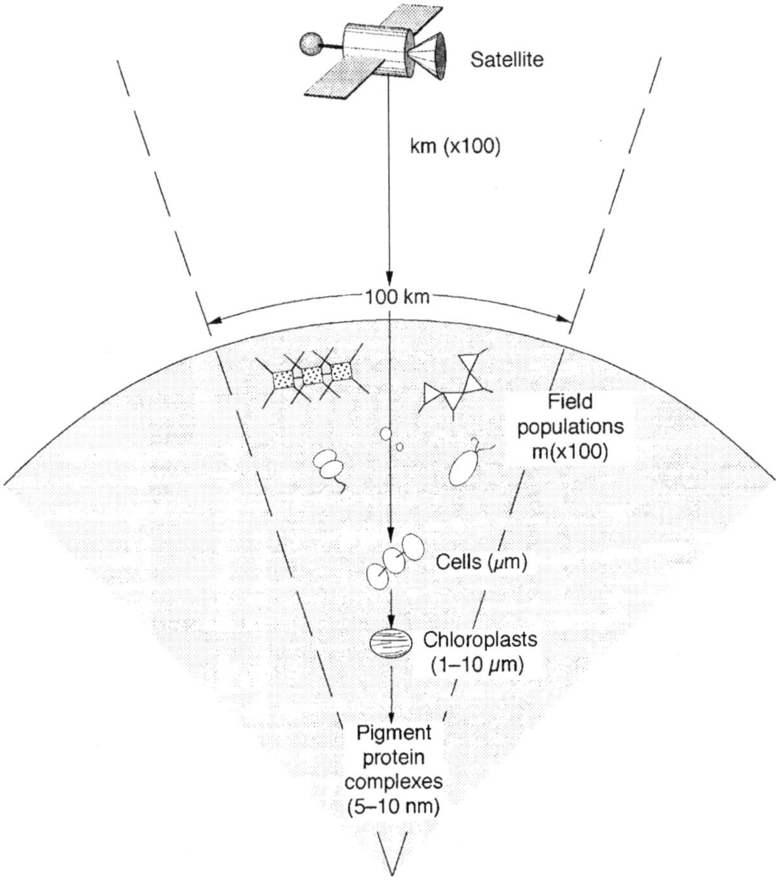

Figure 1.2
Space scales from molecular to global in phytoplankton pigment dynamics.

and ventilation fluxes, and organic sedimentation fluxes. Clearly to fully exploit the unique pigment markers for understanding oceanographic processes, oceanographers must be able to routinely achieve accurate and fully resolved analyses of phytoplankton pigments.

1.1.2 *Milestones in pigment oceanography*

Chlorophyll could not be used to quantify plant biomass in either terrestrial or marine plant systems until pure chlorophylls became available. In the late 1930s, Mackinney (1941), Zscheile (1941) and Zscheile *et al.* (1942) succeeded in purifying and crystallizing chlorophylls *a* and *b* from higher plants. Extinction coefficients which formed the basis for early spectrophotometric measurements, could then be derived. Based on the Mackinney values, Arnon in 1949 and Richards with Thompson in 1952 published the first equations for the spectroscopic determination of chlorophylls *a* and *b*, and chlorophylls *a*, *b* and *c* for terrestrial and marine plants respectively. These classic methods, or improved derivations of them (see Appendix F), are widely used in both marine and terrestrial science to the present day, notwithstanding the tremendous technical advances in analytical procedures that have been made in the meantime. A most important advance was the development of chromatography, which made it possible to rapidly separate and quantitate complex suites of pigments.

The key developments in pigment oceanography since the initial spectrophotometric equations were derived are:

- recognition of chlorophyll as a selective biomarker of phytoplankton in the presence of other biological components such as detritus, bacteria and zooplankton (Richards with Thompson, 1952)
- crystallization of chlorophyll *c* (Jeffrey, 1963) and recognition of two major chlorophyll *c* components (Jeffrey, 1969, 1972), which allowed accurate spectrophotometric equations for chlorophylls *a, b,* c_1 and c_2 to be derived (Jeffrey and Humphrey, 1975)
- fluorometric determination of chlorophyll *a* (Holm-Hansen *et al.* 1965), which was fifty times more sensitive than previous methods; and *in vivo* fluorescence applications (Lorenzen, 1966), which made *in situ* phytoplankton profiling a regular feature of oceanographic studies
- application of thin-layer chromatography to phytoplankton field samples, which enabled routine separation and quantitation of phytoplankton chlorophylls, carotenoids, and their breakdown products to be made (Jeffrey, 1968; 1974; 1976)
- recognition from thin-layer chromatography field applications that pigment signatures were valuable markers for phytoplankton taxa, processes and water masses (Jeffrey, 1974; Hallegraeff, 1981; Jeffrey and Hallegraeff, 1987)
- development of automated high performance liquid chromatography methods for routine shipboard determination of phytoplankton chlorophylls, carotenoids and breakdown products (Mantoura and Llewellyn, 1983; Wright and Shearer, 1984; Gieskes and Kraay, 1983*b*; 1986*a*; Roy, 1987; Zapata *et al.* 1987), which expanded the scale of pigment analysis in oceanography
- satellite remote sensing of ocean colour, which made possible the global mapping of surface chlorophyll (Hovis *et al.* 1980; Robinson, 1983)
- use of epi-fluorescence microscopy (Waterbury *et al.* 1979) and analytical cell flow cytometry (Yentsch *et al.* 1983) to detect chlorophyll and biliprotein autofluorescence in single phytoplankton cells
- application of pigment chemotaxonomy to recognizing classes of microalgae in field samples that had previously been missed because of loss through filtering or preservation damage: (e.g., chlorophyll *b* [chlorophytes and prasinophytes] in the oligotrophic ocean (Jeffrey, 1976; Gieskes *et al.* 1988); alloxanthin [cryptophytes] in the North Sea (Gieskes and Kraay, 1983*a*); zeaxanthin [cyanobacteria] in the North Atlantic (Guillard *et al.* 1985); 19'-hexanoyloxyfucoxanthin [prymnesiophytes] in the Banda Sea (Arpin *et al.* 1976; Gieskes *et al.* 1988); chlorophyll c_3 [prymnesiophytes] in the Atlantic Ocean (Jeffrey and Wright, 1987; Gieskes *et al.* 1988); divinyl chlorophyll *a* and *b* in prochlorophytes in the Sargasso Sea (Gieskes and Kraay, 1983*b*; Chisholm *et al.* 1988; Goericke and Repeta, 1992).

1.1.3 *Basis and scope of this monograph*

High-performance liquid chromatography (HPLC) is the best method for separating, identifying and quantitating chlorophylls and carotenoids, whether from mixed phytoplankton field populations or cultured microalgae. However, for the biological oceanographer embarking on HPLC analysis, there are four outstanding problems:

- the numerous methods reported for the extraction and HPLC analysis of marine plant pigments have not been critically evaluated
- classical spectrophotometric and fluorometric methods have not been adequately evaluated against the new HPLC methods
- standards for identifying marine pigments and their quantitation by HPLC are generally either unstable or unavailable
- physico-chemical data for identifying the many marine pigments are difficult to locate, since they are widely scattered through the literature.

These broad problems were addressed by SCOR Working Group 78 (see Terms of Reference section 1.3.2) at a series of analytical workshops aimed at optimizing and standardizing advanced methods for the analysis of marine pigments by HPLC.

This monograph attempts to compile in one volume all the information required by the oceanographer wanting to use modern pigment methods. The new advances are set in the context of the classical spectrophotometric and fluorometric methods, which are still widely used in oceanography. This monograph:

- reviews recent advances in pigment oceanography (Part I) in the context of previous work
- summarizes the critical comparison of methods undertaken at the three SCOR WG 78 Workshops (see Section 1.3; Part II)
- recommends standard protocols for the reliable use of HPLC in pigment oceanography (Part III)
- compiles, in detailed Data and Graphics sheets (Part IV), spectroscopic and other data needed for identifying marine pigments.

1.2 Previous SCOR pigment working groups

1.2.1 *Spectrophotometric method assessment — UNESCO, Paris, 1964*

SCOR has played a key role in standardizing and developing methods for studying marine plankton (SCOR Handbook, 1991) through its programme of supporting working groups of international experts focussed on particular research topics. SCOR manuals issued for phytoplankton and zooplankton biological methodology (Monographs 2, 3, 4 and 6; see Preface), concentrated on sampling methods, preservation techniques, species identification and productivity measurements. In the case of photosynthetic pigments, SCOR's support for standardizing pigment methodology for the oceanographic community goes back three decades, to the first meeting in Paris in 1964 (Working Group 17, see Table 1.1).

At the UNESCO Headquarters in Paris in the early summer of June 1964, a group of pigment experts met to recommend protocols for spectrophotometric pigment determinations in seawater samples. The basic topics they considered were: type of filters for removing phytoplankton from seawater, filter treatment and storage, extraction solvents, need for grinding, duration and conditions of extraction,

TABLE 1.1. SCOR-funded workshops on pigment methodology

Date	Working group and title	Place	Activity and results	Participants (see footnotes)
1964	Working Group 17 'Determination of Photosynthetic Pigments'	Paris, France	Discussions; UNESCO monograph published 1966.	1
1980	SCOR-funded collaboration: 'Determination of Chlorophyll in Seawater'	Sydney, Australia	Laboratory intercalibration tests: UNESCO Tech. Paper 35 published in 1980.	2
1986	Working Group 78 'Determination of Photosynthetic Pigments in Seawater'	Plymouth, UK	Discussions; plans made for 3 hands-on laboratory Workshops; Report to SCOR.	3
1988 (Nov)	Working Group 78 Chlorophyll intercalibrations	Hobart, Australia	Hands-on laboratory workshop.	4
1989 (Jan-Feb)	Working Group 78 Carotenoid intercalibrations	Hobart, Australia	Hands-on laboratory workshop.	5
1989 (Sept-Oct)	Working Group 78 Field applications	Plymouth, UK	Hands-on laboratory workshop.	6
1992 (May)	Working Group 78 Monograph planning	Hobart, Australia	Discussions; Assembly of data, planning of monograph.	7
1992 (Sept-Oct)	Working Group 78 Monograph writing	Hobart, Australia	Discussions, writing, and data sheets.	8
1993 (Jan-Feb)	Working Group 78 Monograph writing	Hobart, Australia	Discussions, writing and data sheets.	9
1993 (May-June)	Working Group 78 Monograph writing	Plymouth, UK	Discussions, writing and data sheets.	10
1994 (April-May)	Working Group 78 Monograph writing	Hobart, Australia	Writing and discussions	11

Participants

1 K. Banse, H. J. Ceccaldi, V. K. Hansen, G. F. Humphrey, S. W. Jeffrey, J. Krey (Chairman), T. R. Parsons, L. P. Vernon, C. S. Yentsch.

2 S. W. Jeffrey, C. J. Lorenzen.

3 R. Dawson, R. Gagosian, Y. Halim, S. W. Jeffrey, C. Kraay, S. Liaaen-Jensen, R. F. C. Mantoura (Chairman), V. Morcos, D. Repeta, C. Yentsch
(Observers: R. Barlow, C. A. Llewellyn, S. Roy).

4. M. R. Brown, S. W. Jeffrey, C. A. Llewellyn, R. F. C. Mantoura, N. Welschmeyer, S. W. Wright.

5 T. Bjørnland, M. R. Brown, S. W. Jeffrey, D. Repeta, S. W. Wright.

6 R. Barlow, T. Bianchi, H. Claustre, R. Dawson, E. Head, R. F. C. Mantoura, C. A. Llewellyn, D. Repeta, S. W. Wright.

7, 8, 9, 10 S. W. Jeffrey, R. F. C. Mantoura, S. W. Wright (9 included D. Repeta; 10 included T. Bjørnland).

11 S. W. Jeffrey, T. Bjørnland, R. F. C. Mantoura, S. W. Wright.

precision of determinations, and choice of extinction coefficients for chlorophylls *a, b* and *c*. Because chlorophyll *c* had just been crystallized for the first time (Jeffrey, 1963), the SCOR group was able to derive the first quantitative equations for chlorophylls *a, b* and *c* (UNESCO, 1966), updating the Richards with Thompson (1952) equations, which assessed chlorophyll *c* as a non-quantitative milli-specific pigment unit. At the same meeting Dr T. R. Parsons tabled a valuable survey of current pigment methods (1966), which included extractive spectrophotometry and the new tool of fluorometry. Dr G. F. Humphrey also presented data on extractability of pigments from a range of cultured microalgae. These data assisted the SCOR recommendations and were published with the final document (UNESCO, 1966).

1.2.2 *Fluorescence, new spectrophotometric equations and thin-layer chromatography developments*

The period 1960–1975 saw the consolidation of methods, development of new ones, and their use at sea. Because oligotrophic waters are very low in pigments, improved sensitivity for chlorophyll *a* (up to 50 times) was achieved by the use of both *in vivo* (Lorenzen, 1966) and *in vitro* fluorometry (Holm-Hansen *et al.* 1965). Lorenzen (1967) and Holm-Hansen *et al.* (1965) also used an acidification step in spectrophotometry and fluorometry in an attempt to distinguish chlorophyll *a* from degradation products. Vernon (1960) had first introduced this acidification step for pheophytin analysis in higher plant extracts.

New equations appeared as revised extinction coefficients for chlorophylls *a, b* and *c* were published (e.g., Parsons and Strickland, 1963; Jeffrey and Humphrey, 1975). In the late 1960s, chlorophyll *c* was found to consist of two pigments, which could be separated on thin layers of polyethylene (Jeffrey, 1969). Subsequent crystallization of chlorophylls c_1 and c_2 and recrystallization of chlorophylls *a* and *b* allowed new extinction coefficients to be determined (Jeffrey, 1972), which were used for deriving the new spectrophotometric equations for chlorophylls *a, b,* c_1 and c_2 (Jeffrey and Humphrey, 1975).

Another advance was the publication of chromatographic methods for separating microgram quantities of pigments in algal cultures with thin layers of glucose (Madgwick, 1965; 1966), and either sucrose or cellulose or both (Jeffrey, 1968). The two-dimensional cellulose chromatography method separated 18 different chlorophylls and carotenoids from mixed extracts of cultured diatoms, dinoflagellates, green algae and prymnesiophytes. They also separated the degradation products pheophytins *a* and *b*, chlorophyllides *a* and *b*, pheophorbides *a* and *b* and the Mg-free derivatives of chlorophylls c_1 and c_2. When these methods were applied to field phytoplankton samples, the full spectrum of major pigments present in seawater was seen and described for the first time (Jeffrey 1974; 1976). This pigment data also made it possible to map the distribution of algal types and degradation products through the water column, using signature pigments characteristic of algal types and processes (Jeffrey, 1974 and a suite of papers by Jeffrey and Hallegraeff (1980–1987)). In a later development, water types were clearly distinguished by the presence of degradation products in eddy systems (Jeffrey and Hallegraeff, 1987).

1.2.3 *Spectrophotometric and fluorometric methods assessment — Sydney, 1978*

Because of the escalation of pigment methodology and oceanographic effort, the SCOR Executive in 1978 invited Dr C. J. Lorenzen and Dr S. W. Jeffrey to evaluate the spectrophotometric and fluorometric methods in common use. Intercalibration laboratory tests were carried out over 12 days at the CSIRO Division of Fisheries and Oceanography laboratories at Cronulla, Australia, in October 1978 (see Table 1.1). The results, which were quite clear-cut, were published as a UNESCO Technical Paper in Marine Science (Lorenzen and Jeffrey, 1980).

The tests were confined to methodology and procedures subsequent to extraction of the pigments because of time limitations. Technical problems in collecting samples, harvesting phytoplankton, storing filters, choosing extraction solvents and length of extraction time, and homogenization techniques were not addressed. The approach was to prepare the most common chlorophyll pigments found in seawater (chlorophylls *a, b, c,* chlorophyllide *a* and pheophorbide *a*) in spectrally pure form, mix them in a variety of combinations and proportions to simulate different field conditions, and analyze them by a number of spectrophotometric and fluorometric methods.

It was found that:

(a) In mixtures of pure chlorophylls *a, b* and *c*:

- The equations with the best extinction coefficients (see Appendix F) gave the best estimates of chlorophyll *a*. These were Jeffrey-Humphrey (1975); SCOR-UNESCO (1966); Parsons-Strickland (1963) (1–3% error)
- Chlorophyll *b* was measured satisfactorily only with Jeffrey-Humphrey (1975) or SCOR-UNESCO (1966)
- Chlorophyll *c* was satisfactory only with Jeffrey-Humphrey (1975), but could give errors if the concentration of *c* relative to *a* was low (see Appendix G).

(b) In mixtures of pure chlorophylls plus chlorophyllide *a* and pheophorbide *a*:

- All methods erred when chlorophyllide *a* was present, since none differentiated between chlorophyll *a* and chlorophyllide *a*
- Pheophorbide *a* could only be handled by methods designed to account for it (e.g., acidification techniques), and then only with varying degrees of precision. Otherwise chlorophyll *a* was overestimated.
- The presence of chlorophyll *b* interfered with the fluorometric technique since it was partially calculated as 'pheopigment'. Chlorophyll *b* also adversely affected the fluorometric chlorophyll *a* calculation by overestimating it.
- The *in vivo* fluorometric technique (Lorenzen, 1966; Herman and Denman, 1977) was recommended to be used only as a 'search' method and not as a substitute for an accurate analytical method for chlorophyll *a*.
- The selection of a particular chlorophyll method should be based on judgement of the particular marine biological region being sampled (e.g., estuarine, coastal, open ocean or sediments) and the type of chlorophyll and pheo-pigments likely to be present in the samples.

Guidance on how to choose the most appropriate spectrophotometric or fluorometric method for chlorophyll *a* was offered:

1. The trichromatic spectrophotometric equations of Jeffrey and Humphrey (1975) are the most accurate for chlorophylls *a*, *b* and *c* in near-surface euphotic zone

waters. However, errors may occur even with these equations if *b* and *c* are low compared to *a*. If only chlorophyll *a* is required, the three sets of equations — Jeffrey-Humphrey (1975), SCOR-UNESCO (1966) and Parsons-Strickland (1963) — are satisfactory (1–3% error).

2. Samples collected in the near-surface portion of the euphotic zone can be analyzed by either spectrophotometric or fluorometric methods because degradation products are normally absent, or present only in insignificant amounts.
3. Using the trichromatic equations for chlorophyll *a* only, and neglecting corrections for *b* and *c,* will cause errors only if *b* is present in significant quantities. Neglect of these corrections is not recommended, however, since chlorophyll *b* is frequently present.
4. Samples collected further down in the water column should be analysed by an acidification technique, since degradation products from senescent cells, detritus and faecal pellets can be expected.
5. Sediments, or samples containing sediments, could be analyzed by the acidification techniques, but caution should be exercised since these normally contain chlorophyll derivatives and a variety of unidentified pigments.
6. *In vivo* fluorometry (Lorenzen, 1966) should be used only as a 'search' method at sea for locating fronts, defining patch areas, locating chlorophyll maxima, and so on. At present, neither this method nor its DCMU modification (Slovacek and Hannan, 1977) used to maximise *in vivo* fluorescence should be used as a substitute for an accurate chlorophyll *a* measurement. Tests of these techniques show that correlation with chlorophyll *a* concentrations can vary by at least an order of magnitude.

The report also offered four evaluation criteria for international oceanographic chlorophyll data collected over previous decades and analyzed by trichromatic spectrophotometry:

1. Oceanographic data for chlorophyll *a* in surface waters are valid if calculated by Jeffrey-Humphrey (1975), SCOR-UNESCO (1966) or Parsons-Strickland (1963). Data calculated by Richards-Thompson (1952) should be corrected (Wartenberg, 1978; Humphrey, 1978).
2. All oceanographic data for chlorophylls *b* and *c* are suspect in surface waters unless calculated by Jeffrey-Humphrey (1975) equations, or converted to them (Wartenberg, 1978; Humphrey, 1978).
3. Trichromatic spectrophotometric equations are less useful for deeper water layers because of detrital chlorophyll interference (chlorophyllide, pheophorbide). No conversions (Wartenberg, 1978; Humphrey, 1978) can correct these errors.
4. As the Jeffrey-Humphrey (1975) equations are the most accurate, being based on the most accurately determined specific absorption coefficients for crystalline preparations of chlorophylls *a, b,* c_1 and c_2, their use is recommended for surface waters, and for calibration of the *in vitro* and *in vivo* fluorescence methods.

Finally, the report pointed out that if:

> '... precise knowledge of pigment composition is required, chromatography is the only method. Either cellulose thin-layer chromatography, or high-performance liquid chromatography (HPLC) is appropriate. These methods far surpass the simple routine spectrophotometric and fluorometric methods for the complete knowledge of all photosynthetic pigments since they are displayed in a single operation. From the authors' own experience they would like to recom-

mend these techniques for wider use in oceanography, since they are already bringing new insights to our understanding of phytoplankton processes in the sea.' (Lorenzen and Jeffrey, 1980).

1.3 SCOR Working Group 78: 'Determination of Photosynthetic Pigments in Seawater' — scope and objectives

1.3.1 *Introduction*

Between 1975 and 1985, thin-layer chromatography (TLC) methods applied to field phytoplankton samples produced a wealth of pigment information. In the 1980s, more rapid and automated chromatographic methods (high-performance liquid chromatography; HPLC) were applied to field oceanography. Other advances were the resolution of hitherto unknown degradation products, improvements in algal culturing techniques so that more strains of microalgae were available for reference and experimentation, and the discovery, through 'on board' analytical cell-flow cytometry, of new phytoplankton cell types in the sea (e.g., picoplanktonic cyanobacteria and free-living prochlorophytes). The first data from the Coastal Zone Colour Scanner revolutionized the scale at which phytoplankton production could be monitored from orbiting satellites. The question of sea-truthing these satellite images, the need for increasing the depth of the water column that could be sensed for ocean colour, and the possible discrimination of algal types in the plankton both vertically and horizontally, offered new challenges. Further, international concern over climate change and global warming due to greenhouse effects resulted in the setting up of international programs such as JGOFS (Joint Global Ocean Flux Study) to monitor

TABLE 1.2. Typical detection scales of pigments in oceanography

Method	System	Detection scale (pigment/unit)
Satellite ocean colour (CZCS; SeaWIFS)	Field phytoplankton population (≡ meso scale)	2 km^2 (≡ 1 pixel)
Aircraft colour sensors	Field populations	m^2
In situ fluorometers	Vertical or horizontal* profiles	$\mu g\ l^{-1}$
Ship-board flow through fluorometers	Horizontal transects	$\mu g\ l^{-1}$
Spectrophotometry	Discrete extracts (vertical/horizontal)	$\mu g\ l^{-1}$
TLC; HPTLC	Discrete extracts (vertical/horizontal)	$\mu g\ l^{-1}$
HPLC	Discrete extracts (vertical/horizontal)	$ng\ l^{-1}$
Analytical flow cytometry	Discrete *in vivo* particle analysis	$fg\ cell^{-1}$

*Can be attached to horizontally towed bodies.

carbon exchange and phytoplankton abundance in the world oceans. It was imperative that some standardization of methods be recommended to validate this international effort. Typical detection scales for the range of methods now used to measure pigments in oceanography are listed in Table 1.2, and space and time scales involved in phytoplankton pigment dynamics are displayed in Figures 1.2 and 1.3.

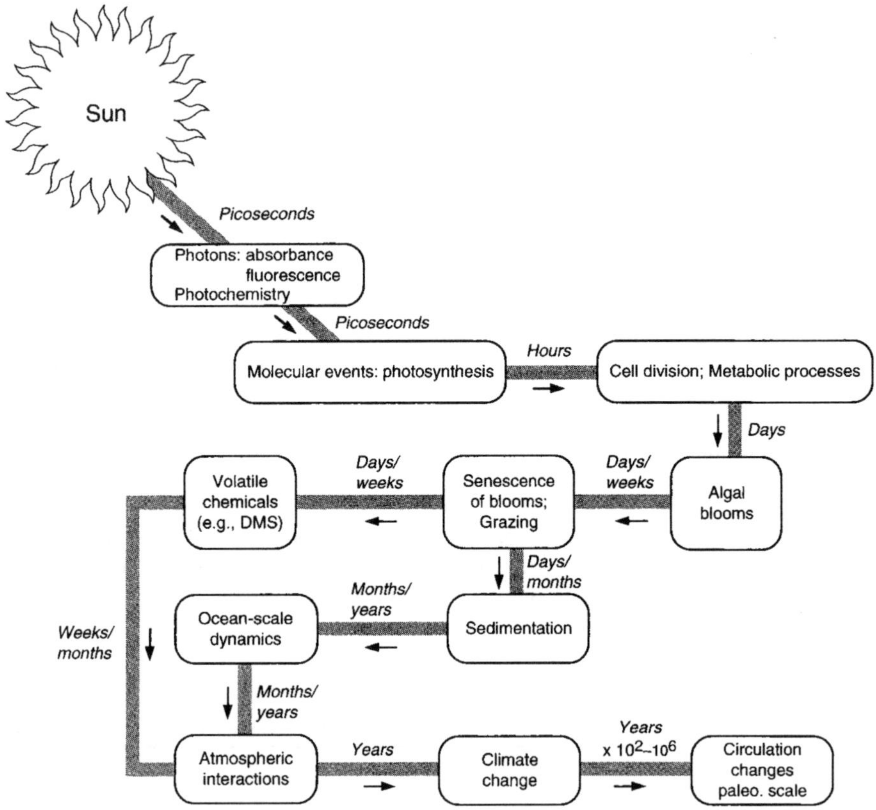

Figure 1.3
Time scales from molecular to global in phytoplankton pigment dynamics.

1.3.2 *Objectives of SCOR Working Group 78*

In 1983, two scientists involved in JGOFS, Dr R. F. C. Mantoura and Dr R. Dawson, proposed to SCOR that a new working group be set up to tackle pigment intercalibration problems. In 1985 the proposal was accepted by the SCOR Executive, and Working Group 78: 'Determination of Photosynthetic Pigments in Seawater' was set up (see Table 1.1). The first meeting was convened at the Institute for Marine and Environmental Research (now the Plymouth Marine Laboratories, UK) in the Spring of 1986 under the chairmanship of Dr R. F. C. Mantoura, with participants as listed in Table 1.1. The terms of reference given by SCOR were modified under Dr Mantoura's leadership when the nature and extent of the work necessary were realized. Six objectives for WG 78 were agreed:

1. To evaluate critically the accuracy of historical data and existing methods for the determination of photosynthetic pigments in seawater, including an appraisal of
   - the SCOR/UNESCO spectrophotometric procedures published in 1966;
   - and the SCOR-supported intercalibrations of Lorenzen and Jeffrey (1980) carried out in 1978;

   for the determination of chlorophylls and their breakdown products in relation to novel chromatographic techniques.
2. To evaluate extraction conditions and storage requirements prior to analysis of individual component chlorophylls, their breakdown products and accessory pigments.
3. To evaluate and recommend the simplest high-performance liquid chromatographic (HPLC) system for ship-board or field-based determination of chlorophylls *a, b, c*, and their principal degradation products.
4. To provide advice on the establishment and provision of reference chloropigments, carotenoid and phycobiliprotein standards for calibration of HPLC, fluorescence and absorbance methods.
5. To evaluate procedures for the extraction and determination of phycobiliproteins, recently identified as important constituents of phytoplankton in seawater.
6. To relate and compare measurements of individual chlorophylls and their breakdown products to *in situ* fluorescence recordings and remotely sensed colour imagery.

To achieve the objectives stated in the terms of reference, it was decided to hold three 'hands-on' workshops: on Chlorophylls, on Carotenoids and on Field Applications. The first two workshops were held at the CSIRO Marine Laboratories in Hobart, Australia, hosted by Dr Jeffrey. The field workshop was held at the Plymouth Marine Laboratories, UK, hosted by Dr Mantoura and assisted by Ms C. Llewellyn. The participants and the dates of the three workshops are shown in Table 1.1. The problem of the commercial non-availability of pigment standards (except for a very few) was solved by deciding to use standard algal cultures, whose pigments had been chemically characterized, grown under standard culture conditions, to provide the range of pigments needed (see Chapters 2, 6 and 13 and Wright *et al.* 1991). The Workshops' objectives are given below.

1.3.3 *Objectives of the Chlorophyll Workshop, Hobart, November 1988*

1. To critically compare and calibrate reverse-phase HPLC (RP–HPLC), absorption and fluorescence systems with standard spectrophotometric and fluorometric methods for the determination of chloropigment composition and concentration, using known mixtures of chlorophyll standards and reference cultures.
2. Using HPLC and diode array spectrophotometric detection, to determine extinction coefficients (E) at absorption maxima (λ_{max}), and at selected HPLC monitoring wavelengths (430, 440 and 450 nm) and in various solvent systems (methanol/H_2O; acetonitrile/H_2O, with and without ion-pairing agent) for all chloropigment standards. Relative molar and gravimetric HPLC fluorescence responses will also be derived for a range of excitation and emission conditions.
3. To fully document all the spectral and chromatographic properties of important chloropigments in the reference standards and reference cultures.

4. To assess pigment extractibility from reference cultures and stability in the extracting solvents using acetone, methanol, grinding, sonication and other appropriate treatments and solvents.
5. To check pigment stability under various storage conditions.
6. To prepare a Draft Report.

1.3.4 *Objectives of the Carotenoid Workshop, Hobart, January–February 1989*

1. To compare the characteristics of RP-HPLC and normal phase HPLC (NP-HPLC) for the separation, identification and quantitation of the principal carotenoids of the 10 reference algal cultures.
2. To document the spectral and chromatographic properties of the principal chemosystematic carotenoid markers in terms of retention times (RT), extinction coefficients, UV-vis spectra in standard and HPLC solvents, as well as simple chemical identification tests where appropriate.
3. To identify and test the suitability of a non-marine carotenoid as an internal standard for quantitative analytical HPLC of marine pigments.
4. To prepare a Draft Report.

1.3.5 *Objectives of the Field Applications Workshop, Plymouth, September–October 1989*

1. To implement 'standardized' protocols derived from the Chlorophyll and Carotenoid Workshops for the quantitative determination of pigments from natural marine samples using HPLC, spectrophotometry and fluorometry. Natural samples included:
   - phytoplankton from the English Channel
   - natural particulates
   - surface sediments
   - sediment trap samples
   - zooplankton faecal material
   - estuarine particulates
   - reference cultures
   - samples provided from other ocean areas by WG members
2. To reproduce HPLC systems and protocols of WG members' own HPLC systems and to compare results in terms of resolution, sensitivity, ease of operation, etc., to enable operational HPLC recommendations to be made.
3. To determine the extractibility and stability of natural phytoplankton pigments from various filters stored under different conditions.
4. To prepare a Draft Report.

In view of the comprehensiveness of the objectives and the short time allowed for each workshop (1–3 weeks), WG members could not prepare a full draft report before the end of each workshop. Several years later (Table 1.1) SCOR gave further support to enable the data to be written up in a comprehensive monograph, although the new recommended HPLC method was actually published in 1991 as a separate scientific paper by the joint authors (Wright, Jeffrey, Mantoura, Llewellyn, Bjørnland, Repeta and Welschmeyer, 1991).

1.4 The present volume

The monograph is arranged in five parts.

Part I groups all the review chapters: the background to this SCOR project (Chapter 1), an introduction to the phytoplankton and their pigment signatures (Chapter 2), metabolism and function of pigments in plants (Chapter 3), pigment methods used in oceanography (Chapter 4), and a bibliography of pigment processes in the sea (Chapter 5).

Part II gives detailed results of the intercalibration experiments carried out at the SCOR 'hands-on' laboratory Workshops at Hobart and Plymouth. This also includes much new experimental work that was carried out *after* the Workshops at both Hobart and Plymouth. The chapters are: procedures for algal culturing (Chapter 6), preparation of chlorophyll and carotenoid standards (Chapters 7 and 8), choice of extraction solvents (Chapter 9), long-term storage effects on pigment stability (Chapter 10), a rapid HPLC method for chlorophylls (Chapter 11), a comprehensive HPLC method for all pigments (Chapter 12), quantitative pigment analyses of the SCOR algal cultures (Chapter 13), and a quantitative comparison of HPLC, spectrophotometric and fluorescence methods (Chapter 14).

Part III gives guidelines for setting up HPLC systems (Chapter 15), HPLC calibration methods (Chapter 16), and recommended analytical protocols for the analysis of field samples (Chapter 17).

Part IV gives detailed Data and Graphics sheets for 47 key chlorophyll and carotenoid pigments to assist oceanographers with identifying pigments.

Part V is an Appendix with information ranging from nomenclature abbreviations, extinction coefficients, and other pigment data, to commercial suppliers of pure pigments.

The members of SCOR Working Group 78 wish to thank SCOR for this opportunity to review, update, intercalibrate, develop and finally to write up their work in the present volume.

References

Aiken, J., Moore, G. F., Holligen, P. M. (1992). Remote sensing of oceanic biology in relation to global climate change. *J. Phycol.* 28: 579–590.

Arnon, D. I. (1949). Copper enzymes in isolated chloroplasts. Polyphenol oxidase in *Beta vulgaris*. *Plant. Physiol.* 24: 1–15.

Arpin, N., Svec, W. A., Liaaen-Jensen, S. (1976). New fucoxanthin-related carotenoids from *Coccolithus huxleyi*. *Phytochem.* 15: 529–532.

Baker, E. W., Palmer, S. E. (1978). Geochemistry of porphyrins. In: Dolphin, D. (ed.) *The Porphyrins*, vol. 1. Academic Press, NY, pp. 485–551.

Barlow, R. G., Mantoura, R. F. C., Gough, M. A., Fileman, T. W. (1993). Pigment signatures of the phytoplankton composition in the north eastern Atlantic during the 1990 spring bloom. *Deep Sea Res.* 40: 459–477.

Buma, A. G. J., Treguer, P., Kraay, G. W., Morvan, J. (1990). Algal pigment patterns in different water masses of the Atlantic sector of the Southern Ocean during Fall 1987. *Polar Biol.* 11: 55–62.

Bunt, J. S. (1975. Primary productivity of marine ecosystems, in Lieth, H., Whittaker, R. H. (eds), *Primary Productivity of the Biosphere*, Springer-Verlag, Berlin, pp. 169–183.

Burkill, P. H., Mantoura, R. F. C., Llewellyn, C. A., Owens, N. J. P. (1987). Microzooplankton grazing and selectivity of phytoplankton in coastal waters. *Mar. Biol.* 93: 581–590.

Callot, H. J. (1991). Geochemistry of chlorophylls. In: Scheer, H. (ed.) *Chlorophylls*. CRC Press, Ann Arbor, pp. 340–364.

Chisholm, S. W., Olson, R. J., Zettler, E. R., Goericke, R., Waterbury, J. B., Welschmeyer, N. A. (1988). A novel free-living prochlorophyte abundant in the oceanic euphotic zone. *Nature*, 334: 340–343.

Conover, R. J., Durvasula, R., Roy, S., Wang, R. (1986). Probable loss of chlorophyll-derived pigments during passage through the guts of zooplankton, and some of the consequences. *Limnol. Oceanogr.* 31: 878–887.

Eckardt, C. B., Carter, J. F., Maxwell, J. R. (1991). Combined liquid chromatography/mass spectrometry of tetrapyrroles of sedimentary significance. *Energy Fuels*, 4: 741–747.

Gieskes, W. W. C., Kraay, G. W. (1983*a*). Dominance of Cryptophyceae during the phytoplankton spring bloom in the central North Sea detected by HPLC analysis of pigments. *Mar. Biol.* 75: 179–185.

Gieskes, W. W. C., Kraay, G. W. (1983*b*). Unknown chlorophyll *a* derivatives in the North Sea and the Atlantic Ocean revealed by HPLC analysis. *Limnol. Oceanogr.* 28: 757–766.

Gieskes, W. W. C., Kraay, G. W. (1986*a*). Analysis of phytoplankton pigments by HPLC before, during and after mass occurrence of the microflagellate *Corymbellus aureus* during the spring bloom in the open northern North Sea in 1983. *Mar. Biol.* 92: 45–52.

Gieskes, W. W. C., Kraay, G. W. (1986*b*). Floristic and physiological differences between the shallow and the deep nanophytoplankton community in the euphotic zone of the open tropical Atlantic revealed by HPLC analysis of pigments. *Mar. Biol.* 91: 567–576.

Gieskes, W. W. C., Kraay, G. W., Nontji, A., Setiapermana, D., Sutomo (1988). Monsoonal alternation of a mixed and a layered structure in the phytoplankton of the euphotic zone of the Banda Sea (Indonesia): a mathematical analysis of algal pigment finger prints. *Neth. J. Sea Res.* 22: 123–137.

Goericke, R., Repeta, D. J. (1992). The pigments of *Prochlorococcus marinus*: the presence of divinyl chlorophyll *a* and *b* in a marine procaryote. *Limnol. Oceanogr.* 37: 425–433.

Guillard, R. R. L., Murphy, L. S., Foss, P., Liaaen-Jensen, S. (1985). *Synechococcus* sp. as likely zeaxanthin-dominant ultraplankton in the North Atlantic. *Limnol. Oceanogr.* 30: 412–414.

Hallegraeff, G. M. (1981). Seasonal study of phytoplankton pigments and species at a coastal station off Sydney; importance of diatoms and nanoplankton. *Mar. Biol.* 61: 107–118.

Hendry, G. A. F., Houghton, J. D., Brown, S. B. B. (1987). The degradation of chlorophyll — a biological enigma. *New Phytol.* 107: 255–302.

Herman, A. W., Denman, K. L. (1977). Rapid underway profiling of chlorophyll with an *in situ* fluorometer mounted on a "Batfish" vehicle. *Deep Sea Res.* 24: 385–397.

Holm-Hansen, O., Lorenzen, C. J., Holmes, R. W., Strickland, J. D. H. (1965). Fluorometric determination of chlorophyll. *J. Conseil.* 301: 3–15.

Hovis, W. A., Clark, D. K., Andersen, F., Austin, R. W., Wilson, W. H., Baker, E. T., Ball, D., Gordon, H. R., Mueller, J. L., El-Sayed, S. Z., Sturm, B., Wrigley, R. C., Yentsch, C. S. (1980). Nimbus-7 Coastal Zone Colour Scanner: System description and initial imagery. *Science*, 210: 60–63.

Humphrey, G. F. (1978). The recalculation of marine chlorophyll concentrations with special reference to Australian waters. *Aust. J. Mar. Freshwat. Res.* 29: 409–416.

Humphrey, G. F., Wootton, M. (1966). Comparison of the techniques used in the determination of phytoplankton pigments. In: *Determination of Photosynthetic Pigments in Seawater.* UNESCO Monographs on Oceanographic Methodology, no. 1, UNESCO, Paris, pp. 37–63.

Jeffrey, S. W. (1963). Purification and properties of chlorophyll *c* from *Sargassum flavicans. Biochem.* J. 86: 313–318.

Jeffrey, S. W. (1968). Quantitative thin-layer chromatography of chlorophylls and carotenoids from marine algae. *Biochim. Biophys. Acta,* 162: 271–285.

Jeffrey, S. W. (1969). Properties of two spectrally different components in chlorophyll *c* preparations. *Biochim. Biophys. Acta*, 177: 456–467.

Jeffrey, S. W. (1972). Preparation and some properties of crystalline chlorophylls c_1 and c_2 from marine algae. *Biochim. Biophys. Acta,* 279: 15–33.
Jeffrey, S. W. (1974). Profiles of photosynthetic pigments in the ocean using thin-layer chromatography. *Mar. Biol.* 26: 101–110.
Jeffrey, S. W. (1976). A report of green algal pigments in the central North Pacific Ocean. *Mar. Biol.* 37: 33–37.
Jeffrey, S. W., Hallegraeff, G. M. (1987). Photosynthetic pigments, species and light climate in a warm core eddy of the East Australian Current. *Deep Sea Res.* 34: 649–673.
Jeffrey, S. W., Hallegraeff, G. M. (1990). Phytoplankton ecology of Australian waters. In: Clayton, M. N. and King, R. J. (eds), *Biology of Marine Plants.* Longman Cheshire, Melbourne, pp. 310–348.
Jeffrey, S. W., Humphrey, G. F. (1975). New spectrophotometric equations for determining chlorophylls *a, b,* c_1 and c_2 in higher plants, algae and natural phytoplankton. *Biochem. Physiol. Pflanzen,* 167: 191–194.
Jeffrey, S. W., Wright, S. W. (1987). A new spectrally distinct component in preparations of chlorophyll *c* from the microalga, *Emiliania huxleyi* (Prymnesiophyceae). *Biochim. Biophys. Acta,* 894: 180–188.
Kleppel, G. S., Frazel, D., Pieper, R. E., Holliday, D. V. (1988). Natural diets of zooplankton off southern California. *Mar. Ecol. Progr. Ser.* 49: 231–241.
Lasker, R. (1975). Field criteria for survival of anchovy larvae: the relation between inshore chlorophyll maximum layers and successful first feeding. *Fish. Bull. U.S.* 73: 453–462.
Lorenzen, C. J. (1966). A method for the continuous measurement of *in vivo* chlorophyll concentrations. *Deep Sea Res.* 13: 223–227.
Lorenzen, C. J. (1967). Determination of chlorophyll and pheopigments: spectrophotometric equations. *Limnol. Oceanogr.* 12: 343–346.
Lorenzen, C. J., Jeffrey, S. W. (1980). Determination of chlorophyll in seawater: report of intercalibration tests. UNESCO Technical Papers in Marine Science 35, 21 pp.
Lovelock, J. E. (1979). *Gaia, New Look at Life on Earth.* Oxford University Press, 157 pp.
Mackinney, G. (1941). Absorption of light by chlorophyll solutions. *J. Biol. Chem.* 140: 315–322.
Madgwick, J. C. (1965). Quantitative chromatography of algal chlorophylls on thin layers of glucose. *Deep Sea Res.* 12: 233–236.
Madgwick, J. C. (1966). Chromatographic determination of chlorophylls in algal cultures and phytoplankton. *Deep Sea Res.* 13: 459–466.
Malin, G., Turner, S. M., Liss, P. S. (1992). Sulfur: the plankton/climate connection. *J. Phycol.* 28: 590–597.
Mantoura, R. F. C., Llewellyn, C. A. (1983). The rapid determination of algal chlorophyll and carotenoid pigments and their breakdown products in natural waters by reverse-phase high-performance liquid chromatography. *Anal. Chim. Acta,* 152: 297–314.
Okaishi, T., Anderson, D. M., Nemoto, T. (1989). *Red Tides: Biology, Environmental Science and Technology.* Elsevier, New York, 489 pp.
Ondrusek, M. E., Bidigare, R. R., Sweet, S. T., Defreitas, D. A., Brooks, J. M. (1991). Distribution of phytoplankton pigments in the North Pacific Ocean in relation to physical and optical variability. *Deep Sea Res.* 38: 243–266.
Parsons, T. R. (1966). The determination of photosynthetic pigments in seawater: a survey of methods. In: *Determination of Photosynthetic Pigments in sea-water,* UNESCO Monographs on Oceanographic Methodology, no. 1, UNESCO, Paris, pp. 19–36.
Parsons, T. R., Strickland, J. D. H. (1963). Discussion of spectrophotometric determination of marine plant pigments, with revised equations for ascertaining chlorophylls and carotenoids. *J. Mar. Res.* 21: 155–163.
Repeta, D. J., Gagosian, R. B. (1987). Carotenoid diagenesis in recent sediments — I. The Peru Continental Shelf (15°S, 75°W). *Geochim. Cosmochim. Acta,* 51: 1001–1009.
Richards, F. A., Thompson, T. G. (1952). The estimation and characterisation of plankton populations by pigment analyses. II. A spectrophotometric method for the estimation of plankton pigments. *J. Mar. Res.* 11: 156–172.

Robinson, I. S. (1983). Satellite observations of ocean colour. *Phil. Trans. R. Soc. Lond. A.* 5)309: 415–432.

Roy, S. (1987). High performance liquid chromatography of chloropigments. *J. Chromatogr.* 391: 19–34.

Sathyendranath, S., Gouveia, A. D., Shetye, S. R., Ravindram, P., Platt, T. (1991*a*). Biological control of surface temperature in the Arabian Sea. *Nature*, 349: 54–56.

Sathyendranath, S., Platt, T., Horne, E. P. W., Harrison, W. G., Ullno, O., Outerbridge, R., Hoepffner, N. (1991*b*). Estimation of new production in the ocean by compound remote sensing. *Nature*, 353: 129–133.

Scheer, H. (ed.) (1991). *Chlorophylls*. CRC Press, Boca Raton, Florida, 1257 pp.

SCOR Handbook, 1991, Halifax, Nova Scotia, Canada, May 1991, 90 pp.

Slovacek, R. I., Hannan, P. J. (1977). *In vivo* fluorescence determinations of phytoplankton chlorophyll *a*. *Limnol. Oceanogr.* 22: 919–925.

Treibs, A. (1936). Porphyrine in Kohle, Liebigs. *Ann.* 520: 144.

Underdal, B., Skulberg, O. M., Dahl, E., Aune, T. (1989). Disastrous bloom of *Chrysochromulina polylepis* (Prymnesiophyceae) in Norwegian coastal waters 1988 – mortality in marine biota. *Ambio,* 18: 265–270.

UNESCO (1966). *Determination of photosynthetic pigments in seawater*. UNESCO Monographs on Oceanographic Methodology, Paris, 69 pp.

Vernon, L. P. (1960). Spectrophotometric determination of chlorophylls and pheophytins in plant extracts. *Anal. Chem.* 32: 1144–1150.

Walker, J. C. G., Klein, C., Schidlowski, M., Schopf, J. W., Stevenson, D. J., Walter, M. R. (1983). Environmental evolution of the archaen-early proterozoic Earth. In: Schopf, J. W. (ed.) *Earth's Earliest Biosphere; its Origin and Evolution*, Princeton University Press, Princeton, N.J. 260 pp.

Wartenberg, D. E. (1978). Spectrophotometric equations: an intercalibration technique. *Limnol. Oceanogr.* 23: 566–570.

Waterbury, J. B., Watson, S. W., Guillard, R. R. L., Brand, L. E. (1979). Widespread occurrence of a unicellular marine planktonic cyanobacterium. *Nature*, 277: 293–294.

Watson, A. J., Robinson, C., Robertson, J. E., Williams, P. J. Le B., Fasham, M. J. R. (1991). Spatial variability in the sink for atmospheric carbon dioxide in the North Atlantic. *Nature*, 350: 50–53.

Williamson, P., Gribbin, J. (1991). How plankton changes the climate. *New Scientist,* 129: 44–48.

Wright, S. W. (1987). Phytoplankton pigment data: Prydz Bay region. Australian Antarctic Research Expeditions. Research Notes 58. Antarctic Division, Kingston, Tasmania, Australia. 102 pp.

Wright, S. W., Shearer, J. D. (1984). Rapid extraction and high-performance liquid chromatography of chlorophylls and carotenoids from marine phytoplankton. *J. Chromatogr.* 294: 281–295.

Wright, S. W., Jeffrey, S. W., Mantoura, R. F. C., Llewellyn, C. A., Bjørnland, T., Repeta, D., Welschmeyer, N. (1991). Improved HPLC method for the analysis of chlorophylls and carotenoids from marine phytoplankton. *Mar. Ecol. Progr. Ser.* 77: 183–196.

Yentsch, C. M., Horan, P. K., Muirhead, K., Dortch, Q., Haugen, E. M., Legendre, L., Murphy, L. S., Perry, M. J., Phinney, D., Pomponi, S. A., Spinrad, R. W., Wood, A. M., Yentsch, C. S., Zahurenec, B. J. (1983). Flow cytometry and sorting: a powerful technique with potential applications in aquatic sciences. *Limnol. Oceanogr.* 28: 1275–1280.

Zapata, M., Ayala, A. M., Franco, J. M., Garrido, J. L. (1987). Separation of chlorophylls and their degradation products in marine phytoplankton by reversed-phase high-performance liquid chromatography. *Chromatographia*, 23: 26–30.

Zscheile, F. P. (1941). Plastid pigments with special reference to their physical and photochemical properties and to analytical methods. *Bot. Rev.* 7: 587–648.

Zscheile, F. P., Comar, C. L., Mackinney, G. (1942). Interlaboratory comparison of absorption spectra by the photoelectric spectrophotometric method; determinations on chlorophyll and Weigert's solutions. *Plant Physiol.* 17: 666–670.

2

Introduction to marine phytoplankton and their pigment signatures

S. W. Jeffrey and M. Vesk

2.1 Introduction

The phytoplankton are the microscopic algae that make up the floating pastures of the world's oceans (*phyton* = plant; *planktos* = wandering). They provide the food base which supports, either directly or indirectly, the entire animal population of the open sea, and they contribute significantly to climatic processes (see Chapter 1). Their diversity is immense and representatives of most algal Divisions (Table 2.1) may be found in these ocean populations. The best known are the exquisite silica-walled diatoms (Bacillariophyta), which account for up to 10,000 living species (Hostetter and Stoermer, 1971; Margulis *et al.* 1990), and many more if fossil species are included. The dinoflagellates (Dinophyta) encompass about 1,200 species, only about half of which are photosynthetic (Taylor 1976*b*). Green algae (Chlorophyceae, Prasinophyceae) and the euglenophytes (Euglenophyta) are common in coastal waters and the former also occur in the open ocean (Jeffrey, 1976*b*; Jeffrey and Hallegraeff 1980*a*, 1980*b*). The polyphyletic golden-brown flagellates of the Haptophyta and

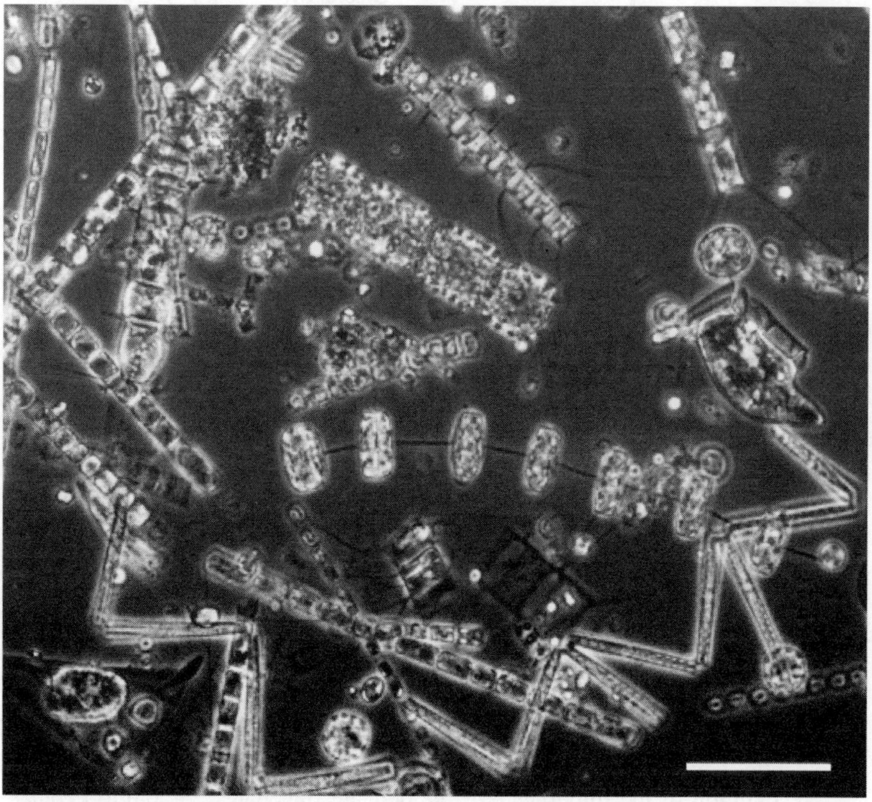

Figure 2.1
Light micrograph of coastal spring bloom of diatoms, East Australian Current. Scale bar = 100 µm.

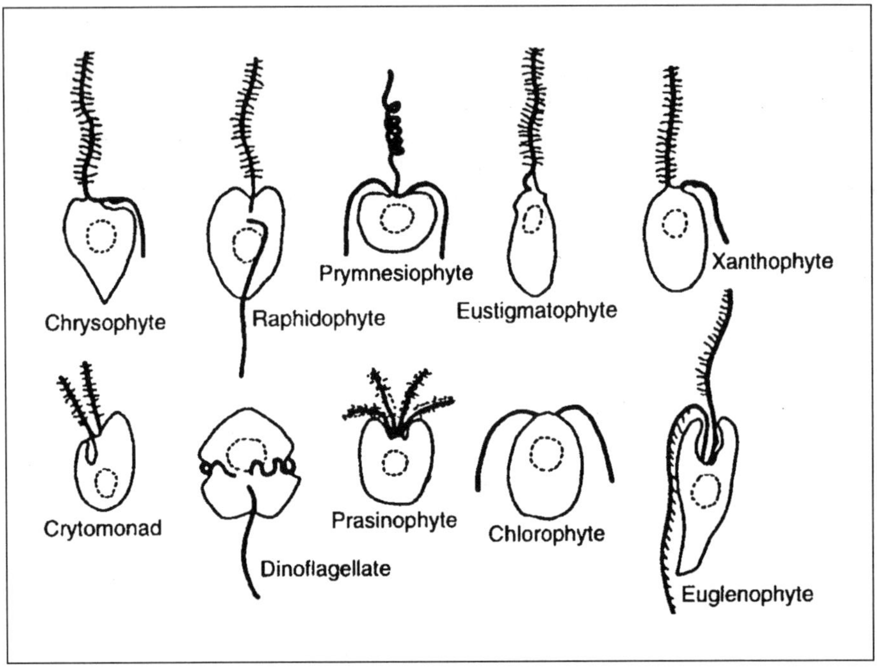

Figure 2.2
Summary of flagellar organization in unicellular algae (after Taylor, 1976*a*).

Chrysophyta (Prymnesiophyceae, Chrysophyceae, Raphidophyceae) are very diverse, and may dominate the phytoplankton in particular regions and at certain times of the year. Picoplanktonic blue-green algae (cyanobacteria; Cyanophyta) and free-living prochlorophytes (Prochlorophyta) are ubiquitous in the world's oceans, often preferring the dimly lit regions at the base of the euphotic zone (Waterbury *et al.* 1979; Chisholm *et al.* 1988). Cryptomonads (Cryptophyta) and eustigmatophytes (Eustigmatophyta) are also widely distributed. The unicellular red algae (Rhodophyta), while common in benthic habitats of tropical reef waters (Jeffrey, unpublished) and Norwegian coastal waters (Paasche and Throndsen, 1970), are not generally detected in oceanic phytoplankton.

In any seawater sample, the silica-walled diatoms (Figs. 2.1, 2.4), the armoured dinoflagellates (Fig. 2.5) and the calcareous coccolithophorids (Fig. 2.6) are the most obviously robust and abundant forms. They are always identifiable microscopically in samples preserved by formalin or Lugol's fixatives. However, the fragile phytoflagellates (Fig. 2.2), which are often equally abundant, are only recognized when gentle fixatives (e.g., buffered glutaraldehyde) are used for their preservation.

Recognition of the importance of the delicate flagellates (5–20 μm) (Tomas, 1993) and minute coccoid forms (0.2–2 μm) has come only in the past two decades when more appropriate techniques for studying the phytoplankton were developed. These included examination of species in the living state after concentrating them in

Table 2.1. Diversity of microalgae found in marine and freshwater phytoplankton[a,b]

Algal Division/Class	Common name	Genera	Approximate of number living species
Golden-brown algal line (chlorophylls *a* and *c*)[c]			
Bacillariophyta	diatoms	210	unknown (5,500–10,000?)
Dinophyta	dinoflagellates	550	4,000
Chrysophyta	golden-brown flagellates		
Chrysophyceae	chrysophytes, silicoflagellates	120	1,000
Raphidophyceae	raphidophytes, (= chloromonads)	4	9
Haptophyta	golden-brown flagellates		
Prymnesiophyceae	coccolithophorids	50	500
Xanthophyta[d]	yellow-green algae	90	600
Cryptophyta[e]	cryptomonads	8	>50
Eustigmatophyta[f]	yellow-green algae	6	12
Green algal line (chlorophylls *a* and *b*)			
Chlorophyta			
Chlorophyceae	green algae	350	2,500
Prasinophyceae[g]	green flagellates	13	120
Euglenophyta	euglenoids	43	650–800 (?)
Red algal line (chlorophyll *a* and biliproteins)			
Rhodophyta[h]	red algae	3	10 (?)
Blue-green algae (cyanobacteria) (chlorophyll *a* and biliproteins)[i]			
Cyanophyta	cyanobacteria	under revision (>10,000 spp.?)[j]	
Prochlorophyta[i]	prochlorophytes	3	3

a. Compiled from Margulis *et al.* (1990); Sournia *et al.* (1991) estimates 474–504 genera and 3444–4375 marine species.
b. The taxonomy of many groups is in a state of flux, and these figures are very tenuous approximations only.
c. Also known as the chromophyte algae, lacking chlorophyll *b*; see Christensen (1980, 1989); Green *et al.* (1989).
d. Few truly planktonic marine species in this group (Hibberd, 1990*b*); contain only traces of chlorophyll *c* (Jeffrey, 1976*a*).
e. Cryptophyta also contain biliproteins.
f. Contain neither chlorophyll *b* or *c* (often included in the chromophyte algae, Green *et al.*, 1989).
g. Includes the Micromonadophyceae (Mattox and Stewart, 1984).
h. While the red macrophytes are prolific (675 genera and 4,100 species, Kraft, 1981) very few unicellular species are known.
i. Urbach *et al.* (1992) recommend that Prochlorophyta should be reclassified in the cyanobacteria. Their pigments are, however, atypical (see Table 2.3).
j. Under revision, based on botanical and bacteriological nomenclature (see Anagnostides and Komárek, 1985, 1986, 1988, 1990; Komárek and Anagnostides, 1989).

a plankton centrifuge, gentler methods of fixation, examination by electron microscopy, size-fractionation of phytoplankton, use of enrichment cultures, onboard cell-flow cytometry and chromatographic pigment analysis (TLC, HPLC) for assessment of algal pigment signatures (see Table 2.3).

TABLE 2.2 Algal types found in phytoplankton size classes (from Jeffrey and Hallegraeff, 1990)

Algal Division/Class	Microplankton 20–200+ μm	Nanoplankton 2–20 μm	Picoplankton 0.2–2 μm
Bacillariophyta	+	+	+
Chlorophyta	+	+	+
Chrysophyceae	+	+	+
Cryptophyta	+	+	–
Cyanophyta	+	+	+
Dinophyta	+	+	–
Euglenophyta	+	–	–
Eustigmatophyta	–	+	–
Prasinophyta	+	+	+
Prochlorophyta	+	+	+
Prymnesiophyceae	–	+	–
Raphidophyceae	+	–	–
Rhodophyta	–	+	–

Phytoplankton can be differentiated into three size classes: the microplankton (20–200 μm), nanoplankton (2–20 μm) and picoplankton (0.2–2 μm) (T able 2.2). The nanoplankton can at times account for up to 90% of the total phytoplankton chlorophyll in coastal and open ocean waters (Jeffrey and Hallegraeff, 1990).

Identification of species composition in field samples requires time and expertise, so routine analyses of phytoplankton are usually limited to measurements of 'chlorophyll biomass' using a variety of techniques (see Chapters 1, 4 and 5). Advances in pigment separation techniques (see Chapters 5, 12 and 13) now allow a great range of chlorophylls and carotenoids to be measured rapidly by chromatographic techniques to give chemotaxonomic signatures of algal types. However, because of the variety of pigment compositions in some algal classes (e.g., the Prymnesiophyceae; see Table 2.3) it is essential that the dominant phytoplankton species are also identified in oceanographic studies.

Taxonomic rankings used in phycology are sometimes confusing to the non-biologist. In this monograph the following hierarchies are used, listed in decreasing rank:

Division	word ending	– phyta
Class		– phyceae
Order		– ales
Family		– aceae
Genus		
Species		

Revisions are constantly appearing in the taxonomic literature as new evidence of relationships comes to light (e.g., the Chrysophyta, section 2.5). For this monograph we have standardized on the classification shown in Table 2.1.

To assist the chemical oceanographer, we provide here a simple introduction to the characteristics of each algal Division (or class), grouped according to classical concepts of algal colour (= pigments, Table 2.1), and point to key literature for further references. Concepts of cell architecture and morphology are described first (to orient the non-biologist). After the descriptions of each Division, what is known of the photosynthetic pigment distribution is summarized, with notes on the many exceptions to the overall scheme. Finally, pigments that have chemotaxonomic usefulness for the study of marine phytoplankton ecology and oceanographic processes are indicated.

2.2 ORGANIZATION OF THE ALGAL CELL

A hypothetical flagellated eukaryotic algal cell is shown in Fig. 2.3A. The outer cell covering may be a dense wall made up of cellulose (some dinoflagellates, some chlorophytes); silica (diatoms, chrysophytes); protein (cryptomonads); or calcite or organic scales (prymnesiophytes). Other forms may be covered by one or more membranes (e.g., raphidophytes, naked dinoflagellates). In the cytoplasm lie the membrane-bound organelles characteristic of all eukaryotic cells. (Prokaryotic forms [e.g., Cyanophyta and Prochlorophyta; see sections 2.11 and 2.12] have no membrane-bound organelles).

The nucleus contains most of the genetic material of the cell. The mitochondria, with either tubular or flattened cristae, are simple or branched elongate organelles, the site of cellular aerobic oxidation processes. The Golgi apparatus consists of stacks of flattened membrane cisternae with such functions as scale formation, packaging of waste materials, formation of polysaccharide and mucous secretions. The endoplasmic reticulum, with attached ribosomes, consists of a network of paired membrane sheets or tubules that permeate through the cell cytoplasm and is the site of protein synthesis.

The chloroplast — the unique organelle characteristic of photosynthetic plant cells — is limited by a two-membraned chloroplast envelope, and in many groups by another two membranes (the chloroplast endoplasmic reticulum). Within the chloroplast (Fig. 2.3B, C) lies a system of membranes (thylakoids). Thylakoids are flattened vesicles consisting of a bilayer membrane (enclosing an aqueous space) in which the photosynthetic pigments are located as macromolecular complexes (see Chapter 3). The pyrenoid, when present, is a proteinaceous area of the chloroplast, the site of the carbon-fixing enzyme of photosynthesis, ribulose *bis*-phosphate carboxylase. The chloroplast DNA is concentrated in nucleoids, which are either scattered within the chloroplast or are ring-shaped structures lying beneath the thylakoid girdle band (Coleman, 1985). Starch grains, the product of photosynthesis in many algal types (e.g., dinoflagellates, chlorophytes), often surround the pyrenoid. Vacuoles containing soluble metabolites are frequent throughout the cytoplasm.

The flagellum (one or several) is usually inserted into the apical region of the cell (prasinophytes, chlorophytes), in a furrow-gullet area (cryptomonads), or two flagella may be inserted — one horizontally and one vertically through pores in a groove — the sulcus (as in the dinoflagellates). The flagella may have hairs (e.g., cryptomonads), scales or both (prasinophytes). They are anchored to the cell by a complex system of roots at their base. Cell shape in the naked flagellates is controlled by a system of cytoskeletal microtubules (not shown).

The organelles form the basic architecture of the eukaryotic algal cell. The following sections highlight their unique characteristics in the different algal groups. These, together with other features, can be used to assist taxonomic identification of

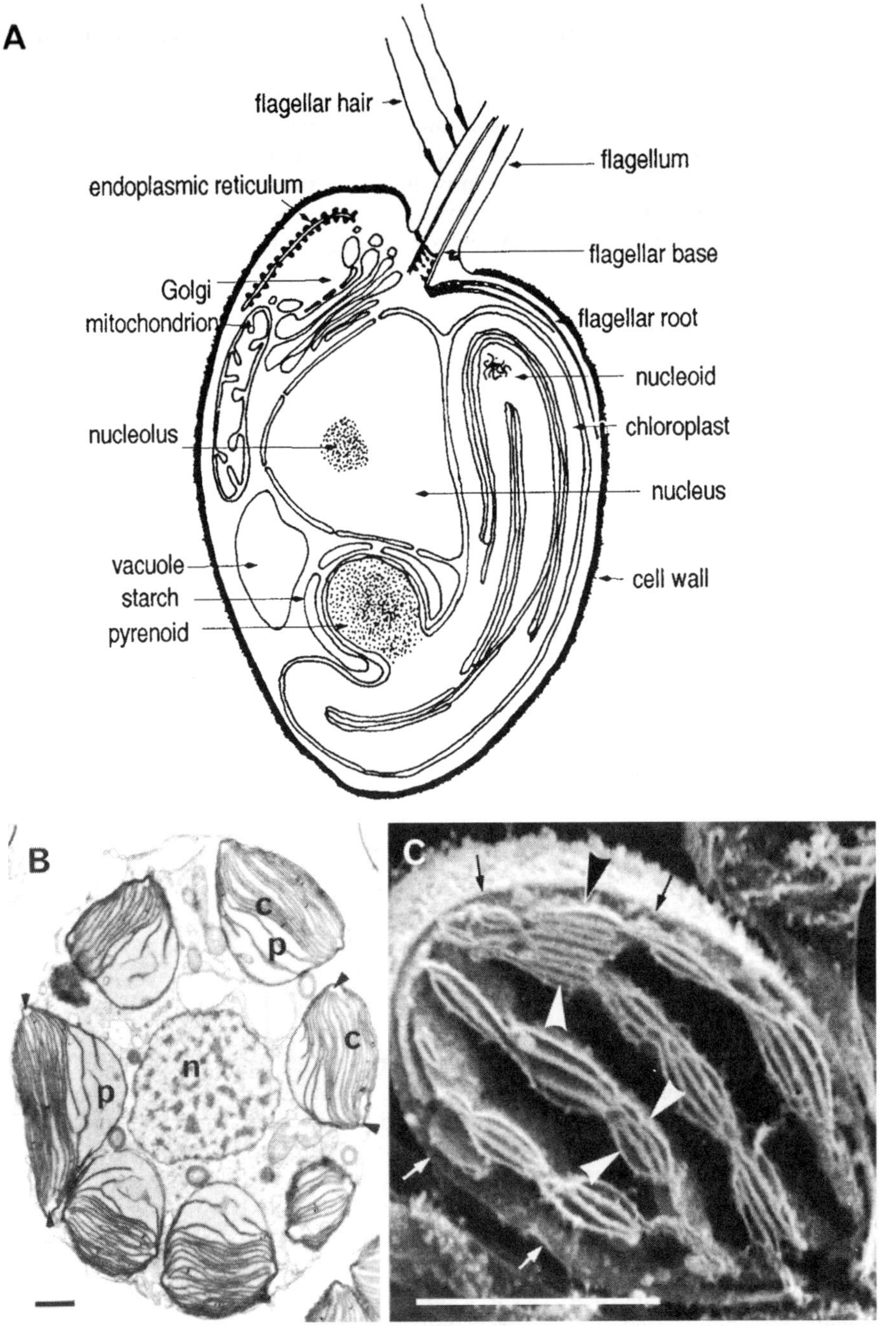

Figure 2.3
A. Longitudinal section through a hypothetical algal cell showing the types of organelles present in most eukaryotes.
B. Transmission electron micrograph (TEM) of a section through the raphidophyte *Heterosigma akashiwo*. Note numerous chloroplasts (c) with pyrenoids (p) and nucleus (n). Arrowheads point to the position of the ring nucloid. Scale bar = 10 μm.
C. High-resolution scanning electron micrograph (SEM) of a fractured spinach chloroplast. Note the stacked thylakoids (arrowheads) forming grana and the surrounding chloroplast envelope (arrows). Scale bar = 1 μm.

algal types. For precise taxonomic identification of species, however, the original taxonomic literature should be consulted. Recent guides to marine phytoplankton genera include Sournia (1986), Ricard (1987) and Chrétiennot-Dinet (1990). Valuable reviews of microalgal ultrastructure include Dodge (1973), Clayton and King (1990) and Berner (1993).

2.3 Diatoms (Bacillariophyta)

Diatoms are the best known of all the unicellular planktonic algae. The variety of silica walls on which diatom taxonomy is based has resulted in more than 10,000 taxa being described (Hostetter and Stoermer, 1971). The number of authentic living species is probably closer to 5,000 (Lipps, 1970) and more than twice that number have been catalogued from the fossil record. This profusion of diatom species may be due to innate prolific speciation in the group, or failure of taxonomists to recognize the inherent variability within each species. New techniques of chloroplast genome analysis are already giving stronger evidence of phylogenetic relationships (Cattolico and Goër, 1989). An excellent modern treatise on diatoms, beautifully illustrated with electron micrographs, has recently been published (Round *et al.* 1990).

2.3.1 *Anatomy and morphology*

Diatoms generally range in size from 5 to 200 μm (Fig. 2.4) with a few reaching 2 mm (e.g., the tropical species *Coscinodiscus gigas*). The most characteristic feature is the box-like silica wall or frustule surrounding the cell which consists of two valves, the smaller valve (hypotheca) fitting into the larger (epitheca) (Fig. 2.4A). The ornamented valves are held together by one or more siliceous girdle bands (Fig. 2.4A, C and D).

Diatoms exhibit a variety of shapes and in valve view may be circular (Fig. 2.4A), boat-shaped (Fig. 2.4F), square, oblong, triangular, elliptical or polygonal. Valve symmetry is used to separate the two orders: Centrales (or centric diatoms) with a concentric radial symmetry around a point or points (Fig. 2.4A-E), and Pennales (or pennate diatoms) with symmetry around a line (Fig. 2.4F).

Diatoms occur either as single cells (Fig. 2.4A, D, F) or in chain-like colonies (Fig. 2.4B, C, E). Individual cells may be held together by interlocking spines or setae as in *Chaetoceros lorenzianum* (Fig. 2.4C), by ridges of interlocking siliceous tubules fused to those of adjacent cells as in *Detonula pumila* (Fig. 2.4B), or by marginal spines or 'teeth' interlocking with adjacent cells. Colonies may also be formed from cells attached to each other by mucuous pads as in *Thalassionema nitzschioides* (Fig. 2.4E).

The diatom frustule has two main components: an outer organic coat and a complex inner wall of polymerized hydrated silica. In many species, the siliceous part is composed of a regular network of hexagonal chambers, which may be closed by a layer of silica with sieve-like perforations. The electron microscope has revealed a number of minute wall structures of various functions (Anon, 1975; Jeffrey *et al.* 1990; Round *et al.* 1990). For more details of wall ultrastructure the above references should be consulted.

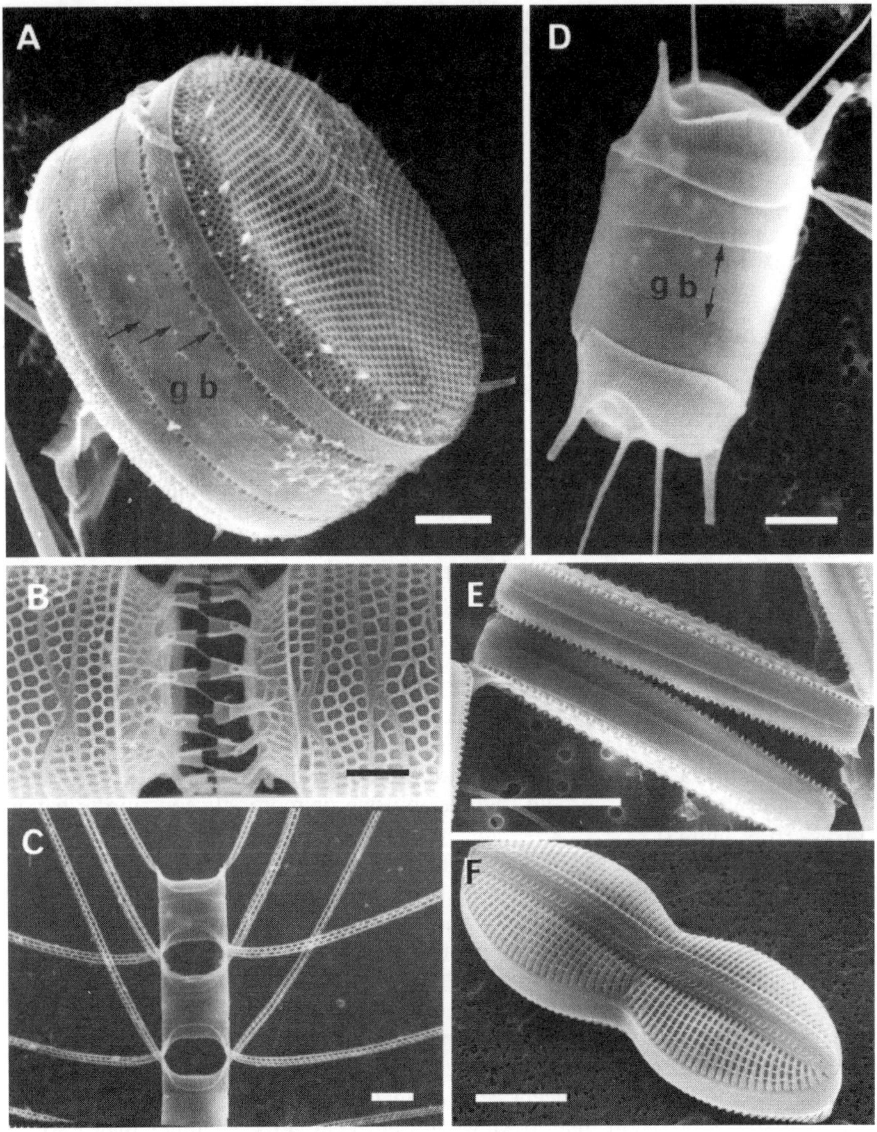

Figure 2.4
SEMs of centric and pennate diatoms showing valve morphology.
A. Centric diatom *Thalassiosira eccentrica* with girdle bands (gb) connecting the epitheca and hypotheca. Scale bar = 10 µm.
B. Detail of the chain-forming *Detonula pumila* showing interlocking tubules that connect the cells. Scale bar = 1 µm.
C. Subtropical *Chaetoceros lorenzianum* with long spines (setae). Scale bar = 10 µm.
D. Tropical *Odontella sinensis* with extended girdle bands (gb). Scale bar = 10 µm.
E. Cells of pennate *Thallasionema nitzschiodes* connected by mucous pads at poles of each cell. Scale bar = 10 µm.
F. Valve view of the benthic pennate *Diploneis* sp. Scale bar = 10 µm.

2.3.2 *Cellular organization*

Diatoms contain the usual plant cell organelles: nucleus, chloroplasts and mitochondria, as well as large vesicles of reserve carbohydrate or oil globules. In centric diatoms there is a large vacuole, often surrounded by streaming cytoplasm. The cell wall is synthesized in special membrane vesicles (the silicalemma) beneath the cell membrane. The unravelling of how the wall is formed was one of the outstanding advances of recent times (Schmidt, 1984).

The gold or golden-brown coloured chloroplasts vary both in number and shape but have a similar ultrastructure. Thylakoids are arranged in bands of three with a girdle band around the whole. The pyrenoid is usually prominent, and the chloroplast DNA occurs as a ring-shaped nucleoid lying beneath the girdle band. The main pigments of diatoms are chlorophylls *a*, c_1 and c_2, and the carotenoids fucoxanthin, diadinoxanthin, diatoxanthin and β,β-carotene. In a small percentage of diatoms, chlorophyll c_3 replaces chlorophyll c_1 (Stauber and Jeffrey, 1988).

2.3.3 *Symbioses*

A special type of symbiosis involves large tropical diatoms — *Rhizosolenia, Hemiaulus* and *Streptotheca* — and the cyanobacterium *Richelia intracellularis*, which occurs as a short nitrogen-fixing filament within the host diatom (Hallegraeff and Jeffrey, 1984).

Several species of the larger foraminifera contain diatom-like endosymbionts (Lee *et al.* 1979). While no frustules are present in the endosymbiotic association, these develop as small pennate diatoms after isolation and culturing. As most invertebrate-algal symbioses involve dinoflagellates (e.g., in tropical clams, corals and foraminifera), a diatom-invertebrate association is unusual. It is obviously important to distinguish free-living from symbiotic forms, when interpreting pigment data.

2.3.4 *Distribution*

Diatoms are ubiquitous throughout the world's oceans and in all kinds of aquatic habitats: from the submerged polar ice-shelf pennate diatom communities to the sparse but exquisite centric diatom communities of tropical waters (Allen and Cupp, 1935). The rapid growth of diatoms generally ensures they are the first to appear in the water column after inorganic nutrient (nitrate) enrichment. The well-known classic species successions ('spring blooms') of northern temperate waters (Round, 1981) result from nutrient enrichment from convective overturn of the water column after winter as well as increased light and temperature. Diatom blooms also occur in the East Australian Current in spring time, although the mechanism of enrichment is different (Hallegraeff and Jeffrey, 1993). The succession patterns are amazingly similar in both situations: small, rapidly growing, chain-forming diatoms bloom first, followed by larger centric diatoms, and then, when the silica for wall formation is depleted, diatoms decline and flagellates take over (Sverdrup, 1953; Margalef, 1960; Jeffrey and Carpenter, 1974; Patrick, 1977; Guillard and Kilham, 1977). Round *et al.* (1990) consider that other aspects of diatom ecology such as sinking rates, cell shape, grazing and resting spore formation (French and Hargraves, 1985) are also key factors in diatom distribution.

2.3.5 *Literature*

Extensive bibliographies and summaries of the classical diatom literature are given in Gran (1912), Lebour (1930), Cupp (1943), Hendey (1964), Hostetter and Stoermer (1971) and Werner (1977). The vast literature on diatom biogeography was collated by Hasle (1976) and Guillard and Kilham (1977), who summarized, region by region, the dominant diatom flora of modern seas. This information now needs to be updated.

The terminology used by diatom taxonomists is published in Anon (1975) and Ross *et al.* (1979), and the ultrastructure of diatoms is summarized in Jeffrey *et al.* (1990). An excellent modern treatise on diatom biology and morphology (Round *et al.* 1990) should be consulted by anyone interested in diatom taxonomy, or by the general reader who would like to be inspired by the beauty of diatom form, so clearly illustrated in this work.

2.3.6 *Characteristics of diatoms (Bacillariophyta)*

Form:	centric or pennate, unicells or in chains
Size:	2–200 μm, some tropical species 2 mm, Antarctic *Thalassiothrix* 4 mm
Major pigments:	chlorophylls *a*, c_2, β,β-carotene, fucoxanthin, diatoxanthin, diadinoxanthin; chlorophyll c_1 usually present (may be replaced by chlorophyll c_3 in a few species)
Energy reserves:	chrysolaminarin (carbohydrate) and oil
Chloroplasts:	one to many; thylakoids in bands of three with girdle band; DNA in ring nucleoid; internal pyrenoid
Cell covering:	silica and organic material (frustule), arranged in two valves, connected by girdle bands
Flagella:	none, except in male gametes
Locomotion:	common in pennate forms via mucilage secretion
Culture colour:	orange to brown, depending on intensity of pigmentation

2.4 DINOFLAGELLATES (DINOPHYTA)

Dinoflagellates (*dinos* = whirling) are a remarkably diverse and complex group of unicellular flagellates, comprising approximately 130 genera and 1200 living species (many more if fossil cyst species are included). About half the species are photosynthetic; others have heterotrophic (animal-like) nutrition.

Some species form chains (e.g., *Gymnodinium catenatum*, Fig. 2.5A) but most are single cells (Fig. 2.5B-E), whose reddish-brown blooms are often referred to as 'red tides'. Dinoflagellates are important components of the phytoplankton of marine and freshwaters, and they also play a key role in symbiotic associations (zooxanthellae) with tropical animals on coral reefs (corals, clams, foraminifera etc.). Some species have attracted attention because of their beautiful bioluminescent properties or their role in producing either toxic or non-toxic blooms (Hallegraeff, 1993).

Dinoflagellates have an impressive fossil history (Sarjeant, 1974); their armoured resting cysts date back to at least 220 million years ago. These fossil cysts are used as stratigraphic markers in petroleum exploration. Modern-day dinoflagellate cysts

Figure 2.5
SEMs of dinoflagellates.
A. Chain-forming, toxic *Gymnodinium catenatum* from Tasmanian estuaries. Note the coiled transverse flagellum lying in the girdle groove (arrows) and the longitudinal flagellum in the sulcus (arrowheads).
B. Tropical *Ceratocorys horrida* with horn-like extensions of the hypotheca.
C. *Amphidinium carterae* showing the two flagella (arrowheads) emerging from the flagellar pores.
D. Divided pair of *Dinophysis tripos*.
E. *Oxytoxum constrictum* with longitudinal ridges on the hypotheca (arrowheads).
Scale bars = 10 µm.

induced to excyst in the laboratory (Wall and Dale, 1968), demonstrated for the first time the relationship between benthic cyst and motile plankton stages.

2.4.1 *Anatomy and morphology*

Dinoflagellates are mostly unicellular and range from about 5 to 2000 µm in diameter. At some stage in their life cycle all species have a motile stage with two dissimilar flagella. The cell is divided by a transverse girdle groove (Fig. 2.5A) into two parts: an upper (epitheca or epicone) and a lower (hypotheca or hypocone) region (see Fig. 2.2; Fig. 2.5A). Within the groove lies the flattened, ribbon-like, transverse flagellum, which encircles the cell. This may be situated centrally (Fig. 2.5A), apically (Fig. 2.5B, C, D and E) or posteriorly. The smooth longitudinal flagellum extends to the base of the cell within a longitudinal furrow (sulcus) (Fig. 2.5A). The transverse flagellum causes the rotational motion of the cell, while the longitudinal flagellum propels the cell forward.

Dinoflagellates, like diatoms, can exhibit extensive morphological variations with spines, horns and wings (Fig. 2.5B, D), especially in tropical species. These morphological extensions are thought to aid flotation, protect against grazing by small zooplankton and enhance nutrient uptake by increasing the surface area and rotational movements.

Dinoflagellates are described as armoured or unarmoured, according to the type of cell covering (theca or amphiesma), which always consists of several layers of membranes. Under the plasma membrane lie flattened vesicles. These may be empty, as in the unarmoured *Gymnodinium* and *Amphidinium* (Fig. 2.5A, C), or may contain cellulose plates of varying thickness as in the armoured forms (Fig. 2.5, B, D, E). The cellulose plates are arranged in a species-specific 'plate tabulation' (Fig. 2.5B, D, E) so they are a most important taxonomic criterion (Taylor, 1980). The number, arrangement and thickness of the thecal plates can vary considerably. They may also be ornamented with pores, depressions, spines, ridges and reticulations (Fig. 2.5B, D, E and Vesk *et al.* 1990*a*).

2.4.2 *Cellular organization*

Dinoflagellates have a complex ultrastructure, with several features unique to this group. The large, prominent nucleus with permanently condensed chromosomes is often visible in the light microscope.

Chloroplasts are present in about half of the dinoflagellate species. Photosynthetic forms can readily be distinguished from non-photosynthetic species (e.g., *Noctiluca* and the large genus *Protoperidinium*) by their brilliant red fluorescence under UV light in the fluorescence microscope (Plate 6.4, Chapter 6; Hallegraeff and Jeffrey, 1984; Hallegraeff and Lucas, 1988). The typical reddish-brown dinoflagellate chloroplasts vary in their shape and in the number per cell. Each chloroplast is surrounded by three membranes and the thylakoids are arranged in bands of two to four, with three being the most common (Dodge, 1975). Girdle-band thylakoids are absent and chloroplast DNA is present in scattered nucleoids (Coleman, 1985). The pyrenoid varies considerably in structure and is of some taxonomic value (Dodge and Crawford, 1971).

Because dinoflagellate chloroplasts are enclosed by three membranes (from two successive endosymbiotic events) it is thought they arose by progressive reduction of a symbiotic eukaryotic alga (Whatley and Whatley, 1981). This idea is enhanced by the finding of pigments characteristic of other algal types within the chloroplast, such as phycobiliproteins in *Dinophysis norvegica* (inferred from fluorescence emission at 580 nm; Geider and Gunter, 1989); green algal pigments in *Lepidodinium viride* (Watanabe *et al.* 1987; 1990) and *Gymnodinium* spp. (Sournia *et al.* 1992); and chrysophyte pigments in *Peridinium foliaceum* (Jeffrey *et al.* 1975; Jeffrey and Vesk, 1976). Most photosynthetic dinoflagellates contain chlorophylls *a*, c_2, β,β-carotene, peridinin, dinoxanthin and diadinoxanthin (Johansen *et al.* 1974; Jeffrey *et al.* 1975). Trace amounts of pigments related to peridinin and neoxanthin are also found (see Chapter 13 and Table 2.3).

Ejectile organelles (trichocysts) are prominent at the cell periphery. These fusiform organelles have a para-crystalline body and a fibrous neck (Vesk *et al.* 1990*a*). In armoured dinoflagellates, the trichocysts lie beneath thecal pores through which they are ejected on stimulation. Their function is unknown, but they might be involved in reduction of sinking rates, osmoregulation, attachment, secretion or defence (Steidinger and Cox, 1980).

Several other unusual organelles occur in the dinoflagellate cell. These include pusules (a complex of vesicles), peduncles and eye-spots. The last resemble the light-perception organs of animals, so may be involved in the well-known phototactic diel vertical migrations of dinoflagellates.

2.4.3 *Symbioses*

Dinoflagellates form extensive endosymbioses in many tropical invertebrates. The best-known associations are the 'zooxanthellae' (*Symbiodinium* spp.) found in corals, clams, radiolarians, acantharians, and foraminiferans. Biochemical, physiological and genetic evidence now indicate that these morphologically similar types belong to at least five different zooxanthellae taxa (Trench and Blank, 1987). These algal 'gardens' contribute significantly to the metabolic economy of the animal host (Muscatine, 1967; Trench *et al.* 1981).

Some tropical heterotrophic members of the Dinophysiales harbour cyanobacterial symbionts of the genera *Synechocystis* and *Synechococcus* (Hallegraeff and Jeffrey, 1984; Lucas, 1991; Kimor *et al.* 1992; Gordon *et al.* 1994). In some forms the symbionts are found extracellularly in the girdle region, in others, in special chambers within the cell. These heterotroph-autotroph consortia are widespread in tropical seas, particularly during periods of nitrogen limitation (Gordon *et al.* 1994).

2.4.4 *Distribution*

Dinoflagellates are widely distributed in the world's oceans, and form specific species assemblages in tropical, subtropical and temperate environments (Jeffrey and Hallegraeff, 1990). Tropical species are typically large heterotrophic forms of exotic shape (Taylor, 1976), often harbouring endosymbionts (see section 2.4.3 above), whereas temperate forms are usually smaller, are photosynthetic and heterotrophic, and are either armoured or unarmoured. Dinoflagellates, like diatoms, can form temporary cysts under unfavourable conditions (Vesk *et al.* 1990*a*). When conditions improve these germinate as vegetative motile cells. Dinoflagellate blooms occur mostly in coastal and inshore waters, often after heavy rain. The humic compounds released from soil run-off are thought to have chelating properties that encourage dinoflagellate growth. For a review of the apparent global increase of harmful algal blooms see Hallegraeff (1993).

2.4.5 *Literature*

Extensive bibliographies and summaries of the classical dinoflagellate literature are given in Schütt (1895), Schiller (1931–1937), Sournia (1973), Taylor (1976*b*) and Dodge (1982). A summary of dinoflagellate morphology and ultrastructure may be found in Vesk *et al.* (1990*a*).

2.4.6 *Characteristics of dinoflagellates (Dinophyta*)*

Form:	unicells of various shapes, including amoeboid
Size:	2–200 μm; some species 2 mm in diameter
Major pigments:	chlorophyll *a*, c_2, β,β-carotene, peridinin, dinoxanthin, diadinoxanthin and peridinin derivatives. Species with endosymbionts have pigments typical of chrysophytes, prymnesiophytes, cryptomonads or chlorophytes.
Energy reserves:	starch and oil
Chloroplasts:	one to many; thylakoids in bands of two or three, with a three-membraned envelope; pyrenoids of various types
Cell covering (theca):	multilayered; armoured forms have cellulose plates within thecal vesicles; girdle divides the cell into epicone and hypocone
Flagella:	two; ribbon-like, transverse flagellum encircles cell in girdle groove; smooth, longitudinal flagellum directed posteriorly
Nucleus:	large, with permanently condensed chromosomes
Trichocysts (ejectile structures):	present
Culture colour:	mostly reddish-brown; can assume the colour of the endosymbiont

* Many species non-photosynthetic

2.5 Golden-brown flagellates (Haptophyta and Chrysophyta)

The golden-brown flagellates comprise a heterogeneous group of unicellular algae. Their taxonomic relationships, ultrastructure and distribution are less well known than many other phytoplankton groups. Several different classifications exist (Christensen, 1980; Bold and Wynne, 1985), and revisions of these have been recommended (Cavalier-Smith, 1993; Green and Leadbeater, 1994; Jordan and Green, 1994). We have used the recent classification of Green and Leadbeater (1994) of two Divisions: Haptophyta and Chrysophyta (see Table 2.1). The former contains the single class Prymnesiophyceae (coccolithophorids and scale-bearing golden-brown flagellates), and the latter, the two classes, Chrysophyceae (including the silicoflagellates) and the Raphidophyceae (the chloromonads).

The marine members of the golden-brown algae in these two Divisions (three classes) range in size from about 2 to100 μm and show considerable variation in pigment composition, cell wall structure and type of flagellated cells. The walls may be of cellulose, silica, calcium carbonate or mucilaginous substances or may be absent altogether. Motile cells may have one or two flagella, equal or unequal in length, smooth, scaly or hairy, and inserted apically or laterally. The characteristics of marine members of the three classes, Prymnesiophyceae, Chrysophyceae and Raphidophyceae are described.

2.5.1 *Prymnesiophyceae*

Most prymnesiophytes are planktonic marine unicells, either flagellate or with flagellate stages in the life history. The best-known are the coccolithophorids (Fig. 2.6A–D; Fig. 2.7A), which have beautiful calcite scales (coccoliths) on the cell surface. Originally found in marine sediments and thought to be of inorganic origin, these tiny crystalline rosettes were eventually discovered on the surface of living cells, and their biological origin was recognized (Black, 1965). The distribution and abundance of coccolithophorids in the world's oceans are receiving increased attention because of their value in dating marine sediments in oil exploration, and their production of volatile sulphur compounds that act as cloud seeding nuclei, influencing weather patterns (Malin *et al.* 1992).

Coccolithophorids are mainly warm-water forms, with the greatest species diversity in tropical and subtropical waters. *Emiliania huxleyi* (Fig. 2.7A) has the widest known geographic and temperature range of any coccolithophorid, with both warm- and cold-water forms, and can exist in waters of 6°C in Subantarctic regions. Several coccolithophorids described from the Arctic (Manton *et al.* 1976, 1977) are found at temperatures as low as -1°C.

Prymnesiophytes with organic scales are less well known than the calcareous coccolithophorids, as they are destroyed by standard preservatives. The pattern of organic scales — too small to be resolved by light microscopy — can be identified by transmission electron microscopy (Moestrup and Thomsen, 1986; Fig. 2.7B).

2.5.1.1 *Anatomy and morphology*

The basic prymnesiophyte cell is bi-flagellate. It is usually not more than 30 µm in diameter; most are in the nanoplankton range (2–20 µm; Table 2.2). They are spherical, oval, elipsoidal, fusiform or flattened (Heimdal, 1993). Their different styles of swimming behaviour are useful aids to species identification (Throndsen, 1993). When swimming rapidly the flagella are at the posterior end of the cell with respect to the direction of motion (Hibberd, 1980). In the case of *Chrysochromulina apheles*, the motion was so rapid (160 µm s^{-1}) as to defy video tape analysis!

Between the two flagella is a third appendage, the haptonema (Fig. 2.7D). This varies in form from a vestigial 'proboscis' in *Isochrysis* to a short appendage in *Pavlova* and a long coiling organelle in *Chrysochromulina*. The long form may reach 160 µm in length, and may function as an attachment organelle, or a collector of food particles for ingestion (Kawachi *et al.* 1991).

In coccolith-forming species, the coccoliths are assembled in the Golgi cisternae from a single-scale base plate and several granular elements, on which the calcite crystals are formed (see Fig. 2.7C). Determination of the site, shape and extent of calcite crystallization all occurs within the Golgi matrix. While the function of coccoliths is still not clearly understood, suggestions include a cytoskeletal function, a protective device for the cell membrane, an easily discarded or synthesized ballast

Figure 2.6
SEMs of coccolithophorids showing the variety of coccolith sizes and shapes.
A. *Scyphosphaera apsteinii*. Scale bar = 10 µm.
B. *Gephyrocapsa oceanica*. Scale bar = 5 µm.
C. *Discosphaera tubifer*. Scale bar = 5 µm.
D. *Umbilicosphaera sibogae*. Scale bar = 10 µm.

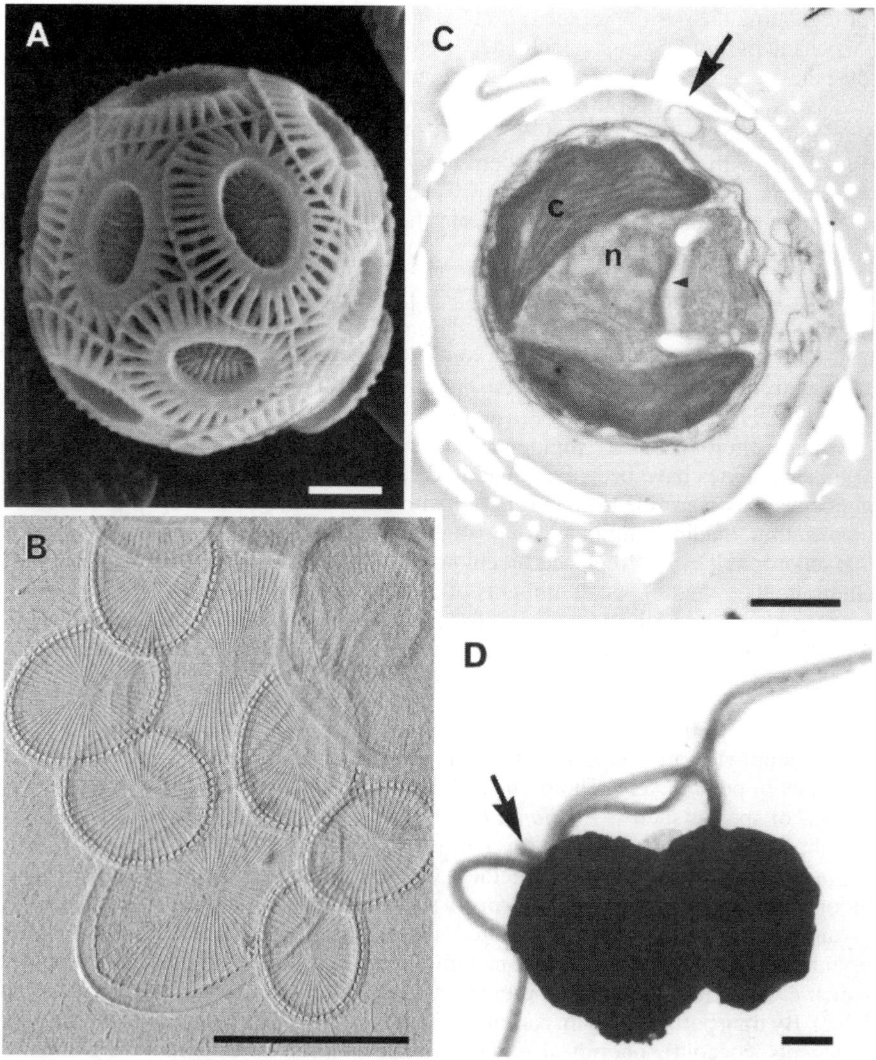

Figure 2.7
A. SEM of the coccolithophorid *Emiliania huxleyi*. Scale bar = 1 μm.
B. Organic body scales of the prymnesiophyte *Chrysochromulina chiton*. Scale bar = 1 μm.
C. TEM of section through *Emiliania huxleyi*, arrow indicates the position of a coccolith covering the cell. Note also the coccolith in a Golgi cisterna (arrowhead), chloroplast (c) and nucleus (n). Scale bar = 1 μm.
D. TEM of shadowed preparation of the prymnesiophyte *Pavlova* sp. with two flagella and a short haptonema (arrow). Scale bar = 1 μm.

for regulating the position of the cell in the water column, a light scattering function, a mechanism for removing calcium deposited in association with a bicarbonate photosynthesis, or simply, God's gift to taxonomists!

Cells without coccoliths, such as *Isochrysis galbana* and *Chrysochromulina chiton*, may be covered with several layers of delicate unmineralized organic scales (Fig. 2.7B) with a characteristic pattern of radiating ridges. These consist of a cellulose matrix and pectic polysaccharides. Like the calcite coccoliths, they are elaborated in the complex Golgi apparatus and then transported to the cell surface (Van der Wal *et al.* 1983).

2.5.1.2 *Cellular organization*

Prymnesiophytes have a cellular organization similar to that shown in Fig. 2.3. Most have one or two discoid peripheral chloroplasts, with internal pyrenoid. The thylakoids are usually in bands of three. There is no girdle band and DNA occurs in scattered nucleoids. The Golgi apparatus is usually prominent at the apical region of the cell.

The photosynthetic pigments in 29 species (50 strains) of cultured prymnesiophytes have been examined in detail by Jeffrey and Wright (1994). Four pigment types were identified. Type 1 had chlorophyll *a*, c_1 and c_2, β,β-carotene, fucoxanthin, diadinoxanthin and diatoxanthin. Type 2 was similar to Type 1, except that chlorophyll c_1 was replaced by chlorophyll c_3. Type 3 contained chlorophyll *a*, chlorophylls c_2 and c_3, small amounts of a phytylated chlorophyll *c*-like derivative (see Chapter 13), with 19'-hexanoyloxyfucoxanthin the dominant carotenoid. In type 4, 19'-butanoyloxyfucoxanthin dominated, with the other chlorophylls and carotenoids present in varying proportions.

2.5.1.3 *Distribution*

Prymnesiophytes are found mostly in tropical and subtropical oceans with a few abundant in polar waters (Thomsen *et al.* 1991). With the advent of remote sensing, blooms of species such as *Emiliania huxleyi* are clearly visible over wide areas, due to light scattered from the coccoliths (Aiken *et al.* 1992). *Phaeocystis antarctica* and *P. pouchetii* are dominant in the Southern Ocean and the North Sea respectively, and blooms of *Chrysochromulina polylepis* in southern Norway have caused widespread disturbance to fish farms. As noted earlier (and in Chapter 1), blooms of prymnesiophytes can affect climate patterns by their production of dimethyl sulphide, which escapes into the atmosphere and acts as cloud-seeding nuclei (Malin *et al.* 1992). By transporting calcium carbonate from the upper oceanic layers to the abyssal sediments, coccolithophorids also play an important role in carbon cycling and the Greenhouse Effect.

2.5.1.4 *Literature*

Some excellent reviews on prymnesiophyte ultrastructure, taxonomy and geographic distribution (e.g., Hibberd, 1980; Green *et al.* 1990) provide the reader with easy access to the literature. A new monograph on the Haptophyta (= the former Prymnesiophyta; Green and Leadbeater, 1994) provides information by experts on all aspects of modern prymnesiophyte biology and ecology. Jordan and Green (1994) provide a modern world check-list of the extant Haptophyta. Heimdal (1993) gives a valuable review of systematic descriptions of the major coccolithophorid genera, and Vesk *et al.* (1990*b*) provide information on prymnesiophyte ultrastructure. An excellent monograph of the coccolithophores links the biology and distribution of living forms with their paleontology (Winter and Slesser, 1994).

2.5.1.5 *Characteristics of prymnesiophytes (Prymnesiophyceae)*

Form:	spherical or ovoid unicell
Size:	5–20 µm
Major pigments:	chlorophyll *a*, c_1 or c_3, c_2, a phytylated chlorophyll *c*-like pigment, β,β-carotene, fucoxanthin and/or 19'-hexanoyloxyfucoxanthin, and 19'-butanoyloxyfucoxanthin, diadinoxanthin and diatoxanthin; four pigment types, depending on the presence of fucoxanthin and chlorophyll *c* derivatives
Energy reserves:	chrysolaminarin
Chloroplasts:	one or two; thylakoids in bands of three; scattered nucleoids
Cell covering:	organic scales; large calcified scales (coccoliths) in many species
Flagella:	two equal or subequal, mostly smooth
Haptonema:	a third flagellum-like appendage present in most species
Culture colour:	gold to golden-brown

2.5.2 *Chrysophyceae*

Most chrysophytes are found in freshwater habitats as unicellular, colonial or filamentous forms. The freshwater members will not be treated here, but the reader may consult the proceedings of a recent symposium (Kristiansen and Andersen, 1986) for further information.

Marine members of the Chrysophyceae are of three types: the well-known and well-defined silicoflagellates (Fig. 2.8A, order Dictyochales); the picoplanktonic *Pelagococcus* (Fig. 2.8D; Lewin *et al.* 1977; Vesk and Jeffrey, 1987), *Aureococcus* (Sieburth *et al.* 1988) and *Pelagomonas*, for which a new class, Pelagophyceae, has been proposed (Andersen *et al.* 1993); and the nanoplanktonic siliceous spheres, for which a new order, Parmales, within the Chrysophyceae has been established (Booth and Marchant, 1987; Fig. 2.8B). These three groups will be briefly described.

2.5.2.1 *Silicoflagellates (order Dictyochales)*

The silicoflagellates form a well-defined group of marine unicells, the cytoplasm of which is contained in a basket-like siliceous skeleton composed of a network of tubular elements (Fig. 2.8A). The cells range in size from 20 to 100 µm and each cell bears a single flagellum. They form a significant biomass in temperate and polar regions (Takahashi, 1987). Naked forms (lacking a skeleton) apparently occur worldwide and have been implicated in a fish-kill bloom event in Denmark (Moestrup and Thomsen, 1990). Although not dominant in the phytoplankton of present-day oceans, their rich fossil record makes them valuable indicators in biostratigraphy and dating of marine sediments. Culture and field studies (Van Valkenburg and Norris, 1970; Van Valkenberg, 1980) show the group to be sensitive to temperature, preferring temperatures below about 15°C.

Under the light microscope, the cytoplasm has a spongy appearance. In the electron microscope, the nucleus is seen in the central dense region, surrounded by a ring of Golgi bodies, while a less dense outer region contains numerous chloroplasts and mitochondria (Vesk *et al.* 1990*b*). The chloroplasts have an internal pyrenoid, and thylakoids are arranged mainly in bands of three.

The taxonomy of the group is based on the shape of the siliceous skeleton. Two genera and three extant species are recognized (Kristiansen, 1990).

2.5.2.2 *Picoplanktonic chrysophytes*

Three new species of 'aberrant' picoplanktonic marine chrysophytes are widespread in the world oceans. *Pelagococcus subviridis* (Lewin *et al.* 1977) was first seen and isolated as 2–3 µm coccoid pale green cells from the North Pacific Ocean. Strains of the same species were subsequently isolated from Norwegian waters (Throndsen and Kristiansen, 1985) and from the East Australian Current (Vesk and Jeffrey, 1987).

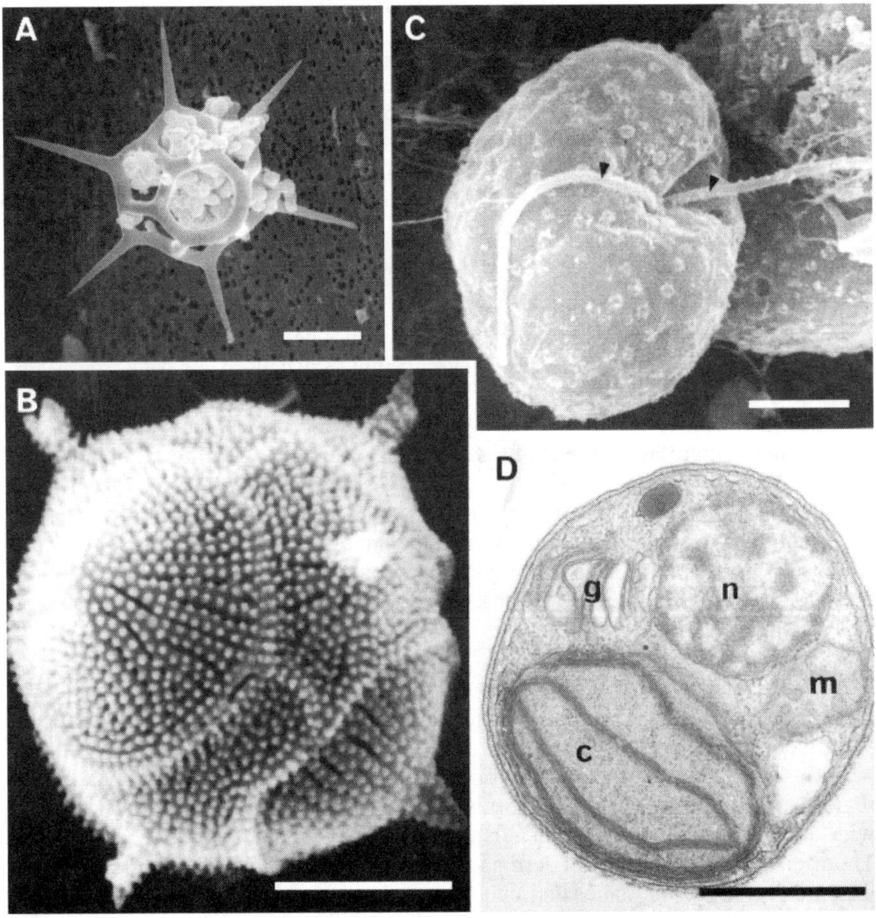

Figure 2.8
A. SEM of the silicoflagellate *Distephanum speculum*. Scale bar = 10 µm.
B. SEM of the chrysophyte *Tetraparma pelagicus* from the Southern Ocean. Scale bar = 1 µm.
C. SEM of the raphidophyte *Fibrocapsa japonica* with two flagella (arrowheads). Scale bar = 10 µm.
D. TEM of section of the chrysophyte *Pelagococcus subviridis* showing single chloroplast (c), nucleus (n), mitochondrion (m) and Golgi body (g). Scale bar = 1 µm.

All strains have been studied in detail, and because of their unique pigment composition (chlorophyll c_3 and 19'-butanoyloxyfucoxanthin), one strain was selected as a SCOR reference culture (see Chapter 6). Another isolate from Narragansett Bay, USA, *Aureococcus anophagefferens* (Sieburth *et al.* 1988; Smayda and Villareal, 1989), which forms 'brown' water under bloom conditions, is closely related, as is the minute flagellate *Pelagomonas calceolata*. A new class, Pelagophyceae, has been proposed to accommodate these forms (Andersen *et al.* 1993).

Ultrastructural features of *Pelagococcus subviridis* (CS-99; Fig. 2.8D), such as the arrangement of thylakoids in the chloroplast and the position of the Golgi apparatus in a depression of the large pear-shaped nucleus, suggest strong affinities with the Chrysophyceae, while the complement of photosynthetic pigments suggests affinities with certain prymnesiophytes (Type 4; Jeffrey and Wright, 1994). Mitosis, however, is atypical of both groups (Vesk and Jeffrey, 1987). Further work is needed to resolve these anomalies.

2.5.2.3 *Nanoplanktonic chrysophytes (Parmales)*

Unique nanoplanktonic spheres (2–5 µm), covered with 5 to 8 ornamented siliceous plates (Fig. 2.8B) were originally found in North Pacific and Antarctic waters (Booth *et al.* 1980, 1982; Marchant and McEldowney, 1986). They have now been found world-wide (Kosman *et al.* 1993). Their algal characteristics, with chloroplasts, and thylakoids arranged in groups of three, were clearly visible in TEM sections of field samples (Marchant and McEldowney, 1986).

A new order, the Parmales, was erected within the Chrysophyceae, to accommodate these microalgae (Booth and Marchant, 1987). More than seven species have been described, based on the morphology of the siliceous plates (Kosman *et al.* 1993). Attempts to culture these cells have so far been unsuccessful, and nothing is therefore known of their life cycles, pigment composition or division processes.

2.5.2.4 *Literature*

For a summary of the present state of knowledge of the Chrysophyceae the reader should consult Kristiansen and Andersen (1986) and Kristiansen (1990).

2.5.2.5 *Characteristics of chrysophytes (Chrysophyceae)*

Form:	coccoid or ovoid
Size:	2–3 µm coccoid, 20–100 µm silicoflagellates
Major pigments:	chlorophyll *a*, c_1, c_2, fucoxanthin, 19'-butanoyloxyfucoxanthin and diadinoxanthin; violaxanthin present in some species (mainly freshwater)
Energy reserves:	chrysolaminarin and lipid
Chloroplasts:	one or two; thylakoids in bands of three with girdle band; DNA usually in a ring nucleoid
Cell covering:	absent, or with silica scales, or within a loose envelope (lorica) of cellulose
Flagella:	absent, or one or two; two are unequal in length, one with tripartite hairs, the other smooth
Culture colour:	gold

2.5.3 *Raphidophyceae*

This small group of flagellates, also known as chloromonads, was previously regarded as predominantly freshwater (Heywood, 1980), but coastal blooms of marine raphidophytes are becoming increasingly common, sometimes causing serious damage to the mariculture industry (Hara and Chihara, 1982). The cells are extremely fragile so are usually not recognizable in preserved phytoplankton field samples (Vesk *et al.* 1990*b*).

2.5.3.1 *Anatomy and morphology*

The motile cells can be 30–100 µm long, with two flagella (Fig. 2.8C). Their shape varies from almost spherical to ovoid, often flattened dorsiventrally. The cells are naked and can change their shape. The colour varies from bright green (in freshwater species) to golden brown (in marine species). Observation of living cells in the light microscope is often complicated by the presence of bullet-like mucocysts, which discharge readily, resulting in cell destruction.

The two flagella are inserted subapically in a slight groove (Fig. 2.8C) and are unequal in length. While swimming, the long hairy flagellum is directed forwards, while the short smooth flagellum trails behind, close to the body.

2.5.3.2 *Cellular organization*

Numerous chloroplasts are present in the peripheral cytoplasm (Fig. 2.3B). The thylakoids tend to be in bands of three, with frequent interconnections. Both a girdle thylakoid and a ring nucleoid are present. Marine species contain a protruding pyrenoid, and photosynthetic pigments are chlorophylls a, c_1 and c_2, fucoxanthin, β,β-carotene and violaxanthin (Fiksdahl *et al.* 1984*b*).

Mucocysts are present in all species, and in the light microscope they can be seen as refractile, bullet-shaped bodies. They discharge explosively to form long threads of mucus (Vesk *et al.* 1990*b*).

There is a single order (Raphidomonadales) with two families. At present, three marine genera — *Chattonella, Fibrocapsa* and *Heterosigma* — are recognized. Heywood (1990) gives an overview of the group.

2.5.3.3 *Characteristics of raphidophytes (Raphidophyceae)*

Form:	coccoid or ovoid
Size:	30–100 µm
Major pigments:	chlorophyll a, c_1, c_2, fucoxanthin, β,β-carotene and violaxanthin are common in marine genera
Energy reserves:	lipid
Chloroplasts:	numerous; thylakoids in bands of three, with a girdle band; DNA in a ring nucleoid
Cell covering:	absent, or a thin mucilaginous layer
Flagella:	two, unequal in length, one with tripartite hairs, the other smooth
Mucocysts (ejectile structures):	present
Culture colour:	gold to golden-brown

2.6 Cryptomonads (Cryptophyta)

The Cryptophyta are a well defined group of mostly photosynthetic nanoplanktonic flagellates, which have affinities with both the Rhodophyta (biliprotein accessory pigments) and the chromophyte algae (chlorophyll *c* accessory pigments). They are common in marine, estuarine and freshwater habitats, but may be missed in field collections unless non-destructive fixation methods are used (Hallegraeff, 1981; Klaveness, 1988–89). Cryptomonads can also be detected in oceanic phytoplankton populations by the presence of the specific cryptomonad carotenoid, alloxanthin (Gieskes and Kraay, 1983; see Chapters 6 and 13). Pigment markers are invaluable for those fragile species that disappear without trace in fixed samples!

2.6.1 *Anatomy and morphology*

Cryptomonads are characteristically ovoid, asymmetrical cells, 6–20 μm in length (Fig. 2.2, 2.9C). Two equal or almost equal flagella are inserted into the apical or subapical gullet or furrow, which can be vertical, horizontal or oblique in orientation, and gives the cell its asymmetric outline (Butcher, 1967; Kugrens *et al.* 1986; Throndsen, 1993). While most cryptomonads are motile, cysts and palmelloid stages are also found. The flagella may vary in length from 4 to 10 μm, and are directed anteriorly when the cell is gyrating. Both flagella have tubular hairs between 1–2 μm long (Jeffrey and Vesk, 1990).

The pellicle or cell covering of cryptomonads is unique (Wetherbee *et al.* 1987). It consists of a ridged periplast (plasma membrane appressed to granular or fibrous material) superimposed on an inner periplast component made up of a regular layer of thin proteinaceous plates or sheets. This pellicle maintains the shape of the cell in the absence of a microtubular cytoskeleton (Jeffrey and Vesk, 1990).

2.6.2 *Cellular organization*

One, or rarely two, large chloroplasts surround the posterior nucleus. Starch grains, ribosomes and a vestigial 'nucleus' (the nucleomorph) are found within a periplastidial compartment enclosed by the endoplasmic reticulum (Jeffrey and Vesk, 1990). The thylakoids of cryptomonad chloroplasts are unique. They occur in pairs and their lumen is filled with electron-dense material. The photosynthetic pigments are chlorophylls *a*, c_2, carotenoids and phycobiliproteins (phycocyanin or phycoerythrin), with the latter located within the thylakoid lumen (Vesk *et al.*, 1992). In this they differ from the red and blue-green algae (cyanobacteria), which have their phycobiliproteins in structures (phycobilisomes) attached to the external face of the thylakoids (see Chapter 3).

The proteinaceous pyrenoid may protrude from the chloroplast and is surrounded by starch plates. If eyespots are present, they consist of a layer of pigmented globules within an extension of the chloroplast stroma.

Ejectile organelles (ejectosomes or trichocysts) are in vesicles around the periphery of the cell or line the furrow-gullet system. They eject membranous ribbons, presumably as a defense mechanism.

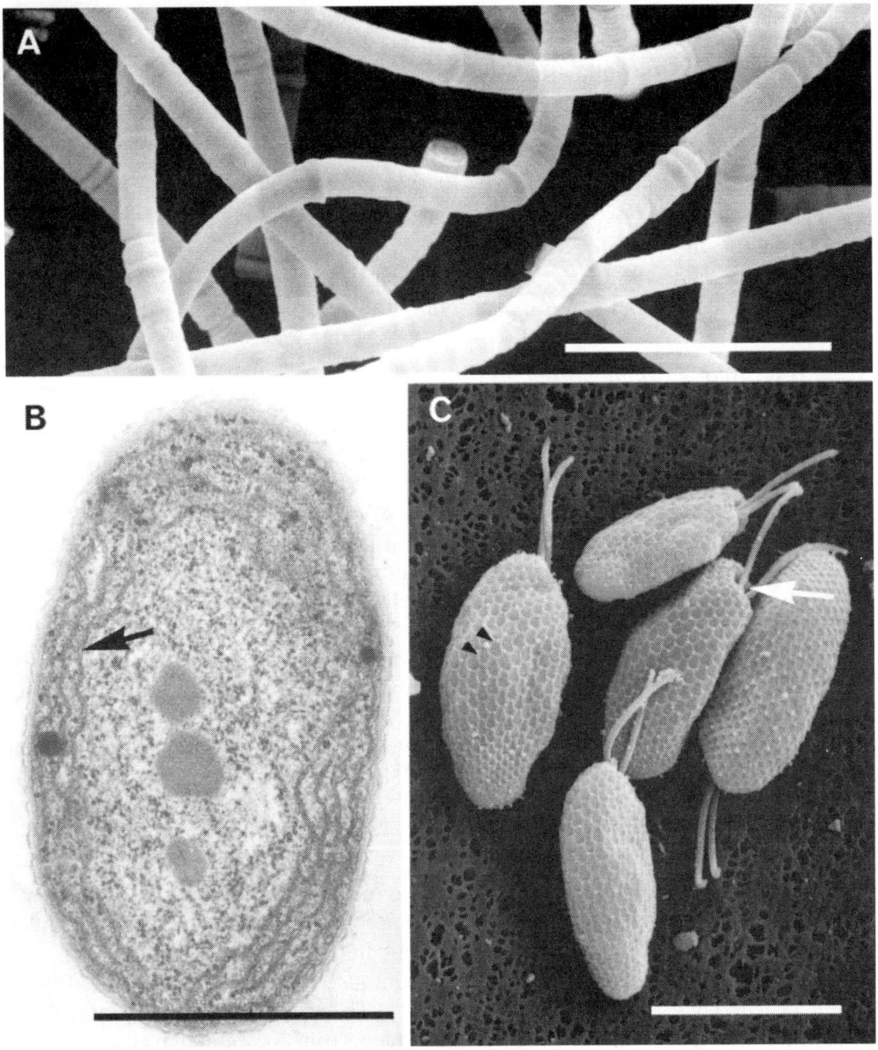

Figure 2.9
A. SEM of the toxic cyanobacterium *Nodularia spumigena* from Australia. Scale bar = 50 µm.
B. TEM of the minute cyanobacterium *Synechococcus* sp. from the East Australian Current. Arrow points to single thylakoids lying in the peripheral cytoplasm. Scale bar = 1 µm.
C. SEM of the cryptophyte *Chroomonas* sp. showing hexagonal pattern of the pellicle (arrowheads) and two flagella emerging from the gullet (arrow). Scale bar = 10 µm.

2.6.3 *Symbioses*

Cryptomonad flagellates are thought to be a primitive group derived from a sequence of two endosymbioses (Whatley and Whatley, 1981). The nucleomorph is a partially degenerate nucleus from a colourless eukaryotic cell, which acquired a photosynthetic prokaryote, to form a red algal-like cell. These events were followed by a further symbiotic association with a colourless eukaryotic flagellate to yield the 'modern' cryptomonad.

Cryptomonads may be found in endosymbiotic association with other organisms. Both the blue-green dinoflagellate *Gymnodinium acidotum* (Wilcox and Wedemayer, 1984) and the red ciliate *Mesodinium rubrum* (Hibberd, 1977), contain cryptomonad-like endosymbionts. A current topic of interest is the pigmentation of the dinoflagellate genus *Dinophysis,* whose thylakoids resemble those of cryptomonads, but for which there is no evidence of endosymbiosis (Vesk *et al.* 1996). *Dinophysis* spp. have not yet been cultured, so until then definitive pigment studies must rely on opportunistic bloom events.

2.6.4 *Literature*

The most recent summaries of the cryptomonad literature may be found in Klaveness (1988–89), Gillott (1990) and Jeffrey and Vesk (1990). The cryptomonads are at present undergoing much taxonomic revision, and papers by D. R. A. Hill are recommended (e.g., Hill, 1991).

2.6.5 *Characteristics of cryptomonads (Cryptophyta)*

Form:	ovoid unicells, often flattened
Size:	6–20 µm
Major pigments:	chlorophyll *a*, c_2, β,ε-carotene, alloxanthin, monadoxanthin; red or blue phycobiliproteins in thylakoid lumen
Energy reserves:	starch
Chloroplasts:	one or occasionally two; thylakoids usually in pairs; phycobiliproteins present in the thylakoid lumen
Nucleomorph:	present in the periplastidial compartment
Cell covering:	proteinaceous pellicle
Flagella:	two equal or nearly equal, with tubular hairs
Ejectosomes (ejectile structures):	present
Culture colour:	red or blue-green, depending on presence of phycoerythrins or phycocyanins

2.7 Green algae (Chlorophyta)

The importance of the small green flagellates in oceanic phytoplankton is now well established from pigment analyses (presence of chlorophyll *b*, Jeffrey, 1976*b*), examination of living samples and enrichment cultures, and gentle fixation followed by examination in the electron microscope (Jeffrey and Vesk, 1981). Small green flagellates and coccoid forms are classified in the Chlorophyceae or Prasinophyceae (which includes Micromonadophyceae, Mattox and Stewart, 1984). They occur in almost any habitat, both marine and freshwater (includes soil and subaerial), and the planktonic forms may be found in both inshore and coastal areas where soluble organics provide nutrient enrichments. Phillips (1990) estimates about 8,000 to 9,000 species are assigned to the Chlorophyta, whereas Melkonian (1990*a*) suggests the range is about twice this number (16,000 species, 500 genera). These estimates include the many attached, filamentous and macrophytic forms not included in Table 2.1.

2.7.1 *Chlorophyceae*

The naked flagellates of the Chlorophyceae are not easily identified in field samples, whereas the Prasinophyceae can be recognized in the electron microscope from their scale patterns. Butcher (1959) gives an introductory account of the green flagellates in British coastal waters, but such a study has not been done for any other coastal or oceanic area. Most of the marine unicells belong in the Volvocales (e.g., *Chlorella, Chlamydomonas, Selenastrum, Dunaliella*). All these genera are easy to cultivate and have been commercialized for various purposes. For example, several thousand tonnes of the edible *Chlorella* are grown each year as a health food supplement, and *Dunaliella salina* is now used for commercial production of the food colouring agent, β,β-carotene, which it synthesizes in massive quantities under conditions of high light, high salinity and low nutrients (Borowitzka and Borowitzka, 1990).

2.7.1.1 *Cellular organization*

The cells of four common genera *Dunaliella*, *Chlamydomonas*, *Chlorella* and *Carteria* are usually small (<20 μm), ovoid, with two or four smooth and equal flagella inserted apically (see *Dunaliella tertiolecta*, Plate 1B, Chapter 6). The wall may be thick as in *Chlamydomonas*, or absent (bounded just by the plasma membrane), as in *Dunaliella*.

In *D. tertiolecta*, the chloroplast is large and cup-shaped and fills the posterior part of the cell (Jeffrey and Vesk, 1981). It contains a number of interconnected and appressed thylakoids, and the large, central pyrenoid is ringed with a starch sheath. The eyespot, which is contained within the chloroplast, consists of a number of highly pigmented orange granules. The nucleus lies at the apical region of the cell, partly within the lobes of the chloroplast cup.

The photosynthetic pigments are chlorophylls *a* and *b*, with major carotenoids, β,β-carotene, lutein, violaxanthin and neoxanthin (see Chapter 13). Minor carotenoids, such as zeaxanthin and tri-hydroxy carotene (loroxanthin), are often present and may have chemotaxonomic associations (Fawley, 1991; 1992).

2.7.2 *Prasinophyceae*

The Prasinophyceae are a heterogeneous collection of primitive green flagellates (Melkonian, 1990*b*), some 2–30 μm in length, and they are found in all size classes of the phytoplankton (see Table 2.2). They have a variety of cell shapes (round, oval, heart or bean-shaped). Both the flagella and the cell body are covered by minute organic scales often highly ornamented. In some genera (e.g., *Tetraselmis*), the body scales have fused to form a rigid theca. Conclusive identification to species level usually requires electron microscopy to evaluate the taxonomically specific patterns of organic scales (McFadden *et al.* 1986*a*).

2.7.2.1 *Cellular organization*

Many prasinophytes (e.g., *Pyramimonas*) have four flagella, which are inserted in an apical pit or depression. In a few genera (e.g., the picoplanktonic *Micromonas*) one or two flagella may be attached laterally (Melkonian, 1990*b*).

A large cup-shaped chloroplast surrounds the central nucleus. The chloroplast contains an eyespot, and many bands of three appressed thylakoids. A starch sheath surrounds the internal pyrenoid. The flagellar scales are formed in the large Golgi bodies situated at the cell apex next to the flagellar bases. In *Pyramimonas* spp., up to five scale types may be produced (McFadden *et al.* 1986*a*). The scales are transported from the scale reservoir *via* a duct to the cell surface (McFadden *et al.* 1986*b*).

The prasinophytes contain chlorophylls *a* and *b*, and some species (e.g., *Tetraselmis*) have carotenoids identical to the Chlorophyceae. Other prasinophytes (e.g., *Pycnococcus provasolii*; Plate 2I, Chapter 6) have a chlorophyll *c*-like pigment (magnesium-2, 4-divinyl pheoporphyrin a_5 monomethyl ester, MgDVP; see section 3.5.5, Chapter 3, for IUPAC nomenclature), in addition to chlorophylls *a* and *b*, and they may have a suite of carotenoids very different from those of the Chlorophyceae. These may include prasinoxanthin, micronone, siphonaxanthin and/or ureolide (see Table 2.3, and Chapter 13).

2.7.3 *Literature*

Melkonian (1990*a, b*) gives an excellent overview of the Chlorophyta, and further information is in Pickett-Heaps (1975), Phillips (1990) and Sym and Pienaar (1993).

2.7.4 *Characteristics of green algae (Chlorophyta: classes Chlorophyceae and Prasinophyceae)*

Form:	coccoid or ovoid unicells
Size:	1–40 µm
Major pigments:	chlorophyll *a* and *b*, β,ε- and/or β,β-carotene, lutein, violaxanthin, zeaxanthin, neoxanthin; some prasinophytes also contain prasinoxanthin or micronone as a major carotenoid and a chlorophyll *c*-like pigment, MgDVP.
Energy reserves:	starch, usually as granules around the pyrenoid
Chloroplasts:	usually one; two to six thylakoids organized into grana; usually one to many pyrenoids; no chloroplast endoplasmic reticulum
Cell covering:	cellulose, xylan, mannan and/or glycoproteins
Flagella:	two, four or eight, or absent
Culture colour:	grass-green or olive green

2.8 EUGLENOIDS (EUGLENOPHYTA)

Euglenophyta are prominent in both freshwaters and marine. Species of *Eutreptiella* are common in temperate inshore and coastal waters (Throndsen, 1973). Euglenoids can utilize soluble organic compounds for growth if they are available, and thus their presence may be an indication of organic pollution. A light micrograph of the freshwater *Euglena gracilis* is shown in Plate 1C, Chapter 6. The number of species of marine and freshwater euglenoids is estimated at about 40–50 genera and 1,000 species (Walne and Kivic, 1990; Phillips, 1990).

2.8.1 *Anatomy and morphology*

The common marine euglenoids are large fusiform flagellates, 40–60 µm long and about 10 µm wide. They are characterized by a frequently changing body outline (metaboly), in which a swelling passes regularly from one end of the cell to the other. Just within the outer plasma membrane lies a proteinaceous layer, the pellicle, which is flexible and allows for extensive body movement. The interlocking proteinaceous strips are visible as helical markings on the cell surface (Jeffrey and Vesk, 1981). Rows of mucilage-producing bodies lying beneath the pellicle open to the exterior by a canal.

Both marine genera — *Eutreptia* and *Eutreptiella* — have two flagella, which are inserted into the anterior gullet. They may be equal or unequal in length, and are thicker than those of other flagellates, because they have an extra structure, the paraflagellar rod.

2.8.2 *Cellular organization*

Marine euglenoids (e.g., *Eutreptia*) have a large nucleus, with a prominent nucleolus, and many condensed chromosomes. The chloroplasts vary in size, shape and pyrenoid structure. The thylakoids may be appressed, as in the Chlorophyta and higher plants. The product of photosynthesis is not starch but a 1–3 glucose polymer, paramylon, which forms discrete granules in the cytoplasm. Some species have an eye-spot, which is a system of pigmented granules located outside the chloroplast.

Photosynthetic euglenoids contain chlorophylls *a* and *b*, as do the Chlorophyta, but their major carotenoid (diadinoxanthin) relates more to the brown algal line than the green. The minor carotenoids include β,β-carotene, zeaxanthin, neoxanthin and antheraxanthin (see Table 2.3 and Chapter 13).

2.8.3 *Literature*

Summaries of euglenoid biology and systematics may be found in Leedale (1967), Jeffrey and Vesk (1981), Phillips (1990), and Walne and Kivic (1990). The last gives a particularly comprehensive reference list to all aspects of euglenoid biology and ecology, including the classic citations.

2.8.4 *Characteristics of euglenoids (Euglenophyta)*

Form:	ovoid unicells, metabolic
Size:	15–500 µm
Major pigments:	chlorophylls *a* and *b*, β,β-carotene, diadinoxanthin, zeaxanthin, neoxanthin
Energy reserves:	paramylon
Chloroplasts:	one to many; thylakoids in extended and interconnected stacks of 2–3; three-membraned envelope
Cell covering:	proteinaceous pellicle
Flagella:	one or two, with single row of fine hairs
Culture colour:	grass-green

2.9 Yellow-green algae (Eustigmatophyta)

The eustigmatophytes are a small group of coccoid microalgae, most of which produce characteristic motile cells (zoospores) containing a prominent red eyespot (stigma) from which the Division gets its name. Many are freshwater organisms found either in soils or running water. Only 6 genera and 12 species are known (Hibberd, 1990*a*). The ecological role of the marine forms is uncertain, although they have been isolated from coastal waters, tide pools and the ocean's surface. One species, *Nannochloropsis occulata* (formerly known as 'marine *Chlorella*'), is widely used in the mariculture industry as a live feed for animal larvae (Volkman *et al.* 1993). Because of the excellent nutritional quality of yellow-green algae and their ease of cultivation, other strains are now being isolated.

The eustigmatophytes were separated from the xanthophytes in classic studies by Hibberd and Leedale (1970, 1971, 1972), based on the unique ultrastructure of the motile cells, and the photosynthetic pigments (Norgård *et al.* 1974). The Xanthophyta are predominantly freshwater forms and are not recognized as members of the marine phytoplankton. The properties of this rather large Division are reviewed by Hibberd (1990*b*).

2.9.1 *Cellular organization*

The eustigmatophytes are spherical or oval, and range in size from 2 to 18 μm. The marine species are usually less than 10 μm (Hibberd, 1990*b*). They are a yellow-green colour and can be confused with the grass-green Chlorophyta. Most representatives contain a large chloroplast, with bands of three thylakoids, and no girdle band. The pyrenoids are present only in the coccoid vegetative cells (not the zoospores) and protrude as a polyhedral structure, connected to the chloroplast by a narrow stalk. Plates of lamellate reserve material surround the pyrenoid.

The structure of the naked motile cell is unique, with usually only one emergent flagellum, with tubular hairs, inserted apically. If a second flagellum is present, it is smooth and short (see Fig. 2.2). The eye-spot, a large orange-red body lying external to the chloroplast, consists of an irregular group of droplets, not bound by a membrane.

2.9.2 *Photosynthetic pigments*

The pigments of the eustigmatophytes are unique, with chlorophyll *a* only (no chlorophylls *b* or *c*), and the carotenoids β,β-carotene, violaxanthin and vaucheriaxanthin ester. In some species canthaxanthin is present (Norgård *et al.* 1974), and in old cultures many minor or trace secondary carotenoids are found (Antia and Cheng, 1982). The Xanthophyta, from which the Eustigmatophyta were separated, have chlorophyll *a*, traces of chlorophyll *c* (Jeffrey, 1976*a*), β,β-carotene, diadinoxanthin (replacing violaxanthin), and vaucheriaxanthin ester (Bjørnland and Liaaen-Jensen, 1989).

2.9.3 *Literature*

Excellent reviews of the Division may be found in Hibberd and Leedale (1972) and Hibberd (1990*a*).

2.9.4 *Characteristics of yellow-green algae (Eustigmatophyta)*

Form:	coccoid unicells or oval flagellates (zoospores)
Size:	2–18 µm
Major pigments:	chlorophyll *a*, β,β-carotene, vaucheriaxanthin ester, violaxanthin
Energy reserves:	chrysolaminarin, lipid
Chloroplasts:	one; thylakoids in bands of three; scattered nucleoids
Cell covering:	wall of unknown composition
Flagella:	lacking in coccoid forms; single in zoospores
Culture colour:	yellow-green

2.10 Red algae (Rhodophyta)

Only a few genera of truly planktonic unicellular red algae are found in the phytoplankton (Chrétiennot-Dinet, 1990). Most Rhodophyta (over 4,000 species; Kraft, 1981) are macrophytes (red seaweeds). *Porphyridium* has been isolated from freshwater and marine habitats, and from soil and sand surfaces (Vonshak, 1988); species of *Rhodella* have been found in Norwegian waters (Paasche and Throndsen, 1970), and of *Rhodosorus* in tropical waters of the Great Barrier Reef (Dibbayawan *et al.* 1990; Jeffrey, unpublished). New methods for identifying the presence of unicellular red algae in the phytoplankton are needed to give a more realistic picture of their abundance. *Porphyridium purpureum* (formerly *cruentum*) is the most studied unicellular red alga; it has been widely used in biotechnology (Vonshak, 1988) and photosynthesis research (Gantt, 1975). It is also a SCOR-recommended culture for pigment standards (see Chapter 6).

2.10.1 *Anatomy and morphology*

Unicellular red algae are usually nanoplanktonic spheres (5–10 µm in size), non-flagellated, and either solitary or found in colonies held together with mucilage (see Plate 2K, Chapter 6). No cell wall is present beneath the continually synthesized sheath of polyanionic sulphated polysaccharide that surrounds the cell (Ramus, 1972).

2.10.2 *Cellular organization*

A comprehensive study of the ultrastructure of two *Rhodella* species is given by Patrone *et al.* (1991).

Porphyridium purpureum has one large stellate chloroplast, with a central pyrenoid. The thylakoids are single, with phycobilisomes on the surface (see section 3.2.2, Chapter 3). Phycoerythrin, the dominant light-harvesting pigment is accompanied by phycocyanin and allophycocyanin in a structured array within the phycobilisomes (see Chapter 3). Chlorophyll *a*, β,ε-carotene and zeaxanthin are associated in macro-molecular pigment-protein complexes within the thylakoids. In most situations, the red colour of the phycoerythrin completely masks that of the other pigments.

2.10.3 *Literature*

The Rhodophyta are reviewed in Gabrielson *et al.* (1990) and Cole and Sheath (1990), while Jeffrey and Vesk (1981) and Patrone *et al.* (1991) describe the ultrastructural features of the unicellular forms and Rowan (1989) reviews the pigmentation of the group. No comprehensive review of unicellular planktonic species has yet been written, despite the need for one.

2.10.4 *Characteristics of red microalgae (Rhodophyta)*

Form:	coccoid unicells or colonies
Size:	5–15 μm
Major pigments:	chlorophyll *a*, β,ε-carotene, β,β-carotene, lutein, zeaxanthin, cryptoxanthin, red or blue phycobiliproteins
Energy reserves:	floridean starch
Chloroplasts:	usually one; single thylakoids with phycobilisomes on the surface; no chloroplast endoplasmic reticulum
Cell covering:	cellulose or xylan with sulphated polysaccharides such as agar and carrageenan
Flagella:	none
Culture colour:	deep red or purple

2.11 Cyanobacteria (blue-green algae) (Cyanophyta)

Cyanobacteria are prokaryotic forms most commonly found in tropical and subtropical phytoplankton. The most abundant filamentous species, *Trichodesmium* (*Oscillatoria*), forms massive blooms both at the sea-surface and at depth (Carpenter and Price, 1977). Coccoid, picoplanktonic species (Fig. 2.9B, Plate 2L, Chapter 6) are also widespread in the world's oceans (Waterbury *et al.* 1979), and are often abundant in the dimly lit regions at the base of the euphotic zone. Symbiotic cyanobacteria are also common as short nitrogen-fixing filaments within large tropical centric diatoms (Hallegraeff and Jeffrey, 1984), or as populations of single cells within special 'culture chambers' formed in girdle list pouches of certain tropical dinoflagellates (Hallegraeff and Jeffrey, 1984; Vesk *et al.* 1990*a*; Gordon *et al.* 1994).

In contrast to the relatively small number of species of cyanobacteria found in the open ocean (Sournia, 1970), a huge range is common at the surface of estuarine muds and in shallow, eutrophic freshwater lakes and rivers (Whitton, 1973). Their ability to flourish in warm temperatures and high light at low carbon dioxide concentrations has enabled them to colonize many terrestrial habitats. Under extreme conditions, blooms of these organisms can cause noxious damage to inland and coastal waterways (e.g., *Nodularia spumigena*, Fig. 2.9A; Hallegraeff, 1993). The taxonomy of the cyanobacteria is being extensively revised, using both botanical and bacteriological criteria (Anagnostidis and Komárek, 1985; and Table 2.1).

2.11.1 *Anatomy and morphology*

Many cyanobacteria are filamentous (Fig. 2.9A) and assume a colonial habit. *Trichodesmium* filaments mass together to form free-floating colonies. The filaments are capable of movement and glide over one another while colony shape is maintained. *Trichodesmium* has densely packed gas vacuoles, regularly arranged within the periphery of the cell. These hollow cylindrical structures are freely permeable to gases, which enable the alga to regulate its position in the water column (Walsby, 1977). They also have high critical collapse pressures and can withstand the hydrostatic pressures of subsurface ocean layers.

Many genera of filamentous cyanobacteria have specialized cells with different functions. The heterocysts are the site of nitrogen fixation, and the akinetes, which are granular resting spores enriched in polyglucan, may play a role in survival strategies (Delaney, 1990).

2.11.2 *Cellular organization*

Cyanobacteria are prokaryotes whose cells contain no membrane-bound organelles. The nuclear material is found as fibrils of DNA within the cytoplasm, not organized into a nucleus bounded by a nuclear membrane, as in the eukaryotes. Similarly, there are no chloroplasts, but thylakoids containing the pigments lie in the peripheral cytoplasm (Fig. 2.9B). The phycobilisomes (containing the phycobiliproteins) are arranged on the outer surface of the thylakoids, similar to those of the red algae (see Chapter 3). The major pigments of the marine coccoid forms are chlorophyll *a*, β,β-carotene, zeaxanthin and phycobiliproteins (phycocyanin, phycoerythrin and allophycocyanin; Rowan, 1989). The marine *Trichodesmium* spp. also have myxoxanthophyll and *cis*-echinenone (Carpenter *et al.* 1993). In freshwater cyanobacteria, the carotenoids echinenone, canthaxanthin, myxoxanthophyll and oscillaxanthin are important (Nicholls, 1973).

The cell covering of cyanobacteria is complex. It is made up of several layers: the cell wall, which is a rigid peptidoglycan layer is surrounded by an outer double-layered membrane and a mucilaginous sheath. A cytoplasmic membrane lies between the wall and the cytoplasm. The composition of these layers is very similar to those of Gram-negative bacteria (Delaney, 1990).

No flagella are present in any members of this Division, but many have inclusions, granules, crystalloids and membranous structures (Jensen, 1993). An important organelle, the carboxysome, is a polyhedral body composed of the enzyme ribulose 1,5-*bis* phosphate carboxylase. Its function may be similar to that of the pyrenoid of eukaryotic algae.

2.11.3 *Literature*

The ecological role of the oceanic picoplanktonic cyanobacteria is reviewed by Waterbury *et al.* (1986), while general references include Carr and Whitton (1982), Carr and Wyman (1986) and Delaney (1990). An excellent review of cyanobacterial ultrastructure is given by Jensen (1993), and an introduction to modern cyanophyte classification may be found in papers by Anagnostidis and Komárek (1985 to 1990).

2.11.4 *Characteristics of the cyanobacteria (Cyanophyta)*

Form:	coccoid unicells, filaments or colonies
Size:	1 µm to 2 mm colonies
Major pigments:	chlorophyll *a*, β,β-carotene, zeaxanthin; (freshwater species may contain echinenone, myxoxanthophyll and oscillaxanthin); phycobiliproteins (C-phycocyanin, allophycocyanin, allophycocyanin-B, C-phycoerythrin, R-phycoerythrin, phycoerythrocyanin)
Energy reserves:	polyglucan, cyanophycin, polyphosphate granules
Chloroplasts:	none; thylakoids in the cytoplasm, phycobilisomes on the surface of thylakoids
Cell covering:	cell wall containing peptidoglycan, lipopolysaccharide and protein
Organization of nuclear material:	prokaryotic; DNA not associated with histone or organised into chromosomes
Flagella:	none
Culture colour:	blue-green, grey-green or red

2.12 Prochlorophytes (Prochlorophyta)

2.12.1 *Discovery and distribution*

The discovery of prokaryotic algal cells with a cell structure resembling that of the Cyanophyta, but with pigments that lacked phycobiliproteins, aroused world-wide interest in the mid-1970s as these algae were considered possible evolutionary ancestors of the Chlorophyta (Lewin, 1977). A new genus *Prochloron* was proposed within the new algal Division Prochlorophyta to accommodate these forms, which were found only in symbiotic association with tropical ascidians. Ten years later, freshwater free-living filamentous prokaryotes with similar ultrastructure and pigment composition were discovered (Burger-Wiersma *et al.*, 1986, 1989), and were included as a second genus, *Prochlorothrix*, in the Prochlorophyta.

More recently great excitement was generated in the oceanographic community by the discovery of free-living picoplanktonic coccoid cells, with procaryotic ultrastructure and pigments similar to, but not identical with, those of *Prochloron*. Detected by ship-board cell-flow cytometry, these cells were described as free-living relatives of the symbiotic *Prochloron* (Chisholm *et al.* 1988), appropriately named *Prochlorococcus marinus* and placed as the third genus in the Prochlorophyta (Chisholm *et al.* 1992). *Prochlorococcus* was originally found in great abundance (1×10^5 cells per ml) at the base of the euphotic zone in the Atlantic and Pacific Oceans, but improved techniques have now shown them throughout the water column, particularly in the Mediterranean Ocean (Li and Wood, 1988; Neveux *et al.* 1989; Olsen *et al.* 1990; Vaulot *et al.* 1990; Veldhuis and Kraay, 1990; Vaulot and Partensky, 1992). Picoplanktonic prokaryotes can now readily be distinguished from picoplanktonic cyanobacteria and eucaryotes by their scatter and fluorescence characteristics (Li, 1994). Their important role in oceanic ecosystems (Stockner, 1988) is demonstrated in part by their substantial contribution to total phytoplankton chlorophyll (Goericke, 1990).

2.12.2 *Cellular organization*

Prochloron cells, always found in association with marine invertebrates, are between 10 and 25 µm in diameter. The cell wall is similar to that of cyanobacteria, with an outer envelope, a central peptidoglycan layer, above a plasma membrane (Cox, 1993). No mucilaginous sheath lies outside the cell wall as it does in cyanobacteria. A band of parallel thylakoids lies at the periphery of the cell. They surround a large central vacuole formed from a dilated thylakoid. *Prochloron* does not have a central DNA-containing nucleoid, but the DNA is scattered as fine filaments throughout the cytoplasm. Carboxysomes (typical of cyanobacteria) are present, as well as a number of other bodies — crystalline and para-crystalline inclusions, pigment granules etc. (Cox, 1993).

Prochlorothrix filaments are unbranched trichomes, 0.5–1.5 µm wide and 3–10 µm long. The cell wall components, while similar to those of *Prochloron*, resemble cyanobacteria more closely. The thylakoids run in an irregular pattern around the periphery of the cell in stacks of two to eight deep, not necessarily parallel to the wall. Unlike *Prochloron*, almost all its DNA is in a central nucleoid. Carboxysomes are present, but are very small (0.2 µm). *Prochlorothrix* cells have gas vacuoles, similar to filamentous cyanobacteria. Most of the biochemical and ultrastructural features of this genus resemble cyanobacteria, except for the thylakoid system, which strongly resembles that of *Prochloron*.

Prochlorococcus cells are picoplanktonic, 0.6 to 0.8 µm in diameter and up to 1.6 µm in length. A sheath surrounds the cell wall. The thylakoids form a broad parallel band at the periphery of the cell. No dilation of thylakoids into vacuoles occurs. The DNA is in a central nucleoid, as in *Prochlorothrix*. Cox (1993) points out that *Prochlorococcus* has few ultrastructural features in common with *Prochloron*, while those it shares with *Prochlorothrix* are common to most cyanobacteria. Chisholm *et al.* (1992) give a formal description of the genus.

2.12.3 *Photosynthetic pigments*

Prochloron, isolated from tropical ascidians has chlorophylls *a* and *b*, no phycobiliproteins (Withers *et al.* 1978), and the carotenoids β,β-carotene (54% of total carotenoids), zeaxanthin (38%), cryptoxanthin (6%) and traces (<1%) of β,β-carotene monoepoxide, mutatochrome and echinenone (Foss *et al.* 1987).

The pigment composition of *Prochlorococcus* determined from field samples (Chisholm *et al.* 1988) and cultured strains (Goericke and Repeta, 1992) is unique. No chlorophylls *a* and *b* or phycobiliproteins are present. Instead, the chlorophylls are divinyl chlorophylls *a* and *b*, and MgDVP (a chlorophyll *c*-like pigment, see section 2.7.2 and the Part IV Data sheets), with the major carotenoids β,β- and β,ε-carotene and zeaxanthin. The availability of cultured *Prochlorococcus* strains (Partensky *et al.* 1993) now allows detailed photophysiology and biochemical studies to be undertaken.

2.12.4 *Phylogeny*

Cladistic analysis of 16S ribosomal RNA sequences of the three genera (Palenik and Haselkorn, 1992; Urbach *et al.* 1992) indicate that the prochlorophytes are a highly diverse polyphyletic group, not monophyletic as previously thought. These authors recommend the taxon Prochlorophyta be abandoned and the group reclassified in the Cyanobacteria. Until this concept has been accepted by the scientific community, we will continue the use of the algal division Prochlorophyta (see Tables 2.1, 2.3).

2.12.5 *Literature*

Cox (1993) gives an excellent review of the status of the Prochlorophyta, particularly ultrastructural characteristics.

2.12.6 *Characteristics of prochlorophytes (Prochlorophyta)*

Form:	coccoid unicells, filaments (freshwater) or endosymbionts within tropical ascidians.
Size:	0.5 μm (unicells), <1 mm (filaments), 10–25 μm (endosymbionts)
Major pigments:	Divinyl chlorophyll *a*, divinyl chlorophyll *b*, MgDVP, β,ε-carotene, zeaxanthin (unicells); chlorophylls *a*, *b*, β,β-carotene, zeaxanthin (endosymbionts)
Energy reserves:	not yet determined
Organization:	prokaryotic
Chloroplasts:	none; parallel thylakoids in peripheral cytoplasm
Cell covering:	cell wall and sheath
Flagella:	none
Culture colour:	pale-green

2.13 Distribution of major pigments in algal divisions/classes

The previous sections gave a brief introduction to the immense diversity of microalgae found in the phytoplankton of the world's oceans. Of these tens of thousands of species (precise number unknown: see Table 2.1 and Sournia *et al.* 1991) probably less than one thousand marine species have been isolated and cultured (again the precise number is unknown, but see Staines *et al.* 1986 for a list of the World Directory of Culture Collections of Microorganisms).

Of those species in culture, only a very few have been examined for pigments by modern techniques, and in some cases, perhaps only two or three species or strains have been examined (e.g., the picoplanktonic chrysophytes). Nevertheless, Table 2.3 lists the major pigments of algal Divisions/Classes, as presently known (mid-1996), and can be used as a guide to the taxonomic usefulness of pigments for marine ecology. Numerous exceptions to pigment distribution trends are noted. This table should be considered only a pigment 'snap-shot' of the 'tip of the iceberg' of all the tens of thousands of species in the world's oceans that have not yet been examined. Pigment analyses of more laboratory-cultured species and strains of all Divisions/Classes will extend our present perceptions.

TABLE 2.3. Distribution of major and taxonomically significant pigments in algal divisions/classes*.

Pigment		Cyanophyta (cyanobacteria)[11]	Prochlorophyta	Rhodophyta	Cryptophyta	Chlorophyceae[14]	Prasinophyceae[7, 8, 9]	Euglenophyta[13]	Eustigmatophyta	Bacillariophyta	Dinophyta[1]	Prymnesiophyceae[5]	Chrysophyceae[6]	Raphidophyceae
		Algal Division/Class												
Chlorophylls														
	a	●	15	●	●	●	●	●	●	●	●	●	●	●
	b		15			●	●	●			1			
	c_1									3 ●	1	5 ●		●
	c_2				●					●	●	●	●	●
	c_3									3		5 ●	●	
Phytylated chlorophyll *c*-like												5 •		
MgDVP			•				●							
Divinyl chlorophyll *a*			●											
Divinyl chlorophyll *b*			●											
Carotenes														
Old terminology	IUPAC													
α	β,ε		●	●	●	•	•					•		
β	β,β	●	●			●	●	●	●	•	•	•	•	●
γ	β,ψ					•								
ε	ε,ε				•								•	
Lycopene	ψ,ψ	11			•		2							
Xanthophylls														
Alloxanthin					●									
Antheraxanthin						•	•	•						
Astaxanthin						2	2							
19'-Butanoyloxyfucoxanthin											1	5 ●	●	
Canthaxanthin		11				2								
Crocoxanthin					•									
Diadinoxanthin								●		●	●	●	●	●
Diatoxanthin								•		•	•	•	•	•
Dinoxanthin											●			
Echinenone		11				2	2	13						
Fucoxanthin										●	1	5 ●	●	●
19'-Hexanoyloxyfucoxanthin											1	5 ●		
Lutein						●	•							
Monadoxanthin					•									
Neoxanthin						●	●	•						
P-457 + P-468 [4]											•			
Peridinin											●			
Peridininol											•			

*Pigment distribution data is taken from recent modern analyses of algal cultures; please note that only very few representatives of each class or division have been examined (e.g., <0.5% of diatoms, Stauber and Jeffrey, 1988).

Pigment	Algal Division/Class												
	Cyanophyta (cyanobacteria)[11]	Prochlorophyta	Rhodophyta	Cryptophyta	Chlorophyceae[14]	Prasinophyceae[7, 8, 9]	Euglenophyta[13]	Eustigmatophyta	Bacillariophyta	Dinophyta[1]	Prymnesiophyceae[5]	Chrysophyceae[6]	Raphidophyceae
Prasinoxanthin						7 ●							
Pyrrhoxanthin										·			
Siphonaxanthin					14	14							
Siphonein					14	14	14						
Vaucheriaxanthin ester								•					
Violaxanthin					●	●		●					
Zeaxanthin	●	●	●		•			·					
Biliproteins													
Allophycocyanins	12 ●		12 ●										
Phycocyanins	12 ●		12 ·	10 ●									
Phycoerythrins	12 ●		12 ●	10 ●									

1. Several dinoflagellates have no trace of peridinin but have pigments characteristic of their endosymbionts, e.g. chrysophyte (chlorophylls $c_1 + c_2$ and fucoxanthin; *Peridinium foliaceum* and *P. balticum*, Jeffrey *et al.* 1975); prymnesiophytes (fucoxanthin derivatives, *Gymnodinium galatheanum, Gyrodinium aureolum, Ptychodiscus brevis*, Bjørnland and Liaaen-Jensen, 1989), or green algae (chlorophyll *b*, *Lepidodinium viride*, Watanabe *et al.*, 1987, 1990; *Gymnodinium* spp., Sournia *et al.*, 1992).
2. Extra-plastidic secondary carotenoids, sometimes found in nitrogen-limited cells.
3. 12% of diatoms tested (71 strains) had chlorophyll c_3 instead of c_1 (Stauber and Jeffrey, 1988).
4. Terminology from Johansen *et al.* (1974).
5. Four pigment types were found in a survey of 50 prymnesiophyte strains (Jeffrey and Wright, 1994). These were separated on the presence/absence of chlorophylls c_1, c_2 and/or c_3 and fucoxanthin, 19'-hexanoyloxy- and/or 19'-butanoyloxyfucoxanthin.
6. Freshwater and brackish-water chrysophytes have major pigments similar to diatoms; occasionally diadinoxanthin is replaced with violaxanthin (Withers *et al.*, 1981).
7. Includes the Micromonadophyceae (Mattox and Stewart, 1984).
8. A significant number of prasinophytes (e.g., *Tetraselmis* spp.) contain normal chlorophyte pigments (Ricketts, 1970) (no prasinoxanthin).
9. A range of new carotenoids have been found in some prasinoxanthin-containing prasinophytes (Egeland and Liaaen-Jensen, 1993).
10. Seven classes of cryptomonad phycobiliproteins are known, containing allophycocyanin, phycocyanin and/or phycoerythrin (see Rowan [1989] for summary).
11. Many freshwater filamentous cyanophytes have a wide range of carotenoids, not found in the picoplanktonic cyanobacteria (e.g., echinenone, myxoxanthophyll, canthaxanthin, see Nichols, 1973).
12. Many types of phycobiliproteins are found in cyanobacteria and rhodophytes, containing R- and C-type phycocyanins, allophycocyanins and/or phycoerythins (see Rowan [1989] for summary).
13. Many trace carotenoids of no known chemosystematic value are found in euglenophytes (Bjørnland, 1987; and Chapter 13).
14. Siphonaxanthin and siphonein, while characteristic of the macrophytic Siphonales (Chlorophyta, Yokohama, 1981) have also been identified in several unicellular members of the Micromonadophyceae (Fawley and Lee, 1990) and a marine planktonic euglenophyte (Fiksdahl *et al.*, 1984*a*; Bjørnland, 1987).
15. The symbiotic *Prochloron* and the freshwater *Prochlorothrix* contain chlorophylls *a* and *b*.

Code ● = major pigment (>10%); • = minor pigment (1–10%); · = trace pigment (<1%) of the total chlorophylls or carotenoids.

Acknowledgements

We thank Mr Chris Puttock for Fig. 2.8C; Dr Gustaaf Hallegraeff for Figs. 2.4A–F, 2.5A, B, D and E, 2.6A–D, 2.7A–B, 2.9A; Dr Harvey Marchant for Fig. 2.8B; Dr A. J. Rees for Figs. 2.8A, 2.9B; and Mr Peter Vesk for Figs. 2.3C and 2.5C.

References

Aiken, J., Moore, G. F., Holligan, P. M. (1992). Remote sensing of oceanic biology in relation to global climate change. *J. Phycol.* 28: 579–590.

Allen, W. E., Cupp, E. E. (1935). Plankton diatoms of the Java Sea. *Annals. Jard. Bot. Buitenz,* 44: 101–174.

Anagnostidis, K., Komárek, J. (1985). Modern approach to the classification system of cyanophytes. 1. Introduction. *Arch. Hydrobiol. Suppl.* 71: 291–302.

Anagnostidis, K., Komárek, J. (1986). Modern approach to the classification system of cyanophytes. 2. Chroococcales. *Arch. Hydrobiol. Suppl.* 73: 157–226.

Anagnostidis, K., Komárek, J. (1988). Modern approach to the classification system of cyanophytes. 3. Oscillatoriales. *Arch. Hydrobiol. Suppl.* 80: 327–472.

Anagnostidis, K., Komárek, J. (1990). Modern approach to the classification system of cyanophytes. 5. Stigonematales. *Arch. Hydrobiol. Suppl.* 86: 1–73.

Andersen, R. A., Saunders, G. W., Paskind, M. P., Sexton, J. P. (1993). Ultrastructure and 18S RNA gene sequence for *Pelagomonas calceolata* gen. et sp. nov. and the description of a new algal class, the Pelagophyceae classis nov. *J. Phycol.* 29: 701–715.

Anon (1975). Proposals for a standardization of diatom terminology and diagnoses. *Nova Hedwigia,* 53: 323–354.

Antia, N. J., Cheng, J. Y. (1982). The keto-carotenoids of two marine coccoid members of the Eustigmatophyceae. *Br. Phycol. J.* 17: 39–50.

Berner, T. (1993). *Ultrastructure of Microalgae*. CRC Press, Ann Arbor, 314 pp.

Bjørnland, T. (1987). Carotenoid structures and lower plant phylogeny. In: Krinsky, N. I., Mathews-Roth, M. M., Taylor, R. F. (eds). *Carotenoids, Chemistry and Biology*, Plenum Press, New York, p. 21–37.

Bjørnland, T., Liaaen-Jensen, S. (1989). Distribution patterns of carotenoids in relation to chromophyte phylogeny and systematics. In: Green, J. C., Leadbeater, B. S. C., Diver, W. L. (eds), *The Chromophyte Algae: Problems and Perspectives.* Clarendon Press, Oxford, pp. 37–61.

Black, M. (1965). Coccoliths. *Endeavour,* 24: 131–137.

Bold, H. C., Wynne, M. J. (1985). *Introduction to the Algae: Structure and Reproduction*, 2nd Edn, Prentice Hall, Engelwood Cliffs, N. J., 720 pp.

Booth, B. C., Marchant, H. (1987). Parmales, a new order of marine chrysophytes, with descriptions of three new genera and seven new species. *J. Phycol.* 23: 245–260.

Booth, B. C., Lewin, J., Norris, R. E. (1980). Siliceous nanoplankton. I. Newly discovered cysts from the Gulf of Alaska. *Mar. Biol.* 58: 205–209.

Booth, B. C., Lewin, J., Norris, R. E. (1982). Nanoplanktonic species predominate in the subarctic Pacific in May and June, 1978. *Deep Sea Res.* 29: 185–200.

Borowitzka, M. A., Borowitzka, L. J. (1990). Algal biotechnology. In: Clayton, M. N., King, R. J. (eds), *Biology of Marine Plants*, Longman Cheshire, Melbourne, p. 385–399.

Burger-Wiersma, T., Stal, L. J., Mur, L. R. (1989). *Prochlorothrix hollandica* gen. nov. sp. nov., a filamentous oxygenic photoautotrophic procaryote containing chlorophylls *a* and *b*: assignment to Prochlorotrichaceae fam. nov. and order Prochlorales Florenzano Balloni and Materassi 1986, with emendation of the original description. *Int. J. Syst. Bacteriol.* 39: 350–360.

Burger-Wiersma, T., Veenhuis, M., Korthals, H. J., Van de Wiel, C. C. M., Mur, L. R. (1986). A new prokaryote containing chlorophylls *a* and *b*. *Nature*, 320: 262–263.
Butcher, R. W. (1959). *An Introductory Account of the Smaller Algae of British Coastal Waters. Part I. Introduction and Chlorophyceae.* Series IV, Min. Ag. Fish. Food, Her Majesty's Stationery Office, London, 17 pp.
Butcher, R. W. (1967). *An introductory account of the smaller algae of British coastal waters. Part IV. Cryptophyceae.* Series IV. Min. Ag. Fish. Food. Her Majesty's Stationery Office, London, 54 pp.
Carpenter, E. J., Price, C. C. (1977). Nitrogen fixation, distribution and production of *Oscillatoria (Trichodesmium)* spp. in the western Sargasso and Caribbean Seas. *Limnol. Oceanogr.* 22: 60–72.
Carpenter, E. J., O'Neil, J. M., Dawson, R., Capone, D. G., Siddiqui, P. J. A., Roenneberg, T., Bergman, B. (1993). The tropical diazotrophic phytoplankter *Trichodesmium*: biological charactersitcs of two common species. *Mar. Ecol. Progr. Ser.* 95: 295–304.
Carr, N. G., Whitton, B. A. (1982). *The Biology of the Cyanobacteria*, Botanical Monographs, vol. 19, University of California Press, Berkeley, 688 pp.
Carr, N. G., Wyman, M. (1986). Cyanobacteria: their biology in relation to the oceanic picoplankton. In: Platt, T., Li, W. K. W. (eds), *Photosynthetic picoplankton. Can. Bull. Fish. Aquat. Sci.* 214: 159–204.
Cattolico, R. A., Goër, S. L. (1989). Analysis of chloroplast evolution and phylogeny: a molecular approach. In: Green, J. C., Leadbeater, B. S. C., Diver, W. L. (eds), *The Chromophyte Algae: Problems and Perspectives*, Clarendon Press, Oxford, p. 85–100.
Cavalier-Smith, T. (1993). Kingdom protozoa and its 18 phyla. *Microbiol. Rev.* 57: 953–994.
Chisholm, S. W., Olsen, R. J., Zettler, E. R., Goericke, R., Waterbury, J. B., Welschmeyer, N. A. (1988). A novel free-living prochlorophyte abundant in the oceanic euphotic zone. *Nature*, 334: 340–343.
Chisholm, S. W., Frankel, S. L., Goericke, R., Olson, R. J., Palenik, B., Waterbury, J. B., West-Johnsrud, L., Zettler, E. R. (1992). *Prochlorococcus marinus* nov. gen. nov. sp.: an oxyphototrophic marine prokaryote containing divinyl chlorophyll *a* and *b*. *Arch. Microbiol.* 157: 297–300.
Chrétiennot-Dinet, M-J. (1990). *Atlas du Phytoplancton Marin* vol. 3: *Chlorarachniophycées, Chlorophycées, Chrysophycées, Cryptophycées, Euglénophycées, Eustigmatophycées, Prasinophycées, Prymnesiophycées, Rhodophycées et Tribophycées*. Éditions du Centre National de la Recherche Scientifique, Paris, 261 pp.
Christensen, T. (1980). *Algae — a Taxonomic Survey. Fasc. 1*, Aio Trykas, Odense, 228 pp.
Christensen, T. (1989). The Chromophyta, past and present. In: Green, J. C., Leadbeater, B. S. C., Diver, W. L. (eds), *The Chromophyte Algae*, Clarendon Press, Oxford, p. 1–12.
Clayton, M. N., King, R. J. (1990). *Biology of Marine Plants*. Longman Cheshire, Melbourne, 501 pp.
Cole, K. M., Sheath, R. G. (1990). *Biology of the Red Algae,* Cambridge University Press, Cambridge, 517 pp.
Coleman, A. W. (1985). Diversity of plastid DNA configuration among classes of eukaryotic algae. *J. Phycol.* 21: 1–16.
Cox, G. (1993). Prochlorophyceae, In: Berner, T. (ed.), *Ultrastructure of Microalgae*, CRC Press, Boca Raton, USA, p. 53–70.
Cupp, E. E. (1943). Marine plankton diatoms of the west coast of North America. *Bull. Scripps Inst. Oceanogr.* 5: 1–237.
Delaney, S. F. (1990). Cyanobacteria, In: Clayton, M. N., King, R. J. (eds), *Biology of Marine Plants*, Longman Cheshire, Melbourne, p. 25–40.
Dibbayawan, T., Vesk, M., Larkum, A. W. D. (1990). Cord-like phycobilisomes of *Rhodosorus marinus* Geitler: fine structure and its functional implications. In: Battsheffsky, M. (ed.), *Current Research in Photosynthesis*, 11: 333–336.
Dodge, J. D. (1973). *The Fine Structure of Algal Cells.* Academic Press, London, 261 pp.

Dodge, J. D. (1975). A survey of chloroplast ultrastructure in the Dinophyceae. *Phycologia,* 14: 253–263.

Dodge, J. D. (1982). *Marine Dinoflagellates of the British Isles.* Her Majesty's Stationery Office, London, 304 pp.

Dodge, J. D., Crawford, R. M. (1971). A fine-structural survey of dinoflagellate pyrenoids and food reserves. *Bot. J. Linn. Soc.* 64: 105–115.

Egeland, E. S., Liaaen-Jensen, S. (1993). New carotenoids and chemosystematics in the Prasinophyceae. Proc. 10th Int. Symp. Carotenoids, Trondheim, Norway, June 1993, pp. 4–6.

Fawley, M. W. (1991). Disjunct distribution of the xanthophyll loroxanthin in the green algae (Chlorophyta). *J. Phycol.* 27: 544–548.

Fawley, M. W. (1992). Photosynthetic pigments of *Pseudoscourfieldia marina* and select green flagellates and coccoid ultraplankton: implications for the systematics of the Micromonadophyceae (Chlorophyta). *J. Phycol.* 28: 26–31.

Fawley, M. W., Lee, C. M. (1990). Pigment composition of the scaly green flagellate, *Mesostigma viride* (Micromonadophyceae) is similar to that of the siphonous green alga, *Bryopsis plumosa* (Ulvophyceae). *J. Phycol.* 26: 666–670.

Fiksdahl, A., Bjørnland, T., Liaaen-Jensen, S. (1984*a*). Algal carotenoids with novel end groups. *Phytochem.* 23: 649–655.

Fiksdahl, A., Withers, N., Guillard, R. R. L., Liaaen-Jensen, S. (1984*b*). Carotenoids in the Raphidophyceae — a chemosystematic contribution. *Comp. Biochem. Physiol.* 78B: 265–271.

Foss, P., Lewin, R. A., Liaaen-Jensen, S. (1987). The carotenoids of *Prochloron* sp. (Prochlorophyta). *Phycologia,* 26: 142–144.

French, F. W., Hargraves, P. E. (1985). Spore formation in the life cycles of the diatoms *Chaetoceros diadema* and *Leptocylindrus danicus. J. Phycol.* 21: 477–483.

Gabrielson, P. W., Garbary, D. J., Sommerfeld, M. R., Townsend, R. A., Tyler, P. L. (1990). Phylum Rhodophyta. In: Margulis, L., Corliss, J. O., Melkonian, M., Chapman, D. J. (eds). *Handbook of Protoctista,* Jones and Bartlett, Boston, p. 102–118.

Gantt, E. (1975). Phycobilisomes: light harvesting pigment complexes. *Bioscience,* 25: 781–788.

Geider, R. J., Gunter, P. A. (1989). Evidence for the presence of phycoerythrin in *Dinophysis norvegica,* a pink dinoflagellate. *Br. Phycol. J.* 24: 195–198.

Gieskes, W. W. C., Kraay, G. W. (1983). Dominance of Cryptophyceae during the phytoplankton spring bloom in the central North Sea detected by HPLC analysis of pigments. *Mar. Biol.* 75: 179–185.

Gillott, M. (1990). Phylum Cryptophyta, In: Margulis, L., Corliss, J. O., Melkonian, M., Chapman, D. J. (eds), *Handbook of Protoctista,* Jones and Bartlett, Boston, p. 139–151.

Goericke, R. (1990). *Pigments as ecological tracers for the abundance and growth of marine phytoplankton.* PhD Thesis, Harvard University, Cambridge, USA, 418 pp.

Goericke, R., Repeta, D. J. (1992). The pigments of *Prochlorococcus marinus*: the presence of divinyl chlorophyll *a* and *b* in a marine procaryote. *Limnol. Oceanogr.* 37: 425–433.

Gordon, N., Angel, D. L., Neori, A., Kress, N., Kimor, B. (1994). Heterotrophic dinoflagellates with symbiotic cyanobacteria and nitrogen limitation in the Gulf of Aqaba. *Mar. Ecol. Progr. Ser.* 107: 83–88.

Gran, H. H. (1912). Pelagic plant life, In: Murray, J., Hjurt, J. (eds), *The Depths of the Oceans,* Macmillan, London, p. 307–386.

Green, J. C., Leadbeater, B. S. C. (1994). *The Haptophyte Algae,* Syst. Assoc. Special Vol. No 51, Clarendon Press, Oxford, 446 pp.

Green, J. C., Leadbeater, B. S. C., Diver, W. L. (1989). *The Chromophyte Algae: Problems and Perspectives,* Clarendon Press, Oxford, 429 pp.

Green, J. C., Perch-Nielsen, K., Westbroek, P. (1990). Phylum Prymnesiophyta, In: Margulis, L., Corliss, J. O., Melkonian, M., Chapman, D. J. (eds), *Handbook of Protoctista,* Jones and Bartlett, Boston, p. 293–317.

Guillard, R. R. L., Kilham, P. (1977). The ecology of marine planktonic diatoms. In: Werner, D. (ed.), *The Biology of Diatoms*. Blackwell Scientific Publications, Oxford, pp. 372–469.

Hallegraeff, G. M. (1981). Seasonal study of phytoplankton pigments and species at a coastal station off Sydney. *Mar. Biol.* 61: 107–118.

Hallegraeff, G. M. (1993). A review of harmful algal blooms and their apparent global increase. *Phycologia,* 32: 79–99.

Hallegraeff, G. M., Jeffrey, S. W. (1984). Tropical phytoplankton species and pigments in continental shelf waters of north and north-west Australia. *Mar. Ecol. Progr. Ser.* 20: 59–74.

Hallegraeff, G. M., Jeffrey, S. W. (1993). Annually recurrent diatom blooms in spring along the New South Wales Coast of Australia. *Aust. J. Mar. Freshw. Res.* 44: 325–334.

Hallegraeff, G. M., Lucas, I. A. N. (1988). The marine dinoflagellate genus *Dinophysis* (Dinophyceae): photosynthetic, neritic and non-photosynthetic ocean species. *Phycologia,* 27: 25–42.

Hara, Y., Chihara, M. (1982). Ultrastructure and taxonomy of *Chattonella* (Class Raphidophyceae) in Japan. *Jpn. J. Phycol.* 30: 47–56.

Hasle, G. (1976). The biogeography of some marine planktonic diatoms. *Deep-Sea Res.* 23: 319–338.

Heimdal, B. R. (1993). Modern coccolithophorids, In: Tomas, C. R. (ed.), *Marine Phytoplankton,* Academic Press, London, pp. 147–247.

Hendey, N. I. (1964). *An Introductory Account of the Smaller Algae of British Coastal Waters. Part V. Bacillariophyceae (diatoms).* Her Majesty's Stationery Office, London, 317 pp.

Heywood, P. (1980). Chloromonads, In: Cox, E. R. (ed.), *Phytoflagellates*, Elsevier North Holland, p. 351–379.

Heywood, P. (1990). Phylum Raphidophyta, In: Margulis, L., Corliss, J. O., Melkonian, M., Chapman, D. J. (eds), *Handbook of Protoctista*, Jones and Bartlett, Boston, p. 318–325.

Hibberd, D. J. (1977). Observations on the ultrastructure of the cryptomonad endosymbiont of the red-water ciliate *Mesodinium rubrum. J. Mar. Biol. Assoc.* U.K., 57: 45–61.

Hibberd, D. J. (1980). Prymnesiophytes (haptophytes), In: Cox, E. R. (ed.) *Phytoflagellates*, Elsevier, North Holland, p. 273–317.

Hibberd, D. J. (1990*a*). Phylum Eustigmatophyta. In: Margulis, L., Corliss, J. O., Melkonian, M., Chapman, D. J. (eds), *Handbook of Protoctista*, Jones and Bartlett, Boston, p. 326–333.

Hibberd, D. J. (1990*b*). Phylum Xanthophyta. In: Margulis, L., Corliss, J. O., Melkonian, M., Chapman, D. J. (eds), *Handbook of Protoctista*, Jones and Bartlett, Boston, p. 686–697.

Hibberd, D. J., Leedale, G. F. (1970). Eustigmatophyceae — a new algal class with unique organization of the motile cell. *Nature*, 225: 758–760.

Hibberd, D. J., Leedale, G. F. (1971). A new algal class — the Eustigmatophyceae. *Taxon*, 20: 523–525.

Hibberd, D. J., Leedale, G. F. (1972). Observations on the cytology and ultrastructure of the new algal class, Eustigmatophyceae. *Ann. Bot.* 36: 49–71.

Hill, D. R. A. (1991). *Chroomonas* and other blue-green cryptomonads. *J. Phycol.* 27: 133–145.

Hostetter, H. P., Stoermer, E. F. (1971). Bibliography of the Bacillariophyceae, In: Roskowski, J. R., Parker, B. C. (eds), *Selected Papers in Phycology*, Dept. Botany, Univ. Nebraska, Lincoln, p. 784–790.

Jeffrey, S. W. (1976*a*). The occurrence of chlorophyll c_1 and c_2 in algae. *J. Phycol.* 12: 349–354.

Jeffrey, S. W. (1976*b*). A report of green algal pigments in the central North Pacific Ocean. *Mar. Biol.* 37: 33–37.

Jeffrey, S. W., Carpenter, S. M. (1974). Seasonal succession of phytoplankton at a coastal station off Sydney. *Aust. J. Mar. Freshw. Res.* 25: 361–369.

Jeffrey, S. W., Hallegraeff, G. M. (1980*a*). Studies of phytoplankton species and photosynthetic pigments in a warm-core eddy of the East Australian Current. I. Summer populations. *Mar. Ecol. Progr. Ser.* 3: 285–294.

Jeffrey, S. W., Hallegraeff, G. M. (1980*b*). Photosynthetic pigment profiles and phytoplankton species in a warm-core eddy of the East Australian Current. II. A note on methodology. *Mar. Ecol. Progr. Ser.* 3: 295–301.

Jeffrey, S. W., Hallegraeff, G. M. (1990). Phytoplankton ecology of Australian waters, In: Clayton, M. N., King, R. J. (eds), *Biology of Marine Plants*, Longman-Cheshire, Melbourne, p. 310–348.

Jeffrey, S. W., Vesk, M. (1976). Further evidence for a membrane-bound endosymbiont within the dinoflagellate *Peridinium foliaceum*. *J. Phycol.* 12: 450–455.

Jeffrey, S. W., Vesk, M. (1981). The phytoplankton — systematics, morphology and ultrastructure. In: Clayton, M. N., King, R. J. (eds), *Marine Botany — an Australasian perspective*. Longman Cheshire, Melbourne, p. 138–179.

Jeffrey, S. W., Vesk, M. (1990). Cryptophyta, In: Clayton, M. N., King, R. J. (eds), *Biology of Marine Plants*, Longman Cheshire, Melbourne, p. 86–95.

Jeffrey, S. W., Wright, S. W. (1994). Photosynthetic pigments in the Prymnesiophyceae. In: Green, J. C., Leadbeater, B. S. C. (eds), *The Haptophyte Algae*. Clarendon Press, Oxford, p. 111–132.

Jeffrey, S. W., Sielicki, M., Haxo, F. T. (1975). Chloroplast pigment patterns in dinoflagellates. *J. Phycol.* 11: 374–384.

Jeffrey, S. W., Vesk, M., Hallegraeff, G. M. (1990). Bacillariophyta, In: Clayton, M. N., King, R. J. (eds), *Biology of Marine Plants*, Longman Cheshire, Melbourne, p. 115–132.

Jensen, T. E. (1993). Cyanobacterial ultrastructure. In: Berner, T. (ed.), *Ultrastructure of Microalgae*, CRC Press, Boca Raton, p. 7–51.

Johansen, J. E., Svec, W. A., Liaaen-Jensen, S., Haxo, F. T. (1974). Carotenoids of the Dinophyceae. *Phytochemistry,* 13: 2261–2271.

Jordan, R. W., Green, J. C. (1994). A check-list of the extant Haptophyta of the world. *J. Mar. Biol. Assoc. U.K.* 74: 149–174.

Kawachi, M., Inouye, I., Maeda, O., Chihara, M. (1991). The haptonema as a food capturing device: observations on *Chrysochromulina hirta* (Prymnesiophyceae). *Phycologia*, 30: 563–573.

Kimor, B., Gordon, N., Neori, A. (1992). Symbiotic associations among the microplankton in oligotrophic marine environments with special reference to the Gulf of Aqaba, Red Sea. *J. Plankton Res.* 14: 1217–1231.

Klaveness, D. (1988–89). Biology and ecology of the Cryptophyceae: status and challenges. *Biol. Oceanogr.* 6: 257–270.

Komárek, J., Anagnostidis, K. (1989). Modern approach to the classification system of cyanophytes. 4. Nostocales. *Arch. Hydrobiol. Suppl.* 82: 247–345.

Kosman, C. A., Thomsen, H. A., Østergaard, J. B. (1993). Parmales (Chrysophyceae) from Mexican, Californian, Baltic, Arctic and Antarctic waters with the description of a new sub-species and several new forms. *Phycologia,* 32: 116–128.

Kraft, G. T. (1981). Rhodophyta: Morphology and Classification. In: Lobban, C. S., Wynne, M. J. (eds), *The Biology of Seaweeds*, University of California Press, Berkeley, p. 6–51.

Kristiansen, J. (1990). Phylum Chrysophyta, In: Margulis, L., Corliss, J. O., Melkonian, M., Chapman, D. J. (eds). *Handbook of Protoctista*. Jones and Bartlett, Boston, 438–453.

Kristiansen, J., Andersen, R. A. (1986). *Chrysophytes: aspects and problems*. Cambridge Univ. Press, Cambridge, 337 pp.

Kugrens, P., Lee, R. E., Andersen, R. A. (1986). Cell form and surface patterns in *Chroomonas* and *Cryptomonas* cells (Cryptophyta) as revealed by scanning electron microscopy. *J. Phycol.* 22: 512–522.

Lebour, M. V. (1930). *The Plankton Diatoms of Northern Seas*, Ray Society, London, 244 pp.

Lee, J. J., McEnery, M. E., Shilo, M., Reiss, Z. (1979). Isolation and cultivation of diatom symbionts from larger Foraminifera (Protozoa). *Nature,* 280: 57–58.

Leedale, G. F. (1967). *Euglenoid Flagellates*. Engelwood Cliffs, New Jersey, 242 pp.

Lewin, J., Norris, R. E., Jeffrey, S. W., Pearson, B. (1977). An aberrant chrysophycean alga, *Pelagococcus subviridis* Norris gen. nov., sp. nov., from the North Pacific. *J. Phycol.* 13: 259–266.

Lewin, R. A. (1977). *Prochloron*, type genus of the Prochlorophyta. *Phycologia*, 16: 217.

Li, W. K. W. (1994). Primary production of prochlorophytes, cyanobacteria and eucaryotic ultraphytoplankton: measurements from flow cytometric sorting. *Limnol. Oceanogr.* 39: 169–175.

Li, W. K. W., Wood, M. (1988). Vertical distribution of North Atlantic ultraplankton: analysis by flow cytometry and epifluorescence microscopy. *Deep-Sea Res.* 35: 1615–1638.

Lipps, J. H. (1970). Plankton evolution, *Evolution,* 24: 1–22.

Lucas, I. A. N. (1991). Symbionts of the tropical Dinophysiales (Dinophyceae). *Ophelia*, 33: 213–224.

Malin, G., Turner, S. M., Liss, P. S. (1992). Sulfur: the plankton/climate connection. *J. Phycol.* 28: 590–597.

Manton, I., Sutherland, J., Oates, K. (1976). Arctic coccolithophorids: two species of *Turrisphaera* gen. nov. from West Greenland, Alaska, and the Northwest Passage. *Proc. R. Soc. Ser. B.* 194: 179–194.

Manton, I., Sutherland, J., Oates, K. (1977). Arctic coccolithophorids: *Wigwamma arctica* gen. et sp. nov. from Greenland and arctic Canada, *W. annulifera* sp. nov. from South Africa and S. Alaska and *Calciarcus alaskensis* gen. et sp. nov. from S. Alaska. *Proc. Roy. Soc. Ser. B.* 197: 145–168.

Marchant, H. J., McEldowney, A. (1986). Nanoplanktonic siliceous cysts from Antarctica are algae. *Mar. Biol.* 92: 53–57.

Margalef, R. (1960). Temporal succession and spatial heterogeneity in phytoplankton. In: Buzzati-Traverso, A. A. (ed.), *Perspectives in Marine Biology*, University of California Press, pp. 332–349.

Margulis, L., Corliss, J. O., Melkonian, M., Chapman, D. J. (1990). *Handbook of Protoctista.* Jones and Bartlett, Boston, 914 pp.

Mattox, K. R., Stewart, K. D. (1984). Classification of the green algae: a concept based on comparative cytology. In: Irvine, D. E. G., John, D. M. (eds), *Systematics of the Green Algae*, Academic Press, London, p. 29–72.

McFadden, G. I., Hill, D. R. A., Wetherbee, R. (1986*a*). A study of the genus *Pyramimonas* (Prasinophyceae, Chlorophyta) from southeastern Australia. *Nordic J. Bot.* 6: 209–234.

McFadden, G. I., Preisig, H. R., Melkonian, R. (1986*b*). Golgi apparatus activity and membrane flow during scale biogenesis in the green flagellate *Scherffelia dubia* (Prasinophyceae). II Cell wall secretion and assembly. *Protoplasma*, 131: 174–184.

Melkonian, M. (1990*a*). Phylum Chlorophyta: Class Chlorophyceae. In: Margulis, L., Corliss, J. O., Melkonian, M., Chapman, D. J. (eds), *Handbook of Protoctista*, Jones and Bartlett, Boston, p. 608–616.

Melkonian, M. (1990*b*). Phylum Chlorophyta: Class Prasinophyceae. In: Margulis, L., Corliss, J. O., Melkonian, M., Chapman, D. J. (eds), *Handbook of Protoctista*, Jones and Bartlett, Boston, p. 600–607.

Moestrup, Ø., Thomsen, H. A. (1986). Ultrastructure and reconstruction of the flagellar apparatus in *Chrysochromulina apheles* sp. nov. (Prymnesiophyceae = Haptophyceae). *Canad. J. Bot.* 64: 593–610.

Moestrup, Ø., Thomsen, H. A. (1990). *Dictyocha speculum* (Silicoflagellata, Dictyochophyceae) studies on armoured and unarmoured stages. *Biologische Skrifter*, 37: 1–57.

Muscatine, L. (1967). Glycerol excretion by symbiotic algae from corals and *Tridacna* and its control by the host. *Science*, 156: 516–519.

Nichols, B. W. (1973). Lipid composition and metabolism. In: Carr, N. G., Whitton, B. A. (eds), *The Biology of Blue-green Algae*, Blackwells, Oxford, p. 144–161.

Norgård, S., Svec, W. A., Liaaen-Jensen, S., Jensen, A., Guillard, R. R. L. (1974). Chloroplast pigments and algal systematics. *Biochem. Syst. and Ecol.* 2: 3–6.

Olson, R. J., Chisholm, S. W., Zettler, E. R., Altabet, M., Dusenberry, J. (1990). Spatial and temporal distributions of prochlorophyte picoplankton in the North Atlantic Ocean. *Deep-Sea Res.* 37: 1033–1051.

Paasche, E., Throndsen, J. (1970). *Rhodella maculata* Evans (Rhodophyceae, Porphiridiales), isolated from the plankton of the Oslo Fjord. *Nytt. Mag. Bot.* 17: 209–212.

Palenik, B., Haselkorn, R. (1992). Multiple evolutionary origins of prochlorophytes, the chlorophyll *b*-containing prokaryotes. *Nature*, 355: 265–267.

Partensky, F., Hoepffner, N., Li, W. K. W., Ulloa, O., Vaulot, D. (1993). Photoacclimation of *Prochlorococcus* sp. (Prochlorophyta) strains isolated from the North Atlantic and the Mediterranean Sea. *Plant Physiol.* 101: 285–296.

Patrick, R. (1977). The ecology of freshwater diatoms and diatom communities. In: Werner, D. (ed.), *The Biology of Diatoms*. Blackwell Scientific Publications, Oxford, pp. 284–332.

Patrone, L. M., Broadwater, S. T., Scott, J. L. (1991). Ultrastructure of vegetative and dividing cells of the unicellular red alga *Rhodella violaceae* and *Rhodella maculata*. *J. Phycol.* 27: 742–753.

Phillips, J. A. (1990). Chlorophyta and Euglenophyta. In: Clayton, M. N., King, R. J. (eds), *Biology of Marine Plants*. Longman Cheshire, Melbourne, p. 183–212.

Pickett-Heaps, J. D. (1975). *Green Algae: Structure, Reproduction and Evolution in Selected Genera*. Sinauer Assoc., Sunderland, Mass., 606 pp.

Ramus, J. (1972). The production of extracellular polysaccharide by the unicellular red alga, *Porphyridium aerugineum*. *J. Phycol.* 8: 97–111.

Ricard, M. (1987). *Atlas du Phytoplancton Marin*, vol. 2, *Diatomophycées*, Éditions du Centre National de la Recherche Scientifique, Paris, 297 pp.

Ricketts, T. R. (1970). The pigments of the Prasinophyceae and related organisms. *Phytochem.* 9: 1835–1842.

Ross, R., Cox, E. J., Karayeva, N. I., Mann, D. G., Paddock, T. B. B., Simonsen, R., Sims, P. A. (1979). An amended terminology for the siliceous components of the diatom cell. *Nova Hedwigia,* 64: 513–533.

Round, F. E. (1981). *The Ecology of Algae*, Cambridge Univ. Press, 653 pp.

Round, F. E., Crawford, R. M., Mann, D. G. (1990). *The Diatoms — Biology and Morphology of the Genera*, Cambridge University Press, 747 pp.

Rowan, K. S. (1989). *Photosynthetic Pigments of Algae.* Cambridge University Press, 334 pp.

Sarjeant, W. A. S. (1974). *Fossil and Living Dinoflagellates*, Academic Press, London, 182 pp.

Schiller, J. (1935). Dinoflagellatae (Peridineae). In: Rabenhorst, L. (ed.), *Kryptogamen-Flora von Deutschland, Osterreich und der Schweiz. Akad. Verlagsges, Leipzig,* 10, (2), 1–589.

Schmidt, A. M. M. (1984). Valve morphogenesis in diatoms, In: Bach, K., Burkhardt, B. (eds), *Diatoms I, Shells in Nature and Technics, IL28*, p. 300–317.

Schütt, F. (1895). Die Peridineen der Plankton Expedition. 1. Studien über die Zellen der Peridineen. *Ergebn. Plankton Exped.* 4: 1–170.

Sieburth, J. M., Johnson, P. W., Hargraves, P. E. (1988). Ultrastructure and ecology of *Aureococcus anophagefferens* gen. et sp. nov. (Chrysophyceae): the dominant picoplankter during a bloom in Narragansett Bay, Rhode Island, summer, 1985. *J. Phycol.* 24: 416–425.

Smayda, T. J., Villareal, T. A. (1989). An extraordinary, noxious brown-tide in Narragansett Bay. I. The organism and its dynamics. In: Okaichi, T., Anderson, D. T., Nemoto, T., *Red Tides: Biology, Environmental Science and Toxicology*, Elsevier North Holland, p. 129–136.

Sournia, A. (1970). Les Cyanophycées dans le plancton marine. *Ann. Biol.* 9: 63–76.

Sournia, A. (1973). Catalogue des espèces et taxons intraspecifiques de Dinoflagellés marins actuels. 1. Dinoflagellés libres. *Beih. Nova Hedwigia.* 48: 1–92.

Sournia, A. (1986). *Atlas du Phytoplancton Marin*, vol. 1, *Cyanophycées, Dictyophycées, Dinophycées, Raphidophycées*, Éditions du Centre National de la Recherche Scientifique, Paris, 219 pp.

Sournia, A., Chrétiennot-Dinet, M-J., Ricard, M. (1991). Marine phytoplankton: how many species in the world ocean? *J. Plankton Res.* 13: 1093–1099.

Sournia, A., Belin, C., Billard, C., Martial, C., Erard-le Denn, E., Fresnel, J., Lassus, P., Pastoureaud, A., Soulard, R. (1992). The repetitive and expanding occurrence of a green, bloom-forming dinoflagellate (Dinophyceae) on the coasts of France. *Cryptogamie Algol.* 13: 1–13.

Staines, J. E., McGowan, V. F., Skerman, V. B. D. (1986). World Directory of Collections of Cultures of Microorganisms, World Data Centre, Univ. Queensland, Australia, 689 pp.

Stauber, J. L., Jeffrey, S. W. (1988). Photosynthetic pigments in fifty-one species of marine diatoms. *J. Phycol.* 24: 158–172.

Steidinger, K. A., Cox, E. R. (1980). Free-living dinoflagellates, In: Cox, E. R. (ed.) *Phytoflagellates*, Elsevier, North Holland, p. 407–432.

Sverdrup, H. U. (1953). On conditions for the vernal blooming of phytoplankton. *J. Conseil, Int. Explor. Mer.* 18: 287–295.

Sym, S. D., Pienaar, R. N. (1993). The class Prasinophyceae. *Progr. Phycol. Res.* 9: 281-376.

Takahashi, K. (1987). Seasonal fluxes of silicoflagellates and *Actiniscus* in the sub-arctic Pacific during 1982 and 1984. *J. Mar. Res.* 44: 397–425.

Taylor, F. J. R. (1976*a*). Flagellate phylogeny: a study in conflicts. *J. Protozool.* 23: 28–40.

Taylor, F. J. R. (1976*b*). Dinoflagellates from the Indian Ocean Expedition. *Bibl. Bot.* 132, 234 pp.

Taylor, F. J. R. (1980). On dinoflagellate evolution. *Biosystems,* 13: 65–108.

Thomsen, H. A., Østergaard, J. B., Hansen, L. E. (1991). Heteromorphic life histories in arctic coccolithophorids (Prymnesiophyceae). *J. Phycol.* 27: 634–642.

Throndsen, J. (1973). Fine structure of *Eutreptiella gymnastica* (Euglenophyceae). *Norw. J. Bot.* 20: 271–280.

Throndsen, J. (1993). The planktonic marine flagellates. In: Tomas, C. R. (ed.), *Marine Phytoplankton: a Guide to Naked Flagellates and Coccolithophorids.* Academic Press, p. 7–145.

Throndsen, J., Kristiansen, S. (1985). *Pelagococcus subviridis* Norris as a major component of the nanoplankton at Haltenbanken, Norwegian Sea in July 1982. *Abst. Int. Phycol. Congr.*, p. 160.

Tomas, C. R. (1993). *Marine Phytoplankton: a Guide to Naked Flagellates and Coccolithophorids.* Academic Press, London, 263 pp.

Trench, R. K., Blank, R. J. (1987). *Symbiodinium microadriaticum* Freudenthal, *S. goreauii* sp. nov., *S. kawagutii* sp. nov. and *S. pilosum* sp. nov.: Gymnodiniod dinoflagellate symbionts of marine invertebrates. *J. Phycol.* 23: 469–481.

Trench, R. K., Wethey, D. S., Porter, J. W. (1981). Observations on the symbiosis with zooxanthellae among the Tridacnidae (Mollusca, Bivalvia). *Biol. Bull.* 161: 180–198.

Urbach, E., Robertson, D. L., Chisholm, S. W. (1992). Multiple evolutionary origins of prochlorophytes within the cyanobacterial radiation. *Nature*, 355: 267–269.

Van der Wal, P., de Jong, E. W., Westbroek, P., de Bruijn, W. C., Molder-Stapel, A. A. (1983). Polysaccharide localization, coccolith formation and Golgi dynamics in the coccolithophorid *Hymenomonas carterae*. *J. Ultrastruct. Res.* 85: 139–158.

Van Valkenburg, S. D. (1980). Silicoflagellates, In: Cox, E. R. (ed.), *Phytoflagellates*, Elsevier, North Holland, p. 335–350.

Van Valkenburg, S. D., Norris, R. E. (1970). The growth and morphology of the silicoflagellate *Dictyocha fibula* Ehrenberg in culture. *J. Phycol.* 6: 48–54.

Vaulot, D., Partensky, F. (1992). Cell cycle distributions of prochlorophytes in the north-western Mediterranean Sea. *Deep-Sea Res.* 39: 727–742.

Vaulot, D., Partensky, F., Neveux, J., Mantoura, R. F. C., Llewellyn, C. (1990). Wintertime presence of prochlorophytes in surface waters of the north-western Mediterranean Sea. *Limnol. Oceanogr.* 35: 1156–1164.

Veldhuis, M. J. W., Kraay, G. W. (1990). Vertical distribution and pigment composition of a

picoplanktonic prochlorophyte in the subtropical North Atlantic: a combined study of HPLC-analysis of pigments and flow cytometry. *Mar. Ecol. Progr. Ser.* 68: 121–127.
Vesk, M., Jeffrey, S. W. (1987). Ultrastructure and pigments of two strains of the picoplanktonic alga *Pelagococcus subviridis* (Chrysophyceae). *J. Phycol.* 23: 322–336.
Vesk, M., Dibbayawan, T. P., Vesk, P. A. (1996). Immunogold localization of phycoerythrin in chloroplasts of *Dinophysis acuminata* and *D. fortii* (Dinophysiales, Dinophyta). *Phycologia,* 35 (in press).
Vesk, M., Hallegraeff, G. M., Jeffrey, S. W. (1990*a*). Dinophyta. In: Clayton, M. N., King, R. J. (eds), *The Biology of Marine Plants*, Longman Cheshire, Melbourne, p. 133–148.
Vesk, M., Jeffrey, S. W., Hallegraeff, G. M. (1990*b*). Golden-brown algae: Prymnesiophyta and Chrysophyta, In: Clayton, M. N., King, R. J. (eds), *Biology of Marine Plants*, Longman Cheshire, Melbourne, p. 97–114.
Vesk, M., Dwarte, D., Fowler, S., Hiller, R. G. (1992). Freeze fracture immunochemistry of light harvesting pigment complexes in a cryptophyte. *Protoplasma*, 170: 166–176.
Volkman, J. K., Brown, M. R., Dunstan, G. A., Jeffrey, S. W. (1993). The biochemical composition of marine microalgae from the class Eustigmatophyceae. *J. Phycol.* 29: 69–78.
Vonshak, A. (1988). *Porphyridium.* In: Borowitzka, M. A., Borowitzka, L. J. (eds), *Microalgal Biotechnology*, Cambridge University Press, p. 122–134.
Wall, D., Dale, B. (1968). Modern dinoflagellate cysts and evolution of the Peridiniales. *Micropaleontol.* 14: 265–304.
Walne, P. L., Kivic, P. A. (1990). Phylum Euglenida. In: Margulis, L., Corliss, J. O., Melkonian, M., Chapman, D. J. (eds), *Handbook of Protoctista*, Jones and Bartlett, Boston, p. 270–287.
Walsby, A. E. (1977). The gas vacuole of blue-green algae. *Sci. Am.* 237: 90–97.
Watanabe, M. M., Suda, S., Inouye, I., Sawaguchi, T., Chihara, M. (1990). *Lepidodinium viride* gen. et sp. nov. (Gymnodiniales, Dinophyta), a green dinoflagellate with a chlorophyll *a* and *b*-containing endosymbiont. *J. Phycol.* 26: 741–751.
Watanabe, M. M., Takeda, Y., Sasa, T., Inouye, I., Suda, S., Sawaguchi, T., Chihara, M. (1987). A green dinoflagellate with chlorophylls *a* and *b*; morphology, fine structure of the chloroplast and chlorophyll composition. *J. Phycol.* 23: 382–389.
Waterbury, J. B., Watson, S. W., Guillard, R. R. L., Brand, L. E. (1979). Widespread occurrence of a unicellular marine planktonic cyanobacterium. *Nature,* 277: 293–294.
Waterbury, J. B., Watson, S. W., Valois, F. W., Franks, D. G. (1986). Biological and ecological characterization of the marine unicellular cyanobacterium *Synechococcus.* In: Platt, T., Li, W. K. W. (eds), *Photosynthetic Picoplankton. Can. Bull. Fish. Aquat. Sci.* 214: 71–120.
Werner, D. (1977). *The Biology of Diatoms*, Blackwell Scientific Publications, Oxford, 498 pp.
Wetherbee, R., Hill, D. R. A., Brett, S. J. (1987). The structure of the periplast components and their association with the plasma membrane in a cryptomonad flagellate. *Can. J. Bot.* 65: 1019–1026.
Whatley, J. M., Whatley, F. R. (1981). Chloroplast evolution. *New Phytol.* 87: 233–247.
Whitton, B. A. (1973). Freshwater plankton, In: Carr, N. G., Whitton, B. A. (eds), *The Biology of Blue-green Algae*, Blackwell, Oxford, p. 353–367.
Wilcox, L. W., Wedemeyer, G. J. (1984). *Gymnodinium acidotum* Nygaard (Pyrrophyta), a dinoflagellate with an endosymbiotic cryptomonad. *J. Phycol.* 20: 236–242.
Winter, A., Siesser, W. G. (1994). *Coccolithophores,* Cambridge Univ. Press, U. K., 242 pp.
Withers, N. W., Alberte, R. S., Lewin, R. A., Thornber, J. P., Britton, G., Goodwin, T. W. (1978). Photosynthetic unit size, carotenoids, and chlorophyll-protein composition of *Prochloron* sp., a prokaryotic green alga. *Proc. Nat. Acad. Sci.* USA, 75: 2301–2305.
Withers, N. W., Fiksdahl, A., Tuttle, R. C., Liaaen-Jensen, S. (1981). Carotenoids of the Chrysophyceae. *Comp. Biochem. Physiol.* 68B: 345–349.
Yokohama, Y. (1981). Distribution of the green light-absorbing pigments siphonaxanthin and siphonein in marine green algae. *Bot. Mar.* 24: 637–640.

3

Metabolism and function of photosynthetic pigments

R. J. Porra, E. E. Pfündel and N. Engel

ABBREVIATIONS

ABA, abscisic acid; ALA, 5-aminolevulinic acid; ATP, adenosine triphosphate; BChl, bacteriochlorophyll; BPhe, bacteriopheophytin; Chl, chlorophyll; Chlide, chlorophyllide; CP, proteins of PSI or PSII core complexes; DMAPP, dimethyl allyl diphosphate; GGPP, geranylgeranyl diphosphate; GSA, glutamate-1-semialdehyde; HMG-CoA, 3-hydroxy-3-methylglutaryl-coenzyme A; IPP, isopentenyl pyrophosphate; LHC I and II, light-harvesting complexes of photosystems I and II; LHCP, light-harvesting Chl *a/b* proteins; NAD^+, nicotine adenine dinucleotide; NADH, reduced nicotine adenine dinucleotide; $NADP^+$, nicotine adenine dinucleotide phosphate; NADPH, reduced nicotine adenine dinucleotide phosphate; PAR, photosynthetically active radiation; PBG, porphobilinogen; PChlide, protochlorophyllide; PS I and PS II, photosystems I and II; RC, reaction centre; TCA, tricarboxylic acid cycle

The tetrapyrrole nomenclature and macrocycle numbering system (Moss, 1987) approved by IUPAC-IUB is used throughout this chapter.

3.1 INTRODUCTION

This chapter provides an introduction to the properties, biosyntheses and functions of the major pigments of photosynthesis. It also discusses the nature, origin and function of the many carotenoid and tetrapyrrolic chromophores and their derivatives, and the degradation products that may be encountered in phytoplankton extracts prepared for pigment assays. The pigments of photosynthetic bacteria are described only briefly.

The sections on the chlorophylls and phycobilins were written by R. J. Porra, the biological degradation of chlorophylls by N. Engel, and the carotenoids by E. E. Pfündel.

3.2 Pigment function in photosynthesis

Photosynthesis is the process whereby light energy is absorbed by photosynthetic pigments in plants, algae and some bacteria, for transduction to chemical energy. While the non-absorbed wavelengths are transmitted, giving these photosynthetic pigments their characteristic colours, the energy of the absorbed wavelengths is converted to chemical energy and reducing power to form carbohydrate from CO_2. Organic matter formed by photosynthesis also fuels living processes in non-photosynthetic organisms. Photosynthesis, therefore, is the basis of almost all life on earth.

The main pigments absorbing photosynthetically active radiation (PAR) include the chlorophylls, carotenoids and phycobilins, which spectacularly colour our terrestrial world during springtime synthesis and autumnal degradation. The greater part of chlorophyll (Chl) is found on land. The global chlorophyll turnover is estimated at about 10^9 tonnes annually (Brown *et al.* 1991), only about 25% of which occurs in terrestrial plants during autumn. The other 75% occurs in oceans, lakes and rivers, where chlorophyll may be turned over about 40 times a year because of the shorter life spans of freshwater and marine algae. The annual synthesis of carotenoids has been estimated at 2×10^8 tonnes (Pfander, 1992; Hendry, 1993), but that of phycobilins is not known.

3.2.1 *Pigment-protein complexes*

Homogeneous solutions of free Chls and carotenoids, as in organic solvents, are not found in photosynthetic cells. The free pigments have no role in photosynthesis: indeed, they are toxic to cells (see section 3.5.2). The pigments function in photosynthesis only when associated with specific *apo*proteins as Chl- and carotenoid-containing pigment-protein complexes.

The properties of pigments are dramatically modified by association with the *apo*proteins, fitting the pigment-proteins for their photosynthetic roles. Thus Chl *a* can function as a light absorber in Light-Harvesting Complexes (LHCs; see section 3.3.1) and also as electron donors and acceptors in Reaction Centres (RCs; see section 3.3.2). Association with proteins modifies Chl spectra: free Chl *a* absorbs at 662 nm in diethylether, but when associated with specific RC I and II *apo*proteins it absorbs at 700 nm (P700) and at 680 nm (P680), respectively (see section 3.3). The modification of Chl spectra by bonding with proteins is also reflected in the spectra of whole cells *(in vivo* spectra) and photosynthetic complexes (Fig. 3.1).

In the chloroplast, the chlorophyll and carotenoid molecules are tightly packed by their *apo*protein bonding sites. In the LHC II complex, for example, the Chl concentration in the total volume of the LHC II trimer is estimated to be about 0.3 M (Kühlbrandt and Wang, 1991), ensuring considerable capacity for light harvesting (see sections 3.3.1 and 3.8.4.3). The carotenoids also have a Chl protection function requiring this tight juxtaposition (see sections 3.8.4.4 and 3.8.4.5). Chls in

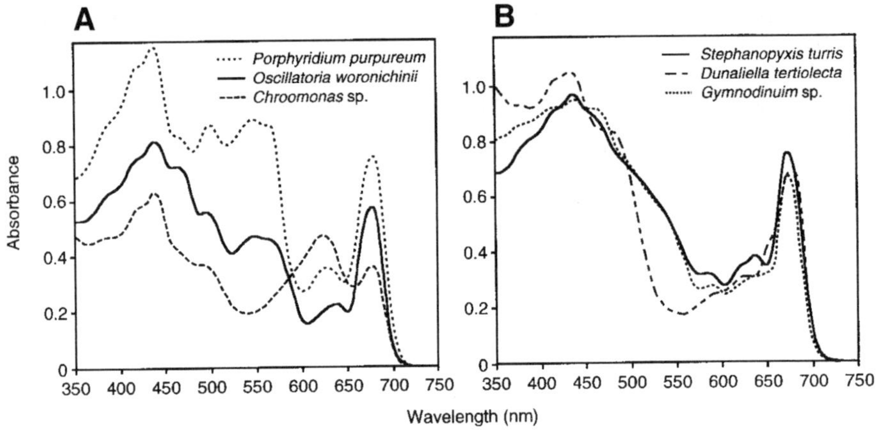

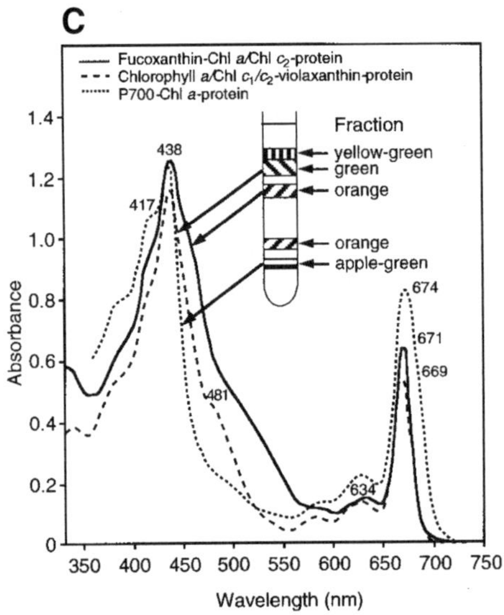

Figure 3.1
In vivo absorption spectra of algal cells from six classes (A, B), and pigment-protein complexes from a brown seaweed (C). A, *Porphyridium purpureum* (red alga); *Oscillatoria woronichinii* (cyanophyte); *Chroomonas* sp. (cryptophyte); B, *Stephanopyxis turris* (diatom); *Dunaliella tertiolecta* (green alga); *Gymnodinium* sp. (dinoflagellate) (after Jeffrey, 1980). C. Absorption spectra of pigment-protein complexes from the brown alga, *Acrocarpia paniculata*: P700-chlorophyll *a*-protein complex and two light-harvesting complexes: chlorophyll a/c_2-fucoxanthin-protein complex, and chlorophyll a/c_1+c_2-violaxanthin-protein complex. Sucrose density gradient separation of pigment fractions from Triton X-100-solubilized thylakoids (after Barrett and Anderson, 1980).

pigment-protein complexes are more stable than free Chls in solution because the carotenoids quench the photo-induced labile triplet-state Chls and the singlet oxygen derived from them (see section 3.6.1).

Pigments are attached to *apo*proteins by several types of bonding, but chlorophylls are further bound by metal-ligand interaction (e.g., Mg-histidine). Phycobilins are covalently bound to their *apo*proteins by thioether bonds arising from adduct formation between cysteine residues of the *apo*protein with a chromophore vinyl group (see section 3.5.4.2).

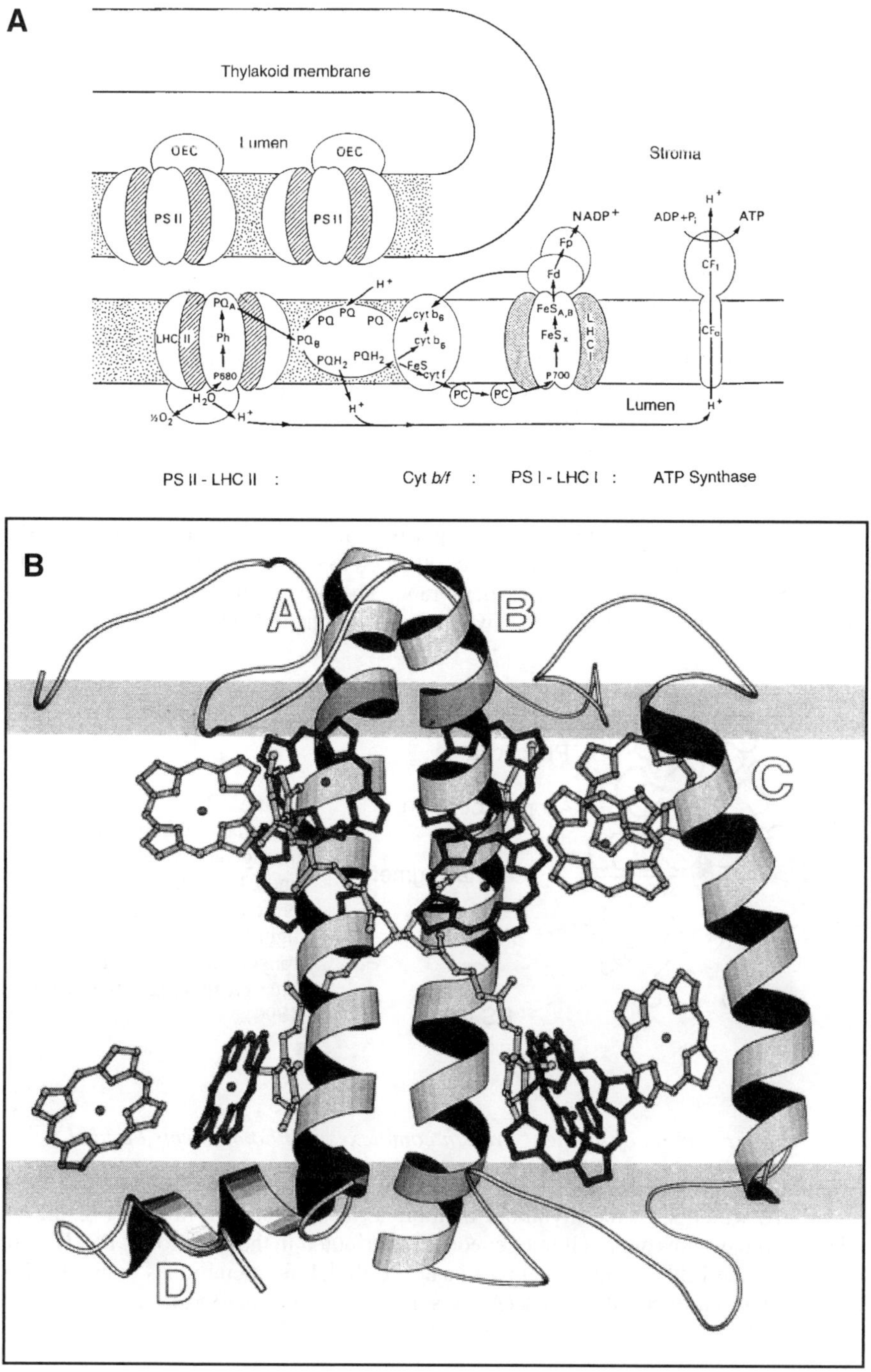

Figure 3.2
Schematic representation of organization of photosystems in the thylakoids and of electron transfer from H_2O to $NADP^+$ *via* PS I and PS II. (A, B): thylakoid membranes of a higher plant or green alga with PS I in the stroma-exposed thylakoids and most of the PS II in the stacked thylakoids. For abbreviations see legend to Fig 3.4 and p. 86 (B). Atomic models of LHC II of higher plants showing the position of the α-helices as ribbons (A to D), and Chl *a* (dark macrocycles), Chl *b* (light macrocycles) and xanthophylls. Horizontal bands indicate the approximate position of the lipid bilayer (from Kühlbrandt *et al.* 1994).

3.2.2 *Assembly of pigment-protein complexes into LHCs and RCs in thylakoids*

The many Chl-carotenoid binding proteins are located in LHCs, Core Proteins (CPs) and RCs (see Boekema *et al.* 1995), which serve different functions in photosynthesis. The roles and arrangement of these complexes are shown in Fig. 3.2.

The pigment-protein complexes of algae and higher plants are within thylakoid membranes of the chloroplast (see Figs. 2.3 and 3.2). Fig. 3.2A shows the thylakoid membranes of a green plant arranged into stacked areas (grana), with the membrane appressed to itself (shaded), and unstacked areas, with the membrane exposed to the stroma (unshaded). The appressed membranes, each about 7 nm thick, are separated by a 3 nm layer of water (Nir and Pease, 1973), which allows protons access to appressed areas. Fig. 3.2B shows an atomic model of the LHCII complex from higher plants, discussed in detail in section 3.3.3.

Photosynthetic bacteria and blue-green algae (cyanobacteria) are prokaryotes (see section 2.11) and contain no chloroplasts. The photosynthetic pigments of the latter are associated with phycobilisomes attached to the outer face of the thylakoids, which lie beneath the cell plasma membrane. The phycobilisomes are poly-aggregates of phycobiliproteins (Fig. 3.3) which harvest light and direct its energy towards the RC.

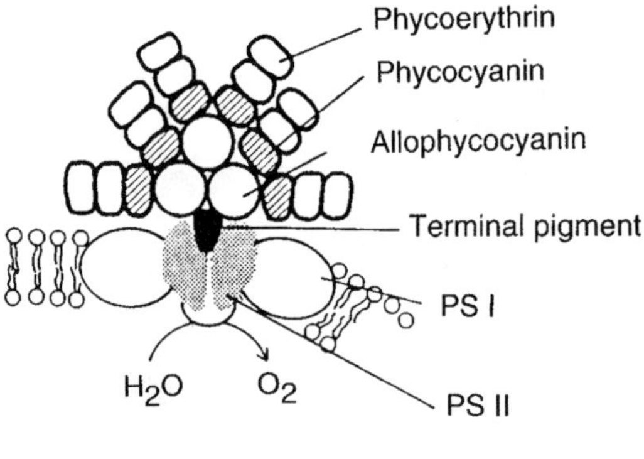

Figure 3.3
Schematic representation of the arrangement of the phycobilin- and Chl-containing pigment-protein complexes in the light-harvesting antennae (phycobilisomes) of red and blue-green algae (after Gantt, 1990).

3.2.3 *Function of pigment-protein complexes in Photosystems I and II*

The pigment-protein complexes are further organized into two photosystems, PS I and PS II, which span the thylakoid membrane from stroma to lumen (Fig. 3.2A). These photosystems are not homogeneously distributed in the thylakoids. PS II, more abundant than PS I, is found mainly in stacked thylakoid membranes, while PS I is found in stroma-exposed regions (Anderson and Andersson, 1988).

3.3 Role of chlorophylls in photosynthetic light transduction to chemical energy

The initial process of photosynthesis is the absorption of light energy (photons) by the Chls and carotenoids of the antenna of the pigment-protein complexes within the

LHC's of PS I and II, respectively (see section 3.3.1). The second process — charge separation — is the utilization of light energy in RC I and II to expel an electron from the Chl *a* of P700* or P680*, respectively, to form P700+• or P680+• cations. The electrons are then passed along an electron transport chain to generate chemical energy (ATP) and reducing power (NADPH) (see section 3.3.2 and Figs. 3.2A and 3.4).

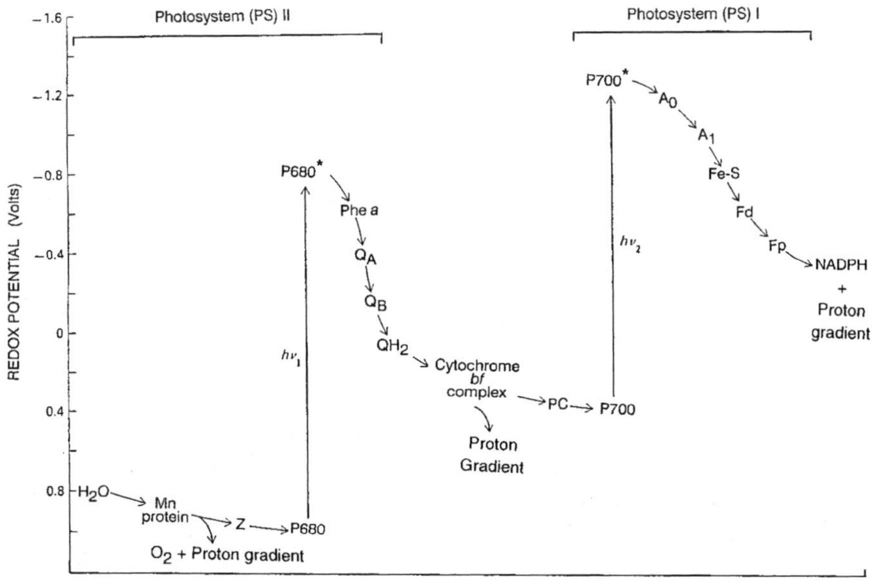

Figure 3.4
The Z scheme, showing PS I and II components in relation to their redox potential. Abbreviations: A_0 and A_1, protein-bound plastoquinones of PS I; Fd, ferredoxin; Fe-S, Riske's iron-sulphur protein; Fp, ferredoxin-NADP+ reductase (a flavoprotein); Mn protein, a manganese protein involved in H_2O splitting; Z, a tyrosine residue of D1 protein; Phe *a*, pheophytin *a*; PC-Cu^{2+}, plastocyanin; Q_A and Q_B, protein-bound plastoquinones of PS II; QH_2, reduced plastoquinone.

3.3.1 *Light-harvesting and energy transfer to Reaction Centres*

Light energy is absorbed by the LHCs of photosynthetic organisms, which contain combinations of Chls *a, b* and/or *c*, biliproteins or bacteriochlorophylls (BChls) together with specific carotenoids (see Table 2.3). The energy of the photon is rapidly and efficiently transferred by electromagnetic interaction to Chl *a* of CPs, which surround both RC I and II. The Chl *a*-binding CPs can also play a significant role in energy absorption. Finally, the energy is transferred to Chl *a* in P700 or P680 of RC I or II, respectively. The arrangement of LHC II and CPs about RC II in PS II is discussed by Boekema *et al.* (1995).

3.3.2 *Charge separation and photosynthetic electron transport*

Fig. 3.4 shows a modern version of the "Z" scheme (Blankenship and Prince, 1985), which explains how the energy of absorbed photons is used to oxidize water to form O_2 and protons, reduce $NADP^+$ to NADPH and provide chemical energy (ATP) from ADP and inorganic phosphate.

The Chl *a* pigment-protein complex of RC II (P680) is converted by light to the active form (P680*) and within *pico*seconds the P680$^{+\bullet}$ cation radical is formed and the expelled electron reduces pheophytin *a* (Phe *a)* in the RC. It is then passed sequentially to a protein-bound plastoquinone (Q_A), another mobile plastoquinone (Q_B) and the cytochrome *b/f* complex to reduce a divalent copper-protein, plastocyanin (PC-Cu^{2+}), to monovalent PC-Cu^+. Meanwhile, the strong oxidant P680$^{+\bullet}$, oxidizes water *via* a Mn-protein, thus reforming P680 and O_2. P700 of PS I is now activated by light to P700*, which also expels an electron to form P700$^{+\bullet}$. The electron sequentially reduces several electron carriers including ferredoxin, and ultimately forms NADPH. Reduced plastocyanin (PC-Cu^+) is oxidized to PC-Cu^{2+}, which donates an electron to regenerate P700 from P700$^+$. Thus, ATP and NADPH are generated and are available to reduce CO_2 to carbohydrate. The simpler bacterial forms of photosynthetic electron transport contain only one photosystem and therefore only one light-boosted reaction (see Foyer, 1984).

The electron transfer reactions of the Z-scheme are shown related to the structures of PS I and PS II in the thylakoid membranes in Fig. 3.2A (Anderson and Andersson, 1988) rather than to redox potential (cf. Fig. 3.4). A proton gradient with high proton density in the lumen is generated across the thylakoid membrane, driving a transmembrane ATP-synthase to form ATP from ADP and phosphate on the stroma side of the thylakoid membrane.

3.3.3 *Molecular composition of the pigment-protein complexes*

Exciting recently-published research is showing the precise molecular structures of some of the various Chl-carotenoid-protein complexes.

The first atomic model of the RC of *Rhodopseudomonas viridis* was obtained using X-ray crystallography (Deisenhofer and Michel, 1989). This RC is composed of 4 mol of BChl *b*, 2 mol of bacteriopheophytin *b*, 1 mol of menaquinone, 1 mol of ubiquinone, 1 mol of carotenoid (dihydroneurosporene) with 4 protein subunits and a *c*-type cytochrome containing 4 haem prosthetic groups. Two of the four BChl *b* molecules were held by protein bonding to form a tightly interacting dimer (i.e. 'special pair'). The dimer is thought to be the electron donor-acceptor of the bacterial reaction centre. P680 of RC II of plant and algal cells may also be a dimer of Chl *a*.

Structural studies on LHC II from PS II in higher-plant chloroplasts have been made by Kühlbrandt and Wang (1991) and Kühlbrandt *et al.* (1994). Kühlbrandt's group developed electron crystallography techniques for their two-dimensional crystals of LHC II and have derived atomic models (Fig. 3.2B).

Each LHC II is a trimer, and within each monomer (Fig. 3.2B) there are a minimum of six molecules of Chl *a,* six of Chl *b* and two xanthophylls (possibly lutein). The four α-helices are shown as spiral bands (Fig. 3.2B). The Chl *b* and carotenoid molecules are very close to Chl *a* molecules for rapid (1 *pico*second) energy transfer. Carotenoids also protect the LHC II from photodamage by quenching toxic singlet oxygen formed aerobically by photodynamic reactions involving triplet Chl (see section 3.8.4).

Recent image analysis studies have enabled Boekema *et al.* (1995) to suggest, for the first time, a tentative molecular structure of PS II isolated from spinach.

3.4 Naturally occurring chlorophylls

The Chls are a group of magnesium coordination complexes of cyclic tetrapyrroles. All Chls contain a fifth *iso*cyclic ring: this 5-ring macrocycle is often referred to as a phorbin (Moss, 1987). The Chls also have a 13^1-oxo substituent (Fig. 3.5) and a 13^2-methylcarboxylate group (Fig. 3.9), the latter not being present in phytoporphyrin, phytochlorin and some BChls. To be consistent with the IUPAC-IUB nomenclature and numbering system (Moss, 1987), most Chls are more correctly referred to as 13^2-methylcarboxylates of Mg-phytoporphyrins and Mg-phytochlorins (Moss, 1987).

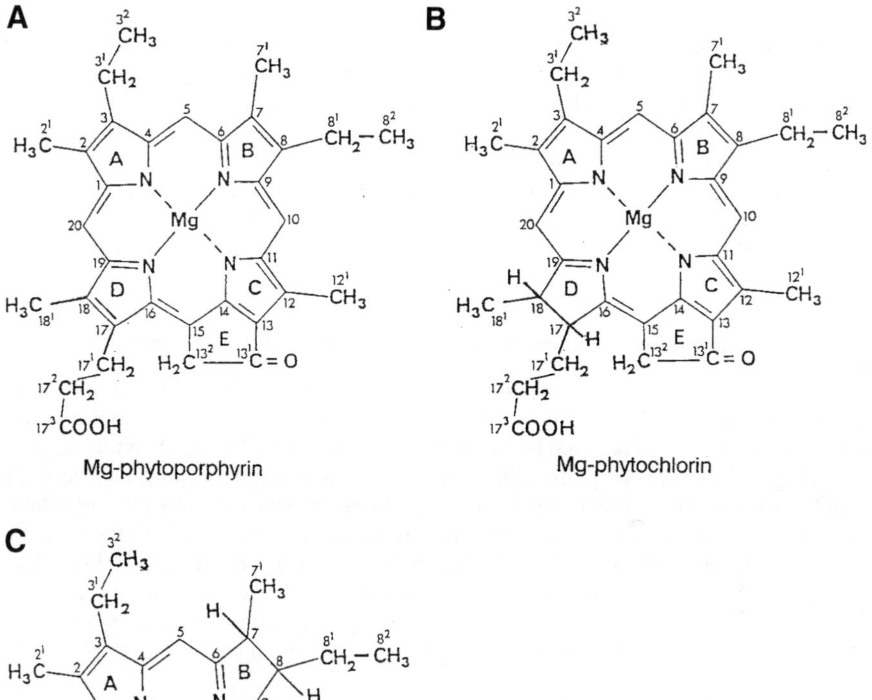

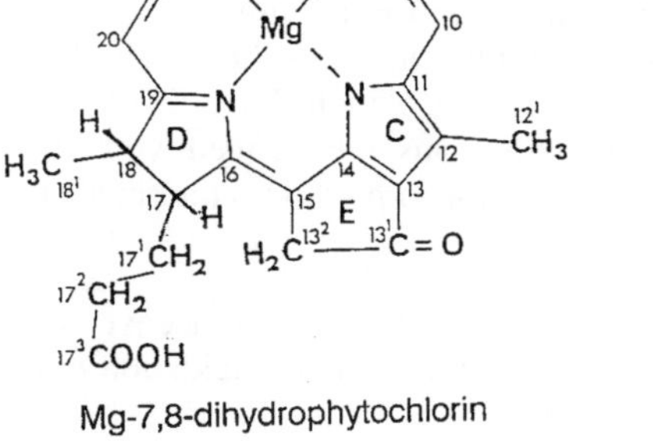

Figure 3.5
Tetrapyrrole nomenclature.
A, Mg-phytoporphyrin;
B, Mg-phytochlorin;
and **C,** Mg-7,8-dihydrophytochlorin.

The Chl *c* family (see section 3.5.6 and Fig. 3.10A), which is widely distributed in the chromophyte algae (Table 2.3), are derived from Mg-phytoporphyrin (Fig. 3.5A) and Chls *a* and *b* from Mg-phytochlorin (Fig. 3.5B). Some bacteriochlorophylls (BChls) are derived from Mg-phytochlorins and others from Mg-7,8-dihydrophytochlorin (Fig. 3.5C; cf. Scheer, 1991).

Many naturally occurring Chls possess a long-chain isoprenoid alcohol ester group (eg. phytyl, farnesyl, geranyl-geranyl) on the C-17 propionic acid side-chain; exceptions are chlorophyllides (Chlides) *a* and *b*, and most of the Chl *c* pigments. Chlides *a* and *b* possess a C-17 propionic acid side-chain but most of the Chl *c* family possess a C-17 acrylic acid side-chain. About 50 photosynthetically active Chls are now known. Their structures and distribution are described by Scheer (1991).

3.5 Biosynthesis of chlorophylls and phycobilins

The formation of haems and phycobilins from their simplest precursors involves about fifteen reaction sequences, while Chl formation from ALA involves well over twenty-five. It is remarkable that such long biosyntheses are achieved so efficiently and economically in healthy cells, with little or no wasteful accumulation of intermediates. This long sequence of enzymic reactions however provides many opportunities for perturbation of the pathway by gene mutation or by toxic chemicals, which may be enzyme-inducing agents or inhibitors. Accumulation of tetrapyrrolic intermediates in pigment extracts of algae might therefore indicate that they were harvested from waters polluted by such toxic agents. Tetrapyrrolic intermediates can also reduce algal numbers by causing cell death by photo-bleaching (see section 3.5.2).

3.5.1 *Significance and mechanisms of 5-aminolevulinate formation*

The first biosynthetic intermediate specifically committed to the formation of all naturally-occurring tetrapyrrole pigments is the C_5 compound, 5-aminolevulinic acid (ALA) (see Fig. 3.6). There are two pathways to form ALA. The ALA-synthase pathway (Fig. 3.6A) mainly operates in haem formation in animals and many bacteria. The C_5-pathway (Fig. 3.6B) operates in algae, higher plants and some photosynthetic and other bacteria. One group of algae, the euglenophytes, possess both the C_5-pathway (Beifuss *et al.* 1988) and the ALA-synthase pathway (Beale *et al.* 1981). Both pathways (cf. Granick, 1966; Porra and Grimme, 1974) are rate-limiting for tetrapyrrole biosynthesis. ALA biosynthesis is susceptible to interference by chemical agents that can increase ALA and tetrapyrrole formation many hundred-fold.

3.5.1.1 *ALA-synthase pathway*

The pyridoxal-phosphate-dependent ALA-synthase pathway requires glycine, activated by pyridoxal phosphate, and succinyl-CoA as substrates (Fig. 3.6A). The ALA-synthase pathway has been intensively studied since its isolation (Laver *et al.* 1958) because it is the sole pathway of ALA formation in humans. It is frequently involved in hereditary diseases (porphyrias) causing accumulation of porphyrin intermediates.

3.5.1.2 *C_5-pathway of ALA formation*

Evidence of ALA formation from glutamate by the C_5-pathway was discovered by Beale and Castelfranco (1973) and Beale *et al.* (1975) long after elucidation of the ALA-synthase route.

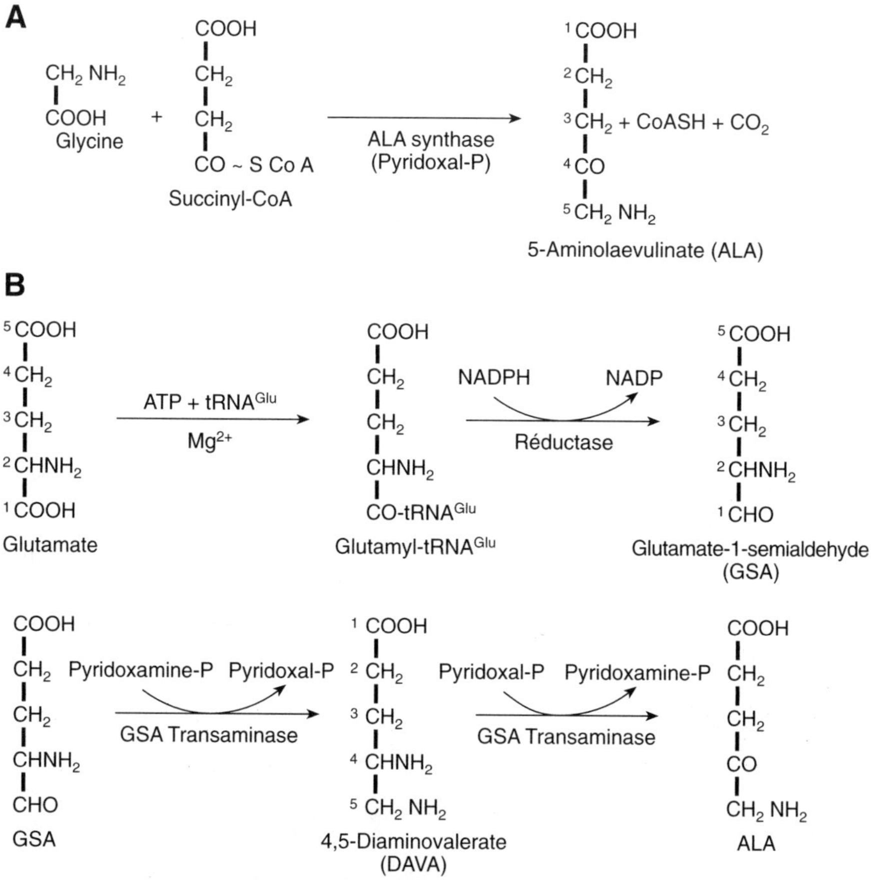

Figure 3.6
Two biosynthetic routes for 5-aminolevulinate (ALA) formation: A, ALA-synthase pathway; B, C_5-pathway

The C_5 pathway incorporates the intact C_5 skeleton of glutamate into ALA *via* glutamate-1-semialdehyde (GSA), 4,5-diaminovalerate (DAVA) (Friedmann *et al.* 1992; Kannangara *et al.* 1994) and involves steps that are dependent on both pyridoxamine-phosphate and pyridoxal-phosphate (Fig. 3.6B). The formation of Chls in green algae (Oh-hama *et al.* 1982) and higher plants (Porra *et al.* 1983) from ALA formed only by the C_5 pathway was unequivocally demonstrated using ^{13}C-NMR-spectroscopic methods.

3.5.2 *Conversion of ALA to protoporphyrin IX*

The sequence of reactions for converting ALA to protoporphyrin IX (Fig. 3.7) is common to the biosynthesis of Chls, haems and phycobilins. Although Fig. 3.7 is self-explanatory some points are worth noting.

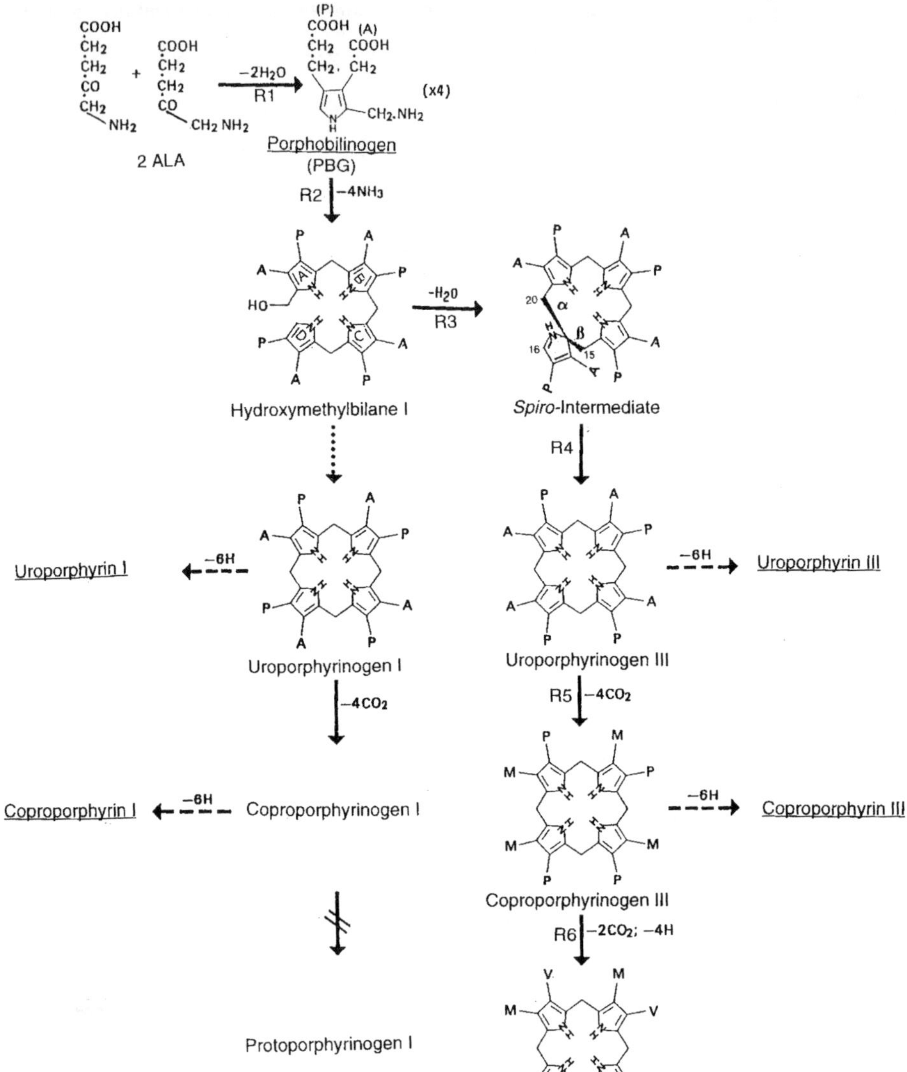

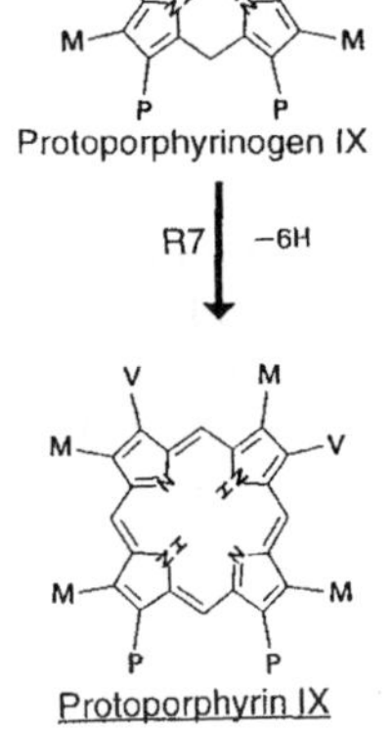

Figure 3.7
The formation of protoporphyrin IX from 5-aminolevulinate (ALA). Underlined compounds may accumulate during abnormal biosynthesis in porphyric conditions (see text). Dotted arrows represent non-enzymic reactions and dashed arrows represent non-enzymic oxidations of porphyrinogens (hexahydroporphyrins) to porphyrins. Uro'gen I can be converted by uro'gen III decarboxylase to copro'gen I, which is not a substrate for copro'gen III oxidative carboxylase. Side chains: A, acetic acid; M, methyl; V, vinyl; P, propionic acid. Reactions R1 to R7 are catalysed by the following enzymes: R1, ALA dehydratase; R2, PBG deaminase; R3 and R4, uro'gen III isomerase; R5, uro'gen III decarboxylase; R6, copro'gen III oxidative decarboxylase; R7, proto'gen IX oxidase.

The closure of the macrocycle to form uroporphyrinogen (uro'gen) III (see R4) occurs at C-20, by breaking bond ß of the *spiro*-intermediate and re-forming it between C-15 and C-16. Re-opening the macrocycle during conversion of protohaem to linear bile pigments and phycobilins (see section 3.5.4.2) occurs at C-5. In the biodegradation of Chls, ring opening generally occurs at C-5 (see sections 3.7.2.1 and 3.7.2.2) but in some special cases occurs at C-20 (see section 3.7.2.3).

The six individual decarboxylation reactions between octa-carboxylic uro'gen III and dicarboxylic protoporphyrin IX occur in a stepwise and ordered fashion (Akhtar, 1991) not shown in Fig. 3.7. The porphyrinogens are hexahydroporphyrins and are readily oxidized aerobically to corresponding octa- to di-carboxylic porphyrins, which may be found in pigment extracts. They are readily identified by high-performance thin-layer chromatography (HPTLC) (Seubert *et al.* 1979).

When ALA formation is non-limiting, uro'gen III isomerase (see R3 and R4) becomes limiting, allowing uro'gen I to be formed non-enzymically from accumulated hydroxymethylbilane I. Uro'gen I may be enzymically decarboxylated as far as copro'gen I. These are readily oxidized to uro- and copro-porphyrins I. If present in pigment extracts, they can be readily distinguished from uro- and coproporphyrins III by HPLC (Sieg and Doss, 1992).

Proto'gen IX oxidase (R7) is inhibited by many herbicides. Massive accumulation of proto'gen IX leads to non-enzymic formation of large amounts of protoporphyrin IX. This oxidation possibly occurs in the cytoplasm because the protoporphyrin IX formed does not readily re-enter the chloroplast where Chl biosynthesis is located. Such herbicides cause cell death in plants and algae, as the Chls are photobleached by toxic singlet oxygen arising from photo-destruction of protoporphyrin IX in the presence of oxygen (Becceril and Duke, 1989).

3.5.3 *Protoporphyrin IX: bifurcation point for iron and magnesium branches of tetrapyrrole formation*

Protoporphyrin IX is the bifurcation point for the formation of iron tetrapyrroles (haems) from which the open-chain phycobilins are derived (see section 3.5.4) and the magnesium tetrapyrroles (Chls) formed (see section 3.5.5).

3.5.4 *Iron branch: formation of phycobilins via protohaem IX*

Protohaem IX is important not only as a prosthetic group of haemoglobins and some cytochromes but also as an intermediate in the formation of phycobilins, the chromophores of the phycobiliproteins of cyanobacteria, red algae and cryptomonads.

3.5.4.1 *Formation of protohaem IX by ferrochelatase*

The insertion of an Fe^{2+} ion into protoporphyrin IX to form protohaem IX (Fig. 3.8) is catalysed by ferrochelatase found in animal mitochondria (Porra and Jones, 1963*a*), in cell membranes of bacteria (Porra & Jones, 1963*b*) and in the chloroplasts of higher-plants and algae (Jones, 1968; Porra and Lascelles, 1968).

3.5.4.2 *Formation of phycobilins*

The cleavage of protohaem IX in the red alga *Cyanidium caldarium* is achieved by a soluble oxygenase requiring O_2, NADPH and ascorbate to form biliverdin IXα (Beale and Cornejo, 1984*a*; Fig. 3.8). The macrocycle is cleaved between rings A and B;

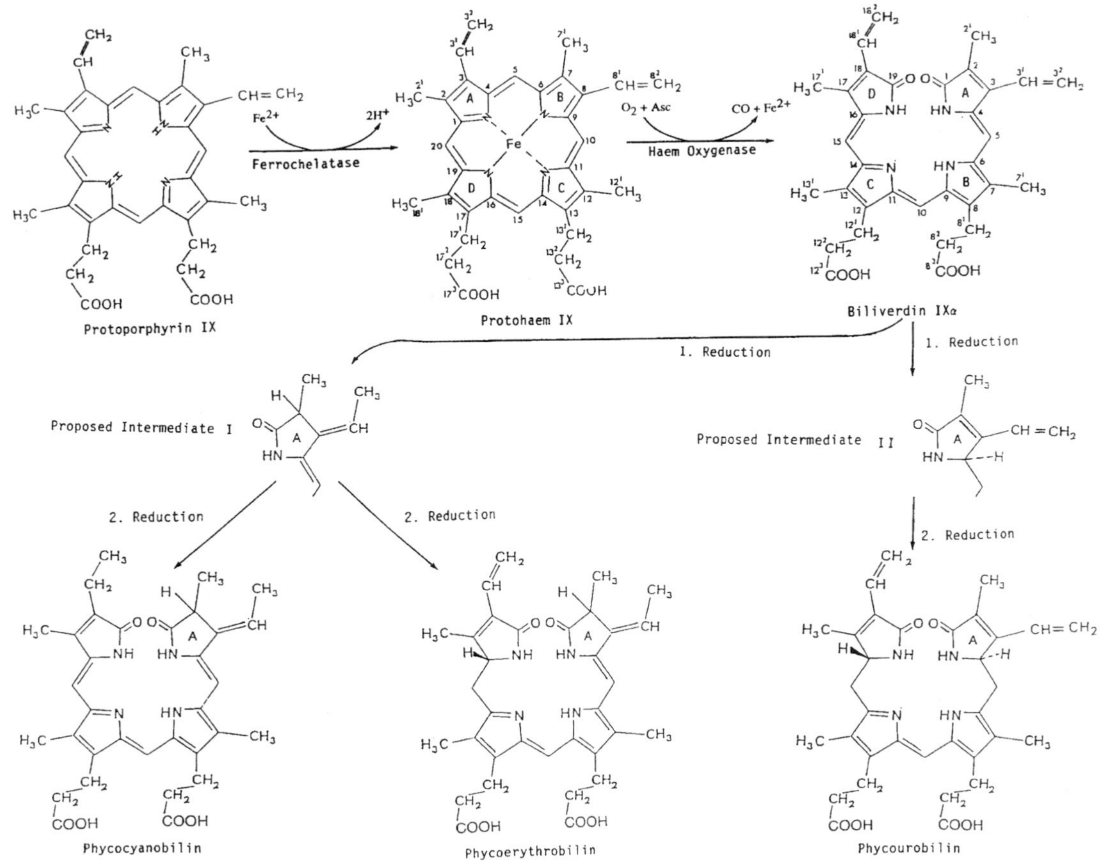

Figure 3.8
The conversion of protoporphyrin IX to protohaem IX, and a hypothetical biosynthetic route for phycobilin formation. Phycocyanobilin is the prosthetic group of C-phycocyanin and allophycocyanin, and is also present in R-phycocyanin together with phycoerythrobilin. Phycoerythrobilin is also the prosthetic group of C-phycoerythrin, which is present in R-phycoerythrin with phycourobilin. C and R represent Cyanophyceae and Rhodophyceae respectively.

note that the numbering and naming of the rings changes in the open-chain tetrapyrroles.

The precise biosynthetic pathway between biliverdin and the phycobilins, which are tetrahydro-biliverdins, is not yet fully understood (cf. Beale and Weinstein, 1991). A possible biosynthetic relationship is shown in Fig. 3.8. Proposed intermediates I and II formed by the first reduction are 2-hydro-3-ethylidene- and 4,5-dihydro-derivatives, respectively. If the second reduction reaction reduces the 18-vinyl or 15–16 double bond of intermediate I, then phycocyanobilin and phycoerythrobilin would be formed, respectively: reduction of the 15–16 double bond of intermediate II would produce phycourobilin.

The phycobilins form phycobiliproteins by covalent bonding through the cysteine residues of their *apo*protein by thioether adduct formation with the 3-ethylidene and 18-vinyl groups of the chromophore (Schoch *et al.* 1976; Beale and Cornejo, 1984*b*). Covalent linkage may also occur by esterification of a chromophore propionic acid sidechain to a serine-hydroxyl of the *apo*protein (Beale and Weinstein, 1991).

3.5.5 *Magnesium branch: conversion of protoporphyrin IX to chlorophylls a and b*

The conversion of protoporphyrin IX to Chls *a* and *b* requires 8 and 10 reaction steps, respectively, which are are shown in Fig. 3.9.

This figure is mainly self-explanatory but some further points are noted.

Mg^{2+}-chelatase (R8) unlike ferrochelatase (see Fig. 3.8) is ATP (energy) dependent.

Formation of *iso*cyclic ring E (Nasrulhaq-Boyce *et al.* 1987; Walker *et al.* 1989) leads to production of Mg-3,8-divinylphytoporphyrin-13^2-methylcarboxylate, known in the old Fischer nomenclature as Mg-2,4-divinylpheoporphyrin a_5 monomethylester (see Appendix A: in this monograph MgDVP serves to abbreviate both names).

Pollutants with metal-chelating properties could inhibit *iso*cyclic ring formation leading to accumulation of protoporphyrin IX-monomethylester and its Mg chelate (cf. Nashrulhaq-Boyce *et al.* 1987).

Rebeiz *et al.* (1983) have demonstrated in higher plants a natural heterogeneity of '3,8-di-vinyl' and '3-vinyl-8-ethyl' forms of Chls *a* and *b*. The divinyl forms are the minor component but can reach 20% of the total Chl. In the prochlorophyte alga, *Prochlorococcus marinus*, the Chls *a* and *b* are mainly in the divinyl form (see Part IV Data sheets). This suggests that reaction R13 (the 8-vinyl reductase step) is missing or defective in the prochlorophyte and, under certain conditions, limiting in higher plants.

Phytylation (R15) is catalysed by the Chl synthase complex after ring E formation and ring D reduction, since PChlide is not used as a substrate (Benz and Rüdiger, 1981). This may explain why most Chl *c* pigments (phytoporphyrins) are not phytylated (see Fig. 3.10A).

Chl synthase catalyses the esterification of Chlide *a* by geranylgeranylpyrophosphate. A reductase also catalyses the stepwise reduction of three of the four geranylgeranyl double bonds to produce a phytyl ester with a single double-bond between C-2 and C-3.

Chl *b* possesses a formyl group at C-7 where Chl *a* has a methyl group. The formyl group is derived solely by oxygenase activity (see Fig. 3.9, R16 and R17) as shown by the formation of a highly enriched ^{18}O-labelled formyl group (Porra *et al.*

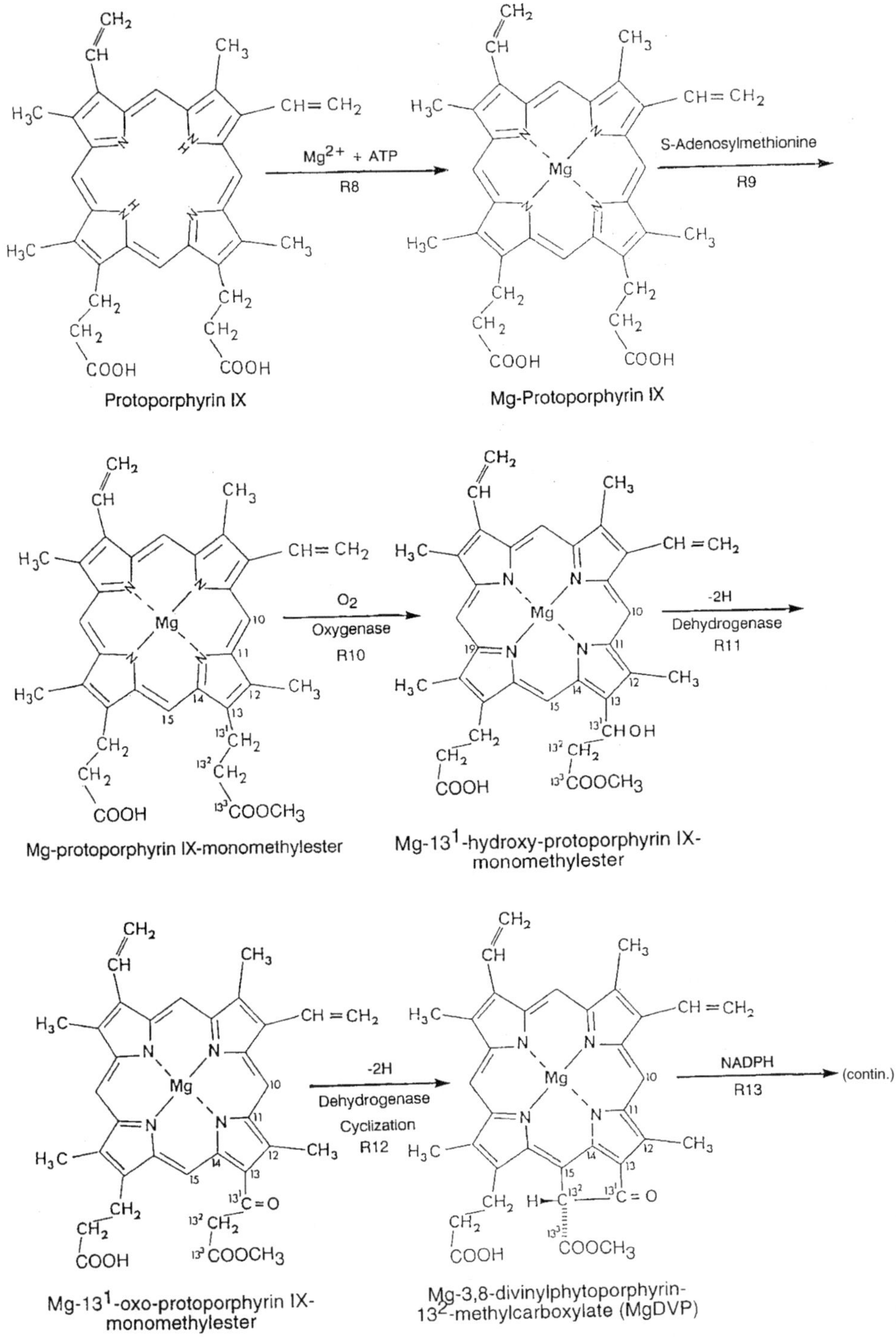
Mg2+ + ATP
R8
S-Adenosylmethionine
R9
Protoporphyrin IX
Mg-Protoporphyrin IX
O2
Oxygenase
R10
-2H
Dehydrogenase
R11
Mg-protoporphyrin IX-monomethylester
Mg-13¹-hydroxy-protoporphyrin IX-monomethylester
-2H
Dehydrogenase
Cyclization
R12
NADPH
R13
(contin.)
Mg-13¹-oxo-protoporphyrin IX-monomethylester
Mg-3,8-divinylphytoporphyrin-13²-methylcarboxylate (MgDVP)

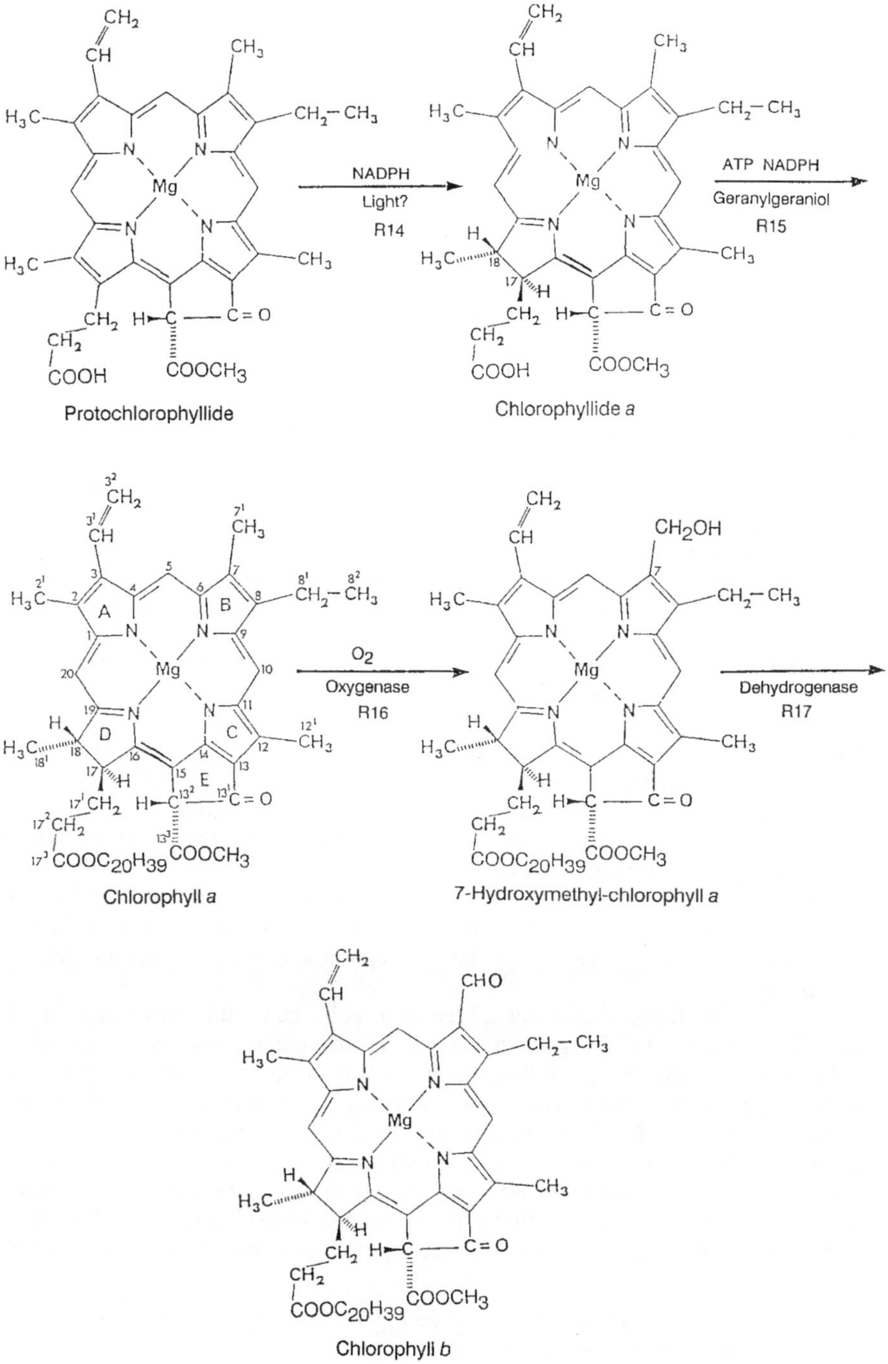

Figure 3.9
The biosynthetic conversion of protoporphyrin IX to Chls *a* and *b*. The nomenclature for the chlorophyll precursors with a five-ring macrocycle is described in section 3.4. Reactions R8 to R17 are catalysed by the following enzymes: R8, Mg-chelatase; R8, Mg-proto IX methyltransferase; R10, R11 and R12, Mg-proto IX monomethylester (oxidative) cyclase; R13, 8-vinyl-reductase; R14, protochlorophyllide reductase; R15, Chl synthetase; R16, Chl *a* oxygenase; R17, 7-hydroxymethyl-Chl *a* dehydrogenase.

1994). This suggests that Chl *b* is formed in a hydrophobic environment preventing isotope exchange to H_2O *via gem*-diol formation. Whether Chl *a*, Chlide *a* or PChlide are substrates for formyl group formation by the oxygenase is still disputed.

3.5.6 *Formation of chlorophyll c pigments*

The chemical structures of three of the six known Chl *c* pigments are fully confirmed: Chls c_1 and c_2 (Dougherty *et al.* 1970), and Chl c_3 (Fookes and Jeffrey, 1989; see Part IV Data sheets). Two further Chl *c* pigments have been isolated from prasinophytes: one, Chl $c_{CS\text{-}170}$, has the tentative structure shown in Fig. 3.10A and the Part IV Data sheets (Fookes and Jeffrey, unpublished); the second, originally suggested by Ricketts (1966) to be MgDVP, has now been tentatively confirmed by Fookes and Jeffrey (unpublished). The structure of a sixth pigment from prymnesiophytes was postulated as a phytylated Chl *c*-like pigment from absorption spectroscopy data, HPLC and ester hydrolysis (Nelson and Wakeham, 1989). Spectrocopic and chromatographic evidence from a variety of sources suggests that more Chl *c*-like pigments exist which need to be further characterized.

From theoretical considerations, the pathway to the Chl *c* group of pigments (Fig. 3.10A) may branch from the Chl *a* biosynthetic route (Fig. 3.9) at MgDVP, or possibly its 3-vinyl-8-ethyl derivative (PChlide). This may account for the predominance of 3,8-divinyl compounds, which also lack the long-chain alcohol ester (see Fig. 3.10A) that is normally introduced at the Chlide stage (Fig. 3.9, R15). The presence of esterified Chl *c* in prymnesiophytes suggests that the esterifying Chl synthase in these algae can permit minor esterification of Chl c_2 (see Table 2.3).

A hypothetical scheme for the formation of the six Chl *c* pigments is shown in Fig. 3.10A. The relationship between Chls c_1, c_2, MgDVP and protochlorophyllide was first proposed by Beale (1984).

Chls c_1, c_2 and c_3 have a 17-acrylate residue on ring D rather than the 17-propionate characteristic of Chls *a* and *b* (Fig. 3.10A); in contrast, Chl $c_{CS\text{-}170}$ and MgDVP (found in some prasinophytes) have the 17-propionate residue (Fookes and Jeffrey, unpublished).

Fig. 3.10B proposes a theoretical reaction sequence for the formation of the 7-methylcarboxylate of Chl c_3 and Chls $c_{CS\text{-}170}$. From studies of oxygenase activity in Chl *b* formation (Porra *et al.* 1994), it appears that the formation of this group may involve a similar oxygenase in a water-containing environment rather than the hydrophobic *milieu* needed for Chl *b* formation (see section 3.5.5). The 7-methyl substituent group could be oxidized to a 7-formyl group (see Fig. 3.9: R16 and R17) but then hydrated to a 7^1-*gem*-diol which on oxidation would form a 7-carboxylate. This carboxylate could then be methylated by a S-adenosylmethionine-dependent methyltransferase, as is the 13^3-carboxyl of Mg-protoporphyrin IX (cf. Fig. 3.9: R9).

3.5.7 *Chlorophyll and carotenoid incorporation into photosynthetic pigment-protein complexes*

The biosynthesis of the *apo*proteins and pigment chromophores in the appropriate ratio must precede formation of the pigment-protein complexes. The process is not well understood: but there must be many regulatory steps for modulating Chl and protein biosynthesis. For example, light regulates reduction of PChlide to Chlide to control Chl biosynthesis. *Apo*protein formation may also be controlled by modulat-

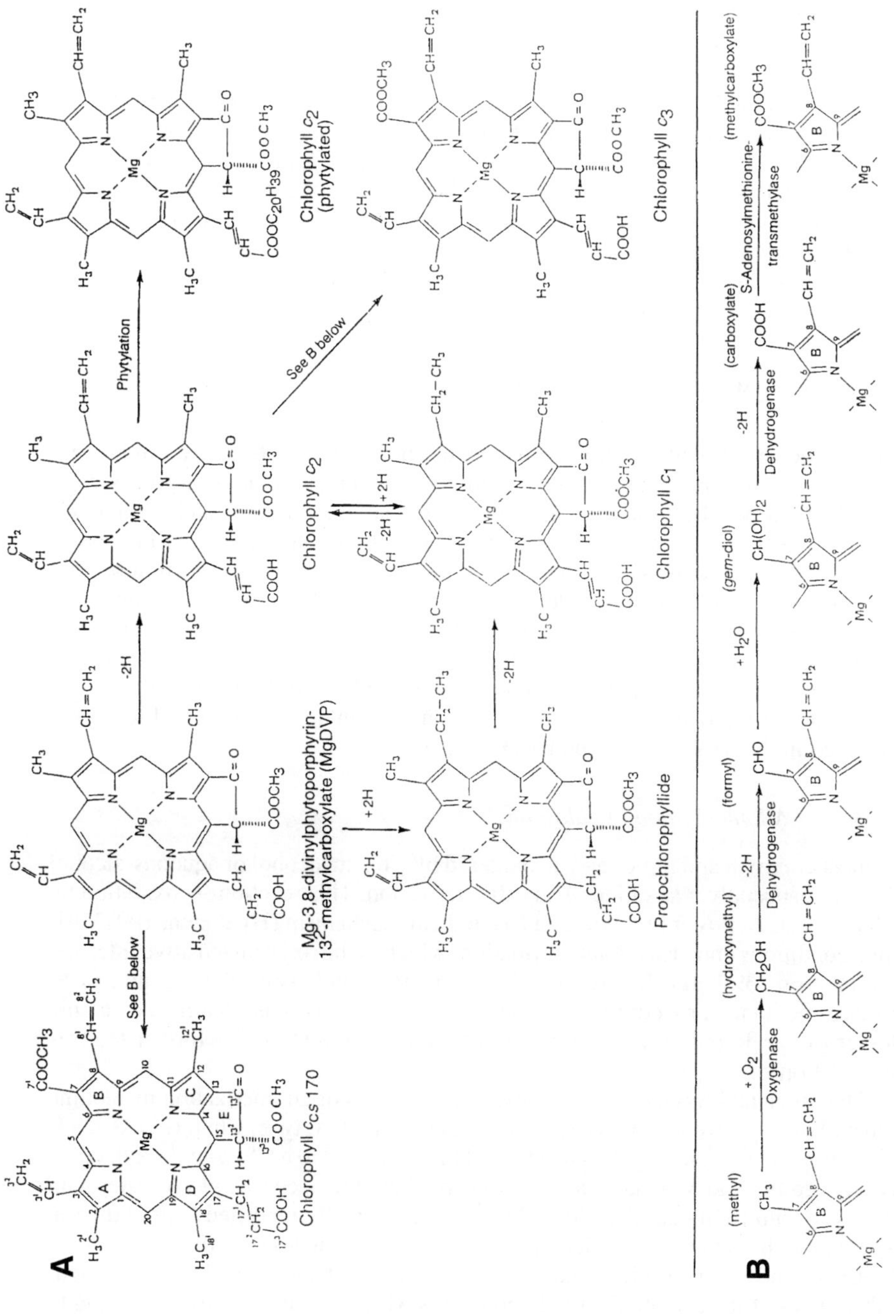

Figure 3.10
Formation of the various members of the Chl *c* family. A, Possible biosynthetic route from Mg-DVP; B, a hypothetical pathway for formation of the 3-methylcarboxylate group of Chl c_3 and Chl $c_{CS\text{-}170}$.

ing the formation of ribonucleic acid (mRNA) required for LHCP II formation (transcriptional control), or by post-transcriptional processes such as specific degradation of mRNAs or, perhaps, by control of protein formation from the mRNA (i.e. translational control).

The formation of pigment-protein complexes has been studied by dis-assembling isolated LHCP II, the major antenna protein of LHC II (Plumley and Schmidt, 1987). After heat-unfolding of the proteins in detergent-sucrose solutions, addition of Chls *a* and *b*, lutein, violaxanthin and neoxanthin in the appropriate ratio reconstituted the LHCP II. Chls are essential, while the omission of carotenoids only diminishes the amount of reconstitution. In more recent experiments, Paulsen *et al.* (1990) took LHCP II, over-expressed in a transgenic *E. coli*, and obtained reconstitution in the presence of appropriate proportions of Chls and xanthophylls.

3.6 Chemically altered chlorophylls

Naturally-occurring Chl-protein complexes in photosynthetic tissues are relatively stable but may be changed by chemical, photochemical and biological processes.

When extracted into an organic solvent, the soluble protein- and carotenoid-free Chl molecules become more labile to chemical and photochemical processes. Many altered Chl products may therefore arise as a result of allomerization (oxidation reactions), epimerization, demetallation, dephytylation, halogenation and decarboxymethylation. Chl lability dictates that extraction, assay and chromatography be performed rapidly under subdued or 'safe' green lights, and that acids and alkalis be avoided (see Chapter 7). When chlorophyll samples are stored, the solvent should be removed under reduced pressure at low temperature and low light and stored in the dark at -15°C under O_2-free N_2.

3.6.1 *Allomerization of chlorophylls*

The most common and rapid fate encountered by Chls in alcohol or aqueous-alcohol solutions, especially at alkaline pH, is allomerization. This reaction — oxidation by triplet oxygen — possibly involves a free radical mechanism (Hynninen, 1991). Alkaline conditions enhance enolate formation, which is the oxygen-sensitive intermediate (see Fig. 3.11, **II**). Extraction and chromatography, using the recommended buffered-methanol procedures (see Chapters 9 and 12), cause barely detectable allomerization. Rapid processing of samples in dim light at pH 7.1 further reduces the oxidation risk.

Hynninen and Assandri (1973) describe two types of allomerization in alkaline methanol in the presence of oxygen (Fig. 3.11). In the first type, Chl *a* (Fig. 3.11; **I**) is converted to 13^2-hydroxy-Chl *a* (**IV**) and in the second to a Chl *a*-hydroxylactone (**VII**) *via* compound **V**. In addition, alkaline methanol replaces the phytyl ester with a methyl group (transesterification). All of the chemically modified Chls shown in Fig. 3.11 can be separated and identified by HPLC (Schaber *et al.* 1984).

Even under anaerobic conditions, alkaline methanol should be avoided; methanolysis of the enolate double-bond occurs which ruptures the *iso*cyclic ring to form a magnesium rhodochlorin (**VI**) with methyl esters of the formic and acetic acid side-chains at C-13 and -15 respectively (Porra, 1991).

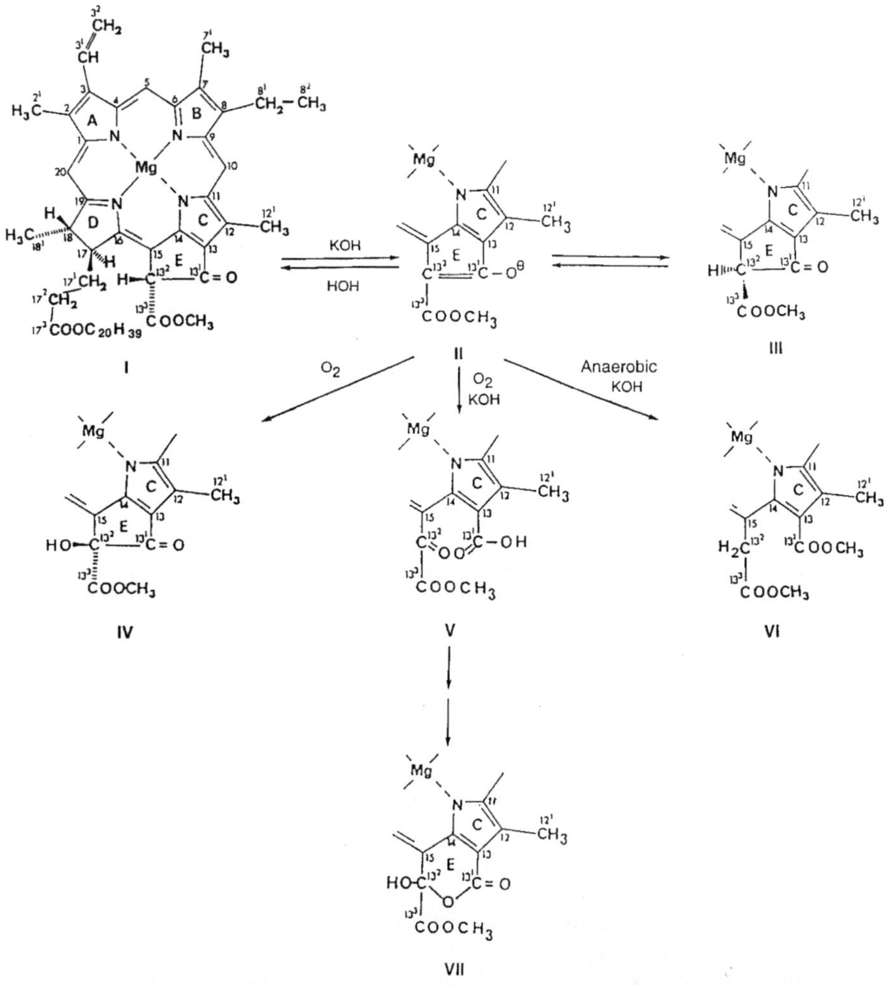

Figure 3.11
A reaction sequence for allomerization and epimerization of Chl *a*. **I**, Chl *a*; **II,** enolate anion of Chl *a*; **III**, Chl *a*', the 13^2-epimer of Chl *a*; **IV**, 13^2-hydroxy-Chl *a*; **V,** Mg-rhodochlorin *a*-13-formic acid-15-methylglycollate; **VI,** Rhodochlorin *a*; **VII**, Chl *a*-hydroxylactone. The numbering of the carbon atoms 13^2 and 13^3 in structures **V**, **VI** and **VII** reflects their biosynthetic origin; correct chemical numbering would be 15^1 and 15^2.

3.6.2 *Epimerization at C-13²*

The Chl enolate ion (Fig. 3.11, **II**) is also an intermediate in the conversion of Chls to their 13^2 epimers (Fig. 3.11, **III**). The epimers, Chls *a*' and *b*', were discovered as fast-running bands just in front of Chls *a* and *b*, respectively, on sucrose columns (Strain and Manning, 1942). The rate of formation of the epimers increases with the basicity of the solvent. Epimerization during chromatography on silica gel, an adsorbant normally not recommended for Chl chromatography, can be prevented by coating with aspartic or ascorbic acids: the latter also prevents oxidation (Brockmann and Risch, 1991). Both epimerization and oxidation are reduced during HPLC on silica gel at low temperature (5–10°C) and increased resolution of pigments may also occur (Watanabe *et al.* 1984; Struck and Scheer, 1990).

3.6.3 *Demetallation reactions*

Magnesium is removed from Chls dissolved in organic solvents in the presence of dilute acids and in high light intensities. They form grey-pink pheophytins that like Chls, absorb in the red region of the spectrum but have a stronger absorption near 530 nm (see Part IV Data sheets). Trace amounts of pheophytins are present in all photosynthetic tissues because they play a role in the Z-scheme of photosynthesis (see section 3.3.2). They can also arise from the postulated activity of the demetallase enzymes.

3.6.4 *Dephytylation*

Chlorophyllides (Chlides) are intermediates in Chl biosynthesis but may also arise by dephytylation of Chls *a* and *b* by the ubiquitous catabolic enzyme, chlorophyllase. Long-chain fatty alcohol esters are transesterified at the C-17 propionic acid to a methyl or ethyl ester when treated with alkaline methanol or ethanol, respectively. Under anaerobic conditions, alkaline alcohols also cleave the *iso*cyclic five-membered ring E non-oxidatively to form a magnesium-rhodochlorin (see section 3.6.1 and Fig. 3.11, **VI**). If oxygen is present during this transesterification, ring E is oxidatively cleaved and recyclized about an oxygen atom to form a six-membered hydroxy-lactone ring (**VII**) *via* a Mg-rhodochlorin derivative (**V**) (see Fig. 3.11).

3.6.5 *Chlorination of chlorophylls*

Chlorination of phytochlorins, such as Chl *a* at the C-20 methine bridge, is now well known (Hynninen and Lötjenen, 1981; Scheer *et al.* 1986; Senge *et al.* 1988).

Chl *a* was chlorinated enzymically by chloroperoxidase in the presence of KCl in phosphate buffer at pH 5.0 (Senge *et al.* 1988) which shifted the red absorption peaks by 7-15 nm to higher wavelengths. Formation of 20-chloro-Chl *a* can also occur during extraction of Chl *a* (Kobayashi *et al.* 1988) from plant tissues, possibly by oxidation of endogenous chloride ions by a chloroplast-stromal peroxidase. A Chl extracted from *Scenedesmus* and named Chl RCI was confused with the prosthetic group of P700 because of its red shifted absorption: it later proved to be the artefact 13^2-hydroxy-20-chloro-Chl *a* (Senge *et al.* 1988).

Chlorination can also occur during silica gel chromatography with the formation of 20-chloro-13^2-hydroxy-Chl *a* (Senge and Senger, 1988; Senge *et al.* 1988). For this reason, silica gel is not recommended for chlorophyll chromatography unless the free negatively-charged silanol groups are 'capped' by treatment with trimethylchlorosilane (Shioi, 1991) to prevent catalysis of chlorination and oxidation reactions (see Chapters 7 and 15). Alternatively such reactions can be minimized by HPLC at low temperatures (Watanabe *et al.* 1984; Struck and Scheer, 1990).

3.6.6 *Decarboxymethylation at C-13²*

The C-13²-carboxymethyl group can be removed by heating anaerobically in pyridine at 100°C for 24 h to form pyrochlorophylls. Decarboxymethylation can also occur enzymically (see section 3.7).

3.7 BIODEGRADATION OF CHLOROPHYLLS

Global Chl degradation is about 10^9 tonnes annually of which about 75% occurs in oceans, lakes and rivers (see section 3.2). Chl derivatives are formed by grazing and digestion by terrestrial and aquatic herbivores, and also by cell senescence and microbial degradation. Until recently, the disappearance of so much Chl without the formation of detectable intermediates or end products was regarded as an enigma (Hendry *et al.* 1987). Useful reviews of Chl biodegradation include those of Simpson *et al.* (1976), Kufner (1980), Rüdiger and Schoch (1989) and Brown *et al.* (1991).

The products of biological degradation are classified here into two categories:

- peripherally modified cyclic tetrapyrroles (see section 3.7.1), and
- linear tetrapyrroles from oxidative cleavage of Chl macrocycles (see section 3.7.2).

3.7.1 *Cyclic tetrapyrrolic catabolites*

In many Chl degradation products, the tetrapyrrole macrocycle remains intact. Enzymes implicated in such degradative processes catalyze pheophytinization, phytyl ester hydrolysis, decarboxymethylation and allomerization. In contrast to the peripheral substituents, the macrocycle is remarkably stable, particularly when oxygen and light are excluded. Some metal complexes of phorbins (five-ring macrocycle; see section 3.4) have survived chemical modification over a long geochemical time scale (diagenesis), losing only the phytyl ester and thus surviving as 'molecular fossils' in a wide range of sediments dating back to the Precambrian period (Callot, 1991). This field of petroporphyrin geochemistry, based on the work of Treibs in the early 1930s, has exploded recently with the availability of sensitive analytical methods for separation and structure determination of petroporphyrins and petrochlorins. While some of these compounds (over 70 are now known; Callot, 1991) can be related to known Chl precursors (e.g., Chls *b*, *c*), others remain an enigma, possibly indicating unknown Chl precursors still awaiting discovery in present-day aquatic ecosystems.

Chlorophyllase, which readily hydrolyses the phytyl ester bond of Chls under physiological conditions, was first discovered by Willstätter and Stoll (1910), and it has been detected in higher plants (Mayer 1930) and phytoplankton (Barrett and Jeffrey, 1964; Jeffrey and Hallegraeff, 1987).

Magnesium dechelatase catalyzes the removal of the magnesium ion (pheophytinization) from all Chls (or Chlides) at neutral pH (Owens and Falkowski, 1982; Ziegler *et al.* 1988).

Pyropheophytins and pyropheophorbides, which arise by removal of the 13²-methylcarboxylate group on *iso*cyclic ring E, have been found in photosynthetic organisms (Ziegler *et al.* 1988; Schoch *et al.* 1981), in marine sediments (Keely and Maxwell, 1991; Sun *et al.* 1993), and during grazing (Hawkins *et al.* 1986).

Phytoporphyrin (cf Fig. 3.4A) has been isolated from the intestines and faeces of some domestic herbivores (Fischer and Stern, 1943), which indicates that, during

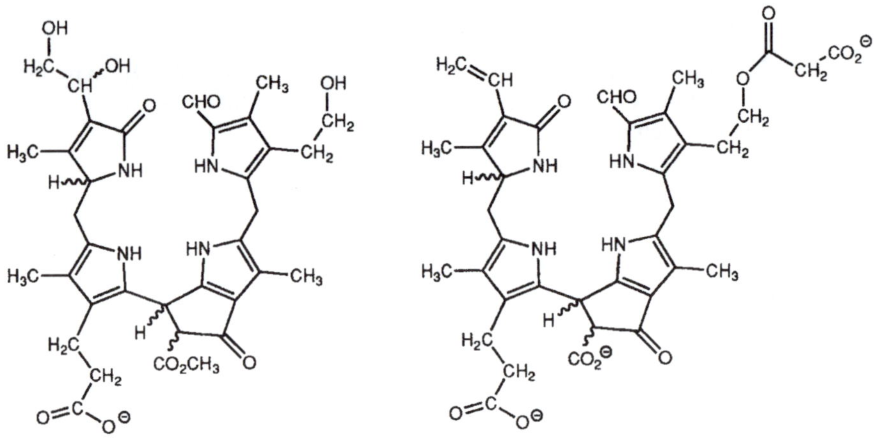

I II

Figure 3.12
Two colourless Chl catabolites, cleaved at C-5, isolated from higher plants I, from *Hordeum vulgare*; II, from *Brassica napus*.

digestion, Chl *a* has been subject to demetallation, dephytylation, reduction of the 3-vinyl substituent, decarboxymethylation and oxidation of ring D. Similar processes occur in marine herbivores (Hawkins *et al.* 1986).

Chl oxidase transforms Chls to their 13^2-hydroxy epimers (Schoch *et al.* 1984).

3.7.2 *Linear tetrapyrrolic catabolites*

Biodegradation of Chls to linear tetrapyrroles in algae and higher plants proceeds mostly by oxidative cleavage of the Chl macrocycle between C-4 and C-5 with the C-5 bridge carbon retained as an aldehyde function. This contrasts with the loss of C-5 as carbon monoxide during phycobilin formation from protohaem IX (see section 3.5.4.2 and compare Figs. 3.8 and 3.12).

Linear tetrapyrroles derived from Chls have been found in higher plants, freshwater chlorophytes and marine organisms. These linear tetrapyrrole catabolites of Chls retain the fifth *iso*cyclic ring. They were the first linear tetrapyrrolic catabolites of Chl to be characterized. No linear pyrrole fragments smaller than a tetrapyrrole have been detected.

3.7.2.1 *Catabolites from higher plants*

Two linear tetrapyrroles (**I**) and (**II**) (Fig. 3.12) derived from the Chl *a* of higher plants have been identified by Kräutler *et al.* (1991) and Mühlecker *et al.* (1993), respectively. In both catabolites, the oxidative cleavage of the Chl macrocycle occurred between C-4 and C-5, with retention of C-5 as a formyl function on ring B.

These colourless, water-soluble catabolites have a UV absorption maximum at 320 nm which has been assigned to the formyl-pyrrole moiety. Exposure to air and light during TLC chromatography converts the colourless catabolites to rust-coloured derivatives.

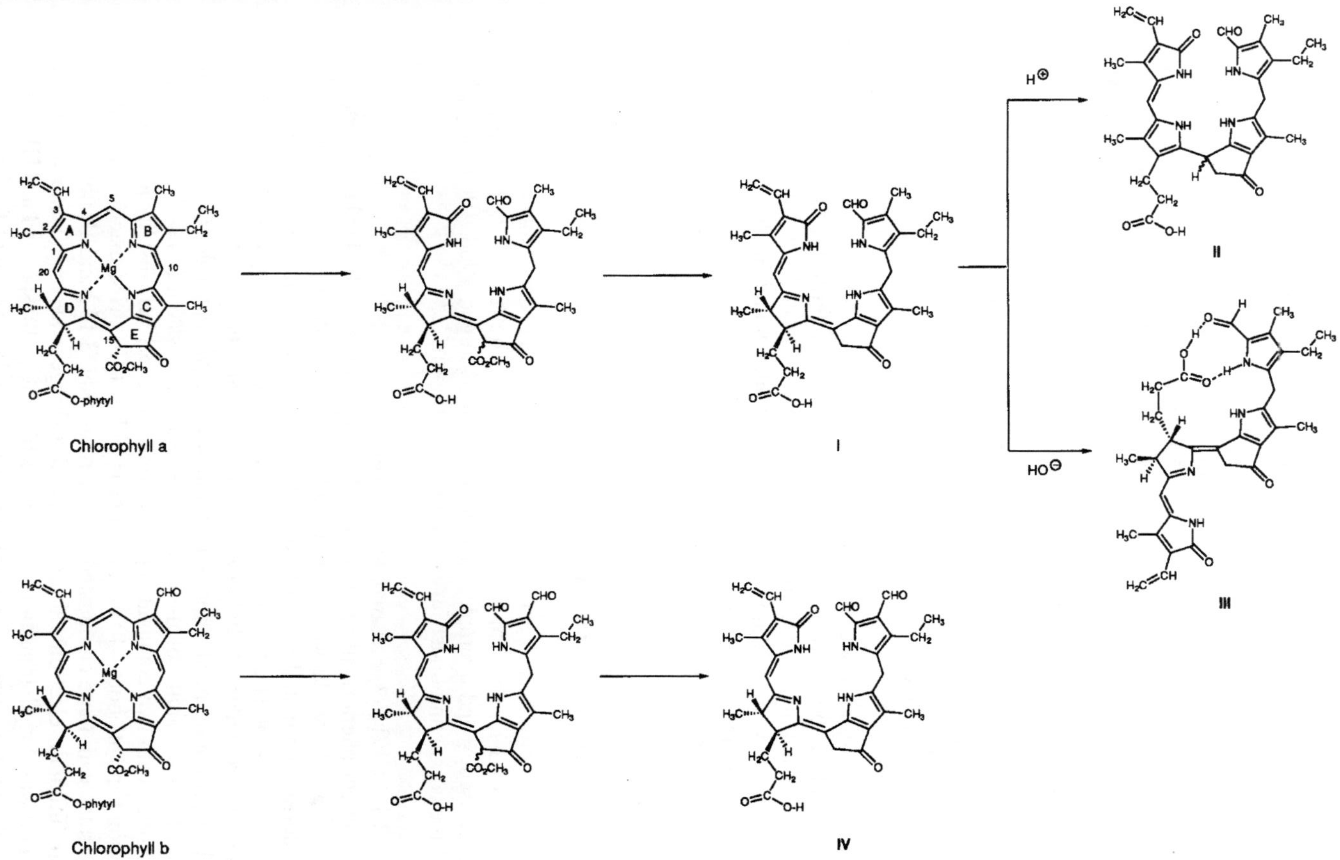

Figure 3.13
A tentative pathway for the degradation of Chls *a* and *b* suggested by the linear tetrapyrrolic catabolites, cleaved at C-5, and isolated from the freshwater alga, *Chlorella protothecoides*. **I** is converted to **II** and **III** by acid and alkaline treatment, respectively, and probably do not represent any enzymic reactions.

3.7.2.2 *Catabolites from freshwater green algae*

The microscopic green alga *Chlorella protothecoides* when grown in media rich in glucose but poor in nitrogen, may excrete several red pigments into the culture medium. These linear tetrapyrrole derivatives of Chls *a* and *b* (Fig. 3.13) are formed by oxidative cleavage between C-4 and C-5, with the 5-formyl group retained on ring B (Engel *et al.* 1991; Iturraspe *et al.* 1993, 1994) as observed in the higher-plant catabolites (see section 3.7.2.1). The red pigment (**I**) from Chl *a* is unstable: it isomerizes under acidic or basic conditions, leading to either yellow (**II**) or red (**III**) pigments, respectively. Being free carboxylic acids, these linear tetrapyrroles are readily soluble in aqueous alkali from which they may be re-extracted into suitable organic solvents by acidification. The purified pigments are readily soluble in alcohols but not in chloroform (Engel *et al.* 1991). Pigment **IV** is the analogous pigment to pigment **I** but derived from Chl *b*.

The spectra I, II and III (Fig. 3.14) are the spectra of the methyl esters of compounds **I, II** and **III** shown in Fig. 3.13 and discussed in section 3.7.3.1.

3.7.2.3 *Unusual Chl catabolites from marine organisms*

Two linear Chl catabolites (Fig. 3.15) — dinoflagellate luciferin (**I**) and krill luciferin (**II**) — are unusual in that they arise from cleavage of Chls at C-20 (Nakamura *et al.* 1988, 1989). They are bioluminescent Chl catabolites found in *Pyrocystis lunula* (dinoflagellates) and *Euphausia pacifica* (krill). As *Euphausia pacifica* is a heterotroph, the linear tetrapyrroles found in these animals must arise from the Chl content of the algal prey.

3.7.3 *Methods of identification of linear catabolites*

NMR-spectroscopy, mass spectrometry and X-ray structure determination have been widely used to determine chlorophyll catabolite structure (Engel *et al.* 1993). UV/VIS spectroscopy and a simple chromic acid test (Rüdiger, 1968, 1969) may also show if the compound under investigation is a tetrapyrrole.

3.7.3.1 *UV/VIS spectroscopic properties*

Native and peripherally altered Chls with an intact macrocycle display the characteristic strong Soret band at about 400 nm. Soret intensity is weaker in the phytochlorins and metallo-phytochlorins than in the porphyrins. All metal-free cyclic tetrapyrroles (other than porphyrinogens) and their Mg-complexes exhibit strong red fluorescence under UV light (see Part IV Data Sheets).

The position of the absorption maxima of linear *oligo*pyrroles (Fig. 3.14) is influenced by the 'length' of the conjugated -electron system (Falk, 1989). Linear tetrapyrroles with two conjugated pyrrole rings (Fig. 3.13 **II**) display just one main absorption band at about 440 nm and appear yellow (Fig. 3.14; spectrum **II**). Linear tetrapyrroles with three conjugated pyrrole rings (Fig. 3.13 **I** and **III**) bring about a 100 nm shift of the electronic absorption bands to longer wavelengths (about 540 nm) and a short wavelength transition becomes visible (Fig. 3.14; spectra **I** and **III**); therefore, solutions of pigments **I** and **III** appear reddish. Possession of four conjugated pyrrole rings would cause the long wavelength absorption band to be shifted bathochromically by 100 nm to 600-650 nm and such pigment solutions would appear blueish (e.g., phycocyanin).

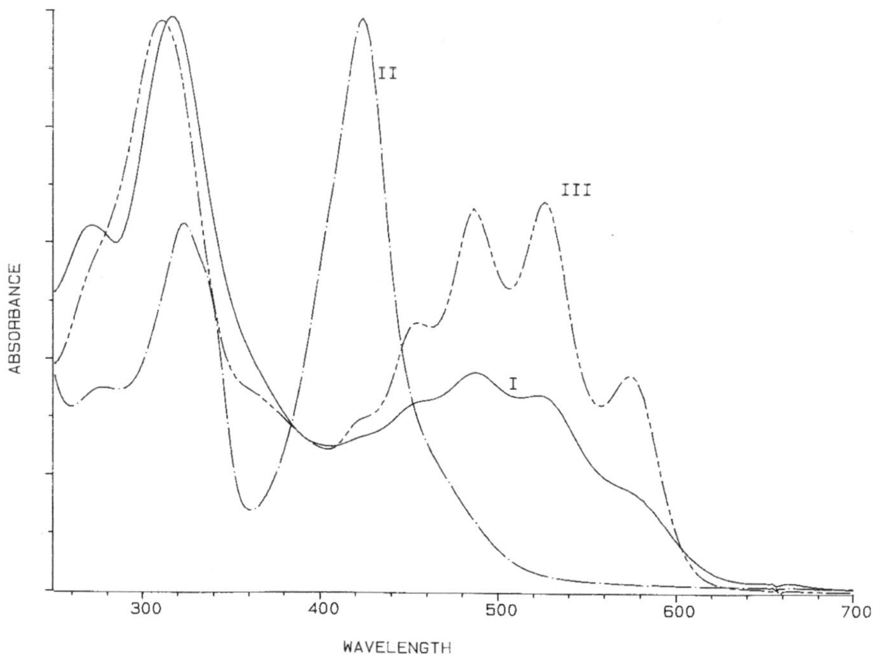

Figure 3.14
Absorption spectra of the methyl esters of algal pigments **I**, **II** and **III**, as designated in Fig 3.13, dissolved in chloroform.

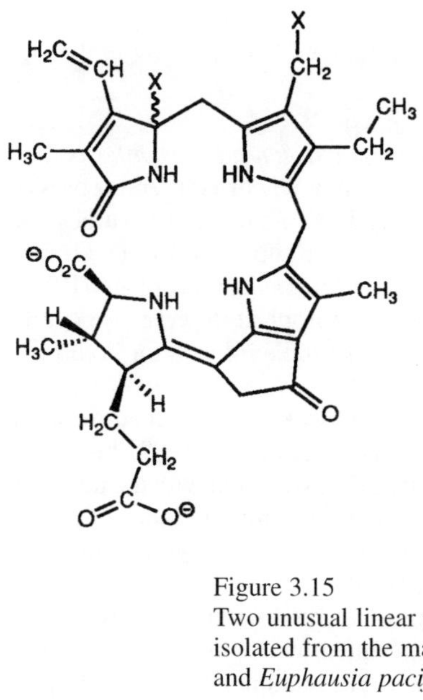

I

Dinoflagellate luciferin
from *Pyrocystis lunula*
(X = H)

II

Krill luciferin
from *Euphausia pacifica*
(X = OH)

Figure 3.15
Two unusual linear tetrapyrrole catabolites of Chl, cleaved at C-20, and isolated from the marine organisms *Pyrocystis lunula* (dinoflagellate; **I**) and *Euphausia pacifica* (krill; **II**).

3.7.3.2 *Chromic acid test*

Besides the non-destructive spectroscopic methods, the chromic acid oxidation test is the method of choice to distinguish Chl- and porphyrin-derived pigments from other chromophores, including anthocyanins and carotenoids. Even non-coloured tetrapyrroles can be made visible by this fast, simple and convenient procedure. Trace amounts (>0.5 µg) of these pigments can be detected on TLC-plates by applying a mixture of chromic and sulphuric acids to the origin spot containing the unknown pigments (Rüdiger, 1968, 1969). After the plate is developed in an appropriate solvent, exposed to a chlorine atmosphere and sprayed with benzidine, the oxidation products are visible as blue spots on a white background. By comparing the pyrrolic fragmentation pattern obtained with suitable standards after applying the chromic acid test, it is sometimes possible to deduce the original structure.

3.8 Biosynthesis and function of carotenoids

De novo synthesis of carotenoids is confined to prokaryotes, fungi, algae and higher plants. Approximately 600 different carotenoids are known. Those found in animals are derived from ingested pigments, often chemically modified after ingestion (Thommen, 1971; Goodwin, 1992; Pfander, 1992; Hendry, 1993).

This section briefly introduces the biosynthetic pathway for C_{40}-carotenoids and their function in photosynthesis. Structural relationships between marine algal carotenoids and visible light spectroscopy of carotenoids are discussed in Appendix C and Appendix D. Young and Britton (1993) review the role of carotenoids in photosynthesis.

3.8.1 *Biosynthesis of carotenoids*

In higher plants, mapping of carotenoid biosynthesis genes confirmed their location in the nucleus (Bartley *et al.* 1994). In bacteria, carotenogenic genes are tightly clustered (Sandmann, 1991) in the DNA nucleoid.

3.8.1.1 *Intracellular distribution of enzymes of carotenoid biosynthesis*

Two models for the intracellular distribution of the enzymes of *iso*prenoid biosynthesis have been suggested (for review see Gray, 1987). The first assumes that chloroplasts are competent to carry out the entire reaction sequence from CO_2 fixation to C_{40}-carotenoids. The second proposes that *iso*pentenyl diphosphate (IPP) is formed in the cytoplasm and then transported to the chloroplast to serve as a carotenoid precursor. As discussed by Gray (1987), experimental evidence for both types of distribution exist.

The inner envelope membrane of the chloroplast was long assumed to be the site of conversion of IPP to carotenoids (Lütke-Brinkhaus *et al.* 1982) but recent work has shown that most phytoene desaturase activity is associated with the thylakoid membranes (Linden *et al.* 1993). The epoxidation of zeaxanthin to violaxanthin was also demonstrated in osmotically-shocked thylakoid preparations, which are envelope-free (Siefermann and Yamamoto, 1975*a*; 1975*b*).

Little is known about the distribution of most of the enzymes catalyzing the biosynthesis of C_{40}-carotenoids, because they are integrated within the membrane and difficult to isolate and characterize (Britton, 1990; Sandmann, 1991). They occur *in situ* as multi-enzyme complexes, and are easily perturbed during isolation procedures (Britton, 1982; Bramley, 1985; Britton, 1988).

3.8.1.2 *Early enzymes of carotenoid biosynthesis and the formation of* iso*pentenyl diphosphate*

The initial steps of carotenoid biosynthesis from acetyl-CoA to geranylgeranyl diphosphate (GGPP, see Fig. 3.16) are also involved in the formation of the *iso*prenoid side chains of Chls (e.g., phytol) and quinones. The enzymes of geranylgeranyl diphosphate formation have been recently reviewed (Gray, 1987; Goodwin 1993).

In eukaryotic photosynthetic organisms, 3-hydroxy-3-methylglutaryl-CoA (HMG-CoA) is formed by the reaction of one molecule of acetyl-CoA with one molecule of acetoacetyl-CoA. HMG-CoA is also formed from the degradation of leucine in bacteria, animals and fungi; whether HMG-CoA is also formed by leucine degradation in algae and plants is unclear.

The subsequent conversion of HMG-CoA into mevalonate is catalyzed by the HMG-CoA reductase in a two-step reaction where two molecules of NADPH are used to reduce the thioester group of HMG-CoA to the 5'-hydroxyl of mevalonate. This conversion occurs *via* the enzyme-bound intermediate, mevaldate. HMG-CoA reductase, which is the rate-limiting enzyme in *iso*prenoid formation in animals, may also be rate-determining in plants (Gray, 1987).

Mevalonate is converted to *iso*pentenyl pyrophosphate (IPP) by a sequence of three ATP-consuming enzymes: mevalonate kinase, mevalonate 5-phosphate kinase and mevalonate 5-diphosphate carboxylase.

3.8.1.3 *Later enzymes converting IPP to carotenes*

The enzymic isomerization of IPP yields dimethyl-allyl-diphosphate (DMAPP) which condenses with a second molecule of IPP to form geranyl diphosphate, a C_{10} allylic diphosphate. This in turn accepts another one or two molecules of IPP to form farnesyl- or geranylgeranyl-diphosphate (GGPP), respectively. These reactions are mediated by a group of prenyl transferases that differ in their substrate specificity.

Phytoene, the first C_{40}-compound in carotenoid biosynthesis, results from the head-to-head condensation of GGPP (Fig. 3.16) *via* prephytoene diphosphate (Bramley and Mackenzie, 1988; Britton, 1988). This two-step reaction is probably catalyzed by a single enzyme and, in eukaryotic plants, forms predominantly the 15-*cis* isomer of phytoene (Bramley and Mackenzie, 1988; Sandmann, 1991).

The three conjugated double-bonds in the colourless phytoene are increased to eleven in the red-coloured lycopene by four desaturation steps. The extension of the -electron system occurs alternately on either side of the central chromophore, with the intermediates being phytofluene, ζ-carotene (7,8, 7',8'-tetrahydrolycopene) and neurosporene. These desaturase reactions are catalysed by one or two enzymes (Bramley and Mackenzie, 1988; Sandmann, 1991). Phytoene desaturation is enhanced by either NAD or NADP and requires O_2, which suggests an oxidative dehydrogenation mechanism (Sandmann, 1991).

In contrast to phytoene, cyclic carotenoids usually occur in the all-*trans* form. Thus, the presence of a *cis-trans*-isomerase acting on phytoene or phytofluene has been suggested (Bramley and Mackenzie, 1988).

The cyclization of lycopene to form carotenes in higher plants yields either a ß-ionone ring, with the double-bond in the 5,6-position, or an ε-ionone ring, with the double-bond in the 4,5-position (for carbon atom numbering see Fig. IV.1, Part IV

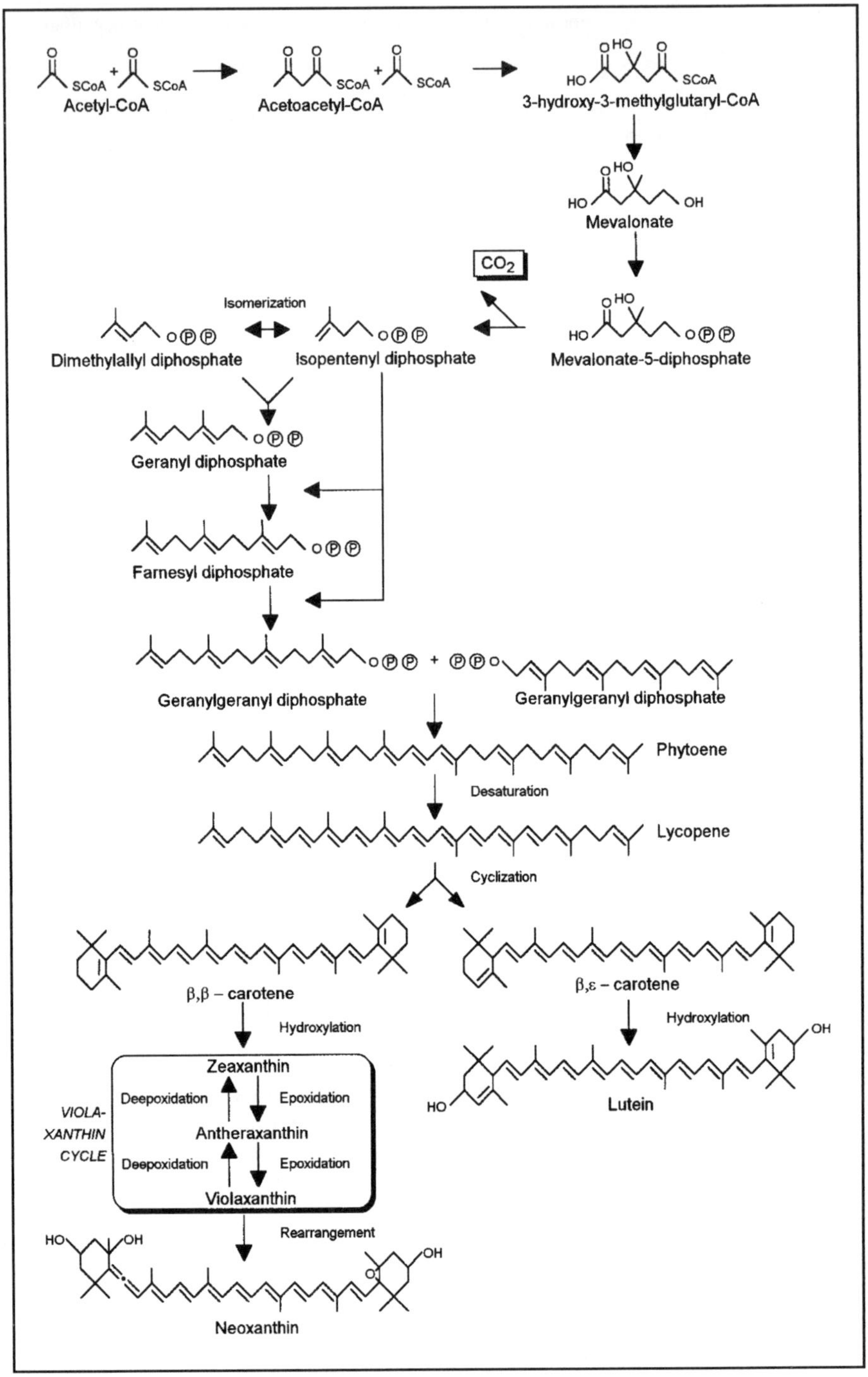

Figure 3.16
The biosynthetic route for the formation of β,β-, β,ε-carotene and some xanthophylls. Cofactors and some intermediates in phytoene to lycopene transformation are not shown. Boxed intermediates are shown in detail in Fig 3.17.

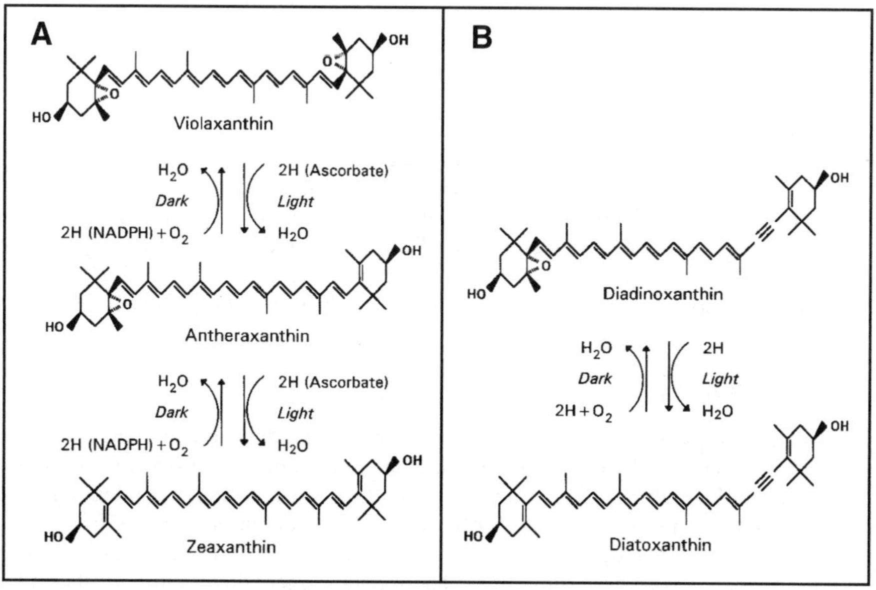

Figure 3.17
The violaxanthin (higher plants and green algae) and diadinoxanthin (chromophyte algae) cycles (adapted from Siefermann-Harms, 1977; Hager, 1980). The configurations of violaxanthin, zeaxanthin and diatoxanthin are taken from Weedon (1971*b*). For diadinoxanthin, the stereochemistry of antheraxanthin was assumed.

Data sheets). The formation of the two ß-ionone rings of ß,ß-carotene is probably catalyzed by a single enzyme, cyclase ß. However, ß,ε-carotene (one ß-ionone ring and one ε-ionone ring) is formed by two enzymes: the cyclase ß and cyclase ε (Sandmann, 1991).

3.8.1.4 *Formation of xanthophylls*

Xanthophylls, containing C, H and O atoms, are oxidation products of carotenes, which contain only C and H atoms. Algae contain more than 30 major carotenoids which are listed in Appendix C and Chapter 2. Biosynthetic pathways, where known, are reviewed by Britton (1993). Here only the formation of the major carotenoids of chlorophyte (green) algae and higher plants are described.

The xanthophylls, zeaxanthin and lutein, are formed by the hydroxylation of the 3 and 3' carbon atoms of ß,ß-carotene or ß,ε-carotene, respectively (Fig. 3.16). In ß,ß-carotene, presumably both hydroxylation steps are catalyzed by a single enzyme (Sandmann, 1991).

The introduction of an epoxy group in the 5,6-position of zeaxanthin forms antheraxanthin, and in the 5',6'-position yields violaxanthin (Figs. 3.16 and 3.17). Both the hydroxylation of ß,ß-carotene (Sandmann and Bramley, 1985) and the epoxidation of zeaxanthin require NADPH and O_2 (Hager, 1975; Siefermann and Yamamoto, 1975*a*; 1975*b*). The enzymes involved in these reactions are assumed to be mixed-function oxygenases. Recently, zeaxanthin epoxidation activity was attributed to a protein in the LHC II complex of higher plants (Gruszecki and Krupa, 1993*a*; 1993*b*) but this is in conflict with the work of Siefermann-Harms *et al.* (1980), who claimed zeaxanthin epoxidation in intact leaves devoid of LHC II.

Acetylenic and allenic xanthophylls (see Appendix C) may be synthesized at the latter stages of xanthophyll biosynthesis (Britton, 1988). Allenic neoxanthin (Fig. 3.16) may be formed by the rearrangement of one of the ionone rings of violaxanthin (Goodwin, 1971). The acetylenic diadinoxanthin (Fig. 3.17) is probably formed by the conversion of the allenic end-group of neoxanthin (Britton, 1988). The C_{37}-carotenoid peridinin (see Part IV Data sheets and Appendix C) is probably derived from neoxanthin by removal of the carbons 13, 14 and 20. Bjørnland (Appendix C) points out the lack of knowledge of biosynthetic pathways of highly oxidized algal xanthophylls.

3.8.2 *Regulation of carotenoid biosynthesis*

Although most plants can synthesize carotenoids in the dark, the amount produced is strongly influenced by environmental factors. For example, the size and composition of the violaxanthin cycle pigment pool is highly variable and dependent on light intensity. Thayer and Björkman (1990) observed that the pool size was four times greater in sun-tolerant species grown under sun-exposed conditions than in shade-tolerant species grown under low light. These data suggest that, within limits set by genetic disposition, environmental factors significantly influence the carotenoid content of plants and algae.

The control of carotenoid biosynthesis by either transcription or feed-back regulation has been suggested (Gray, 1987; Bramley and Mackenzie, 1988; Sandmann, 1991). The photoreceptors that modulate carotenoid biosynthesis are chlorophyll, chlorophyllide, cryptochrome and phytochrome (Harding and Shropshire, 1980; Gray, 1987; Bramley and Mackenzie, 1988; Humbeck and Krupinska, 1993) but the mechanism of these regulation processes is largely unknown.

3.8.3 *Biological degradation of carotenoids*

3.8.3.1 *Senescence processes*

During senescence, nitrogen-containing compounds in plants, including chlorophylls, are mostly converted to, and salvaged as, biologically useful amino acids because such forms of nitrogen are scarce in most natural environments. Carotenoids, however, contain no nitrogen, and have no enzymic degradation pathway of comparable importance to that of the chlorophylls (G. Sandmann, private communication). Conspicuous evidence of salvaging chlorophyll nitrogen and the relative stability of carotenoids is the loss of green and retention of red and yellow carotenoid colours in autumn. Senescent autumn leaves accumulate carotenoid acyl esters (Goodwin, 1958) but the physiological significance of this is not clear. The relative stability of carotenoids may also be associated with their ability to quench singlet oxygen to form triplet-state carotenoids, which rapidly lose energy as heat and return to the ground state (Foyer and Harbinson, 1994). Under extreme stress conditions, carotenoids can undergo oxidative and free-radical-mediated processes to form ß,ß-carotene epoxides (Young *et al.* 1989, Liebler and Kennedy, 1992).

3.8.3.2 *Examples of chemical modification of carotenoids in animals*

Animals cannot synthesize carotenoids *de novo* but can modify carotenoids ingested in their diet. Some are deposited unchanged in animal tissues; for example, lutein and zeaxanthin are the major macular pigments of the human retina (Bone and Landrum, 1992).

Carotenoids containing at least one ß-ionone ring can be converted to vitamin A by oxidative cleavage of the 15-15'-double bond to form retinal, which is then reduced to retinol (Pitt, 1971). Vitamin A is essential to animal growth and vision.

Crustaceans oxidize ß,ß-carotene to astaxanthin (3,3'-dihydroxy-ß,ß-carotene-4,4'-dione), which accumulates in the carapace (cf. Schiedt *et al.* 1993) and is bound non-covalently to a protein to form a blue-green pigment. Heating releases the free red pigment, which accounts for the blue-green to red shift when cooking crustaceans. In salmonids the red astaxanthin accumulates in the muscular tissue to give the characteristic pink flesh. Numerous carotenoids are deposited in the feathers and skin of birds (Hudon and Brush, 1992; Weedon. 1971*a*): two notable and colourful examples are the canary and flamingo.

3.8.4 *Function of carotenoids*

3.8.4.1 *Free carotenoids and protein complexes binding carotenoids exclusively*

In chloro- and chromo-plasts of photosynthetic eukaryotes, carotenoids can be found both free and complexed to proteins. These compounds are probably not linked to photosynthesis but may be involved in photomorphogenesis and phototropism or in photoprotection by screening against harmful radiation (Bidigare *et al.* 1989; Rau, 1988). In cyanobacteria and prochlorophytes, membrane-associated protein complexes that exclusively bind carotenoids have been characterized by Reddy *et al.* (1993). It is thought they might play a role in ion-uptake processes.

3.8.4.2 *Carotenoids in chlorophyll pigment-protein complexes*

In most photosynthetic organisms, the major fraction of carotenoids is non-covalently bound and highly ordered in pigment-protein complexes of the photosynthetic apparatus. They serve as light-harvesting pigments (see 3.8.4.3) or as protectors of the photosynthetic apparatus (see 3.8.4.4 and 3.8.4.5). In early evolutionary times, the main function of carotenoids was probably light-harvesting and they later assumed photoprotective functions during the formation of the global aerobic atmosphere (Siefermann-Harms, 1987; Gust and Moore, 1991).

3.8.4.3 *Light-harvesting properties*

Carotenoids in the Chl-containing pigment-protein complexes of the photosynthetic apparatus absorb light energy and transfer it to Chls. Carotenoid absorption in the visible wavelengths is due to transition from the ground state to the second excited singlet (1^1B_u) state (see Fig. 3.18).

Energy transfer from carotenoid to Chl probably occurs *via* the Dexter electron exchange mechanism (Dexter, 1953), which requires spectral overlap between both pigments (Gillbro *et al.* 1993). Relative to the Chl first excited singlet state, the carotenoid 1^1B_u singlet state energy is too high to meet this requirement; consequently it is probably the 2^1A_g singlet state that is involved in the light-harvesting function of carotenoids (Fig. 3.18). The carotenoid first excited singlet state (2^1A_g) can be populated from the 1^1B_u state by internal conversion (Fig. 3.18).

The excited carotenoid singlet state is extremely short-lived, so the highly efficient energy transfer from carotenoids to Chl, observed in photosynthesis, would require the permanent contact of both -electron systems (Siefermann-Harms, 1985, 1987; Cogdell *et al.* 1994). This is achieved by the highly ordered arrangement of pigments in the pigment-protein complexes of relatively fixed stoichiometries (Siefermann-Harms, 1985; 1987; Wilhelm, 1990). Recently, Kühlbrandt *et al.* (1994) demonstrated the close vicinity of xanthophyll and chlorophyll in the LHC II of higher plants by electron crystallography (see section 3.3.3 and Figs. 3.2 B, C).

3.8.4.4 *Photoprotection by carotenoids*

In stressful high-light conditions, damaging excited singlet state oxygen ($^1O_2$) can be formed by energy transfer from triplet-state chlorophyll to ground-state oxygen (Siefermann-Harms, 1985; 1987; Young, 1991; Foyer and Harbinson, 1994). Triplet state Chl can be created by intersystem crossing from the lowest singlet state Chl.

Carotenoids can reduce the formation of $^1O_2$ by quenching: (1) the excited triplet state of Chl, and (2) the excited singlet state of Chl. In addition, carotenoids can directly deactivate $^1O_2$. In all cases, carotenoids can dissipate the accepted energy and return chemically unchanged to their ground state.

Quenching of triplet chlorophyll and singlet oxygen probably occurs *via* the formation of triplet state carotenoid, which has about half the energy of the 2^1A_g state (Cogdell and Frank, 1987; not shown in Fig. 3.18). Quenching of excited singlet Chl probably involves the 2^1A_g carotenoid state. Since the excited states of chlorophyll and of singlet oxygen are relatively long-lived, it is possible that carotenoids quench these excited states in a temporarily formed collision complex.

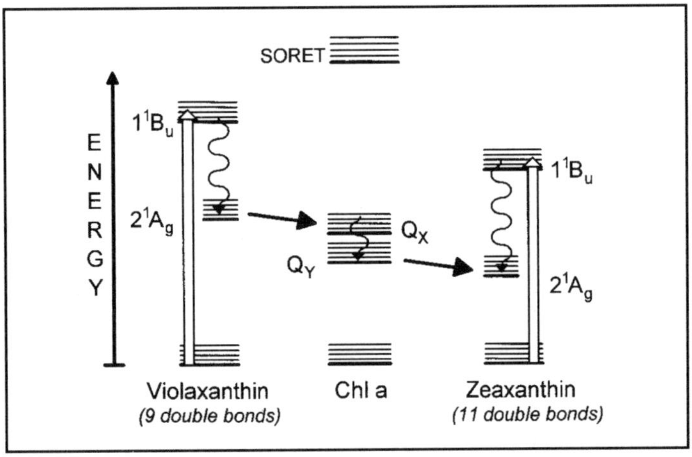

Figure 3.18
Possible pathways for singlet energy-transfer between carotenoids and Chls (adapted from Siefermann-Harms, 1987; Owens *et al.* 1992). 2^1A_g and 1^1B_u represent the first and second excited carotenoid singlet state, respectively. Q_x and Q_y correspond to singlet states of Chl *a* giving rise to *in vivo* absorption bands around 625 and 670 nm. Indicated energy levels are not proportional to the actual electronic states. Straight open arrows: excitation by visible light; straight solid arrows: singlet energy transfer; wavy arrows: internal conversion. The 2^1A_g state energy of zeaxanthin with its 11 conjugated double-bonds is assumed to be lower than the Q_y state of chlorophyll *a*, thus permitting chlorophyll-to-carotenoid energy transfer. Violaxanthin with nine conjugated double bonds is capable of transferring singlet excitation energy to chlorophyll (Owens *et al.* 1992; Owens, 1994).

3.8.4.5 *Protection against photoinhibition by xanthophyll cycles*

The epoxidation of zeaxanthin together with the de-epoxidation of violaxanthin (Fig. 3.17) form the three-component violaxanthin cycle of higher plants, green and brown algae (Stransky and Hager, 1970; Hager, 1980). A two-component xanthophyll cycle — the diadinoxanthin cycle (Fig. 3.17) — is present in all chromophyte algal groups (see Tables 2.1, 2.3). In the dark-adapted state, most of the xanthophyll-cycle pigments are in the epoxide form. In higher plants de-epoxidation and epoxidation occur with half-times of about 1 and 10 min, respectively, which is significantly shorter than that of gross carotenoid biosynthesis (see Siefermann-Harms, 1977; Yamamoto, 1979; Hager, 1980; Pfündel and Bilger, 1994).

The isolated de-epoxidase has a pH optimum of about 5 (Hager, 1969). *In situ*, de-epoxidase activity is regulated within an extremely narrow pH interval and shows a plateau of maximum activity at pH values <pH 5.8 (Pfündel and Dilley, 1993). De-epoxidation *in vivo* is triggered by high light or, more generally, when light absorption exceeds light utilization by photosynthesis. Hager (1969) theorized that under such conditions considerable proton accumulation (acidification) is likely in the thylakoid lumen, and this would activate violaxanthin de-epoxidase.

Demmig-Adams (1990) suggested that zeaxanthin protects the photosynthetic apparatus against photoinhibition by quenching the excited Chl singlet state (see 3.8.4.4). Under high light conditions, this would prevent over-reduction of the chloroplast electron transport chain and decrease the formation of triplet Chl in the RC. Recent data indicate that the diadinoxanthin cycle in diatoms could have a similar function (Fig. 3.17; Arsalane *et al.* 1994).

The excited chlorophyll singlet state might be quenched by structural changes of the LHC II mediated by zeaxanthin (Horton *et al.* 1991; Horton and Ruban, 1994). Such changes could bring about Chl species with a high capacity for radiationless energy dissipation.

Another hypothesis on the role of the violaxanthin cycle in energy quenching was formulated by Owens and co-workers (Owens *et al.* 1992; Owens, 1994). They considered the increase in the number of conjugated double-bonds that occurs during de-epoxidation of violaxanthin to zeaxanthin (Fig. 3.17) and suggested a corresponding decrease in the 2^1A_g state energy. In the case of zeaxanthin, the 2^1A_g state energy was assumed to be below the lowest excited Chl singlet state (Fig. 3.18). Thus, zeaxanthin was expected to accept singlet energy from Chl (quenching), and violaxanthin (with high 2^1A_g state) to transfer energy to Chl (light-harvesting). Recently, Frank *et al.* (1994) determined the 2^1A_g state energy of the pigments of the violaxanthin cycle and confirmed the prediction of Owens.

3.9 Summary

This chapter introduces the oceanographer to the biosynthetic pathways and degradative processes of chlorophylls, phycobilins and carotenoids. It provides an introductory account of photosynthesis and the function and location of pigment-protein complexes in the cell. Pigment-proteins may serve several functions: metabolic regulators, light harvesters (in antennae), electron donors or acceptors (in reaction centres) and photoprotection of nearby pigment molecules. We have attempted to explain these multi-purpose roles and the ingenious way nature has combined them for maximum efficiency and economy. Their role in the marine environment is introduced in the next two chapters.

ACKNOWLEDGEMENTS

RJP thanks the Deutsche Forschungsgemeinschaft, Sonderforschungsbereich 143 (Projekt A1) for financial support and the CSIRO Division of Plant Industry, Canberra, for the use of communication, library and other facilities. EP thanks his wife, Anette, whose help made his participation in the writing of this chapter possible. NE thanks the Swiss National Science Foundation for financial support and Professor A. Gossauer for his continued and expert advice. SWJ thanks the authors for their special co-operation in the development of this chapter.

REFERENCES

Akhtar, M. (1991). Mechanism and stereochemistry of the enzymes involved in the conversion of uroporphyrinogen III to haem. In: Jordan, P. M. (ed.), *Biosynthesis of Tetrapyrroles*. Elsevier, Amsterdam, p. 67–100.

Anderson, J. M., Andersson, B. (1988). The dynamic photosynthetic membrane and regulation of solar energy conversion. *Trends Biochem. Sci.* 13: 351–355.

Arsalane, W., Rousseau, B., Duval, J.-C. (1994). Influence of the pool size of the xanthophyll cycle on the effects of light stress in a diatom: competition between photoprotection and photoinhibition. *Photochem. Photobiol.* 60: 237–243.

Barrett, J., Anderson, J. M. (1980). The P700-Chlorophyll *a*-protein complex and two major light-harvesting complexes of *Acrocarpia paniculata* and other brown seaweeds. *Biochim. Biophys. Acta,* 590: 309–323.

Barrett, J., Jeffrey, S. W. (1964). Chlorophyllase and formation of an atypical chlorophyllide in marine algae. *Plant Physiol.* 39: 44–47.

Bartley, G. E., Scolnik, P. A., Giuliano, G. (1994). Molecular biology of carotenoid biosynthesis in plants. *Ann. Rev. Plant Physiol. Plant Mol. Biol.* 45: 287–301.

Beale, S. I. (1984). Biosynthesis of photosynthetic pigments. In: Baker, N. R., Barber, J. (eds), *Chloroplast Biogenesis*. Elsevier, Amsterdam, p. 135–205.

Beale, S. I., Castelfranco, P. A. (1973). ^{14}C incorporation from exogenous compounds into δ-aminolevulinic acid by greening cucumber cotyledons. *Biochem. Biophys. Res. Commun.* 52: 143–149.

Beale, S. I., Cornejo, J. (1984*a*). Enzymatic heme oxygenase activity in soluble extracts of the unicellular red alga, *Cyanidium caldarium*. *Arch. Biochem. Biophys.* 235: 371–384.

Beale, S. I., Cornejo, J. (1984*b*). Enzymic transformation of biliverdin to phycocyanobilin by extracts of the unicellular red alga, *Cyanidium caldarium*. *Plant Physiol.* 76: 7–15.

Beale, S. I., Foley, T., Dzelzkalns, V. (1981). δ-Aminolaevulinic acid synthase from *Euglena gracilis*. *Proc Nat. Acad. Sci. USA*, 78: 1666–1669.

Beale, S. I., Gough, S. P., Granick, S. (1975). Biosynthesis of δ-aminolevulinic acid from the intact carbon skeleton of glutamic acid in greening barley. *Proc. Nat. Acad. Sci. USA,* 72: 2719–2723.

Beale, S. I., Weinstein, J. D. (1991). Biochemistry and regulation of photosynthetic pigment formation in plants and algae. In: Jordan, P. M. (ed.), *Biosynthesis of Tetrapyrroles*. Elsevier, Amsterdam, p. 155–223.

Becerril, J. M., Duke, S. O. (1989). Protoporphyrin IX content correlates with activity of photobleaching herbicides. *Plant Physiol.* 90: 1175–1181.

Beifuss, U., Hart, G. J., Miller, A. D., Battersby, A. R. (1988). ^{13}C-NMR studies on the pyrromethane cofactor of hydroxymethylbilane synthase. *Tetrahedron Lett.* 29: 2591–2594.

Benz, J., Rüdiger, W. (1981). Incorporation of 1-^{14}C-isopentenyl diphosphate, geraniol and farnesol into chlorophyll in plastid membrane fractions of *Avena sativa* L. *Z. Pflanzenphysiol.* 102: 95–100.

Bidigare, R. R., Schofield, O., Prézelin, B. B. (1989) Influence of zeaxanthin on quantum yield of *Synechococcus* clone WH7803 (DC2). *Mar. Ecol. Prog. Ser.* 56: 177–188.

Blankenship, R. E., Prince, R. C. (1985). Excited-state redox potentials and the Z scheme of photosynthesis. *Trends Biochem. Sci.* 10: 382–383.

Boekema, E. J., Hankamer, B., Bald, D., Kruip, J., Nield, J., Boomstra, A. F., Barber, J., Rögner, M. (1995). Supramolecular structure of the photosynthetic complex from green plants and cyanobacteria. *Proc. Natl. Acad. Sci. USA*, 92: 175–179.

Bone, R. A., Landrum, J. T. (1992). Distribution of macular pigment components, zeaxanthin and lutein, in the human retina. *Methods Enzymol.* 213: 360–366.

Bramley, P. M. (1985). The *in vitro* biosynthesis of carotenoids. *Adv. Lipid. Res.* 21: 243–279.

Bramley, P. M., Mackenzie, A. (1988). Regulation of carotenoid biosynthesis. *Curr. Top. Cellul. Regul.* 29: 291–343.

Britton, G. (1982). Carotenoid biosynthesis in higher plants. *Physiol. Vég.* 20: 735–755.

Britton, G. (1988). Biosynthesis of carotenoids. In: Goodwin, T. W. (ed.), *Plant Pigments*, Academic Press, London, p. 133–182.

Britton, G. (1990). Carotenoid biosynthesis: an overview. In: Krinsky, N. I., Mathews-Roth, M. M., Taylor, R. F. (eds), *Carotenoids: Chemistry and Biology*, Plenum Press, New York, p. 167–194.

Britton, G. (1993). Biosynthesis of carotenoids. In: Young, A., Britton, G. (eds). *Carotenoids in Photosynthesis*. Chapman and Hall, London, p. 96–126.

Brockman, H., Risch, N. (1991). Preparative chromatography. In: Scheer, H. (ed.), *Chlorophylls*, CRC Press Boca Raton, USA, p. 104–114.

Brown, S. B., Houghton, J. D., Hendry, G. A. F. (1991). Chlorophyll breakdown. In: Scheer, H. (ed.), *Chlorophylls*. CRC Press, Boca Raton, USA, p.465–489.

Callot, H. J. (1991). Geochemistry of chlorophylls. In: Scheer, H. (ed.), *Chlorophylls*. CRC Press, Boca Raton, USA, p. 339–364.

Cogdell, R. J., Frank, H. A. (1987). How carotenoids function in photosynthetic bacteria. *Biochim. Biophys. Acta*, 895: 63–79.

Cogdell, R. J., Gillbro, T., Andersson, P. O., Liu, R. S. H., Asato A. E. (1994). How carotenoids function as accessory light-harvesting pigments in photosynthesis. *Pure Appl. Chem.* 66: 1041–1046.

Deisenhofer, J., Michel, H. (1989). The photosynthetic reaction center from the purple bacterium *Rhodopseudomonas viridis*. *Science*, 245: 1463–1473.

Demmig-Adams, B. (1990). Carotenoids and photoprotection in plants: a role for the xanthophyll zeaxanthin. *Biochim. Biophys. Acta*, 1020: 1–24.

Dexter, D. L. (1953). A theory of sensitized luminescence in solids. *J. Chem. Phys.* 21: 836–850.

Dougherty, R. C., Strain, H. H., Svec, W. A., Uphaus, R. A., Katz, J. J. (1970). The structure, properties and distribution of chlorophyll *c*. *J. Amer. Chem. Soc.* 92: 2826–2833.

Engel, N., Gossauer, A., Gruber, K., Kratky, C. (1993). X-ray molecular structure of a red bilin derivative from *Chlorella protothecoides*. *Helv. Chim. Acta*, 76: 2236–2238.

Engel, N., Jenny, T. A., Mooser, V., Gossauer, A. (1991). Chlorophyll catabolism in *Chlorella protothecoides*. Isolation and structure elucidation of a red bilin derivative. *FEBS Lett.* 293: 131–133.

Falk, H. (1989). *The Chemistry of Linear Oligopyrroles and Bile Pigments*. Springer Verlag, Vienna, 621 pp.

Fischer, H., Stern, A. (1943). *Die Chemie des Pyrroles*, vol. II, 2. Hälfte. Leipzig Acad. Verlagsgesellschaft, p.390–400.

Fookes, C. J. R., Jeffrey, S. W. (1989). The structure of chlorophyll c_3, a novel marine photosynthetic pigment. *J. Chem. Soc. Chem. Comm.* 23: 1827–1828.

Foyer, C. H. (1984). *Photosynthesis,* Wiley Interscience Publications: New York, Chichester, p. 22–48.

Foyer, C. H., Harbinson, J. (1994). Oxygen metabolism and the regulation of photosynthetic electron transport. In: Foyer, C. H., Mullineaux, P. M. (eds), *Causes of Photooxidative*

Stress and Amelioration of Defense Systems in Plants, CRC Press, Boca Raton, USA, p. 1–42.

Frank, H. A., Cua, A., Chynwat, V., Young, A., Gosztola, D., Wasielewski, M. R. (1994). Photophysics of the carotenoids associated with the xanthophyll cycle in photosynthesis. *Photosynth. Res.* 41: 389–395.

Friedmann, H. C., Duban, M. E., Valasinas, A., Frydman, B. (1992). The enantioselective participation of (S) and (R)-diaminovaleric acids in the formation of δ-aminolevulinic acid in cyanobacteria. *Biochem. Biophys. Res. Commun.* 185: 60–68.

Gantt, E. (1990). Pigmentation and photoacclimation. In: Cole, K. M., Sheath, R. G. (eds). *Biology of the Red Algae*. Cambridge Univ. Press, p. 203–220.

Gillbro, T., Andersson, P. O., Liu, R. S. H., Asato, A. E., Takaishi, S., Cogdell, R. J. (1993). Location of the carotenoid 2Ag-state and its role in photosynthesis. *Photochem. Photobiol.* 57: 44–48.

Goodwin, T. W. (1958). Carotenogenesis. 24. Changes in carotenoid and chlorophyll pigments in leaves of deciduous trees during autumn necrosis. *Biochem. J.* 68: 503–511.

Goodwin, T. W. (1971). Biosynthesis. In: Isler, O. (ed.), *Carotenoids*. Birkhäuser Verlag, Basel, p. 577–636.

Goodwin, T. W. (1992). Distribution of carotenoids. *Methods Enzymol.* 213: 167–172.

Goodwin, T. W. (1993). Biosynthesis of carotenoids. An overview. *Methods Enzymol.* 214: 330–340.

Granick, S. (1966). The induction *in vitro* of the synthesis of 5-aminolevulinic acid synthetase in chemical porphyria: a response to certain drugs, sex hormones and foreign chemicals. *J. Biol. Chem.* 241: 1359–1375.

Gray, J. C. (1987). Control of isoprenoid biosynthesis in higher plants. *Adv. Bot. Res.* 14: 25–91.

Gruszecki, W. I., Krupa, Z. (1993*a*). LHCII preparation exhibits properties of a zeaxanthin epoxidase. *J. Photochem. Photobiol. B: Biol.* 17: 291–292.

Gruszecki, W. I., Krupa, Z. (1993*b*). LHCII, the major light-harvesting pigment-protein complex is a zeaxanthin epoxidase. *Biochim. Biophys. Acta*, 1144: 97–101.

Gust, D., Moore, T. A. (1991). Mimicking photosynthetic electron transport and energy transfer. In: Volman, D., Hammond, G., Neckers, D. (eds), *Advances in Photochemistry*. John Wiley and Sons, New York, pp. 1–65.

Hager, A. (1969). Lichtbedingte pH-Erniedrigung in einem Chloroplasten-Kompartiment als Ursache der enzymatischen Violaxanthin → Zeaxanthin-Umwandlung; Beziehungen zur Photophosphorylierung. *Planta*, 89: 224–243.

Hager, A. (1975). Die reversiblen, lichtabhängigen Xanthophyllumwandlungen im Chloroplasten. *Ber. Deutsch. Bot. Ges.* 88: 27–44.

Hager, A. (1980). The reversible light-induced conversions of xanthophylls in the chloroplast. In: Czygan, F.-C. (ed.) *Pigments in Plants*, Fischer, Stuttgart, p. 57–79.

Harding, R. W., Shropshire, W. (1980). Photocontrol of carotenoid biosynthesis. *Ann. Rev. Plant Physiol.* 31: 217–238.

Hawkins, A. J. S., Bayne, B. L., Mantoura, R. F. C., Llewellyn, C. A. (1986). Chlorophyll degradation and absorption through the digestive system of the blue mussel *Mytilus edulis*. *J. Exp. Mar. Biol. Ecol.* 96: 213–223.

Hendry, G. A. F. (1993). Plant Pigments. In: Lea, P. J., Leegood, R. C. (eds), *Plant Biochemistry and Molecular Biology*, John Wiley & Sons, New York, p. 181–196.

Hendry, G. A. F., Houghton, J. D., Brown, S. B. (1987). The degradation of chlorophyll: a biological enigma. *New Phytol.* 107: 255–302.

Horton, P., Ruban, A. V. (1994). The role of light-harvesting complex II in energy quenching. In: Baker, N. R., Bowyer, J. R. (eds), *Photoinhibition of Photosynthesis*. Bios Scientific Publishers, Oxford, pp. 111–128.

Horton, P., Ruban, A. V., Rees, D., Pascal, A. A., Noctor, G., Young, A. J. (1991). Control of the light-harvesting function of chloroplast membranes by aggregation of the LHCII chlorophyll-protein complex. *FEBS Lett.* 292: 1–4.

Hudon, J., Brush, A. H. (1992). Identification of carotenoids in birds. *Methods Enzymol.* 213: 312–321.
Humbeck, K., Krupinska, K. (1993). Control of carotenoid biosynthesis by light. *Methods Enzymol.* 214: 175–184.
Hynninen, , P. H. (1991). Chemistry of chlorophylls: modifications. In: Scheer, H (ed.), *Chlorophylls,* CRC Press, Boca Raton, USA, p.145–206.
Hynninen, P. H. (1980). Mechanism of the allomerization of chlorophyll: inhibition of the allomerization by carotenoid pigments. *Z. Naturforsch.* 86b, 1010-1016
Hynninen, P. H., Assandri, S. (1973). Chlorophylls II. Allomerization of chlorophylls *a* and *b*. *Acta Chem. Scand.* 27: 1478–1486.
Hynninen, P. H., Lötjenen, S. (1981) Electrophilic substitution at the δ-methine bridge of phaeophorbide *a*. *Tetrahedron Lett.* 22: 1846–1847.
Itturaspe, J., Engel, N., Gossauer, A. (1994). Chlorophyll catabolism. Isolation and structure elucidation of chlorophyll *b* catabolites in *Chlorella protothecoides*. *Phytochemistry*, 35: 1387–1390.
Itturaspe, J., Engel, N., Matzinger, P., Mooser, V., Gossauer, A. (1993). Chlorophyll catabolism. Part 3. Structure elucidation and partial synthesis of a new red bilin derivative from *Chlorella protothecoides*. *Photochem. Photobiol.* 58: 116–119.
Jeffrey, S. W. (1980). Algal pigment systems. In: Falkowski, P. G. (ed.), *Primary Productivity of the Sea.* Plenum Press, New York, p. 33–58.
Jeffrey, S. W., Hallegraeff, G. M. (1987). Chlorophyllase distribution in 10 classes of phytoplankton — a problem for chlorophyll analysis. *Mar. Ecol. Progr. Ser.* 35: 293–304.
Jones, O. T. G. (1968). Ferrochelatase of spinach chloroplasts. *Biochem. J.* 107: 113–119.
Kannangara, C. G., Anderson, R.V., Pontoppidan, B., Willows, R., von Wettstein, D. (1994). Enzyme and mechanistic studies on the conversion of glutamate to 5-aminolevulinate. In: *The Biosynthesis of the Tetrapyrrole Pigments*, CIBA Foundation Symp. No. 180, John Wiley and Sons, Chichester, p. 3–25.
Keely, B. J., Maxwell, J. R. (1991). Structural characterisation of the major chlorins in a Recent sediment. *Organic Geochem.* 17: 663–669.
Kobayashi, M., Watanabe, T., Struck, A., Scheer, H. (1988). *Meso*-chlorination of chlorophyll *a* in the course of pigment extraction. *FEBS. Lett.* 235: 293–297.
Kräutler, B., Jaun, B., Bortlik, K., Schellenberg, M., Matile, P. (1991). On the enigma of chlorophyll degradation: the constitution of a secoporphinoiden catabolite. *Angew. Chem. Int. Ed. Engl.* 30: 1315–1318.
Kufner, R. B. (1980). The biological degradation of chlorophyll in senescent tissues. In: Czygan, F.-C. (ed.), *Pigments in Plants.* Gustav Fischer, Stuttgart, p. 308–313.
Kühlbrandt, W., Wang, D. N. (1991). Three-dimensional structure of plant light-harvesting complex determined by electron crystallography. *Nature*, 350: 130–134.
Kühlbrandt, W., Wang, D. N., Fujiyoshi, A. (1994). Atomic model of plant light-harvesting complex by electron crystallography. *Nature*, 367: 614–621.
Laver, W. G., Neuberger, A., Udenfriend, S. (1958). The initial stages in the biosynthesis of porphyrins. 1. The formation of δ-aminolaevulic acid by particles obtained from chicken erythrocytes. *Biochem. J.* 70: 4–14.
Liebler, D. C., Kennedy, T. A. (1992). Epoxide products of ß-carotene antioxidant reactions. *Methods Enzymol.* 213: 472–479.
Linden, H., Lucas, M. M., de Filipe, M. R., Sandmann, G. (1993). Immunogold localization of phytoene desaturase in higher plant chloroplasts. *Physiol. Plantarum,* 88: 229–236.
Lütke-Brinkhaus, F., Liedvogel, B., Kreuz, K., Kleinig, H. (1982). Phytoene synthase and phytoene dehydrogenase associated with envelope membranes from spinach chloroplasts. *Planta*, 156: 176–180.
Mayer, H. (1930). Untersuchungen über die Chlorophyllase. *Planta,* 11: 294–330.
Moss, G. P. (1987). Nomenclature of tetrapyrroles. *Pure Appl. Chem.* 59: 779–832; (a) p. 790, 794, 800, 825 and 826.

Mühlecker, W., Kräutler, B., Ginsburg, S., Matile, P. (1993). Breakdown of chlorophyll: a tetrapyrrolic chlorophyll catabolite from senescent rape leaves. *Helv. Chim. Acta*, 76: 2976–2980.

Nakamura, H., Kishi, J., Shimomura, O., Morse, D., Hastings, J. W. (1989). Structure of dinoflagellate luciferin and its enzymatic and non-enzymatic air-oxidation products. *J. Am. Chem. Soc.* 111: 7607–7611.

Nakamura, H., Musicki, B., Kishi, Y., Shimomura, O. (1988). Structure of the light emitter in krill *(Euphausia pacifica)* bioluminescence. *J. Am. Chem. Soc.* 110: 2683–2685.

Nashrulhaq-Boyce, A., Griffith, W. T., Jones, O. T. G. (1987). The use of continuous assays to characterize the oxidative cyclase that synthesizes the chlorophyll isocyclic ring. *Biochem. J.* 243: 23–29.

Nelson, J. R., Wakeham, S. G. (1989). A phytol-substituted chlorophyll *c* from *Emiliania huxleyi* (Prymnesiophyceae). *J. Phycol.* 25: 761–766.

Nir, I., Pease, D. C. (1973). Chloroplast organization and the ultrastructural localization of photosystems I and II. *J. Ultrastruct. Res.* 42: 534–550.

Oh-hama, T., Seto, H., Otake, N., Myachi, S. (1982). ^{13}C-NMR evidence for the pathway of chlorophyll biosynthesis in green algae. *Biochem. Biophys. Res. Commun.* 105: 647–652.

Owens, T. G. (1994). Excitation energy transfer between chlorophylls and carotenoids. A proposed molecular mechanism for non-photochemical quenching. In: Baker, N. R., Bowyer, J. R. (eds), *Photoinhibition of Photosynthesis,* Bios Scientific Publishers, Oxford, p. 95–107.

Owens, T. G., Falkowsky, P. G. (1982). Enzymatic degradation of chlorophyll *a* by marine phytoplankton *in vitro*. *Phytochem.* 21: 979–984.

Owens, T. G., Shreve, A. P., Albrecht, A. C. (1992). Dynamics and mechanism of singlet energy transfer between carotenoids and chlorophylls: light harvesting and non-photochemical fluorescence quenching. In: Murata, N. (ed.), *Research in Photosynthesis*, vol. 1. Kluwer Academic Press, The Netherlands, p. 171–178.

Paulsen, H., Rümler, U., Rüdiger, W. (1990). Reconstitution of pigment-containing complexes from light-harvesting chlorophyll *a/b*-binding protein overexpressed in *Escherichia coli. Planta*, 181: 204–211.

Pfander, H. (1992). Carotenoids: an overview. *Methods Enzymol.* 231: 3–13.

Pfündel, E., Bilger, W. (1994). Regulation and possible function of the violaxanthin cycle. *Photosynth. Res.* 42: 89–109.

Pfündel, E., Dilley, R. A. (1993). The pH dependence of violaxanthin deepoxidation in isolated pea chloroplasts. *Plant Physiol.* 101: 65–71.

Pitt, G. A. J. (1971). Vitamin A. In: Isler, O. (ed.), *Carotenoids*. Birkhäuser, Basel. p. 717–742.

Plumley, F. G., Schmidt, G. W. (1987). Reconstitution of chlorophyll *a/b* light harvesting complexes: xanthophyll-dependent assembly and energy transfer. *Proc. Natl. Acad. Sci. USA*, 84: 146–150.

Porra, R. J. (1991). Recent advances and re-assessments in chlorophyll extraction and assay procedures for terrestrial, aquatic and marine organisms including recalcitrant algae. In: Scheer, H. (ed.), *Chlorophylls*. CRC Press, Boca Raton, USA, p. 32–57.

Porra, R. J., Grimme, L. H. (1974). Chlorophyll synthesis and intracellular fluctuations of 5-aminolaevulinate formation during regreening of nitrogen-deficient *Chlorella fusca*. *Arch. Biochem. Biophys.* 148: 37–43.

Porra, R. J., Jones, O. T. G. (1963*a*). Studies on ferrochelatase. 1. Assay and properties of ferrochelatase from a pig-liver mitochondrial extract. *Biochem. J.* 87: 181–185.

Porra, R. J., Jones, O. T. G. (1963*b*). Studies on ferrochelatase. 2. An investigation of the role of ferrochelatase in the biosynthesis of various haem prosthetic groups. *Biochem. J.* 87: 186–192.

Porra, R. J., Klein, O., Wright, P. E. (1983). The proof by ^{13}C-NMR spectroscopy of the predominance of the C_5 pathway over the Shemin pathway in chlorophyll biosynthesis in higher plants and of the formation of the methyl ester group of chlorophyll from glycine. *Eur. J. Biochem.* 130: 509–516.

Porra, R. J., Lascelles, (1968). Studies on ferrochelatase. The enzymic formation of haem in proplastids, chloroplasts and plant mitochondria. *Biochem. J.* 108: 343–348.
Porra, R. J., Schäfer, W., Cmiel, E., Katheder, I., Scheer (1994). The derivation of the formyl-group oxygen of chlorophyll *b* in higher plants from molecular oxygen. *Eur. J. Biochem.* 219: 671–679.
Rau, W. (1988). Functions of carotenoids other than in photosynthesis. In: Goodwin, T.W. (ed.), *Plant Pigments*. Academic Press, London, p. 231–255.
Rebeiz, C. A., Wu, S. M., Kuhadja, M., Daniell, H., Perkins, E. J. (1983). Chlorophyll *a* biosynthetic routes and chlorophyll *a* chemical heterogeneity in plants. *Mol. Cell. Biochem.* 57: 97–125.
Reddy, K. J., Bullerjahn, G. S., Sherman, L. A. (1993). Characteristics of membrane-associated carotenoid-binding proteins in cyanobacteria and prochlorophytes. *Methods Enzymol.* 214: 390–401.
Ricketts, T. R. (1966). Magnesium 2,4-divinyl pheoporphyrin a_5 monomethyl ester, a protochlorophyll-like pigment present in some unicellular flagellates. *Phytochem.* 5: 223–229.
Rüdiger, W. (1968) In: Goodwin, T. W. (ed.), *Porphyrins and Related Components*. Academic Press, London, p. 121–130.
Rüdiger, W. (1969). Chromatabbau von Gallenfarbstoffen. *Hoppe-Seyler's Z. Physiol. Chem.* 350: 1291–1300.
Rüdiger, W., Schoch, S. (1989). Abbau von Chlorophyll. *Naturwis.* 76: 453–457.
Sandmann, G. (1991). Biosynthesis of cyclic carotenoids: Biochemistry and molecular genetics of the reaction sequence. *Physiol. Plantarum,* 83: 186–193.
Sandmann, G., Bramley, P. M. (1985). The *in vitro* biosynthesis of ß-cryptoxanthin and related xanthophylls with *Aphanocapsa* membranes. *Biochim. Biophys. Acta,* 843: 73–77.
Schaber, P. M., Hunt, J. E., Fries, R., Katz, J. J. (1984). High-performance liquid chromatographic study of the chlorophyll allomerization reaction. *J. Chromatog.* 316: 25–41.
Scheer, H. (1991). Chemistry of chlorophylls. In: Scheer, H. (ed.), *Chlorophylls*. CRC Press, Boca Raton, USA, p. 3–30.
Scheer, H., Gross, E., Nitsche, B., Cmiel, E, Schneider, S., Schäfer, W., Schiebel, H.-M., Schuten, H.-R. (1986). Structure of methylphaeophorbide-RC1. *Photochem. Photobiol.* 43: 559–571.
Schiedt, K., Bischof, S., Glinz, E. (1993). Metabolism of carotenoids *in vivo*. Racemization of (3S,3'S)-astaxanthin in the crustacean *Penaeus*. In: Packer, L. (ed.), *Carotenoids*, Methods in Enzymol. vol. 214, 148–168.
Schoch, S., Klein, G., Linsenmeier, U., Rüdiger, W. (1976). Über die Bindungen zwischen Chromophor und Protein in Biliproteiden, III Eliminieriungs-reaktionen an Modellsubstanzen. *Liebigs Ann. Chem.* 115: 549–558.
Schoch, S., Rüdiger, W., Lüthy, B., Matile, P. (1984). 13^2-Hydroxychlorophyll *a*, the first product of the reaction of chlorophyll oxidase. *J. Plant Physiol.* 115: 85–89.
Schoch, S., Scheer, H., Schiff, J. A., Rüdiger, W., Siegelman, H. W. (1981). Pyropheophytin *a* accompanies pheophytin *a* in dark and light grown cells of *Euglena. Z. Naturforsch.* 36c: 827–833.
Senge, M., Senger, H. (1988). Chlorination of isolated chlorophylls *in vitro*. *Photochem. Photobiol.* 48: 711–717.
Senge, M., Struck, A, Dörnemann, D., Scheer, H., Senger, H. (1988). Hydroxylation of chlorinated and unchlorinated chlorophylls *in vitro*. *Z. Naturforsch.* 43c: 515–518.
Seubert, A., Seubert, S., Ippen, H. (1979). Eine neue schnelle Methode zur Trennung von Porphyrinen mittels HPTLC. *Dtsch. Med. Wochenschr.* 104: 1459.
Shioi, Y. (1991) Analytical chromatography of chlorophylls In: Scheer, H. (ed.), *Chlorophylls*. CRC Press, Boca Raton, p. 59–88.
Siefermann, D., Yamamoto, H. Y. (1975*a*). Properties of NADPH and oxygen-dependent zeaxanthin epoxidation in isolated chloroplasts. *Arch. Biochem. Biophys.* 171: 70–77.
Siefermann, D., Yamamoto, H. Y. (1975*b*). NADPH and oxygen-dependent epoxidation of zeaxanthin in isolated chloroplasts. *Biochem. Biophys. Res. Commun.* 62: 456–461.

Siefermann-Harms, D. (1977). The xanthophyll cycle in higher plants. In: Tevini, M., Lichtenthaler, H. K. (eds), *Lipids and Lipid Polymers in Higher Plants.* Springer, Berlin, p. 218–230.

Siefermann-Harms, D. (1985). Carotenoids in photosynthesis. I. Location in photosynthetic membranes and light-harvesting function. *Biochim. Biophys. Acta*, 811: 325–355.

Siefermann-Harms, D. (1987). The light-harvesting and protective functions of carotenoids in photosynthetic membranes. *Physiol. Plantarum*, 69: 561–568.

Siefermann-Harms, D., Michel, J.-M., Collard, F. (1980). Carotenoid transformation underlying the blue absorbance change in flashed bean leaves during the induction of oxygen evolution. *Biochim. Biophys. Acta*, 589: 315–323.

Sieg, I., Doss, M. O. (1992). HPLC der Koproporphyrin-I/III-Isomere bei hepatischen Porphyrinopathien. *Lab. Med.* 16: 89–96

Simpson, K. L., Rodriguez, T. G., Chichester, C .D. (1976). Metabolism in senescent and stored tissues. In: Goodwin, T. W. (ed.), *Chemistry and Biochemistry of Plant Pigments,* Vol. 1, Academic Press, London, p. 779–824.

Strain, H. H., Manning, W. M. (1942). Isomerization of chlorophylls *a* and *b*. *J. Biol. Chem.* 146: 275–276.

Stransky, H., Hager, A. (1970). Das Carotinoidmuster und die Verbreitung des lichtinduzierten Xanthophyllcyclus in verschiedenen Algenklassen. *Arch. Mikrobiol.* 73: 315–323.

Struck, A., Scheer, H. (1990). Modified reaction centers from *Rhodobacter sphaeroides* R26. 1. Exchange of monomeric bacteriochlorophyll with 13^2-hydroxy-bacteriochlorophyll. *FEBS Lett.* 261: 385–388.

Sun, M-Y., Lee, C., Aller, R. C. (1993). Anoxic and oxic degradation of 14C-labelled chloropigments and 14C-labelled diatoms in Long Island Sound Sediments. *Limnol. Oceanogr.* 38: 1438–1451.

Thayer, S. S., Björkman, O. (1990). Leaf xanthophyll content and composition in sun and shade determined by HPLC. *Photosynth. Res.* 23: 331–343.

Thommen, H. (1971). Metabolism. In: Isler, O. (ed.), *Carotenoids*, Birkhäuser, Basel, p. 637–668.

Walker, C. J., Mansfield, K. E., Smith, K. M., Castelfranco, P. A. (1989). Incorporation of atmospheric oxygen into the carbonyl functionality of the protochlorophyllide isocyclic ring. *Biochem. J.* 257: 599–602.

Watanabe, T., Hongu, A., Honda, H., Nakazato, M., Konno, M., Saitoh, S. (1984). Preparation of chlorophylls and pheophytins by isocratic liquid chromatography. *Analyt. Chem.* 56: 251–256.

Weedon, B. C. L. (1971*a*). Occurrence. In: Isler, O. (ed.), *Carotenoids*, Birkhäuser, Basel, p. 29–59.

Weedon, B. C. L. (1971*b*). Stereochemistry. In: Isler, O. (ed.), *Carotenoids*, Birkhäuser, Basel, p. 267–323.

Wilhelm, C. (1990). The biochemistry and physiology of light-harvesting processes in chlorophyll *b*- and chlorophyll *c*-containing algae. *Plant Physiol. Biochem.* 28: 293–306.

Willstätter, R., Stoll, A. (1910). Über Chlorophyllase. *Ann. d. Chem.* 378: 18–72.

Yamamoto, H. Y. (1979). Biochemistry of the violaxanthin cycle in higher plants. *Pure Appl. Chem.* 51: 639–648.

Young, A. J. (1991). The photoprotective role of carotenoids in higher plants. *Physiol. Plantarum,* 83: 702–708.

Young, A. J., Barry, P., Britton, G. (1989). The occurrence of ß-carotene-5,6-epoxide in the photosynthetic apparatus of higher plants. *Z. Naturforsch.* 44c: 959–965.

Young, A. J., Britton, G. (1993). *Carotenoids in Photosynthesis*, Chapman and Hall, London, 498 pp.

Ziegler, R., Blaheta, A., Guha, N., Schönegge, B. (1988). Enzymatic formation of pheophorbide and pyropheophorbide during chlorophyll degradation in a mutant of *Chlorella fusca* Shihira et Kraus. *J. Plant Physiol.* 132: 327–332.

4

Application of pigment methods to oceanography

S. W. Jeffrey

4.1 Introduction

Deep in the sea are riches beyond compare;
But if you seek safety it is on the shore.
Saadi, 13th Century poet.

A brief overview of milestones in pigment oceanography (Chapter 1) sets the scene for the present chapter. Here examples of the use of different methods in field situations show how method development has opened new windows on the phytoplankton and their role in the global oceans. Details of the various methods are found in the chapters and appendices and will not be treated here. Each new advance created a 'population explosion' of new papers and it was difficult to select from the wealth of data available from the vast oceanographic literature. This Chapter is dedicated to those whose work contributed to the advances outlined in this brief review.

Detection of pigments relies on the absorption and/or fluorescence properties of the pigment molecules — chlorophylls, carotenoids and biliproteins — whether *in situ*, as pigment-protein complexes in the phytoplankton cell, or extracted in a monomeric state in a solvent. Pigments may be detected using a spectrophotometer, fluorometer or microscope in the laboratory, ashore or at sea. They may also be detected from sensors attached to aircraft or satellites (see Fig. 1.2, Chapter 1; Table 17.1, Chapter 17). Combined with separation techniques of increasing selectivity and resolution, oceanographers now have powerful tools available for phytoplankton exploration in the sea.

4.2 Spectrophotometry

The first spectrophotometric method described for the estimation of the three chlorophylls *a*, *b* and *c* in phytoplankton extracts was that of Richards with Thompson (1952). Because the extinction coefficients used were low, this method over-estimates chlorophyll *a*, especially if chlorophyll degradation products are also present in the sample (see Appendix F; Chapter 14). The chlorophyll *c* values are in 'millispecific pigment units', because no extinction coefficients were available at the time. Methods for recalculation, assuming the absence of degradation products, are given by Humphrey (1978) and Wartenberg (1978).

An example of this widely used method is shown in Figure 4.1 for the south-eastern Indian Ocean (Humphrey, 1966). In this paper Humphrey used the first extinction coefficients for chlorophyll *c* (Jeffrey, 1963), resulting in chlorophyll *c* values (in $\mu g\ l^{-1}$) which are too high (*c.f.* Jeffrey's (1972) revised extinction coefficients for crystalline chlorophylls c_1 and c_2). Phytoplankton samples were collected from 1959 to 1962 in both summer and winter. Summer depth profiles of chlorophylls *a* and *c* are averaged for each depth over a given square. At most stations sub-surface maxima at about 75 m depth are clearly seen, with chlorophyll *a* and *c* values of 0.14 to 0.19 and 0.24 to 0.35 $\mu g\ l^{-1}$ respectively at the maximum. While these values are both over-estimated by the method (chlorophyll *a* by about 25%; chlorophyll *c* by about 100%; see Appendix F; and Lorenzen and Jeffrey, 1980), this and related studies (Humphrey, 1960; Humphrey and Kerr, 1969; Humphrey, 1970) were the first extensive records of phytoplankton biomass data (chlorophyll measurements) to be published for both the Indian and Pacific Oceans. Later, a comprehensive atlas of data from the Indian Ocean gave both chlorophyll and productivity profiles (Krey and Babenerd, 1976).

Improvements in the trichromatic spectrophotometric technique using more accurate extinction coefficients were published in the next decade by Parsons and Strickland (1963), UNESCO (1966), and Jeffrey and Humphrey (1975). These equations are all accurate for chlorophyll *a*, but only those of Jeffrey and Humphrey (1975) have terms which are also accurate for chlorophylls *b* and *c* (see Appendix F, G). All of these equations were derived for mixtures of pure chlorophylls, and none allow for the presence of similarly absorbing chlorophyll degradation products. They are

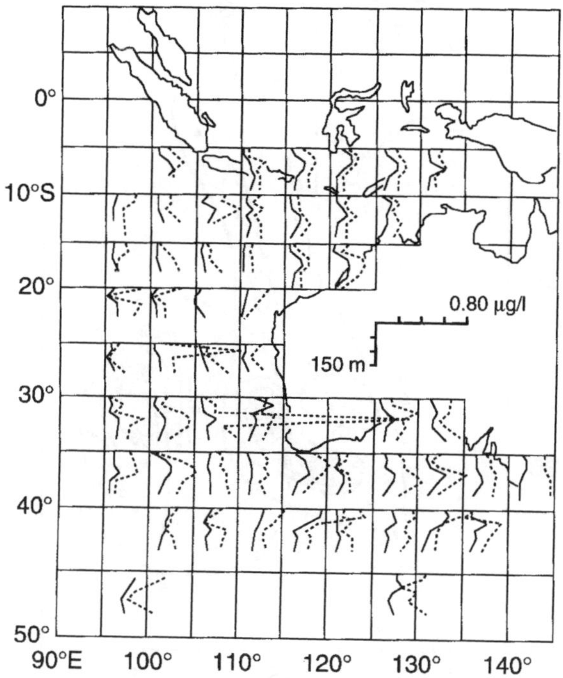

Figure 4.1
Summer depth profiles of chlorophylls *a* (——) and *c* (········) in µgl^{-1} in the south-east Indian Ocean. The data was collected over the period 1959–1962. The mean value at each depth for a given square is plotted. Note that chlorophyll *a* is over-estimated by about 25%, chlorophyll *c* by about 100% — see text section 4.2 (from Humphrey, 1966; chlorophyll method used, Richards with Thompson, 1952).

therefore in error to varying degrees if interfering pigments are present (see Lorenzen and Jeffrey, 1980; and Chapter 14). However valuable data continues to be obtained with these techniques if used appropriately (see Chapter 14).

The Jeffrey and Humphrey (1975) method was used to measure chlorophyll contours in and around a coalesced eddy system (Eddy Mario) of the East Australian Current (Figure 4.2; Jeffrey and Hallegraeff, 1987*a*). Thirty-five stations were occupied, and chlorophylls were determined in depth profiles and horizontal transects from discrete, filtered and extracted samples. Areas rich in chlorophyll resulting from convective overturn and upwelling at the eddy centre and the western margins respectively were identified, with highest chlorophyll *a* values of 60.1±10.5 mg m^{-2} at the eddy centre, and lowest values of 26.0±2.8 mg m^{-2} north-west of centre. The accuracy of the equations in the upper water column is suggested by inspection of UV-visible absorption spectra of the phytoplankton extracts both above and below 100 m (Fig. 4.3A, B). The Soret (blue):red band ratios of 2.50 to 2.69 above 100 m are indicative of healthy algal cells (diatom bloom), whereas the very high ratios of 5.61 to 7.31 in the eddy centre below 100 m (Fig. 4.3B), indicate the presence of

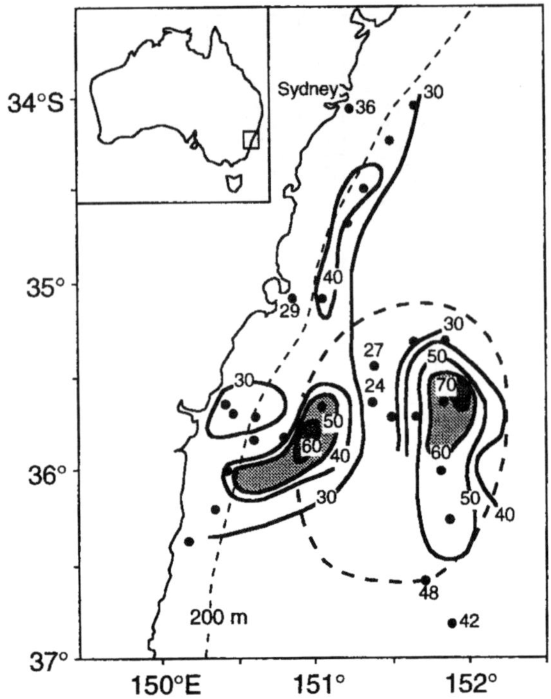

Figure 4.2
Water column chlorophyll contours (mg m^{-2}) in and around Eddy Mario in the East Australian Current, May, 1981. Dashed line (- - - -) indicates the approximate edge of Eddy Mario (from Jeffrey and Hallegraeff, 1987; chlorophyll method used, Jeffrey and Humphrey, 1975).

degradation products from senescent cells or detrital material. Indeed, temperature-depth profiles (not shown, see Jeffrey and Hallegraeff, 1987*a*) indicated that the detritus-rich waters below 100 m were the remains of an 'old' submerged eddy system (Leo). This data was a clear case of a pigment signature combined with a temperature profile indicating a specific water mass.

The acidification method of Vernon (1960) was used to correct analyses of chlorophyll *a* in higher plant tissues from interference from Mg-free chlorophyll derivatives (pheophytins and pheophorbides, collectively termed 'pheopigments'). This method was further developed for spectrophotometry and fluorometry of phytoplankton (Lorenzen, 1967; Yentsch and Menzel, 1963; Holm-Hansen *et al.* 1965). The accuracy of these methods was tested by Lorenzen and Jeffrey (1980) and participants of the SCOR WG 78 (see Chapter 14). The corrections improve the assessments of chlorophyll *a*, but the wide-range of degradation products now known to be included in the 'pheopigment' fraction (see Chapter 14), makes this measurement less useful. Values were reliable when chlorophyll *a* and pheopigments were the only pigments present. For further details see Chapter 14 and Appendix F.

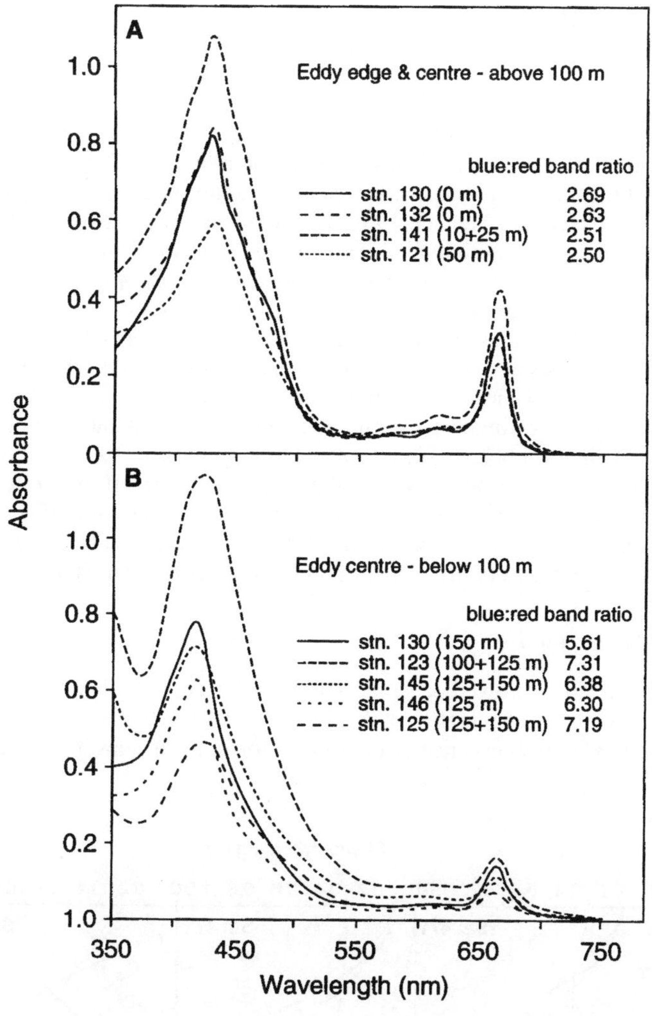

Figure 4.3
Absorption spectra of phytoplankton pigments in diethyl ether from selected stations and depths in and around Eddy Mario, East Australian Current, May, 1981. These spectra display characteristics associated with living phytoplankton or detritial material. A, Blue:red ratios, 2.50 to 2.69, eddy edge and centre stations, above 100 m depth; B, blue:red ratios, 5.61 to 7.31, eddy centre stations, 100–200 m depth (from Jeffrey and Hallegraeff, 1987*a*).

4.3 Fluorometry

4.3.1 *Extracted fluorometry*

Fluorometric measurement of chlorophyll *a* is up to 50 times more sensitive than spectrophotometry, allowing analyses of much smaller seawater (phytoplankton) samples. For this and other reasons, fluorometric analysis of chlorophylls became very popular (Yentsch and Menzel, 1963; Holm-Hansen *et al.* 1965), and is widely used by the oceanographic community. Fig. 4.4 shows the use of this method (Yentsch and Menzel, 1963) in depth profiles of chlorophyll *a* and 'pheopigments' at selected stations in the southern Atlantic Ocean (Yentsch, 1965). Two stations (471 and 472) showed chlorophyll *a* values greater than those for pheopigments, but at station 484, the pheopigment values at depth exceeded those of chlorophyll *a*. This fluorometric acidification technique is now known to be inaccurate when chlorophyll *b* is present (Fig. 4.5), grossly overestimating the pheopigment component and limiting the reliability of the method (Gibbs, 1979; Lorenzen, 1981; Gieskes, 1991). Chlorophyll *b* (green algae) is widely distributed even in oligotrophic waters (Jeffrey, 1976*a*; Gieskes and Kraay, 1986*a*, *b*; Vernet and Lorenzen, 1987) and a new modification of the method has been introduced (Welschmeyer, 1994) to help correct for it. Quantitative confirmation of these observations is presented in detail in Chapter 14.

4.3.2 *In vivo fluorometry*

A spectacular advance allowed oceanographers to detect phytoplankton chlorophyll fluoresence *in situ* (Lorenzen, 1966). A Turner model III fluorometer was adapted for recording *in vivo* fluorescence of phytoplankton cells in seawater, using a flow-

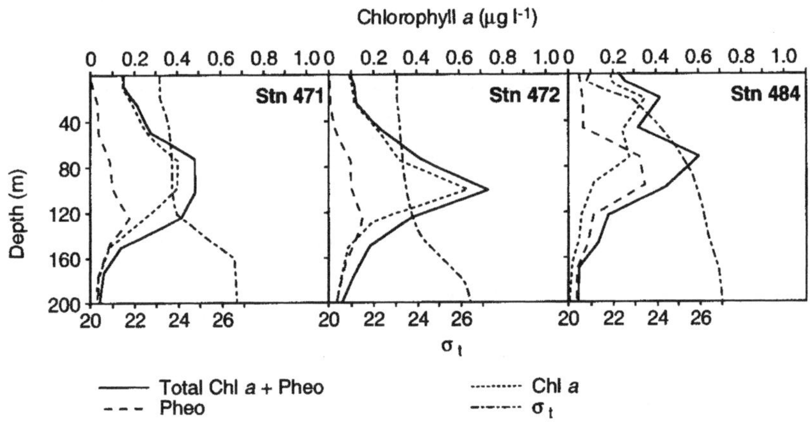

Figure 4.4
Depth distribution of chlorophyll *a* + pheopigment (——), chlorophyll *a* (········), pheopigment (- - - -) and density expressed as sigma t (-·-·-), from selected stations in the southern Atlantic Ocean off Brazil, December, 1964 (adapted from Yentsch, 1965; method used, Yentsch and Menzel, 1963).

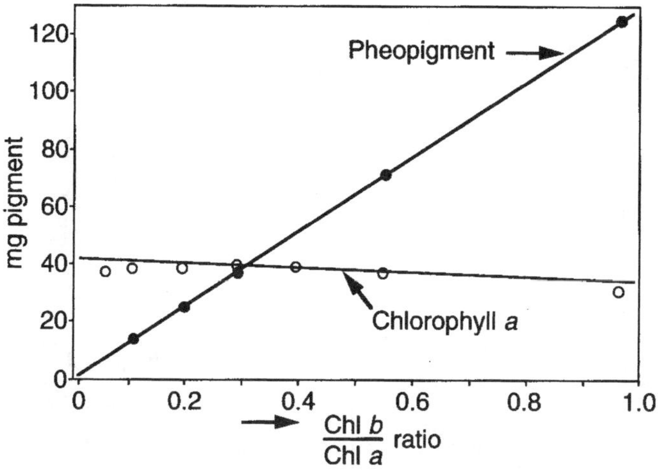

Figure 4.5
The effect of chlorophyll *b* on the estimation of chlorophyll *a* and the over-estimation of 'pheopigments' using the fluorometric acidification technique of Yentsch and Menzell (1963) and Holm-Hansen *et al.* (1965) (from Gieskes, 1991).

through cuvette, to give a continuous fluorescence record. The seawater was pumped into the shipboard laboratory either from the engine-cooling water or from a discrete intake line. The chlorophyll *a* concentration was checked at regular intervals by spectrophotometric or fluorometric sample analysis, allowing the *in vivo* fluorescence signal to be calibrated frequently. Underway horizontal profiles of phytoplankton abundance in surface waters could thus readily be obtained using this technique. It was not long before *in situ* instruments were also developed for vertical and horizontal fluorescence profiles — VarioSens (Früngel and Koch, 1976; Herman and Denmann, 1977), AquaTracka (Fasham *et al.* 1983) and SeaTech Inc. (Guo and Dunstan, 1995) being the most successful.

The use of two of these instruments is illustrated in Figs 4.6 and 4.7. *In vivo* fluorescence measurements taken with a SeaTech fluorometer in the Sargasso Sea showed a deep chlorophyll maximum just above 100 m depth (Guo and Dunstan, 1995; Fig. 4.6A). A VarioSens depth profile of chlorophyll fluorescence at a station in the Baltic Sea (Derenbach *et al.* 1979; Fig. 4.6B), shows a number of 'chlorophyll peaks' in and below the thermocline at 30 m depth. These were interpreted as reflecting an interaction between the microscale physical structure of the thermocline and the sinking of phytoplankton cells (see Cullen, 1982). However, Dr D. J. Mackey has shown in unpublished experiments with the VarioSens fluorometer (Mark III) that well-mixed algal cultures (e.g., *Lauderia annulata*, *Skeletonema costatum* (diatoms), *Pavlova lutheri* (prymnesiophyte) and *Micromonas pusilla* (prasinophyte)) all give sharp and variable fluorescence signals (noise) at low chlorophyll concentrations (Fig. 4.6C), similar to those often encountered in the field. He suggests that the sharp signals seen in data such as that in Fig. 4.6B may be artefacts due to the low sensitivity of the VarioSens and its approximately logarithmic output. Other in-

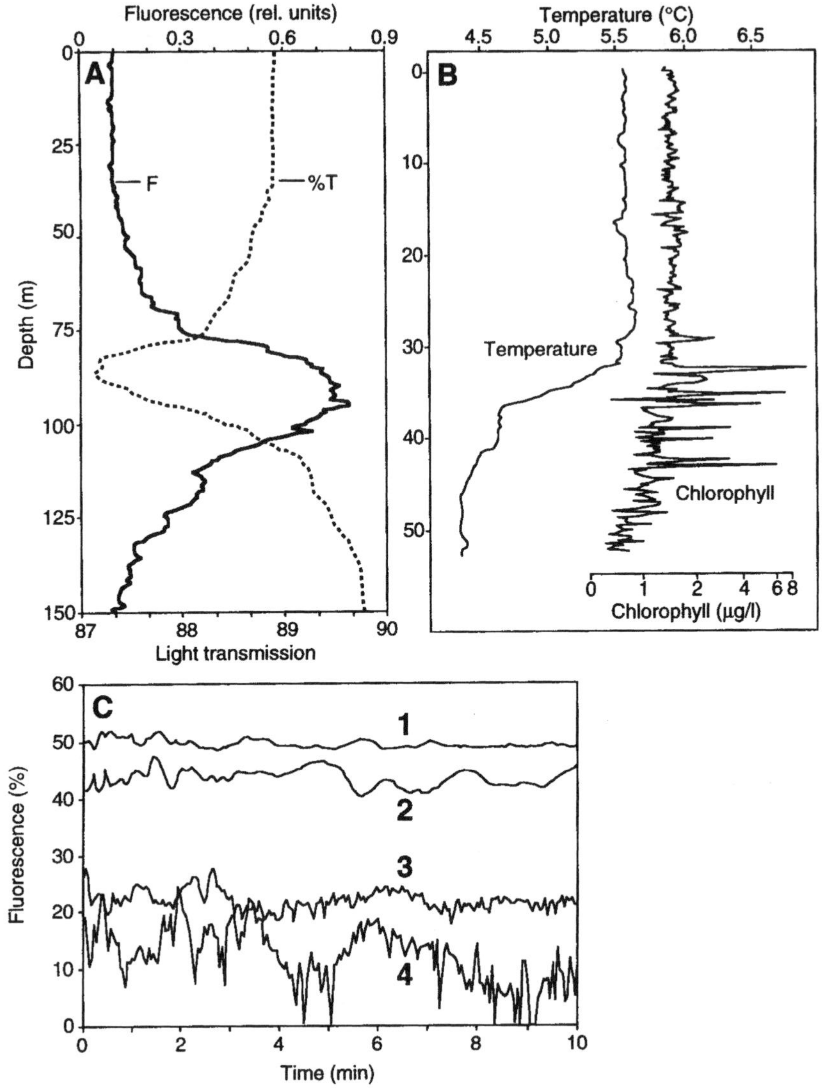

Figure 4.6
In vivo chlorophyll fluorescence profiles in the sea. A. Depth profile of *in situ* fluorescence and % light transmission using SeaTech Inc. instruments at a station in the Sargasso Sea, April, 1989 (from Guo and Dunstan, 1995); B. Vertical profile of temperature and chlorophyll *a* using a VarioSens fluorometer at a station in the Baltic Sea, May, 1975 (from Derenbach *et al.* 1979); C. *In vivo* fluorescence traces of well-mixed cultures of the diatom *Lauderia annulata* at four chlorophyll concentrations. 1, 8.89 µg Chl *a* l^{-1}; 2, 4.45 µg Chl *a* l^{-1}; 3, 0.56 µg Chl *a* l^{-1}; 4, 0.28 µg Chl *a* l^{-1}. (Dr D. Mackey, unpublished observations).

struments, such as the SeaTech *in situ* fluorometer, have a linear response to chlorophyll, are more sensitive and do not show such 'noisy' fluorescence responses at very low phytoplankton densities (Dr D. J. Mackey, unpublished data). Falkowski and Kolber (1995), however, interpret similar signals to indicate real changes in phytoplankton patch-scale dynamics, due to turbulence, light fluctuations and intrapatch heterogeneities. The relative importance of these effects requires further investigation.

In addition to instrument problems, it is well known that *in vivo* fluorescence is only a semi-quantitative measure of chlorophyll concentration due to the variability of the chlorophyll fluorescence signal (see Appendix F). Factors contributing to this variability include the non-uniformity of cells (shapes and sizes) in field phytoplankton populations, their fluctuating nutrient status, their previous light history, the presence of senescent cells and 'inactive' chlorophyll, their location, temperature history (season), time of day the measurements are taken, and the variable fluorescence of chlorophyll degradation products (Jeffrey and Hallegraeff, 1990). Strickland (1968) tested the *in vivo* fluorescence response of seven algal cultures from four algal classes and showed that the relative fluoresence response varied from 0.13 to 0.67 fluorescence units for 1 µg chlorophyll *a* per litre. Nevertheless, *in situ* fluorescence profiling of chlorophyll, both vertically (Fig. 4.6) and horizontally (Fig. 4.7), is excellent when used at sea for obtaining rapid indications of phytoplankton distribution, and particularly for locating the deep chlorophyll maximum (see Venrick *et al.* 1973). Appropriate locations can then be targetted for more detailed and accurate study by other methods.

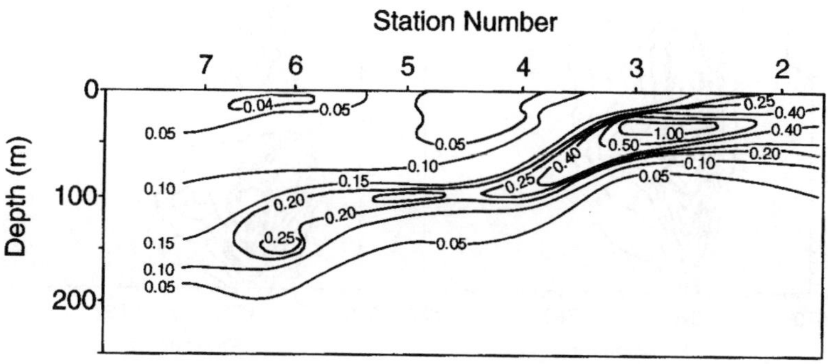

Figure 4.7
Horizontal and vertical distribution of chlorophyll *a* obtained by an *in situ* VarioSens fluorometer at seven stations in the eastern tropical North Atlantic Ocean, off the West Coast of Africa, September-October, 1977 (from Gieskes *et al.* 1978). Units, µg Chl *a* l^{-1}.

4.3.3 *Fluorescence spectral signatures*

Attempts have been made to identify algal types from their *in vivo* excitation and emission fluorescence characteristics. These techniques (see Fig. 4.8) rely on either pigment complement (Fig. 4.8A, Yentsch and Yentsch, 1979; Yentsch and Phinney, 1985) or total fluorescence spectral matrices (Fig. 4.8B, Oldham *et al.* 1985). Four basic phytoplankton types with specific excitation – emission spectra were distin-

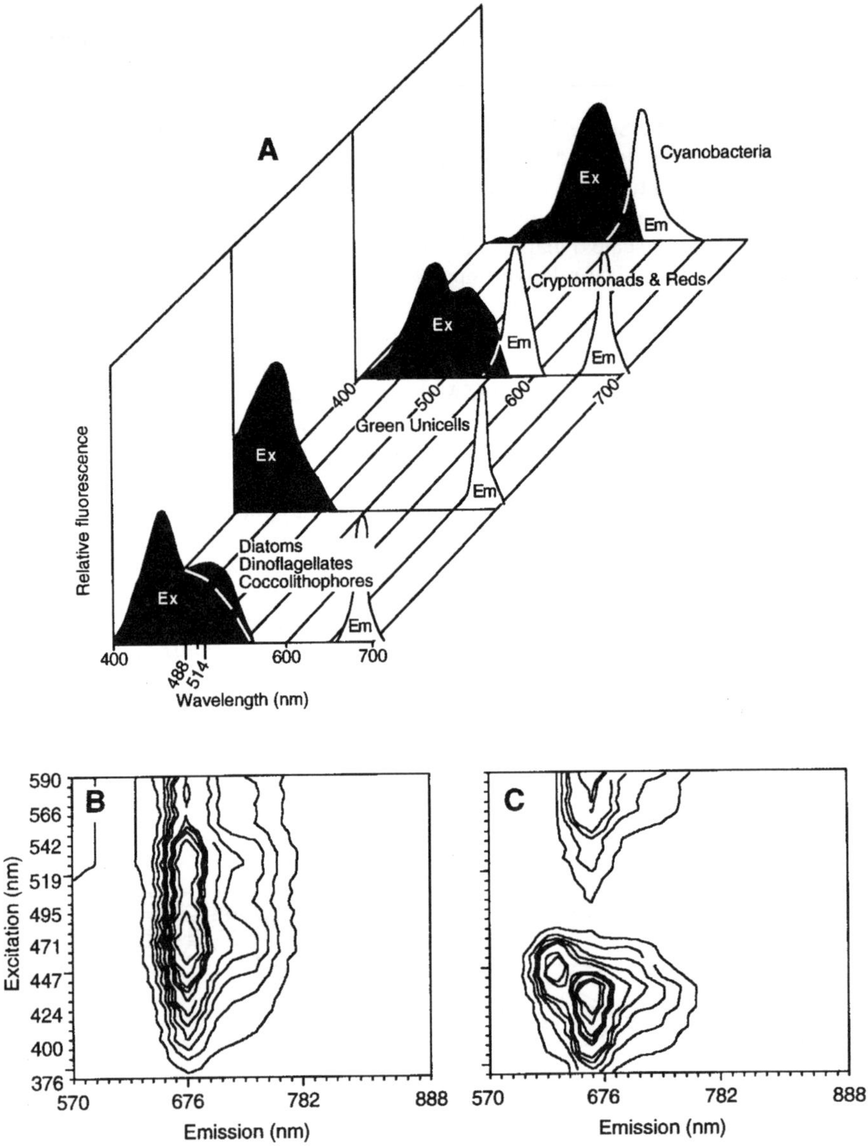

Figure 4.8
Fluorescence characteristics of algal types. A, Spectral fluorescence signatures of major phytoplankton groups (from Yentsch and Phinney, 1985). B, C, Two-dimensional contour plots of *Amphidinium carterae* pigment fluorescence, B, *in vivo* and C, in 90% acetone extracts. Note the decoupling of energy transfer after pigments are extracted into solvent (from Oldham *et al.* 1985).

guished by Yentsch and Phinney (1985): chromophyte algae (chlorophylls *a*, *c* and light-harvesting carotenoids), green algae (chlorophylls *a* and *b*), cryptomonads and red algae (chlorophyll *a* and phycoerythrin) and cyanobacteria (chlorophyll *a* and phycocyanin).

These groupings were confirmed and extended by studies of cultured microalgae from eight classes (Poryvkina *et al.* 1994) in which excitation and emission signals were displayed in 2-dimensional matrix plots (Fig. 4.8B, C; Oldham *et al.* 1985). However Soo Hoo *et al.* (1986), and Hilton *et al.* (1989) showed in careful studies of algal cultures that photoadaptation can dramatically change pigment content and fluorescence responses, the latter being as great as the taxonomic differences in algal pigmentation. Photoadaptive effects must therefore always be considered when interpreting vertical profiles of fluorescence signatures, especially where light gradients are large. On present evidence it seems that *in vivo* fluorescence matrix spectroscopy is unlikely by itself to adequately identify algal classes or species in mixed phytoplankton field populations, unless the cells are examined individually, as in cell flow cytometry (Chisholm *et al.* 1988; Olson *et al.* 1988, 1989). Further work will obviously be needed before the gross *in vivo* spectral fluorescence signatures described above can be applied to the full characterization of natural phytoplankton communities.

4.3.4 *'Active' fluorescence as a measure of phytoplankton photosynthetic quantum efficiency*

An innovative advance in *in situ* fluorescence technology has been made by Kolber and Falkowski (1993). This method estimates the maximum quantum efficiency of photosystem II (PSII, see Chapter 3) in natural phytoplankton communities by a 'pump and probe' fluorometric technique. Both ship-based laboratory and *in situ* instruments were developed (Falkowski and Kolber, 1995). The change in the quantum yield of fluorescence ($\Delta\phi_{max}$) of a weak 'probe' flash, preceding and succeeding an actinic 'pump' flash, gave measurements of the kinetics of phytoplankton photosynthetic electron transfer, which could (from biophysical models) be related to ocean depth, irradiance and time. Profiles of dark-adapted $\Delta\phi_{max}$ and the effective absorption cross-section of PSII, δ_{PSII}, measured at many ocean sites and areas, showed marked variability.

Areas of upwelling, such as off the north-west coast of Africa, have variable fluorescence (F_v) to maximum florescence (F_m) ratios of 0.65, close to the maximum achievable in nutrient-replete cultures. However, in nutrient-deficient areas of central ocean basins, quantum efficiency may be reduced by more than 50%, suggesting a loss of functional PSII reaction centres. Both nitrogen and iron-limitation have been implicated.

From knowledge of the incident spectral irradiance, $\Delta\phi_{max}$, δ_{PSII} and photochemical quenching, the absolute photosynthetic electron transport rate can be obtained non-destructively in real time. Falkowski and Kolber (1995) are using untended, moored, *in situ* fluorometers to obtain continuous measurements of *in vivo* fluorescence parameters, so that satellite observations of the global ocean may be related to basin-scale seasonal estimates of phytoplankton carbon fixation. Chlorophyll fluorescence, used in this mode, may thus link biophysical responses to ocean dynamics.

4.4 Chromatographic techniques

The classical methods of column chromatography used by Strain and colleagues to separate algal pigments (Strain, 1958; see Chapter 7) guided the development of quantitative micromethods for phytoplankton pigments in the mid-sixties. It was becoming evident that complex mixtures of pigments, such as those likely to be found in field samples, could not be assayed accurately by simple spectrophotometric and fluorometric methods for chlorophyll *a*, even when steps to distinguish the 'pheopigment' component were included (Jeffrey, 1968; see Appendix F; Chapter 14). The search for greater precision demanded that separation techniques be sought on the microscale level for the wide range of algal pigments likely to be encountered in seawater samples, allowing for more accurate assessments of chlorophylls.

4.4.1 *Thin-layer chromatography*

For a chromatographic technique to be useful in oceanography, it must be rapid and simple to execute, it must resolve all major components quantitatively (both chlorophylls and carotenoids), the adsorbants must be innocuous and cause no chemical change to the pigments, and it must be able to resolve the small (<1 µg) quantities of pigment available from a 'normal' seawater sample (e.g., 5–10 l). The excellent TLC methods used by Liaaen-Jensen and her group for carotenoid analysis (see Liaaen-Jensen and Jensen, 1971; Kjøsen and Liaaen-Jensen, 1972; and related papers) were not suitable for samples containing chlorophylls, which were damaged by the inorganic adsorbants used. The systems of Hager and Stransky (1970), which used oil-impregnated silica gel, were also not adopted, because solvent elution procedures caused unwanted oil-contamination of purified pigments. Appropriate resolution of algal pigments was finally obtained with organic adsorbants such as glucose (Madgwick, 1965), sucrose and cellulose (Jeffrey, 1968). A rapid (10 min) 2-dimensional system separated over 25 major algal chlorophylls, carotenoids and chlorophyll degradation products with greater than 90% quantitative recovery (Jeffrey, 1968). The method, originally developed using algal cultures, was later successfully adapted for phytoplankton field samples (Jeffrey, 1974; 1976*a*).

A depth profile of phytoplankton pigments separated by cellulose thin-layer chromatography (TLC) is given in Figure 4.9A (Jeffrey, 1974). The full complement of diatom pigments occurred throughout the 100 m water column at this coastal station off Sydney, Australia. Chlorophyll *b* (green algae) was found at 8 to 28 m depth, astaxanthin (a copepod carotenoid) occurred at three depths from 8 to 75 m, while the degradation products, chlorophyllide *a* (senescent diatoms or filtering artefact; Jeffrey and Hallegraeff, 1987*a*), pheophytin *a* and pheophorbide *a* (zooplankton faecal pellets) were distributed through the water column. The results are summarized in Figure 4.9B to show the distribution of living algal populations and detrital material.

Chlorophyll *b* was also identified by TLC in the oligotrophic waters of the central North Pacific Gyre (Jeffrey, 1976*a*). This unexpected observation showed that green algae were present in the oligotrophic ocean far from the coastal environment they were thought to occupy. Their widespread occurrence necessitated modification of the fluorescence technique for chlorophyll *a* and 'pheopigments' (see section 4.3), recently developed by Welschmeyer (1994) (see Appendix F). In hindsight, the then unknown picoplanktonic prochlorophyte, *Prochlorococcus marinus*, discovered in the 1980s, may also have contributed some divinyl chlorophyll *b* (DVchl *b*) to these samples.

An improved and more robust cellulose TLC method (Jeffrey, 1981) was used to investigate new chlorophyll *a* degradation products in field samples (see Fig. 4.10; Hallegraeff and Jeffrey, 1985). Phytoplankton populations in the East Australian Current and its warm core eddies (Hallegraeff, 1981; Jeffrey and Hallegraeff, 1980*a*,

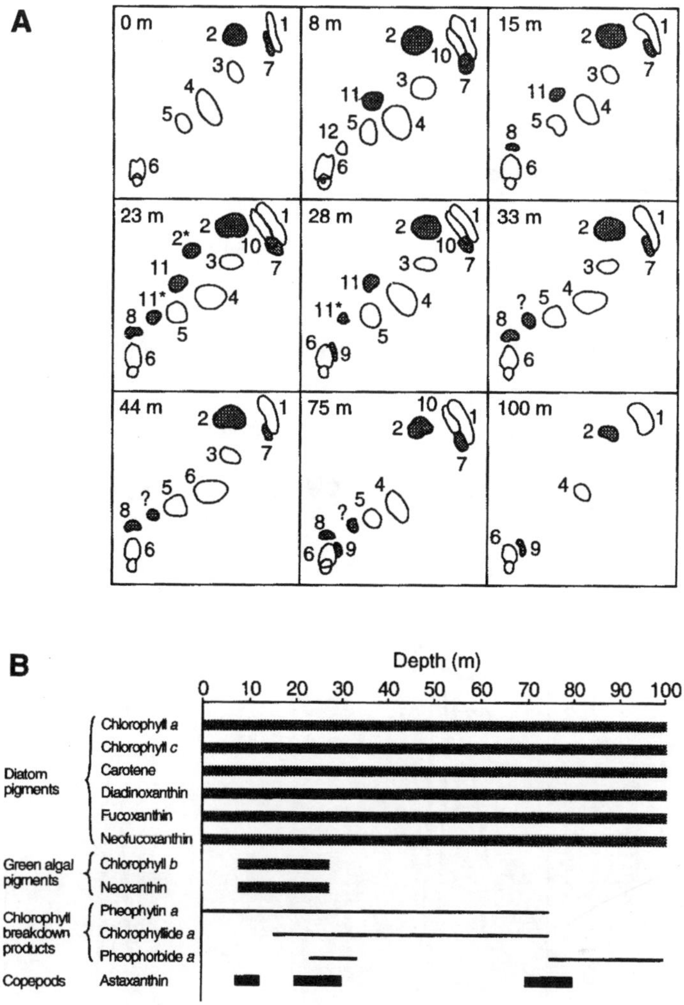

Figure 4.9
Phytoplankton pigments at a coastal station off Sydney, October 1970 (34°05'30"S; 151°15'30"E). A. TLC of photosynthetic pigments through the water column at 9 depths. 1, carotene (orange); 2, chlorophyll *a* (blue-green); 2*, chlorophyll *a* derivative (blue-green); 3, diadinoxanthin (yellow); 4, fucoxanthin (orange); 5, fucoxanthin derivative (orange); 6, chlorophyll *c* (light-green); 7, pheophytin *a* (grey); 8, chlorophyllide *a* (blue-green); 9, pheophorbide *a* (grey); 10, astaxanthin (pink); 11, chlorophyll *b* (olive-green); 11*, chlorophyll *b* derivatives (olive-green); 12, neoxanthin (yellow); 13, peridinin (red); 14, peridinin derivative (red). B. Profiles of pigments with depth from chromatograms shown in A. Solid bars: pigments associated with living cells; narrow bars: pigments associated with detrital material (from Jeffrey, 1974).

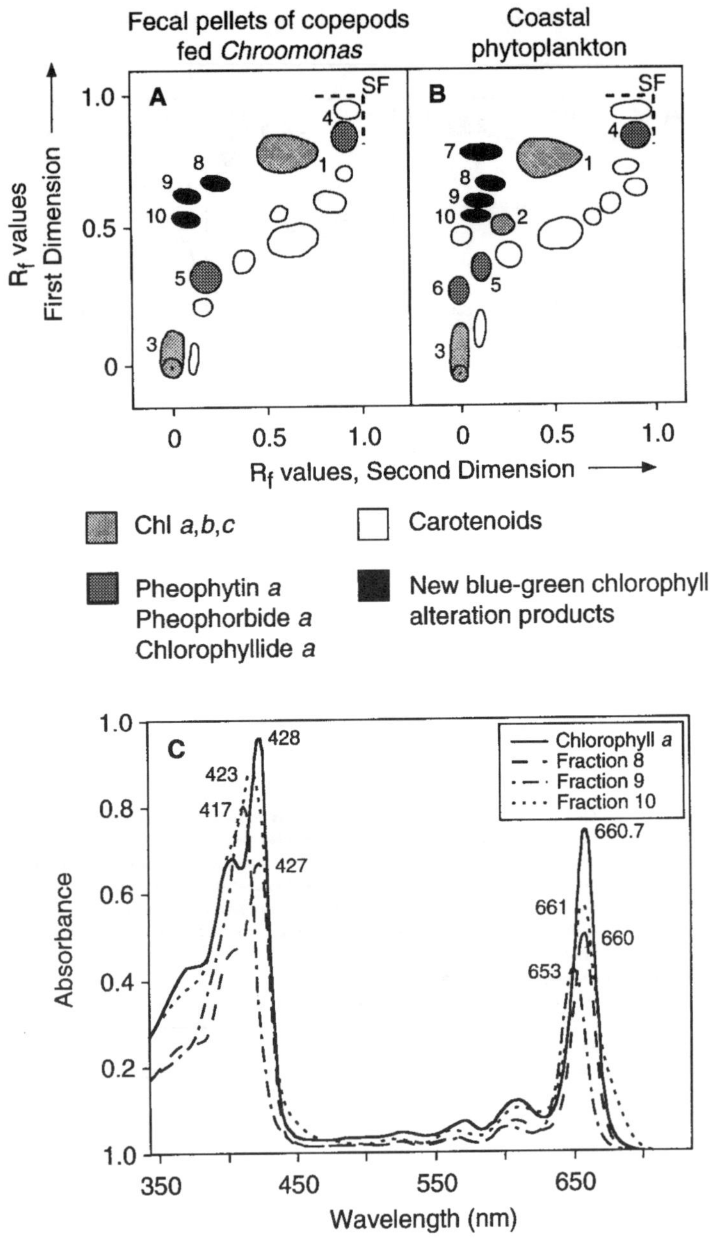

Figure 4.10
Thin-layer chromatograms of chlorophylls and carotenoids in A, faecal pellets of copepods fed cultures of the cryptomonad, *Chroomonas* sp. and B, phytoplankton from 40 m depth at a coastal station off Sydney, February, 1979. Pigment fractions: 1, chlorophyll *a*; 2, chlorophyll *b*; 3, chlorophyll *c*; 4, pheophytin *a*; 5, pheophorbide *a*; 6, chlorophyllide *a*; 7–10, new blue-green chlorophyll *a* alteration products. C. Absorption spectra of chlorophyll *a* and three blue-green chlorophyll *a* derivatives in diethyl ether. Fractions 8, 9 and 10 are similar to 10-methoxylactone derivatives of chlorophyll *a*, 10-hydroxy chlorophylls or isomerization products respectively (from Hallegraeff and Jeffrey, 1985).

b; Jeffrey and Hallegraeff, 1987*a*), and tropical waters of north and north western Australia (Hallegraeff and Jeffrey, 1984) were also studied by this method. Phytoplankton populations in the Australian region were dominated by a fucoxanthin-chlorophyll *c* photosynthetic system (diatoms, prymnesiophytes). More than 50% of the dinoflagellates were non-photosynthetic (lack of peridinin), green microalgae (chlorophyll *b*) were abundant even in off-shore phytoplankton populations, and pheophytin *a*, pheophorbide *a*, chlorophyllide *a* and other blue-green chlorophyll derivatives (Fig. 4.10) were important chlorophyll breakdown products in these waters. A number of signature pigments were clear indicators of algal types and processes (Jeffrey, 1974). In spite of the value of these observations to phytoplankton ecology, chromatographic techniques were slow to gain acceptance in oceanography, until fully automated liquid chromatography techniques became available. Then the chromatographic explosion really took off! — see Fig. 5.1, Chapter 5.

4.4.2 *High-performance liquid chromatography*

4.4.2.1 *Classical methods*

High-performance liquid chromatographic systems were developed for higher plant pigments in the late 1970's and early 1980's (e.g., Eskins *et al.* 1977; reviewed by Schwartz and von Elbe, 1982; Roy, 1987). Subsequently HPLC systems of increasing complexity were published for chlorophylls and carotenoids from microalgae and natural phytoplankton populations (Abaychi and Riley, 1979; Mantoura and Llewellyn, 1983; Wright and Shearer, 1984; Gieskes and Kraay, 1983*a*, *b*; 1986*a*, *b*). Both reversed-phase and normal phase systems were used.

No one method was ideal for the 50 or so chlorophylls, carotenoids and degradation products likely to be important in aquatic systems (see Table 2.3, Chapter 2). Not only were selective HPLC methods needed for accurate quantitative analysis of chlorophylls free from degradation products but pigments unique to certain algal classes, taxa or processes needed to be unequivocally separated. The value of such signature pigments in field oceanography was becoming apparent.

One of the objectives of SCOR WG 78 was to evaluate gradient HPLC systems for phytoplankton pigments. Up to 1988, the best methods for the separation of phytoplankton carotenoids was that of Wright and Shearer (1984), and for the full range of polar and non-polar chlorophylls and derivatives was that of Mantoura and Llewellyn (1983). At the SCOR WG 78 chlorophyll workshop, Dr N. A. Welschmeyer suggested changes to the initial solvent system of Wright and Shearer (1984) which allowed complete separation of chlorophyllide *a*, chlorophyllide *b*, chlorophyll c_3, and chlorophyll $c_1 + c_2$ + Mg3,8DVP (one fraction).

This method also separated two pheophorbide *a* pigments, pheophytin *a* and *b* and chlorophyll *a* allomer and epimer fractions. While over 30 carotenoids were separated, including the important lutein and zeaxanthin (to base-line), two carotenoid pairs were imperfectly separated: 19'-hexanoyloxyfucoxanthin and 9'-*cis* neoxanthin, and 19'-butanoyloxyfucoxanthin and siphonaxanthin, respectively. Diode array detection allowed these pigments to be resolved spectroscopically if not chromatographically. The method was published as Wright *et al.* (1991) and is described in detail for mixed phytoplankton pigments in Chapter 12 and algal cultures in Chapter 13. Typical resolution of a suite of algal pigments (Fig. 4.11B) is compared with a chromatogram of leaf pigments from a higher plant (Fig. 4.11A).

Several investigators concentrated mainly on chlorophylls and derivatives.

Mantoura and Llewellyn (1983) included a simple system which separated chlorophyll *a* from nine known and several unknown chlorophyll pigments found in field samples. This greatly improved quantitative chlorophyll *a* assays. Sartory (1985) and Zapata *et al.* (1987) (see Fig. 4.11C) increased this resolution by separating 14 chlorophylls and 18 derivatives in a single run. HPLC fluorescence detection is extremely sensitive for chlorophylls and excludes the carotenoids, while absorbance detection around 440 nm can be used for both chlorophylls and carotenoids (see Fig. 4.11).

While normal phase HPLC systems are in general less selective for algal pig-

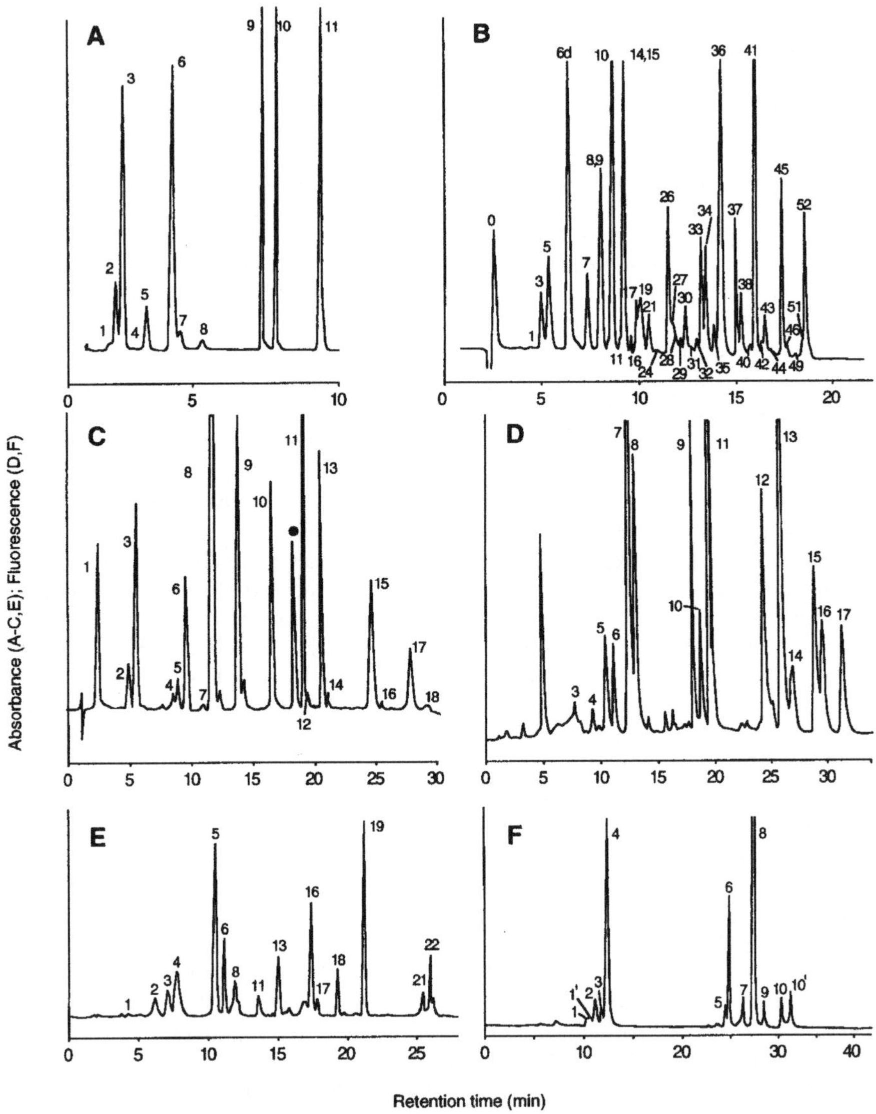

ments than reversed-phase methods, new advances were nevertheless made by using them. Gieskes and Kraay (1983*a*) detected a new chlorophyll *a* derivative, subsequently identified in reversed-phase systems as divinyl chlorophyll *a* (Goericke and Repeta, 1993). A new chlorophyll *c* derivative was also detected in blooms of the prymnesiophyte *Corymbellus aureus* (Gieskes and Kraay, 1986*a*). This pigment was studied as chlorophyll c_3 (Jeffrey and Wright, 1987), and identified as the methoxycarbonyl derivative of chlorophyll c_2 (Fookes and Jeffrey, 1989; Appendix B).

Figure 4.11
Various HPLC separation systems for chlorophylls and carotenoids from higher plants and algal cultures. Note detection by absorbance in panels A–C, E and by fluorescence in panels D and F.
A, Separation of leaf pigments (*Populus nigra*) by a C_{18} reversed-phase system. Pigment fractions: 1, neoxanthin; 2, *cis*-neoxanthin; 3, violaxanthin; 4, taraxanthin; 5, antheraxanthin; 6, lutein; 7, zeaxanthin; 8, *cis*-lutein; 9, chlorophyll *b*; 10, chlorophyll *a*; 11, β,β-carotene (from de las Rivas *et al.* 1989).
B, Separation of mixed algal pigments from a green alga (*Dunaliella tertiolecta*), a diatom (*Phaeodactylum tricornutum*), a cyanophyte (*Synechococcus* sp.), a prasinophyte (*Pycnococcus provasolii*), a cryptomonad (*Chroomonas* sp.), a prymnesiophyte (*Emiliania huxleyi*) and a chrysophyte (*Pelagococcus subviridis*). Several other authentic carotenoids were added to the mixture. For the identity of pigment fractions see Fig. 12.1, and Table 12.2, Chapter 12.
C, Reversed-phase HPLC separation of chlorophylls and derivatives by Zapata *et al.* (1987). Pigment fractions: 1, chlorophyllide *b*; 2, methyl chlorophyllide *b*; 3, chlorophyllide *a*; 4, chlorophyll c_1+c_2; 5, pheophorbide *b*; 6, methyl chlorophyllide *a*; 7, methyl chlorophyllide c_1+c_2; 8, pheophorbide *a*; 9, methyl pheophorbide *b*; 10, methyl pheophorbide *a*; 11, chlorophyll *b*; 12, chlorophyll *b*' (epimer); 13, chlorophyll *a*; 14, chlorophyll *a*' (epimer); 15, pheophytin *b*; 16; pheophytin b' (epimer); 17, pheophytin *a*; 18, pheophytin a' (epimer); • = unknown.
D, Reversed-phase HPLC separation of chlorophyll derivatives in faecal pellets of the protozoan *Strobilidium* fed on cultures of *Isochrysis galbana* and *Micromonas pusilla* (from Strom, 1993). Pigment fractions (author's designations): 3, pheophytin *c*-like pigments; 4, blue-green chlorophyll *a*-like pigment; 5, pheophorbide b_1; 6, pheophorbide b_2; 7, pheophorbide a_4; 8, pheophorbide a_5; 9, unknown carotenoid; 10, pheophytin a_0; 11, pheophytin a_0'; 12, pheophytin b_1; 13, pheophytin a_1; 14, pheophytin a_1'; 15, pheophytin a_2; 16, pheophytin b_2; 17, pheophytin a_3.
E, Reversed-phase HPLC separation of chlorophyll *c* pigments in addition to chlorophylls *a*, *b* and carotenoids on a high carbon-loaded HPLC column (from Kraay *et al.* 1992). Pigment fractions: 1, chlorophyllide *a*; 2, chlorophyll c_3; 3, chlorophyll c_1 (+Mg3,8 DVP); 4, chlorophyll c_2; 5, fucoxanthin; 6, 19'-hexanoyloxyfucoxanthin; 8, unknown; 11, violaxanthin; 13, diadinoxanthin; 16, lutein; 17, zeaxanthin; 18, chlorophyll *b*; 19, chlorophyll *a*; 21, β,ε-carotene; 22, β,β-carotene.
F, Reversed-phase HPLC separation of chlorophylls, particularly chlorophyll *c* pigments, on a wide-pore polymeric C_{18} column (from Garrido and Zapata, 1993*a*). Pigment fractions: 1 and 1', chlorophyll c_3; 2, chlorophyll c_1; 3, Mg3,8 DVP; 4, chlorophyll c_2; 5, chlorophyll *b* allomer; 6, chlorophyll *b*; 7, chlorophyll *a* allomer; 8, chlorophyll *a*; 9, chlorophyll *a*' (epimer); 10 and 10', phytylated chlorophyll *c*-like. Note splitting of the chlorophyll c_3 fraction and the phytylated chlorophyll *c*-like fraction into 2 peaks. For more information on these 'split' fractions see Garrido *et al.* (1995).

4.4.2.2 *New advances*

Since the SCOR Workshops, further advances in HPLC pigment methods have been published. One is the improvement in 'pheopigment' resolution. Strom (1991; 1993) separated a number of pheophytin *a*-like and pheophorbide *a*-like fractions from open ocean samples and microalgae/zooplankton feeding experiments (see Fig. 4.11D). Five pigments designated as pheophorbide *a*-like (a_1 to a_5) and four as pheophytin *a*-like (a_0 to a_3) of unknown chemistry and function have been recorded.

Another advance is the introduction of C_8 columns for separating divinyl chlorophylls *a* and *b* (Goericke and Repeta, 1992). Modified C_{18} systems and polymeric silica columns were used to improve resolution of the chlorophyll *c* family (Kraay *et al.* 1992; Garrido and Zapata, 1993*a*, *b*; Garrido *et al.* 1995), and difficult carotenoid pairs (van Heukelem *et al.* 1992; 1994).

Kraay *et al.* (1992) separated chlorophylls c_1, c_2 and c_3 for the first time on an HPLC system using a high ionic strength solvent in combination with a high carbon-loaded C_{18} column (see Fig. 4.11E). Garrido and Zapata (1993*a*, *b*) also showed that improved resolution of polar and other chlorophylls on polymeric C_{18} columns was achievable (Fig. 4.11F). In this system, not only were chlorophylls c_1, c_2, c_3 and MgDVP separated, but chlorophyll c_3 and phytylated chlorophyll *c*-like fractions each split into two peaks. A new paper gives the spectral properties of these new chlorophyll *c* pigment pairs (Garrido *et al.* 1995), suggested to be divinyl derivatives.

Van Heukelem *et al.* (1992) used polymeric columns to separate several difficult carotenoid pairs: lutein and zeaxanthin, 19'-hexanoyloxyfucoxanthin and 9'-*cis* neoxanthin and β,ε- and β,β-carotene. 19'-Hexanoyloxyfucoxanthin and 9'-*cis* neoxanthin can now also be resolved to baseline in the system of Wright *et al.* (1991) (Dr S. W. Wright, unpublished data) using an Alltech Alltima C18 column instead of a spherisorb ODS2. The value of varying the column temperature for enhancing pigment resolution was also described (Van Heukelem *et al.* 1994).

It is obvious from these examples that HPLC methodology is steadily improving, and we can expect the discovery of new pigments and improved resolution of 'old' ones with further developments. The hope that one HPLC system may eventually separate all important pigment components is still elusive at the present time.

4.4.2.3 *Signature pigments*

Many authors have listed marker pigments and pigment suites which indicate algal types and processes in the sea (e.g., Bidigare *et al.* 1987; Hooks *et al.* 1988; Jeffrey and Hallegraeff, 1990; Williams and Claustre, 1991; Wright *et al.* 1991; Bidigare *et al.* 1992; Millie *et al.* 1993*a*, *b*). Other authors have studied pigment distribution in more detail within algal classes and taxa, e.g., carotenoids and chlorophyll *c* in chromophyte algae (Bjørnland and Liaaen-Jensen, 1989; Jeffrey, 1989); chlorophylls and carotenoids in prymnesiophytes (Jeffrey and Wright, 1994).

One problem with designating marker pigments for algal types in field studies is that some pigment 'finger prints' are recommended on the basis of examination of only a few algal species from the group. Table 2.3 (Chapter 2) lists the distribution of major and minor pigments across algal classes, and many exceptions to these listings are noted. One example is the analysis of 50 prymnesiophyte strains (29 species) by Jeffrey and Wright (1994) which clearly shows four types of pigment distribution within the class (see Fig. 4.12). Various combinations of fucoxanthin, 19'-hexanoyloxyfucoxanthin, 19'-butanoyloxyfucoxanthin and chlorophylls c_1, c_2, c_3,

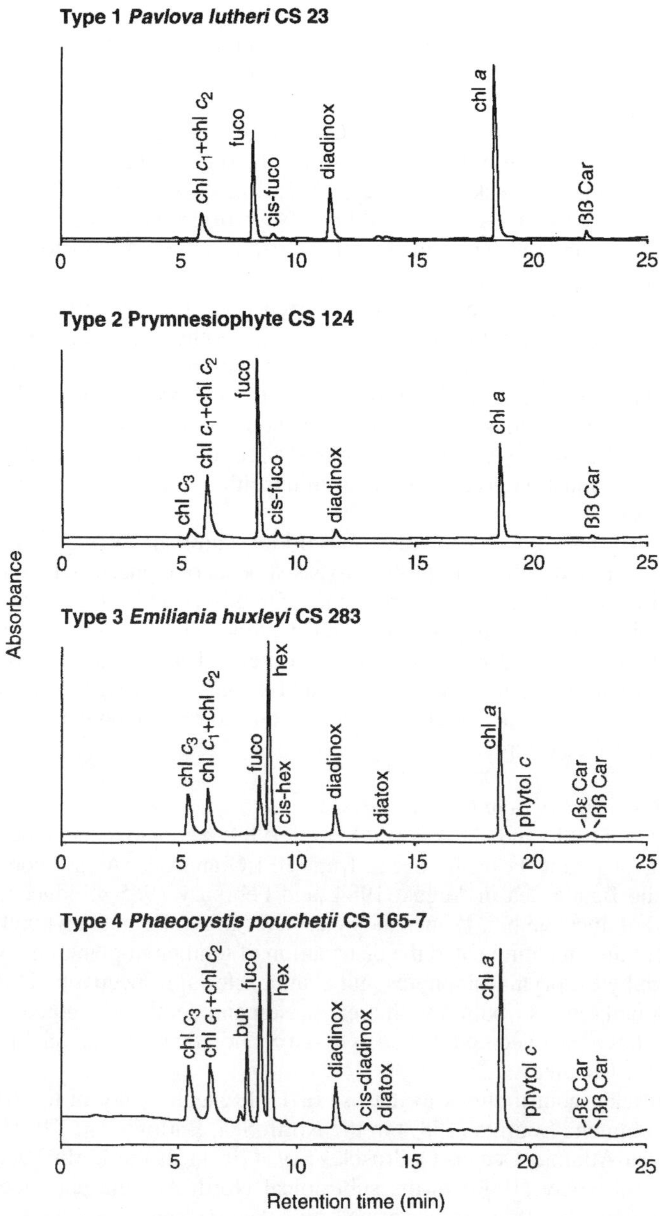

Figure 4.12
Representative HPLC chromatograms of each prymnesiophyte pigment group: Type 1, *Pavlova lutheri* (CS-23); Type 2, Prymnesiophyte (CS-124); Type 3, *Emiliania huxleyi* (CS-283); Type 4, *Phaeocystis pouchetii* (CS-165-7) (from Jeffrey and Wright, 1994).

and the phytylated chlorophyll *c*-like pigment distinguished the four prymnesiophyte groups. This data provides a cautionary note to authors and users of marker pigment schemes, not to ascribe a pigment characteristic to an algal class on the basis of examination of only one or a few species; or if this is done, to know the limitations. The carotenoid, 19'-butanoyloxyfucoxanthin is often suggested as a marker for marine chrysophytes — but only three strains of one species, *Pelagococcus subviridis*, have been examined! (Vesk and Jeffrey, 1987; Bjørnland *et al.* 1989). This situation will surely change with our improved knowledge of growing microalgae in culture (see Chapter 6), and the very wide variety of strains now available for experimentation in culture collections across the world.

Another problem is the 'atypical' occurrence of pigment suites, normally characteristic of one class, within representatives of another class. For example, green algal pigments, prymnesiophyte and diatom pigments have been found in dinoflagellates and reflect endosymbiotic events (see footnotes to Table 2.3, Chapter 2). A similar problem is the occurrence of endosymbiotic cyanobacteria (*Richelia*) within tropical diatoms (e.g., *Rhizosolenia*; see section 2.3.3, Chapter 2). Obviously, some attention must be given to species composition when interpreting HPLC pigment profiles.

Table 4.1 provides yet another (!) working summary of signature pigments, upgrading that of Wright *et al.* (1991), to assist oceanographers in the interpretation of HPLC field data. Only minimal referencing to this Table is given here, since more extensive bibliographies may be found in Chapters 5 and 13. The citations in Millie *et al.* (1993*a* and *b*) are also recommended. Perusal of Table 4.1 shows the poor taxon specificity of many key pigments, and suggests that HPLC analysis should often be supported by species composition studies to resolve potential 'pigment-origin' problems (see Chapter 2).

4.4.2.4 *Pigment contours in the sea*

Gieskes *et al.* (1988) and Gieskes and Kraay (1989) were pioneers in the geographic contouring of pigments in the ocean from HPLC analysis. A first comprehensive study of the Banda Sea in August 1984, and February 1985, produced profiles of chlorophyll *a*, fucoxanthin, 19'-hexanoyloxyfucoxanthin, 19'-butanoyloxyfucoxanthin and zeaxanthin which indicated the distribution of total phytoplankton, diatoms plus prymnesiophytes, prymnesiophytes and cyanobacteria, respectively. Differences in the phytoplankton distribution with area, season and depth were noted. Many similar studies have now been published as JGOFS and other international programmes develop (see Chapter 5).

Significant contributions to the use of HPLC technology in the field include those of Klein and Sournia (1987) in the North Sea, Barlow *et al.* (1993*a*, *b*) in the northeastern Atlantic Ocean, Ondrusek *et al.* (1991) in the north Pacific Ocean, Veldhuis and Kraay (1990) in the subtropical North Atlantic and Hoepffner and Sathyendranath (1992) in the western North Atlantic Ocean (see Chapter 5 for more extensive citations). Here just two examples of oceanic pigment contours are given (Figs. 4.13; 4.14).

Seasonal variations in algal chlorophylls and carotenoids from the central North Pacific gyre (station ALOHA) were recorded monthly over a 3 year period (Letelier *et al.* 1993). The distribution of density plots, chlorophyll *a*, chlorophyll *b*, zeaxanthin, 19'-hexanoyloxyfucoxanthin, 19'-butanoyloxyfucoxanthin for 18 months of the study period are shown in Fig. 4.13A–F. High chlorophyll *a* concentrations in the

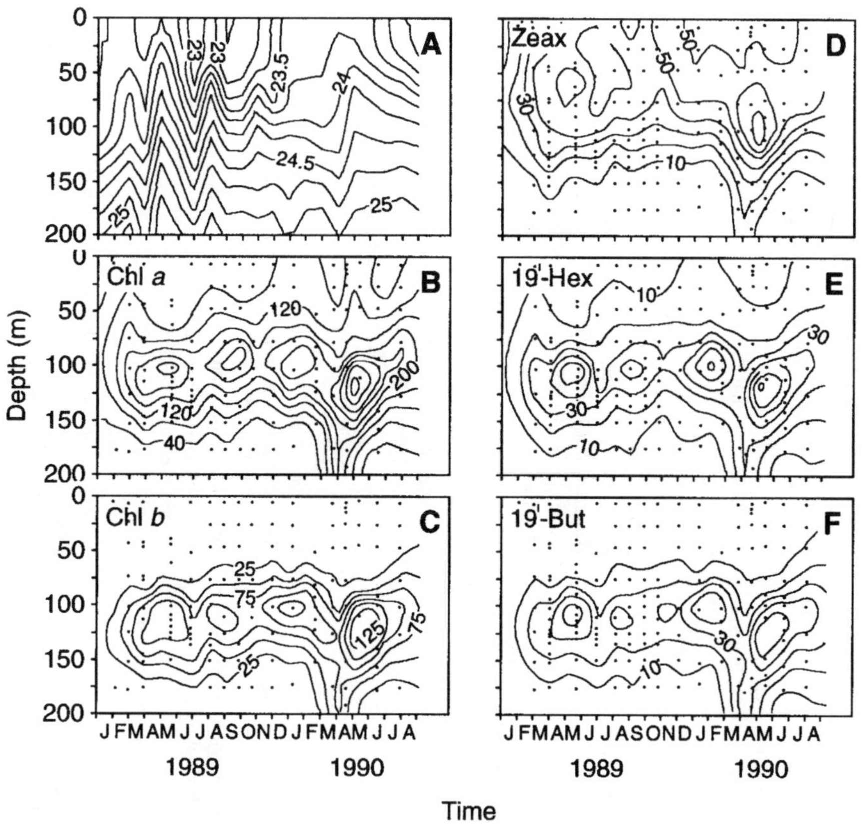

Figure 4.13
Distribution of density, chlorophylls and major accessory pigments in the water column, February 1989 to August 1990, at station ALOHA (22°45'N, 158°00'W) in the central North Pacific gyre. A, density; B, chlorophyll *a*; C, chlorophyll *b*; D, zeaxanthin; E, 19'-hexanoyloxyfucoxanthin; F, 19'-butanoyloxyfucoxanthin. All pigment concentrations in ng.l^{-1} (from Letelier *et al.* 1993).

mixed layer were the result of photoadaptation as the mixed layer deepened in winter. The deep chlorophyll maximum layer was made up of *Prochlorococcus* (39%), cyanobacteria (24%), prymnesiophytes (22%), and chrysophytes (13%). These percentages were derived from algorithms relating ratios of chlorophyll *a* to marker pigments present in algal taxa (see sections 4.4.2.3; 4.2.4.5).

The second example shows pigment contours in the equatorial Pacific in October 1990 along 155°E (see Fig. 4.14). HPLC-derived chlorophyll *a* contours (A) are compared with those obtained *in situ* with a SeaTech fluorometer (C). HPLC derived contours for chlorophyll *b* (B) and zeaxanthin (D) are also compared (see Mackey *et al.* 1995). The distribution of chlorophyll *a* was bimodal, with more detail in the contours of the *in vivo* fluorescence data (C). The vertical distribution of

TABLE 4.1. Summary of signature pigments useful as markers of algal groups and processes in the sea (see also Table 2.3, Chapter 2; and Chapter 13)[a,b]

Pigment	Algal group or process	Key references
A. Chlorophylls (see Table 2.3, Chapter 2)		
Chl *a*	All photosynthetic microalgae (except prochlorophytes)	See Table 2.3, Chapter 2
Divinyl chl *a*	Prochlorophytes	Goericke & Repeta (1992)
Chl *b*	Green algae: chlorophytes, prasinophytes, euglenophytes	See Table 2.3, Chapter 2
Divinyl chl *b*	Prochlorophytes	Goericke & Repeta (1992)
Chl *c* family	Chromophyte algae	Jeffrey (1989)
Chl c_1	Diatoms, some prymnesiophytes, some freshwater chrysophytes, raphidophytes	Jeffrey (1976*b*); Stauber & Jeffrey (1988); Jeffrey (1989); Andersen & Mulkey (1983)
Chl c_2	Most diatoms, dinoflagellates, prymnesiophytes, raphidophytes, cryptophytes	Jeffrey *et al.* (1975); Stauber & Jeffrey (1988); Andersen & Mulkey (1983)
Chl c_3[c]	Some prymnesiophytes, one chrysophyte, several diatoms and dinoflagellates	Jeffrey & Wright (1987); Vesk & Jeffrey (1987); Jeffrey (1989); Johnsen and Sakshaug (1993)
Chl $c_{CS\text{-}170}$	One prasinophyte	Jeffrey (1989)
Phytylated chl *c*-like[c]	Some prymnesiophytes	Nelson & Wakeham (1989); Jeffrey & Wright (1994)
Mg3,8 DVP	Some prasinophytes	Ricketts (1966); Jeffrey (1989)
Bacteriochlorophylls	Anoxic sediments	Repeta *et al.* (1989); Repeta and Simpson (1991)
B. Carotenoids (see Table 2.3, Chapter 2)		
Alloxanthin	Cryptophytes	Chapman (1966); Pennington *et al.* (1985)
19'-Butanoyloxyfucoxanthin	Some prymnesiophytes, one chrysophyte, several dinoflagellates	Bjørnland & Liaaen-Jensen (1989); Bjørnland *et al.* (1989); Jeffrey & Wright (1994)
β,ε-carotene	Cryptophytes, prochlorophytes, rhodophytes, green algae	See Table 2.3, Chapter 2
β,β-carotene	All algae except cryptophytes and rhodophytes	See Table 2.3, Chapter 2
Crocoxanthin	Cryptophytes (minor pigment)	Pennington *et al.* (1985)
Diadinoxanthin	Diatoms, dinoflagellates, prymnesiophytes, chrysophytes, raphidophytes, euglenophytes	See Table 2.3, Chapter 2
Dinoxanthin	Dinoflagellates	Johansen *et al.* (1974); Jeffrey *et al.* (1975)
Echinenone	Cyanophytes (but see Table 2.3, Chapter 2)	Foss *et al.* (1987)
Fucoxanthin	Diatoms, prymnesiophytes, chrysophytes, raphidophytes, several dinoflagellates	Stauber & Jeffrey (1988); Bjørnland & Liaaen-Jensen (1989)
19'-Hexanoyloxyfucoxanthin	Prymnesiophytes, several dinoflagellates	Arpin *et al.* (1976); Bjørnland & Liaaen-Jensen (1989)
Lutein	Green algae: chlorophytes, prasinophytes	See Table 2.3, Chapter 2

Micromonal	Some prasinophytes	Egeland & Liaaen-Jensen (1992, 1993)
Monadoxanthin	Cryptophytes (minor pigment)	Pennington *et al.* (1985)
9'-*cis* Neoxanthin	Green algae: chlorophytes, prasinophytes, euglenophytes	See Table 2.3, Chapter 2
Peridinin	Dinoflagellates	Johansen *et al.* (1974); Jeffrey *et al.* (1975)
Peridininol, P-457, P-468	Dinoflagellates (minor pigments)	Bjørnland & Liaaen-Jensen (1989)
Prasinoxanthin	Some prasinophytes	Foss *et al.* (1984)
Pyrrhoxanthin	Dinoflagellates (minor pigment)	Bjørnland & Liaaen-Jensen (1989)
Siphonaxanthin	Several prasinophytes; one euglenophyte	Fawley & Lee (1990); Bjørnland (1989)
Vaucheriaxanthin ester	Eustigmatophytes	Bjørnland & Liaaen-Jensen (1989)
Violaxanthin	Green algae: chlorophytes, prasinophytes; eustigmatophytes	See Table 2.3, Chapter 2
Zeaxanthin	Cyanophytes, prochlorophytes, rhodophytes, chlorophytes, eustigmatophytes (minor pigment)	Guillard *et al.* (1985); Gieskes *et al.* (1988); Goericke & Repeta (1992); Table 2.3, Chapter 2
C. Biliproteins		
Allophycocyanins	Cyanophytes, rhodophytes	Rowan (1989)
Phycocyanin	Cyanophytes, cryptophytes, rhodophytes (minor pigment)	Rowan (1989)
Phycoerythrin	Cyanophytes, cryptophytes, rhodophytes	Rowan (1989)
D. Chlorophyll degradation products (see Bibliographies, Chapter 5)		
Pheophytin *a*[d]	Zooplankton faecal pellets, sediments	Vernet & Lorenzen (1987)
Pheophytin *b*[d]	Protozoan faecal pellets	Strom (1991; 1993)
Pheophytin *c*[d]	Protozoan faecal pellets	Strom (1991; 1993)
Pheophorbide *a*[d]	Protozoan faecal pellets	Strom (1991; 1993); Head *et al.* (1994); Welschmeyer & Lorenzen (1985)
Pheophorbide *b*[d]	Protozoan faecal pellets	Strom (1991; 1993)
Chlorophyllide *a*	Senescent diatoms; extraction artefact	Jeffrey & Hallegraeff (1987*b*)
Blue-green chl *a* derivatives (lactones, 10-hydroxy chlorophylls)	Senescent microalgae	Hallegraeff & Jeffrey (1985)
Pyrochlorophyll *a*	Sediments	See Bibliography Table 5.4, Chapter 5
Pyropheophytin *a*	Sediments	See Bibliography Table 5.4, Chapter 5
Pyropheophorbide *a*	Copepod grazing; faecal pellets	Head *et al.* (1994); Table 5.4, Chapter 5
Mesopheophorbide *a*	Sediments	See Bibliography Table 5.4, Chapter 5

a. Note that many carotenoids in trace quantities have wider distributions than those shown (Bjørnland and Liaaen-Jensen, 1989; and Chapter 13).
b. Note that some 'atypical' pigment contents reflect endosymbiotic events (see footnotes to Table 2.3, Chapter 2).
c. Two spectrally distinct pigments have been found in these fractions (Garrido *et al.* 1995)
d. Multiple zones found in each of these fractions (Strom, 1993).

cyanobacteria (zeaxanthin) varied markedly (D), while the chlorophyll *b* profiles (B) were more diffuse (green algae; *Prochlorococcus*). This brief presentation illustrates the value of HPLC marker pigment contours to indicate phytoplankton distributions. For more information on pigment processes in the sea the original publications and the many citations in Chapter 5 should be consulted.

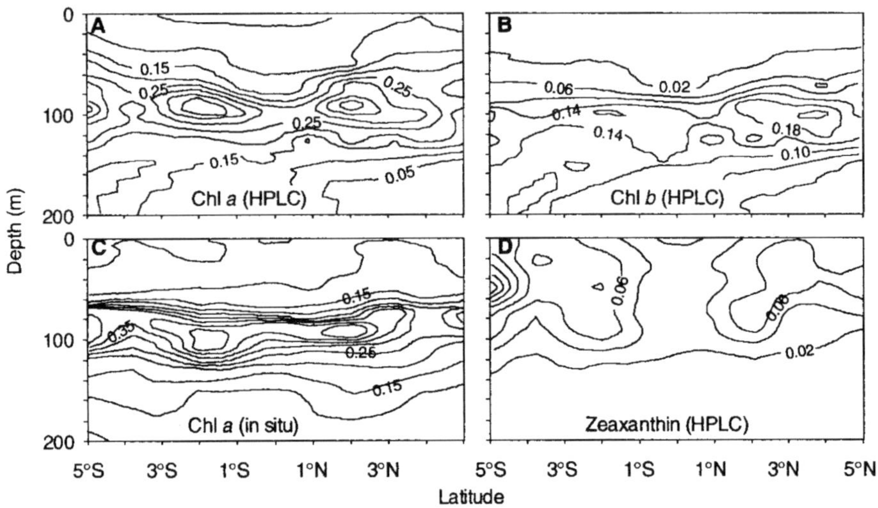

Figure 4.14
Distribution of pigment contours in the Equatorial Pacific along 155°E in October 1990. A, chlorophyll *a* (HPLC); C, *in situ* chlorophyll *a* (SeaTech fluorometer); B, chlorophyll *b* (HPLC);
D, zeaxanthin (HPLC). All pigment concentrations in $\mu g.l^{-1}$ (from Mackey *et al.* 1995).

4.4.2.5 *Phytoplankton class (taxa) abundances estimated from marker pigments*

The information obtained from pigment TLC and HPLC studies of field phytoplankton populations has been very informative, but it was often used only as qualitative data (e.g., Jeffrey and Hallegraeff, 1980*a*, *b*; 1987*a*; Hallegraeff and Jeffrey, 1984; Ridout and Morris, 1985; Klein and Sournia, 1987; Gieskes and Kraay, 1983*b*; 1986*a*, *b*). More recently quantitative estimates of phytoplankton class abundances were made from marker pigment concentrations (Gieskes *et al.* 1986*b*; 1988). A robust method for such estimations would be invaluable, as it could lead to the development of fast semi-automated algal class identification using HPLC data, allowing for more wide-spread investigation of phytoplankton abundances and distributions than is now possible from cell counts and flow cytometry (see section 4.5).

The technique used by Gieskes *et al.* (1988) assumed that pigment ratios are constant within an algal group, and required a large data set for statistical validity. The method also included only the contributions of pigment-related groups to the phytoplankton population. A pigment with taxon-poor specificity such as fucoxanthin, could have contributed 50% of the chlorophyll biomass to a given sample, but diatoms, prymnesiophytes, chrysophytes and raphidophytes, which all contain this pigment, were not differentiated. Minor groups tended to get lost in the data variance and the technique failed if the concentration of marker pigments co-varied. Changes in algal pigment ratios with variations in water column light intensity were not accounted for by the method.

An alternative approach (Everitt *et al.* 1990) involved dividing the plankton into classes on the basis of pigment types, and then determining the contribution of each class to the total chlorophyll *a* in the sample from measured pigment abundances and chlorophyll *a* : marker pigment ratios derived from the experimental literature. Quantitative measures of those classes without unique marker pigments were calculated by difference. Discrepancies between the calculated and observed concentration of chlorophyll *a* indicated how well the model predictions matched the experimental results. An iterative procedure was used to minimize this difference by varying the chlorophyll *a* : marker pigment ratios. One serious drawback was that the process of calculation by difference for those classes without clear marker pigments sometimes led to unrealistic or even negative abundances for these classes. A more robust procedure was obviously required to make full use of the data from HPLC analyses.

Most of these difficulties are overcome in a new PC based computer program (written in Pascal and MATLAB) based on factor analysis. The method requires an initial estimate of the pigment ratios appropriate for the algal classes likely to be in the sample. From this, an initial value for the phytoplankton class abundances is derived. The pigment ratio matrix is then varied by a predetermined amount and the abundances recalculated. A steepest descent algorithm is then used to obtain a better estimate of both class abundances and pigment ratios. The process is iterated until the least squares fit satisifies predetermined limits.

The method was evaluated using a series of mathematically constructed data sets of HPLC pigment concentrations and corresponding algal class abundances. The addition of known and random errors to the pigment ratios and to class abundances allowed the importance of variations in pigment ratios and experimental accuracy to be assessed. A manuscript on this new work is in preparation (Dr D. J. Mackey, pers. comm.).

4.5 Additional methods: biliproteins and bacteriochlorophylls

The quantitative determination of biliproteins and bacteriochlorophylls were not addressed at the SCOR WG 78 Workshops. The water-soluble red phycoerythrins and the blue phycocyanins and allophycocyanins are found in cryptomonads, cyanobacteria and several planktonic rhodophytes (see Table 2.3, Chapter 2). Methods for extraction, identification and determination by spectroscopy, fluorometry and *in vivo* fluorometry include those of Exton *et al.* (1983), Stewart and Farmer (1984), Beer and Eshel (1985), and Wyman (1992). Extinction coefficients of biliproteins from many sources are listed in Rowan (1989), and equations are recommended which are particularly useful for macrophytes and algal cultures. These need to be evaluated for use with phytoplankton field samples. Although macro-scale separation techniques for biliproteins such as hydroxyapatite columns and electrophoresis are well documented (Alberte *et al.* 1984), no micro-technique, suitable for phytoplankton biliprotein separation and analysis, has yet been developed, comparable to the highly selective HPLC techniques now available for chlorophylls and carotenoids (see sections 4.4.2.1 and 4.4.2.2). Experts in biliprotein macro-scale analysis are urgently requested to develop such methods!

Analytical techniques for bacteriochlorophylls have been available for some time, particularly HPLC methods (see review by Shioi, 1990). These pigments are seldom detected in oceanography since photosynthetic bacteria usually inhabit anoxic environments. However, Repeta *et al.* (1989) and Repeta and Simpson (1991) found bacteriochlorophylls in their studies of anoxic environments in the Black Sea (see Table 5.4, Chapter 5). Bacteriochlorophylls and bacteriopheophytins have also been found on HPLC chromatograms of Tasmanian lake sediments by the method of Wright *et al.* (1991) (see Table 4.2; D. Hodgson and S. W. Wright, unpublished data).

TABLE 4.2. Pigments identified in anoxic sediments from Lake Fidler, Tasmania. HPLC method used: Wright *et al.* (1991).

Retention time	Pigment	Mol. weight†
10.21	Bacteriochlorophyll *c* 1*	
10.63	Bacteriochlorophyll *c* 2*	
11.19	Bacteriochlorophyll *c* 3*	
12.00	Pheophorbide *a* derivative	592.7
12.51	Alloxanthin	564
14.28	Diatoxanthin	566.9
14.61	Lutein	568.7
15.22	Bacteriopheophytin *c* 1^1*	770.5
15.36	Bacteriopheophytin *c* 1^2*	770.5
16.09	Bacteriopheophytin *c* 2*	784.5
16.23	Chlorophyll *a*	892.4
17.12	Isorenieratine	528.7
17.55	Bacteriopheophytin *c* 7^1*	788.5
17.65	β,β-Carotene	536.5
18.28	Bacteriopheophytin *c* 7^2*	776.4
20.89	Pyropheophytin *a*	812.5

* Nomenclature of bacteriochlorophylls and bacteriopheophytins from Caple *et al.* (1978). (D. Hodgson and S. W. Wright, unpublished data).
† Data from mass spectroscopy of fractions.

4.6 Cell-flow cytometry and epifluorescence microscopy

The fluorescence properties of chlorophylls and biliproteins are utilized in cell-flow cytometry and epi-fluorescence microscopy of phytoplankton. In the former technique, cells are passed rapidly in single file through a laser beam, and their optical properties are characterized (Ormerod, 1994). Forward angle light scatter (related to cell size), 90° side scatter (related to cell shape) and three fluorescence emission bands (related to pigmentation) are usually measured. This allows sub-populations of algal cultures and natural phytoplankton populations to be distinguished and quantified. With appropriate equipment, sub-populations can be sorted for further study (e.g., microscopy, cell culture). Fluorescent stains specific for biochemicals such as nucleic acids or proteins can also be used (see Yentsch *et al.* 1983). Fast and objective analysis of large numbers of measured particles is thus achieved, easily yielding statistically sound data.

Spectacular advances have resulted from the use of cell-flow cytometry at sea, particularly in differentiating picoplankton populations that are difficult to assess by conventional microscopy (see Platt and Li, 1985; and Chapter 5). A now classic paper using this method (Chisholm *et al.* 1988) reported the discovery of free-living coccoid picoplanktonic procaryotes in the North Atlantic and Pacific Oceans. These 0.6–0.8 µm cells (now known as *Prochlorococcus marinus*; Chisholm *et al.* 1992) fluoresced red, contained divinyl chlorophylls *a* and *b*, and were abundant in the deep euphotic zone at concentrations greater than 10^5 cells per ml. They were readily distinguished from the slightly larger orange-fluorescing picoplanktonic cyanobacteria (*Synechococcus* spp.; see Waterbury *et al.* 1979), and other red-fluorescing small eucaryotes (Chisholm *et al.* 1988). In the depth profile shown in Fig. 4.15, cyanobacteria were more abundant in the upper water column (from 0 to 75 m), whereas the prochlorophytes dominated the deeper layers (45 to 120 m). *Prochlorococcus marinus* is a free-living relative of *Prochloron* . It has a wide distribution in temperate and tropical oceans, where it is found throughout the euphotic zone (see section 2.12, Chapter 2).

Quantitative epifluorescence microscopy has also added to our knowledge of the picoplankton. Initially this technique was used to distinguish and quantify the orange fluorescence of phycoerythrin in small coccoid cyanobacteria from the red fluorescence of other autotrophic cells (see Plate 4.A, Chapter 6). Improved techniques, combined with size fractionation, have now shown the presence of five different autotrophic picoplankton groups (see Ishizaka *et al.* 1994). These were designated: autotrophic microflagellates, red-fluorescing picoplankton, *Synechococcus* spp., *Prochloroccocus*, and orange-fluorescing picoplankton. Their distribution along a section in the central north Pacific at 175°E from 45°N to 10°N (which included sub-arctic, central gyre and tropical waters) during the late northern summer of 1990 is shown in Fig. 4.16 (from Ishizaka *et al.* 1994). The distribution of autotrophic microflagellates and red-fluorescing picoplankton matched that of chlorophyll *a*, most of which was present in cells less than 3 µm diameter. *Synechococcus* sp. was found mostly in surface waters of the Kuroshio extension, while *Prochloroccocus* was found at depth south of the Kuroshio. Orange-fluorescing picoplankton, larger than typical *Synechococcus*, were found in the oligotrophic surface waters of the central gyre.

This excellent work is an example from the growing literature of the detail now achievable in picoplankton analysis using epifluorescence microscopy and cell-flow cytometry. Further information on the species composition of this group may come from new picoplankton cell isolations, electron microscopy of mixed picoplankton and isolated cells, and HPLC analysis of the differently pigmented groups. Analyses such as those of Simon *et al.* (1994) and Moore *et al.* (1995) on cultured picoplankton, may help to discriminate further sub-populations of picoplankton in the sea.

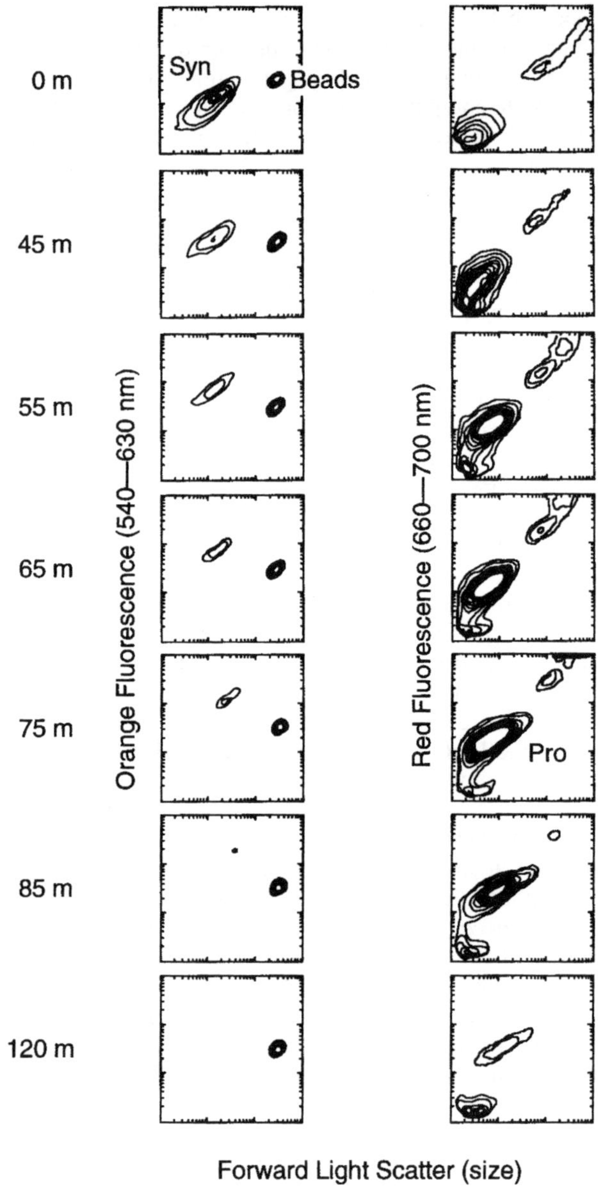

Figure 4.15
Flow cytometry signatures of *Synechococcus* spp. compared to those of prochlorophytes from a depth profile off Southern California, December 1987. Left panels, forward angle light scatter versus orange fluorescence from phycoerythrin shows *Synechococcus* abundant down to 75 m. Fluorescent (0.9 µm) beads act as internal standards. Right panels, forward angle light scatter versus red fluorescence shows prochlorophytes dominating deep samples. The larger cells with a brighter fluorescence are eukaryotic cells 2–5 µm in diameter (from Chisholm *et al.* 1988).

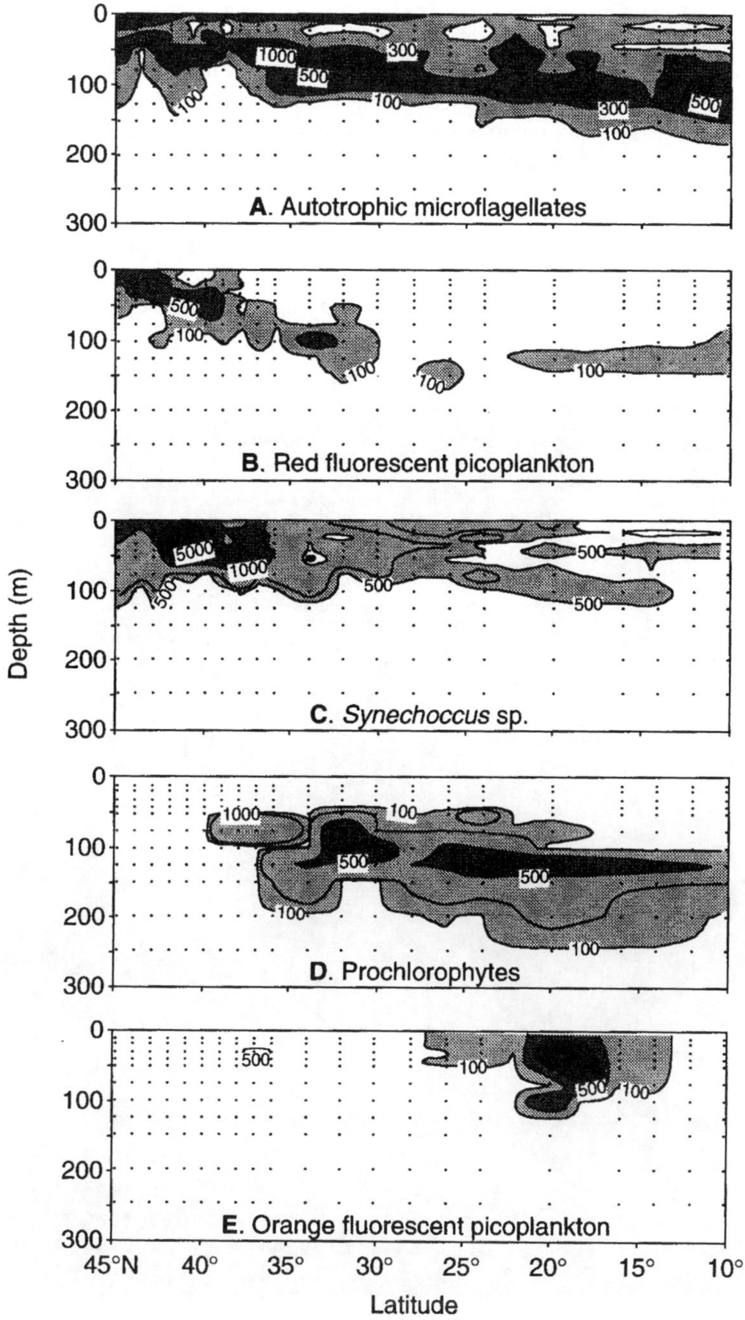

Figure 4.16
Distribution of picoplankton in the subarctic, central gyre and equatorial regions of the North Pacific Ocean (175°E; 45°N to 10°N) during August and September 1990. A, Autotrophic microflagellates; B, red fluorescent picoplankton; C, *Synechococcus* sp.; D, prochlorophytes; E, orange fluorescent picoplankton (from Ishizaka *et al.* 1994).

4.7 Remotely-sensed ocean colour

Remote sensing of ocean colour is a powerful tool for estimating algal biomass from space. Spectral variations in light leaving the sea surface (differential absorption and backscatter of irradiance), corrected for atmospheric distortion, are detected by satellite-mounted spectroradiometers at wavelengths appropriate to the detection of phytoplankton pigments (see Sathyendranath (1986) for discussion of the concepts and techniques involved). Synoptic images of chlorophyll distribution in the upper ocean can be obtained at time scales relevant to phytoplankton processes (Sathyendranath, 1986; Lewis, 1989). These images provide detailed data over broad areas with resolution impossible to achieve by ship-based sampling, and this has

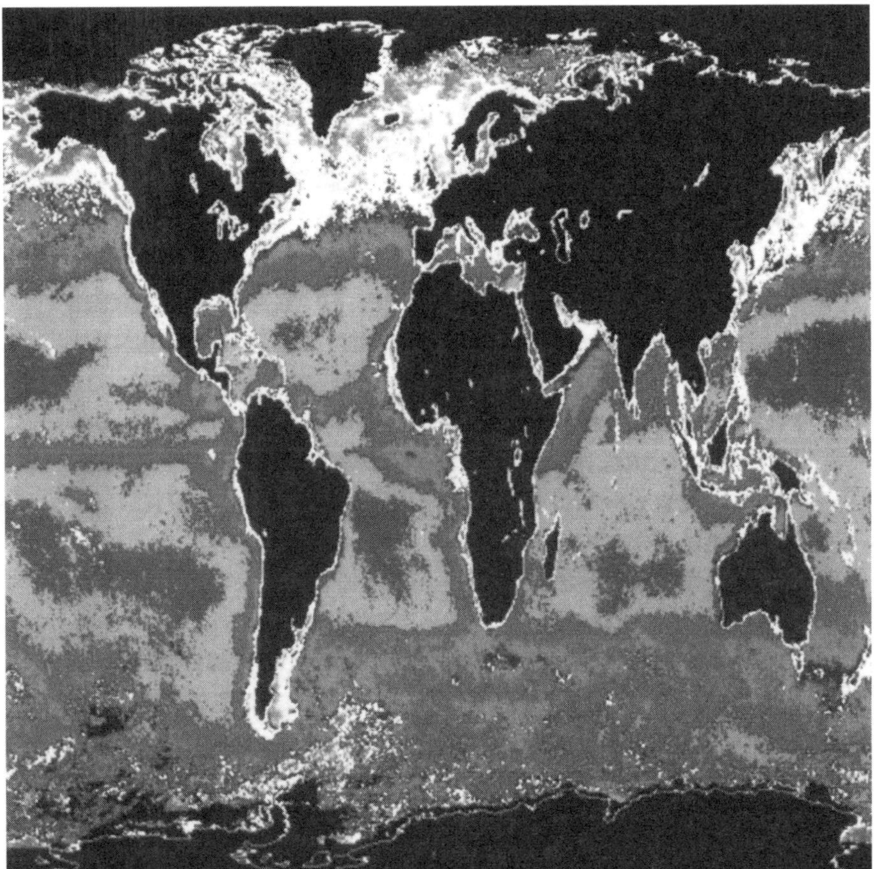

Figure 4.17
Seven year mean pigment fields derived from the Coastal Zone Colour Scanner. The phytoplankton spring blooms are indicated by the white areas (adapted from Mitchell *et al.* 1991).

revolutionized the oceanographer's scale of perception. Fig. 4.17 shows the extent of the northern hemisphere phytoplankton spring-blooms (white areas, adapted from Mitchell *et al.* 1991), which have been called the largest remotely-sensed biological signal of the planet (Lewis, 1989).

Unfortunately the images are limited to the upper water column — about 25 m in clear oligotrophic waters, decreasing to less than 5 m in chlorophyll-rich coastal areas. Sathyendranath (1986) calculated that the top 22% of the euphotic zone accounts for 90% of the signal received by the satellite sensor. The widespread deep chlorophyll maximum layer is therefore not detected by remote sensing techniques, which cannot simply replace ship's observations. Indeed, ground-truthing must be provided by ship or *in situ* observations to determine the depth structure of phytoplankton populations and to calibrate the surface measurements. An accurate picture of chlorophyll distribution from satellite sensors can thus be achieved.

Reflectance spectra are invaluable for showing blooms of oceanic coccolithophorids (Aiken *et al.* 1992). The calcite coccoliths (see Chapter 2) are powerful light reflectors; as much as 30% of the incident sunlight penetrating the sea may be reflected from the sea surface by concentrations of 4×10^5 cells per ml (Balch *et al.* 1991). These blooms are clearly visible as 'milky' patches over large areas (Holligan and Balch, 1991).

All currently available remotely-sensed data on ocean colour has come from the Coastal Zone Colour Scanner (CZCS), which transmitted data from the NIMBUS-7 satellite from 1979 to 1986. The CZCS had five wavelength detection channels at 443, 520, 550, 670 and 750 nm, which were used to produce imagery of pigment biomass (chlorophyll *a* plus 'pheopigments'). In order to gain more information from satellite colour images, new generations of colour imagers are being designed (see Aiken *et al.* 1992). The improved and eagerly awaited scanner, the Sea-Viewing Wide-Field-of-View Sensor (SeaWIFS), which is due to be launched by NASA in the near future, has eight narrow bands across the visible spectrum and much improved resolution, with a pixel size of only 1.13 km (Aiken *et al.* 1992).

Interpretation of satellite ocean colour data requires information on the *in vivo* absorption spectra of algae (see Fig. 3.1, Chapter 3). Hoepffner and Sathyendranath (1991) determined the contribution of individual pigments to the *in vivo* absorption spectrum of algal cultures by using Gaussian curve deconvolution techniques. Although the specific absorption coefficients are in error due to the use of inappropriate spectrophotometric equations for chlorophylls and carotenoids (see Appendix F),

TABLE 4.3. *In vivo* pigment absorption maxima from Gaussian deconvolution curve analysis of algal cultures (from Hoepffner and Sathyendranath, 1991).

Pigment	*In vivo* maxima (nm)
Chlorophyll *a*	435; 675
Chlorophyll *b*	470–490; 650
Chlorophyll *c*	450; 630
Carotenoids	475 — 540
Phycocyanin	580—600
Phycoerythrobilin	540—565
Phycourobilin	480–500

the broad placement of the most taxonomically-useful *in vivo* pigment absorption peaks are valid (see Table 4.3).

Key questions now are: what can satellite ocean colour images see of near-surface particulate absorption (Garver *et al.* 1994); and can further detail on the composition and structure of phytoplankton communities be detected from satellite sensors in addition to chlorophyll estimates? Millie and Schofield (1995) even suggest that *in vivo* absorption spectra may add more information to the bio-optical properties of the water column than HPLC pigment signatures. However, two opposing views are current.

Garver *et al.* (1994) found that remotely-sensed ocean colour spectra reflected only three statistically significant components — the total amount of particulate material, the relative amounts of chlorophyll-containing biomass, and detrital materials. They analyzed the spectral variability of 400 *in situ* particulate absorption spectra from contrasting oceans to obtain this result. Because of the similarity of *in vivo* spectral shapes, these authors believe that it will be difficult to retrieve anything more from ocean colour spectra beyond a simple description of algal and detrital properties. Even this will be a valuable advance, and is supported by the modelling studies of Cleveland and Perry (1994), who clearly distinguished algal from detrital spectra (see Fig. 4.3). Additional information on algal speciation and accessory pigment concentrations in field samples may still be achieved (Hoepffner and Sathyendranath, 1992) if problems associated with variability and measurement noise, and the spectral similarity of many key pigments, can be overcome.

Johnsen *et al.* (1994) have a more positive view based on *in vivo* spectral absorption characteristics of 31 species of bloom-forming microalgae. By using a combination of three to five wavelengths (481, 535, 586, 628 and 649 nm), four main groups of phytoplankton were clearly separated, based on the presence of accessory chlorophylls: chlorophyll *b* (chlorophytes, prasinophytes, euglenophytes), chlorophylls $c_1 + c_2$ (chromophyte algae), chlorophyll c_3 (prymnesiophytes and several dinoflagellates), and no accessory chlorophylls (cyanophytes, eustigmatophytes and rhodophytes). These authors recognize that because they worked only with algal cultures, their approach may be most relevant to *in situ* multi-wavelength attenuation sensors suspended from buoys, where monospecific blooms of known species composition may develop.

It has been suggested that oceans may be partitioned into various provinces based on regional absorption characteristics which may be 'quasi-stable' for each season (Hoepffner and Sathyendranath, 1992; Claustre, 1994; Lutz *et al.* 1996). Such characteristics could then be used to improve remote-sensing algorithms for each oceanographic province. These ideas can be tested when the new generation of satellites (e.g., SeaWIFS) is launched. We expect that in the future further differentiation of phytoplankton populations will result from remotely-sensed ocean colour.

Acknowledgements

Thanks are due to Dr D. J. Mackey, CSIRO Division of Oceanography, Hobart, for contributing section 4.4.2.5 and Fig. 4.6C, and Dr S. W. Wright for contributing Table 4.2. Both colleagues are thanked for their helpful comments on the manuscript.

REFERENCES

Abaychi, J. K., Riley, J. P. (1979). The determination of phytoplankton pigments by high performance liquid chromatography. *Anal. Chim. Acta*, 107: 1–11.

Aiken, J., Moore, G. F., Holligan, P. M. (1992). Remote sensing of ocean biology in relation to global climate change. *J. Phycol.* 28: 579–590.

Alberte, R. S., Wood, A. M., Kursar, T. A., Guillard, R. R. L. (1984). Novel phycoerythrins in marine *Synechococcus* spp. *Plant Physiol.* 75: 732–739.

Andersen, R. A., Mulkey, T. J. (1983). The occurrence of chlorophylls c_1 and c_2 in the Chrysophyceae. *J. Phycol.* 19: 289–294.

Arpin, N., Svec, W. A., Liaaen-Jensen, S. (1976). A new fucoxanthin-related carotenoid from *Coccolithus huxleyi*. *Phytochemistry*, 15: 529–532.

Balch, W. M., Holligan, P. M., Ackleson, S. G., Voss, K. J. (1991). Biological and optical properties of mesoscale coccolithophorid blooms in the Gulf of Maine. *Limnol. Oceanogr.* 36: 629–643.

Barlow, R. G., Mantoura, R. F. C., Gough, M. A., Fileman, T. W. (1993*a*). Pigment signatures of the phytoplankton composition in the northeastern Atlantic during the 1990 spring bloom. *Deep-Sea Res.* 40: 459–477.

Barlow, R. G., Mantoura, R. F. C., Gough, M. A., Fileman, T. W. (1993*b*). Phaeopigment distribution during the 1990 spring bloom in the northeastern Atlantic. *Deep-Sea Res.* 40: 2229–2242.

Beer, S., Eshel, A. (1985). Determining phycoerythrin and phycocyanin concentrations in aqueous crude extracts of red algae. *Aust. J. Mar. Freshw. Res.* 36: 785–792.

Bidigare, R. R., Prézelin, B. B., Smith, R. C. (1992). Bio-optical models and the problems of scaling, In: Falkowski, P. G., Woodhead, A. D. (eds). *Primary Productivity and Biogeochemical Cycles in the Sea*, Plenum Press, New York, p. 175–212.

Bidigare, R. R., Smith, R. C., Baker, K. S., Marra, J. (1987). Oceanic primary production estimates from measurements of spectral irradiance and pigment concentrations. *Global Biogeochem. Cycles*, 1: 171–186.

Bjørnland, T. (1989). Carotenoid structure and lower plant phylogeny. In: Krinsky, N. I., Mathews-Roth, M. M., Taylor, R. F. (eds). *Carotenoids: Chemistry and Biology*. Plenum Press, New York, p. 21–32.

Bjørnland, T., Liaaen-Jensen, S. (1989). Distribution patterns of carotenoids in relation to chromophyte phylogeny and systematics. In: Green, J. C., Leadbeater, B. S. C., Diver, W. L. (eds), *The Chromophyte Algae: Problems and Perspectives*. Clarendon Press, Oxford, pp. 37–60.

Bjørnland, T., Liaaen-Jensen, S., Throndsen, J. (1989). Carotenoids of the marine chrysophyte *Pelagococcus subviridis*. *Phytochemistry*, 28: 3347–3353.

Caple, M. B., Chow, H., Strouse, C. E. (1978). Photosynthetic pigments of green sulphur bacteria: the esterifying alcohols of bacteriochlorophylls *c* from *Chlorobium limicola*. *J. Biol. Chem.* 253: 6730–6737.

Chapman, D. J. (1966). Three new carotenoids isolated from algae. *Phytochemistry*, 5: 1331–1333.

Chisholm, S. W., Olson, R. J., Zettler, E. R., Goericke, R., Waterbury, J. B., Welschmeyer, N. A. (1988). A novel free-living prochlorophyte abundant in the oceanic euphotic zone. *Nature*, 334: 340–343.

Chisholm, S. W., Frankel, S., Goericke, R., Olson, R., Palenik, B., Urbach, E., Waterbury, J., Zettler, E. (1992). *Prochlorococcus marinus* nov. gen. nov. sp.: an oxyphototrophic marine prokaryote containing divinyl chlorophyll *a* and *b*. *Arch. Microbiol.* 157: 297–300.

Claustre, H. (1994). The trophic status of various oceanic provinces as revealed by phytoplankton pigment signatures. *Limnol. Oceanogr.* 39: 1206–1210.

Cleveland, J. S., Perry, M. J. (1994). A model for partitioning particulate absorption into phytoplanktonic and detrital components. *Deep-Sea Res.* 41: 197–221.

Cullen, J. L. (1982). The deep chlorophyll maximum: comparing vertical profiles of chlorophyll *a*. *Canad. J. Fish. Aquat. Sci.* 39: 791–803.

De las Rivas, J., Abadia, A., Abadia, J. (1989). A new reversed-phase HPLC method resolving all major higher plant photosynthetic pigments. *Plant Physiol.* 91: 190–192.

Derenbach, J. B., Astheimer, H., Hansen, H. P., Leach, H. (1979). Vertical microscale distribution of phytoplankton in relation to the thermocline. *Mar. Ecol. Progr. Ser.* 1: 187–193.

Egeland, E. S., Liaaen-Jensen, S. (1992). Eight new carotenoids from a chemosystematic evaluation of Prasinophyceae. *Proc. 7th Int. Symp. on Marine Natural Products*, Capri, July 1992, p. 2.

Egeland, E. S., Liaaen-Jensen, S. (1993). New carotenoids and chemosystematics in the Prasinophyceae. *Proc. 10th Int. Symp. Carotenoids*, Norway, June, 1993, pp. 4–6.

Eskins, K., Scholfield, C. R., Dutton, H. J. (1977). High-performance liquid chromatography of plant pigments. *J. Chromatogr.* 135: 217–220.

Everitt, D. A., Wright, S. W., Volkman, J. K., Thomas, D. P., Lindstrom, E. (1990). Phytoplankton community compositions in the western equatorial Pacific determined from chlorophyll and carotenoid pigment distributions. *Deep-Sea Res.* 37: 975–997.

Exton, R. J., Houghton, W. M., Esaias, W., Haas, L. W., Hayward, D. (1983). Spectral differences and temporal stability of phycoerythrin fluoresence in estuarine and coastal waters due to the domination of labile cryptophytes and stable cyanobacteria. *Limnol. Oceanogr.* 28: 1225–1231.

Falkowski, P. G., Kolber, Z. (1995). Variations in chlorophyll fluorescence yields in phytoplankton in the world oceans. *Aust. J. Plant Physiol.* 22: 341–355.

Fasham, M. J. R., Holligan, P. M., Pugh, P. R. (1983). The spatial and temporal development of the spring phytoplankton bloom in the Celtic Sea, April 1979. *Prog. Oceanogr.* 12: 87–145.

Fawley, M. N., Lee, C. M. (1990). Pigment composition of the scaly green flagellate *Mesostigma viride* (Micromonadophyceae) is similar to that of the siphonous green alga *Bryopsis plumosa* (Ulvophyceae). *J. Phycol.* 26: 666–670.

Fookes, C. J. R., Jeffrey, S. W. (1989). The structure of chlorophyll c_3, a novel marine photosynthetic pigment. *J. Chem. Soc. Chem. Comm.* 23: 1827–1828.

Foss, P., Guillard, R. R. L., Liaaen-Jensen, S. (1984). Prasinoxanthin — a chemosystematic marker for algae. *Phytochemistry*, 23: 1629–1633.

Foss, P., Lewin, R. A., Liaaen-Jensen, S. (1987). Carotenoids of *Prochloron* sp. (Prochlorophyta). *Phycologia*, 26: 142–144.

Früngel, F., Koch, C. (1976). Practical experience with the VarioSens equipment in measuring chlorophyll concentrations and fluorescent substances like rhodamine, fluorescein and some new substances. *IEEE J. Ocean. Eng.* vol. OE-1, p. 21–32.

Garrido, J. L., Zapata, M. (1993*a*). High performance liquid chromatographic separation of polar and non-polar chlorophyll pigments from algae using a wide pore polymeric octadecyl silica column. *J. High Res. Chromatogr.* 16: 229–233.

Garrido, J. L., Zapata, M. (1993*b*). High performance liquid chromatography of chlorophylls c_3, c_1, c_2 and *a* and of carotenoids of chromophyte algae on a polymeric octadecyl silica column. *Chromatographia*, 35: 543–547.

Garrido, J. L., Zapata, M., Muniz, S. (1995). Spectral characterization of new chlorophyll *c* pigments isolated from *Emiliania huxleyi* (Prymnesiophyceae) by high performance liquid chromatography. *J. Phycol.* 31: 761–768.

Garver, S. A., Siegel, D. A., Mitchell, B. G. (1994). Variability in near-surface particulate absorption spectra: what can a satellite ocean colour imager see? *Limnol. Oceanogr.* 39: 1349–1367.

Gibbs, C. F. (1979). Chlorophyll *b* interference in the fluorometric determination of chlorophyll *a* and 'phaeo-pigments'. *Aust. J. Mar. Freshw. Res.* 30: 597–606.

Gieskes, W. W. C. (1991). Algal pigment fingerprints: clue to taxon-specific abundance, productivity and degradation of phytoplankton in seas and oceans. In: Demers, S. (ed.). *Particle Analysis in Oceanography*, NATO ASI Ser. vol. G27, p. 61–99.

Gieskes, W. W. C., Kraay, G. W. (1983*a*). Unknown chlorophyll *a* derivatives in the North Sea and the tropical Atlantic Ocean revealed by HPLC analysis. *Limnol. Oceanogr.* 28: 757–766.

Gieskes, W. W. C., Kraay, G. W. (1983*b*). Dominance of Cryptophyceae during the phytoplankton spring bloom in the central North Sea detected by HPLC analysis of pigments. *Mar. Biol.* 75: 179–185.

Gieskes, W. W. C., Kraay, G. W. (1986*a*). Analysis of phytoplankton pigments by HPLC before, during and after mass occurrence of the microflagellate *Corymbellus aureus* during the spring bloom in the open northern North Sea in 1983. *Mar. Biol.* 92: 45–52.

Gieskes, W. W. C., Kraay, G. W. (1986*b*). Floristic and physiological differences between the shallow and the deep nanophytoplankton community in the euphotic zone of the open tropical Atlantic revealed by HPLC analysis of pigments. *Mar. Biol.* 91: 567–576.

Gieskes, W. W. C., Kraay, G. W. (1989). Estimating the carbon-specific growth rate of the major algal species groups in eastern Indonesian waters by ^{14}C labelling of taxon-specific carotenoids. *Deep-Sea Res.* 36: 1127–1139.

Gieskes, W. W. C., Kraay, G. W., Nontji, A., Setiapermana, D., Sutomo (1988). Monsoonal alteration of a mixed and layered structure in the phytoplankton of the euphotic zone of the Banda Sea (Indonesia): a mathematical analysis of algal pigment fingerprints. *Neth. J. Sea Res.* 22: 123–137.

Gieskes, W. W. C., Kraay, G. W., Tijssen, S. B. (1978). Chlorophylls and their degradation products in the deep pigment maximum layer of the tropical North Atlantic. *Neth. J. Sea-Res.* 12: 195–204.

Goericke, R., Repeta, D. J. (1992). The pigments of *Prochlorococcus marinus*: the presence of divinyl chlorophyll *a* and *b* in a marine procaryote. *Limnol. Oceanogr.* 37: 425–433.

Goericke, R., Repeta, D. J. (1993). Chlorophylls *a* and *b* and divinyl chlorophylls *a* and *b* in the open sub-tropical North Atlantic Ocean. *Mar. Ecol. Progr. Ser.* 101: 307–313.

Guillard, R. R. L., Murphy, L. S., Foss, P., Liaaen-Jensen, S. (1985). *Synechococcus* spp. as likely zeaxanthin-dominant ultraphytoplankton in the north Atlantic. *Limnol. Oceanogr.* 30: 412–414.

Guo, C., Dunstan, W. M. (1995). Depth-dependent changes in chlorophyll fluorescence number at a Sargasso Sea station. *Mar. Biol.* 122: 333–339.

Hager, A., Stransky, H. (1970). Das Carotenoidmuster und die Verbreitung des lichtinduzierten Xanthophyll-cyclus in Verschiedenen Algen Klassen. V. Einzelne Vertreter der Cryptophyceae, Euglenophyceae, Bacillariophyceae, Chrysophyceae und Phaeophyceae. *Arch. Mikrobiol.* 73: 77–89.

Hallegraeff, G. M. (1981). Seasonal study of phytoplankton pigments and species at a coastal station off Sydney: importance of diatoms and the nanoplankton. *Mar. Biol.* 61: 107–118.

Hallegraeff, G. M., Jeffrey, S. W. (1984). Tropical phytoplankton species and pigments of continental shelf waters of north and north-west Australia. *Mar. Ecol. Progr. Ser.* 20: 59–74.

Hallegraeff, G. M., Jeffrey, S. W. (1985). Description of new chlorophyll *a* alteration products in marine phytoplankton. *Deep-Sea Res.* 32: 697–705.

Head, E. J. H., Hargrave, B. T., Subba Rao, D. V. (1994). Accumulation of a pheophorbide *a*-like pigment in sediment traps during late stages of a spring bloom: a product of dying algae? *Limnol. Oceanogr.* 39: 176–181.

Herman, A. W., Denman, K. L. (1977). Rapid underwater profiling of chlorophyll with an *in situ* fluorometer mounted on a 'Batfish' vehicle. *Deep-Sea Res.* 24: 385–397.

Hilton, J., Rigg, E., Jaworski, G. (1989). Algal identification using *in vivo* fluorescence spectra. *J. Plankton Res.* 11: 65–74.

Hoepffner, N., Sathyendranath, S. (1991). Effect of pigment composition on absorption properties of phytoplankton. *Mar. Ecol. Progr. Ser.* 73: 11–23.

Hoepffner, N., Sathyendranath, S. (1992). Bio-optical characteristics of coastal waters: absorption spectra of phytoplankton and pigment distribution in the western North Atlantic. *Limnol. Oceanogr.* 37: 1660–1679.

Holligan, P. M., Balch, W. M. (1991). From oceans to cells: coccolithophore optics and biogeochemistry. In: Demers, S. (ed.). *Particle Analysis in Oceanography*, NATO ASI Series, vol. G27, Springer, pp. 301–324.

Holm-Hansen, O., Lorenzen, C.J., Holmes, R. W., Strickland, J. D. H. (1965). Fluorometric determination of chlorophyll. *J. Conseil.* 301: 3–15.

Hooks, C. E., Bidigare, R. R., Keller, M. D., Guillard, R. R. L. (1988). Coccoid eukaryotic marine ultraplankters with four different HPLC pigment signatures. *J. Phycol.* 24: 571–580.

Humphrey, G. F. (1960). The concentration of plankton pigments in Australian waters. *CSIRO Div. Fish. Oceanogr. Tech. Paper* No. 9, 27 pp.

Humphrey, G. F. (1966). The concentration of chlorophylls *a* and *c* in the south-eastern Indian Ocean. *Aust. J. Mar. Freshwat. Res.* 17: 135–145.

Humphrey, G. F. (1970). The concentration of chlorophylls *a* and *c* in the south-west Pacific Ocean. *Aust. J. Mar. Freshwat. Res.* 21: 1–10.

Humphrey, G. F. (1978). The recalculation of marine chlorophyll concentrations with special reference to Australian waters. *Aust. J. Mar. Freshwat. Res.* 29: 409–416.

Humphrey, G. F., Kerr, J. D. (1969). Seasonal variations in the Indian Ocean along 110°E. III. Chlorophylls *a* and *c*. *Aust. J. Mar. Freshwat. Res.* 20: 55–64.

Ishizaka, J., Kiyosawa, H., Kazunori, I., Ishikawa, K., Takahashi, M. (1994). Meridional distribution and carbon biomass of autotrophic picoplankton in the Central North Pacific Ocean during late northern summer 1990. *Deep-Sea Res.* 41: 1745–1766.

Jeffrey, S. W. (1963). Purification and properties of chlorophyll *c* from *Sargassum flavicans. Biochem. J.* 86: 313–318.

Jeffrey, S. W. (1968). Quantitative thin-layer chromatography of chlorophylls and carotenoids from marine algae. *Biochim. Biophys. Acta*, 162: 271–285.

Jeffrey, S. W. (1972). Preparation and some properties of crystalline chlorophylls c_1 and c_2 from marine algae. *Biochim. Biophys. Acta*, 279: 15–33.

Jeffrey, S. W. (1974). Profiles of photosynthetic pigments in the ocean using thin-layer chromatography. *Mar. Biol.* 26: 101–110.

Jeffrey, S. W. (1976*a*). A report of green algal pigments in the central North Pacific Ocean. *Mar. Biol.* 37: 33–37.

Jeffrey, S. W. (1976*b*). The occurrence of chlorophyll c_1 and c_2 in algae. *J. Phycol.* 12: 349–354.

Jeffrey, S. W. (1981). An improved thin-layer chromatographic technique for marine phytoplankton pigments. *Limnol. Oceanogr.* 26: 191–197.

Jeffrey, S. W. (1989). Chlorophyll *c* pigments and their distribution in the chromophyte algae. In: Green, J. C., Leadbeater, B. S. C., Diver, W. L. (eds), *The Chromophyte Algae: Problems and Perspectives*, Clarendon Press, Oxford, p. 13–36.

Jeffrey, S. W., Hallegraeff, G. M. (1980*a*). Studies of phytoplankton species and photosynthetic pigments in a warm-core eddy of the East Australian Current. I. Summer populations. *Mar. Ecol. Progr. Ser.* 3: 285–294.

Jeffrey, S. W., Hallegraeff, G. M. (1980*b*). Studies of phytoplankton species and photosynthetic pigments in a warm-core eddy of the East Australian Current. II. A note on pigment methodology. *Mar. Ecol. Progr. Ser.* 3: 295–301.

Jeffrey, S. W., Hallegraeff, G. M. (1987*a*). Phytoplankton pigments; species and light climate in a complex warm-core eddy of the East Australian Current. *Deep-Sea Res.* 34: 649–673.

Jeffrey, S. W., Hallegraeff, G. M. (1987*b*). Chlorophyllase distribution in 10 classes of phytoplankton — a problem for chlorophyll analysis. *Mar. Ecol. Progr. Ser.* 35: 293–304.

Jeffrey, S. W., Hallgraeff, G. M. (1990). Phytoplankton ecology of Australasian waters. In: Clayton, M. N., King, R. J. (eds), *The Biology of Marine Plants*, Longman Cheshire, Melbourne, p. 310–348.

Jeffrey, S. W., Humphrey, G. F. (1975). New spectrophotometric equations for determining chlorophylls *a, b,* c_1 and c_2 in higher plants, algae and natural phytoplankton. *Biochem. Physiol. Pflanzen*, 167: 191–194.

Jeffrey, S. W., Sielicki, M., Haxo, F. T. (1975). Chloroplast pigment patterns in dinoflagellates. *J. Phycol.* 11: 374–385.

Jeffrey, S. W., Wright, S. W. (1987). A new spectrally distinct component in preparations of chlorophyll *c* from the microalga *Emiliania huxleyi* (Prymnesiophyceae). *Biochim. Biophys. Acta*, 894: 180–188.

Jeffrey, S. W., Wright, S. W. (1994). Photosynthetic pigments in the Haptophyta, In: Green, J. C., Leadbeater, B. S. C. (eds), *The Haptophyte Algae*, Clarendon Press, Oxford, p. 111–132.

Johansen, J. E., Svec, W. A., Liaaen-Jensen, S., Haxo, F. T. (1974). Carotenoids of the Dinophyceae. *Phytochemistry*, 13: 2261–2271.

Johnsen, G., Sakshaug, E. (1993). Bio-optical characteristics and photoadaptive responses in the toxic and bloom-forming dinoflagellates *Gyrodinium aureolum, Gymnodinium galatheanum*, and two strains of *Prorocentrum minimum*. *J. Phycol.* 29: 627–642.

Johnsen, G., Samset, O., Granskog, L., Sakshaug, E. (1994). *In vivo* absorption characteristics in 10 classes of bloom-forming phytoplankton: taxonomic characteristics and responses to photoadaptation by means of discriminant and HPLC analysis. *Mar. Ecol. Progr. Ser.* 105: 149–157.

Kjøsen, H., Liaaen-Jensen, S. (1972). Carotenoids of higher plants. Total synthesis of lycoxanthin and lycophyll. *Acta Chem. Scand.* 26: 4121–4129.

Klein, B., Sournia, A. (1987). A daily study of the diatom spring bloom at Roscoff (France) in 1985. II. Phytoplankton pigment composition studied by HPLC analysis. *Mar. Ecol. Progr. Ser.* 37: 265–275.

Kolber, Z., Falkowski, P. G. (1993). Use of active fluorescence to estimate phytoplankton photosynthesis *in situ*. *Limnol. Oceanogr.* 38: 1646–1665.

Kraay, G. W., Zapata, M., Veldhuis, M. J. W. (1992). Separation of chlorophylls c_1, c_2 and c_3 of marine phytoplankton by reversed-phase C_{18} high-performance liquid chromatography. *J. Phycol.* 28: 708–712.

Krey, J., Babenerd, B. (1976). *Phytoplankton production: Atlas of the International Indian Ocean Expedition*. Univ. Kiel, Germany, 70 pp.

Letelier, R. M., Bidigare, R. R., Hebel, D. V., Ondrusek, M., Winn, C. D., Karl, D. M. (1993). Temporal variability of phytoplankton community structure based on pigment analysis. *Limnol. Oceanogr.* 38: 1420–1437.

Lewis, M. (1989). The varigated ocean: a view from space. *New Scientist*, 124: 19–22.

Liaaen-Jensen, S., Jensen, A. (1971). Quantitative determination of carotenoids in photosynthetic tissues. *Meth. Enzymol.* 23: 586–602.

Lorenzen, C. J. (1966). A method for the continuous measurement of *in vivo* chlorophyll concentrations. *Deep-Sea Res.* 13: 223–227.

Lorenzen, C. J. (1967). Determination of chlorophyll and pheopigments: spectrophotometric equations. *Limnol. Oceanogr.* 12: 343–346.

Lorenzen, C. J. (1981). Chlorophyll *b* in the ocean. *Deep-Sea Res.* 28: 1049–1056.

Lorenzen, C. J., Jeffrey, S. W. (1980). Determination of chlorophyll in seawater: report of intercalibration tests. *UNESCO Technical Papers in Marine Science*, 35, 21 pp.

Lutz, V. A., Sathyendranath, S., Head, E. J. H. (1996). Absorption coefficient of phytoplankton: regional variations in the North Atlantic. *Mar. Ecol. Progr. Ser.* (in press).

Mackey, D. J., Parslow, J., Higgins, H. W., Griffiths, F. B., O'Sullivan, J. E. (1995). Plankton productivity and biomass in the western Equatorial Pacific: biological and physical controls. *Deep-Sea Res.* 42: 499–533.

Madgwick, J. C. (1965). Quantitative chromatography of algal chlorophylls on thin layers of glucose. *Deep-Sea Res.* 12: 233–236.

Mantoura, R. F. C., Llewellyn, C. A. (1983). The rapid determination of algal chlorophyll and carotenoid pigments and their breakdown products in natural waters by reverse-phase high-performance liquid chromatography. *Anal. Chim. Acta*, 151: 297–314.

Millie, D. F., Paerl, H. W., Hurley, J. P., Kirkpatrick, G. J. (1993*a*). Algal pigment determinations in aquatic ecosystems: analytical evaluations, applications and recommendations. *Current Topics Bot. Res.* 1: 1–13.

Millie, D. F., Paerl, H. W., Hurley, J. P. (1993*b*). Microalgal pigment assessments using high-performance liquid chromatography: a synopsis of organismal and ecological applications. *Canad. J. Fish. Aquat. Sci.* 50: 2513–2527.

Millie, D. F., Schofield, O. M. (1995). What are the requirements for bio-optical characterization of harmful algal blooms? *J. Phycol.* Suppl. 31: 12.

Mitchell, B. G., Brody, E. A., Holm-Hansen, O., McClain, C., Bishop, J. (1991). Light limitation of phytoplankton biomass and macronutrient utilization in the Southern Ocean. *Limnol. Oceanogr.* 36: 1662–1677.

Moore, L. R., Goericke, R., Chisholm, S. W. (1995). Comparative physiology of *Synechococcus* and *Prochlorococcus*: influence of light and temperature on growth, pigments, fluorescence and absorptive properties. *Mar. Ecol. Progr. Ser.* 116: 259–275.

Nelson, J. R., Wakeham, S. G. (1989). A phytol-substituted chlorophyll *c* from *Emiliania huxleyi* (Prymnesiophyceae). *J. Phycol.* 25: 761–766.

Oldham, P. B., Zillioux, E. J., Warner, I. M. (1985). Spectral fingerprinting of phytoplankton populations by two-dimensional fluorescence and Fourier-transform-based pattern recognition. *J. Mar. Res.* 43: 893–906.

Olson, R. J., Chisholm, S. W., Zettler, E. R., Armbrusty, E. V. (1988). Analysis of *Synechococcus* pigment types in the sea using single and dual beam flow cytometry. *Deep-Sea Res.* 35: 425–440.

Olson, R. J., Zettler, E. R., Anderson, O. K. (1989). Discrimination of eukaryotic phytoplankton cell types from light scatter and autofluorescence properties measured by flow cytometry. *Cytometry*, 10: 636–643.

Ondrusek, M. E., Bidigare, R. R., Sweet, S. T., Defreitas, D. A., Brooks, J. M. (1991). Distribution of phytoplankton pigments in the North Pacific Ocean in relation to physical and optical variability. *Deep-Sea Res.* 38: 243–266.

Ormerod, M. G. (1994). *Flow Cytometry*. Royal Microscopical Society Microscopy Handbooks, No. 29, BIOS Scientific Publishers, Oxford, 77 pp.

Parsons, T. R., Strickland, J. D. H. (1963). Discussion of spectrophotometric determination of marine plant pigments, with revised equations for ascertaining chlorophylls and carotenoids. *J. Mar. Res.* 21: 155–163.

Pennington, F. C., Haxo, F. T., Borch, G., Liaaen-Jensen, S. (1985). Carotenoids of Cryptophyceae. *Biochem. Syst. Ecol.* 13: 215–219.

Platt, T., Li, W. K. W. (eds) (1986). *Photosynthetic Picoplankton. Can. Bull. Fish. Aquat. Sci.* 214: 583 pp.

Poryvkina, L., Babichenko, S., Kaitala, S., Kuosa, H., Shalapjonok, A. (1994). Spectral fluorescence signatures in the characterization of phytoplankton community composition. *J. Plankt. Res.* 16: 1315–1327.

Repeta, D. J., Simpson, D. J. (1991). The distribution and recycling of chlorophyll, bacteriochlorophyll and carotenoids in the Black Sea. *Deep-Sea Res.* 39: 969–984.

Repeta, D. J., Simpson, D. J., Jorgensen, B. B., Jannash, H. W. (1989). Evidence for anoxygenic photosynthesis from the distribution of bacteriochlorophylls in the Black Sea. *Nature*, 342: 69–72.

Richards, F. A., Thompson, T. G. (1952). The estimation and characterization of plankton populations by pigment analyses. II. A spectrophotometric method for the estimation of plankton pigments. *J. Mar. Res.* 11: 156–172.

Ricketts, T. R. (1966). Magnesium 2,4-divinyl phaeoporphyrin a_5 monomethyl ester, a protochlorophyll-like pigment present in some unicellular flagellates. *Phytochemistry*, 5: 223–229.

Ridout, P. S., Morris, R. J. (1985). Short-term variations in the pigment composition of a spring phytoplankton bloom from an enclosed experimental ecosystem. *Mar. Biol.* 87: 7–11.

Rowan, K. S. (1989). *Photosynthetic Pigments of Algae*. Cambridge Univ. Press, 334 pp.

Roy, S. (1987). High-performance liquid chromatographic analysis of chloropigments. *J. Chromatog.* 391: 19–34.

Sartory, D. P. (1985). The determination of algal chlorophyllous pigments by high performance liquid chromatography and spectrophotometry. *Water Res.* 19: 605–610.
Sathyendranath, S. (1986). Remote sensing of phytoplankton: a review, with special reference to picoplankton. In: Platt, T., Li, W. K. W. (eds), *Photosynthetic Picoplankton. Can. Bull. Fish. Aquat. Sci.* 214: 561–583.
Shioi, Y. (1990). Analytical chromatography of chlorophylls. In: Scheer, H. (ed.), *Chlorophylls*, CRC Press, Boca Raton, Florida, p. 60–88.
Simon, N., Barlow, R. G., Marie, D., Partensky, F., Vaulot, D. (1994). Characterization of oceanic photsynthetic picoeukaryotes by flow cytometry. *J. Phycol.* 30: 922–935.
SooHoo, J. B., Kiefer, D. A., Collins, D. J., McDermid, I. S. (1986). *In vivo* fluorescence excitation and absorption spectra of marine phytoplankton. I. Taxonomic characteristics and responses to photoadaptation. *J. Plankton Res.* 8: 197–214.
Stauber, J. L., Jeffrey, S. W. (1988). Photosynthetic pigments in fifty-one species of marine diatoms. *J. Phycol.* 24: 158–172.
Stewart, D. E., Farmer, F. H. (1984). Extraction, identification and quantitation of phycobiliprotein pigments from phototrophic plankton. *Limnol. Oceanogr.* 29: 392–397.
Strain, H. H. (1958). *Chloroplast pigments and chromatographic analysis.* 32nd Annual Priestly Lecture, Penn. State University, USA, 180 pp.
Strickland, J. D. H. (1968). Continuous measurement of *in vivo* chlorophyll: a precautionary note. *Deep-Sea Res.* 15: 225–227.
Strom, S. L. (1991). Growth and grazing rates of an herbivorous dinoflagellate (*Gymnodinium* sp.) from the open subarctic Pacific Ocean. *Mar. Ecol. Progr. Ser.* 78: 103–113.
Strom, S. L. (1993). Production of pheopigments by marine protozoa: results of laboratory experiments analysed by HPLC. *Deep-Sea Res.* 40: 57–80.
Swartz, S. J., von Elbe, J. H. (1982). High performance liquid chromatography of plant pigments — a review. *J. Liquid Chromatog.* 5: 43–73.
UNESCO (1966). *Determination of photosynthetic pigments in seawater.* UNESCO Monographs on Oceanographic Methodology, Paris, 69 pp.
Van Heukelem, L., Lewitus, A. J., Kana, T. M., Craft, N. E. (1992). High performance liquid chromatography of phytoplankton pigments using a polymeric reversed-phase C_{18} column. *J. Phycol.* 28: 867–872.
Van Heukelem, L., Lewitus, A. J., Kana, T. M., Craft, N. E. (1994). Improved separations of phytoplankton pigments using temperature-controlled high performance liquid chromatography. *Mar. Ecol. Progr. Ser.* 114: 303–313.
Veldhuis, M. J. W., Kraay, G. W. (1990). Vertical distribution and pigment composition of a picoplankton prochlorophyte in the subtropical North Atlantic: a combined study of HPLC analysis of pigments and flow cytometry. *Mar. Ecol. Progr. Ser.* 68: 121–127.
Venrick, E. L., McGowan, J. A., Mantyla, A. W. (1973). Deep maxima of photosynthetic chlorophyll in the Pacific Ocean. *Fishery Bulletin*, 71: 41–52.
Vernet, M., Lorenzen, C. J. (1987). The presence of chlorophyll *b* and the estimation of pheopigments in marine phytoplankton. *J. Plankton Res.* 9: 255–265.
Vernon, L. P. (1960). Spectrophotometric determination of chlorophylls and pheophytins in plant extracts. *Anal. Chem.* 32: 1144–1150.
Vesk, M., Jeffrey, S. W. (1987). Ultrastructure and pigments of two strains of the picoplanktonic alga *Pelagococcus subviridis* (Chrysophyceae). *J. Phycol.* 23: 322–336.
Wartenberg, D. E. (1978). Spectrophotometric equations: an intercalibration technique. *Limnol. Oceanogr.* 23: 566–570.
Waterbury, J. B., Watson, S. W., Guillard, R. R. L., Brand, L. E. (1979). Widespread occurrence of a unicellular, marine, planktonic, cyanobacterium. *Nature*, 277: 293–294.
Welschmeyer, N. A. (1994). Fluorometric analysis of chlorophyll *a* in the presence of chlorophyll *b* and pheopigments. *Limnol. Oceanogr.* 39: 1985–1992.
Welschmeyer, N. A., Lorenzen, C. J. (1985). Chlorophyll budgets: zooplankton grazing and phytoplankton growth in a temperate fjord and the Central Pacific Gyre. *Limnol. Oceanogr.* 30: 1–21.

Williams, R., Claustre, H. (1991). Photosynthetic pigments as biomarkers of phytoplankton populations and processes involved in the transformation of particulate organic matter at the Biotrans site (47°N, 20°W). *Deep-Sea Res.* 38: 347–355.

Wright, S. W., Jeffrey, S. W., Mantoura, R. F. C., Llewellyn, C. A., Bjørnland, T., Repeta, D., Welschmeyer, N. A. (1991). Improved HPLC method for the analysis of chlorophylls and carotenoids from marine phytoplankton. *Mar. Ecol. Progr. Ser.* 77: 183–196.

Wright, S. W., Shearer, J. D. (1984). Rapid extraction and high-performance liquid chromatography of chlorophylls and carotenoids from marine phytoplankton. *J. Chromatog.* 294: 281–295.

Wyman, M. (1992). An *in vivo* method for the estimation of phycoerythrin concentrations in cyanobacteria. *Limnol. Oceanogr.* 37: 1300–1306.

Yentsch, C. M., Horan, P. K., Muirhead, K., Dortch, Q., Haugen, E., Legendre, L., Murphy, L. S., Perry, M. J., Phinney, D. A., Pomponi, S. A., Spinrad, R. W., Wood, M., Yentsch, C. S., Zahuranec, B. J. (1983). Flow cytometry and cell sorting: a technique for analysis and sorting of aquatic particles. *Limnol. Oceanogr.* 28: 1275–1280.

Yentsch, C. S. (1965). The relationship between chlorophyll and photosynthetic carbon production with reference to the measurement of decomposition products and chloroplastic pigments. Proc. IBP Symp. *'Primary Production in Aquatic Environments'*, Pallanza, Italy, p. 323–346.

Yentsch, C. S., Menzel, D. W. (1963). A method for the determination of phytoplankton chlorophyll and phaeophytin by fluorescence. *Deep-Sea Res.* 10: 221–231.

Yentsch, C. S., Phinney, D. A. (1985). Spectral fluorescence: an ataxonomic tool for studying the structure of phytoplankton populations. *J. Plankt. Res.* 7: 617–632.

Yentsch, C. S., Yentsch, C. M. (1979). Fluorescence spectral signatures: the characterization of phytoplankton populations by the use of excitation and emission spectra. *J. Mar. Res.* 37: 471–483.

Zapata, M., Ayala, A. M., Franco, J. M., Garrido, J. L. (1987). Separation of chlorophylls and their degradation products in marine phytoplankton by reversed-phase high-performance liquid chromatography. *Chromatographia*, 23: 26–30.

5

Pigment processes in the sea: a selected bibliography

S. W. Jeffrey, C. A. Llewellyn, R. G. Barlow and R. F. C. Mantoura

5.1 Introduction

This chapter was to review concepts of phytoplankton processes in the sea derived from the use of pigment separation methods. The field is however large and multifaceted. Major topics which have been studied include HPLC contributions to chemotaxonomy, phytoplankton distributions and successions, quantitation of biomass, carbon and pigment budgets, chromatic signatures and ocean optics, zooplankton grazing processes, the digestive fate of pigments, phytodetritus and sedimentation processes, and pigment diagenesis in sediments. Vertical and horizontal pigment distributions have been measured in contrasting oceans: from oligotrophic environments (e.g., the Pacific Ocean), to cold productive systems (e.g., the Southern Ocean), stratified water columns (e.g., the Indian Ocean), deep mixing systems (e.g., the North Atlantic), blooms in shelf environments (e.g., Chesapeake Bay) and ocean basins (e.g., the North Atlantic) and degradation processes in inshore eutrophic estuaries (e.g., the Tamar estuary). New concepts in phytoplankton dynamics have resulted from the application of pigment separation methods.

It soon became apparent that a full treatment of this topic could not be justified in a monograph devoted to method evaluation and development. Indeed, an in depth treatment of phytoplankton pigment processes in the sea would require its own monograph (not written by us!), which would be of major impact when the results of present international studies of the global ocean were completed. Two such reports have al-

TABLE 5.1. Phytoplankton pigments: chemotaxonomy, biomass budgets and contrasting seas

Reference	Topics studied															Ocean
	Chlorophylls	Carotenoids	Pheopigments	^{14}C labelling of pigments	Multiple regression	Spring bloom	Chlorophyll budgets	Sediment traps	Biochemical parameters	Chromatic adaptation/photoadaptation	Phytoplankton species/successions	Water masses/fronts	Seasonal changes	Carbon budgets	Cell flow cytometry	
Barlow *et al.* (1993*a*)	+	+			+	+										N. E. Atlantic
Barlow *et al.* (1993*b*)	+		+			+										N. E. Atlantic
Barlow *et al.* (1995)	+	+	+		+	+	+	+								W. Alboran Sea
Bidigare *et al.* (1990)	+	+				+				+	+					Sargasso Sea
Bidigare *et al.* (1993)	+	+														Caribbean Sea
Brunet *et al.* (1992)	+	+										+				English Channel
Buma *et al.* (1990)	+	+										+	+			Southern Ocean
Burford *et al.* (1995)	+	+									+		+			Gulf of Carpentaria
Claustre *et al.* (1990)	+	+							+		+					Irish Sea
Claustre *et al.* (1994*a*)	+	+										+		+		E. Alboran Sea
Claustre *et al.* (1994*b*)	+	+								+		+				E. Alboran Sea
Denant *et al.* (1990)	+	+														E. China Sea
Everitt *et al.* (1990)	+	+														W. Equatorial Pacific
Fiala *et al.* (1994)	+	+									+	+			+	E. Alboran Sea
Gieskes & Kraay (1983*a*)	+	+														Tropical Atlantic; North Sea
Gieskes & Kraay (1983*b*)	+	+			+	+					+					Central North Sea
Gieskes & Kraay (1986*a*)	+	+				+					+					N. North Sea
Gieskes & Kraay (1986*b*)	+	+		+												Tropical Atlantic
Gieskes & Kraay (1989)	+	+		+												E. Indonesian waters

Gieskes *et al.* (1988)	+	+			+											Banda Sea
Goericke & Repeta (1993)	+	+														Subtropical N. Atlantic
Goericke & Welschmeyer (1993*a*)	+	+		+						+	+					Sargasso Sea
Goericke & Welschmeyer (1993*b*)	+	+		+						+	+					Subarctic Pacific; Chesapeake Bay
Goericke & Welschmeyer (1993*c*)	+	+		+							+					Subarctic Pacific; Chesapeake Bay
Hallegraeff (1981)	+	+	+			+					+		+			East Australian Current
Hallegraeff & Jeffrey (1984)	+	+	+								+					North West Shelf, Australia
Harrison *et al.* (1993)	+	+				+		+			+	+		+		W. North Atlantic
Jeffrey (1976)	+	+														Central North Pacific
Jeffrey & Hallegraeff (1980)	+	+	+								+		+			East Australian Current
Jeffrey & Hallegraeff (1987)	+	+	+								+	+				East Australian Current
Klein & Sournia (1987)	+	+	+			+										French coastal waters
Latasa *et al.* (1992)	+	+	+									+				N. W. Mediterranean Sea
Letelier *et al.* (1993)	+	+											+			Pacific Ocean, Hawaii
Mackey *et al.* (1995)	+	+			+						+	+				Equatorial Pacific
Malej *et al.* (1995)	+	+											+			Gulf of Trieste
Marchant *et al.* (1987)	+	+									+	+				Southern Ocean
McManus & Dawson (1994)	+	+									+				+	Caribbean Sea; W. trop. Atlantic
Mejanelle *et al.* (1995)	+	+											+			N. E. Atlantic
Ondrusek *et al.* (1991)	+	+														N. Pacific
Pollehne *et al.* (1993)	+	+					+									N. Indian Ocean
Roy (1989)	+		+			+										Bedford Basin, Canada
Siegal *et al.* (1990)	+	+				+					+					Sargasso Sea
Trees *et al.* (1986)	+		+													Gulf of Strean; N. Sargasso Sea
Vaulot *et al.* (1990)	+	+									+		+		+	N. W. Mediterranean
Veldhuis & Kraay (1990)	+	+									+				+	Subtrop. N. Atlantic
Veldhuis & Kraay (1993)	+	+									+				+	S. Red Sea
Weeks *et al.* (1993)	+	+	+													N. E. Atlantic
Williams & Claustre (1991)	+	+	+													N. E. Atlantic
Wright (1985)	+	+	+													Prydz Bay, Southern Ocean

ready appeared as a collection of papers in *Limnology and Oceanography*, Chisholm and Morel (eds) (1991): 'Control of phytoplankton production in nutrient-rich areas of the open sea', and in *Deep-Sea Research*, Ducklow and Harris (eds) (1993): 'The JGOFS North Atlantic Bloom Experiment'.

To assist the oceanographer with the literature, we offer here a selected bibliography of 83 recent papers which have used pigment separation techniques (HPLC or TLC) to study phytoplankton and ocean processes. The bibliography was selected on the basis that pigment separation techniques (e.g., HPLC) were an essential part of the methodology, and each paper was an important contribution to our knowledge of phytoplankton processes.

The bibliography is grouped into four major topics (see Tables 5.1 to 5.4) with no paper being cited more than once. Where overlap of topics within papers occurs, the major thrust is given the priority in the Tables. The listing is not meant to be exhaustive and we regret any inadvertant omissions. We hope this bibliography may provide the nucleus for a future review of phytoplankton processes in the sea, perhaps encouraged by the methods recommended and the data listings given in the present monograph.

5.2 Bibliography

5.2.1 *Phytoplankton pigments: chemotaxonomy, biomass budgets and contrasting seas*

Recent papers that have used HPLC or TLC in general phytoplankton studies in the world oceans are given in Table 5.1. Of the 49 papers listed, 44 used HPLC and five used TLC techniques. Fig. 5.1 shows the location of these and other studies in the world oceans (see Tables 5.2–5.4). The major effort has been in the northern hemisphere in the Atlantic Ocean, the Mediterranean Sea and the North Sea, with some focus in the western equatorial Pacific.

All authors studied chlorophyll and carotenoid patterns, and some included chlorophyll degradation products. Many papers highlighted spring blooms with phytoplankton species and successions also included. Phytoplankton distributions across fronts and water masses, seasonal changes, and chlorophyll and carbon budgets were studied in some areas. Overall, a wide range of topics was included in these general studies of phytoplankton populations in contrasting seas.

5.2.2 *Phytoplankton pigments and optical processes*

A small number of papers focused on the impact of optical processes and phytoplankton populations (Table 5.2). While there is a vast literature on optical oceanography not cited here, the six papers in Table 5.2 all used HPLC techniques for pigment analysis. Three papers addressed the exciting topic of light-induced epoxide (xanthophyll) cycles in the sea, to determine whether the light history of populations could be inferred from diadinoxanthin:diatoxanthin pigment ratios (see Chapter 3, section 3.8.4.5). Other studies distinguished optical differences in water masses/fronts, one paper proposed a bio-optical model, while another focused on pigment photoadaptation in the sea. The control of phytoplankton pigment processes by light requires further study, so that oceanographers may interpret depth profiles of pigments from a physiological, as well as a taxonomic viewpoint.

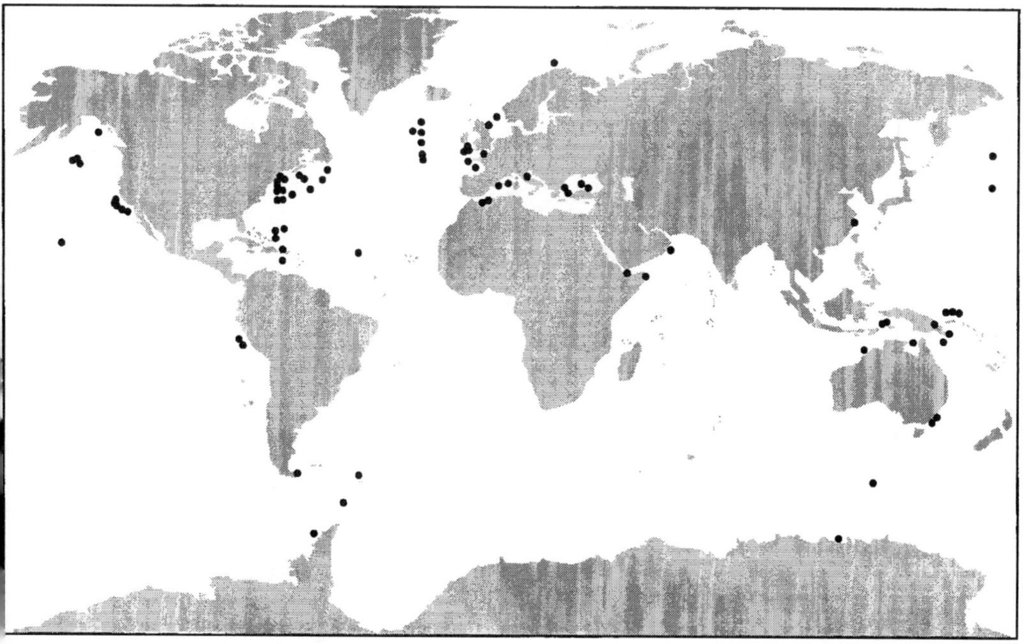

Figure 5.1
Geographic location of study sites of papers listed in Tables 5.1 to 5.4. All used pigment separation techniques to study phytoplankton processes.

TABLE 5.2. Phytoplankton pigments and optical processes

Reference	Topics studied										Ocean
	Chlorophylls	Carotenoids	Absorption spectra	Photosynthetic quantum yield	Xanthophyll cycling	Photoadaptation	Water masses/fronts	Bio-optical model	*In situ* production	Spring bloom	
Bidigare *et al.* (1987)	+	+						+	+		Sargasso Sea
Brunet *et al.* (1993)	+	+			+	+	+				Eastern English Channel
Nelson *et al.* (1993)	+	+	+								Southern California Bight
Olaizola *et al.* (1992)	+	+			+					+	Auke Bay, Alaska
Schofield *et al.* (1993)	+	+	+	+							Southern California Bight
Welschmeyer & Hoepffner (1986)	+	+			+		+				Coastal Atlantic (Massachussetts)

TABLE 5.3. Phytoplankton pigments and grazing processes

Reference	Topics studied									Zooplankton animals	Ocean
	Chlorophylls	Carotenoids	Pheopigments and degradation products	Growth rates from ^{14}C labelling	Animal tissues	Water column	Sediment traps	Faecal material	Phytoplankton species		
Abele-Oeschger *et al.* (1992)	+	+	+		+					Giant isopods	Gulf of Gdansk, Poland
Burkill *et al.* (1986)	+	+				+			+	Microzooplankton	Celtic Sea; Camarthan Bay
Hawkins *et al.* (1986)	+	+	+		+					Mussels	Mesocosms
Head & Harris (1992)	+	+						+	+	Copepods	East coast, Newfoundland
Head & Horne (1993)	+	+	+				+	+	+	Copepods	North Atlantic
Kleppel & Pieper (1984)	+	+			+					Copepods	Southern California coast
Kleppel *et al.* (1991)	+	+							+	Copepods, microplankton	California coastal waters; Irish Sea
McManus & Ederington-Cantrell (1992)	+	+							+	Microzooplankton	Chesapeake Bay, USA
Nelson (1989)	+	+	+		+			+		Macrozooplankton, copepods, salps	Southern California Bight
Strom & Welschmeyer (1991)	+	+		+		+			+	Microzooplankton	Subarctic Pacific Ocean
Verity & Vernet (1992)	+	+	+						+	Microzooplankton	Norwegian fiords
Vernet & Lorenzen (1987*a*)	+		+			+					Central N. Pacific; Oregon coast; Dabob Bay, Washington
Vernet & Lorenzen (1987*b*)			+			+	+	+			Dabob Bay
Wassman *et al.* (1990)	+	+	+				+	+	+	Macrozooplankton	Barents Sea
Welschmeyer *et al.* (1991)	+	+	+	+		+				Microzooplankton	Subarctic Pacific Ocean

5.2.3 *Phytoplankton pigments and grazing processes*

Fourteen papers concentrated on phytoplankton pigments and grazing processes (Table 5.3). A range of zooplankton animals were studied, from giant isopods to protozoa. Some papers used experimental animals grazing on phytoplankton cultures, but many used faecal material from field studies. A number of key 'pheopigments' are indicators of grazing processes and senescent algae, pyropheophorbide *a* and pheophorbide *a* being two of the most important (see Table 4.1, Chapter 4). The bibliography in this section overlaps somewhat with that of section 5.2.4, since grazing and sedimentation processes in the water column are intimately linked.

5.2.4 *Phytoplankton pigments, sediments and sedimentation processes*

Thirteen papers show the diagenetic changes occurring in phytoplankton pigments in sediments, and during sedimentation processes (Table 5.4). Some papers focused on transformations in carotenoids, others on chlorophylls. Most studies used sediments or sediment traps as a source of biological material. In one paper the presence of bacteriochlorophylls in sediments indicated anoxic conditions. The use of HPLC methods has uncovered a complex web of diagenetic transformations in sediments,

TABLE 5.4. Phytoplankton pigments, sediments and sedimentation processes

Reference	Topics studied									Ocean
	Chlorophylls	Carotenoids	Pheopigments; degradation products	Bacteriochlorophylls	Sediments	Sediment traps	Water column	Salt-marsh pond	Spectrophotometry	
Barlow *et al.* (1990)	+	+	+					+	+	Mid-Atlantic coast: France
Bianchi *et al.* (1993)	+	+			+		+			Hudson River estuary
Eckardt *et al.* (1992)	+				+					Baltic Sea; Black Sea
Head *et al.* (1994)	+		+			+				Bedford Basin, Canada
King & Repeta (1991)			+		+					Black Sea
Repeta (1989)		+	+		+					Peru continental shelf
Repeta & Gagosian (1982)		+	+							Peru continental shelf
Repeta & Gagosian (1984)	+	+	+			+				Peru upwelling
Repeta & Gagosian (1987)		+	+		+					Peru continental shelf
Repeta *et al.* (1989)				+	+		+			Black Sea
Repeta & Simpson (1991)	+	+		+	+		+			Black Sea
Riaux-Gobin *et al.* (1987)	+	+	+		+					Brittany coast: France
Sun *et al.* (1991)	+		+		+					Long Island Sound
Sun *et al.* (1993)	+		+		+					Long Island Sound

both in chlorophylls and carotenoids. The conditions for, and the precise pathways of, these changes in aquatic systems must be known before accurate pigment budgets can be determined in aquatic systems.

5.3 Conclusions

The bibliography listed above shows that HPLC pigment methods are now a regular part of many oceanographic programmes. New pigments are being discovered in natural phytoplankton populations and in marine sediments, as well as in sedimentation, grazing and diagenesis processes. It is now essential to ensure that pigments are identified securely, that marker pigments are correctly assigned to algal types or processes (see Table 4.1, Chapter 4), and that physiological changes in pigment ratios induced by light intensity and colour are recognized in the interpretation of field data. We hope the methods and data listings in this monograph (Parts II–V) will contribute to these goals.

Acknowledgements

We thank Mr T. Fileman for assistance with Fig. 5.1.

References

Abele-Oeschger, D., Szaniawska, A., Theede, H. (1992). Do pigments reflect the turnover of plant material in food chain studies? Analysis of plant pigments in the intestine of Saduria *(Mesidotea)* entomon. *Mar. Ecol. Prog. Ser.* 81: 43–49.

Barlow, R. G., Collos, Y., Maestrini, S. Y., Roy, S. (1990). Microphytobenthic pigments in a salt marsh pond determined by HPLC and spectrophotometry. *Mar. Microbial Food Webs* 4: 117–128.

Barlow, R. G., Mantoura, R. F. C., Gough, M. A., Fileman, T. W. (1993*a*). Pigment signatures of the phytoplankton composition in the northeastern Atlantic during the 1990 spring bloom. *Deep-Sea Res.* II. 40: 459–477.

Barlow, R. G., Mantoura, R. F. C., Gough, M. A., Fileman, T. W. (1993*b*). Phaeopigment distribution during the 1990 spring bloom in the northeastern Atlantic. *Deep-Sea Res.* 40: 2229–2242.

Barlow, R. G., Mantoura, R. F. C., Peinert, R. D., Miller, A. E. J., Fileman, T. W. (1995). The distribution, sedimentation and fate of pigment biomarkers following thermal stratification in the Western Alboran Sea. *Mar. Ecol. Prog. Ser.* 125: 279–291.

Bianchi, T. S., Findlay, S., Dawson, R. (1993). Organic matter sources in the water column and sediments of the Hudson River estuary: the use of plant pigments as tracers. *Estuarine, Coastal and Shelf Science*. 36: 359–376.

Bidigare, R. R., Marra, J., Dickey, T. D., Iturriaga, R., Baker, K. S., Smith, R. C., Pak, H. (1990). Evidence for phytoplankton succession and chromatic adaptation in the Sargasso Sea during spring 1985. *Mar. Ecol. Prog. Ser.* 60: 113–122.

Bidigare, R. R., Ondrusek, M. E., Brooks, J. M. (1993). Influence of the Orinoco River outflow on distributions of algal pigments in the Caribbean Sea. *J. Geophys. Res.*, 98: 2259–2269.

Bidigare, R. R., Smith, R. C., Baker, K. S., Marra, J. (1987). Oceanic primary production estimates from measurements of spectral irradiance and pigment concentrations. *Global Biogeochem. Cycles.* 1: 171–186.

Brunet, C., Brylinski, J. M., Frontier, S. (1992). Productivity, photosynthetic pigments and hydrology in the coastal front of the eastern English Channel. *J. Plankt. Res.* 14: 1541–1552.

Brunet, C., Brylinski, J. M., Lemoine, Y. (1993). *In situ* variations of the xanthophylls diatoxanthin and diadinoxanthin: photoadaptation and relationships with a hydrodynamical system in the eastern English Channel. *Mar. Ecol. Prog. Ser.* 102: 69–77.

Buma, A. G. J., Treguer, P., Kraay, G. W., Morvan, J. (1990). Algal pigment patterns in different water masses of the Atlantic sector of the Southern Ocean during fall 1987. *Polar Biol.* 11: 55–62.

Burford, M. A., Rothlisberg, P. C., Wang, Y-G. (1995). Spatial and temporal distribution of tropical phytoplankton species and biomass in the Gulf of Carpentaria, Australia. *Mar. Ecol. Prog. Ser.* 118: 255–266.

Burkill, P. H., Mantoura, R. F. C., Llewellyn, C. A., Owens, N. J. P. (1987). Microzooplankton grazing and selectivity of phytoplankton in coastal waters. *Mar. Biol.* 93: 581–590.

Chisholm, S. W., Morel, F. M. M. (eds) (1991). What controls phytoplankton production in the nutrient-rich areas of the open sea? *Limnol. Oceanogr.* vol. 36(8), 1965 pp.

Claustre, H., Kerhervé, P., Marty, J.-C., Prieur, L. (1994*a*). Phytoplankton photoadaptation related to some frontal physical processes. *J. Mar. Sys.* 5: 251–265.

Claustre, H., Kerhervé, P., Marty, J-C., Prieur, L., Videau, C., Hecq, J-H. (1994*b*). Phytoplankton dynamics associated with a geostrophic front: ecological and biogeochemical implications. *J. Mar. Res.* 52: 711–742.

Claustre, H., Poulet, S. A., Williams, R., Marty, J-C., Coombs, S., BenMlih, F., Hapette, A. M., Martin-Jezequel, V. (1990). A biochemical investigation of a *Phaeocystis* sp. bloom in the Irish Sea. *J. Mar. Biol. Ass. U.K.* 70: 197–207.

Denant, V., Saliot, A., Mantoura, R. F. C., Llewellyn, C. A. (1990). Algal chlorophyll and carotenoid pigments in the Changjiang estuary and the adjacent east China Sea. In: Yu, G., Martin, J. M., Zhou, J., (eds), *Biogeochemical Study of the Chanjiang Estuary*, China Ocean Press, Beijing, p. 716–737.

Ducklow, H. W., Harris, R. P. (eds) (1993). *Tropical Studies in Oceanography. JGOFs: the North Atlantic Bloom Experiment. Deep-Sea Res.* II, vol. 40, 641 pp.

Eckardt, C. B., Pearce, G. E. S., Keely, B. J., Kowalewska, G., Jaffé, R., Maxwell, J. R. (1992). A widespread chlorophyll transformation pathway in the aquatic environment. *Org. Geochem.* 19: 217–227.

Everitt, D. A., Wright, S. W., Volkman, J. K., Thomas, D. P., Lindstrom, E. J. (1990). Phytoplankton community compositions in the western equatorial Pacific determined from chlorophyll and carotenoid pigment distributions. *Deep-Sea Res.* 37: 975–997.

Fiala, M., Sournia, A., Claustre, H., Marty, J-C., Prieur, L., Vétion, G. (1994). Gradients of phytoplankton abundance, composition and photosynthetic pigments across the Almeria-Oran front (SW Mediterranean Sea). *J. Mar. Sys.* 5: 223–233.

Gieskes, W. W., Kraay, G. W. (1983*a*). Unknown chlorophyll *a* derivatives in the North Sea and the tropical Atlantic Ocean revealed by HPLC analysis. *Limnol. Oceanogr.* 28: 757–766.

Gieskes, W. W., Kraay, G. W. (1983*b*). Dominance of Cryptophyceae during the phytoplankton spring bloom in the central North Sea detected by HPLC analysis of pigments. *Mar. Biol.* 75: 179–185.

Gieskes, W. W., Kraay, G. W. (1986*a*). Analysis of phytoplankton pigments by HPLC before, during and after mass occurrence of the microflagellate *Corymbellus aureus* during the spring bloom in the open northern North Sea in 1983. *Mar. Biol.* 92: 45–52.

Gieskes, W. W., Kraay, G. W. (1986*b*). Floristic and physiological differences between the shallow and the deep nanophytoplankton community in the euphotic zone of the open tropical Atlantic revealed by HPLC analysis of pigments. *Mar. Biol.* 91: 567–576.

Gieskes, W. W., Kraay, G. W. (1989). Estimating the carbon-specific growth rate of the major algal species groups in eastern Indonesian waters by ^{14}C labeling of taxon-specific carotenoids. *Deep-Sea Res.* 36: 1127–1139.

Gieskes, W. W. C., Kraay, G. W., Nontji, A., Setiapermana, D., Sutomo. (1988). Monsoonal alteration of a mixed and a layered structure in the phytoplankton of the euphotic zone on the Banda Sea (Indonesia): A mathematical analysis of algal pigment fingerprints. *Netherlands J. Sea Res.* 22: 123–137.

Goericke, R., Repeta, D. J. (1993). Chlorophylls *a* and *b* and divinyl chlorophylls *a* and *b* in the open subtropical North Atlantic Ocean. *Mar. Ecol. Prog. Ser.* 101: 307–313.

Goericke, R., Welschmeyer, N. A. (1993*a*). The marine prochlorophyte *Prochlorococcus* contributes significantly to phytoplankton biomass and primary production in the Sargasso Sea. *Deep-Sea Res.* I. 40: 2283–2294.

Goericke, R., Welschmeyer, N. A. (1993*b*). The chlorophyll-labelling method: measuring specific rates of chlorophyll *a* synthesis in cultures and in the open ocean. *Limnol. Oceanogr.* 38: 80–95.

Goericke, R., Welschmeyer, N. A. (1993*c*). The carotenoid-labeling method: measuring specific rates of carotenoid synthesis in natural phytoplankton communities. *Mar. Ecol. Prog. Ser.* 98: 157–171.

Hallegraeff, G. M. (1981). Seasonal study of phytoplankton pigments and species at a coastal station off Sydney: importance of diatoms and the nanoplankton. *Mar. Biol.* 61: 107–118.

Hallegraeff, G. M., Jeffrey, S. W. (1984). Tropical phytoplankton species and pigments of continental shelf waters of north and north-west Australia. *Mar. Ecol. Progr. Ser.* 20: 59–74.

Harrison, W. G., Head, E. J. H., Horne, E. P. W., Irwin, B., Li, W. K. W., Longhurst, A. R., Paranjape, M. A., Platt, T. (1993). The western North Atlantic Bloom Experiment. *Deep-Sea Res.* II, 40: 279–305.

Hawkins, A. J. S., Bayne, B. L., Mantoura, R. F. C., Llewellyn, C. A. (1986). Chlorophyll degradation and absorption throughout the digestive system of the blue mussel *Mytilus edulis* L. *J. Exp. Mar. Biol. Ecol.* 96: 213–223.

Head, E. J. H., Hargrave, B. T., Subba Rao, D. V. (1994). Accumulation of a pheophorbide *a*-like pigment in sediment traps during late stages of a spring bloom: a product of dying algae? *Limnol. Oceanogr.* 39: 176–181.

Head, E.J . H., Harris, L. R. (1992). Chlorophyll and carotenoid transformation and destruction by *Calanus* spp. grazing on diatoms. *Mar. Ecol. Prog. Ser.* 86: 229–238.

Head, E. J. H., Horne, E. P. W. (1993). Pigment transformation and vertical flux in an area of convergence in the North Atlantic. *Deep-Sea Res.* II, 40: 329–346.

Jeffrey, S. W. (1976). A report of green algal pigments in the central North Pacific Ocean. *Mar. Biol.* 37: 33–37.

Jeffrey, S. W., Hallegraeff, G. M. (1980). Studies of phytoplankton species and photosynthetic pigments in a warm-core eddy of the East Australian Current. I. Summer populations. *Mar. Ecol. Progr. Ser.* 3: 285–294.

Jeffrey, S. W., Hallegraeff, G. M. (1987). Phytoplankton pigments, species and light climate in a complex warm-core eddy of the East Australian Current. *Deep-Sea Res.* 34: 649–673.

King, L. L., Repeta, D. J. (1991). Novel pyropheophorbide sterylesters in Black Sea sediments. *Geochim .Cosmochim. Acta*, 55: 2067–2074.

Klein, B., Sournia, A. (1987). A daily study of the diatom spring bloom at Roscoff (France) in 1985. II. Phytoplankton pigment composition studied by HPLC analysis. *Mar. Ecol. Prog. Ser.* 37: 265–275.

Kleppel, G. S., Pieper, R. E. (1984). Phytoplankton pigments in the gut contents of planktonic copepods from coastal waters off southern California. *Mar. Biol.* 78: 193–198.

Kleppel, J. S., Holliday, D. V., Pieper, R. E. (1991). Trophic interactions between copepods and microplankton: a question about the role of diatoms. *Limnol. Oceanogr.* 36: 172–178.

Latasa, M., Estrada, M., Delgado, M. (1992). Plankton-pigment relationships in the Northwestern Mediterranean during stratification. *Mar. Ecol. Prog. Ser.* 88: 61–73.

Letelier, R. M., Bidigare, R. R., Hebel, D. V., Ondrusek, M., Winn, C. D., Karl, D. M. (1993). Temporal variability of phytoplankton community structure based on pigment analysis. *Limnol. Oceanogr.* 38: 1420–1437.

Mackey, D. J., Parslow, J., Higgins, H. W., Griffiths, F. B., O'Sullivan, J. E. (1995). Plankton productivity and biomass in the western Equatorial Pacific: biological and physical controls. *Deep-Sea Res.* 42: 499–533.

Malej, A., Mozetic, P., Malicic, V., Terzic, S., Ahel, M. (1995). Phytoplankton reponses to fresh water inputs in a small semi-enclosed gulf (Gulf of Trieste, Adriatic Sea). *Mar. Ecol. Prog. Ser.* 120: 111–121.

Marchant, H. J., Davidson, A., Wright, S. W. (1987). The distribution and abundance of chroococcoid cyanobacteria in the Southern Ocean. *Proc. MPR. Symp. Polar Biol.* 1: 1–9.

McManus, G. B., Dawson, R. (1994). Phytoplankton pigments in the deep chlorophyll maximum of the Caribbean Sea and the western tropical Atlantic Ocean. *Mar. Ecol. Prog. Ser.* 113: 199–206.

McManus, G. B., Ederington-Cantrell, M. C. (1992). Phytoplankton pigments and growth rates, and microzooplankton grazing in a large temperate estuary. *Mar. Ecol. Prog. Ser.* 87: 77–85.

Méjanelle, L., Laureillard, J., Fillaux, J., Saliot, A., Lambert, C. (1995). Winter distribution of algal pigments in small- and large-size particles in the northeastern Atlantic. *Deep-Sea Res.* I. 42: 117–133.

Nelson, J. R. (1989). Phytoplankton pigments in macrozooplankton feces: variability in carotenoid alterations. *Mar. Ecol. Prog. Ser.* 52: 129–144.

Nelson, N. B., Prézelin, B. B., Bidigare, R. R. (1993). Phytoplankton light absorption and the package effect in California coastal waters. *Mar. Ecol. Prog. Ser.* 94: 217–227.

Olaizola, M., Bienfang, P. K., Ziemann, D. A. (1992). Pigment analysis of phytoplankton during a Subarctic spring bloom: xanthophyll cycling. *J. Exp. Mar. Biol. Ecol.* 158: 59–74.

Ondrusek, M. E., Bidigare, R. R., Sweet, S. T., Defreitas, D. A., Brooks, J. M. (1991). Distribution of phytoplankton pigments in the North Pacific Ocean in relation to physical and optical variability. *Deep-Sea Res.* 38: 243–266.

Pollehne, F., Klein, B., Zeitzschel, B. (1993). Low light adaptation and export production in the deep chlorophyll maximum layer in the northern Indian Ocean. *Deep-Sea Res.* II. 40: 737–752.

Repeta, D. J. (1989). Carotenoid diagenesis in recent marine sediments: II. Degradation of fucoxanthin to loliolide. *Geochim. Cosmochim. Acta,* 53: 699–707.

Repeta, D. J., Gagosian, R. B. (1982). Carotenoid transformations in coastal marine waters. *Nature,* 295: 51–54.

Repeta, D. J., Gagosian, R. B. (1984). Tranformation reactions and recycling of carotenoids and chlorins in the Peru upwelling region (15°S, 75°W). *Geochim. Cosmochim. Acta,* 48: 1265–1277.

Repeta, D. J., Gagosian, R. B. (1987). Carotenoid diagenesis in recent marine sediments. I. The Peru continental shelf (15°S, 75°W). *Geochim. Cosmochim. Acta,* 51: 1001–1009.

Repeta, D. J., Simpson, D. J. (1991). The distribution and recycling of chlorophyll, bacteriochlorophyll and carotenoids in the Black Sea. *Deep-Sea Res.* 39: 969–984.

Repeta, D. J., Simpson, D. J., Jorgensen, B. B., Jannasch, H. W. (1989). Evidence for anoxygenic photosynthesis from the distribution of bacteriochlorophylls in the Black Sea. *Nature,* 342: 69–72.

Riaux-Gobin, C., Llewellyn, C. A., Klein, B. (1987). Microphytobenthos from two subtidal sediments from North Brittany. II. Variations of pigment compositions and concentrations determined by HPLC and conventional techniques. *Mar. Ecol. Prog. Ser.* 40: 275–283.

Roy, S. (1989). HPLC-measured chlorophyll-type pigments during a phytoplankton spring bloom in Bedford Basin (Canada). *Mar. Ecol. Prog. Ser.* 55: 279–290.

Schofield, O., Prézelin, B. B., Bidigare, R. R., Smith, R. C. (1993). *In situ* photosynthetic quantum yield. Correspondence to hydrographic and optical variability within the Southern California Bight. *Mar. Ecol. Prog. Ser.* 93: 25–37.

Siegel, D. A., Iturriaga, R., Bidigare, R. R., Smith R. C., Pak, H., Dickey, T. D., Marra, J., Baker, K. S. (1990). Meridional variations of the springtime phytoplankton community in the Sargasso Sea. *J. Mar. Res.* 48: 379–412.

Strom, S. L., Welschmeyer, N. A. (1991). Pigment-specific rates of phytoplankton growth and microzooplankton grazing in the open subarctic Pacific Ocean. *Limnol. Oceanogr.* 36: 50–63.

Sun, M., Aller, R. C., Lee, C. (1991). Early diagenesis of chlorophyll *a* in Long Island Sound sediments: a measure of carbon flux and particle reworking. *J. Mar. Res.* 49: 379–401.

Sun, M-Y., Lee, C., Aller, R. C. (1993). Anoxic and oxic degradation of ^{14}C-labeled chloropigments and a ^{14}C-labeled diatom in Long Island Sound sediments. *Limnol. Oceanogr.* 38: 1438–1451.

Trees, C. C., Bidigare, R. R., Brooks, J. M. (1986). Distribution of chlorophylls and phaeopigments in the Northwestern Atlantic Ocean. *J. Plankton Res.* 3: 447–458.

Vaulot, D., Partensky, F., Neveux, J., Mantoura, R. F. C., Llewellyn, C. A. (1990). Winter presence of prochlorophytes in surface waters of the northwestern Mediterranean Sea. *Limnol. Oceanogr.* 35: 1156–1164.

Veldhuis, M. J. W., Kraay, G. W. (1990). Vertical distribution and pigment composition of a picoplanktonic prochlorophyte in the subtropical North Atlantic: a combined study of HPLC-analysis of pigments and flow cytometry. *Mar. Ecol. Prog. Ser.* 68: 121–127.

Veldhuis, M. J. W., Kraay, G. W. (1993). Cell abundance and fluorescence of picoplankton in relation to growth irradiance and nitrogen availability in the Red Sea. *Netherlands J. Sea Res.* 31: 135–145.

Verity, P. G., Vernet, M. (1992). Microzooplankton grazing, pigments, and composition of plankton communities during late spring in two Norwegian Fjords. *Sarsia*, 77: 263–274.

Vernet, M., Lorenzen, C. J. (1987*a*). The presence of chlorophyll *b* and the estimation of phaeopigments in marine phytoplankton. *J. Plankton Res.* 9: 255–265.

Vernet, M., Lorenzen, C. J. (1987*b*). The relative abundance of pheophorbide *a* and pheophytin *a* in temperate marine waters. *Limnol. Oceanogr.* 32: 352–358.

Wassmann, P., Vernet, M., Mitchell, B. G., Rey, F. (1990). Mass sedimentation of *Phaeocystis pouchetii* in the Barents Sea. *Mar. Ecol. Prog. Ser.* 66: 183–195.

Weeks, A., Conte, M. H., Harris, R. P., Bedo, A., Bellan, I., Burkill, P. H., Edwards, E. S., Harbour, D. S., Kennedy, H., Llewellyn, C. A., Mantoura, R. F. C., Morales, C. E., Pomroy, A. J., Turley, C. M. (1993). The physical and chemical environment and changes in community structure associated with bloom evolution: the Joint Global Flux Study North Atlantic Bloom Experiment. *Deep-Sea Res.* II. 40: 347–368.

Welschmeyer, N. A., Hoepffner, N. A. (1986). Rapid xanthophyll cycling: an *in situ* tracer mixing in the upper ocean. *Eos, Trans. Am. Geophys. Union*, 67: 969.

Welschmeyer, N., Goericke, R., Strom, S., Peterson, W. (1991). Phytoplankton growth and herbivory in the subarctic Pacific: a chemotaxonomic analysis. *Limnol. Oceanogr.* 36: 1631–1649.

Williams, R., Claustre, H. (1991). Photosynthetic pigments as biomarkers of phytoplankton populations and processes involved in the transformation of particulate organic matter at the Biotrans site (47°N, 20°W). *Deep-Sea Res.* 38: 347–355.

Wright, S. W. (1985). Phytoplankton pigment data: Prydz Bay region — SIBEX II, M. V. Nella Dan, January, 1985. ANARE Res. Notes, No. 58, 50 pp.

Part II

Experimental results of SCOR Working Group 78 workshops

6

Simple procedures for growing SCOR reference microalgal cultures

S. W. Jeffrey and J.-M. LeRoi

6.1 Introduction

Because of the commercial non-availability of all but a few pure algal pigment standards (see Appendix J), SCOR Working Group 78 recommended that the source of such standards should be well characterized strains of microalgae whose pigments had been securely identified by modern chemical methods (Bjørnland and Liaaen-Jensen, 1986). A small number of species were selected from a range of algal classes (Table 6.1) that together contained most of the significant chlorophylls and carotenoids known to be found in marine phytoplankton. To reproduce exactly the pigment analy-

TABLE 6.1. Microalgal species recommended as authentic sources of pigments, and culture conditions used for their growth.

Species	Class	Clonal designation and origin	CSIRO culture code	Culture medium[a]	Axenic	Growth temperature °C	Light intensity $\mu Em^{-2}sec^{-1}$	Culture colour (log phase)
Amphidinium carterae	Dinophyceae	AMPH1, Plymouth 450	CS-212	G	+	20	80	reddish-brown
Chroomonas salina	Cryptophyceae	CCMP 56/00/00	CS-174	fE	+	20	80	reddish-brown
Dunaliella tertiolecta	Chlorophyceae	CCMP	CS-175	f2	+	20	80	green
Emiliania huxleyi	Prymnesiophyceae	BT6	CS-57	f2	–	20	80	gold
Euglena gracilis	Euglenophyceae	Univ. NSW, strain Z	CS-66	MBL/NB_2	+	20	80	green
Micromonas pusilla	Prasinophyceae	UTEX LB991	CS-86	G	–	20	80	olive-green
Pavlova lutheri	Prymnesiophyceae	CCMP	CS-182	f2	+	20	80	gold
Pelagococcus subviridis	Chrysophyceae	East Aust. Current	CS-99	f2	–	20	80	green-gold
Phaeodactylum tricornutum	Bacillariophyceae	Plymouth (1052/1)	CS-29	f2	+	20	80	orange-brown
Porphyridium cruentum[b]	Rhodophyceae	Halifax, Canada	CS-25	f2	+	20	80	red
Pycnococcus provasolii	Prasinophyceae	CCMP, omega 48-23AX	CS-185	f2	+	20	80	olive-green
Synechococcus sp.	Cyanophyceae	DC2	CS-197	f2	+	17.5	20	red

a. Media code: f = Guillard & Ryther (1962); f2 = half concentration of medium f; MBL/NB_2 = Guillard's MBL in Nichols (1973); fE = f + EDTA (Jeffrey 1980); G = Loeblich & Smith (1968). All cultures were grown on 12:12 hour light:dark cycles.
b. This species is now known as *Porphyridium purpureum*, but we are using the more familiar name for this volume.

Cultures may be obtained from either the CSIRO Algal Culture Collection or the Guillard-Provasoli Culture Collection (see Appendix L).

ses of the standard species (see Chapter 13), both the strains and culturing conditions must be identical. The present Chapter is a brief introduction for the chemical oceanographer to the principles of culturing microalgae and the procedures for replicating the culturing conditions used in our CSIRO Algal Culture Laboratory and the SCOR Workshops.

6.2 Principles of microalgal culture: an introduction

6.2.1 *Historical perspective*

Although microalgae have been cultivated for over 100 years, success was largely limited to fast-growing 'weed' species. Most early attempts at cultivating unicellular algae were based on the work of Miquel, who achieved in 1890, growth of a few diatoms under artificial conditions. The early culture medium was seawater, to which a few nutrients (nitrate, phosphate and silicate) had been added. In the 1930s and 1940s, the medium was improved by adding soil extract and trace metals (Pringsheim, 1946), which allowed more species, both marine and freshwater, to be successfully cultivated.

A breakthrough came in the 1950s with the discovery by Dr L. Provasoli and co-workers that many microalgae needed vitamins for growth: B_{12}, thiamine and biotin being the most important. Their next discovery (Provasoli *et al.* 1957; Provasoli, 1958), that addition of chelating compounds successfully controlled metal availability in growth media, greatly increased the number of microalgae that could be cultured axenically. The partial replacement of the multifunctional soil extract — which provided various trace elements, vitamins and metal-chelating organic compounds, particularly those solubilizing iron — with specific compounds, has also been a step forward. The design of new media for ultra- and pico-plankton from oligotrophic environments, and the recognition that many species require selenium are recent advances (Brand, 1986; Keller *et al.* 1987; Harrison *et al.* 1988).

Further improvements in culture techniques have come from identifying toxic trace-metal contamination from various sources in culture media: culture flasks, sterilization techniques, chemicals used to prepare culture media and natural seawater (Brand, 1986).

Consolidation of knowledge of algal culturing has been assisted by two excellent publications — one for phycologists (Stein, 1973) and one for aquaculturists (Guillard, 1975). Together with Brand (1986), who reviews advances in the cultivation of pico-planktonic oligotrophic species, these books and articles are highly recommended sources of information on culturing algae. Table 6.2 lists the optimal characteristics of an algal culture laboratory for general information.

6.2.2 *Chemical and physical factors: approximating the marine environment*

Overview

Isolating an alga from its natural environment, free from competitors and predators, removes both good and bad biotic influences. In addition, the high biomass that can be achieved in artificial culture can change the chemical composition of the medium

TABLE 6.2. Optimal characteristics of an algal culture laboratory

- Constant-temperature rooms or cabinets with back-up (duplicate) compressor systems and alarm systems
- Range of temperatures available for growing polar to tropical species (-2 to 40°C)
- Range of light intensities, colour and day:night cycles available
- No toxic plastics or metals in the culture environment
- Regular preventative maintenance of the growth rooms and cabinets
- Autoclave, hot-air oven for dry sterilization, laminar flow cabinet, good distilled water supply.
- Microscopes
- Clean environment (working benches, light shelves etc.)
- Glass-washing facility separate from the algal culture laboratory (no detergents in the same room as the algal cultures)
- Purified seawater, high-purity chemicals, range of culture vessels (sizes, shapes and materials)
- Adequate cold storage for purified seawater, stock nutrients and culture media
- Frequent testing for contaminating bacteria in axenic cultures
- Constant supervision of the Collection by devoted staff!

in which the algae are growing, and autoinhibitory substances, which are dilute in nature, may accumulate in high concentration and adversely affect growth. For physiological and nutritional studies, both chemical and physical variables must be controlled. These include inorganic and organic nutrients, toxins (organic or metallic), type and geometry of culture flasks, light (intensity, day-length, light quality), temperature, dissolved gases and aeration. The biological factors include the origin and strain type of the clone, the presence of contaminants (bacteria, viruses and fungi), culture age and time of harvest. Many of these factors can affect the pigment composition of algal cells, so in the SCOR studies we have recommended specific algal strains and standard growth conditions that rigorously control all the variables likely to affect pigment composition. If these guidelines are followed, the pigment composition of the standard algal strains described in Chapter 13 can be readily achieved.

Culture Media

The basic problem in establishing algal cultures is the design of isolation (culture) media (see Table 6.3) that imitate the natural environment sufficiently closely (McLaughlin, 1973). Usually, purified off-shore seawater (Jeffrey, 1980) or artificial seawater (Provasoli *et al.* 1957) provide the basis of the media, to which precisely defined quantities of major and minor nutrients are added: nitrate, phosphate, silicate, magnesium, potassium, sodium, chloride and sulphate, with minor or trace metals (copper, zinc, cobalt, manganese, molybdenum, iron and selenium), and vitamins (B_{12}, thiamine and biotin) added in varying proportions. Trace elements are chelated with EDTA (ethylene diamine tetraacetic acid) to form a non-toxic reservoir of complexed cations, and to ensure a constant supply of low concentrations of trace elements. Tris (hydroxymethylamino methane) is sometimes used as a culture buffer. Soil extract, prepared by autoclaving fresh uncultivated soil, free from herbicides and fertilizers, is used to add humic acids, trace metals, chelating compounds and other unknown nutrients to some culture media (e.g., G medium, Loeblich and Smith, 1968; Loeblich, 1975).

TABLE 6.3. Micro-molar concentrations of nutrients used in algal culture media[a]

	fE	f2[b]	G	MBL/NB$_2$
Major nutrients				
KNO_3	–	–	1978	–
$NaNO_3$	1760	880	–	880
K_2HPO_4	–	–	200	40
$NaH_2PO_4.2H_2O$	72.4	36.2	–	–
$Na_2SiO_3.5H_2O$	107.0	53.5	–	482
$CaCl_2.2H_2O$	–	–	–	250
$MgSO_4.7H_2O$	–	–	–	150
$NaHCO_3$	–	–	–	149
Tris buffer	–	–	–	4120
Trace metals				
$CoCl_2.6H_2O$	0.093	0.047	0.11	0.047
$CuSO_4.5H_2O$	0.079	0.04	–	0.04
H_3BO_3	–	–	111.0	–
$MnCl_2.4H_2O$	1.82	0.91	4.35	0.91
$NaMoO_4.2H_2O$	0.052	0.026	–	0.026
$ZnCl_2$	–	–	0.44	–
$ZnSO_4.7H_2O$	0.15	0.075	–	0.075
Chelators				
Citric acid	42.8	21.4	–	42.8
Ferric citrate	26.9	13.45	–	26.9
$FeCl_3.6H_2O$	–	–	1.07	–
$Na_2EDTA.2H_2O$	80.6	–	16.10	12.09
Vitamins				
Biotin	0.004	0.002	0.0082	0.004
Cyanocobalamin (B_{12})	0.0007	0.0004	0.0007	0.0007
Thiamine HCl	0.6	0.3	2.96	0.6

a. All media have been slightly modified from the original recipes.
b. Half strength of the original f medium of Guillard and Ryther (1962) was designated as f/2. Our modification is here designated f2.

The composition of a variety of marine and freshwater media may be found in Stein (1973), while Keller *et al.* (1987) and Harrison *et al.* (1988) recommend new formulations both for general purpose use and difficult-to-cultivate species.

Chemicals

All chemicals used in the preparation of culture media should be at least Analytical Reagent Grade to reduce trace contaminants possibly harmful to microalgae. Stock solutions are usually prepared with highly purified distilled water (see Tables 6.4, 6.5) which are then added in small quantities to large volumes of seawater, to prevent precipitation problems. If artificial seawater is used instead of natural seawater as a base, large quantities of salts are required, which can add excessive quantities of unknown and unwanted contaminants to the final media. This could be a problem in the cultivation of delicate species.

TABLE 6.4. Composition of stock solutions used for preparing culture media

A. MAJOR NUTRIENTS	g/100 ml distilled water	B. TRACE METALS	g/l distilled water
KNO_3	10.0	**Trace metal solution 1**	
$NaNO_3$	15.0	$CuSO_4.5H_2O$	0.020
K_2HPO_4	3.48	$ZnSO_4.7H_2O$	0.044
$NaH_2PO_4.2H_2O$	1.13	$CoCl_2.6H_2O$	0.022
$Na_2SiO_3.5H_2O$[a]	2.27	$MnCl_2.4H_2O$	0.360
$NaHCO_3$	1.26	$NaMoO_4.2H_2O$	0.013
$CaCl_2.2H_2O$	3.68	**Trace metal solution 2**[c]	
$MgSO_4.7H_2O$	3.7	Na_2EDTA	6.0
Na_2EDTA	3.0	$FeCl_3$	0.29
Ferric citrate[b]	0.9	H_3BO_3	6.85
Citric acid[b]	0.9	$MnCl_2.4H_2O$	0.86
Tris buffer	25.0	$ZnCl_2$	0.06
		$CoCl_2.6H_2O$	0.026

C. VITAMINS

Thiamine, biotin and vitamin B_{12} are added to all culture media.

Primary stock solutions

Biotin	50 mg in 500 ml distilled water
B_{12}	10 mg in 100 ml distilled water

Dispense in 10 ml aliquots and store frozen (-18°C)[d]

Working stock solution I — for all f media and MBL/NB$_2$

Thiamine HCl	20 mg in 100 ml distilled water
To this solution add	
Biotin (primary stock)	1 ml
B_{12} (primary stock)	1 ml

Working stock solution II — for G medium

Thiamine HCl	100 mg in 100 ml distilled water
To this solution add	
Biotin (primary stock)	2 ml
B_{12} (primary stock)	1 ml

Prepare working stock solutions every 3 months and store in the dark at 5°C

a. Heat to dissolve.
b. To dissolve, autoclave both chemicals together in 100 ml of distilled H_2O.
c. Ensure Na_2EDTA is fully dissolved before adding other chemicals. When all chemicals have dissolved, adjust the pH of **trace metal solution 2** to 7.8–8.0 with NaOH pellets.
d. We recommend that thawed primary stocks are not refrozen or reused.

Seawater

Natural seawater is the preferred base for algal culture media. It should be obtained off-shore to minimize the concentration of metal and organic pollutants and fine sediments. (Inshore water can be highly variable in salinity, and organic and inorganic loads, and can produce highly variable algal growth rates if used for culture media). Seawater from any location should be purified for the routine maintenance of algal stocks, although for the initial isolation of new strains from particular loca-

TABLE 6.5. Volumes of stock solutions used for preparing fE, f2, G and MBL/NB_2 media

fE medium	Seawater	1000 ml
	$NaNO_3$	1.0 ml
	$NaH_2PO_4.2H_2O$[a]	a
	$Na_2SiO_3.5H_2O$	1.0 ml
	Trace metal **solution 1**	1.0 ml
	Ferric citrate/citric acid	1.0 ml
	Na_2EDTA	1.0 ml
	Vitamins – **working stock solution I**	1.0 ml
f2 medium	Seawater	1000 ml
	$NaNO_3$	0.5 ml
	$NaH_2PO_4.2H_2O$[a]	a
	$Na_2SiO_3.5H_2O$	0.5 ml
	Trace metal **solution 1**	0.5 ml
	Ferric citrate/citric acid	0.5 ml
	Vitamins – **working stock solution I**	0.5 ml
G medium	Seawater	750 ml
	Distilled water	250 ml
	KNO_3	2.0 ml
	K_2HPO_4[a]	a
	Trace metal **solution 2**	5.0 ml
	Vitamins – **working stock solution II**	1.0 ml
	Soil extract[b]	5.0 ml
MBL/NB_2[c]	Distilled water	1000 ml
	$NaNO_3$	0.5 ml
	K_2HPO_4	0.2 ml
	$Na_2SiO_3.5H_2O$	4.5 ml
	$NaHCO_3$	1.0 ml
	$CaCl_2.2H_2O$	0.8 ml
	$MgSO_4.7H_2O$	0.5 ml
	Trace metal **solution 1**	0.5 ml
	Ferric citrate/citric acid	1.0 ml
	Vitamins – **working stock solution I**	1.0 ml
	Tris buffer	2.0 ml
	Nutrient broth No. 2 (Oxoid)	2.5 g

a. Phosphate is added separately to seawater media to prevent precipitation on autoclaving – see separate protocol, Table 6.6
b. See Table 6.7 for preparation of soil extract
c. Adjust pH of MBL/NB_2 medium to 7.2, using HCl, before dispensing and autoclaving

tions it is often better initially to use the original natural seawater in the growth medium (see Section 6.2.3).

At the CSIRO Algal Culture Laboratory, seawater is collected from oceanic waters, 12 km off-shore, far removed from river influences, at 10 m depth and purified by filtering sequentially through three filters (see Section 6.3.2; Fig. 6.3C). The purified seawater is stored at 5°C in 10 l glass carboys in the dark. Other laboratories recommend aging seawater at 15–20°C for several months to decrease the or-

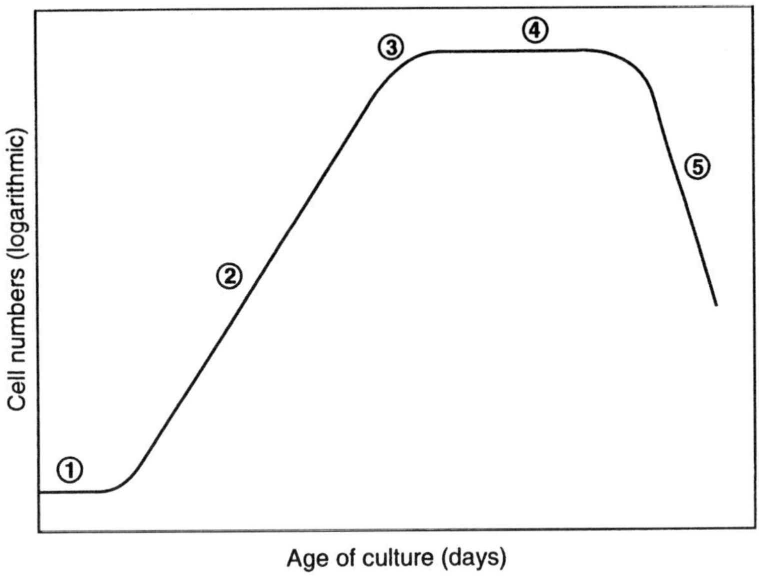

Figure 6.1
The characteristic pattern of growth shown by a unicellular alga in a culture of limited volume; (1) lag phase, (2) logarithmic (exponential) phase, (3) phase of declining relative growth rate, (4) stationary phase, (5) death phase.

ganic load, and then filtering. Seawater purified by the CSIRO method has never failed to give excellent results.

Distilled Water

Highly purified distilled water or Millipore Milli-Q water (equivalent in purity to four times distilled water), or double glass-distilled water are used for making nutrient, trace metal and vitamin stocks. Stills with copper or other metal elements are not recommended.

Culture Vessels

Straight-necked pyrex glass 125 ml Erlenmeyer flasks with tooled rims are used for the algal culture stocks. The flasks are plugged with Steristoppers* or cotton wool stoppers that fit firmly against the tooled rim to prevent aerial contamination (Plate 3). The use of inverted beakers instead of regular stoppers is not recommended if cultures are to remain axenic. For larger volumes, 250 ml or 1–5 l Erlenmeyer flasks may be used. Many delicate oligotrophic algal species (*not* the robust SCOR cultures) cannot tolerate glass flasks, and may require polyethylene petri-dishes, or flasks made of high-quality inert plastics such as teflon or polycarbonate (see Brand (1986) for further information). For such organisms, filter sterilization (0.22 μm filters) of media, rather than autoclave sterilization, is recommended. Plastic tissue-culture

* Heinz Herenz, Hamburg, Germany

flasks, although convenient to use at sea, are also not recommended; in our experience, most microalgal species grow poorly in such flasks.

Culture Mode

Batch, semi-continuous or continuous culture techniques may be used (Stein, 1973). Batch cultures of given volume are used for all stock cultures and were employed for all SCOR culture experiments. Such cultures are physiologically healthy provided they are harvested for experimentation well before the end of the logarithmic-phase of growth (see Fig. 6.1) when nutrients, including CO_2, may be depleted or toxic exudates accumulated. Semi-continuous culture (in which part of the culture volume is withdrawn and replaced with an equal volume of fresh medium every few days) or continuous culture (in which culture harvesting and fresh medium replacement are ongoing) are often used for physiological experiments or when a continuous supply of algal biomass is required (Ukeles, 1973).

Aeration

The SCOR cultures grown in 125 or 250 ml flasks were not aerated, although this is often necessary in larger volumes of culture, both to keep the cultures mixed and to increase the CO_2 supply. Filtered air may be used, or air enriched with CO_2 (0.5–3% v/v). In the SCOR experiments all cultures were gently swirled once daily until harvested.

Temperature

Most algal cells have a broad temperature optimum of at least 10°C for growth, but growth often declines sharply 2–3°C higher or lower than this optimal range. For the different oceanic temperature climates at the sea-surface, these ranges may be:

tropical	30–40°C
sub-tropical	15–30°C
warm temperate	8–25°C
temperate	5–15°C
polar	-2 to 5°C

In tropical and temperate oceans, temperature can drop 8–10°C from the surface to the base of the euphotic zone (approximately 200 m depth), but as this can vary seasonally, algae being carried with the ocean currents must tolerate significant temperature variations. This may not be the case with species that inhabit specific habitats (e.g., hot springs or slush ice). Most of the SCOR strains, which are robust species, can tolerate a range of temperatures from 15–23°C. As shown in Table 6.1, we standardized growth temperatures at 20°C, except for *Synechococcus* sp. (CS-197), which originated from the base of the euphotic zone in the Atlantic Ocean and needed a lower temperature (17.5°C). Temperature *per se* is not known to affect pigment composition, but because of its effects on growth rates and enhancement of the effects of light stress, it may contribute to pigment variations.

Light

Light intensity, light quality and day-length can have profound effects on both algal growth and pigment composition. Light intensities in the ocean range from 2,400 μE m^{-2} s^{-1} at the sea-surface in the tropics at midday, to 10–500 μE m^{-2} s^{-1} in the oceanic euphotic zone (down to 200 m depth). Most microalgae are adapted to a relatively low light intensity of between 50 and 300 μE m^{-2} s^{-1} (Jeffrey, 1984). The cell pigment content is usually inversely related to light intensity: more pigment is produced at lower light intensities to capture the available light more effectively (see Fig. 6.2). At very high light intensities, extra carotenoids may be synthesized. They filter light to protect the cell from harmful radiation (e.g., synthesis of β,β-carotene at high light and low nutrients in *Dunaliella salina*; Borowitzka and Borowitzka, 1988). For the SCOR reference cultures, we standardized light intensity at 80 $\mu E\ m^{-2}\ s^{-1}$, except for the low-light species *Synechococcus* sp. (CS-197), which received 20 $\mu E\ m^{-2}s^{-1}$.

Light quality (colour) also affects pigment composition. It varies considerably between oceanic habitats: the full spectrum of white light (400–700 nm) is found only within 1 metre of the sea surface. Most of the oceanic photosynthetic zone exhibits a blue-green spectrum (450–500 nm), the deep ocean has a blue-violet spectrum (350–450 nm), and shallow coastal waters receive mostly green (550 nm) to yellow wavelengths (580–590 nm) (Jerlov, 1976). Blue-green light increases pigment synthesis in a number of species (Jeffrey and Vesk, 1977; Vesk and Jeffrey, 1977), and also prolongs viability in cultured tropical species (Jeffrey, unpublished). In the CSIRO Algal Culture Collection all tropical species are maintained on blue-green light. The SCOR reference cultures were standardized on low-intensity white light (80 $\mu E\ m^{-2}\ s^{-1}$) provided by Philips 'white' or 'cool white' fluorescent tubes.

Day-length also affects algal growth rates and pigment synthesis cycles. Some algal culturists use continuous light, but this is only experienced in nature by plants in polar regions in summer. Most algae prefer a day:night cycle of 16:8, 14:10 or 12:12 hours. Because pigments increase before cell division, the content of pigment in each cell can vary throughout the daily growth cycle. The SCOR reference cultures were standardized on a 12:12 hour light:dark regime and cells were harvested 4–6 hours into the light cycle.

6.2.3 *Isolation of new species*

To obtain new isolates in culture, several methods can be used. The most common is to place a mixed phytoplankton field sample under an inverted microscope or a stereomicroscope, and pick out the algal cells with a fine-tipped sterile capillary pipette (five times the diameter of the cell to prevent capillary forces from damaging the cell). After washing the cell several times in drops of sterile seawater, it is placed in a small volume (0.5–1.0 ml) of dilute culture medium in small tubes, petri-dishes, tissue-culture compartments or cavities made in agar slopes. After several weeks' incubation under suitable conditions of light and temperature, growth may be evident.

A method for species that can tolerate growth on an agar surface (not large flagellates) is to use a wire loop to streak a small volume of a dilute plankton sample over the surface of a nutrient agar plate or slope. After 7–10 days algal colonies may appear. They can be aseptically removed with a wire loop and placed in a liquid medium.

When a particular algal species is dominant in a field sample, a serial dilution technique can be used. A set of about 10 culture tubes is prepared, each tube containing 9 ml culture medium. One ml of dilute mixed phytoplankton from the field sample is added to the first tube, and gently mixed. Then 1 ml of this mixture is transferred to the next tube, and so on throughout the set. Often a unialgal culture will be started in one of the last tubes.

Differential filtration using polycarbonate Nucleopore filters of fixed pore-size can often be used to separate algal cells of different size classes e.g., picoplankton

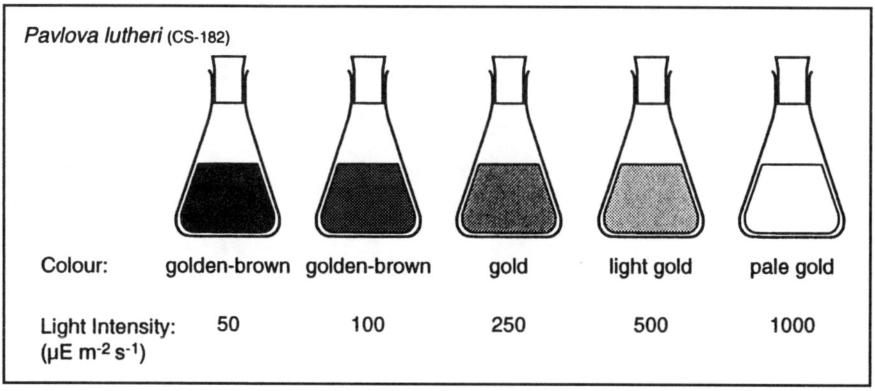

Figure 6.2
The effect of light intensity on culture colour of the prymnesiophyte *Pavlova lutheri*, grown at 20°C for 14 days on a 12:12 light:dark cycle at different light intensities of white fluorescent light (50–1000 µE m^{-2} s^{-1}).

(0.5 µm) from nanoplankton (5–20 µm) or net plankton (>30–100 µm). Some laboratories recommend a 35 µm prefilter to remove protozoa before processing.

Another isolation method is to inoculate a drop of a phytoplankton field sample into various types of dilute nutrient media (e.g., different dilutions of f, G or K media). To isolate non-diatom species, the diatom inhibitor germanium dioxide can be added to the medium. This is a competitive inhibitor of silica, which is needed to form diatom cell walls (Lewin, 1966; Markham and Hagmeier, 1982). After a few days' growth, algal cells can be picked out from these cultures, sometimes more easily than from the original field samples.

Several well-illustrated articles on algal isolation techniques may be found in Hoshaw and Roscowski (1973), Guillard (1973*a, b*) and Throndsen (1973).

6.2.4 *Aseptic techniques*

Microalgal cultures are susceptible to bacterial and fungal contamination, so strict aseptic techniques must be used. As soon as the medium is prepared it must be sterilized (see Section 6.3.4), and aseptic techniques must be used for all transfers of microalgae (Fig. 6.3 and Section 6.3.5). Before discarding, cultures should always be sterilized by autoclaving or adding a disinfectant to prevent contamination elsewhere.

The following references give excellent outlines on the theory and use of aseptic technique in microbiology: Paul (1970), Freshney (1987) and Brock and Madigan (1991). Protocols recommended for aseptic transfer of the SCOR cultures will be found in Section 6.3.5 and Table 6.9.

6.2.5 *Basic requirements for an algal culture laboratory*

The CSIRO Algal Culture Laboratory holds approximately 700 living strains of microalgae (as at 1996). Wherever possible, stocks are maintained in axenic (bacteria-free) condition in small-volume (50, 75 or 125 ml) Erlenmyer flasks or on agar

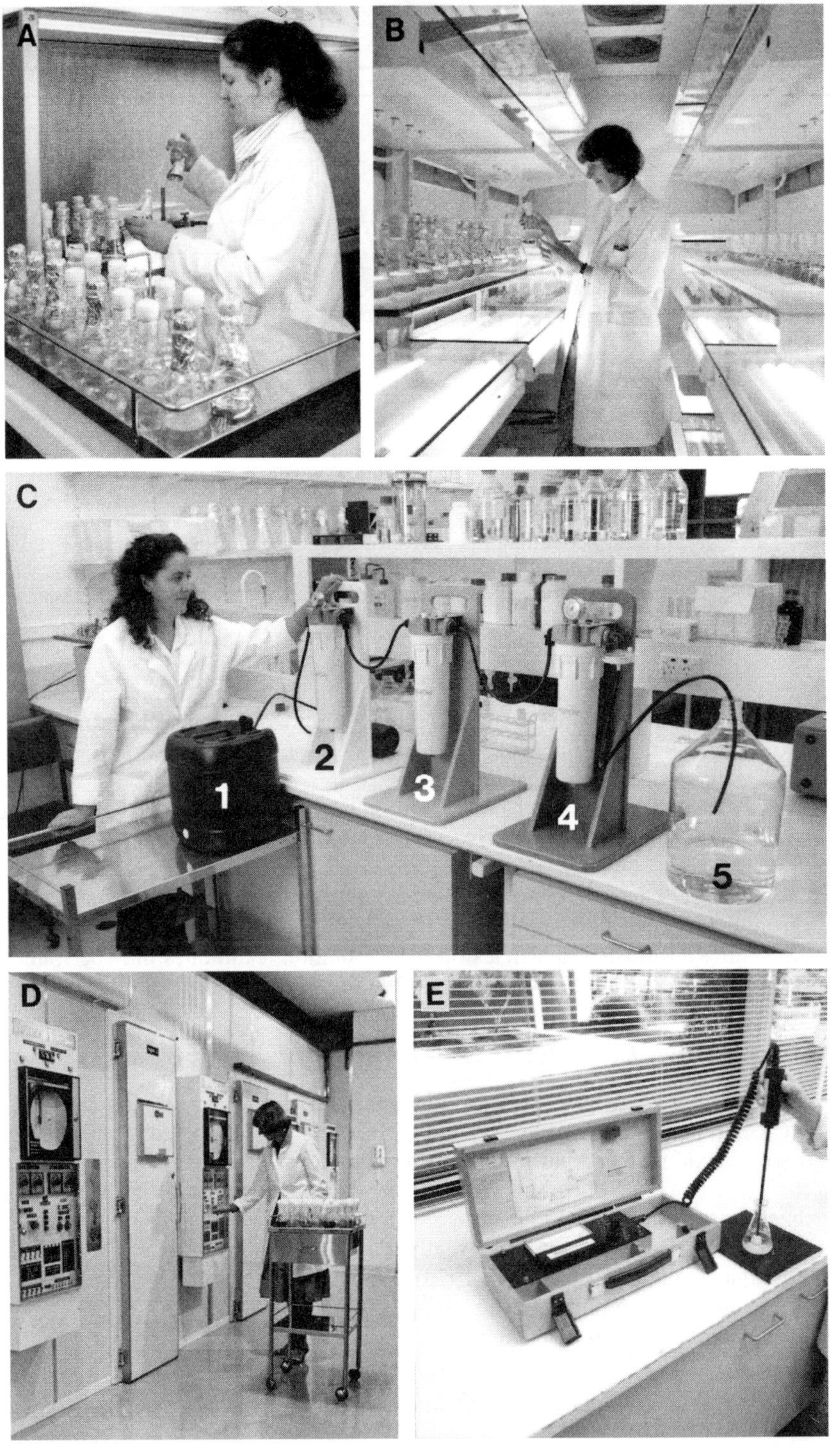

Figure 6.3
Algal culture procedures and facilities at the CSIRO Algal Culture Laboratory.
A. Jeannie-Marie LeRoi transfers algal cultures in laminar-flow hood.
B. Shirley Jeffrey examines SCOR cultures in 20°C growth room.
C. Catherine Johnston adjusts seawater purification system. 1, raw seawater; 2, prefilter; 3, activated charcoal filter; 4, 0.45 μm filter; 5, purified seawater.
D. Walk-in algal culture growth rooms, showing control panels.
E. Biospherical Instruments QSL100 light meter inserted into culture flask containing seawater medium.

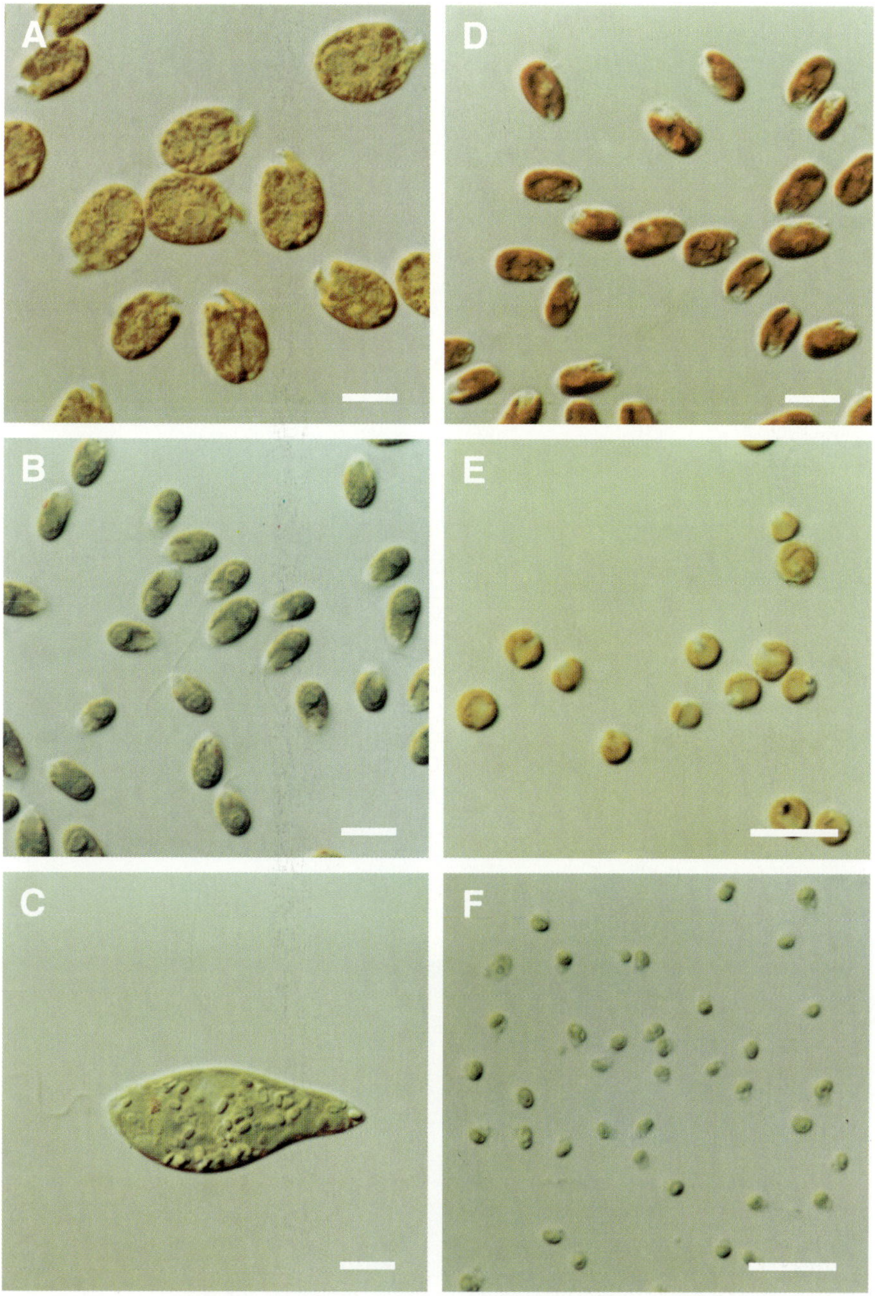

Plate 1
Light micrographs of microalgae (taken under Nomarski optics) recommended as sources of algal pigments.
A, *Amphidinium carterae*; B, *Dunaliella tertiolecta*; C, *Euglena gracilis*; D, *Chroomonas salina*; E, *Emiliania huxleyi*; F, *Micromonas pusilla*.
All culture code numbers as in Table 6.1. Scale bars = 10 µm.
Photographs by Dr A. Rees.

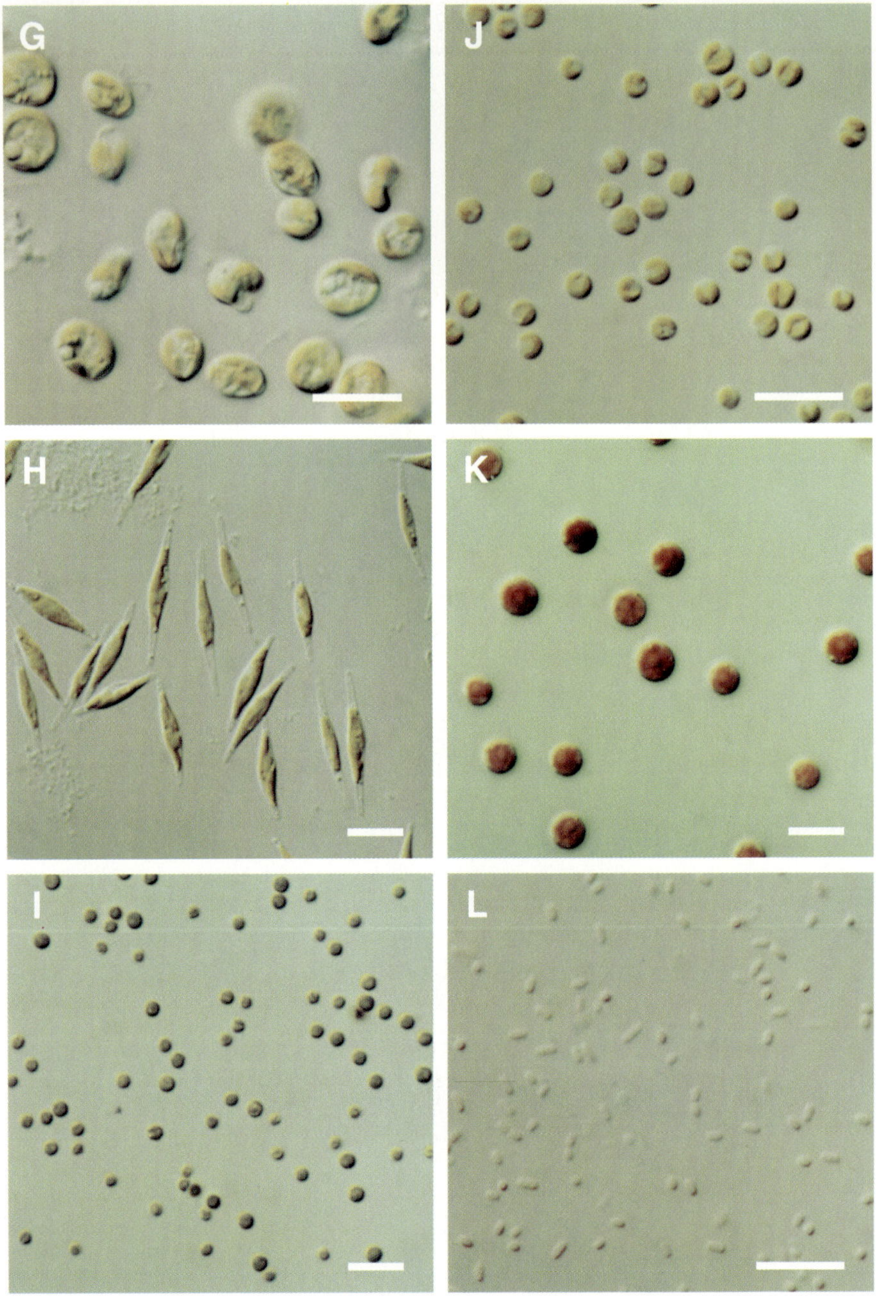

Plate 2
Light micrographs of microalgae (taken under Nomarski optics) recommended as sources of algal pigments.
G, *Pavlova lutheri*; H, *Phaeodactylum tricornutum*; I, *Pycnococcus provasolii*; J, *Pelagococcus subviridis*; K, *Porphyridium cruentum*; L. *Synechococcus* sp.
All culture code numbers as in Table 6.1. Scale bars = 10 µm.
Photographs by Dr A. Rees.

Plate 3
Logarithmic phase cultures of some SCOR reference algae grown under standard conditions and ready for harvest. All cultures in panels A–D are listed from left to right.
A. *Pavlova lutheri; Chroomonas salina; Micromonas pusilla*
B. *Pleurochrysis* sp. (*not* SCOR culture); *Synechococcus* sp.; *Emiliania huxleyi*
C. *Lauderia annulata* (*not* SCOR culture); *Porphyridium cruentum*; *Amphidinium carterae*
D. *Pelagococcus subviridis; Dunaliella tertiolecta; Pycnococcus provasolii.*
Photographs by Mr G. Johnson.

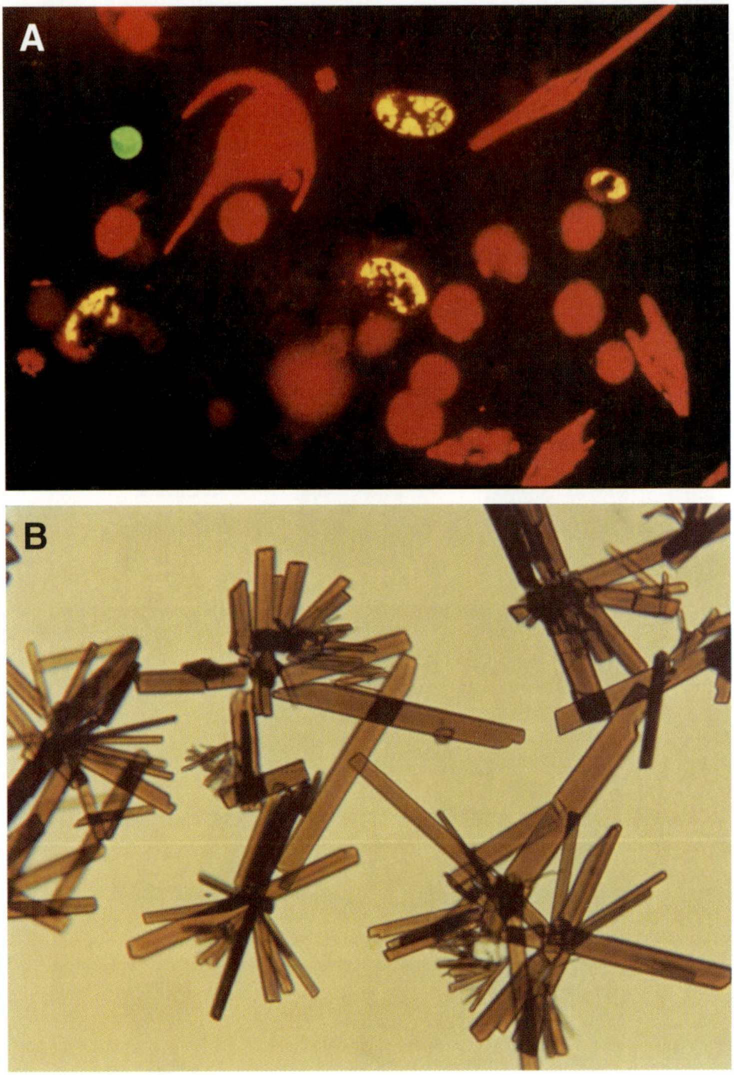

Plate 4

A. Light micrograph of a field sample of phytoplankton taken under UV light to show red chlorophyll fluorescence of pigmented cells. Some heterotrophic cells fluoresce green or yellow.
Micrograph courtesy of Dr A. Rees.

B. Crystalline preparations of chlorophyll c_2 (S. W. Jeffrey, unpublished). Photograph by the late Dr A. C. Heron.

slopes. Inocula are successfully dispatched world-wide in 20 ml or 200 ml volumes (see Appendix L).

The optimal characteristics of an algal culture laboratory are listed in Table 6.2, and illustrated in Fig. 6.3. For researchers wishing to hold just a few algal strains, an extensive facility is not necessary, so long as a few basic principles are observed. Working areas should be clean and free from dust; drafts should be excluded while transferring (keep doors and windows shut); facilities for washing glassware used for algae should be kept separate from those used for chemicals and away from the algal growth area; and an even temperature and lighting for the algal cultures should be maintained (see Section 6.3). Further details on algal culture facilities and their general maintenance may be found in Starr (1973).

6.3 Practical protocols for growing SCOR reference cultures

In this section, simple protocols are given for growing the SCOR reference cultures. These procedures are basically those used for the maintenance of microalgal cultures in the CSIRO Algal Culture Collection (Jeffrey, 1980), but they have been simplified for oceanographers wishing to maintain just a few robust strains for pigment standards. The topics and protocols cover selection of the SCOR reference cultures, choice of seawater or artificial seawater, preparation of culture media and soil extract, transfer techniques, light and temperature regimes, harvest times, measurement of algal growth rates, and glassware cleanliness.

6.3.1 *SCOR reference microalgal strains*

The 12 reference cultures recommended by SCOR Working Group 78 as a source of authentic pigment standards for oceanographic studies are listed in Table 6.1. The origin of each culture, the CSIRO Culture Code Number and the culture medium, growth temperature, light intensity, culture colour and absence of bacteria (axenic culture) of each strain are also listed. The two laboratories supplying these strains world-wide are found in Appendix L.

Other cultures were also used for SCOR experiments. The eustigmatophyte *Nannochloropsis oculata* (CS-216) was used as a source of vaucheriaxanthin for pigment identification studies (see Part IV Data and Graphics sheets). If the pigments in this species had been rigorously identified chemically, it would have been included in the SCOR reference cultures. *Prochlorococcus marinus*, a ubiquitous prochlorophyte now in culture (Partensky *et al.* 1993) was not available to SCOR WG 78 at the time of the Workshops. It is hoped that an easy-to-grow strain will soon be available as a source for oceanographers of DV chlorophyll *a* and DV chlorophyll *b*.

Nannochloris atomus (CS-183) and *Tetraselmis suecica* (CS-187) were two 'difficult-to-extract' chlorophytes that were used for the solvent extractability experiments (Chapter 9). These species, and *N. oculata* above, were grown under the same conditions as the SCOR reference algal cultures (Table 6.1), viz., f2 medium, 75 ml cultures in 125 ml Erlenmeyer flasks, 20°C, 80 $\mu E\ m^{-2}\ s^{-1}$ light intensity on 12:12 light:dark cycles. Cultures were harvested 4–6 hours into the light cycle.

Light micrographs of the 12 species (Table 6.1) photographed under a Zeiss Axioplan microscope using Nomarski optics are shown in Plates 1 and 2. The species range in size (largest diameter) from 0.5 µm (*Synechococcus* sp., CS-197) to 20 µm (*Chroomonas salina*, CS-174) as shown by the scale bars on each micrograph. These micrographs are of logarithmic phase cells, grown in the media listed in Table 6.1. Plate 3, which shows the culture colours of 12 cultures in late-log phase, should provide a guide to the expected density of culture colour of the differently pigmented algal strains. Chapter 13 details the pigment composition of each culture grown under the standard conditions described here. The culture colour will, of course, reflect the pigment content of the algal cells.

6.3.2 *Choice of seawater or artificial seawater*

Purified natural or artificial seawater (see Section 6.2.2) is the basis for all marine culture media. The CSIRO system uses clean off-shore seawater from 10 m depth, which is purified by pumping sequentially through three separate filter cartridges (see Fig. 6.3C) in a system that can process 100 l in 2–3 hours. Fig. 6.3C shows the purification sequence:

1. raw seawater
2. a prefilter to remove particulate matter (Millipore Rogard, Catalogue No. CD PRM 1206)
3. an activated charcoal filter to eliminate soluble organics (Cuno-Pacific, Aquapure AP117)
4. a 0.45 µm filter for final purification (Millipore Durapore, Catalogue No. JMAS 463).
5. purified seawater

Smaller volumes of seawater can be filtered sequentially through

a) a prefilter (Whatman GF/C 90 mm diameter or Schleicher and Schüll 90 mm diameter Glass-fibre filter, No. 8) in a filter holder and filtered under suction (laboratory vacuum line or pump)
b) approximately 1 g activated charcoal powder in 20 l seawater to remove organics, stirring for 4 hours and allowing to settle overnight
c) a prefilter (Millipore 90 mm AW) placed on top of a 0.45 µm Millipore or Sartorius 90 mm membrane filter in a filter holder to remove both charcoal and fine particulates.

This latter system was used by CSIRO for many years when the Culture Collection was small, and only small volumes of seawater were required. It was entirely successful, but more time-consuming than the present system.

In one set of experiments (see Chapter 13) an artificial seawater made from Sigma Sea Salts was used for a set of SCOR cultures. While the microalgae usually grew at half the rate of those microalgae grown in a purified natural seawater medium, in most species, the pigment composition (µg/10^6 cells) was similar. For laboratories that have no access to clean seawater, artificial seawater may be used, but with caution!

6.3.3 *Preparation of culture media*

Three marine media and one freshwater medium were used for the SCOR-recommended strains (Table 6.1). The final composition of the four culture media in µmoles are compared in Table 6.3.

Media are always prepared from stock solutions in highly purified Milli-Q water or distilled water (Table 6.4), small aliquots of which (Table 6.5) are added to a given large volume of seawater for marine media, or distilled water for freshwater media. This method is to prevent the precipitation problems that often occur with seawater media. Table 6.4 gives the concentrations of the main nutrients, chelating agents, culture buffers, trace metals and vitamins of the stock solutions used for the four culture media. If fungal contamination is a problem, stocks should be autoclaved or filter-sterilized (0.22 µm filter) and aseptically dispensed. Table 6.5 gives volumes of stock nutrients used per litre of seawater or distilled water in the different media. These solutions are routinely stored at 5°C.

Phosphate causes precipitation problems when autoclaved in seawater. We therefore recommend autoclaving the phosphate in separate tubes and then adding appropriate aliquots aseptically to the cooled 'sterile medium minus phosphate' in the 125 ml Erlenmeyer flasks (Table 6.6).

G medium contains soil extract as a source of extra trace elements, metal-chelating organic compounds and vitamins. A protocol for preparing soil extract is given in Table 6.7. The original soil-extract medium, denoted 'Erdschreiber', originated from Schreiber in 1927 (cited by McLachlan, 1973). It has been highly successful in its various forms.

A general protocol for media preparation, which is applicable to all media, is given in Table 6.8.

TABLE 6.6. Protocol for preparing phosphate solutions for culture media

Phosphate is autoclaved separately from the seawater medium, to prevent precipitation problems.

- Use $NaH_2PO_4.2H_2O$ stock solution for all f media.
  Use K_2HPO_4 stock solution for G and MBL/NB_2 media.
- Dilute the concentrated phosphate stock (Table 6.5) with distilled water such that 1 ml added to each flask of sterile medium will give the required concentration of phosphate in the medium.
  For example:
    for 10 × 125 ml Erlenmeyer flasks, each containing 75 ml medium, prepare dilute phosphate stock for full-strength f or G medium as follows:
    Take 0.75 ml of original phosphate stock and make up to 10 ml with distilled water. Pour into a 20 ml test tube and plug with cotton wool or Steristopper.*
  Scale up in the same proportion for larger volumes.
- Autoclave dilute phosphate stock at 121°C (15 min at 15 PSI). Allow to cool before use.
- Using a sterile dispenser or sterile 1 ml pipettes, add 1 ml of dilute phosphate stock *aseptically* to each flask of sterile medium.
  Ensure that the correct phosphate is added to the correct medium (e.g., $NaH_2PO_4.2H_2O$ for f media; K_2HPO_4 for G and MBL/NB_2).
- To reduce the risk of contamination, work in a laminar flow cabinet and flame neck of flask before and after adding phosphate (see Table 6.9 for aseptic techniques).

* In our experience some screw caps contain materials toxic to algae, so we do not recommend their use.

6.3.4 *Sterilization procedures*

Glassware and media used for algal culture are sterilized to prevent bacterial or fungal contamination. They are either autoclaved (sterilized by steam under pressure),

TABLE 6.7. Protocol for the preparation of soil extract

Soil must be collected from a natural uncultivated environment, though a rich garden loam may be suitable. No fungicides, insecticides or garden fertilizers should be present. At CSIRO, topsoil from a local sandy-loam bushland environment has proved to have particularly beneficial growth-promoting properties. Soil from clay or other soil types are less suitable in our experience.

1. Soil should not be stored or processed in the algal culture laboratory, since it is a potent source of unwanted microorganisms.
2. Sift dry soil (not recently treated with fertilizer or herbicide) once through a coarse sieve (4 or 2 mm square mesh), and twice through a finer sieve (1 mm mesh).
3. Mix 1 kg of soil into 2 l of distilled water (Milli-Q).
4. Autoclave for 60 minutes at 121°C, at 15 PSI, and cool overnight.
5. Depending on the amount of debris, either filter through cotton wool loosely packed into the top of the stem of a glass filter funnel, or centrifuge at 5000 rpm for 20 minutes in 250 ml polyethylene centrifuge tubes, and collect the deep brown supernatant.
6. Dispense the supernatant (50 ml aliquots) into 100 ml Schott reagent bottles or 100 ml media bottles.
7. Autoclave for 15 minutes at 121°C and 15 PSI.
8. After cooling, wrap caps with parafilm to prevent airborne contamination from fungal spores or bacteria.
9. Store sterile soil extract at 4°C (cold room or refrigerator).

TABLE 6.8. General protocol for media preparation

1. Assemble all glassware (e.g., Erlenmeyer flasks; glass carboys for seawater; Schott bottles and test tubes for dilute phosphate solutions).
2. Label with media type and date (use a label that survives autoclaving).
3. Assemble stock solutions and seawater.
4. Measure out seawater required in a large flask or carboy (e.g., 1–10 l).
5. Dispense small volumes of stock solutions one by one into the seawater, stirring after each addition to avoid precipitation problems (see Tables 6.3–6.5).
6. Dispense 75 ml media into 125 ml Erlenmeyer flasks (or other volumes and flasks as appropriate).
7. Plug with Steristoppers or cotton wool, and cap with foil (see Plate 3B) to prevent dust or bacterial contamination.
8. Dilute phosphate stock and dispense into tubes (see Table 6.6).
9. Autoclave flasks of media and tubes of dilute phosphate at 121°C and 15 PSI for 15 min.
10. Cool.
11. When cold, add phosphate to media aseptically in a laminar flow hood (see Tables 6.6 and 6.9).
12. Leave 2–3 days before using the media to allow CO_2 equilibration. (Autoclaving drives off CO_2 needed for algal photosynthesis and raises the pH to undesirable levels.)
13. Store media at 4°C in the dark until ready for use.

dry-heat sterilized, or media may be filter-sterilized through a bacteriological 0.22 μm membrane filter.

In the CSIRO Algal Culture Laboratory all culture media are autoclaved in a horizontal-loading Atherton autoclave. Sterilization is dependent on time, temperature, steam pressure and the volume being autoclaved. Flasks containing 75–150 ml of media are autoclaved for 15 minutes at 121°C and 15 PSI (105 kPa), while flasks containing larger volumes are autoclaved for longer periods and cooled slowly.

To prevent spillage during autoclaving, flasks are never more than two-thirds full and stoppers or screwcaps are slightly loosened (but tightened immediately afterwards).

Pipettes used for culture transfers are usually dry-sterilized in an oven with forced-air circulation. Pipettes are first plugged with non-absorbent cotton wool and packed into glass cylinders (tips downwards). They are then placed into a preheated oven at 160°C for 1–1½ hours. Sterilization is checked with an indicator tape or Browne's tubes, both of which change colour when the desired temperature is reached and maintained long enough for sterilization to be effected.

Pipettes and other glassware can also be sterilized in autoclave bags (paper or autoclavable plastic).

6.3.5 *Transfer techniques*

In the CSIRO Algal Culture Laboratory, algal stocks (including the SCOR cultures) are routinely transferred (subcultured) to fresh medium every 6 weeks, although delicate species may need to be transferred every 2–3 weeks. For example, 0.1–1 ml of 'parent' culture may be used to inoculate 75 ml fresh medium in a 125 ml Erlenmeyer flask ('daughter' culture). After logarithmic growth has been established in the new daughter cultures, the light intensity is reduced from 80 $\mu E\ m^{-2}\ s^{-1}$ to approximately 20 $\mu E\ m^{-2}\ s^{-1}$ to slow growth and minimize the need for frequent transfers. Stock parent algal cultures, being 'inviolate', are opened only once, to inoculate the daughter stock culture. All subsequent inocula are then taken from the parent rather than the daughter culture.

Experimental cultures are conditioned before experimentation by inoculating a small volume (e.g., 0.5 ml) of stock culture into fresh medium and incubating it under the growth conditions required for the experiment. When this 'pre-experiment' culture is in logarithmic phase, it is used to inoculate the 'real' experimental culture.

Transfers are carried out under strictly sterile conditions in a UV-sterilized cabinet provided with sterile laminar air flow (Fig. 6.3A). If these facilities are not available, a clean bench, disinfected first with ethanol, may be used. All air drafts in the laboratory should be eliminated (keep doors and windows closed and no 'people interruptions') to prevent air-borne contamination when transferring. General protocols for successful aseptic transfers are given in Table 6.9.

6.3.6 *Light, day length, temperature and harvest times*

Light intensities are critical to both growth and pigment content of algae. At the CSIRO Algal Culture Laboratory, a Biospherical Instruments light meter QSL-100* is used for all light measurements (see Fig. 6.3E). As the spherical (4 π) light sensor collects light from 360°, not 180° as does a 2 π or a flat bed cosine collector (e.g., a LICOR meter), it more accurately measures the light intensity received by the algal culture. Light intensities of photosynthetically-active radiation (400–700 nm) can be measured with the QSL-100 within the culture flask or on the shelf in either wet or

* Biospherical Instruments Inc., 4901 Morena Blvde, Suite 1003, San Diego, California 92117, USA

TABLE 6.9. General protocols for aseptic algal transfers

1. Sterilize laminar flow cabinet with UV radiation (10 min), and swab bench surfaces with technical-grade alcohol or disinfectant before use.
2. Assemble stock cultures and culture flasks with fresh medium, and label new flasks with culture name, code and date of transfer. Ensure culture medium is at 20°C before transferring, to avoid temperature shock to the inoculum.
3. For inoculating new media with stock cultures, ensure a sufficient supply of sterile plugged 1–5 ml graduated pipettes, or sterile, plugged Pasteur pipettes.
4. Effect the sterile transfer of each algal culture to its new flask by following this sequence:
   - carefully remove sterile pipette from container (see section 6.3.4). Do not allow the pipette tip to touch any surface. Always bring flasks to pipette to minimize movement within the laminar flow cabinet.
   - withdraw cotton stopper from the stock culture
   - flame the neck of the flask (1–2 s) with a Bunsen burner flame
   - carefully withdraw 0.1–1 ml stock into the sterile pipette, using a silicone bulb or rubber teat
   - flame the neck of the flask of stock culture
   - insert the cotton stopper into the stock flask and set aside
   - withdraw the cotton stopper from the new flask
   - flame the neck of the new flask as described above
   - carefully insert the pipette tip below the surface of the new culture medium and gently expel inoculum (do not blow air bubbles into the medium)
   - flame the neck of the new flask
   - insert stopper into the inoculated flask
   - discard used pipette into disinfectant
5. Place new flasks and stocks on culture shelf at appropriate light and temperature (see Table 6.1).

dry modes. For our experiments we use 20 W Philips 'White deluxe' (TL3694) fluorescent tubes beneath glass culture shelves (Fig. 6.3B) at a light intensity of 80 or 100 µE m^{-2} s^{-1} on 12:12 light:dark cycles. 'Day' always begins at 6 am, 'night' at 6 pm, and we usually harvest 4–6 hours into the light cycle.

Some tropical isolates (e.g., *Micromonas pusilla* CS-170) grow better and last longer on a blue-green light regime (Jeffrey and Vesk, 1977; Vesk and Jeffrey, 1977). This is obtained by placing sheets of Belgian Signal Glass on the glass shelves beneath the culture flasks, which are illuminated from below (Fig. 6.3B). The transmittance spectrum of this blue-green filter has a maximum at 480 nm, with no red light above 600 nm (Jeffrey and Vesk, 1977), which simulates the light climate of oligotrophic water at 10 m depth. This blue-green light culture regime is referred to elsewhere (e.g., Part IV Data and Graphics sheets).

The SCOR cultures are maintained at a temperature of 20°±1°C, although most species can tolerate 15–23°C (except *Synechococcus* sp., CS-197, which requires low light and temperature, Table 6.1). The CSIRO facility has culture rooms (Fig. 6.3B, D) and cabinets that can be set at any temperature between 5 and 30°C, so that simulated polar through to tropical conditions can be maintained.

Harvest time can vary according to inoculum size, cell density of the inoculum, culture volume, environmental growth conditions, and age at which the cells are to be harvested. If the inoculum consists of logarithmic phase cells at a density of 1 x

10^6 cells/ml, and the ratio of inoculum to culture volume is about 1 to 75, then cultures would be expected to reach mid to late logarithmic phase in about 7–10 days. Cell densities obtained with the SCOR cultures in the present work under our standard growth conditions are listed in Table 13.3, Chapter 13.

6.3.7 *Cell counts*

Growth rates are measured by cell counts over time (Fig. 6.1). Haemocytometers (Neubauer) are used at the CSIRO Algal Culture Laboratory for algal cells within the size range 1–30 μm diameter, but Coulter counters (Parsons, 1973), calibrated for each individual cell type, may also be used. We prefer the haemocytometer method as cell counts are done microscopically and the health of each culture is observed while counting. Three excellent references on counting techniques may be found in Guillard (1973, 1978) and Schoen (1988).

6.3.8 *Glassware cleanliness*

All glassware to be used for culturing algae or media preparation must be dedicated to 'culturing only' purposes, and must be scrupulously clean and free from toxic organic or metallic contaminants. Glassware should be washed away from areas of the laboratory where media is prepared or algae cultured, to ensure no traces of detergent or acids are in the same air-space as the microalgae. While different laboratories have different protocols for washing procedures (e.g., McLachlan, 1973), Table 6.10 gives the simple but rigorous procedures used for glassware washing in the CSIRO Algal Culture Laboratory.

6.4 Conclusion

It is hoped that this introduction to the principles of algal culture, and the detailed protocols we have provided, will give the oceanographer enough information to cul-

Table 6.10. Recommended protocol for washing glassware

- Ensure all glassware is washed in a room separate from the algal culture facilities.
- Before disposal, autoclave all microalgal cultures, to ensure local waterways will not become contaminated.
- Scrub test-tubes, flasks and Schott bottles with appropriate brush.
- Soak glassware in a biodegradable detergent suitable for tissue culture and bacteriological work (e.g., Pyroneg) *or* wash glassware in dishwasher (e.g., with Deconex detergent).*
- Rinse all washed glassware thoroughly. Include at least 5 hot and 2 cold rinses and at least 5 distilled water rinses (use Milli-Q water or equivalent).
- Dry glassware in oven at 60–70°C.
- Store glassware in a clean cupboard, and cover neck of open flasks with foil, to prevent dust particles settling.

* Pyroneg is marketed by Diversey (A/Asia) Pty Ltd, Sydney, Australia. Deconex is marketed by Borer Chemie, CH-4528, Zuchwil, Switzerland.

ture successfully all of the algal strains required for pigment standards (see Chapter 13). If the practitioner attempts to mimic the marine environment as far as possible in the growth conditions outlined above (e.g., no direct sunlight on culture flasks, appropriate media, temperatures, glassware) then success can be assured. We hope that this Chapter has dispelled the myth that algal cultures are difficult to grow; instead, we hope that, armed with these procedures, researchers will isolate and study more algae from the world's oceans, with further beneficial insights into global phytoplankton dynamics.

REFERENCES

Bjørnland, T., Liaaen-Jensen, S. (1986). Suggested algal cultures as pigment references for HPLC analysis of seawater samples, Unpublished Report prepared for SCOR WG 78, 6 pp.

Borowitzka, M. A., Borowitzska, L. J. (1988). Dunaliella. In: Borowitzka, M. A. and Borowitzska, L. J. (eds), *Microalgal Biotechnology*, Cambridge University Press, Cambridge, U.K., p. 27–58.

Brand, L. E. (1986). Nutrition and culture of autotrophic ultraplankton and picoplankton. In: Platt, T. and Li, W. E. (eds), *Photosynthetic Picoplankton. Canad. Bull. Fish. Aquat. Sci.* 214: 205–233.

Brock, T. D., Madigan, M. T. (1991). *Biology of Microorganisms*, 6th edn, Prentice Hall Inc., Englewood Cliffs, New Jersey, 874 pp.

Freshney, R. I. (1987). *Culture of Animal Cells — a Manual of Basic Techniques*, 2nd edn, Alan R. Liss, Inc., N. Y., 295 pp.

Guillard, R. R. L. (1973*a*). Division rates. In: Stein, J. (ed.), *Handbook of Phycological Methods: Culture Methods and Growth Measurements*. Cambridge University Press, U.K., p. 289–311.

Guillard, R. R. L. (1973*b*). Methods for microflagellates and nanoplankton. In: Stein, J. (ed.), *Handbook of Phycological Methods: Culture Methods and Growth Measurements*. Cambridge University Press, U.K., p. 69–85.

Guillard, R. R. L. (1975). Culture of phytoplankton for feeding marine invertebrates. In: Smith, W. L. and Chanley, M. (eds), *Culture of Marine Invertebrate Animals*. Plenum Press, New York, p. 29–60.

Guillard, R. R. L. (1978). Counting slides, In: *Phytoplankton Manual*, UNESCO *Monograph in Oceanographic Methodology*, UNESCO, Paris, p. 182–189.

Guillard, R. R. L., Ryther, J. H. (1962). Studies of marine plankton diatoms. I. *Cyclotella nana* Hustedt and *Detonula confervacea* (Cleve) Gran. *Can. J. Microbiol.* 8: 229–239.

Harrison, P. J., Yu, P. W., Thompson, P. A., Price, N. M., Phillips, D. J. (1988). Survey of selenium requirements in marine phytoplankton. *Mar. Ecol. Progr. Ser.* 47: 89–96.

Hoshaw, R. W., Roscowski, J. R. (1973). Methods for microscopic algae. In: Stein, J. (ed.), *Handbook of Phycological Methods: Culture Methods and Growth Measurements*. Cambridge University Press, U.K., p. 53–67.

Jeffrey, S. W. (1980). Cultivating unicellular marine plants. CSIRO (Hobart, Tasmania) *Fish. and Oceanogr. Ann. Rep.* 1977–1979, p. 22–43.

Jeffrey, S. W. (1984). Responses of unicellular marine plants to natural blue-green light environments. In: Senger, H. (ed.), *Blue Light Effects in Biological Systems*. Springer-Verlag Berlin, p. 497–508.

Jeffrey, S. W., Vesk, M. (1977). Effect of blue-green light on photosynthetic pigments and chloroplast structure in the marine diatom, *Stephanopyxis turris*. *J. Phycol.* 13: 271–279.

Jerlov, N. G. (1976). *Marine Optics,* Elsevier Oceanography Series 14, Elsevier, Amsterdam, 231 pp.

Lewin, J. (1966). Silicon metabolism in diatoms. V. Germanium dioxide, a specific inhibitor of diatom growth. *Phycologia*, 6: 1–12.

Keller, M. D., Selvin, R. C., Claus, W., Guillard, R. R. L. (1987). Media for the culture of oceanic ultraphytoplankton. *J. Phycol.* 23: 633–638.

Loeblich, A. R. (1975). A seawater medium for dinoflagellates and the nutrition of *Cachonina niei*. *J. Phycol.* 11: 80–86.

Loeblich, A. R., Smith, V. E. (1968). Chloroplast pigments of the marine dinoflagellate *Gyrodinium resplendens*. *Lipids*, 3: 5–13.

Markham, J. W., Hagmeier, E. (1982). Observations on the effects of germanium dioxide on the growth of macroalgae and diatoms. *Phycologia*, 21: 125–130.

McLachlan, J. (1973). Growth media — marine. In: Stein, J. R. (ed.), *Handbook of Phycological Methods: Culture Methods and Growth Measurements*. Cambridge University Press, Cambridge, U.K., p. 25–51.

Nichols, H. W. (1973). Growth media—freshwater. In: Stein, J. R. (ed.), *Handbook of Phycological Methods: Culture Methods and Growth Measurements*. Cambridge University Press, Cambridge, U.K., pp. 7–24.

Parsons, T. R. (1973). Coulter counter for phytoplankton. In: Stein, J. (ed.), *Handbook of Phycological Methods: Culture Methods and Growth Measurements*. Cambridge University Press, U.K., p. 345–358.

Partensky, F., Hoepffner, N., Li, W. K. W., Ulloa, O., Vaulot, D. (1993). Photoacclimation of *Prochlorococcus* sp. (Prochlorophyta) strains isolated from the north Atlantic and the Mediterranean Sea. *Plant Physiol.* 101: 285–296.

Paul, J. (1970). *Cell and Tissue Culture*, E. and S. Livingstone, Edinburgh, 430 pp.

Pringsheim, E. G. (1946). *Pure Cultures of Algae*. Cambridge University Press, London, 119 pp.

Provasoli, L. (1958). Nutrition and ecology of protozoa and algae. *Ann. Rev. Microbiol.* 12: 279–308.

Provasoli, L., McLaughlin, J. J. A., Droop, M. R. (1957). The development of artificial media for marine algae. *Arch. Mikrobiol.* 25: 392–428.

Schoen, S. (1988). Cell counting, In: Lobban, C. S., Chapman, D. J., Kremer, B. P. (eds), *Experimental Phycology—a Laboratory Manual*. Cambridge University Press, U.K., p. 16–22.

Starr, R. C. (1973). Apparatus and maintenance. In: Stein, J. (ed.), *Handbook of Phycological Methods: Culture Methods and Growth Measurements*. Cambridge University Press, U.K., p. 171–180.

Stein, J. R. (1973). *Handbook of Phycological Methods: Culture Methods and Growth Measurements*, Cambridge University Press, Cambridge, U.K., 448 pp.

Throndsen, J. (1973). Special methods—micromanipulators. In: Stein, J. (ed.), *Handbook of Phycological Methods: Culture Methods and Growth Measurements*. Cambridge University Press, U.K., p. 139–144.

Ukeles, R. (1973). Continuous culture—a method for the production of unicellular algal foods. In: Stein, J. (ed.), *Handbook of Phycological Methods: Culture Methods and Growth Measurements*. Cambridge University Press, Cambridge, U.K., p. 233–254.

Vesk, M., Jeffrey, S. W. (1977). Effect of blue-green light on photosynthetic pigments and chloroplast structure in unicellular marine algae from six classes. *J. Phycol.* 13: 280–288.

7

Preparation of chlorophyll standards

S. W. Jeffrey

7.1 INTRODUCTION

Chlorophylls and associated carotenoids are invaluable chemosystematic markers of phytoplankton in the oceans. Any method of pigment measurement — whether spectroscopic, fluorometric or chromatographic — is calibrated with pure pigment standards. Of the 47 pigments listed in the Part IV Data sheets as important in phytoplankton ecology, only a handful are available commercially (see Appendix J).

Therefore SCOR Working Group 78 recommended that the source of pure pigments should be reference microalgal cultures that together could supply all the pigments identified as important in present-day oceanography. If these strains are grown under standard conditions (see Chapter 6), extracted and chromatographed as outlined in Chapter 13, then the full range of pigments needed for standardization should be readily available to practitioners.

In this chapter, methods for obtaining large-scale (mg to gm) and small-scale (μg) quantities of pure chlorophyll standards are described; methods for isolating carotenoids are described in Chapter 8. These methods are based on those used at the SCOR 'hands-on' Workshops (see Chapter 1), supplemented by well-tested methods from the literature and the experience of working group members.

7.2 Historical background

To isolate pure compounds from biological materials requires sophisticated separation techniques. The Russian scientist M. S. Tswett (1872–1919), who pioneered the study of leaf pigments, was the first to develop such a method. As described by Ettre (1991), Tswett succeeded in separating coloured zones of chlorophylls and carotenoids from leaf extracts placed on small columns of calcium carbonate in glass tubes and washed with solvent. Later, he used the more inert inulin and sucrose to minimize

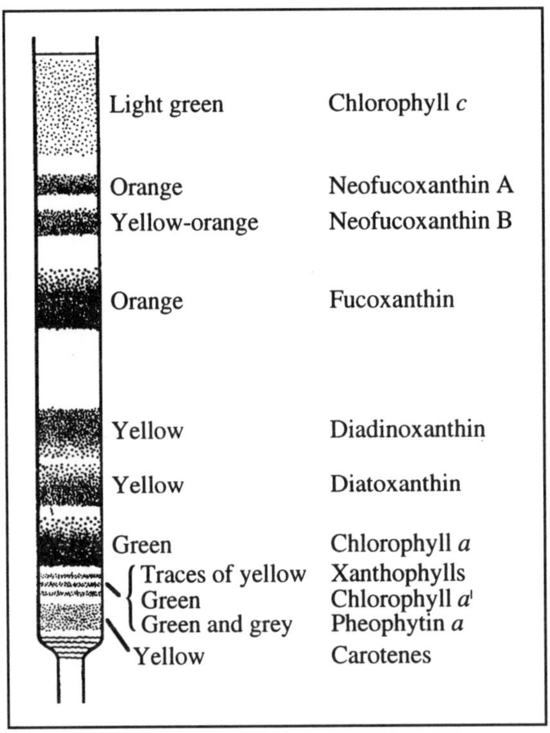

Figure 7.1
Column chromatography of diatom pigments on powdered sucrose. Solvent 0.5% *n*-propanol in petroleum ether (60–80°C fraction). Original, after Strain *et al.* (1944): neofucoxanthin A and B was Strain's terminology for oxidized fucoxanthins.

chemical damage to the pigments. The adsorbent and the coloured zones were named the 'chromatogram' (*chroma* = colour; *graphein* = to write). To examine the pigments, the calcium carbonate containing the green and yellow zones was extruded from the tubes, the zones collected, and the pigment fractions eluted with solvent. This technique, first published in 1906 by Tswett, is the basis of all sophisticated separation techniques (TLC, HPTLC, HPLC, GC) now practised in modern chromatography.

When partition (liquid/liquid) chromatography for amino acids was developed in 1941 by A. J. P. Martin and R. L. M. Synge, the way was open for H. H. Strain and colleagues to apply it to plant and algal pigments. Large columns of powdered sucrose were packed dry (Strain *et al.* 1944 and Fig. 7.1), washed with a non-polar solvent (petroleum ether), and the concentrated pigment extract was applied to the top of the column. The chromatogram was developed by increasing the solvent polarity with small volumes of *n*-propanol. Strain's classic column chromatography of plant pigments (for review, see Strain, 1958) became the model for all subsequent chlorophyll and carotenoid separation techniques used in aquatic science (e.g., TLC, HPTLC and HPLC). Indeed, for large-scale pigment purifications, column chromatography based on Strain's original method is still used today (see Hynninen and Lötjönen, 1983; Porra *et al.* 1989; and Section 7.5).

In the late 1930s, spectroscopic characterization of pigment fractions was made possible with the development of manual spectrophotometers (Zscheile, 1934). In careful, painstaking studies, the spectroscopic properties of purified chlorophylls and their derivatives were gradually established (e.g., Harris and Zscheile, 1943; see Fig. 7.2). Even today, with our sophisticated automatic recording spectrophotometers and elegant HPLC separation techniques, the careful studies of the early pigment chemists remain a model for all who follow. Their work led to rapid advances in pigment purification, crystallization and structure determination, with applications to terrestrial agriculture, food technology and, eventually, to aquatic science.

7.3 Sources of standard chlorophylls

Of the 17 chlorophylls and derivatives important in oceanography (see Part IV Data and Graphics sheets), only two — chlorophylls *a* and *b* — are available commercially (see Appendix J). All commercial preparations of chlorophylls *a* and *b* require purification as they contain epimers, 13^2 hydroxy chlorophylls and other allomers (oxidation products). Other chlorophylls required as standards must be isolated by the researcher from biological materials. The 12 microalgal strains recommended by SCOR WG 78 as standard pigment source materials (see Chapter 6) were used for the small-scale (μg) preparations at the SCOR Workshops (see Table 7.1).

Sometimes large-scale (>10 mg) preparations of chlorophylls are needed for identification or other purposes (e.g., determination of extinction coefficients, elucidation of new pigment structures), and in these cases leaves of higher plants or seaweed fronds or blades may be a more convenient source of large amounts of chlorophylls than huge volumes of microalgae culture. For example, large-scale preparations of chlorophylls *a* and *b* are usually derived from leaf tissue such as stinging nettle (Smith and Benitez, 1955), spinach (Strain *et al.* 1963), maize (Porra *et al.* 1989) or a green seaweed (e.g., *Ulva lactuca*; Jeffrey, unpublished), and large-scale preparations of chlorophylls c_1 and c_2 are readily isolated from the brown seaweed *Sargassum flavicans* (Jeffrey, 1969, 1972). Table 7.1 lists some alternative sources of pigments,

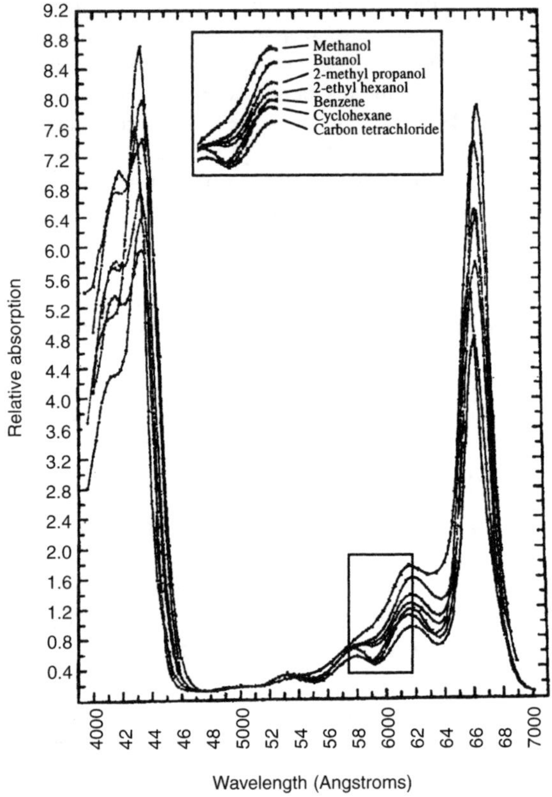

Figure 7.2
Absorption spectra of highly purified chlorophyll *a* in different solvents. Original, after Harris and Zscheile (1943).

and also recommends chromatographic methods for purifying small quantities of various chlorophylls (see also section 7.6).

Further guidance in the choice of a suitable biological source for chlorophylls and derivatives is given in Table 7.2. These hints take account of the biochemical properties of the plant tissues. For example, if isolating chlorophylls, the researcher should choose plant tissue or species that lack active chlorophyllase, oxidase or demetallase enzyme activity, to prevent the formation of degradation products. Similarly, if isolating chlorophyll *c*, an alga rich in chlorophyllase should not be used (e.g., the diatom *Phaeodactylum tricornutum*), since the chlorophyllide *a* formed as an unwanted side reaction will need an additional separation step to remove it from the chlorophyll *c* fraction. Further, if isolating chlorophyll c_2, the procedure is very easy if the source material does not contain chlorophyll c_1 or c_3 (e.g., dinoflagellates or cryptomonads). Table 2.3 in Chapter 2 gives details of pigment distribution among algal classes that can assist the choice of source material.

TABLE 7.1. Biological sources and methods used for the preparation of chlorophyll standards

Chlorophyll	Microalgal source	CSIRO culture code no.	Purification methods[a] (cultures)	Alternative sources	Purification methods (alternative sources)
Chlorophyll *a*	*Dunaliella tertiolecta*	CS–14	HPLC–1	Sigma chlorophyll *a*[b];	HPLC–1; TLC–1;
	Chroomonas salina	CS–174	HPLC–1	Higher plants: spinach, maize	CC–1
	Synechococcus sp.	CS–94	HPLC–1	Seaweed: *Ulva lactuca*	
Chlorophyllide *a*	*Phaeodactylum tricornutum*	CS–29	Hydrolysed extract; HPLC–1; HPLC–3		
Pheophytin *a*				Acidification of chlorophyll *a*	
Pheophorbide *a*				Acidification of chlorophyllide *a*	
Chlorophyll *b*	*Dunaliella tertiolecta*	CS–14	HPLC–1	Sigma chlorophyll *b*[b];	HPLC–1; TLC–1;
	Micromonas pusilla	CS–86	HPLC–1	Higher plants: spinach, maize	CC–1
				Seaweed: *Ulva lactuca*	
Chlorophyllide *b*	*Dunaliella tertiolecta*	CS–14	Hydrolysed extract; HPLC–1; HPLC–3		
Pheophytin *b*				Acidification of chlorophyll *b*	
Pheophorbide *b*				Acidification of chlorophyllide *b*	
Pyropheophytin *a*				Reflux pheophytin *a* in pyridine	
Chlorophyll $c_1 + c_2$	*Lauderia annulata*	CS–30	HPLC–1; CC; TLC–1	Seaweed: *Sargassum* spp.	HPLC–1; CC–1; TLC–1
Chlorophyll c_1	*Lauderia annulata*	CS–30	TLC–2; HPLC–2	Seaweed: *Sargassum* spp.	CC–1; CC–2; TLC–1; TLC–2; HPLC–2
Chlorophyll c_2	*Amphidinium carterae*	CS–21	HPTLC–1; HPLC–1		
	Chroomonas salina	CS–174	HPTLC–1; HPLC–1		
Chlorophyll c_3	*Emiliania huxleyi*	CS–57	HPTLC–1; HPLC–1		
	Pycnococcus provasoli	CS–185	HPTLC–1; HPLC–1		
Chlorophyll c_{CS-170}	*Micromonas pusilla*	CS–170	HPTLC–1		
Mg 3,8 DVP	*Micromonas pusilla*	CS–86; CS–98	HPTLC–1		
Phytylated Chl *c*-like	Many prymnesiophytes[c]		HPLC–1		

a. HPLC–1 Reversed phase C18 column (Wright *et al.* 1991; see Chapter 12)
HPLC–2 Polyethylene column (Jeffrey, 1989)
HPLC–3 Mantoura *et al.* (1996) — (see Chapter 11)
HPTLC–1 Jeffrey and Wright (1987)
TLC–1 Cellulose TLC (Jeffrey, 1981)
TLC–2 Polyethylene TLC (Jeffrey, 1969)
CC–1 Sugar or cellulose column chromatography (Strain *et al.* 1963; Jeffrey, 1972)
CC–2 Polyethylene column chromatography (Jeffrey, 1972)

b. Epimers usually present (up to 20%)

c. See Jeffrey and Wright (1994)

TABLE 7.2. Guide to the choice of algal sources for chlorophyll preparations

Chlorophyll	Comments
Chlorophyll *a*	Preferably no chlorophyll *b* present; use chromophyte alga with chlorophyll *a* + *c*, or chlorophyll *a* only (e.g., cyanobacteria, rhodophyte, eustigmatophyte; see Table 2.3, Chapter 2); if possible no highly active chlorophyllase present (see Jeffrey and Hallegraeff, 1987); no oxidases or de-metallase enzymes (see Owens and Falkowski, 1982); no acid cytoplasm to form 'pheopigments'.
Chlorophyll *b*	Use chlorophyll *b*-rich alga (e.g., prasinophytes or chlorophytes grown in low light); if possible no active chlorophyllase, oxidase or de-metallase enzymes.
Chlorophyllide *a*	Use alga with highly-active chlorophyllase (e.g., *Phaeodactylum tricornutum*), but check simultaneous oxidative degradation and formation of atypical chlorophyllide (see Barrett and Jeffrey, 1964; Fig. 7.3B).
Chlorophyllide *b*	As for chlorophyllide *a*, but use chlorophyll *b*-rich chlorophyte (e.g., *Dunaliella tertiolecta*).
Chlorophyll $c_1 + c_2$	Use any easy-to-extract chromophyte alga containing both $c_1 + c_2$ (see Table 2.3, Chapter 2).
Chlorophyll c_1	Choose diatom/prymnesiophyte without active chlorophyllase, to avoid contaminating chlorophyll c_1 fraction with chlorophyllide *a*.
Chlorophyll c_2	Use a chlorophyll c_2-only alga (e.g., dinoflagellate or cryptomonad).
Chlorophyll c_3	Use an easy-to-extract prymnesiophyte rich in chlorophyll c_3 (see Jeffrey and Wright, 1994). *Emiliania huxleyi* was the original source (see Jeffrey and Wright, 1987), but new chlorophyll *c* derivatives have recently been detected in this species (Garrido *et al.*, 1995).
Chlorophyll c_{CS-170}	Only known source is the tropical prasinophyte *Micromonas pusilla* (CS–170), which also contains Mg 3,8 DVP.
Phytylated chlorophyll *c*-like	*Emiliania huxleyi* was the original source but many prymnesiophytes are now available (see Jeffrey and Wright, 1994).
Mg 3,8 DVP	Common in prasinophytes such as *Micromonas pusilla* (CS-86 or CS–98).

7.4 CHLOROPHYLL ISOLATION

Classic treatises on chlorophyll purification include those of Willstätter and Stoll (1913), Smith and Benitez (1955), French (1960) and Strain and Svec (1966), which contain a wealth of material for the chlorophyll practitioner. These should be consulted as background information. Modern references to analytical and preparative chromatography of chlorophylls may be found in Shioi (1991) and Brockmann and Risch (1991) respectively. Excellent modern reviews by Scheer (1991) and Hynninen (1991) give the current state of knowledge of chlorophyll structure, occurrence and chemistry. In the plant and bacterial kingdoms more than 50 chlorophyll pigments are now known to have some functional role in photosynthesis (Scheer, 1991). Luckily for the oceanographer, the key functional chlorophylls known for phytoplankton are less than ten.

7.4.1 *General precautions*

Chlorophylls and carotenoids, being highly reactive molecules, need to be handled with extreme care during isolation procedures. Chlorophylls in particular are susceptible to degradation by light (photo-oxidation), heat (isomerization), air/oxygen (oxidation), acids (de-metallation) and bases (macro-cycle opening). Solvents, chromatographic adsorbents and the plant tissues themselves are also potentially destructive agents to chlorophylls (as they are to carotenoids; see Chapter 8).

Degradation reactions of chlorophylls *in situ* or during purification are described in section 3.6, Chapter 3 (and Hynninen, 1991). They are:

- allomerization (oxidation)
- epimerization at C–13^2
- de-metallation
- de-phytylation and trans-esterification
- chlorination at the C–20 methine bridge
- decarboxymethylation at C–13^2 (formation of 'pyro' derivatives)

The derivatives from the above reactions may occur in chlorophyll preparations and must be removed before final purification (see later sections). Dim light must be used for all procedures to prevent photo-degradation, and because of the solvents used for extraction, the work should be carried out in a fume hood. Evaporation of solvents is done *in vacuo* or under N_2, below 25°C, and 'pure' pigments should be stored under N_2 at -18°C to -20°C or in liquid nitrogen. Solvent purity is essential, particularly diethyl ether, which must be free from peroxide (e.g., Mallinckrodt is recommended). Further precautions, such as cleanliness of glassware and chromatographic adsorbent purity are essential, and are discussed in detail in Chapter 8 (section 8.3.1).

7.4.2 *Spectroscopic purity of chlorophyll preparations*

Because of the extreme lability of chlorophylls, it is essential to monitor purity by chromatographic and spectroscopic tests as the chlorophyll preparation progresses, particularly if the preparation is on a large scale. Both the position of absorption maxima and ratios of the absorption bands are important spectrophotometric parameters of pigment purity; these are listed for seven chlorophylls and three de-metallated chlorophyll *a* derivatives ('pheopigments') in Table 7.3. Further details are given in the Part IV Data and Graphics sheets.

The full spectra of several chlorophylls and derivatives are shown for comparison in Fig. 7.3A–D. Panels A and B show the spectral differences of chlorophyll *b* and the atypical (oxidized) chlorophyllide *a*, to chlorophyll *a*, respectively. Panels C and D show clearly shifts in the Soret (blue) bands of 428 to 410 nm, when Mg is removed from chlorophyll *a* and chlorophyllide *a* to form pheophytin *a* and pheophorbide *a* respectively. Both pheopigments also have a higher absorbance at 506 and 535 nm than the parent chlorophylls. These are indicative spectral characteristics of pheopigment contamination, which can readily appear in chlorophyll preparations. The spectral identities of chlorophyll *a* and chlorophyllide *a*, and of pheophytin *a* and pheophorbide *a*, should also be noted; they emphasize that phytol has no effect on the spectral characteristics of the macrocycle, whereas that of Mg is significant (see Scheer, 1991).

Absorption spectra of six chlorophyll *c* pigments, in diethyl ether or acetone, are compared in Fig. 7.4A–F, and their band ratios are given in Table 7.3. Details are given in the Part IV Data sheets.

TABLE 7.3. Band ratios of pure chlorophylls and derivatives useful in assessing purity of preparations (λ_{max} in brackets)

Chlorophyll	Solvent	Soret/red	Soret/violet	Soret/505nm	Orange/red	Reference
Chlorophyll *a*	diethyl ether	1.29 (428:660 nm)	1.63 (428:408 nm)	58.5 (428:505 nm)		Jeffrey (unpubl., see Fig. 7.3A)
Chlorophyll *b*	diethyl ether	2.84 (453:643 nm)	2.79 (453:428 nm)	42.5 (428:505 nm)		Jeffrey (unpubl., see Fig. 7.3A)
Chlorophyll c_1	diethyl ether	9.93 (444:628 nm)			0.67 (580:628 nm)	Jeffrey (1969)
Chlorophyll c_2	diethyl ether	14.12 (448.7:629.3 nm)			1.15 (580:629.3 nm)	Jeffrey (1969)
Chlorophyll c_3	diethyl ether	32.1 (451.3:625.9 nm)			3.79 (584.5:625.9 nm)	Jeffrey and Wright (1987)
Chlorophyll c_{CS-170}	diethyl ether	22.1 (448.1:620.3 nm)			2.59 (580.1:620.3 nm)	Jeffrey (1989)
Mg 3,8 DVP	diethyl ether	9.72 (437.1:623.5 nm)			0.54 (573.9:623.5 nm)	Jeffrey and Wright (1987)
Pheophytin *a*	acetone (100%)	2.25 (410:665 nm)		8.2 (410:506 nm)		Jeffrey and Lorenzen (unpubl., see Fig. 7.3C)
Pheophorbide *a*	acetone (100%)	2.32 (409:666 nm)		8.4 (409:505 nm)		Jeffrey and Lorenzen (unpubl. see Fig. 7.3D)
Pyropheophytin *a*	acetone (100%)	2.26 (410.2:666.7 nm)		9.46 (410.2:506.7 nm)		SCOR WG 78 (see Part IV Graphics sheets)

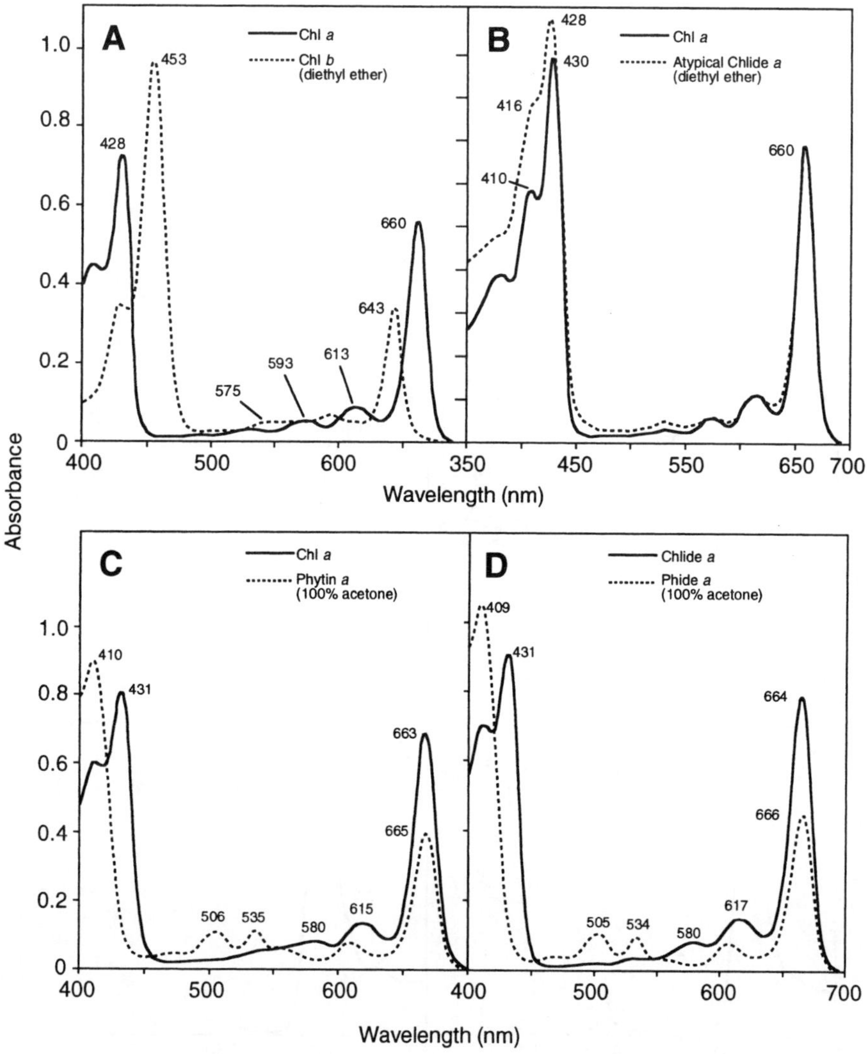

Figure 7.3
Absorption spectra of highly purified chlorophylls and derivatives.
A. Chlorophylls *a* and *b* in diethyl ether (Jeffrey, unpublished);
B. Chlorophyll *a* and the atypical (oxidized) chlorophyllide *a* in diethyl ether from the diatom *Phaeodactylum tricornutum* (from Barrett and Jeffrey, 1964). Note the atypical chlorophyllide has a higher Soret:red band ratio of 1.42–1.46, and an inflection at 415–417 nm rather than a minor peak at 410 nm. C. Chlorophyll *a* and pheophytin *a*; and
D. Chlorophyllide *a* and pheophorbide *a* in 100% acetone. (C and D, Jeffrey and Lorenzen, unpublished).

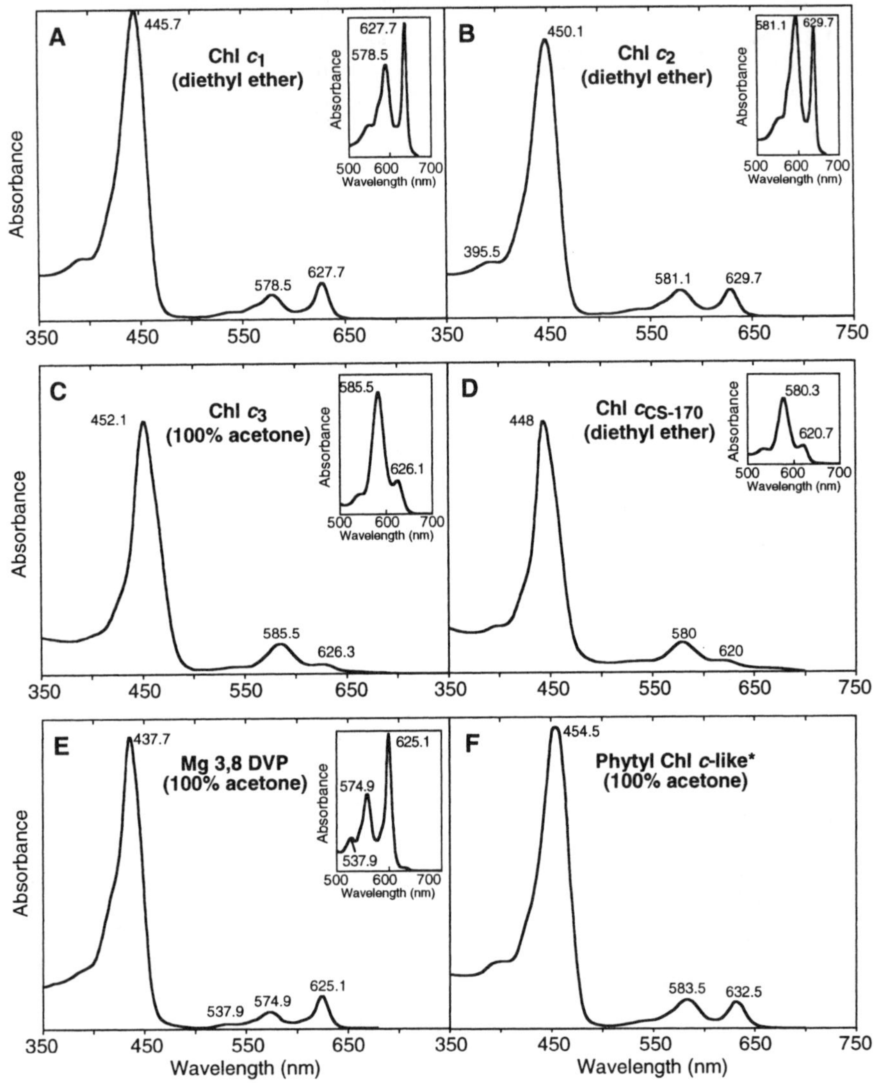

Figure 7.4
Absorption spectra of six chlorophyll *c* pigments in diethyl ether or acetone. A, B, Chlorophyll c_1 and c_2 (Jeffrey, 1969); C, E, Chlorophyll c_3 and Mg 3,8 DVP (Jeffrey and Wright, 1987); D, Chlorophyll $c_{CS\text{-}170}$ (Jeffrey, 1989); F, Phytylated chlorophyll *c*-like, (Nelson and Wakeham, 1989).

7.4.3 *Chromatographic purity of chlorophyll preparations*

The purity of chlorophyll preparations must be checked by small-scale chromatographic tests which can detect contamination by chlorophylls, chlorophyll degradation products and carotenoids. Three chromatographic systems — TLC, HPTLC and HPLC — are all suitable for this purpose. Shioi (1991) lists published TLC and HPLC methods for chlorophylls, but here only those systems that have proved useful for phytoplankton pigments will be discussed.

Thin-layer chromatography.

TLC and HPTLC are ideally suited for a monitoring function: the procedures are simple, rapidly executed (10–20 min) and inexpensive. They need only μg quantities of pigment and are especially useful when HPLC equipment is not available. Thin-layer adsorbents that do not cause pigment degradation, but which have good resolving power for both chlorophylls and carotenoids, are sucrose (Jeffrey, 1968*a*; Sahlberg and Hynninen, 1984), cellulose (Sievers and Hynninen, 1977; Jeffrey, 1968*a*, 1981), polyethylene (Jeffrey, 1969) and silica bonded with C–8 and C–18 silanes (e.g., Merck RP–8 HPTLC plates, Jeffrey and Wright, 1987; and Merck RP–18 HPTLC plates, Suzuki *et al.* 1987, Ziegler *et al.* 1988). Kieselgur and silica gel with unprotected end-groups are not recommended as they cause serious degradation of chlorophylls (see Shioi, 1991), although they are suitable for chromatography of metal-free derivatives (see Hynninen and Lötjönen, 1980). Indeed, chlorination of chlorophyll *a* can occur during silica gel chromatography with the formation of artefacts such as 20-chloro-13^2-hydroxy-chlorophyll *a* (Senge and Senger, 1988; see section 3.6.5, Chapter 3).

Relevant examples of some of these techniques are shown in Fig. 7.5. Panel A shows chlorophyll *a*, *b* and *c* separations on washed-cellulose layers (Macherey-Nagel MN300; Jeffrey, 1968*a*, 1981); Panel B shows chlorophyll *c* pigments separated on polyethylene layers (Jeffrey, 1969, 1989); Panel C shows chlorophyll *c* separations on RP–8 plates (Jeffrey, 1989); and Panel D shows chlorophyll *a* and derivatives on RP–18 plates (Ziegler *et al.* 1988). The RP–8 HPTLC system is often used for small-scale high-purity preparations of chlorophyll *c* pigments (see Jeffrey, 1989), particularly when available HPLC methods are not selective enough to achieve the same resolution (cf. Wright *et al.* 1991, and Fig. 7.7).

High-performance liquid chromatography.

Many HPLC systems suitable for monitoring chlorophyll preparations are available (see Shioi, 1991). The one developed at the SCOR Workshops (see Chapter 11) for separating chlorophylls from degradation products is recommended. It uses a reversed-phase C-18 column with a 25 min step-isocratic development system (Fig. 7.6A), with fluorescence detection of pigments. Carotenoids, which do not fluoresce, can also be detected in this system if absorbance is monitored (e.g., at 440 nm).

HPLC columns of polyethylene are not only useful for separating chlorophyll *c* pigments (Fig. 7.6B; Jeffrey, 1989), but also for resolving mono- and divinyl chlorophyllides and other chlorophyll mixtures (Shioi and Beale, 1987).

The reversed-phase C-18 HPLC method described in Chapter 12 (Wright *et al.* 1991) gives excellent separation of chlorophylls *a*, *b*, $c_1 + c_2$ + Mg 3,8 DVP (one fraction), c_3, pheophytins *a* and *b*, pheophorbides *a* and *b* and chlorophyllides *a* and *b* (Fig. 7.7). The lack of resolution of chlorophyll c_1, c_2, $c_{CS\text{-}170}$ and Mg 3,8 DVP is

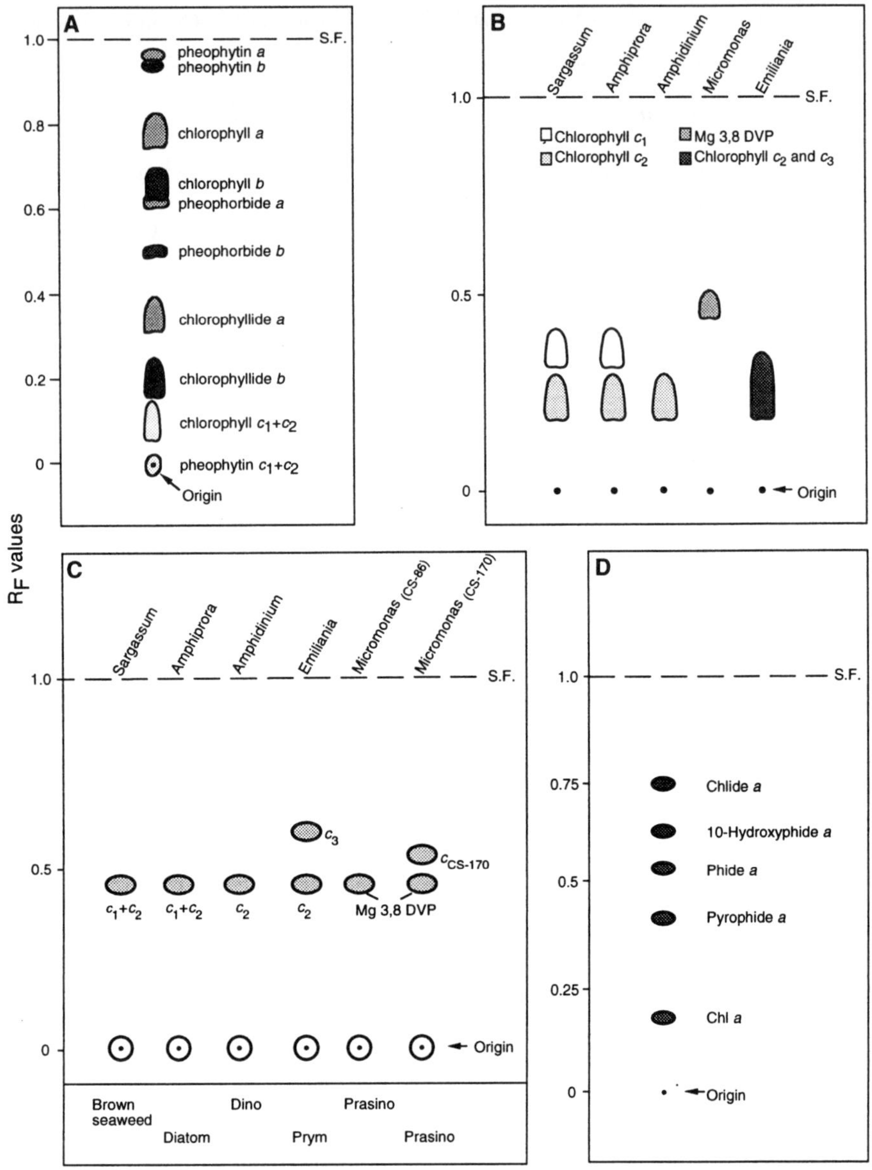

Figure 7.5
TLC and HPTLC methods for chlorophylls and derivatives.
A. TLC separation of chlorophyll *a*, *b* and *c* and derivatives on thin-layers of sucrose (Macherey-Nagel MN200). Solvent system, acetone: water = 9:1 (v/v) (from Jeffrey, 1968); B. TLC separation of chlorophyll *c* pigments on thin layers of polyethylene (PRX-1025, Polysciences Inc.); Solvent system, 100% acetone (from Jeffrey, 1969, 1989); C. HPTLC separation of chlorophyll *c* pigments on Merck RP-8 plates. Solvent system, methanol: water = 9:1 (v/v) (from Jeffrey, 1989); D. HPTLC separation of chlorophyll *a* and derivatives on Merck RP-18 plates. Solvent system, methanol: acetone: water = 30:20:2 (v/v/v) (after Ziegler *et al.* 1988).

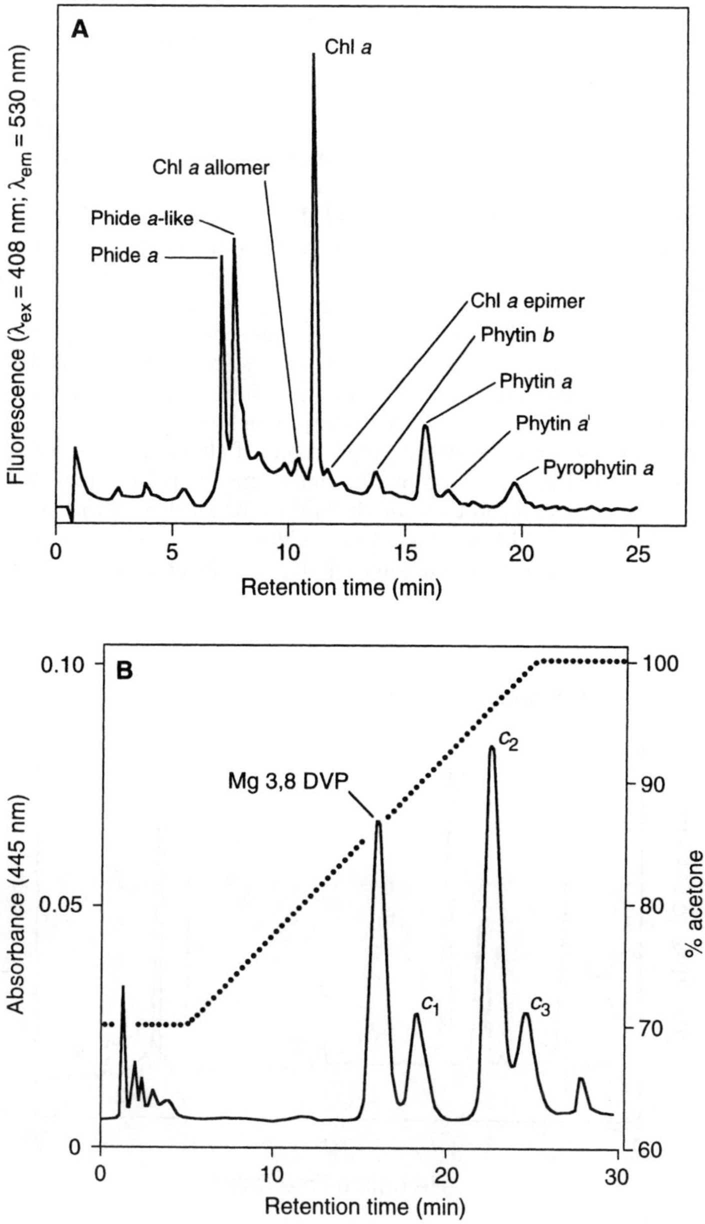

Figure 7.6
HPLC systems for chlorophylls and derivatives. A. Step-isocratic HPLC separation of chlorophyll *a* and derivatives on a C18 column (see Chapter 11). B. HPLC separation of four chlorophyll *c* pigments on polyethylene columns (PRX-1025, Polysciences Inc.). Solvent system: 70% (v/v) aqueous acetone (0–5 min); linear gradient 70–100% acetone (5–25 min), 100% acetone (25–30 min; from Jeffrey, 1989).

a problem compared to the excellent resolution obtained by the thin-layer methods described above. However, Garrido and colleagues, using polymeric octadecyl silica columns, have now improved HPLC chlorophyll *c* resolution, and they have identified several new chlorophyll *c* derivatives (Garrido and Zapata, 1993; Garrido *et al.* 1995). These exciting new observations extend the number of known chlorophyll *c* pigments to at least eight, and further work should establish their chemistry, distribution and function.

Judicious use of these TLC, HPTLC and HPLC methods can rapidly determine whether chlorophyll preparations are free from contamination by other chlorophylls, chlorophyll derivatives and carotenoids. Two types of ubiquitous alteration products — chlorophyll *a* epimers and allomers — are also readily separated by reversed-phase HPLC (see Fig. 7.6, 7.7 and Chapter 12). Chlorophyll *a* and *b* epimers (a' and b'), which show a configurational change at C–13^2 (stereoisomers), are also readily resolved on sucrose and cellulose TLC plates, running just ahead of chlorophylls *a* and *b* in the system shown in Fig. 7.5A (Jeffrey, unpublished). This particular separation is important, since commercial preparations of chlorophylls *a* and *b* usually contain up to 20% of the total chlorophyll as the epimer and also some 13^2-hydroxy chlorophylls and allomers, and these products must be purified before use (Jeffrey, 1981). The small spectral differences between chlorophyll *a* and its epimer, and the lack of spectral difference between chlorophyll *b* and its epimer, are clearly illus-

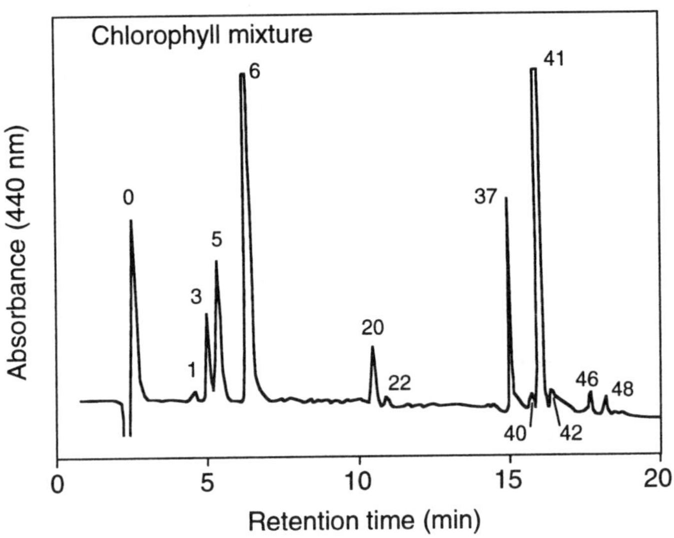

Figure 7.7
HPLC separation of chlorophylls and derivatives on a reversed-phase C-18 column (see Wright *et al.* 1991 and Chapter 12; pigment numbering system as in these references). Fraction 1, chlorophyllide *b*; 3, chlorophyllide *a*; 5, chlorophyll c_3; 6, chlorophyll c_1 + c_2 + Mg 3,8 DVP; 20, pheophorbide *a*; 22, pheophorbide *a*-like; 37, chlorophyll *b*; 40, chlorophyll *a* allomers; 41, chlorophyll *a*; 42, chlorophyll *a* epimer; 46, pheophytin *b*; 48, pheophytin *a*.

trated in Hynninen (1973). Additional properties of the epimers, and methods for their large-scale preparation, are described by Hynninen and Lötjönen (1983).

The allomerization reaction, first observed by Willstätter and Stoll (1913), includes a complicated series of oxidation reactions at C–13^2 by triplet oxygen in alcohol solutions (see Chapter 3). These reactions, which include the formation of 13^2-hydroxy chlorophylls, can occur during isolation of chlorophyll, and these products can then contaminate the preparation. Seven allomerization products of chlorophyll *a* were resolved by normal-phase HPLC (Kuronen *et al.* 1993). Chemical structures, the possible mechanisms of formation and UV-visible spectra are given in this paper.

The next section presents methods for the large-scale preparation of chlorophylls *a*, *b*, c_1 and c_2, while the following section describes preparation of 17 chlorophylls and derivatives on a small scale. As shown in all the preparation schemes (Figs. 7.8 to 7.14) both spectral and chromatographic accuracy are essential for establishing 'purity' of chlorophyll calibration standards.

7.5 Large-scale preparations of chlorophylls

Large-scale (>10 mg) preparations of chlorophylls are needed to

- determine extinction coefficients
- analyse the structure of new pigments
- prepare derivatives
- identify unknowns

Classical column chromatography of extracted leaf or algal pigments followed by crystallization procedures (based on Strain *et al.* 1944; Strain, 1958; and Strain *et al.* 1963) are the usual methods employed. They were adapted by Hynninen and Lötjönen (1983) for preparing mg quantities of chlorophyll epimers and by Porra *et al.* (1989) for re-determining extinction coefficients of chlorophylls *a* and *b*.

7.5.1 *Chlorophylls a and b*

The method outlined below (and summarized in Fig. 7.8) is based on that of Strain *et al.* (1963). It was adapted by the present author originally for purifying chlorophylls *a* and *b* for extinction coefficient measurements needed for deriving trichromatic spectrophotometric equations for chlorophyll analysis (Jeffrey and Humphrey, 1975). While Strain *et al.* (1963) used spinach (200 gm portions) as source material and sucrose columns, the present author used 100 gm portions of green seaweed (*Ulva lactuca*, sea lettuce) as source material, and cellulose columns for easier handling. *Ulva lactuca* lacks an acid cytoplasm, which avoids pheophytin formation during extraction.

All operations were performed speedily in darkened fume hoods. When storage was necessary during procedures, pigments were placed in the dark at -20°C. The plant material (in this case seaweed blades) was initially dipped for a few seconds in boiling seawater buffered with 1.5% ammonium acetate, then removed rapidly and chilled in ice-cold seawater, damp-dried on filter paper and extracted for several minutes with methanol:petroleum ether (b.p. <40°C), 5:1 (v/v). The 'black' extract was collected, and the pale-green fronds were re-extracted in acetone until colourless (not more than 5 min). Some authors prefer to use 80% aqueous acetone for

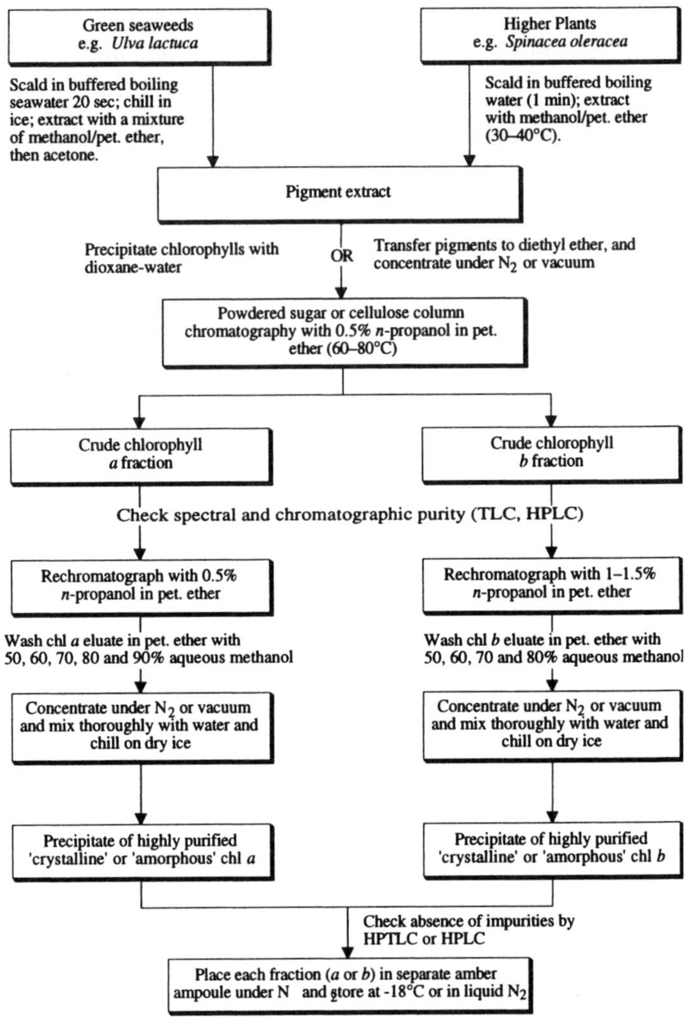

Figure 7.8
Preparation of highly purified 'crystalline' chlorophylls *a* and *b* from algae and higher plants.

extraction, buffered at pH7.8 with 2.5 mM sodium phosphate, to minimize conversion of chlorophylls to pheophytins (Porra, 1990).

The extracts were combined, filtered, and transferred to petroleum ether by gently mixing with an excess of 10% cold NaCl solution. The petroleum ether upper layer was collected, and concentrated to a small volume under N_2 or vacuum (rotary evaporator), ready for application to the sucrose or cellulose column. For all details of solvent and 'wash' volumes, column sizes etc. the original literature should be consulted (e.g., Strain *et al.* 1963; Jeffrey, 1972).

Alternatively, the chlorophylls can be precipitated by adding dioxane and water to pigment extracts in aqueous methanol or acetone solution (Iriyama *et al.* 1974), leaving most of the unwanted carotenoids in solution (Porra *et al.* 1989). Omata and Murata (1983) used columns of DEAE-Sepharose CL-6B and Sepharose CL-6B for obtaining 'clean' preparations of chlorophylls *a* and *b* from spinach, free from carotenoids, pheophytins, chlorophyllides and plastoquinones.

The concentrated pigment extract was then placed on the top of either a sucrose column (packed dry and washed with a little petroleum ether, b.p. 60–80°C) or a cellulose column (Whatman CC31; packed as a slurry in petroleum ether; b.p. 60–80°C). The pigments were washed into the column with a little petroleum ether, and the column developed with 0.5% *n*-propanol in petroleum ether. The carotenes, non-polar lipid, pheophytin *a* and chlorophyll *a*' eluted off the column before chlorophyll *a* (see Fig. 7.1) and were discarded so a 'clean' fraction of chlorophyll *a* could be collected. The chlorophyll *b*, which remained on the column adsorbed to the cellulose or sucrose, was removed by allowing the column to run dry and the powder extruded by gentle pressure from N_2 from a gas cylinder. The chlorophyll *b* zone was collected, packed into a fresh column, and eluted with 4% ethanol in petroleum ether. After each column purification, spectroscopic and TLC checks were performed to monitor the purity of the preparations (see sections 7.4.2, 7.4.3). Both 'crude' fractions of chlorophyll *a* and *b* were separately rechromatographed as above to re-purify and remove contaminating carotenoids. Colourless impurities that had eluted from the sucrose and the cellulose, as well as remaining traces of carotenoids, were removed by extracting the separate petroleum ether solutions of chlorophylls *a* and *b* with successive portions of aqueous methanol: 50, 60, 70, 80 and 90% aqueous methanol for chlorophyll *a*; and 50, 60, 70 and 80% aqueous methanol for chlorophyll *b*. After concentrating under N_2 or vacuum, the purified petroleum ether solutions were deep blue (chlorophyll *a*) or deep emerald green (chlorophyll *b*), with a red fluorescence. These were gently shaken with water and chilled on dry ice, and the chlorophylls precipitated as semi-crystalline water-adducts. The spectroscopic and chromatographic purity of the products were again checked. The 'crystalline' preparations, after careful drying, were stored in amber ampoules in the dark at -20°C under nitrogen. For further details of these methods, the original papers should be consulted (e.g., Strain *et al.* 1963; Strain and Svec, 1966).

7.5.2 *Chlorophylls c_1 and c_2*

The method outlined below is that of Jeffrey (1969, 1972), developed for crystallizing chlorophylls c_1 and c_2 for measuring extinction coefficients (Jeffrey and Humphrey, 1975). Chlorophyll *c* was originally detected in brown seaweed and diatom extracts by Strain and Manning (1942), and was first separated by chromatography on thin-layers of polyethylene (Jeffrey, 1968*b*) into two components with characteristic spectral properties (Jeffrey, 1969; see Fig. 7.4). Large-scale purification of chlorophylls c_1 and c_2 from brown seaweed pigment extracts was subsequently achieved on columns of cellulose and polyethylene (PRX-1025; Polysciences Inc.).

The stages of purification of chlorophylls c_1 and c_2 are summarized in Fig. 7.9, and commented on briefly here. The same conditions used for chlorophylls *a* and *b* — speedy operations in a darkened fume hood, and storage at -20°C in the dark — should be adhered to for the chlorophyll *c* pigments (see section 7.5.1). If both chlorophylls c_1 and c_2 are needed, then the source material can be brown seaweeds

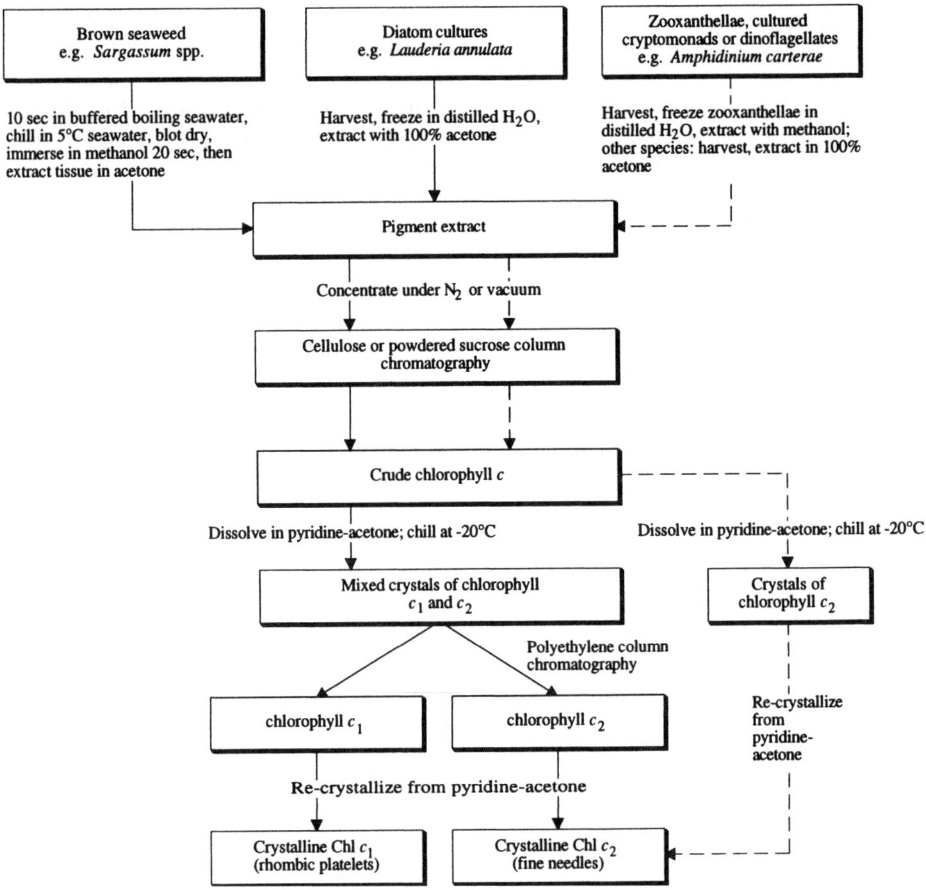

Figure 7.9
Preparation of crystalline chlorophylls c_1 and c_2 from marine algae.

with thin fronds (e.g., *Sargassum* spp.) or diatom cultures (e.g., *Lauderia annulata*). If only chlorophyll c_2 is required then an alga containing only this pigment (see Jeffrey, 1976) is recommended — a cryptomonad (e.g., *Chroomonas salina*) or a free-living or symbiotic dinoflagellate (e.g., *Amphidinium carterae* cultures, or *Symbiodinium microadriaticum* isolated from corals or clams; see Jeffrey, 1972). Tropical clam mantle tissue (e.g., *Tridacna* spp.) provides a rich source of dinoflagellates; these must be freed from animal tissue before extraction (Jeffrey and Haxo, 1968).

Preparation of extracts

Seaweed fronds (300–700 gm) were immersed for 10 sec in boiling seawater containing 1.5% ammonium acetate, chilled immediately in ice-cold seawater, damp-dried on filter paper, and extracted with acetone, or a mixture of methanol:diethyl ether:petroleum ether (b.p. <40°C; 5:2:1, v/v/v). The pale green fronds were extracted once or twice more in acetone until almost colourless (5 min). The black crude pigment extracts were combined, filtered, and the pigments transferred to diethyl ether by gently mixing 500 ml portions of extract, *plus* 300 ml diethyl ether with 4 l cold (5°C) 10% NaCl in a 5 l separating funnel. The ether extracts were collected, concentrated under N_2 or *in vacuo* to about 100 ml and stored in the dark at 5°C before cellulose or sucrose column chromatography.

Microalgal cultures (e.g., 3 × 40 l *Lauderia annulata*, *Amphidinium carterae* or *Chroomonas salina*) were harvested by continuous centrifugation; the diatom pellet was frozen for several hours to weaken the cell walls before extraction. Pigments were extracted with 100% acetone (with sonication or grinding), and transferred to diethyl ether as described above for *Sargassum*.

Symbiotic dinoflagellates were harvested from clam (*Tridacna crocea*) mantle tissue (Jeffrey and Haxo, 1968) and frozen in distilled water for several hours to weaken the tough cell wall. Pigments were then readily extracted in 100% methanol, and were transferred to diethyl ether as described above.

Separation of 'crude' chlorophyll c fraction

The chlorophyll *c* fraction was readily obtained by chromatogaphy on either cellulose or sucrose columns (see Fig. 7.1). Being very polar, it remains at the top of the column, and can easily be washed free from other pigments. Cellulose powder (Whatman CC21) was packed as a slurry in petroleum ether (60–80°C fraction). A suitable portion of the concentrated pigment extract in diethyl ether was diluted with an equal volume of petroleum ether, and applied to the top of the column. The column was developed with 0.5% *n*-propanol in petroleum ether (60–80°C), which washed the carotenes, chlorophyll *a* and most of the xanthophylls from the light-green zone of chlorophyll *c*. Traces of fucoxanthin (diatoms, seaweeds) or peridinin (dinoflagellates) were removed from the pale green zone by washing with small volumes of 2% *n*-propanol in light petroleum. If traces of chlorophyllide *a* (blue-green zone) contaminated the chlorophyll *c*, this could be readily removed by washing with 2% *n*-propanol in petroleum ether (b.p. <40°C). The chlorophyll *c* was then concentrated to a narrow band with 5% methanol in diethyl ether, the chlorophyll *c* zone collected, packed into a fresh column and eluted with the same solvent.

The 'crude' emerald green chlorophyll *c* solution was reduced to a small volume under nitrogen (or *in vacuo*), and the chlorophyll *c* was precipitated by adding 3–4 volumes of petroleum ether (b.p. 60–80°C) and leaving overnight at 5°C. The red or black precipitate was collected, dissolved in the minimum volume (e.g., 1 ml or less) of pyridine (black solution with an intense red fluorescence), diluted with 5–10 volumes of acetone (deep emerald-green solution with a red fluorescence), and left overnight at -20°C for the chlorophyll *c* to crystallize. In the case of seaweed and diatom extracts, mixed crystals of chlorophylls c_1 and c_2 resulted, and in the case of dinoflagellate and cryptomonad extracts, fine needles of chlorophyll c_2 were obtained. The latter, after being checked for spectral and chromatographic purity (see sections 7.4.2 and 7.4.3), usually needed no further purification other than re-crystallization. Crystals were either black or red to the naked eye, the latter being due to their intense red fluorescence. Under the light microscope the crystals were green by transmitted light and had a red fluorescence (see Plate 4B, Chapter 6).

Separation of chlorophyll c_1 and c_2

The mixed crystals of chlorophylls c_1 and c_2 from the seaweed or diatom extracts were dissolved in the minimum volume of pyridine and diluted with about 10 volumes of acetone. This solution was applied to the top of a dry-packed polyethylene column (PRX-1025, Polysciences Inc.), which was then developed with acetone until two light green zones separated. Chlorophyll c_1 was the faster moving zone. Development was stopped, the column sucked dry, and the polyethylene extruded as a column under N_2 pressure. The two fractions were 'cut' and collected, dried in the dark, repacked into fresh columns and eluted with pyridine. After drying each solution *in vacuo* (rotary evaporator), the residues were dissolved in the minimum volume of pyridine and crystallized from acetone at -20°C as above. Chlorophyll c_1 crystallized as rhombic platelets; chlorophyll c_2 as fine needles. The crystals were collected by centrifugation, washed with aqueous 50% acetone, and dried in the dark at room temperature over silica gel.

Because highly purified chlorophylls c_1 and c_2 are insoluble in diethyl ether, acetone or methanol, extinction coefficients and spectral properties were taken by dissolving crystals in a drop of pyridine and making up to volume with appropriate solvents (see Jeffrey, 1972).

Large-scale preparations of the recently described chlorophyll c_3 (Jeffrey and Wright, 1987), chlorophyll $c_{CS\text{-}170}$ (Jeffrey, 1989), and the phytylated chlorophyll *c* (Nelson and Wakeham, 1989) have not yet been published. Adaptations of the above methods should be appropriate for chlorophylls c_3 and $c_{CS\text{-}170}$, while the phytylated chlorophyll *c*, which elutes near to chlorophyll *a* on HPLC columns (see Jeffrey and Wright, 1994), would probably be amenable to large-scale purification by modified methods for chlorophylls *a* and *b* (section 7.5.1). It is hoped that large-scale preparation of these chlorophyll *c* pigments will soon be attempted, so that extinction coefficients for accurate quantitative analysis become available.

7.6 Small-scale preparations

Small-scale (µg) preparations of spectrally and chromatographically pure chlorophylls and derivatives are described in this section. These were prepared at the SCOR Workshops for calibrating the HPLC instruments, obtaining accurate spectral and chromatographic data (see the Part IV Data and Graphics sheets), and testing the accuracy of spectrophotometric, fluorometric and HPLC methods from accurately prepared pigment mixtures of known concentration (see Chapter 14). The methods used were a combination of TLC, HPTLC and semi-preparative HPLC, and the source materials were either commercially available pigments that were repurified, logarithmic-phase microalgal cultures (see Chapter 6), or other plant tissues (see Table 7.1).

In this section, purification schemes for chlorophylls and derivatives are outlined (Figs. 7.10–7.14), supplemented with brief descriptions of the procedures.

7.6.1 *Chlorophylls a and b*

Dunaliella tertiolecta (CS-14) was the source of chlorophylls *a* and *b* (Fig. 7.10), but when chlorophyll *a* alone was required, then a red alga (e.g., *Porphyridium purpureum*), a cyanophyte (e.g., *Synechococcus* sp.), or a eustigmatophyte (e.g., *Nannochloropsis occulata*) were all good source material (see Table 2.3, Chapter 2). Healthy logarithmic-phase cultures were used to minimize the chlorophyll degradation products. Cultures (10–50 ml, depending on cell density) were harvested by fil-

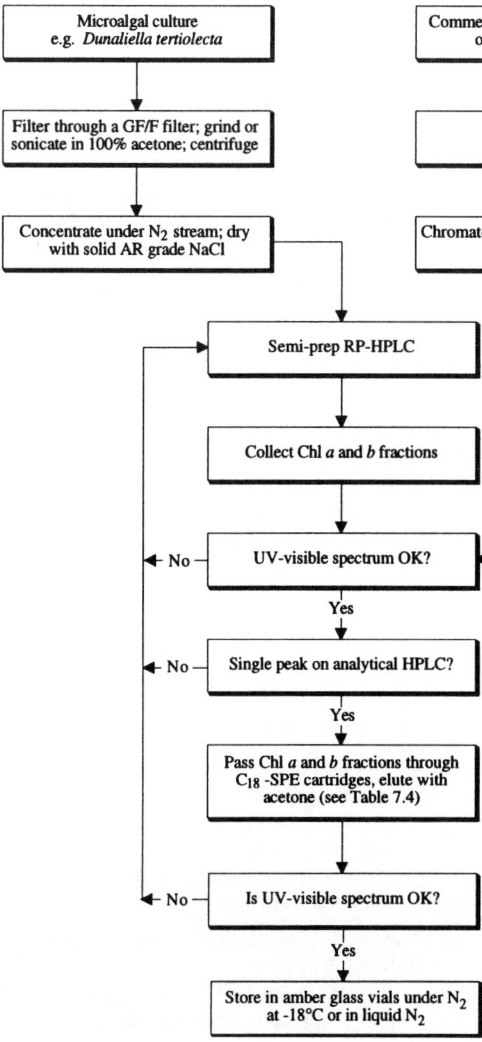

Figure 7.10
Small-scale preparation of microgram quantities of chlorophylls *a* and *b* for calibration standards.

tering on a GF/F filter, and immediately extracted by sonicating or grinding in 100% acetone (see Chapter 9). The extract was centrifuged, concentrated under a stream of nitrogen, and dried with solid AR grade NaCl (Na_2SO_4 is not recommended, since it sometimes adsorbs pigments). Alternatively, the ether solution can be dried by placing in a deep freeze (at -15° to -20°C) for 30 minutes, the ether decanted, leaving the water as ice on the sides of the vessel (Dr R. Porra, pers. comm.). An appropriate sample (e.g., 100 µl, see Chapter 13) in a suitable solvent was injected on to a C-18 column, developed according to Wright *et al.* (1991) (see Chapter 12), and the chlorophyll *a* and *b* fractions collected (see Fig. 7.7). These should be free of epimers and allomers that run just before and behind the main chlorophyll peak, respectively, in this HPLC system. The visible spectra in eluant were checked to ensure the purity of the fractions (see Part IV Data and Graphics sheets and Fig. 7.3A), which were rechromatographed if necessary. To transfer the chlorophyll *a* and *b* solutions from HPLC eluant to a standard solvent (e.g., acetone) they were passed through a C_{18} Solid Phase Extraction (SPE) cartridge, according to the protocols in Table 7.4, and eluted with acetone. If the UV-visible spectra matched those in the Part IV Data and Graphics sheets and Table 7.3 (to within ±1 nm for λ_{max} and 5% for band ratios), the chlorophylls were deemed chromatographically and spectrally 'pure', and were stored in amber glass vials under nitrogen at -20°C or in liquid nitrogen. It should be noted that pigments obtained by these methods may also contain colourless substances (e.g., lipids), and so are not chemically 'pure' in the same way crystalline samples are. Nevertheless they are perfectly suitable for HPLC calibration (see Chapter 16).

If commercial sources of chlorophylls *a* or *b* are used (e.g., Sigma chlorophylls), they should be purified by either HPLC, TLC (Jeffrey, 1981) or column chromatography (Strain *et al.* 1963) before use, to free the chlorophylls from epimer and 13^2-hydroxy chlorophyll contamination (see section 7.4.3 and Table 7.1). Both these alteration products which form so readily are hard to distinguish from the parent chlorophylls by spectra alone.

TABLE 7.4. Protocols for the use of C_{18} Solid Phase Extraction (SPE) cartridges for the transfer of pigments in HPLC eluate to reference solvent (e.g., acetone).

	Protocols
A. *Cartridge preparation*	Wash dry cartridge with • 1 cartridge volume of 100% methanol • 2 cartridge volumes of 100% acetone • 1 cartridge volume of distilled H_2O Do not allow the cartridge to run dry between washes
B. *Transfer of pigments from HPLC eluate to acetone* (a) *Non-polar pigments* (e.g., fucoxanthin to carotenes, see HPLC separation, Chapter 12)	1. Transfer eluant with pigment to graduated centrifuge tube (0.5 ml). 2. Fill to 2 ml with 50% acetone; mix with vortex. 3. Apply to top of washed SPE cartridge already prepared (see A above). 4. Pigment retained at top of column; dry with nitrogen. 5. Elute with 1 ml 100% HPLC grade acetone, accurately pipetted; flush with nitrogen. 6. Transfer to amber vial; store at -20°C.
(b) *Polar pigments* (e.g., chlorophyllides *a* and *b*, chlorophyll c_1+c_2, c_3)	1. Transfer eluant with pigment to graduated centrifuge tube (0.5 ml). 2. Fill to 2 ml with 30% acetone; mix with vortex. 3. Apply to top of washed SPE cartridge, already prepared (see A above). 4–6. As (a) above.

Note: Chlorophyll c_2 does not elute from the cartridge in acetone (B(a) 5), so diethyl ether was used. It was then evaporated under nitrogen and made up to 1 ml with HPLC grade acetone.

7.6.2 *Chlorophyllides a and b*

Chlorophyllides *a* and *b* are readily obtained from log-phase cultures of *Dunaliella tertiolecta* (CS-14), which contain an active chlorophyllase (Fig. 7.11). If only chlorophyllide *a* is required, then a good source is the chlorophyllase-rich diatom, *Phaeodactylum tricornutum* (see Barrett and Jeffrey, 1964; Jeffrey and Hallegraeff, 1987). The cultures (about 50 ml) are centrifuged, the pellet re-suspended in several ml 50% acetone, and incubated in the dark at 20°C for 30 min to allow de-phytylation to occur. In *Phaeodactylum*, all the chlorophyll *a* is converted to chlorophyllide *a* during this period. The reaction is terminated by adding acetone to a final concentration of 90%, the extract is centrifuged and the supernatant concentrated to a small

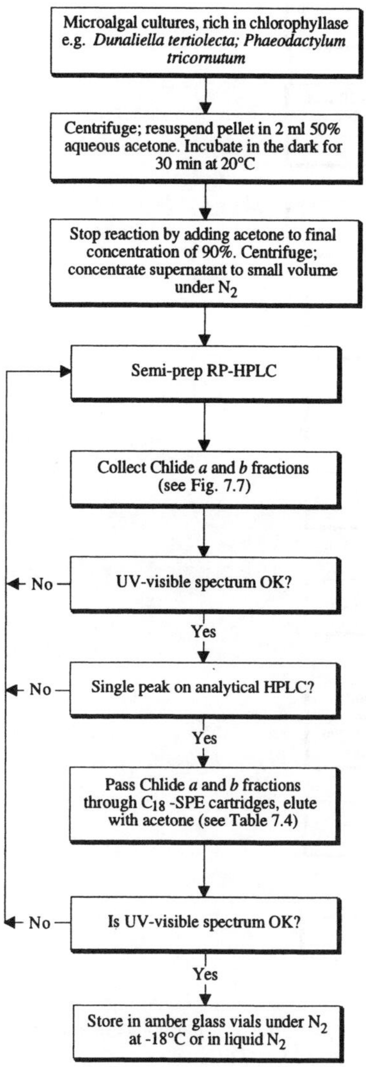

Figure 7.11
Small-scale preparation of microgram quantities of chlorophyllides *a* and *b* for calibration standards.

volume under N_2. After drying with solid Analytical Reagent grade NaCl, the chlorophyllides *a* and *b* are collected by semi-preparative HPLC (see Fig. 7.7, Table 7.1).

Chlorophyllide *a* can sometimes oxidize during the chlorophyllase reaction in *Phaeodactylum* preparations (see Fig. 7.3B). It is essential to ensure that this enzymatic reaction, thought to produce hydroxy chlorophyllides, does not occur if 'normal' chlorophyllide *a* is required (see Barrett and Jeffrey, 1964).

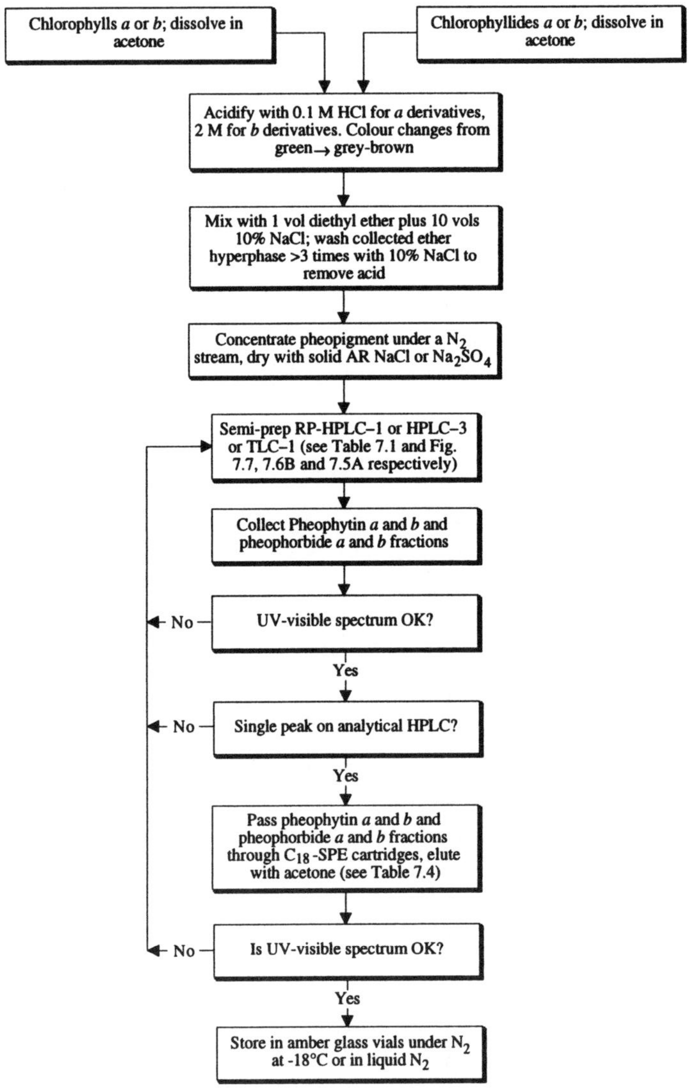

Figure 7.12
Small-scale preparation of microgram quantities of pheophytins *a* and *b* and pheophorbides *a* and *b* for calibration standards.

7.6.3 *Pheophytins a and b; Pheophorbides a and b*

Pheophytins *a* and *b* and pheophorbides *a* and *b* were prepared at the SCOR Workshops by acidification of chlorophylls *a* and *b* and chlorophyllides *a* and *b*, respectively (Fig. 7.12). The *a* series are de-metallated more readily than the *b* series, so different concentrations of HCl are required. The method will first be described for preparing pheophytin *a* from chlorophyll *a*.

Chlorophyll *a* (about 10–20 µg) was dissolved in acetone (about 5 ml) and 1 ml 0.1 M HCl added. After about 5 min the colour changed from deep blue-green to pale grey-brown indicating formation of pheophytin *a*. The reaction was stopped by adding 1 volume of diethyl ether. After 10 volumes of 10% cold NaCl solution were added and gently mixed, the pheophytin migrated to the diethyl ether hyperphase. The aqueous hypophase was discarded, and the ether extract was washed several more times until the pH of the aqueous phase was greater than 6. The ether layer containing the pheophytin *a* was then collected, concentrated under N_2 (or *in vacuo*) and dried over solid NaCl. The pheophytin *a* was chromatographed on either HPLC–1 (see Table 7.1; Fig. 7.7), HPLC–3 (Fig. 7.6B) or cellulose TLC–1 (Fig. 7.5A), to obtain a chromatographically 'pure' fraction. The UV-visible spectrum was recorded. The cycle of chromatography and checking of spectral properties was repeated until the pheophytin *a* was 'pure' by the criteria outlined in sections 7.4.2 and 7.4.3. The product was then passed through a C_{18} SPE cartridge (see Table 7.4) to transfer the pheophytin from HPLC eluant to acetone, the spectrum re-checked and the 'pure' product stored in an amber vial under N_2 at -20°C or in liquid N_2 (see Fig. 7.12). The same procedure was used to produce pheophorbide *a* from chlorophyllide *a*.

The methods for obtaining pheophytin *b* from chlorophyll *b* and pheophorbide *b* from chlorophyllide *b*, were similar, except that a longer incubation time (about 1 hour) and a higher acid concentration (2M HCl) were needed for de-metallation. More rigorous washing procedures were also needed to ensure the pheophytin *b* and pheophorbide *b* solutions were free from acid. Otherwise the preparative procedures were the same (Fig. 7.12).

Pheophytins *a* and *b* can also be prepared by dissolving the parent chlorophylls in glacial acetic acid, and then blowing the acid off with a jet of N_2 (Dr R. Porra, pers. comm.). Evaporation is rapid, leaving a dry film of the pheophytin, which can then be checked for chromatographic and spectroscopic purity.

Pheophytins *a* and *b* and pheophorbides *a* and *b* can also be prepared directly from unresolved mixtures of chlorophylls *a* and *b*, utilizing only the difference in the hydrolyzability of the *a* and *b* phytyl esters and differential liquid/liquid partition (Hynninen and Lötjönen, 1980). This method was not used at the SCOR Workshops, but since the products have high purity and yield, the method is recommended.

7.6.4 *Chlorophylls c_1 and c_2*

Two sources of biological material can be used for small-scale preparation of chlorophylls c_1 and c_2: microalgae (such as diatoms and prymnesiophytes) that have c_1 and c_2, and dinoflagellates or cryptomonads that contain only c_2 (see Table 2.3, Chapter 2). Both sources were used at the SCOR Workshops (see Table 7.1 and Fig. 7.13).

Logarithmic-phase cultures of the diatom *Lauderia annulata* and of the flagellates *Amphidinium carterae* and *Chroomonas salina* were harvested by filtration (GF/

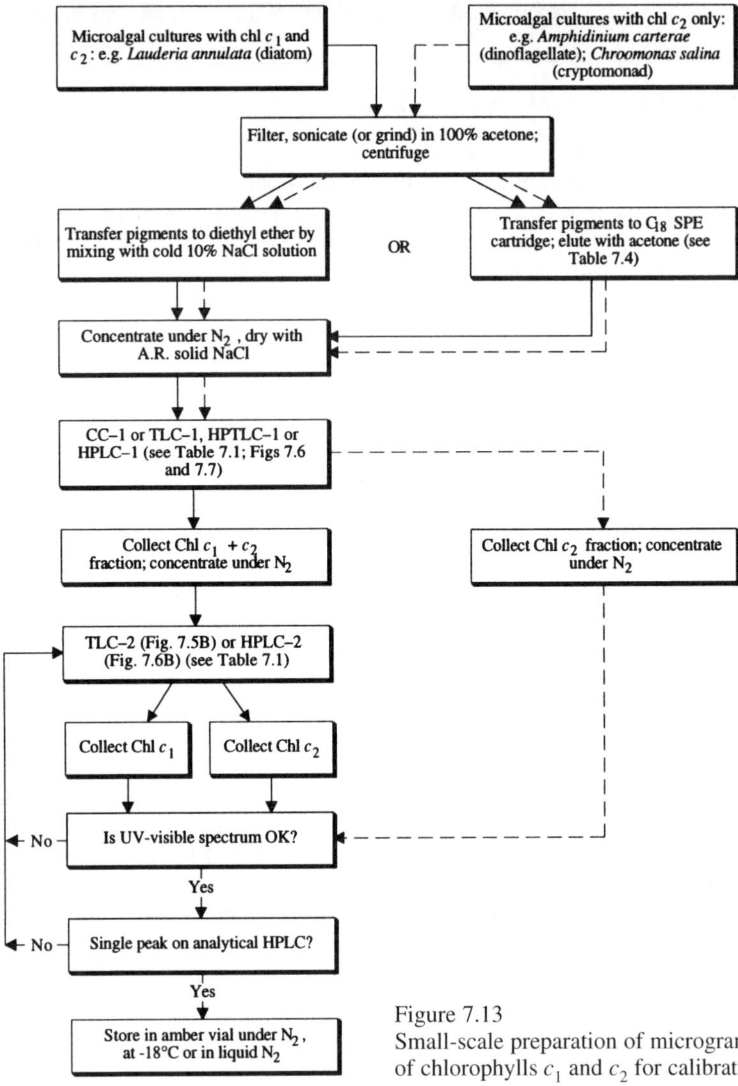

Figure 7.13
Small-scale preparation of microgram quantities of chlorophylls c_1 and c_2 for calibration standards.

F filters), and the pigments extracted in acetone, with sonication or grinding. After centrifugation, diethyl ether was added to the acetone supernatant and the pigments were transferred to diethyl ether by mixing with cold 10% NaCl solution. Alternatively, the acetone extract was passed through a C_{18} SPE cartridge as described in Table 7.4. The extracts were concentrated under N_2, dried with solid AR NaCl and chromatographed on either a cellulose or sucrose column (CC–1), thin-layer plates (TLC–1 or HPTLC–1) or HPLC (see Table 7.1 and Figs. 7.1, 7.5A, 7.5C or 7.7 respectively). The chlorophyll $c_1 + c_2$ zone was collected (from the *Lauderia* extract), or the chlorophyll c_2 zone (from the *Amphidinium* or *Chroomonas* extracts). The former required another chromatographic step to separate the chlorophylls c_1 and c_2 (e.g., polyethylene TLC–2 or polyethylene HPLC–2, Fig. 7.5B or 7.6B, respectively). The separated pigments (c_1 and c_2) were eluted with acetone or diethyl ether, concentrated under N_2 and dried with solid AR NaCl. The preparations were given final chromatographic and UV-visible spectroscopy checks (Fig. 7.13), before being stored in amber vials under N_2 at -20°C.

7.6.5 *Chlorophylls c_3, $c_{CS\text{-}170}$ and Mg 3,8 DVP*

Chlorophyll c_3 is now known to be quite widely distributed in prymnesiophytes (Jeffrey and Wright, 1994) so a number of sources are available for its isolation. Chlorophyll $c_{CS\text{-}170}$, which is known only from the tropical prasinophyte, *Micromonas pusilla* (CS-170), is thought to be a close derivative of chlorophyll c_3 (see Appendix B; Fookes and Jeffrey, unpublished). Mg 3,8 DVP is found in many prasinophytes (see Table 2.3, Chapter 2), which are suitable biological sources for its isolation. In the SCOR Workshops, cultures of *Emiliania huxleyi* and *Pycnococcus provasolii* were the sources of chlorophyll c_3; *Micromonas pusilla* (CS-170) was the source of chlorophyll $c_{CS\text{-}170}$; and *Micromonas pusilla* (CS-86) of Mg 3,8 DVP (see Table 7.1 and Fig. 7.14).

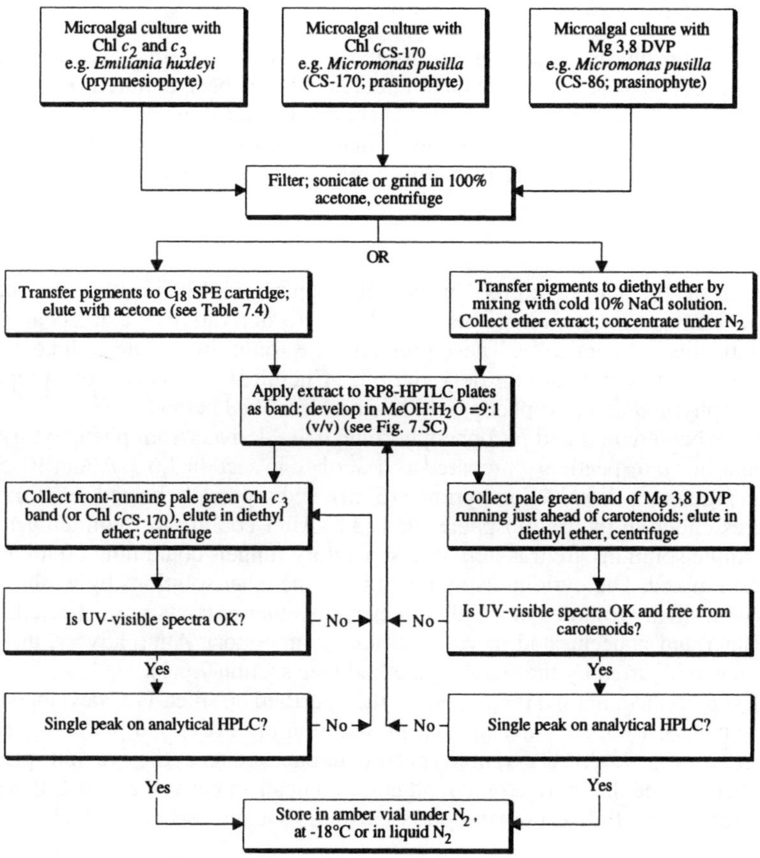

Figure 7.14
Small-scale preparation of microgram quantities of chlorophyll c_3, $c_{CS\text{-}170}$ and Mg 3,8 DVP for calibration standards.

Logarithmic-phase cultures were harvested by filtration on GF/F filters (50–100 ml, depending on cell density), and pigments were extracted by sonicating or grinding in acetone. After centrifugation, pigments were transferred to diethyl ether by adding 1 volume of diethyl ether and washing with 10 volumes of 10% NaCl solution. Alternatively, the extract was passed through a C_{18} SPE cartridge (see Table 7.4) and eluted with acetone. The extracts were concentrated under nitrogen, dried with solid NaCl, and streaked as a band on HPTLC Merck RP–8 plates, which were developed with methanol:H_2O = 9:1 (v/v) (Fig. 7.5C and Table 7.1). The front running pale-green zones of the three extracts (though with different R_F values, see Fig. 7.5C) were collected: chlorophyll c_3 (*Emiliania*), chlorophyll $c_{CS\text{-}170}$ (*Micromonas*, CS-170) and Mg 3,8 DVP (*Micromonas* CS-86).

The zones were scraped off the plates with a small spatula into centrifuge tubes and eluted with small volumes of acetone or diethyl ether. After centrifugation, the spectrum of each chlorophyll *c* fraction was checked, and if chromatographically and spectrally 'pure', they were stored in amber vials under N_2 at -20°C (see Fig. 7.14).

7.6.6 *Divinyl chlorophylls a and b*

Divinyl chlorophylls *a* and *b* are found in the free-living prochlorophyte, *Prochlorococcus marinus*. While the microalga has now been cultured (Partensky *et al.* 1993), and the chlorophylls identified (Goericke and Repeta, 1992), no culture was available to SCOR Working Group 78 in time for isolation details to be included in this monograph. The reader is therefore referred to the two papers above.

7.6.7 *Pyro-derivatives*

Removal of the carbomethoxy group at position C–13^2 on the chlorophyll cyclopentanone ring to form the so-called pyro-derivatives* can occur either enzymatically or chemically. These pigments are found in marine sediments, and result from phytoplankton degradation. Chemical methods of preparing pyropheophytin *a* and pyropheophorbide *a* are described below.

Pyropheophytin *a* and pyropheophorbide *a* are derived from pheophytin *a* and pheophorbide *a* respectively, prepared as described in section 7.6.3. About 10–50 µg of each pigment is dissolved in pyridine (5 ml), and separately heated to reflux temperatures under an inert atmosphere for 1–2 h. After cooling to room temperature, the pyridine solutions are transferred to separatory funnels containing 20 ml diethyl ether and mixed. The pyridine is removed from the ether solutions by washing 4–5 times with 80 ml volumes of distilled water, the ether extracts are collected, dried over NaCl and concentrated *in vacuo* (rotary evaporator). Alternatively, the ether solution can be dried by the freezing method (see section 7.6.1).

The pyropheophytin *a* preparation is then purified on silica TLC developed with acetone:hexane (5:95, v/v), with final purification on a reversed-phase C_{18} HPLC column (see Fig. 7.6A). UV-visible spectra are checked as in the previous preparation schemes, and the 'pure' compound is stored in an amber vial at -20°C under N_2. The pyropheophorbide *a* preparation is purified by cellulose TLC developed with

* This nomenclature is not approved by IUPAC but is widely used in the aquatic literature (see Part IV Data sheets for pyropheophytin *a*).

acetone:hexane = 3:7 (v/v), with final purification on a reversed-phase C_{18} HPLC column (solvent system 0.5 N ammonium acetate:methanol = 1:5 (v/v)). UV-visible spectra are checked for purity as in previous reaction schemes before storing the 'pure' compound in an amber vial at -20°C under N_2.

7.7 Extinction coefficients of chlorophylls and derivatives in 90% acetone

With all the above chlorophylls and derivatives carefully prepared and stored under N_2 at -20°C, they are now ready to use for HPLC calibration (see Chapter 16), or designing a range of pigment mixtures to check the accuracy of analytical methods (see Chapter 14). The 'best' extinction coefficients from the literature were assembled to accurately quantitate the pigments purified above. These are presented in a common solvent — 90% acetone — in Table 7.5. Extinction coefficients of chlorophylls and carotenoids in other solvents are listed in Appendix E.

Conclusion. Guidelines to the purification of various chlorophylls and derivatives are given in this chapter, both on a large scale to obtain crystalline products, and on a small scale to provide adequately purified pigments for HPLC calibration. Readers are also recommended to consult the original literature for further details.

Table 7.5. Extinction coefficients of chlorophylls and derivatives at the red band in 90% acetone used in the SCOR Workshops

Pigment	Molecular weight	Extinction coefficient in 90% acetone $l.g^{-1}\,cm^{-1}$	Source of extinction coefficients
Chlorophyll *a*	893.50	87.7	Jeffrey and Humphrey (1975)
Chlorophyllide *a*	614.97	127.0	Lorenzen and Jeffrey (1980); calculated
Pheophytin *a*	871.21	51.2	Lorenzen and Jeffrey (1980); calculated
Pheophorbide *a*	592.67	74.2	Lorenzen and Jeffrey (1980); calculated
Pyrochlorophyll *a*	835.48	87.5	B. Keely, Univ. Bristol (pers. comm.)
Pyropheophytin *a*	813.17	52.3	B. Keely, Univ. Bristol (pers. comm.)
Chlorophyll *b*	907.49	51.4	Jeffrey and Humphrey (1975)
Chlorophyllide *b*	628.97	74.1	Derived from Chl *b*, assuming constant molar extinction coefficients
Pheophytin *b*	885.20	31.8	N. Welschmeyer: calculated
Pheophorbide *b*	606.67	46.4	Derived from pheophytin *b*, assuming constant molar extinction coefficients
Pyrochlorophyll *b*	849.47	37.3	B. Keely, Univ. Bristol (pers. comm.)
Pyropheophytin *b*	827.18	28.6	B. Keely, Univ. Bristol (pers. comm.)
Chlorophyll $c_1 + c_2$		42.6	Jeffrey (1972)
Chlorophyll c_1	610.95	44.8	Jeffrey (1972)
Chlorophyll c_2	608.95	40.4	Jeffrey (1972)
Chlorophyll c_3	652.95	42.6	Not known; assumed similar to Chl $c_1 + c_2$
Mg 3,8 DVP	610.94	42.6	Not known; assumed similar to Chl $c_1 + c_2$

ACKNOWLEDGEMENTS

Thanks are due to Dr N. A. Welschmeyer for initiating the semi-preparative HPLC method for chlorophyll isolation at the SCOR Workshops, Dr D. J. Repeta for the method of pyro-derivative formation, and Dr R. F. C. Mantoura for contributing to the isolation schemes (Figs. 7.8–7.13). Dr P. Saenger assisted with the preparation of crystalline chlorophyll *b* from *Ulva lactuca* (section 7.5.1).

REFERENCES

Barrett, J., Jeffrey, S. W. (1964). Chlorophyllase and formation of an atypical chlorophyllide in marine algae. *Plant Physiol.* 39: 44–47.

Brockmann, H., Risch, N. (1991). Preparative chromatography. In: Scheer, H. (ed.), *Chlorophylls*, CRC Press, Boca Raton, Florida, p. 103–114.

Ettre, L. S. (1991). From Tswett to partition chromatography. *Int. Lab.* 21: 18–24.

French, C. S. (1960). The chlorophylls *in vivo* and *in vitro*, In: Ruhland, W. (ed.), *Encyclopedia of Plant Physiology*, vol. 5, Springer, Heidelberg, p. 252–297.

Garrido, J. L., Zapata, M. (1993). High performance liquid chromatography of chlorophylls c_3, c_1, c_2 and *a*, and of carotenoids of chromophyte algae on a polymeric octadecyl silica column. *Chromatographia*, 35: 543–547.

Garrido, J. L., Zapata, M., Muniz, S. (1995). Spectral characterization of new chlorophylls isolated from *Emiliania huxleyi* (Prymnesiophyceae) by high performance liquid chromatography. *J. Phycol.* 31: 761–768.

Goericke, R., Repeta, D. J. (1992). The pigments of *Prochlorococcus marinus*: the presence of divinyl chlorophyll *a* and *b* in a marine procaryote. *Limnol. Oceanogr.* 37: 425–433.

Harris, D. G., Zscheile, F. P. (1943). Effects of solvent upon absorption spectra of chlorophylls *a* and *b*; their ultraviolet absorption spectra in ether solution. *Bot. Gaz.* 104: 515–527.

Hynninen, P. H. (1973). Chlorophylls. III Keto-enol tautomerisin of chlorophylls *a* and *b*: the nature of chlorophylls *a'* and *b'*. *Acta Chemica Scandinavica* 27: 1487–1495.

Hynninen, P. H. (1991). Chemistry of chlorophylls: modifications, In: Scheer, H. (ed.), *Chlorophylls*, CRC Press, Boca Raton, Florida, p. 145–209.

Hynninen, P. H., Lötjönen, S. (1980). Preparation of phorbin derivatives from chlorophyll mixture utilizing the principle of selective hydrolysis. *Synthesis*, 7: 539–541.

Hynninen, P. H., Lötjönen, S. (1983). Large-scale preparation of crystalline (10*S*) - chlorophylls *a* and *b*. *Synthesis*, 9: 705–708.

Iriyama, K., Ogura, N., Takamiya, A. (1974). A simple method for extraction and partial purification of chlorophyll from plant material, using dioxane. *J. Biochem.* 76: 901–904.

Jeffrey, S. W. (1968*a*). Quantitative thin-layer chromatography of chlorophylls and carotenoids from marine algae. *Biochim. Biophys. Acta*, 162: 271–285.

Jeffrey, S. W. (1968*b*). Two spectrally distinct components in preparations of chlorophyll *c*. *Nature*, 220: 1032–1033.

Jeffrey, S. W. (1969). Properties of two spectrally different components in chlorophyll *c* preparations. *Biochim. Biophys. Acta*, 177: 456–467.

Jeffrey, S. W. (1972). Preparation and some properties of crystalline chlorophyll c_1 and c_2 from marine algae. *Biochim. Biophys. Acta*, 279: 15–33.

Jeffrey, S. W. (1976). The occurrence of chlorophylls c_1 and c_2 in algae. *J. Phycol.* 12: 349–354.

Jeffrey, S. W. (1981). An improved thin-layer chromatographic technique for marine phytoplankton pigments. *Limnol. Oceanogr.* 26: 191–197.

Jeffrey, S. W. (1989). Chlorophyll *c* pigments and their distribution in the chromophyte algae. In: Green, J. C., Leadbeater, B. S. C., and Diver, W. L. (eds), *The Chromophyte Algae: Problems and Perspectives*, Clarendon Press, Oxford, p. 13–36.

Jeffrey, S. W., Haxo, F. T. (1968). Photosynthetic pigments of symbiotic dinoflagellates (zooxanthellae) from corals and clams. *Biol. Bull.* 135: 149–165.

Jeffrey, S. W., Hallegraeff, G. M. (1987). Chlorophyllase distribution in 10 classes of phytoplankton: a problem for chlorophyll analysis. *Mar. Ecol. Progr. Ser.* 35: 293–304.

Jeffrey, S. W., Humphrey, G. F. (1975). New spectrophotometric equations for determining chlorophylls *a*, *b*, c_1 and c_2 in higher plants, algae and natural phytoplankton. *Biochem. Physiol. Pflanzen*, 167: 191–194.

Jeffrey, S. W., Wright, S. W. (1987). A new spectrally-distinct component in preparations of chlorophyll *c* from the micro-alga *Emiliania huxleyi* (Prymnesiophyceae). *Biochim. Biophys. Acta*, 894: 180–188.

Jeffrey, S. W., Wright, S. W. (1994). Photosynthetic pigments in the Haptophyta. In: Green, J. C., Leadbeater, B. S. C. (eds), *The Haptophyte Algae*, Oxford University Press, p. 111–132.

Kuronen, P., Hyvärinen, K., Hynninen, P. H. (1993). High-performance liquid chromatographic separation and isolation of the methanolic allomerization products of chlorophyll *a*. *J. Chromatog.* 654: 93–104.

Lorenzen, C. J., Jeffrey, S. W. (1980). Determination of chlorophyll in seawater: report of intercalibration tests. *UNESCO Tech. Pap. Mar. Sci.* 35, 20 pp.

Martin, A. J. P., Synge, R. L. M. (1941). A new form of chromatogram employing two liquid phases, 1. A theory of chromatography; 2. Application to the micro-determinations of the higher monoamino acids in proteins. *Biochem. J.* 35: 1358–1368.

Nelson, J. R., Wakeham, S. G. (1989). A phytol-substituted chlorophyll *c* from *Emiliania huxleyi* (Prymnesiophyceae). *J. Phycol.* 25: 761–766.

Omata, T., Murata, N. (1983). Preparation of chlorophyll *a*, chlorophyll *b* and bacteriochlorophyll *a* by column chromatography with DEAE-Sepharose CL-6B and Sepharose CL-6B. *Plant and Cell Physiol.* 24: 1093–1100.

Owens, T. G., Falkowski, P. G. (1982). Enzymatic degradation of chlorophyll *a* by marine phytoplankton *in vitro*. *Phytochem.* 21: 979–984.

Partensky, F., Hoepffner, N., Li, W. K. W., Ulloa, O., Vaulot, D. (1993). Photoacclimation of *Prochlorococcus* sp. (Prochlorophyta) strains isolated from the North Atlantic and the Mediterranean Sea. *Plant Physiol.* 101: 285–296.

Porra, R. J. (1990). The assay of chlorophylls *a* and *b* converted to their respective magnesium-rhodochlorin derivatives by extraction from recalcitrant algal cells with aqueous alkaline methanol: prevention of allomerization with reductants. *Biochim. Biophys. Acta*, 1015: 493–502.

Porra, R. J., Thompson, W. A., Kriedemann, P. E. (1989). Determination of accurate extinction coefficients and simultaneous equations for assaying chlorophylls *a* and *b* extracted with four different solvents: verification of the concentration of chlorophyll standards by atomic absorption spectroscopy. *Biochim. Biophys. Acta*, 975: 384–394.

Sahlberg, I., Hynninen, P. H. (1984). Thin-layer chromatography of chlorophylls and their derivatives on sucrose layers. *J. Chromatog.* 291: 331–338.

Scheer, H. (1991). Structure and occurrence of chlorophylls. In: Scheer, H. (ed.), *Chlorophylls*, CRC Press, Boca Raton, Florida, p. 3–30.

Senge, M., Senger, H. (1988). Chlorination of isolated chlorophylls *in vitro*. *Photochem. Photobiol.* 48: 711–717.

Shioi, Y. (1991). Analytical chromatography of chlorophylls. In: Scheer, H. (ed.), *Chlorophylls*, CRC Press, Boca Raton, Florida, p. 59–88.

Shioi, Y., Beale, S. I. (1987). Polyethylene-based high performance liquid chromatography of chloroplast pigments: resolution of mono- and divinyl chlorophyllides and other pigment mixtures. *Anal. Biochem.* 162: 493–499.

Sievers, G., Hynninen, P. H. (1977). Thin-layer chromatography of chlorophylls and their derivatives on cellulose layers. *J. Chromatog.* 134: 359–364.

Smith, J. H. C., Benitez, A. (1955). Chlorophylls: Analysis in plant materials. In: Paech, K., Tracey, M. V. (eds). *Modern Methods of Plant Analysis*, vol. 4, Springer, Heidelberg, p. 142–196.

Strain, H. H. (1958). *Chloroplast pigments and chromatographic analysis*. 32nd Annual Priestley Lectures, Pennsylvania State University Press, USA, 180 pp.

Strain, H. H., Manning, W. M. (1942). Chlorofucine (chlorophyll γ), a green pigment of diatoms and brown algae. *J. Biol. Chem.* 144: 625–636.

Strain, H. H., Manning, W. M., Hardin, G. (1944). Xanthophylls and carotenes of diatoms, brown algae, dinoflagellates and sea anemones. *Biol. Bull.* 86: 169–191.

Strain, H. H., Svec, W. A. (1966). Extraction, separation, estimation and isolation of the chlorophylls. In: Vernon, L. P., Seely, G. R. (eds), *The Chlorophylls*, Academic Press, London, p. 21–66.

Strain, H. H., Thomas, M. R., Katz, J. J. (1963). Spectral absorption properties of ordinary and fully deuteriated chlorophylls *a* and *b*. *Biochim. Biophys. Acta*, 75: 306–311.

Suzuki, N., Saitoh, K., Adachi, K. (1987). Reversed-phase high-performance thin-layer chromatography and column liquid chromatography of chlorophylls and their derivatives. *J. Chromatog.* 408: 181–190.

Tswett, M. (1906). Adsorptionanalyse und chromatographische Methode. Anwendung auf die Chemie des Chlorophylls. *Bert. Bot. Ges.* 24: 384–393.

Willstätter, R., Stöll, A. (1913). *Untersuchungen über Chlorophyll, Methoden, Ergebnisse*. Berlin, 385 pp.

Wright, S. W., Jeffrey, S. W., Mantoura, R. F. C., Llewellyn, C. A., Repeta, D. J., Bjørnland, T., Welschmeyer, N. A. (1991). An improved HPLC method for the analysis of chlorophylls and carotenoids from marine phytoplankton. *Mar. Ecol. Progr. Ser.* 17: 183–196.

Ziegler, R., Blaheta, A., Guha, N., Schönegge, B. (1988). Enzymatic formation of pheophorbide and pyropheophorbide during chlorophyll degradation in a mutant of *Chlorella fusca* Shihira et Kraus. *J. Plant Physiol.* 132: 327–332.

Zscheile, F. P. (1934). A quantitative spectro-photoelectric analytical method applied to solutions of chlorophylls *a* and *b*. *J. Phys. Chem.* 38: 95–102.

8

Preparation of carotenoid standards

D. J. Repeta and T. Bjørnland

8.1 INTRODUCTION

One of the objectives of SCOR WG 78 was to 'provide advice on the establishment and provision of reference chlorophyll, carotenoid and phycobiliprotein standards for calibration of HPLC, fluorescence and absorbance methods' (see section 1.3.2). Only a few algal carotenoids are commercially available, either purified from biological sources or synthesized *de novo* (see Appendix J). Most reference carotenoids must therefore be obtained from biological sources, usually by the practitioner.

About 600 carotenoids have been isolated and structurally characterized (Straub, 1987). Much of this work was done by an exceptional group of chemists, who have been publishing in this field since the 1930s (Karrer and Jucker, 1950; Zechmeister, 1962; Isler, 1971; Davies, 1976; Britton and Goodwin, 1982; Krinsky *et al.* 1989). Approximately 50 carotenoids may be present in marine phytoplankton samples in varying amounts and frequencies. Progress in structure elucidation of algal carotenoids

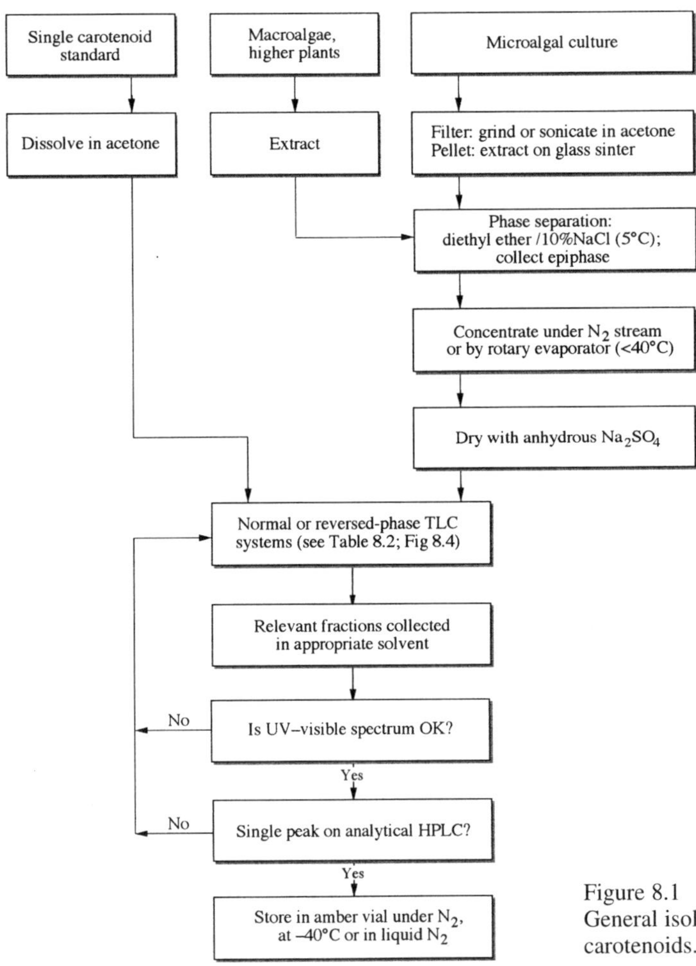

Figure 8.1
General isolation scheme for carotenoids.

TABLE 8.1. Biological sources and methods used for preparation of carotenoid pigment standards

Carotenoid	Microalgal source	Alternative source	Purification methods[a]
Alloxanthin	*Chroomonas salina* (CS-174)		TLC-1, RPTLC-1
Antheraxanthin	*Dunaliella tertiolecta* (CS-175)	*Spinacia oleracea*, leaves	RPTLC-2
Astaxanthin		synthetic (Hoffman La Roche)	none
19'-Butanoyloxyfucoxanthin	*Pelagococcus subviridis* (CS-99)		RPTLC-4
Canthaxanthin		synthetic (Fluka Chemical)	none
β,β-Carotene	*Pavlova lutheri* (CS-182)	*Daucus carota*, tubers; synthetic	TLC-2, TLC-3
β,ε-Carotene	*Porphyridium cruentum* (CS-25)	*Daucus carota*, tubers	TLC-2, TLC-3
β,ψ-Carotene	*Dunaliella tertiolecta* (CS-175)	synthetic	TLC-2
ε,ε-Carotene	*Pelagococcus subviridis* (CS-99)		TLC-2, TLC-3
Crocoxanthin	*Chroomonas salina* (CS-174)		HPLC-1
Diadinochrome	*Amphidinium carterae* (CS-212)		TLC-1, RPTLC-1
Diadinoxanthin	*Amphidinium carterae* (CS-212)		TLC-1, RPTLC-2
Diatoxanthin	*Amphidinium carterae* (CS-212)		TLC-1, RPTLC-2
Dinoxanthin	*Amphidinium carterae* (CS-212)		TLC-1, RPTLC-2
Echinenone		synthetic (Hoffman La Roche)	none
Ethyl 8'-apo-carotenoate		synthetic (Fluka Chemical)	TLC-1, RPTLC-2
Fucoxanthin	*Phaeodactylum tricornutum* (CS-29)	*Fucus vesiculosus*, fronds	RPTLC-4
19'-Hexanoyloxyfucoxanthin	*Emiliania huxleyi* (CS-57)		RPTLC-3
Lutein	*Dunaliella tertiolecta* (CS-175)	*Spinacia oleracea*, leaves	TLC-1, RPTLC-2
Lycopene	*Micromonas pusilla* (CS-86)	*Lycopersicon esculentum*, fruits	TLC-2
Monadoxanthin	*Chroomonas salina* (CS-174)		HPLC-1
9'-*cis*-Neoxanthin	*Dunaliella tertiolecta* (CS-175)	*Spinacia oleracea*, leaves	RPTLC-3
all-*trans*-Neoxanthin[b]	*Dunaliella tertiolecta* (CS-175)	*Spinacia oleracea*, leaves	CC, HPLC-2
P-457	*Amphidinium carterae* (CS-212)		RPTLC-2
Peridinin	*Amphidinium carterae* (CS-212)		TLC-1, RPTLC-2
Prasinoxanthin	*Pycnococcus provasolii* (CS-185)		RPTLC-2, HPLC-1
Siphonaxanthin		*Codium fragile*, fronds	RPTLC-4
Siphonein		*Codium fragile*, fronds	RPTLC-4
Vaucheriaxanthin	*Nannochloropsis oculata* (CS-216)		HPLC-1
Vaucheriaxanthin ester	*Nannochloropsis oculata* (CS-216)		HPLC-1
Violaxanthin	*Dunaliella tertiolecta* (CS-175)	*Spinacia oleracea*, leaves	RPTLC-3
Zeaxanthin	*Synechococcus* sp. (CS-197)	*Zea mays*	TLC-1, RPTLC-1

a. **Purification methods:** TLC-1: 30:70 acetone:*n*–hexane on activated silica plate; TLC-2: 100% *n*–hexane on activated silica plate; TLC-3: 100% *n*–hexane, on silica G: Kieselguhr: $Ca(OH)_2$:MgO=14:16:9:9; RPTLC-1: 100% methanol, Merck RP-8 HPTLC plate; RPTLC-2: 95% methanol (aq), Merck RP-8 HPTLC plate; RPTLC-3: 92% methanol (aq), Merck RP-8 HPTLC plate; RPTLC-4: 90% methanol (aq), Merck RP-8 HPTLC plate; CC:open column chromatography; 4% acetone in benzene on $CaCO_3$; HPLC-1: see Chapter 13; HPLC-2: *n*-hexane:*iso*-propylacetate: acetone=76:17:7 on a 250 mm analytical nitrile column (cyano-5).

b. Obtained by iodine-catalyzed stereomutation of 9'-*cis*-neoxanthin *ex S. oleracea*, and subsequent chromatography.

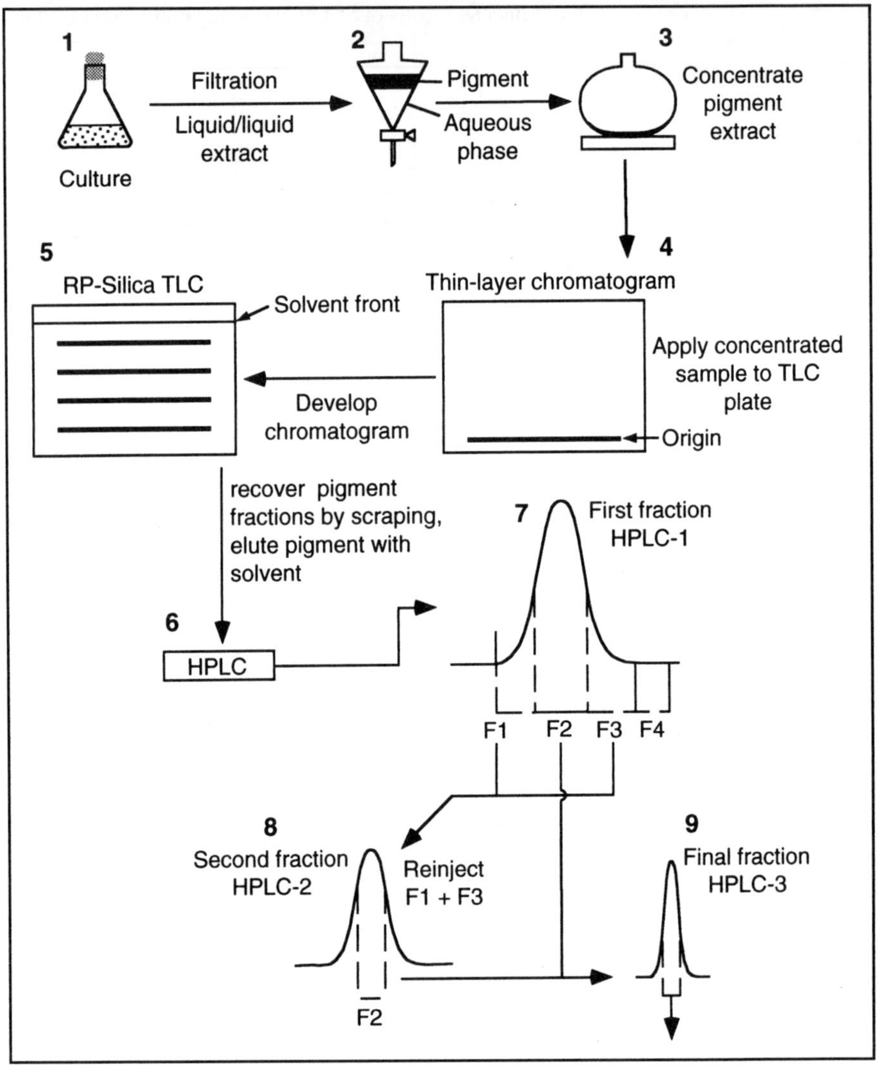

Figure 8.2
Isolation of standard carotenoids from microalgal cultures.

and knowledge of their taxonomic distribution have been reviewed frequently (Goodwin, 1971; Liaaen-Jensen, 1977, 1978 and 1985; Bjørnland and Liaaen-Jensen, 1989; Bjørnland, 1989; Egeland and Liaaen-Jensen, 1993). *International Symposia on Carotenoids* may also be consulted (e.g., Pure and Applied Chemistry, vols 14, 20, 47, 51, 57 and 66). Although modern analytical techniques and instrumentation have supplanted many of the methods described in the early literature, the general approach remains the same: extraction from biological tissues by organic solvents, chromatographic purification, crystallization and finally spectrometric characterization (Fig. 8.1). The original literature is a valuable introduction when formulating plans for preparation of reference carotenoids.

This chapter summarizes the methods used in the SCOR WG 78 carotenoid workshop to purify mg quantities of 29 carotenoids of chemosystematic importance in phytoplankton ecology, and three carotenoids of interest as internal standards. These carotenoids were used both as calibration standards for HPLC, for testing the resolving capacity of recommended HPLC systems (e.g., Chapter 12), and as single authentic pigments to confirm the identity of carotenoids in samples. Five carotenoids were synthetic, the rest were isolated from microalgae, macroalgae and higher plants. The carotenoids, their origin and the methods used for their isolation, are listed in Table 8.1. Properties of 30 carotenoids important in phytoplankton ecology are detailed in the Part IV Data and Graphics sheets.

The choice of methods for carotenoid isolation depends both on how much pigment is needed and what it is for. If carotenoids are to be used simply as chromato-

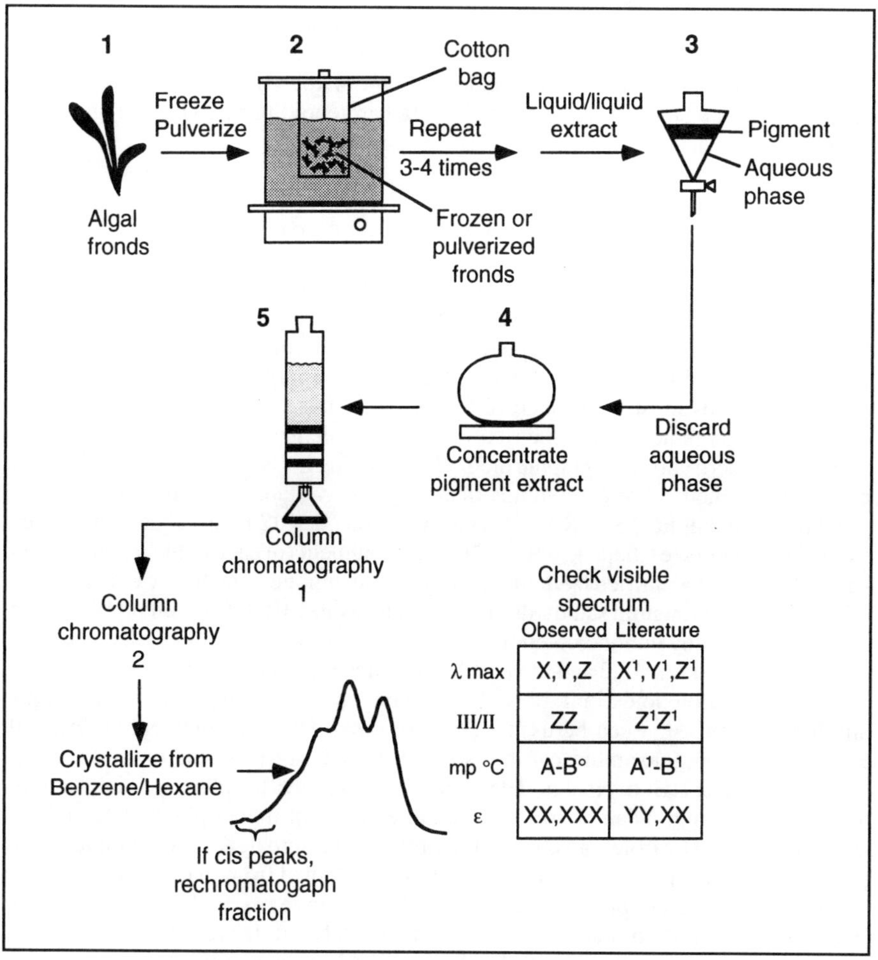

Figure 8.3
Isolation of standard carotenoids from macroalgae.

graphic standards, then the final product need only be free from other pigments ('spectrally pure'), and the small amounts (μg to mg) of material needed can be conveniently prepared by analytical HPLC (see Chapters 12, 13). Chapter 13 details the carotenoids that can be obtained in μg quantities from HPLC runs of extracts of the 12 SCOR reference algal cultures. If, however, the carotenoids are to be used for nuclear magnetic resonance spectrometry, for stable isotope analysis, or for the partial synthesis of degradation products, then the final isolate must be both 'spectrally pure' and free from all lipids. In such cases more material is needed, and much more rigorous purification is required. This chapter recommends general methods for isolation of

- small to moderate (1–3 mg) amounts of carotenoids from microalgae suitable as chromatographic standards (Fig. 8.2)
- larger amounts (>10 mg) of carotenoids from macroalgae and higher plants suitable for the applications listed above (Fig. 8.3).

Specific carotenoid isolations will obviously vary in detail depending on the biological source material, the amount of tissue extracted and the carotenoid to be isolated, as well as available laboratory equipment.

Selected general methods for obtaining chromatographically pure carotenoids in the crystalline state, criteria for evaluating carotenoid purity, and storage conditions are also discussed.

8.2 Sources of standard carotenoids

Only a limited number of synthetic carotenoids are commercially available: β,β-carotene, canthaxanthin, astaxanthin, ethyl 8'-apo-carotenoate, and 8'-apo-carotenal. Lycopene, lutein, zeaxanthin, and β,ε-carotene isolated from biological source materials are also available through specialty biochemical supply houses (Appendix J). All other carotenoids must be isolated from biological materials. These should be well characterized botanically and their carotenoid complement should have been determined by modern chromatographic methods, including mass or nuclear magnetic resonance spectrometry. For carotenoids that are only available from microalgae (e.g., peridinin, alloxanthin), SCOR WG 78 has recommended 12 microalgae as biological source materials (see Chapters 6, 13). The major pigments of each of the recommended strains have been identified by modern analytical techniques, and they were the source of many of the carotenoid standards prepared during the SCOR Carotenoid Workshop. All major and many minor carotenoids in these algae have been structurally characterized (see Tables 13.1, 13.2 and the Part IV Data and Graphics sheets).

Several carotenoids in Table 8.1 also have convenient sources in macroalgae and higher plants that can be used when more than 10 mg of pigment are needed. Fucoxanthin and zeaxanthin can be isolated in quantity from the brown seaweed *Fucus vesiculosus* (Bonnett *et al.* 1969; Haugan *et al.* 1992), and lutein, violaxanthin, and 9'-*cis*-neoxanthin can be isolated from the leaves of higher plants (Cholnoky *et al.* 1969). Several carotenoids, available only from microalgae, can be obtained by partial chemical synthesis of related carotenoids obtained from macroalgae or higher plant tissues. For example, all-*trans* neoxanthin can give rise to dinoxanthin by partial acetylation, and to diadinochrome by acid-catalyzed dehydration and furanoid rearrangement (Egger *et al.* 1969; Buchecker and Liaaen-Jensen, 1975).

When selecting biological source material for pigment isolation, it is important to recognize that carotenoids may occur naturally in a variety of optical (*R,S*) and geometrical (*cis/trans*) isomeric forms. Many of these isomers can be separated by modern chromatographic methods, particularly HPLC (Vecchi and Müller, 1979).

Optical isomerism may occur at any chiral carbon, most importantly at C-3, C-3' in carotenols, C-6' in lutein and β,ε-carotene, at the allenic position in fucoxanthin and peridinin, at C-5, C-6 in carotenoid epoxides and at C-8 in carotenoid furanoids. All theoretically possible optical isomers for lutein (8 in total) and astaxanthin (3 in total) have been isolated from biological material (Müller *et al.* 1982; Bernhard, 1990). However, in microalgae, only one optical isomer of each carotenoid is known with certainty, and this is the one drawn in the Graphics sheets, Part IV. Optical isomers are generally not interconvertable, e.g., by long term storage, and they are therefore considered different chemical compounds.

Geometrical isomerism results from rotation of conjugated double bonds, a reversible process that results in *cis/trans* isomers and may occur during long-term storage. Most naturally occurring carotenoids are all-*trans*; but a particular carotenoid may be present as a *cis* isomer (e.g., 9'-*cis*-neoxanthin (Bjørnland *et al.* 1987; Baumeler *et al.* 1994) in chlorophytes, higher plants and two euglenophytes). Carotenoids to be used as standards in oceanography and phytoplankton ecology must be in the relevant isomeric form.

SCOR reference cultures are recommended as the biological source material for all carotenoid standards used in oceanographic and phytoplankton ecology applications (see Tables 6.1, 13.1). These strains may be obtained from the Algal Culture Collections listed in Appendix L. If non-algal biological source materials are selected, then great care must be exercised to avoid the isolation of non-relevant isomers. The most recent literature should be consulted, and in all cases the chromatographic and spectroscopic properties of the carotenoid should be compared to standards isolated and characterized by SCOR WG 78 from the relevant SCOR algal culture (see Chapter 13, and Part IV Data sheets).

8.3 Carotenoid isolation

8.3.1 *General precautions*

Excellent discussions of analytical strategies and problems with isolation of carotenoids from higher plant tissues are given by Davies (1976), Rüedi (1985), Goodwin and Britton (1988), Packer (1992) and Britton and Young (1993).

Carotenoids vary in their chemical properties, but they should always be considered very labile. They are susceptible to degradation by light, heat, oxygen (air), acids and bases. Solvents, adsorbents and the plant tissues themselves are all potentially destructive agents. Carotenoids should always be handled in subdued light, preferably in fume hoods; evaporation of solvents should be done *in vacuo* below 40°C and pigments should be stored under N_2 at -20°C (or in liquid nitrogen). Equipment must be clean, and some solvents and reagents freshly purified. HPLC and specialty grade solvents are sufficiently free of contaminating lipids and nonvolatile residues to be used without redistillation. Even the best grades of solvents may contain trace amounts of acids (e.g., methylene chloride, chloroform, acetonitrile) or peroxides (e.g., tetrahydrofuran, diethyl ether) that may cause extensive degradation of the isolate when concentrated during rotary evaporation or during storage. These solvents must be redistilled and tested for residual acids and peroxides before use.

Many solvents also contain appreciable amounts of water, which must be removed when highly reproducible normal-phase chromatography or anhydrous derivatization reactions are required (see Perrin *et al.* (1988) for guidelines on purifying solvents and laboratory chemicals). Silica gel, alumina, and reversed-phase (RP) packing materials used in column and thin-layer chromatography must be cleaned, neutralized and (except RP packing) activated before use.

As with any natural product isolation or chemical derivatization, all new procedures should be tested on a small sample of the tissue or extract before scaling up to preparatory levels. Unless the specific chromatographic or other purification technique has been well established, it is recommended that the entire sample should not be run in a single step. This approach lengthens the analysis time, but it guards against catastrophic loss of the entire sample. Carotenoid isolation protocols for microalgal cultures and macroalgae are summarized in Figs. 8.1 to 8.3.

8.3.2 *Extraction*

a) Cultures

Cultures are grown according to procedures outlined in Chapter 6. Microalgae from 10–100 l cultures are harvested by continuous flow centrifugation or by filtration onto large-diameter (90 mm) glass fibre filters. Smaller -volume cultures (75 ml–2 l) are filtered as described in Chapter 12. The resulting pellet or filter is extracted several times with acetone or buffered methanol (Chapter 9) with grinding or sonication, until the extract and pellet are colourless (see Fig. 8.2).

b) Plant tissues

Macroalgae and higher plant tissues are handled somewhat differently from microalgae and there is no single, generally applicable protocol. Sometimes fresh leaves or seaweed fronds are used, but other methods start with frozen tissues (see Fig. 8.3). One such method will be described.

The leaves or fronds of spinach or *Fucus* (brown seaweed) are frozen with dry ice or liquid nitrogen, then pulverized with either a mortar and pestle, high speed laboratory blender, or ball mill (Fig. 8.3). The powdered sample is placed in a cotton bag and soaked in acetone or methanol overnight and mildly agitated. Three or more serial extractions are needed for optimum recovery of pigment. If the target compound is susceptible to acid-catalyzed degradation (epoxide rearrangement, ester hydrolysis), several grams of solid $NaHCO_3$ are added to the mixture to serve as a buffer. When sufficient pigment has been extracted, the cotton bag is removed and allowed to drain. If no bag is used, the sample may be filtered after extraction is complete. The solvent containing the pigment is then concentrated to 10% of the original volume by vacuum rotary evaporation.

The combined pigment extracts from microalgae or macroalgae sources are partitioned in a separatory funnel by adding an equal volume of peroxide-free diethyl ether to the sample concentrate and then diluting with ten volumes of cold (5°C) 10% (w/v aqueous) NaCl. The mixture is gently shaken (2 min), with frequent venting to release the pressure build-up, and the two layers allowed to separate. Pigments migrate to the ether (epiphase, top layer), which is collected. The aqueous phase (bottom layer) should be extracted at least once more with diethyl ether to ensure that all pigments have been transferred to the epiphase. The combined ether extract is washed twice with 5% $NaHCO_3$ to remove residual acids, followed by two additional washes with 5% NaCl, then dried over anhydrous Na_2SO_4 and filtered. The sample may be concentrated under nitrogen, and stored at -20°C (or liquid nitrogen) before use.

8.3.3 *Chromatography*

Great advances continue to be made in chromatographic supports and instrumentation, such that a wide variety of techniques are available to the analyst. The choice of chromatographic technique to be used in a carotenoid preparation depends on the purity and amount of pigment required. Classical column chromatography (Strain, 1958; Strain and Svec, 1966) is suitable for µg to g scale preparations. The method is most frequently used for a first 'rough' separation of an extract on adsorbant particles of relatively large diameter and size variation. The resulting fractions are often of inferior purity, and need to be purified further. Thin-layer chromatography has a limited sample capacity, but gives products of high purity. The technique is inexpensive, easy to use, and is suitable for µg to mg amounts of material. High-performance liquid chromatography (HPLC) achieves very high resolution, and therefore purity, but quantities are usually limited to less than 1 mg of material per HPLC run. Serial injections (Haugan *et al.* 1992) may be done on analytical columns and can yield 10 mg pigment per day, even with difficult separations. Preparative-scale HPLC systems suitable for mg to g purifications are available at a greater cost.

A large number of carotenoid purifications have been reported by TLC and HPLC on both reversed and normal phases (see Table 11.1). Although most discussions in this monograph are confined to reversed-phase chromatographic systems using mixtures of methanol, acetone and water as solvents, the analyst should be aware that a large number of stationary and mobile phases have been successfully exploited for different applications.

8.3.3.1 *Column chromatography*

Column chromatography is a classic and effective method for isolating pigments from biological samples (Strain, 1958) and is the method of choice for processing samples with over 10 mg of material. Chemically bonded reversed-phase adsorbents are limited to C-18, but all other packing materials common to TLC and HPLC are available for column chromatography. Columns can be eluted isocratically, or with a gradient of solvent polarity.

To pack a column, a 1–5 cm diameter glass column (depending on the amount of sample to be processed) is filled to one half the total volume with the initial mobile phase (solvent). A slurry containing 10–100 g of adsorbant mixed with the mobile phase is added and allowed to settle in a uniform bed (the ratio of packing:sample should be >100 by weight). Depending on the difficulty of separation, columns are packed to a height 5–10 times the column diameter. The column is then flushed with three column volumes of solvent to expel trapped air and to pack the adsorbent into a firm, stable bed. Finally, a 1–2 cm cap of glass microbeads (or sand) is added to guard against disturbing the bed during sample application and elution. The sample, dissolved in a minimum amount of mobile phase, is applied to the top of the column, and allowed to drain to the top of the adsorbant bed. This process is repeated with a rinse from the sample flask and with fresh mobile phase (solvent), which moves the sample fully into the column. The pigment is then eluted with fresh mobile phase, according to the relevant protocol.

Step-wise gradient elution is also satisfactory. Gradient chromatography on normal-phase adsorbents (silica, alumina) begins with a nonpolar solvent and proceeds to progressively more polar solvent mixtures. Columns with RP adsorbents are packed and eluted with a polar solvent and developed in progressively less polar solvent mixtures. Changes in solvent composition during a gradient separation sometimes result in the generation of heat at the solvent front due to exothermic displacement of nonpolar solvent from the stationary phase, and in columns more than 2 cm diameter this can be significant. The problem is particularly acute when mixtures of hexane and acetone are used, and can result in poor recovery of the pigment. In these situations, the heat must be dissipated by slow elution and cooling of the column.

Flash chromatography is a variation of conventional column chromatography that uses moderate pressure (5–10 psi) to push the mobile phase down the column. The application of pressure allows smaller and more uniformly sized silica particles (20–50 µm) to be used as the packing material, thereby increasing pigment resolution. When used properly, the technique achieves separations comparable to thin-layer chromatography, but can accommodate much higher loadings of sample (10 mg to 2 g). Best separations are achieved by isocratic elution of the pigment with a solvent system that yields an R_f value (see section 8.3.3.2) of 0.3–0.4 by TLC for the target pigment. Step-wise gradient elution may also be applied with satisfactory results. For large samples (>100 mg) repeated flash chromatographic purification of the target pigment may be carried out, using different mobile-phase combinations on progressively smaller diameter columns to obtain a pigment concentrate that can be further purified by crystallization. Speciality columns suitable for flash chromatography are available at moderate cost. This technique is discussed by Still *et al.* (1978).

8.3.3.2 *Thin-layer chromatography*

Thin-layer chromatography (TLC) is a technique that uses a stationary phase (adsorbant) applied as a thin film coating (0.2–1 mm) on a glass, plastic or aluminum plate. After the sample is applied, the plate is placed in a developing tank, and components in the sample are separated as the mobile phase migrates up the plate by capillary action (Figs. 8.2, 8.4). Thin-layer plates coated with silica, alumina, cellulose, polyamide, and a variety of reversed phases (C-18, C-8, C-2, cyano, amino, phenyl) are commercially available. Some researchers prefer to make their own plates if certain plate sizes or adsorbants are not available commercially. The TLC technique is described in a number of excellent monographs on chromatography (Stahl and Mangold, 1975; Poole and Schuete, 1984; Hamilton and Hamilton, 1987; Poole and Poole, 1991). Systems adapted for the separation of carotenoids are given by Hager and Meyer-Bertenrath (1966), Davies (1976) and Britton and Young (1993).

The TLC systems used by SCOR WG 78 to separate and purify 1–3 mg quantities of various carotenoid standards from the SCOR reference cultures are listed in Table 8.1. Two TLC systems were used to purify each pigment. The greatest selectivity was achieved by combining normal- and reversed-phase systems. In some instances where the pigment complement of the source organism was well characterized, one chromatographic system was sufficient (e.g., 19'-hexanoyloxyfucoxanthin from *Emiliania huxleyi*; Merck RP-8 HPTLC plate; solvent methanol: H_2O = 9:1 (v/v))

All TLC plates should be cleaned before use. Silica plates are washed by developing with 100% acetone or acetone:methanol (1:1, v/v), air-dried, then activated at 140°C for 4 h. Contaminants that migrate to the top of the plate are isolated from the sample by removing a band of the adsorbent (= scoring) approximately 1 cm

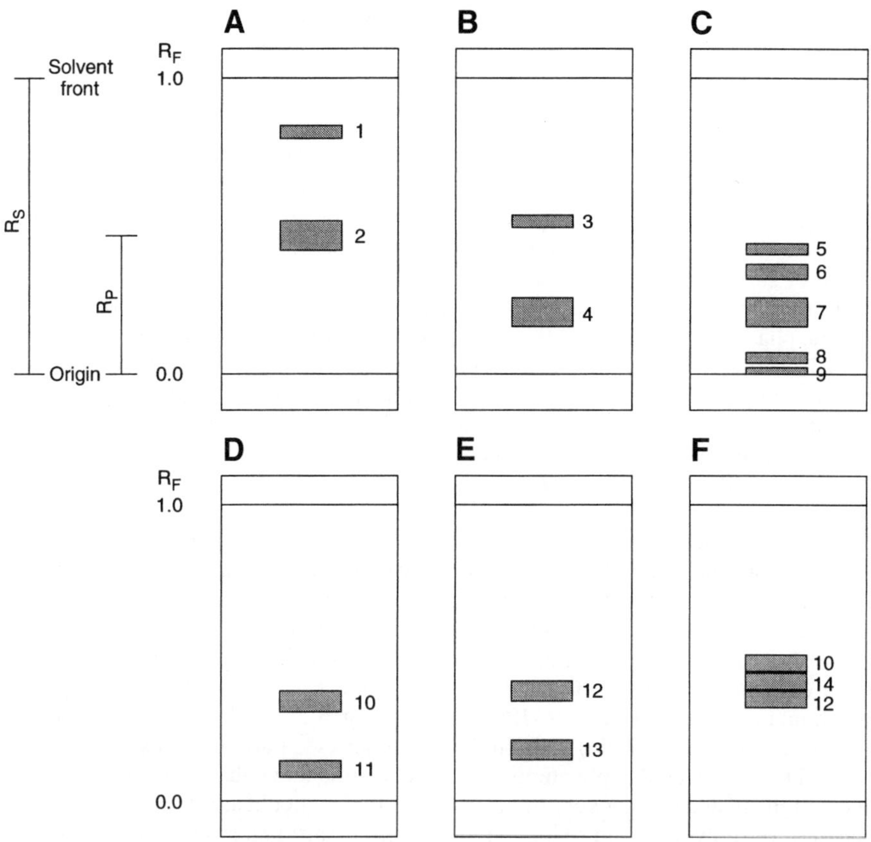

Figure 8.4
TLC systems for separating closely related carotenoids.
A. The structurally related β,β- and β,ε-carotenes (1, 2 respectively) Stationary phase: silica G:Kieselguhr:$Ca(OH)_2$:MgO = 16:14:9:9 (w/w). Mobile phase: *n*-hexane:acetone:*iso*-propanol = 69:30:1 (v/v/v). Dimensions for calculating R_f (R_s, R_p) are shown for β,ε-carotene (refer section 8.3.3.2).
B. Lutein (3) and zeaxanthin (4). Stationary phase: as in A; Mobile phase: *n*-hexane:acetone:*iso*-propanol = 69:30:1 (v/v/v).
C. Dinoflagellate carotenoids of intermediate polarity: dinoxanthin (5), pyrrhoxanthin (6), diadinoxanthin (7), diatoxanthin (8), diadinochromes (9). TLC system as in B.
D. 19'-Butanoyloxyfucoxanthin (10) and siphonaxanthin (11). TLC system as in B.
E. 19'-Hexanoyloxyfucoxanthin (12) and 9'-*cis*-neoxanthin (13). TLC system as in B.
F. Prymnesiophyte carotenoids: 19'-butanoyloxyfucoxanthin (10), fucoxanthin (14), and 19'-hexanoyloxyfucoxanthin (12). Stationary phase: RP8-HPTLC (Merck); mobile phase: methanol:water = 9:1 (v/v) (Wright and Jeffrey, 1987).

below the top of the plate. Plates are best cleaned and activated immediately before use, but they may be stored in a desiccator after cleaning for 1 or 2 days. Reversed-phase C-18 and C-8 plates are washed with acetone or acetone:methylene chloride (1:1, v/v), scored, and allowed to air dry. RPTLC plates do not need heat activation before use.

The sample is applied to the plate after dissolving the concentrated extract (see section 8.3.2) in a minimum volume of solvent (acetone, diethyl ether, or methylene chloride). Application is usually from a capillary tube, and the sample is spotted or streaked along a line 1–2 cm above the bottom edge of the plate, well above the depth of solvent in the tank. The TLC chamber should be pre-equilibrated for 5 to 10 min before use. The sides are lined with adsorbant paper, which dips into the mobile phase and speeds equilibration of the vapour of the mobile phase. The loaded plate is carefully placed into the TLC chamber and allowed to develop until the solvent front reaches the scored line at the top of the plate. After separation is complete, the plate is removed from the chamber and the R_f value (a measure of pigment retention by the adsorbant, see Fig. 8.4) is recorded, where:

$R_f = R_p/R_s$

and

R_p = distance from the origin to the middle of pigment band

R_s = distance from the origin to the solvent front.

The band containing the desired pigment is scraped with a fine spatula on to glassine paper or aluminium foil and transferred to a filter funnel or centrifuge tube. The pigment is recovered by washing the adsorbent with small volumes of acetone, diethyl ether or other appropriate solvent. Pigment purity is checked by chromatography of an aliquot on a second TLC or HPLC system (Fig. 8.2) and by comparing spectral properties (λ_{max}, III/II ratio) with published values (see Part IV Data sheets).

In nearly all TLC applications, exposure to air is inevitable during the development of the plate, and recovery of the pigment. Plates should always be developed in the dark (to prevent photo-oxidation), and if necessary under an inert atmosphere. When handling in air is unavoidable, manipulations should be completed in dim light as rapidly as possible. Excessive exposure of the adsorbed pigment to air may lead to photo-bleaching, *cis/trans* isomerization, epoxide rearrangement, and irreversible adsorption of the pigment. With care during handling, these problems can all be minimized. In special cases developed plates can be wrapped in foil and stored in the dark at -20°C for no more than 30 min before degradation becomes significant.

The retention of carotenoids on silica and RPTLC plates is primarily determined by their differences in relative polarity (see Table 8.2 and Appendix C). Polar pigments are strongly retained by silica; nonpolar pigments by reversed-phase adsorbents. Carotenes (β-/ε-)are separated efficiently by alkaline adsorbents ($CaCO_3$, MgO, $Ca(OH)_2$) (Hager and Meyer-Bertenrath, 1966; Hager and Stransky, 1970; Davies, 1976; and Fig. 8.4A). The nature and position of the double bonds are important in determining a pigment's relative retention on alkaline plates, often to such an extent that polarity differences are obscured (Table 8.2). For the SCOR WG 78 Carotenoid Workshop the alkaline plates of Bjørnland (Bjørnland, 1990; Bjørnland and Liaaen-Jensen, 1989) were used to purify all-*trans* carotenoids and separate isomeric carotenes (see Table 8.2, TLC systems III and IV). Some experience is necessary to make high-quality alkaline TLC plates. Alkaline adsorbents do not adhere strongly to the glass support plate, and are prone to cracking during drying. Bjørnland's adsorbent mixtures were formulated to minimize these problems, but careful attention to the

TABLE 8.2. R_f values of 29 algal carotenoids on four TLC systems

Carotenoids	R_f values				Occurence of carotenoid structural features[a]									
	I	II	III	IV	A	B	C	D	E	F	G	H	I	J
ε,ε-Carotene	0.83	1.00	1.00	~0.90		2								
β,ε-Carotene	0.82	1.00	1.00	0.84	1	1								
β,β-Carotene	0.82	1.00	1.00	0.52		2								
β,ψ-Carotene	0.68	0.99	0.75	0.01		1		1						
ψ,ψ-Carotene	0.58	0.99	0.24	0.00				2						
Echinenone		0.83	0.88		2								1	
β-Cryptoxanthin		0.72	0.66		2					1				
Canthaxanthin		0.70	0.62		2								2	
Siphonein		0.63	0.48		1	1				2			1	1
Vaucheriaxanthin diester		0.58	0.57		2				1	2	1			2
Lutein		0.58	0.53		1	1				2				
Zeaxanthin		0.58	0.25		2					2				
Diatoxanthin		0.58	0.07		2			1		2				
Alloxanthin		0.58	0.02		2			2		2				
Dinoxanthin		0.58	~0.43		2				1	2	1			1
Pyrrhoxanthin		0.58	~0.38		2			1		1	1			2
Diadinochrome		0.58	0.02		2			1		2		1		
Astaxanthin		0.54			2					2			2	
Antheraxanthin		0.50	0.35		2					2	1			
Diadinoxanthin		0.50	0.26		2			1		2	1			
Fucoxanthin		0.46	0.40		2				1	2	1		1	1
19'-Hexanoyloxyfucoxanthin		0.46	0.40		2				1	2	1		1	2
Violaxanthin		0.44	0.46		2					2	2			
19'-Butanoyloxyfucoxanthin		0.41	0.36		2				1	2	1		1	2
Peridinin		0.38	0.37		2				1	2	1			2
all-*trans*-Neoxanthin		0.27	0.22		2				1	3	1			
9'-*cis*-Neoxanthin		0.25	0.20		2				1	3	1			
Siphonaxanthin		0.22	0.14		1	1				3			1	
Vaucheriaxanthin		0.11	0.08		2				1	4	1			

TLC systems

I: Adsorbant = silica G: $CaCO_3$, 2:1 (w/w); Solvent = light petroleum (40–60°C): acetone, 99.5: 0.5 (v/v)

II: Adsorbant = silica G: $CaCO_3$, 2:1 (w/w); Solvent = light petroleum (40–60°C): acetone: *i*-propanol, 71:28:1 (v/v/v)

III: Adsorbant = silica G: Kieselguhr: $Ca(OH)_2$: MgO, 14:16:9:9 (w/w/w/w); Solvent = light petroleum (40–60°C): acetone: *i*-propanol, 73:25:2 (v/v/v)

IV: Adsorbant = silica G: Kieselguhr: $Ca(OH)_2$: MgO, 14:16:9:9 (w/w/w/w); Solvent = light petroleum (40–60°C): acetone, 97:3 (v/v)

a. Structural characteristics. A: β-ring end-group; B: ε-ring end-group; C: aliphatic end-group; D: acetylenic bond; E: allenic bond; F: hydroxyl group; G: 5,6-epoxide; H: 5,8-furanoid; I: keto group; J: ester function (for full structures, see Appendix C).

drying process (pre-drying for 10 to 15 min at room temperature followed by 4 h in a convection oven at 110–120°C) is necessary to produce high-quality plates.

The chromatographic behaviour of carotenoids on alkaline adsorbents is more complex than on silica and reversed-phase supports. Many separations traditionally achieved on alkaline TLC can be duplicated by reversed-phase and specialty HPLC columns, or by HPLC columns packed in-house with alkaline adsorbents. Thus the baseline resolution of β,ε-carotene, β,β-carotene, β,ψ-carotene, and lycopene have been widely reported (Rhodes *et al.* 1988; Craft *et al.* 1992), as has the separation of lutein from zeaxanthin (Juhler and Cox, 1990; Stalcup *et al.* 1990; Gilmore and Yamamoto, 1991; Craft, 1992; Epler *et al.* 1992). The separation of all-*trans* from *cis* isomers of carotenoids was recently reviewed by O'Neil and Schwartz (1992).

8.3.3.3 *High-performance liquid chromatography*

For difficult separations, or when apparatus for TLC purification of pigments is not readily available, carotenoids may be purified by high performance liquid chromatography (HPLC). Analytical HPLC columns have maximum sample loading limits of less than 1 mg, so it is often advantageous to separate the target pigment from other lipids (and pigments) by column chromatography or TLC before final HPLC purification.

Preparatory HPLC-scale purification can be made on wide bore (2–5 cm) columns that have sample loadings of up to 100 mg (Scalia and Francis, 1989; Isaksen and Francis, 1989; Francis, 1992). Normal-phase or nonaqueous reversed-phase systems are preferred, as water is excluded from the mobile phase. As with column chromatography, single injection preparatory HPLC can be operated in either isocratic (see Chapter 11) or gradient mode. For isocratic separations, optimal resolution occurs at 3 to 5 times the dead volume of the column (Appendix H).

Thin-layer chromatography is a useful exploratory technique for developing preparatory HPLC separations. However, separation conditions developed for TLC cannot be directly transferred to HPLC, and the final separation should be optimized on the preparatory HPLC system with a small aliquot of the sample before scaling to high-column loadings. The need to inject very high concentrations of pigment in relatively small sample volumes often requires a sample solution close to saturation of the pigment. Peak distortion or spurious peaks in the chromatogram due to solvent-pigment interactions have been observed at even moderate pigment concentrations in the injection solution (Khachik *et al.* 1988, Zapata and Garrido, 1991; and Appendix H). To scale up a preparatory HPLC system, 1% of the (final) target volume of sample is injected first to test the separation. Injections are then scaled up to 10%, 50% and finally 100% of the target injection volume. Any deterioration due to column overloading or injection volume is monitored (see Appendix H).

If carotenoids are to be purified by reversed-phase methods such as those of Wright *et al.* 1991 (see Chapters 12 and 13) no buffer or ion-pairing reagent (such as that used by Mantoura and Llewellyn, 1983) should be present in the mobile phase. The buffer has little effect on the separation of carotenoids, but it will contaminate the final product.

Many carotenoids can be isolated free from spectral and chromatographic impurities by a combination of reversed and normal phase HPLC. The target carotenoid from the first separation (e.g., by reversed-phase HPLC) is dried, and rechromatographed on a second system (e.g., normal phase HPLC). For such purifications it is absolutely essential that the carotenoid be isolated from a well characterized biological source material known to be free of co-eluting pigments (see Chapter 13 for HPLC separations of pigments from the SCOR reference cultures), that the peak elutes with good peak shape (free from overlapping peaks or shoulders), and that the visible spectrum be in close agreement with published spectra from pure compounds, including the absence of any *cis* peaks (see Part IV Data and Graphics sheets).

For SCOR WG 78 carotenoid standards, only the middle fraction of the HPLC peak was collected (see Fig. 8.2). Leading and trailing portions of the peak were collected as one fraction, recombined, then chromatographed a second time to collect a second middle fraction of the peak. After the pigment was collected and its purity established by analytical HPLC, the solvent was removed by vacuum rotary evaporation and the pigment prepared for storage, crystallization or further analysis.

8.4 Two procedures for carotenoid purification and identification

Two examples of protocols for isolating carotenoids are described. First, we discuss the use of microalgal cultures for isolating peridinin and diadinoxanthin, which typifies many of the small-scale isolations performed for the SCOR Carotenoid Workshop (Fig. 8.2). Second, we describe the isolation of lutein from a higher plant (spinach). This method uses techniques appropriate for preparatory scale (10–100 mg) isolation of carotenoids from higher plant tissues (Fig. 8.3) and macroalgae.

8.4.1 *Source material, algal cultures: isolation of peridinin and diadinoxanthin from the dinoflagellate* Amphidinium carterae

Cultures of the dinoflagellate *Amphidinium carterae* (CS-212) were grown according to procedures outlined in Chapter 6. Cells were harvested from 20 l of culture (see Chapter 12) and filtered by gravity or gentle vacuum on to a 29 cm glass fibre (Whatman GF/F) filter. The filter was stored at -40°C until extraction (up to 1 month; see Chapter 10). From the size of the culture and its cell density, the pigment yield was expected to be in the mg to sub-mg range. Thin-layer chromatography and HPLC are both appropriate techniques to use for this scale of preparation. A protocol that used normal-phase TLC followed by reversed-phase HPLC was therefore adopted.

The frozen filter was cut into small strips, placed in a 500 ml Erlenmeyer flask, and 150 ml acetone/methanol (3:7, v/v) was added. The sample was extracted with sonication, the solvent decanted, and replaced several times with fresh solvent (150 ml). The combined extracts were reduced by rotary vacuum evaporation to a small volume (50 ml) and then transferred with ether rinses to a 250 ml separatory funnel. Diethyl ether (peroxide-free, 100 ml) and 5% NaCl (100 ml) were added and the pigments were partitioned into ether with gentle shaking. The upper ether layer was collected, dried with anhydrous Na_2SO_4 (1.5 h, with stirring) and evaporated to dryness under nitrogen.

Amphidinium carterae contains a suite of major and minor carotenoids (see Chapter 13). Peridinin and diadinoxanthin (both major fractions) were first separated by TLC. A TLC plate (Whatman PLK5, 150 Å silica, 20 × 20 cm, 1000 μm coating) was washed as described in section 8.3.3.2. The *Amphidinium* extract was dissolved in 300 µl of methylene chloride and applied as a line extending across the plate approximately 1 cm from the bottom and sides (Fig. 8.4). The pigments were separated with acetone/hexane (40:60, v/v) to yield relatively pure peridinin ($R_f = 0.53$) and diadinoxanthin ($R_f = 0.62$). The two bands (peridinin, brick-red; diadinoxanthin, yellow), were scraped from the plate into beakers, and the pigments eluted from the silica with acetone.

Final purification was achieved by reversed-phase HPLC on a 150 × 4.6 mm 3 µm Adsorbosphere C-18 HS column, with methanol as eluting solvent. The sample was sufficiently small (several mg) for all the pigment to be loaded on to the column in one injection, but to guard against loss of the entire sample through equipment or column failure, the sample was applied in two portions. A trial separation established the chromatographic conditions, and then one half of the sample was injected. The all-*trans* peak was collected in three fractions (F_1–F_3, Figure 8.2) with the second fraction (centre of the peak) retained as the reference carotenoid standard. Beginning and trailing portions of the peak were recombined, injected, and collected in the same fashion to yield a second portion of the pure pigment. In this manner 1.4 mg peridinin with λ_{max} (acetone) 471, 457 nm, and 0.8 mg diadinoxanthin with λ_{max} (acetone) 478, 448, 429 nm (III/II = 0.69) were isolated. The pigments eluted as single compounds with good peak shape on the reversed-phase HPLC system (see Part IV, Preamble to Data sheets, section IV.1.7.1), and on a second normal-phase HPLC system (silica, 250 × 3.6 mm column packed with 3 µm adsorbosphere, eluted with *iso*-propanol:hexane = 40:60, v/v).

The RP-HPLC system above is an alternative to the alkaline TLC system illustrated for dinoflagellate xanthophylls in Fig. 8.4C.

8.4.2 *Source material, leaf tissues: isolation of lutein from* Spinacia oleracea

Spinach leaves (120 g wet weight) were frozen overnight at -20°C, thawed, and squeezed to remove excess water. The leaves were hardened by submersion in liquid nitrogen and pulverized by grinding with a large mortar and pestle. The pulverized tissue was extracted by soaking in degassed methanol (500 ml for 3 h) followed by two soaks in acetone (300 ml for 3 h). The combined extracts were filtered to remove finely divided tissue debris, and concentrated to 250 ml by vacuum rotary evaporation (<40°C). The pigments were partitioned into diethyl ether by adding 300 ml peroxide-free diethyl ether to the extract, followed by 1 l of 5% NaCl in a 2 l separatory funnel. After gentle shaking for 2 min with venting, the two layers were allowed to separate. The ether epiphase containing the pigments was collected, dried by stirring over anhydrous Na_2SO_4 (4 h), and evaporated to dryness.

Lutein was separated from the other pigments and lipids by column chromatography. A 2 cm diameter chromatography column was packed with 35 g of alumina (neutral, 100–200 mesh, cleaned and activated at 150°C overnight) in acetone/hexane (1:4, v/v). The sample was applied to the column in 5 ml of mobile phase (see section 8.3.3.1) and the carotenes were eluted first with 3 column volumes of acetone/hexane (1:4, v/v). To elute chlorophyll and lutein, the column was then washed with 4 column volumes of acetone/hexane (2:3, v/v), which eluted chlorophyll immediately ahead of lutein, without, however, fully separating the two pigments. Final purification of the lutein-rich fraction resulted from flash chromatography on a 1 cm diameter column of activated silica (20–45 µm particle size), with acetone/hexane

(3:10, v/v) as mobile phase. This column achieved excellent separation of chlorophyll from lutein, which eluted as a narrow, deep-red band in 3 column volumes of eluant. The lutein fraction was collected, evaporated to dryness, and the remaining oil containing the pigment redissolved in 1 ml methanol. Crystallization occurred overnight to yield 5 mg lutein of mp 180–181°C (uncorrected mp 183°C, Buchecker *et al.* 1974), λ_{max} (acetone) 473, 445, 423 nm, and III/II of 0.70 (see Part IV Data sheets). The crystalline lutein was at least 99% pure, and if traces of zeaxanthin were present it was less than 1%.

Figure 8.3 describes a similar large-scale isolation of fucoxanthin from the brown seaweed *Fucus vesiculosus*.

8.5 Microchemical tests for functional group determination

A summary of microchemical tests used in carotenoid identifications is given in Davies (1976). Here two useful microchemical tests to detect the presence of 5,6-epoxide groups and primary and secondary alcohols in carotenoids are described.

8.5.1 *Acid-catalyzed rearrangements of 5,6-epoxides*

In most cases, the presence of a 5,6-epoxide can be detected by acid-catalyzed epoxide rearrangement to the 5,8-furanoid. This transformation is characterized by a hypsochromic shift of 20 nm in λ_{max} (i.e. λ_{max} moves to shorter wavelengths), a reduction in absorbance, and a reduction of fine structure in the visible spectrum (reduction in III/II ratio). The principal exception is fucoxanthin, which has a ketone at the C-8 position and therefore does not undergo such rearrangements.

To confirm the presence of a 5,6-epoxide in, for example, diadinoxanthin isolated from *Amphidinium carterae* (section 8.4.1) several µg of diadinoxanthin were dissolved in 1 ml of acetone and the visible spectrum recorded (λ_{max} = 478, 448, 429 nm; $\text{absorbance}_{448nm}$ = 0.98, and III/II = 0.69). Ten µl of 1N HCl were added to the cuvette, and the spectral changes at 448 were monitored until the absorption was stable (approximately 5–10 min). The visible spectrum had the following properties: λ_{max} = 455, 428, 411 nm; $\text{absorbance}_{428nm}$ = 0.57, and III/II = 0.44.

8.5.2 *Acetylation of primary and secondary alcohols*

Acetylation of carotenols with acetic anhydride to yield acetates is useful for detecting the presence of primary and secondary alcohol groups in a carotenoid by the reduced polarity of the pigment (increased retention time in RP-HPLC). Tertiary alcohols, such as the 5'-hydroxyl group in fucoxanthin, peridinin and neoxanthin, are not acetylated by this procedure.

To acetylate lutein, 50 µl of acetic anhydride and 50 µl of dry pyridine were added to 5 µg of the dried purified pigment. The reaction vial was capped, and allowed to react at room temperature in the dark overnight. The contents of the vial were transferred to a small (30 ml) separatory funnel to which 10 ml of methylene chloride and 10 ml of 0.1 mM HCl were added. The sample was shaken to transfer the pigment into the methylene chloride (bottom layer) and neutralize the pyridine. The aqueous epiphase was then discarded. The acid wash was repeated twice, followed by a final wash with distilled water. The organic phase containing the pigment was dried over anhydrous Na_2SO_4 and the solvent removed by rotary evaporation. Acetylation resulted in a decrease in the polarity of the pigment, which was monitored as a change in relative retention (R_f) on silica TLC, developed with acetone/hexane (3:7, v/v). The R_f value increased from 0.31 (lutein) to 0.82 (lutein diacetate). Small amounts (<10%) of lutein monoacetate were also observed ($R_f = 0.35$). A decrease in polarity (increase in R_f) can be interpreted as successful acetylation and therefore indicates the presence of primary or secondary alcohol groups or both.

Acetylation was used in the preliminary identification of fucoxanthinol in oceanic sediment-trap samples, and in the partial identification of other carotenoid degradation products (Repeta and Gagosian, 1987). Acetylation of the putative fucoxanthinol yielded fucoxanthin acetate, which was compared by HPLC to fucoxanthin acetate prepared by the acetylation of authentic fucoxanthin (Repeta and Gagosian, 1982).

8.6 CRYSTALLIZATION

8.6.1 *Background*

Crystallization is a very effective purification step: it yields mg to g amounts of carotenoids in their chemically most stable condition. Highly symmetric carotenoids (e.g., lycopene, β,β- and β,ε-carotene, lutein, zeaxanthin) crystallize readily, often after only a single chromatographic purification procedure. Crystalline preparations of asymmetric carotenoids (fucoxanthin, peridinin) are more difficult to obtain.

The ease with which a carotenoid can be crystallized also depends upon the amount and nature of accompanying colourless contaminants, the degree of *cis* isomerization, and the method of isolating the pigment. Biological source materials with low lipid content are best suited for the isolation of crystalline pigments. Several methods of crystallization may be used depending on the investigators' personal preferences and the carotenoid in question. Crystallization procedures should be done in clean tubes or small conical flasks which can be fitted to a vacuum rotary evaporator so that solvents can be changed easily if necessary. A concentration of 1–5 mg of purified carotenoid per ml is suitable for the three types of crystallization procedures outlined below.

8.6.2 *Crystallization by cooling of saturated solutions*

The pigment-solvent mixture is gently warmed (<40°C) in a N_2-atmosphere until all the pigment is in solution. Crystallization occurs during cooling. As the best crystal quality is obtained when cooling proceeds slowly, crystallization tubes or flasks should not be placed directly in a freezer, but placed in a Dewar and then transferred to a freezer (-20°C). Disturbances and mechanical vibrations should be kept to a minimum.

Nonpolar xanthophylls and carotenes can be crystallized from acetone by slow cooling. Methanol should be used for xanthophylls of intermediate polarity (lutein, zeaxanthin, diatoxanthin, diadinoxanthin, etc.). If necessary, the xanthophyll is first dissolved in a minimum volume of an intermediate polarity solvent (acetone, benzene, methylene dichloride, diethyl ether, etc.), and methanol added in small portions to near-saturation, as observed by increasing cloudiness after each addition of methanol. More polar xanthophylls (violaxanthin, fucoxanthin, 19'-acyloxy-fucoxanthins, peridinin, neoxanthin, siphonaxanthin, vaucheriaxanthin, etc.) are crystallized by adding a nonpolar solvent (hexane) to a solution of the pigment in a minimum volume of acetone.

8.6.3 *Crystallization by liquid-solid diffusion*

The carotenoid is dissolved in an appropriate volume of benzene and cooled to a solid phase at -20°C. (Benzene should be used with caution because of its hazards to human health). One to three volumes of pre-cooled *n*-hexane (for polar xanthophylls) or methanol (for carotenes and nonpolar xanthophylls) are added on the top of the solid benzene layer and the biphasic system is left at -20°C. Over 1 to 3 days, the epiphasic hexane or methanol layer will slowly dissolve the solid benzene and the carotenoid will crystallize continuously at the interface between the layers.

This is probably the most reliable method of obtaining crystalline samples of carotenoids. Crystals can be obtained from about 0.5 mg pigment, but they are generally very small — either microcrystalline or semicrystalline. If further purification is needed, the process should be repeated or the carotenoid recrystallized by the techniques described in sections 8.6.2 or 8.6.4.

8.6.4 *Crystallization by liquid-liquid diffusion*

If crystals of very high purity are required, crystallization is allowed to proceed slowly in a small-bore (diffusion) tube protected from vibration and held at constant temperature. For this technique the pigment is dissolved in a small amount of a very dense polar solvent, such as methylene chloride or chloroform. *n*-Hexane is then gently added on the top of the hypo-layer without mixing. Diffusive mixing of the two solvents over several days to weeks often yields crystals of >99.5% purity. The crystals are collected by filtration, washed with a cold organic solvent in which they are insoluble, and dried under vacuum before storage.

8.7 PURITY CRITERIA

Calibration of an HPLC detector and quantitative carotenoid analysis are highly dependent on the purity of isolated pigment standards. Carotenoid standards prepared as crystalline compounds should have melting points of the same sharpness and temperature as those reported for authentic pigments. Lower melting points, or melting over a wide temperature range, generally suggest the presence of impurities. Extinction coefficients should also agree with literature values determined on authentic carotenoids. For carotenoid standards that have not been crystallized, the purified product should elute as a single peak with good peak shape on two different HPLC systems. Spectral properties must be determined off-line (spectrophotometrically) in a reference solvent. On properly calibrated spectrophotometers, values for λ_{max} should agree with literature values within the precision of the instrument (±1–2 nm). The III/II ratio should also be determined and agree with published ratios to within ±10% (see Part IV Data sheets). One of the principal difficulties in obtaining pure carotenoids is the persistence of *cis*-isomers in many isolates. The UV-vis spectra of all-*trans* preparations should not display any *cis*-peak in the near UV region (cf. Appendix D).

8.8 Storage

Crystalline carotenoids can be stored under an inert atmosphere at -40°C for 5–8 years with little degradation. Noncrystalline carotenoids and mixtures (>0.1 mg) are stored *in vacuo* or under N_2 (or He) in a sealed ampoule. Very small amounts of carotenoid (<0.1 mg) are dissolved in degassed benzene and sealed in an ampoule under an inert atmosphere. The ampoule is then stored at -40°C or in liquid nitrogen in the dark. The benzene freezes at +5.5°C and the solid phase promotes stability of the pigment. However, some degradation and isomerization of carotenoids occurs with all forms of storage, regardless of the precautions taken.

Acknowledgement

We thank Dr S. W. Jeffrey for expertise contributed to the SCOR Carotenoid Workshop, and for assistance in the preparation of this chapter.

References

Baumeler, A., Zerbe, O., Kunz, R., Eugster, C. H. (1994). (6*R*, 9'Z)– Neoxanthin: Synthese, Schmelzverhalten, Spektren und Konformationsberechnungen. *Helv. Chim. Acta,* 77: 909–930.

Bernhard, K. (1990). Synthetic astaxanthin. The route of a carotenoid from research to commercialisation. In: Krinsky, N. I., Mathews-Roth, M. M., Taylor R. F. (eds.), *Carotenoids. Chemistry and Biology*. Plenum Press, New York, p. 337–363.

Bjørnland, T. (1989). Carotenoid structures and lower plant phylogeny. In: Krinsky, N. I., Mathews-Roth, M. M., Taylor, R. F. (eds), *Carotenoids. Chemistry and Biology*. Plenum Press, New York, p. 21–37.

Bjørnland, T. (1990). Chromatographic and spectrometric characterization of native carotenoids of the marine dinoflagellate *Thoracosphaera heimii. Biochem. Syst. Ecol.* 18: 307–316.

Bjørnland, T., Liaaen-Jensen, S. (1989). Distribution patterns of carotenoids in relation to chromophyte phylogeny and systematics, In: Green, J. C., Leadbeater, B. S. C., Diver, W. L. (eds), *The Chromophyte Algae: Problems and Perspectives.* Clarendon Press, Oxford, p. 37–61.

Bjørnland, T., Skjetne, T., Liaaen-Jensen, S. (1987). Studies on the stereoisomeric set of neoxanthin — non-existence of *R/S* allene isomerism in carotenoids? Abstr. 8th Intl. Symp. Carotenoids, Boston, p. 3.

Bonnett, R., Mallams, A. K., Spark, A. A., Tee, J. L., Weedon, B. C. L., McCormick, A. (1969). Carotenoids and related compounds. Part XX. Structure and reactions of fucoxanthin. *J. Chem. Soc.*, 429–453.

Britton, G., Goodwin, T. W. (1982). *Carotenoid Chemistry and Biochemistry*. Pergamon Press, New York, 397 pp.

Britton, G., Young, A. J. (1993). Methods for the isolation and analysis of carotenoids, In: Young, A. J., Britton, G. (eds), *Carotenoids in Photosynthesis*. Chapman and Hall, Bury St. Edmunds, UK, p. 409–457.

Buchecker, R., Liaaen-Jensen, S. (1975). Reaktionen an Allencarotinoiden 2. *Helv. Chim. Acta*, 58: 89–95.

Buchecker, R., Hamm, P., Eugster, C. H. (1974). Absolute Konfiguration von Xanthophyll (lutein). *Helv. Chim. Acta*, 57: 631–635.

Cholnoky, L., Györgyfy, K., Rónai, A., Szabolcs, J., Tóth, G., Galasko, G., Mallams, A. K., Waight, E. S., Weedon, B. C. L. (1969). Carotenoids and related compounds. Part XXI. Structure of neoxanthin (foliaxanthin). *J. Chem. Soc.* 1256–1263.

Craft, N. E. (1992). Carotenoid reversed-phase high-performance liquid chromatography methods: reference compendium, In: Packer, L. (ed.), *Methods in Enzymology,* vol 213, Academic Press, London, p. 185–205.

Craft, N. E., Wise, S. A., Soares, J. H. Jr. (1992). Optimization of an isocratic high-performance liquid chromatographic separation of carotenoids. *J. Chromatogr.* 589: 171–176.

Davies, B. H. (1976). Carotenoids, In: Goodwin, T. W. (ed.), *Chemistry and Biochemistry of Plant Pigments*, Academic Press, New York, p. 38–165.

Egeland, E. S., Liaaen-Jensen, S. (1993). New carotenoids and chemosystematics in Prasinophyceae. Abstr. 10th Intl. Symp. Carotenoids, Trondheim, p. 4–6.

Egger, K., Dabbagh, A., Nitsche, H. (1969). Überführung von Neoxanthin in Diadinochrom. *Tet. Lett.* 2995–2998.

Epler, K. S., Sander, L. C., Ziegler, R. G., Wise, S. A., Craft, N. E. (1992). Evaluation of reversed-phase liquid chromatographic columns for recovery and selectivity of selected carotenoids. *J. Chromatogr.* 595: 89–101.

Francis, G. W. (1992). Preparative high-performance liquid chromatography of carotenoids, In: Packer, L. (ed.), *Methods in Enzymology*, vol 213, Academic Press, London, p. 246–250.

Gilmore, A. M., Yamamoto, H. Y. (1991). Resolution of lutein and zeaxanthin using a non-endcapped, lightly carbon-loaded C_{18} high-performance liquid chromatographic column. *J. Chromatogr.* 543: 137–145.

Goodwin, T. W. (1971). Algal Carotenoids. In: Goodwin, T. W. (ed.), *Aspects of Terpenoid Chemistry and Biochemistry*, Academic Press, London, p. 315–356.

Goodwin, T. W., Britton, G. (1988). Distribution and analysis of carotenoids, In: Goodwin, T. W. (ed.), *Plant Pigments*. Academic Press, London, p. 61–132.

Hager, A., Meyer-Bertenrath, T. (1966). Die Isolierung und quantitative Bestimmung der Carotinoide und Chlorophylle von Blättern, Algen und isolierten Chloroplasten mit Hilfe Dünnschichtchromatographischer Methoden. *Planta*, 69: 198–217.

Hager, A., Stransky, H. (1970) Das Carotinoid Muster und die Verbreitung des lichtinduzierten Xanthophyllcyclus in verschiedenen Algenklassen. *Arch. Microbiol.* 73: 77–89.

Hamilton, R., Hamilton, S. (1987). *Thin-layer Chromatography*. John Wiley and Sons, London, 129 pp.

Haugan, J. A., Aakermann, T., Liaaen-Jensen, S. (1992). Isolation of fucoxanthin and peridinin, In: Packer, L. (ed.), *Methods in Enzymology*, vol 213, Academic Press, London, p. 231–245.

Isaksen, M., Francis, G. W. (1989). Preparative HPLC of carotenoids. *Chromatographia*, 27: 325–327.

Isler, O. (1971). *Carotenoids*. Birkhaüser Verlag, Basel, 932 pp.

Juhler, R. K., Cox, R. P. (1990). High-performance liquid chromatographic determination of chloroplast pigments with optimized separation of lutein and zeaxanthin. *J. Chromatogr.* 508: 232–235.

Karrer, P., Jucker, E. (1950). *Carotenoids*. Elsevier, Amsterdam, 384 pp.

Khachik, F., Beecher, G. R., Vanderslice, J. T., Furrow, G. (1988). Liquid chromatographic artifacts and peak distortion: sample solvent interactions in the separation of carotenoids. *Anal. Chem.* 60: 807–815.

Krinsky, N. I., Mathews-Roth, M.M., Taylor, R. F. (1989). *Carotenoids. Chemistry and Biology*, Plenum Press, New York, 382 pp.

Liaaen-Jensen, S. (1977). Algal carotenoids and chemosystematics. In: Faulkner, D. J., Fenical, W. H. (eds), *Marine Natural Products Chemistry*. Plenum Press, New York, p. 239–259.

Liaaen-Jensen, S. (1978). Marine Carotenoids. In: Scheuer, P. (ed.), *Marine Natural Products*. Academic Press, New York, p. 2–73.

Liaaen-Jensen, S. (1985). Carotenoids of lower plants — recent progress. *Pure Appl. Chem.* 57: 649–658.

Mantoura, R. F. C., Llewellyn, C. (1983). The rapid determination of algal chlorophyll and carotenoid pigments and their breakdown products in natural waters by reverse-phase high-performance liquid chromatography. *Anal. Chim. Acta,* 151: 297–314.

Müller, R.K., Bernhard, K., Vecchi, M. (1982). Recent advances in the synthesis and analysis of 3,4-oxygenated xanthophylls, In: Britton, G., Goodwin, T. W. (eds), *Carotenoid Chemistry and Biochemistry*. Pergamon Press, New York, p. 27–54.

O'Neil, C.A., Schwartz, S. J. (1992). Chromatographic analysis of *cis/trans* carotenoid isomers. *J. Chromatogr.* 624: 235–252.

Packer, L. (1992). *Carotenoids, Methods in Enzymology*, vol 213, Academic Press, London, 538 pp.

Perrin, D. D., Armarego, W. L. F., Perrin, D. R. (1988). *The Purification of Laboratory Chemicals*, Pergamon Press, New York, 568 pp.

Poole, C. F., Poole, S. K. (1991). *Chromatography Today*. Elsevier, Amsterdam, 1026 pp.

Poole, C. F., Schuete, S. A. (1984). *Contemporary Practice of Chromatography*, Elsevier, Amsterdam, 708 pp.

Repeta, D. J., Gagosian, R. B. (1982). Carotenoid transformations in coastal marine waters. *Nature*, 295: 51–54.

Repeta, D. J., Gagosian, R. B. (1987). Carotenoid diagenesis in recent marine sediments I. The Peru continental shelf (15°S, 75°W). *Geochim. Cosmochim. Acta*, 51: 1001–1009.

Rhodes, S. H., Netting, A. G., Milborrow, B. V. (1988). Normal-phase high-performance liquid chromatography of carotenes. *J. Chromatogr.* 442: 412–419.

Rüedi, P. (1985). HPLC: a powerful tool in carotenoid research. *Pure Appl. Chem.* 57: 793–800.

Scalia, S., Francis, G. W. (1989). Preparative scale reversed-phase HPLC method for simultaneous separation of carotenoids and carotenoid esters. *Chromatographia*, 28: 129–132.

Stahl, E., Mangold, H. K. (1975). Techniques of thin-layer chromatography, In: Heftmann, E. (ed.), *Chromatography: a Laboratory Handbook of Chromatographic and Electrophoretic Methods*.Van Nostrand Reinhold Co., New York, p. 164–186.

Stalcup, A. M., Jin, H. L., Armstrong, D. W., Mazur, P., Derguini, F., Nakanishi, K. (1990). Separation of carotenes on cyclodextrin-bonded phases. *J. Chromatogr.* 499: 627–635.

Still, W. C., Kahn, M., Mitra, A. (1978). Rapid chromatographic technique for preparative separations with moderate resolution. *J. Org. Chem.* 43: 2923–2925.

Strain, H. H. (1958). *Chloroplast Pigments and Chromatographic Analysis.* 32nd Annual Priestly Lecture. Pennsylvania State University Press, USA, 180 pp.

Strain, H. H., Svec, W. A. (1966). Extraction, separation, estimation and isolation of chlorophylls. In: Vernon, L. P., Seely, G. R. (eds), *The Chlorophylls*, Academic Press, London, p. 21–66.

Straub, O. (1987). Carotenoids. In: Pfander, H., Gerspacher, M., Rychener, M., Schwabe, R. (eds), *Key to Carotenoids*, Birkhaüser, Basel, 296 pp.

Vecchi, M., Müller, R. K. (1979). Separation of (3*S*,3'*S*)–, (3*R*,3'*R*)– and (3*S*,3'*R*)– astaxanthin via(-)-camphanic acid esters. *J. High Res. Chromatog.* 2: 195–196.

Wright, S. W., Jeffrey, S. W. (1987). Fucoxanthin pigment markers of marine phytoplankton analysed by HPLC and HPTLC. *Mar. Ecol. Progr. Ser.* 38: 259–266.

Wright, S. W., Jeffrey, S. W., Mantoura, R. F. C., Llewellyn, C. A., Bjørnland, T., Repeta, D., Welschmeyer, N. (1991). *Mar. Ecol. Progr. Ser.* 77: 183–196.

Zapata, M., Garrido, J. L. (1991). Influence of injection conditions in reversed-phase high-performance liquid chromatography of chlorophylls and carotenoids. *Chromatographia*, 31: 589–594.

Zechmeister, L. (1962) *Cis-trans Isomeric Carotenoids, Vitamins A, and Arylpolyenes*. Springer, Vienna, 251 pp.

9

Evaluation of methods and solvents for pigment extraction

S. W. Wright, S. W. Jeffrey and R. F. C. Mantoura

9.1 Introduction

The first step in chromatographic analysis is to extract the pigments of a phytoplankton sample into a compatible solvent. The properties of common solvents and the processes of solvation are reviewed in Hayes (1985), but extraction of pigments from an algal cell is more complex than a direct solvent–solute interaction. The solvent must first penetrate the algal cell wall, and then solubilize the lipids and lipoproteins of chloroplast membranes in an aqueous matrix. The efficiency of the extraction depends on the resistance of the cell wall, the penetration power and the solvation properties of the solvent, the duration of the extraction, and whether mechanical disruption is used.

The process is complicated by algal cell types having cell walls with widely differing mechanical and chemical properties and containing a range of pigments with very different polarities (see Table 2.3). Many algae are notoriously difficult to extract — notably armoured dinoflagellates, heavily silicified benthic diatoms, cyanobacteria, and heavily walled green algae (Porra, 1991), whilst planktonic diatoms and naked flagellates usually yield their pigments very easily. The literature reports several extraction techniques involving soaking, grinding, or ultrasonic disruption in various solvents, which have varying degrees of success. This chapter reviews the efficacy of these techniques and presents the SCOR workshop evaluations of the more promising procedures.

Several parameters were considered in assessing the extraction techniques:

1. *Extractability*. All pigments should be completely extracted from phytoplankton samples, irrespective of the types of algae present.
2. *Fidelity*. The pigments should be extracted without alteration products being formed. The method should prevent the activation of chlorophyllase enzymes, de-metallation, allomerization or epimerization of chlorophylls, and degradation or isomerization of the carotenoids. Pigment degradation should be minimized by using a fast extraction technique with cold solvents of neutral pH in the dark. The extracted pigments should be stable in the extraction solvent for at least one day to allow for delays in injection while samples are queued in an autosampler.
3. *Compatibility*. If the samples are to be analysed by HPLC, the solvents used must be compatible with the HPLC system (solvents and column) into which the sample will be injected.
4. *Precision*. The technique should allow good replication with high precision.
5. *Simplicity*. The technique should be as simple as possible for the operator, with very few handling steps, to enable large numbers of field samples to be processed rapidly.
6. *Safety*. The solvents used should have low toxicity and flammability. As many solvents capable of dissolving lipid-soluble pigments are irritating to human skin, contact must be avoided. The extraction technique should minimize transfers of solvents or manipulation of solvent-soaked filters. Safety is especially important when extractions are done at sea, where the ship's motion and often less-than-ideal laboratory conditions increase both the risk and consequences of an accident.

While many evaluations of the efficacy of various extraction techniques for spectrophotometric analysis of chlorophyll *a* have been published, there is very little on either the alteration of chlorophylls to spectrally similar compounds (chlorophyllides, their esters, allomers, epimers) during extraction or the extractability or alteration of carotenoids. The next section briefly reviews extraction techniques and solvents from the literature, while Section 9.3 evaluates SCOR workshop trials of extraction methods which used HPLC analysis to determine their success.

9.2 Review of common extraction techniques and solvents

9.2.1 *Extraction techniques*

The main methods used to enable solvents to penetrate into algal cells are soaking, grinding and sonication.

Soaking

Extraction by soaking alone is the most common method. It is also the easiest, and is suitable for preparing large numbers of samples for spectrophotometry. Soaking requires extended extraction times — from 10 min at 65°C in dimethylsulphoxide (Burnison, 1980), to overnight in 90% acetone (SCOR-UNESCO, 1966), to about 40 hours at 4°C in various solvents (Stauffer *et al.* , 1979), which provide ample opportunity for pigments to degrade or isomerize. Storage of flammable volatile solvents in a refrigerator is a safety hazard, as the vapours can readily be ignited by electric contacts inside the cabinet, unless special precautions are taken (acetone flash point = -17°C, Table 9.1).

Grinding

An early method (SCOR-UNESCO, 1966) used a motor-driven teflon pestle rotating at 500 r.p.m. in a glass homogenization tube to break up the algal cells and allow solvent penetration. This method is still widely used. The clearance between the pestle and the tube greatly affects the efficiency of extraction. A pestle that fits too tightly prevents the filter from entering the gap, causing delays and sometimes overheating. The resultant slurry must be transferred to a centrifuge tube, an awkward operation that can cause losses and also puts the operator at risk of solvent contact. The eyes and hands must be protected (by safety glasses and stout leather gloves) in case the glass tube shatters.

Shaking for 20 sec in a glass-pearl homogenizer (Gieskes *et al.* 1988) has also been used for pigment extraction.

Sonication.

Ultrasonic waves disrupt cells by forming and collapsing cavities in the solvent. Local pressure changes of up to 20 000 atmospheres produce shock waves that are transmitted for a distance of a few micrometres and can produce instantaneous temperatures of 10 000°K (Alliger, 1975).

Comparisons between methods are complicated by the different designs and geometry of the sonicator, which can greatly influence its effectiveness. The probe sonicator that Chang and Rossman (1982) found to be inferior to grinding delivered 73.5 W.cm^{-2}, whereas that of Wright and Shearer (1984), which produced better results than grinding, delivered 398 W.cm^{-2}. Bath sonicators, as used by Sartory and Grobbelaar (1984), usually deliver only 60 W total power into the bath, and less would be transferred through the walls of the extraction tube into the solvent and sample. The differences in power strongly affect extraction times. The 10 minutes' sonication of Sartory and Grobbelaar (1984) could lead to considerable heating of the solvent and degradation of the pigments in a high-powered sonicator. For safety, the operator should be perform sonication in a sound-proof cabinet or wear ear muffs. The sample tube must not be touched during sonication. A suitable set-up is described in Chapters 12 and 17.

9.2.2 *Solvents*

The physical and safety data on the seven solvents discussed below are summarized in Table 9.1. Laboratory safety manuals should be consulted before working with these solvents.

Acetone

80% acetone was used by Arnon (1949) to extract chlorophylls for spectrophotometry. To extract chlorophyll *a* from marine phytoplankton, Richards with Thompson (1952) soaked samples for 9 h in 90% acetone, which proved an excellent extraction solvent for marine phytoplankton (Humphrey and Wooton, 1966). This method was recommended by SCOR-UNESCO (1966), Strickland and Parsons (1972), Parsons *et al.* (1984) and others. Spectral data for chlorophylls *a, b* and *c* in diethyl ether, acetone, and 90% acetone are well known (see Part IV), and accurate spectrophotometric equations (Jeffrey and Humphrey, 1975) have been derived (see Appendix F).

Because chlorophyllase activity in acetone solutions increases with the water content (Barrett and Jeffrey, 1964), Jeffrey (1974) recommended that, for chromato-

TABLE 9.1. Physical characteristics and safety considerations for seven extraction solvents.

Solvent	MP (°C)	BP (°C)	FP (°C)	LD_{50}	Safety considerations
Acetone[(a)]	-94	56	-17	5.8	Extremely flammable; severe eye irritant; skin irritant.
Chloroform[(a)]	-63	61	–	0.91	Carcinogen; irritant; highly toxic; damage to liver, kidneys, heart; toxicity increased by alcohol; toxic decomposition products; usually contains HCl.
Dimethyl acetamide[(a)]	-20	165	70	4.9	Harmful; irritant; may cause birth defects; damage to liver; reacts with highly halogenated compounds.
Dimethyl formamide[(a)]	-61	153	57	2.8	Harmful; irritant; intolerance for ingested alcohol for up to 4 days after exposure; potent liver toxin; readily absorbed through skin.
Dimethyl sulfoxide[(b)(d)]	18.6	189	88	>15	Irritant; readily absorbed through skin; enhances skin absorption of other toxins; not compatible with contact lenses; moderately flammable, produces toxic fumes.
Ethanol[(b)]	-130	78	13	13.7	Eye and respiratory irritant, mild skin irritant; highly flammable
Methanol[(b)(c)]	-98	65	12	5-15	Highly flammable; highly toxic; vapours harmful to respiratory tract.

Abbreviations: MP: melting point, BP: boiling point, FP: flash point at atmospheric pressure, LD_{50}: lethal dose for 50% of cases (oral, rat, g/kg body weight).

Sources: (a) Lenga (1988) (b) Chemwatch (1993) (c) Weiss (1986) (d) De Renzo (1986) graphic analysis, algae should first be extracted with 100% acetone (which prevents chlorophyllase activity) and then diluted to 90% acetone for spectrophotometric analysis. From a safety perspective, the biggest problem with acetone is its high flammability, although this is lessened when 10% water is added.

Methanol

Methanol is a more efficient extraction solvent than 90% acetone for many green algae (Marker, 1972; Rai, 1973) and cyanobacteria (Stauffer *et al.* 1979; Sand-Jensen, 1976; Holm-Hansen and Riemann, 1978; Riemann, 1980).

On the other hand, 90% and 100% methanol are less effective extractants than 100% or 90% acetone, or 50:50 acetone–methanol, for the freshwater cyanobacteria *Microcystis aeruginosa* and *Anabaena oscillaroides*, the chlorophyte *Chlorella vulgaris*, and the diatom *Cyclotella meneghiniana* (Bowles *et al.* 1985). The latter authors found that algal pigments were extracted differentially by various solvents, and there was no single combination of solvent and extraction time that was best for all pigments. They recommended 50:50 acetone–methanol or 90% acetone for general use. Undiluted (100%) methanol was superior to 90% acetone or 90% acetone:methanol (5:1 v/v) for extraction of freshwater blooms dominated by chrysophytes and diatoms, Chlorococcales and Cryptophyceae, or the cyanobacterium *Aphanizomenon flos-aquae*, as well as cultures of *Scenedesmus abundans, Chlamydomonas geitleri* and *Chlorella minutissima* (Pechar 1987). For the field samples, Pechar found little difference between hot or cold methanol with homogenization, or hot methanol without homogenization, but for robust cultured algae, hot methanol with homogenization was best overall.

The stability of chlorophyll *a* is reportedly lower in methanol than in acetone (Tett *et al.*. 1975, Bowles *et al.* 1985), since methanol promotes the formation of allomerization products (Diehn and Seely, 1968). However, in a recent survey of prymnesiophyte pigments where buffered methanol was used as an extractant (98% methanol + 2% ammonium acetate, 0.5 M, pH 7.1), Jeffrey and Wright (1994) found less than 1.5% of chlorophyll *a* present as chlorophyllides, allomers and epimers. When chlorophyllase activity persists during methanol extraction, methyl chlorophyllides are formed (Barrett and Jeffrey, 1971). This could be advantageous, since it distinguishes chlorophyllide formed as an artefact of methanol extraction from that which may have been present in the original sample (provided that this does not itself esterify with methanol).

In terms of chromatographic compatibility, Zapata and Garrido (1991) found that methanol as extractant produced sharper HPLC peaks for the more polar pigments than did acetone. Buffered methanol (98% methanol + 2% ammonium acetate, 0.5 M) was used as the extractant by Wright *et al.* (1991) and Kraay *et al.* (1992), since it improved chromatographic resolution of the early-eluting peaks over that obtained with 100% methanol.

Dimethyl sulphoxide

Dimethyl sulphoxide (DMSO) is a strong penetrating agent, which has either been used alone or as a supplement to acetone. DMSO and 90% acetone extracted equal quantities of chlorophyll from diatoms and blue-green algae, but DMSO was superior to 90% acetone for all green algae tested, extracting between 2 and 60 times more chlorophyll, depending on the species (Shoaf and Lium, 1976). Burnison (1980) used DMSO for 10 min at 65° C, then rinsed with 90% acetone prior to spectrophotometric analysis. He found 20% more chlorophyll *a* was extracted from *Scenedesmus* and *Selenastrum* by this technique than by grinding in either 90% acetone or DMSO alone. Pre-treating *Nannochloris atomus* with DMSO extracted ten-fold more chlorophyll *a* than did 90% acetone alone, and a further three-fold increase was obtained when the mixture was sonicated (Heyward, 1991). Other authors have found DMSO/acetone mixtures to be superior to acetone for thick-walled algal species (Stauffer *et al.* 1979, Speziale *et al.* 1984, Hains, 1985).

Dimethyl formamide

Dimethyl formamide (DMF) was used by Volk and Bishop (1968) after they found methanol extracted only part of the chlorophyll from the cyanobacterium *Cyanidium*. Methanol:DMF was significantly better than 90% acetone in extracting chlorophyll from freshwater phytoplankton, although similar in power to acetone–DMSO or methanol (Stauffer *et al.*1979, Hains, 1985). DMF extracted chlorophyll *a* more effectively than 80% acetone from higher plant tissue, and it could be used without grinding, unlike acetone (Moran and Porath, 1980). DMF used at 20°C, 5°C and -25°C without grinding was significantly better for extraction of chlorophyll from marine field samples, than grinding in 90% acetone (Suzuki and Ishimaru, 1990).

With 14 out of 16 marine phytoplankton cultures (Neveux, 1988), 90% aqueous DMF was not superior to acetone and the mean chlorophyll concentrations were slightly lower, although with the chlorophyte *Chlamydomonas magnusii* and the cyanobacterium *Synechocystis* sp., 10 to 300% more chlorophyll was extracted with 90% DMF than with 90% acetone. In comparing the effectiveness of 90% acetone, 90% DMF, 90% methanol, 90% ethanol, 100% ethanol, and 100% DMF for extraction of chlorophyll *a* from *Synechocystis* sp., Neveux (1988) found that both DMF extractants were better than the others, and that grinding significantly improved extractability in all cases. Without grinding, unusually high proportions of pheopigments (determined spectrophotometrically) were sometimes found in acetone extracts of exponentially growing diatoms (perhaps due to chlorophyllase enzymes being activated and chlorophyllide subsequently degrading to pheophorbide, as noted by Suzuki and Fujita (1986)). Neveux recommended 90% DMF as an extraction solvent for samples containing cyanobacteria or coccoid green algae.

DMF proved slightly better than DMSO–acetone for chlorococcalean species of green algae; neither required grinding, unlike 90% acetone (Speziale *et al.* 1984). Similarly, DMF extracted 60% more chlorophyll than acetone–DMSO from the recalcitrant chlorophyte *Nannochloris atomus* and, though less dramatically, from *Tetraselmis suecica* (Heyward, 1991). Heyward considered DMF to be more useful for marine microalgae than acetone, as pigments were extracted completely after soaking for only 30 minutes.

Chlorophyll solutions in DMF appear to be quite stable. In spectrophotometric studies, DMF extracts stored in the dark at 4°C were stable for 20 days (Moran and Porath, 1980), while chlorophyll *a* was stable in DMF for at least one month at 5°C (Heyward, 1991). HPLC checks were not done in either case. Chlorophyll concentrations were stable in DMF for at least 24 days at 5°C or -25°C, as analysed by thin-layer chromatography (Suzuki and Ishimaru, 1990).

DMF is quite toxic to humans (Table 9.1), but has the advantage of being much less flammable than acetone or alcohol. It is also much less volatile than other solvents, and evaporative loss of DMF during handling and storage is almost negligible at, or below, room temperature (Suzuki and Ishimaru, 1990).

Dimethyl acetamide

To our knowledge, dimethylacetamide (DMA) has not been used previously for extracting pigments from marine phytoplankton. It was included in the SCOR trials as a possible alternative to DMF, since it is a somewhat less aggressive solvent, is less toxic, and has an even higher flash point.

Ethanol

Ethanol appears to be the safest solvent available for pigment extraction, but while it is frequently used for freshwater phytoplankton it is rarely used for marine phytoplankton. Moed and Hallegraeff (1978) and Nusch (1980) showed that the efficiency of ethanol is similar to methanol, although Arvola (1981) found that methanol was slightly superior. Sartory and Grobbelaar (1984) found boiling 100% methanol and 90% ethanol to be superior to acetone and 90% acetone–DMSO (1:1 v/v) for chlorophyll *a* extraction from *Scenedesmus quadricauda* and *Microcystis aeruginosa* (freshwater species). They tested homogenization and sonication (in a bath) but these did not improve extractability. Jespersen and Christoffersen (1987) also found that ethanol was as efficient as methanol for extracting chlorophyll from freshwater chlorophyte cultures (mostly desmids) and natural freshwater phytoplankton, when soaked for 6 hours at room temperature, but longer extraction or heating was needed if green algae were present. The extraction efficiency dropped by 10–25% when the aqueous ethanol concentration was reduced from 96 to 83.5%. Webb *et al.*(1992) found no significant difference between 95% ethanol, 90% ethanol and DMSO/90% acetone (2:3 v/v) in the efficiency with which chlorophyll was extracted from natural lake samples dominated by cyanobacteria and diatoms. Pigments were determined both by spectrophotometry and HPLC.

Other solvents

The lipid solvent chloroform-methanol (2:1 v/v, Bligh and Dyer, 1959) extracted significantly more chlorophyll from algal cultures and marine and freshwater phytoplankton than DMSO, methanol, methanol + H_2S, or 90% acetone (Wood, 1985). Soaking for 4 hours in chloroform-methanol yielded 87% more chlorophyll *a* than other treatments.

In a series of papers, Porra studied the extraction of chlorophylls from the recalcitrant marine alga *Nannochloris atomus*. He found that buffered 80% acetone extracted <10% and methanol <75% of chlorophylls *a* and *b* (Porra, 1990*a*). Repeated extraction (4 times) with 33% petroleum ether (b.p. 40–60°C) in methanol extracted less than 10% (Porra, 1990*b*). As the difficulty appeared to be the impermeability of the cell wall to the solvent, Porra tried several very aggressive solvents: 2.1 M pyridine in 0.35M NaOH (Porra and Grimme, 1974), 2% KOH + 1.5 mM sodium dithionite in 85% aqueous methanol (Porra, 1990*a*), and 1.5 mM sodium dithionite in 85% aqueous methanol (Porra, 1990*b*). The first two solvent mixtures extracted chlorophylls as their hydroxylacetones and magnesium–rhodochlorins respectively (and would also have degraded epoxy-carotenoids due to the alkaline conditions; Liaaen-Jensen, 1990), thus rendering them unsuitable for HPLC, but the third extracted undegraded chlorophylls (determined spectrophotometrically). Porra (1991) postulated that the reducing agent in these mixtures may weaken the cell wall of *N. atomus* by disrupting the protein disulphide bridges, enabling the solvent to penetrate.

9.2.3 *Summary of extraction methods in the literature*

The above literature on extraction of algal pigments for spectrophotometry presents a rather confusing picture, but some generalizations can be made:

- Pigments are readily extracted from algae such as marine planktonic diatoms, prymnesiophytes or unarmoured dinoflagellates. Soaking overnight in acetone or methanol is usually sufficient for complete extraction of pigments, if the samples are to be analysed only by spectrophotometry, which does not distinguish between chlorophylls and their de-esterified, epimerized, or allomerized alteration products.
- For more resistant algae, such as armoured dinoflagellates, cyanobacteria and picoplankton, or freshwater species with heavily walled cells, the extra penetrating power of solvents such as DMSO or DMF is required.
- Mechanical disruption of the cell wall greatly speeds extraction and reduces the action of chlorophyllase, but the reports conflict as to whether grinding or sonication is superior. The ease of operation, lower risk of contamination and accuracy of manipulation favour sonication, provided the samples are kept cool and the sonicator has sufficient power to disrupt the cells.
- For physiological experiments with algal cultures, a simple protocol may be sufficient if it gives quantitative extraction of all pigments (e.g., colourless residues). However in field studies, where the exact composition of the phytoplankton population is seldom known at the time of extraction, a rigorous technique is required to ensure that all pigments from all species are extracted.

9.3 Experimental evaluation of extraction protocols

Solvents and disruption techniques were evaluated at the SCOR Workshops by extracting a number of algal cultures and using HPLC to determine the recoveries of pigments and the extent of pigment degradation. The algae were deliberately chosen because they were either difficult to extract or possessed chlorophyllase enzymes.

9.3.1 *Extraction protocols tested*

A screening test was performed to compare the efficacy of cell-disruption techniques and solvents. The protocols were:

- sonication, grinding, and soaking for 24 hours then grinding, in each of:
- 90% acetone, 100% acetone, or methanol.

Each protocol was performed both with and without pre-freezing the filtered sample (which was thought might aid extraction by disrupting the cells). Each test was performed on each of three algae: *Nannochloris atomus* and *Tetraselmis suecica* (both of which are notoriously difficult to extract) and *Phaeodactylum tricornutum* (which has a highly active chlorophyllase; Jeffrey and Hallegraeff, 1987). No replicates were done for the initial screening due to the large matrix of treatments (54 in all).

Once the most promising procedures were found from this screening, further replicated tests were carried out to rigorously compare these and other potentially useful procedures. Four algae, used as a mixture, reduced the number of treatments and allowed replication. Each alga had definitive marker pigments, and uniform growth in culture (no cell clumping), but their ease of extraction varied.

The algae were:

- *Nannochloris atomus* (a difficult-to-extract chlorophyte containing chlorophyll *b*, violaxanthin, and lutein as marker pigments)
- *Synechococcus* sp. (a difficult-to-extract cyanobacterium containing zeaxanthin with abundant ß,ß-carotene relative to other chromophyte algae)
- *Phaeodactylum tricornutum* (an easily extracted diatom with high chlorophyllase activity, containing fucoxanthin, chlorophylls c_1+c_2 and diadinoxanthin)
- *Emiliania huxleyi* (an easily extracted coccolithophorid containing chl c_3 and 19'-hexanoyloxyfucoxanthin).

The following protocols (with 4 replicates) were used:

- sonication in methanol, DMF, DMA, or mixtures of methanol and DMF
- grinding in acetone
- soaking 24 hr in DMSO then grinding the filter in acetone
- soaking 24 hr in acetone, DMF, or hot methanol + 1.5mM sodium dithionite.

9.3.2 *Materials and Methods*

Microalgae: growth and harvest

Cultures of *Nannochloris atomus* (CS-183), *Tetraselmis suecica* (CS-187), *Phaeodactylum tricornutum* (CS-29), *Synechococcus* sp. (CS-197), and *Emiliania huxleyi* (CS-57) were grown in 80 µE m^{-2} s^{-1} light, on 12:12 light:dark cycles at 20°C, as described in Chapter 6. Where mixtures of the algae were used, the individual cultures were first analysed separately by HPLC, then mixed in appropriate proportions for peaks of the marker pigments for each species to be clearly visible.

Algal cultures were harvested at middle to late log-phase (see Chapter 6). They were always carefully decanted into clean flasks at harvest to leave any cell residues behind. Replicate aliquots (5–15 ml) of decanted cultures containing about 10 µg chlorophyll *a* per aliquot were filtered onto 25 mm diameter Whatman GF/F filters, using 0.5 atmospheres of vacuum or less. The filters were folded once (algae inside) then quickly blotted between layers of filter paper to reduce variability due to water content.

In the screening experiment, the filters were either extracted immediately ('Fresh' treatment) or after freezing at -18°C for about 2 hours ('Frozen' treatment). In the replicated experiment, all samples were frozen immediately after filtration by dipping them in liquid nitrogen before being placed in containers and stored in a freezer at -18°C. The entire process from beginning of filtration to freezing took no more than 30 seconds.

Extraction

Pigments were extracted by one of four methods:

- *Sonication*. Sonication was by Wright and Shearer's (1984) method described in Chapters 12 and 17. Filters were cut and sonicated in modified centrifuge tubes (detailed in Chapter 17) with 3.5 ml of solvent for 30 s at 50 W using a Braun Labsonic 1510 sonicator equipped with a 4 mm diameter probe inserted directly into the solvent. The extract was collected by centrifugal filtration (700 g, 3min) and the pellet washed with a further 0.5 ml of solvent before re-centrifugation. This procedure usually took 9 to 10 min, from cutting filters to obtaining the clarified extract.
- *Grinding*. The grinding method was modified from that of SCOR-UNESCO

(1966). Filters were ground in 3 ml of solvent in a glass homogenizer (25 ml capacity, held in an ice-bath) with a motor-driven teflon pestle, until the filter was completely disrupted (about 1 min). The resultant slurry was quantitatively transferred to the modified centrifuge tubes, and centrifuged and washed as described above. This procedure took about 12 min.

- *Soaking*. Filters were placed in 3 ml of solvent for 24 hours at 4°C in the dark . They were thoroughly mixed before and after storage with a Labline Supermixer.
- *Soaking and grinding*. Filters were placed in 3 ml of solvent in a centrifuge tube for 24 hours at 4°C in the dark (or 20 min at 60°C, in hot 85% methanol + 1.5 mM sodium dithionite) before being ground and centrifuged as described above. For the DMSO/acetone treatment, filters were soaked in 1.5 ml DMSO for 24 hours at 4°C in the dark. The DMSO was removed and the filter was ground in 2.0 ml of acetone as described above. Acetone was used for rinsing and adjusting the extract volume to 5.0 ml, reducing the final concentration of DMSO to 30%.
- *Exhaustive extraction for nominal 100% extraction values*. After all treatments were evaluated, the two methods found to be most effective (sonication in methanol and sonication in dimethylformamide) were each used twice in sequence on the same sample (in quadruplicate) to provide an estimate of 100% extractability. This procedure took about 20 minutes. The recovery from this exhaustive extraction was sometimes less than a particular treatment (e.g., chlorophyll *a* recovery was 2.06 ± 0.01 µg/ml for sonication in DMF compared to 1.99 ± 0.01 µg/ml for the exhaustive extraction). In such cases, the highest value was used as the 100% estimate.

Pigment stability in solvents

To test the stability of pigments in the various solvents, three replicate extracts from sonication in methanol, sonication in DMF, and grinding in acetone were re-injected into the HPLC after standing in the autoinjector at -10°C for about 24 hours. The recovery and degradation of the major pigments were then compared with those of the fresh extracts.

HPLC analysis

In all treatments, the final centrifuged extract was made up to 5.0 ml in a volumetric flask before a final filtration through a 0.45 µm pore-size teflon filter into an amber glass autosampler vial.

The extracts were then placed in a refrigerated (-10°C) autosampler stage in a darkened room for up to 8 hours until HPLC analysis was completed. HPLC was by the SCOR-recommended method (Wright *et al.* 1991; Chapter 12). Immediately before injection, the extract was diluted with water (20% by volume) which improves peak-shape and detectability (Wright *et al.* 1991). The absorbance chromatograms (λ = 435 nm, see Fig. 9.2) were quantitated using Waters Maxima software after peak integration was checked manually.

Calculation of results

In the initial screening experiment (see Section 9.3.1), the absolute extractability of pigments was not determined. Instead, the amount of each major pigment extracted in each sample was divided by the greatest amount extracted by any treatment for the same species to obtain a relative-extractibility figure for each pigment. The extractabilities for all pigments in a treatment were then averaged to produce a single relative-extractibility figure for each treatment.

In the replicated experiment, the absolute amount of each pigment in an algal mixture was estimated after exhaustive extraction, as described above. The percentage extractibility of each pigment could be then be calculated for each treatment.

In both experiments, contributions from pigment degradation products (mainly *cis*-isomers of carotenoid and allomers and epimers of chlorophylls) were included in the extractibility figures for each pigment. The percentages (w/w) of degradation products are presented separately.

9.4 RESULTS

9.4.1 *Extractability*

Preliminary screening experiment

The extractability of pigments from the three algal species for each of the 18 treatments are shown in Table 9.2A. The percentages of significant pigment degradation products are tabulated separately (Table 9.2B).

TABLE 9.2. Preliminary screening extraction experiment. A. Relative pigment extraction. Figures as percentage of best value (summed for all pigments per species). B. Percentage (w/w) of major degradation products in each treatment.

	100% Acetone			90% Acetone			Methanol			Range
	Sonic	Grind	Soak/G	Sonic	Grind	Soak/G	Sonic	Grind	Soak/G	
A. Percentage extractibility (all pigments)										
Nannochloris atomus										
Fresh	15	68	85	19	61	46	99	87	89	15–99
Frozen	12	60	74	19	49	60	75	84	72	12–84
Tetraselmis suecica										
Fresh	64	81	84	65	60	75	98	65	79	60–98
Frozen	80	80	69	79	68	83	100	81	73	68–100
Phaeodactylum tricornutum										
Fresh	66	84	87	64	60	69	88	80	95	60–95
Frozen	71	71	72	64	67	75	84	84	84	71–84
B. Percentage degradation										
Chlorophyll *a* degradation (% of total chlorophyll *a*)										
Nannochloris atomus										
Fresh	0	1	2	0	1	45	5	3	13	0–45
Frozen	5	1	7	0	1	9	6	5	7	0–9
Tetraselmis suecica										
Fresh	1	2	1	2	1	1	6	2	4	1–6
Frozen	5	0	2	1	1	3	3	2	9	0–9
Phaeodactylum tricornutum										
Fresh	17	2	11	14	4	42	13	2	15	2–42
Frozen	12	21	6	18	21	29	8	22	4	4–29
Carotenoid isomers (% of total carotenoids)										
Tetraselmis suecica										
Fresh	56	42	40	39	47	25	21	29	22	21–56
Frozen	30	30	60	27	24	49	13	18	45	13–60

Methanol was the best solvent for each species with either sonication or 'soaking and grinding' extracting the most pigment. There was little difference between 90% and 100% acetone, but both were inferior to methanol in each species.

The extractibility of fresh or frozen filters were similar except in *Nannochloris atomus* where fresh filters yielded slightly more pigment.

Overall, there was a wide range of effectiveness between the different solvents and treatments. In *N. atomus*, the range was 15–99% for fresh filters and 12–84% for frozen; in *Tetraselmis suecica* it was 60–98% for fresh and 68–100% for frozen; and in *Phaeodactylum tricornutum*, it was 60–95% for fresh and 71–84% for frozen filters.

Degradation of chlorophyll *a* to chlorophyllide was only substantial when 90% acetone was used with the 'soaking and grinding' treatment in *N. atomus* and *P. tricornutum* (45% and 42% degradation respectively). Other treatments caused 0 to 22% degradation in these species, and always less than 9% in *T. suecica*. In *T. suecica*—the only species in which carotenoid isomerization was significant—methanol always produced less isomers than 90% or 100% acetone.

Three observations showed the complexity of the interactions of solvents and treatments in the extraction process. First, when using either 90% or 100% acetone, grinding gave much better extraction than sonication, yet when using methanol, sonication was better than grinding. Second, freezing inhibited chlorophyllase with the 'soak and grind' treatment, and had little effect on chlorophyllase in sonicated treatments, but promoted chlorophyllase activity when grinding alone was used. Third, freezing consistently inhibited carotenoid isomerization using sonication or grinding, but produced the greatest isomerization when used with the 'soak and grind' treatment.

As the complex interactions precluded a simple step-by-step optimization of the extraction process, a more extensive replicated experiment was done to compare promising combinations of extraction parameters, with the goal of finding the best extraction procedure.

Replicated experiment

The extractability of the major pigments (grouped by algal source) and the degradation found for each technique are shown in Figure 9.1. The results, starting with the most successful treatments are described below.

Sonication in DMF or methanol

The highest pigment concentrations were obtained by sonication in DMF (Fig. 9.1A); this gave better than 98% extractability for every major pigment. Reproducibility of extraction was excellent: standard deviations for the major pigments ranged from 0.4 to 6.0%. However, there was 10% degradation of chlorophyll *a* (producing chlorophyllide *a*, chlorophyll *a* allomer and epimer), 15% degradation of chlorophyll *b* (mainly epimer) and 4% formation of carotenoid *cis*-isomers.

Sonication in methanol (Fig. 9.1B) gave the second highest pigment extractions. Chlorophyll *a* and the cyanobacterial pigments (zeaxanthin and ß,ß-carotene) were 15 - 20% less than those from sonication in DMF, but the extractions of other marker pigments were similar. The standard deviations were ±5%, with an upper limit of 13% for ß,ß-carotene (a minor pigment). Pigment degradation was less in methanol than in DMF: only 8% degradation of chlorophyll *a* (to methyl chlorophyllide *a* and chlorophyll *a* epimer), negligible degradation of chlorophyll *b*, and 4% of carotenoids recovered as *cis*-isomers.

Sonication in DMF:methanol combinations

The results obtained for sonication in 90%, 50%, and 10% DMF in methanol are shown in Figure 9.1C–E. Apart from the extractibility of chlorophyll *a* and ß,ß-carotene, which increased slightly with increasing concentrations of DMF, the results of the three treatments were similar, although more variable in the third. Chlorophyll *a* degradation fell from 11% to 6% as the DMF concentration was reduced (the main derivatives were epimer and allomer, and in the two higher concentrations, chlorophyllide). Chlorophyll *b* degradation ranged from 0 to 12% in the three treatments. Carotenoid isomerization was constant at about 3%.

Acetone treatments

Grinding in 100% acetone without prior soaking (Fig. 9.1F) produced substantially lower recoveries of most pigments: about 90% of fucoxanthin and 19'-hexanoyloxyfucoxanthin (reflecting the ease with which *Phaeodactylum tricornutum* and *Emiliania huxleyi*, respectively, yield their pigments); only about 60% of chlorophylls c_1, c_2 and c_3; 74% of chlorophyll *a*; and less than 48% of the green algal pigments (chlorophyll *b*, violaxanthin, and lutein). There was 5% degradation of chlorophyll *a* (to allomer and epimer), 7% degradation of chlorophyll *b* (to epimer), and 8% isomerization of carotenoids.

Soaking for 24 hours in DMSO before grinding in acetone (Fig. 9.1G) produced a similar result to grinding in acetone, except that the green algal pigments were better extracted (with an corresponding 10% increase in extraction of chlorophyll *a*). However, there was 21% degradation of chlorophyll *a* (predominantly epimer, with some chlorophyllide and allomer) and 31% degradation of chlorophyll *b* (epimer and unknown derivative). Surprisingly, only 5% of the carotenoids were isomerized.

Soaking for 24 hours in 100% acetone (Fig. 9.1H) extracted only 56% of chlorophyll *a* and less than 13% of the green algal pigments. Only 6% of the chlorophyll *b* was extracted, and 66% of that was recovered as the epimer. Degradation of chlorophyll *a* was 8% (mainly to a derivative eluting immediately after the epimer, plus some allomer), whilst 13% of carotenoids were recovered as *cis*-isomers.

Miscellaneous treatments

Soaking in DMF (Fig. 9.1I) produced good average recoveries for some carotenoids (most greater than 85%), and chlorophyll *a* (90%), but chlorophylls *b*, c_1+c_2, and c_3 averaged less than 72%. There was considerable variability in extraction: the standard deviations for most pigments were 10 to 48% (chlorophyll *a* was 2%). Moreover, 20% of chlorophyll *a* was degraded (chlorophyllide, epimer, and a derivative preceding the allomer), 25% of chlorophyll *b* was epimerized, and 12% of carotenoids were isomerized.

Sonication in dimethylacetamide (DMA, Fig. 9.1J) was clearly inferior to DMF: average extraction figures were 22% lower. There was 20% degradation of chlorophyll *a* (predominantly epimer, with chlorophyllide and allomer), 34% degradation of chlorophyll *b* (epimer and unknown derivative), and 8% isomerisation of carotenoids.

Soaking in hot 85% methanol plus 1.5 mM sodium dithionite or sonication in the same (cold) solvent produced the results shown in Figures 9.1K, L. Poor extractability of green algal and cyanobacterial pigments occurred in the two treatments, with only 44% and 22% extraction of chlorophyll *a*, respectively. In the hot-soak treatment, 8% of chlorophyll *a* was degraded, chiefly to the epimer, but in the soni-

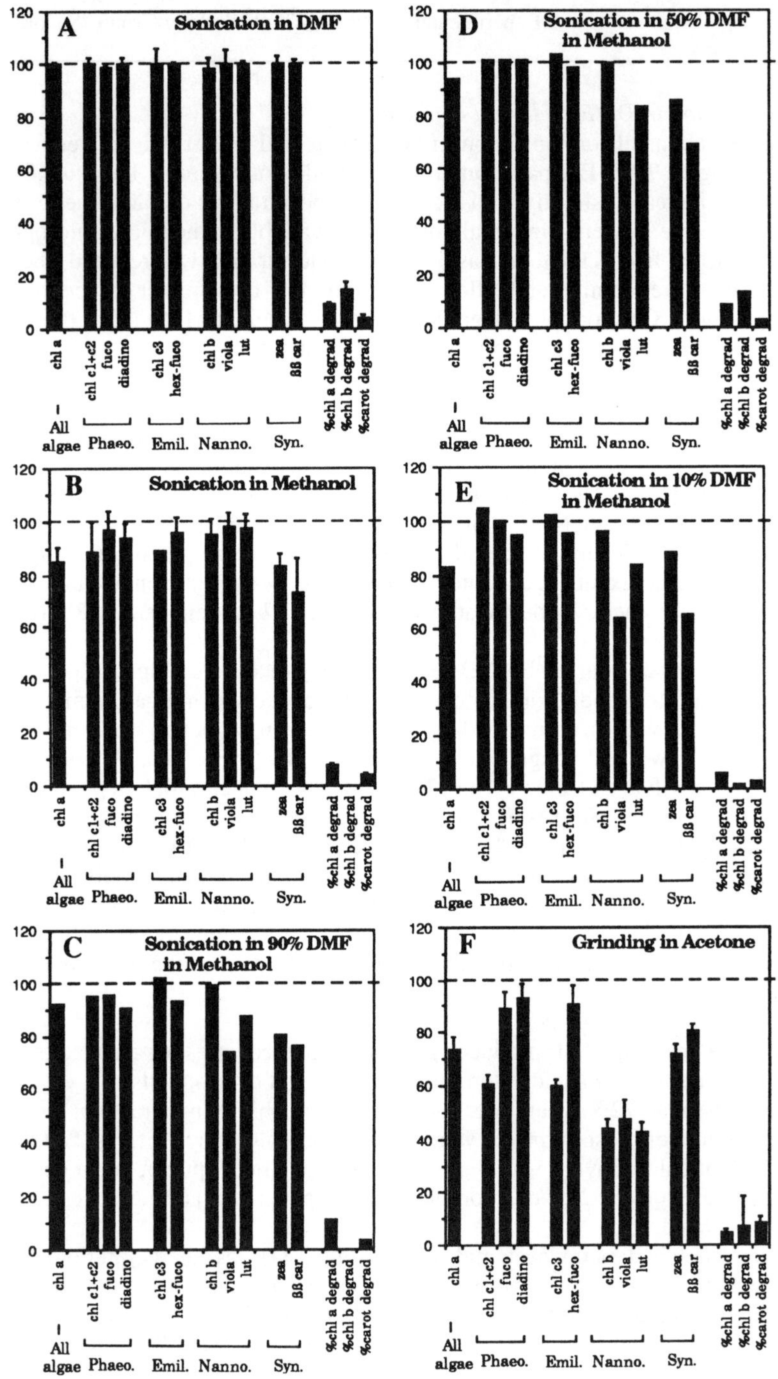

Figure 9.1
Extractability of marker pigments from a mixture of cultures of *Phaeodactylum tricornutum* (diatom), *Emiliania huxleyi* (coccolithophorid), *Nannochloris atomus* (chlorophyte), and *Synechococcus* sp. (cyanobacterium), and the percentage (w/w) degradation of pigments. Pigments are grouped by their algal source. The 100% extraction efficiency figure for each pigment was obtained from exhaustive extraction of the cultures, using repeated sonication in dimethyl formamide and in methanol. These were (in nanograms per 100 µl injected): chlorophyll *a*, 186.4; chlorophyll c_1+c_2, 18.3; fucoxanthin, 29.3; diadinoxanthin, 9.0; chlorophyll c_3, 8.5; 19'-hexanoyloxyfucoxanthin, 43.4; chlorophyll *b*, 11.4; violaxanthin,

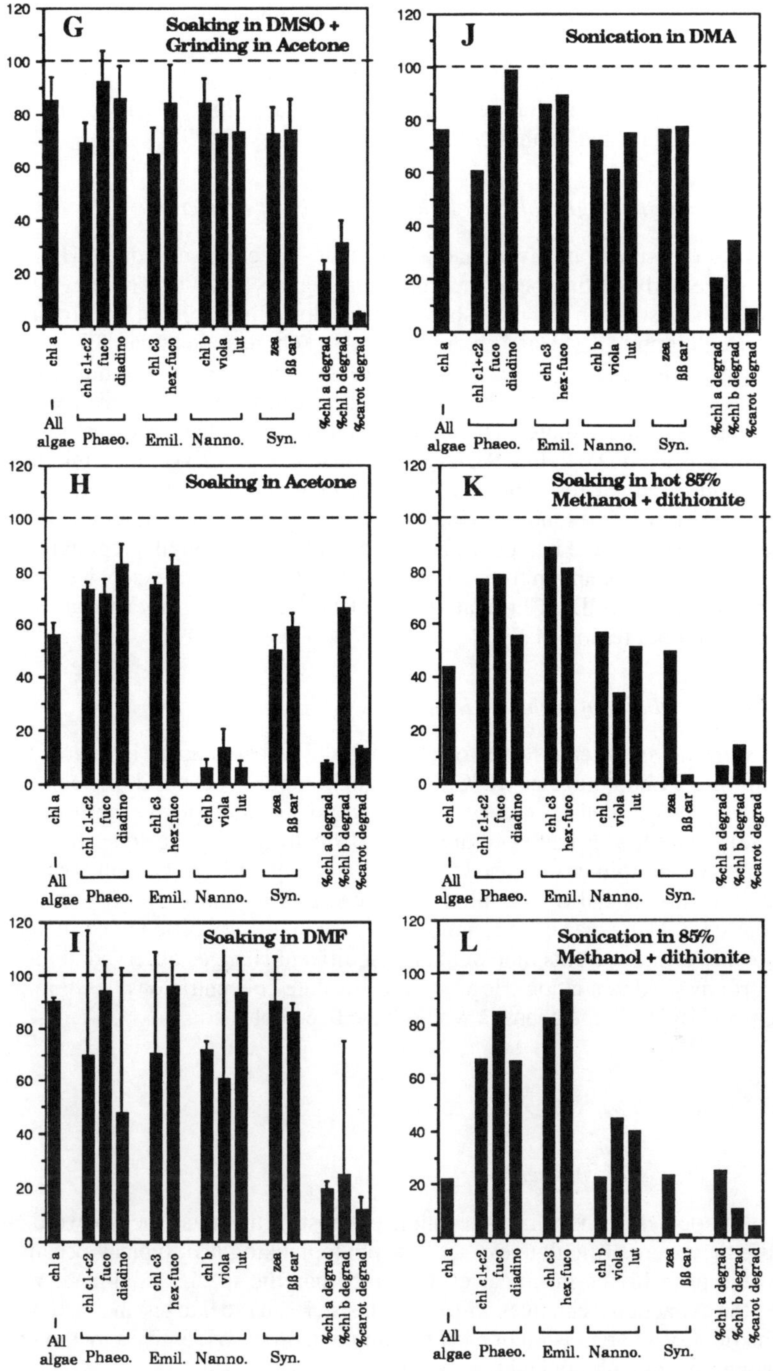

2.2; lutein, 9.2; zeaxanthin, 5.4; ß,ß-carotene, 10.7. Error bars represent 1 standard deviation in the treatments where triplicate extractions were performed. A, sonication in DMF; B, sonication in methanol; C, sonication in DMF:methanol = 9:1 (v/v); D, sonication in DMF:methanol = 1:1 (v/v); E, sonication in DMF:methanol = 1:9 (v/v); F, grinding in acetone; G, soaking in DMSO (24 hr) then grinding in acetone; H, soaking in acetone (24 h); I, soaking in DMF (24 h); J, sonication in DMA; K, soaking (20 min) in hot (60°C) 85% methanol plus 1.5 mM sodium dithionite; L, sonication in cold 85% methanol plus 1.5 mM sodium dithionite.

cated treatment, chlorophyll *a* degradation increased to 25%, chiefly by production of chlorophyllide *a* and methyl chlorophyllide *a*.

9.4.2 *Chromatographic resolution after solvent extraction*

The choice of extraction solvent had a noticeable effect on subsequent HPLC resolution of the early-eluting peaks, as shown in Figures 9.2A–D. The biggest effects were with the resolution and peak shape of the chlorophyll *c* region (chlorophyllide *a* (peak 3), chlorophyll c_3 (5), and chlorophyll $c_1 + c_2$ (6b; retention times 3.7 to 7.1 min.) and the resolution of neoxanthin (14) from 19'-hexanoyloxyfucoxanthin (15) (retention times 8.8 to 9.4 min.). The differences between treatments, which were quite reproducible, depended more on the solvent than on the mechanical extraction method used. Extracts from DMF (Fig. 9.2A) and methanol (Fig. 9.2B) or their mixtures (not shown) had excellent resolution in the chlorophyll *c* region, but the resolution of neoxanthin from 19'-hexanoyloxyfucoxanthin was poor. Chromatograms from acetone extracts (Fig. 9.2C) had poor peak shape in the chlorophyll *c* region but excellent resolution of neoxanthin from 19'-hexanoyloxyfucoxanthin (except where DMSO was included, Fig. 9.2D). Chromatograms of DMA extracts (not shown) had poor resolution in both regions.

9.4.3 *Stability of pigments in solvents*

There were no significant losses of chlorophylls or carotenoids in any of the solvents during 24 hr storage at -10°C, but a small increase occurred in the proportion of chlorophyll degradation products. The proportion of chlorophyll *a* epimer increased from 2.0% to 3.8% in methanol and from 0.7% to 6.0% in acetone. In DMF, the proportion of the epimer increased from 0.9% to only 1.0%, but an unknown derivative eluting just ahead of the allomer increased from 0.8% to 4.2% of the total chlorophyll *a*. There was no measurable increase in the proportion of carotenoid isomers in any of the solvents, nor were any significant changes observed in the extent of chlorophyll *b* degradation. However the absolute concentrations of chlorophyll *b* were low and only large changes would have been detected.

9.5 Discussion

9.5.1 *General comments*

To test the power of the various techniques, most of the algal species were chosen because they were difficult to extract. The phytoplankton in a 'typical' oceanic sample would probably be much easier to extract than the species used here. We challenged the extraction techniques with difficult species in order to test their extractibility and the fidelity and precision of each technique in order to identify a method powerful enough for any phytoplankton sample.

9.5.2 *Extractability*

Sonication in DMF was clearly the best procedure, with nearly 100% extractability for all pigments. Reproducibility was excellent, with a standard deviation of 0.7%

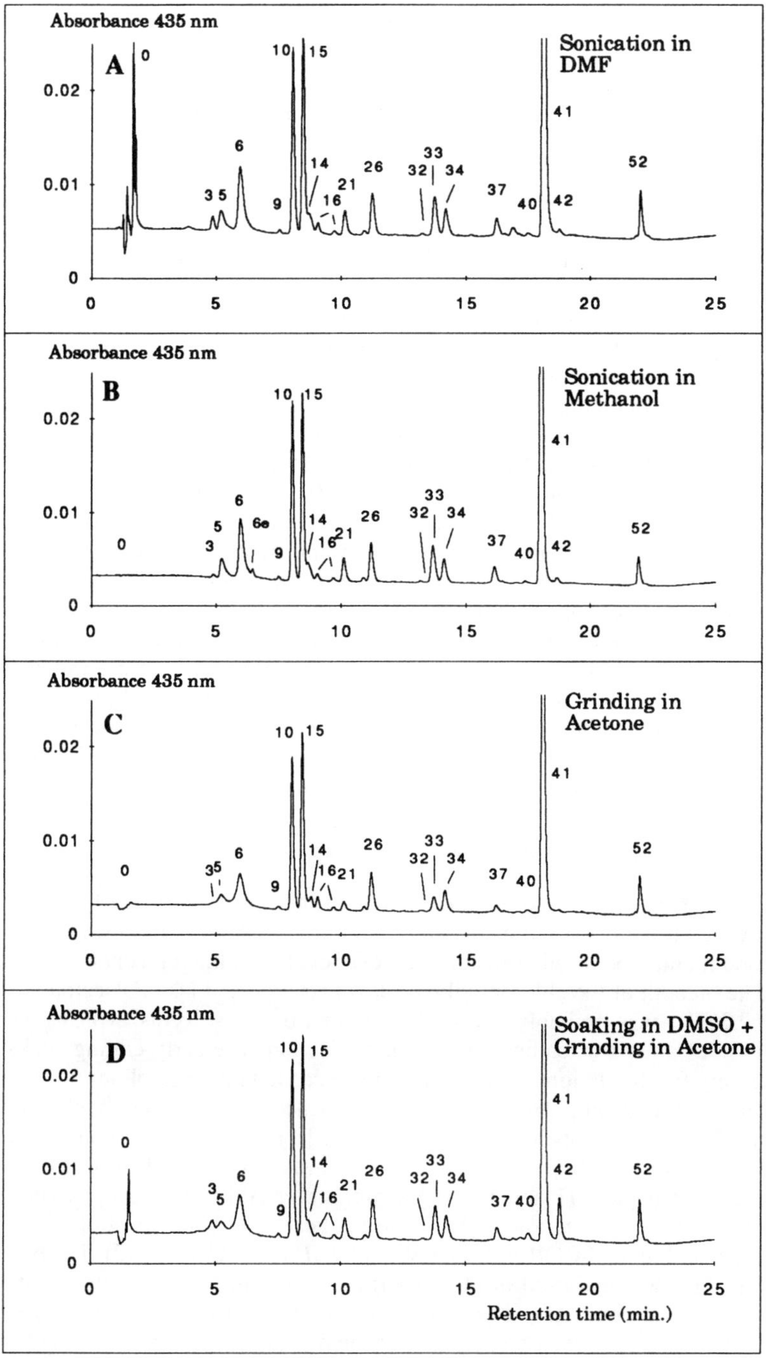

Figure 9.2
Representative HPLC chromatograms for some of the extraction protocols tested on a mixture of *Phaeodactylum tricornutum*, *Emiliania huxleyi*, *Nannochloris atomus* and *Synechococcus* sp.: A, sonication in DMF; B, sonication in methanol; C, grinding in acetone; D, soaking in DMSO (24h) then grinding in acetone. Peak numbers match those of Table 12.2 and are as follows: 3, chlorophyllide *a*; 5, chlorophyll c_3; 6, chlorophyll $c_1 + c_2$; 6e, methyl chlorophyllide *a*; 9, 19'-butanoyloxyfucoxanthin; 10, fucoxanthin; 14, neoxanthin; 15, 19'-hexanoyloxyfucoxanthin; 16, *cis*-fucoxanthin; 21, violaxanthin; 26, diadinoxanthin; 32, diatoxanthin; 33, lutein; 34, zeaxanthin; 37, chlorophyll *b*; 40, chlorophyll *a* allomer; 41, chlorophyll *a*; 42, chlorophyll *a* epimer; 52, ß,ß-carotene.

for chlorophyll *a* for the entire procedure from dispensing the algal cultures, analytical HPLC and integration, and the extraction itself. Degradation of chlorophylls *a* and *b* was slightly higher than with other solvents, but acceptable at 9.6% and 15% respectively, while only 4.4% of carotenoids were isomerized. The chromatographic resolution of the most polar pigments was excellent, although that of the 19'-hexanoyloxyfucoxanthin-neoxanthin pair (Fig. 9.2A, peaks 15 and 14) was not as good as with acetone (Fig. 9.2C).

Soaking for 24 hours in DMF, on the other hand, did not achieve the same extractability, was much more variable, and produced 19.6% and 25.0% degradation of chlorophylls *a* and *b* respectively.

Sonication in methanol was the next best procedure, and gave good recoveries of most marker pigments, with about 16% underestimation of cyanobacterial pigments. Combinations of methanol and DMF increased the extraction of chlorophyll *a* as the DMF concentration increased, but even 90% DMF in methanol did not improve the extractability of zeaxanthin or ß,ß carotene over that of methanol.

Soaking or grinding in acetone seriously underestimated the green algal pigments from the recalcitrant *N. atomus* and *T. suecica*. Soaking for 24 hours produced considerable degradation of chlorophyll *b*. Even fucoxanthin and 19'-hexanoyloxyfucoxanthin were incompletely extracted from *Phaeodactylum tricornutum* and *Emiliania huxleyi*, respectively. Addition of a DMSO soak before grinding in acetone increased the extraction of chlorophyll *a* and cyanobacterial pigments but at the expense of degrading chlorophyll *a* and *b* by 21% and 31%, respectively.

Neither dimethylacetamide nor 85% methanol/dithionite appeared to have any promise as extractants for HPLC of phytoplankon pigments.

9.5.3 *HPLC peak resolution*

Differences between the extraction solvents in the resolution of early-eluting pigments were probably caused by differences in the polarity of the solvents as well as the amount and type of other material co-extracted with the pigments.

The breadth of the chlorophyll *c* peaks in acetone and DMSO/acetone extracts (Figs. 9.2C,D) was probably due to the strength of the solvents, which prevented formation of a narrow loading band during injection. The early-eluting peaks have less affinity for the stationary phase and can be carried into the column by the injection solvent, broadening the loading band and hence the subsequently-eluted peaks. Perhaps the resolution of chlorophyll *c* pigments from these extracts could have been improved by further diluting the extract with water or buffer before injection, as discussed in Chapters 11 and 12. The samples were diluted by 20% water, added immediately before injection, which gave the best results for injection of buffered 98% methanol in the SCOR method (Wright *et al.* 1991; Chapter 12). The best dilution for other solvents used in this experiment were not individually sought. The stronger solvents could have been diluted further to improve the shape of the initial peaks, but this would be at the expense of the overall sensitivity of the method.

The differences in resolution between the later-emerging peaks, notably neoxanthin (peak 14) and 19'-hexanoyloxyfucoxanthin (15), in extracts from different solvents were probably caused by a mechanism other than the injection effect noted above, since the effects were not general and affected the retention time of only some pigments. These effects could have been produced by compounds, such as lipids or lipoproteins, being differentially extracted with the pigments, and subsequently affecting the affinity of pigments for the column by binding to the stationary phase or complexing the pigments in the mobile phase. The extent to which this occurred was not investigated, although it has been noted (Wright, unpublished) that DMF seems to extract more contaminating material than either methanol or acetone. This material can precipitate during concentration of extracts for semi-preparative work and may reduce the life of the column.

9.6 Conclusions and recommendations

Sonication in dimethyl formamide clearly gave the best pigment extraction of all protocols tested and may be thus regarded as the 'Reference' extraction method. Close to 100% extractability was obtained for all pigments, even from the recalcitrant chlorophyte algae used. This technique fulfils five of the six criteria required of a good extractant (Section 9.1): extractability, acceptable fidelity, compatibility with the HPLC solvents, precision and simplicity of the method. However, because of its toxicity, it does not fulfil the sixth criterion: safety for the operator (Table 9.1). We are therefore reluctant to recommend DMF for routine work, particularly at sea, although if safety procedures are fully observed, its low flammability and low volatility, which minimizes evaporative losses during handling and storage (Suzuki and Ishimaru, 1990), may offset its toxicity.

Sonication in methanol appears to be a practical alternative to DMF. Even with the most recalcitrant algae tested here, there was only 15–20% underestimation of the pigments. In natural field samples, such algae would seldom be major components (except in freshwater environments) so underestimation of the total population would be negligible. The fidelity of extraction with methanol is excellent, as is the compatibility with HPLC solvents.

We therefore recommend that routine field samples should be sonicated in methanol. The absolute extractability of the methanol should be monitored at intervals against the DMF 'Reference' method to ensure there is no significant underestimation of a particular component. For physiological studies with cultures, this may mean a single comparison to ensure that the species is completely extracted by methanol, but in field studies, occasional duplicate filters should be taken in different zones and depths to guard against unrecognized changes in the composition of algal populations.

We strongly recommend that soaking without mechanical disruption should not be used as an extraction method since recovery is low, variable, and always accompanied by degradation products. Pre-soaking before grinding in acetone generally increases pigment extractibility, but often at the expense of chlorophyll degradation.

Freezing of the sample before extraction did not appear to increase extractability, and the effect on chlorophyllase activity depended markedly on the subsequent extraction technique. It is recommended that unless samples are to be extracted immediately, they should be frozen and stored in the dark. The recommended methods for storage are described in Chapters 10 and 17.

For spectrophotometric analysis, acetone may still be used to extract pigments from diatoms and naked flagellates (in cultures or field samples), since there are accurate simultaneous equations for chlorophylls *a, b,* and *c* in this solvent (Jeffrey and Humphrey, 1975). Porra *et al.* (1989) have recently published equations for chlorophylls *a* and *b* in both methanol and DMF allowing use of these solvents for green algae or other samples where chlorophyll *c* is not present.

Acknowledgements

We thank Ms. C. Llewellyn for HPLC analyses of samples from the screening experiments and Ms. J. M. Leroi for growing the algal cultures.

References

Alliger, H. (1975). Ultrasonic Disruption. American Laboratory, Oct. 1975.

Arnon, D. I. (1949). Copper enzymes in isolated chloroplasts: polyphenoloxidase in *Beta vulgaris*. *Plant Physiol.* 24: 1–15.

Arvola, L. (1981). Spectrophotometric determination of chlorophyll *a* and pheopigments in ethanol extractions. *Ann. Bot. Fennici,* 18: 221–227.

Barrett, J., Jeffrey, S. W. (1964). Chlorophyllase and formation of an atypical chlorophyllide in marine algae. *Plant Physiol.* 33: 44–47.

Barrett, J., Jeffrey, S. W. (1971). A note on the occurrence of chlorophyllase in marine algae. *J. Exp. Mar. Biol. Ecol.* 7: 255–262.

Bligh, E. G., Dyer, W. J. (1959). A rapid method of total lipid extraction and purification. *Can. J. Biochem. Physiol.* 37: 911–917.

Bowles, N. D., Paerl, H. W., Tucker, J. (1985). Effective solvents and extraction periods employed in phytoplankton carotenoid and chlorophyll determinations. *Can. J. Fish. Aquat. Sci.* 42: 1127–1131.

Burnison, B. K. (1980). Modified dimethyl sulfoxide (DMSO) extraction for chlorophyll analysis of phytoplankton. *Can. J. Fish. Aquat. Sci.* 37: 729–733.

Chemwatch (1993). Chemical Management Systems, Australia.

Chang, W. Y. B., Rossmann, R. (1982). The influence of phytoplankton composition on the relative effectiveness of grinding and sonication for chlorophyll extraction. *Hydrobiol.* 88: 245–249.

De Renzo, D. J. (1986). *Solvents Safety Handbook.* Noyes Publications, Park Ridge, U. S. A., 696 pp.

Diehn, B., Seely, G. R. (1968). The oxidation of chlorophyll *a* in alcohols. *Biochim. Biophys. Acta,* 153: 862–867.

Gieskes, W. W. C., Kraay, G. W., Nontji, A., Setiapermana, D., Sutomo (1988). Monsoonal alteration of a mixed and a layered structure in the phytoplankton of the euphotic zone on the Banda Sea (Indonesia): a mathematical analysis of algal pigment fingerprints. *Neth. J. Sea Res.* 22: 123–137.

Hains, J. J. (1985). Practical considerations for routine chlorophyll measurements: precautions and comparisons of extraction methods. *J. Freshw. Ecol.* 3: 175–180.

Hayes, M. H. B. (1985). Extraction of humic substances from soil. In: Aiken, G. R., McKnight, D. M., Wershaw, R. L, McCarthy, P. (eds), *Humic Substances in Soil, Sediment, and Water*, John Wiley and Sons, New York, pp. 329–362.

Heyward, A. J. (1991). Chlorophyll extraction from marine microalgae: verification and extension of an improved technique. *Proc. Austr. Soc. Phycol. Aquat. Bot.* p. 9.

Holm-Hansen, O., Riemann, B. (1978). Chlorophyll *a* determination: improvements in methodology. *Oikos*, 30: 438–447.

Humphrey, G. F., Wooton, M. (1966). Comparison of techniques used in the determination of phytoplankton pigments. In: *Determination of Photosynthetic Pigments in Seawater*, UNESCO Monograph on Oceanographic Methodology, No. 1, Paris, p. 37–63.

Jeffrey, S. W. (1974). Profiles of photosynthetic pigments in the ocean using thin-layer chromatography. *Mar. Biol.* 26: 101–110.

Jeffrey, S. W., Hallegraeff, G. M. (1987). Chlorophyllase distribution in ten classes of phytoplankton: a problem for chlorophyll analysis. *Mar. Ecol. Prog. Ser.* 35: 293–304.

Jeffrey, S. W., Humphrey, G. F. (1975). New spectrophotometric equations for determining chlorophylls *a, b,* c_1 and c_2 in higher plants, algae and natural phytoplankton. *Biochem. Physiol. Pflanzen,* 167: 191–194.

Jeffrey, S. W., Wright, S. W. (1994). Photosynthetic pigments in the Prymnesiophyceae. In: Green, J. C., Leadbeater, B. S. C. (eds), *The Haptophyte Algae*, Oxford University Press, Oxford, p. 111–132.

Jespersen, A. -M., Christoffersen, K. (1987). Measurements of chlorophyll *a* from phytoplankton using ethanol as an extraction solvent. *Arch. Hydrobiol.* 109: 445–454.

Kraay, G. W., Zapata, M., Veldhuis, M. J. W. (1992). Separation of chlorophylls c_1, c_2, and c_3 of marine phytoplankton by reversed-phase C18 high-performance liquid chromatography. *J. Phycol.* 28: 708–712.

Lenga, R. E. (1988). *The Sigma-Aldrich Library of Chemical Safety Data*, vol. 1, 2nd ed., Sigma-Aldrich Corp., Milwaukee, 2175 pp.

Liaaen-Jensen, S. (1990). Artifacts of natural carotenoids - unintended carotenoid synthesis. In: Krinsky, N. I., Mathews-Roth, M. M., Taylor, R. F. (eds), *Carotenoids. Chemistry and Biology*. Plenum Press, New York, pp. 149–165.

Marker, A. F. H. (1972). The use of acetone and methanol in the estimation of chlorophyll in the presence of pheophytin. *Freshw. Biol.* 2: 361–385.

Moed, J. R., Hallegraeff, G. M. (1978). Some problems in the estimation of chlorophyll *a* and pheopigments from pre- and post-acidification spectrophotometric measurements. *Int. Rev. Ges. Hydrobiol.* 63: 787–800.

Moran, R., Porath. D. (1980). Chlorophyll determination in intact tissues using N,N-dimethylformamide. *Plant Physiol.* 65: 478–479.

Neveux, J. (1988). Extraction of chlorophylls from marine phytoplankton. *Verh. Internat. Verein. Limnol.* 23: 928–932.

Nusch, E. A. (1980). Comparison on different methods for chlorophyll and pheopigment determination. *Arch. Hydrobiol. Beih. Ergebn. Limnol.* 14: 14–36.

Parsons, T. R., Maita, Y., Lalli, C. M. (1984). *A Manual of Chemical and Biological Methods for Seawater Analysis*. Pergamon Press: Oxford. 173 pp.

Pechar, L. (1987). Use of an acetone:methanol mixture for the extraction and spectrophotometric determination of chlorophyll *a* in phytoplankton. *Arch. Hydrobiol. Suppl.* 78: 99–117.

Porra, R. J. (1990*a*). The assay of chlorophylls *a* and *b* converted to their respective magnesium–rhodochlorin derivatives by extraction from recalcitrant cells with aqueous methanol: prevention of allomerisation with reductants. *Biochim. Biophys. Acta*, 1015: 493–502.

Porra, R. J. (1990*b*). A simple method for extracting chlorophylls from the recalcitrant alga, *Nannochloris atomus*, without formation of spectroscopically different magnesium–rhodochlorin derivatives. *Biochim. Biophys. Acta*, 1019: 137–141.

Porra, R. J. (1991). Recent advances and re-assessments in chlorophyll extraction and assay procedures for terrestrial, aquatic, and marine organisms, including recalcitrant algae. In: Scheer, H. (ed.), *Chlorophylls*. CRC Press, Boca Raton, pp 31–57.

Porra, R. J., Grimme, L. H. (1974). A new procedure for the determination of chlorophylls *a* and *b* and its application to normal and regreening *Chlorella. Anal. Biochem.* 57: 255–262.

Porra, R. J., Thompson, W. A., Kriedmann, P. E. (1989). Determination of accurate extinction coefficients and simultaneous equations for assaying chlorophylls *a* and *b* extracted with four different solvents: verification of the concentration of chlorophyll standards by atomic absorption spectroscopy. *Biochim. Biophys. Acta*, 975: 384–394.

Rai, H. (1973). Methods involving the determination of photosynthetic pigments using spectrophotometry. *Verh. Internat. Verein. Limnol.* 18: 1864–1875.

Richards, F. A., Thompson, T. G. (1952). The estimation and characterisation of plankton populations by pigment analysis. II. A spectrophotometric method for the estimation of plankton pigments. *J. Mar. Res.* 11: 156–172.

Riemann, B. (1980). A note on the use of methanol as an extraction solvent for chlorophyll *a* determination. *Arch. Hydrobiol. Beih. Ergebn. Limnol.* 14: 70–78.

Sand-Jensen, K. (1976). A comparison of chlorophyll *a* determinations of unstored and stored plankton filters extracted by methanol and acetone. *Vatten*, 4: 337–341

Sartory, D. P., Grobbelaar, J. U. (1984). Extraction of chlorophyll *a* from freshwater phytoplankton for spectrophotometric analysis. *Hydrobiologia*, 114: 177–187.

SCOR-UNESCO Working Group 17 (1966). *Determination of Photosynthetic Pigments in Seawater*. UNESCO Monographs on Oceanographic Methodology, 1, Paris, 69 pp.

Shoaf, W. T., Lium, B. W. (1976). Improved extraction of chlorophyll *a* and *b* from algae using dimethyl sulfoxide. *Limnol. Oceanogr.* 21: 926–928.

Speziale, B. J., Schreiner, S. P., Giammatteo, P. A., Schindler, J. E. (1984). Comparison of N, N-dimethylformamide, dimethyl sulfoxide and acetone for the extraction of phytoplankton chlorophyll. *Can. J. Fish. Aquat. Sci.* 41: 1519–1522.

Stauffer, R. E., Lee, G. F., Armstrong, D. E. (1979). Estimating chlorophyll extraction biases. *J. Fish. Res. Board Can.* 36: 152–157.

Strickland, J. D. H., Parsons, T. R. (1972). *A Practical Handbook of Seawater Analysis*. 2nd ed. *Bull. Fish. Res. Bd. Can.* No. 167, 310 pp.

Suzuki, R., Fujita, Y. (1986). Chlorophyll decomposition in *Skeletonema costatum*: a problem in chlorophyll determination of water samples. *Mar. Ecol. Prog. Ser.* 28: 81–85.

Suzuki, R., Ishimaru, T. (1990). An improved method for the determination of phytoplankton chlorophyll using N,N-dimethylformamide. *J. Oceanogr. Soc. Jpn.* 46: 190–194.

Tett, P., Kelly, M. G., Hornberger, G. M. (1975). A method for the spectrophotometric measurement of chlorophyll *a* and pheophytin *a* in benthic microalgae. *Limnol. Oceanogr.* 20: 887–896.

Volk, S. L., Bishop, N. E. (1968). Photosynthetic efficiency of a phycocyanin-less mutant of *Cyanidium. Photochem. Photobiol.* 8: 213–221.

Webb, D. J., Burnison, B. K., Trimbee, A. M., Prepas, E. E. (1992). Comparisons of chlorophyll *a* extractions with ethanol and dimethyl sulfoxide/acetone, and a concern about spectrophotometric pheopigment correction. *Can. J. Fish. Aquat. Sci.* 49: 2331–2336.

Weiss, G. (1986). *Hazardous Chemicals Data Book*. 2nd ed. Noyes Data Corporation, Park Ridge, U. S. A., 1069 pp.

Wood, L. W. (1985). Chloroform–methanol extraction of chlorophyll *a*. *Can. J. Fish. Aquat. Sci.* 42: 38–43.

Wright, S. W., Shearer, J. D. (1984). Rapid extraction and high-performance liquid chromatography of chlorophylls and carotenoids from marine phytoplankton. *J. Chromatogr.* 294: 281–295.

Wright, S. W., Jeffrey, S. W., Mantoura, R. F. C., Llewellyn, C. A., Bjørnland, T., Repeta, D., Welshmeyer, N. (1991). Improved HPLC method for the analysis of chlorophylls and carotenoids from marine phytoplankton. *Mar. Ecol. Progr. Ser.* 77: 183–196.

Zapata, M., Garrido, J. J. (1991). Influence of injection conditions in reversed-phase high-performance liquid chromatography of chlorophylls and carotenoids. *Chromatographia*, 31: 589–94.

10

Filtration and storage of pigments from microalgae

R. F. C. Mantoura, S. W. Wright, S. W. Jeffrey, R. G. Barlow and D. E. Cummings

10.1 Introduction

As phytoplankton pigments are biochemically reactive, labile compounds they should preferably be analysed immediately after sampling. However, during intensive oceanographic or geochemical surveys large numbers of filtered samples can accumulate. It may be necessary to store the filters for periods of weeks to months, depending on the analytical capacity of the laboratory. Reliable storage methods are also needed for preserving and transporting filtered samples of both phytoplankton and standard SCOR cultures between ship and laboratory, and between collaborating scientists.

In this chapter, a brief review of filters and storage methods is presented first, followed by the results of the SCOR WG 78 storage experiments. Filtered samples of cultured microalgae and natural phytoplankton were stored under five different conditions for periods of up to one year, and HPLC was used to follow the recovery and degradation of pigments. Recommendations for optimal storage of filtered samples are made following assessment of the experiments.

10.2 Review of filtration and storage of phytoplankton pigments

10.2.1 *Filters for sampling phytoplankton pigments*

Filtration is the first step in the analysis of phytoplankton and detrital pigments following collection of seawater samples. Table 10.1 summarizes the filters and pigment methods oceanographers use, and lists their advantages and disadvantages. This information was compiled from a survey of 104 papers published between 1980 to 1993 in the refereed international literature.

Two classes of filters are commonly used: glass-fibre or 'depth' filters and membrane or 'surface' filters (see Strickland and Parsons, 1968; Parsons *et al.* 1984). It is striking that glass-fibre filters were selected in 88% of HPLC and 100% of TLC applications in oceanography. Membrane filters were used extensively with fluorometric methods, since they conveniently dissolve in the extracting solvent (Strickland and Parsons, 1968). There are important differences between glass-fibre and membrane filters in their particle retention properties and compatibility with chromatographic methods (see Table 10.1).

Glass-fibre filters

Glass-fibre filters retain particles on their surface and throughout the internal spaces of the filter and are the most common form of 'depth' filter. However, the size range of particles retained on glass-fibre filters is less precise than those of membrane filters, and is usually expressed as the size of particles trapped with 98% efficiency. The most widely used glass-fibre filters in pigment oceanography are the Whatman GF/F (0.7 µm) and GF/C (1.2 µm) and the Gelman A/E (1 µm) filters. Their principal advantages are: (1) large filtration capacity, (2) fast filtration rates, (3) inertness in organic solvents, (4) ability to be ashed for elemental (C, N, H) analysis, (5) potential aid to cell disruption during extraction, and (6) lower cost than membrane filters.

Early studies of the chlorophyll retention efficiencies of various filters (Humphrey and Wooton, 1966; SCOR–UNESCO Working Group 17, 1966; Yentsch, 1983) found that Whatman GF/C (1.2 µm) filters recovered only 80–95% of the

TABLE 10.1 Survey of filters used for sampling phytoplankton pigments in oceanography[a]

Filter type	Commercial suppliers	Filter usage (%)					Pore size (μm)	Advantages	Disadvantages
		HPLC	TLC	SP	SF	Total			
Glass-fibre									
GF/F	Whatman	42	0	2	3	47	0.7	High filtration capacity and rates. Insoluble in pigment-extracting solvents. GF/F retains >94% of picoplankton[b], *Synechococcus* sp.[c], *Prochlorococcus* sp.[d], and picoeukaryots[e]. Aids cell disruption. Less expensive than membrane filters. Compatible with HPLC, TLC and elemental (C, N, H) analyses.	Broad size cut-off. Unsuitable for size-fractionation.
GF/C	Whatman	12	8	9	6	35	1.2		
Other	Sartorius, Gelman, Schleicher and Schüll	2	0	3	1	6	0.7-5.0		
Total		56	8	14	10	88			
Membrane									
Cellulose esters	Millipore (HA), Sartorius	0	0	0	5	5	0.02-8.0	Available in wide range of narrow pore sizes (0.01–10 μm). Suitable for size-fractionation of phyto-biomass and production[f]. Compatible with SP, SF, fluorometry and gravimetry[g].	Low filtration capacity and rates. Cellulose-ester filters dissolve in pigment-extracting solvents; therefore unsuitable for HPLC, TLC. Nuclepore polycarbonate filters release dyes that interfere with HPLC[h]. More expensive than glass-fibre filters.
Polyester	Nuclepore	4	0	0	0	4	0.4		
Other	Gelman, Anapore, Millipore	1	0	0	2	3	0.02-10.0		
Total		5	0	0	7	12			
Filter total		61	8	14	17	100			
$MgCO_3$ addition (%)		4	4	5	4	17		Buffers pH, improves particle retention[g].	Adsorbs Chlides and Phides[i]. Retards filtration.

a. Survey of 104 publications (1980–1993) in the refereed literature that used high performance liquid chromatography (HPLC) thin-layer chromatography (TLC) spectrophotometry (SP) and spectrofluorometry (SF).
b. Li *et al.* (1983), Phinney and Yentsch (1985), Taguchi and Laws (1988).
c. Guillard *et al.* (1985), Li (1986), Kana *et al.* (1988).
d. Goericke and Repeta (1992, 1993), Partensky *et al.* (1993).
e. Hooks *et al.* (1988), Vaulot *et al.* (1990).
f. Joint *et al.* (1993), Li *et al.* (1983).
g. Strickland and Parsons (1968).
h. Mantoura (unpublished results).
i. Daly *et al.* (1973).

chlorophyll *a* from coastal seawater compared to that recovered by Millipore HA (0.45 μm) filters. Nevertheless, Holm-Hansen (1978) and Marker *et al.* (1980) preferred GF/C filters because of their many practical advantages (Table 10.1). Venrick (1987) has since noted that HA filters yield a significant blank in the fluorometric determination of chlorophyll *a*. With the more recent availability of the GF/F (0.7 μm) filters, Phinney and Yentsch (1985) re-evaluated the chlorophyll *a* retention characteristics of these and seven glass-fibre filters, using contrasting samples from the north Atlantic Ocean. The GF/F filters consistently yielded the highest concentration of chlorophyll *a* and were statistically indistinguishable from concentrations obtained with Millipore HA (0.45 μm) membrane filters.

It is now recognised that the phytoplankton biomass in the open ocean is often dominated by small 'picoplanktonic' cells (0.2–2.0 μm) including cyanobacteria (e.g., *Synechococcus* sp.; Waterbury *et al.* 1979), prochlorophytes (Chisholm *et al.* 1988) and the 'picoeukaryots' (Hooks *et al.* 1988; Vaulot *et al.* 1990). Significant errors in pigment measurement result if these small cells are not retained by the filters.

GF/F filters recovered >94% of the picoplanktonic chlorophyll *a* (or fixed $^{14}CO_2$) recovered by 0.2 μm Nucleopore filters in studies in the tropical Caribbean (Li *et al.* 1983) and the Pacific Ocean off Hawaii (Taguchi and Laws, 1988). Chlorophyll *a* particles that passed through GF/F filters could be recovered by re-filtering through another filter of either type (Taguchi and Laws, 1988).

GF/F filters are now routinely used to harvest cultures of small (0.8–1.2 μm) cyanobacteria (e.g., *Synechococcus* sp.; Guillard *et al.* 1985; Li, 1986; Kana *et al.* 1988) and the even smaller (0.4–0.8 μm) prochlorophytes (e.g., *Prochlorococcus* sp.; Goericke and Repeta, 1992, 1993; Partensky *et al.* 1993). GF/F filters have been used in almost all recent HPLC studies of pigment distribution in oceans and lakes (e.g., Repeta *et al.* 1989; Hurley and Armstrong, 1990; Hurley and Watras, 1991; Wright *et al.* 1991; Millie *et al.* 1992; Barlow *et al.* 1993; Letelier *et al.* 1993).

The addition of $MgCO_3$ to filters to inhibit acid-degradation of pigments and act as a filter aid (SCOR-UNESCO, 1966; Strickland and Parsons,1968) is now not recommended (see Table 10.1). It does not increase the yield of chlorophyll *a* or decrease the formation of pheopigments (Holm-Hansen, 1978; Marker *et al.* 1980) and may cause adsorptive losses of chlorophyllides and pheophorbides during filtration and extraction (Daly *et al.* 1973).

Given the overwhelming evidence that the GF/F filter retains close to 100% of phytoplankton pigments in oceanographic samples of all types, SCOR WG 78 concluded that no further re-evaluation of filtration methods was necessary. Consequently, the GF/F filter was used for all the SCOR pigment experiments described in Chapters 9 to 14.

Membrane filters

Membrane filters are 'surface' filters: particles larger than the pore size of the filter are retained wholly on the surface of the filter. They are manufactured from materials ranging from the traditional cellulose-acetate and -nitrate esters (e.g., Millipore HA, Sartorius SM) to the more recently introduced polycarbonate and polyester (e.g., Nuclepore, Microporetics), polysulfone (e.g., Gelman), nylon and polytetrafluorethylene 'teflon' polymers (Millipore, Gelman), and inorganic filters (Anapore). Polycarbonate membrane filters are available with well-defined pore sizes (0.02–10 μm), which has allowed size-fractionation of bacteria and phytoplankton into pico-, nano- and micro-planktonic components (Li *et al.* 1983; Joint *et al.* 1993).

Compared to glass fibre filters, the disadvantages of membrane filters are: (1) they have lower flow rates and filtration capacity, (2) they are more expensive, (3) they may dissolve in pigment-extracting solvents (e.g., acetone, methanol, dimethylformamide) and block up the analytical columns on the HPLC, and (4) the Nuclepore polycarbonate filters while insoluble in methanol, contain extractable yellow dye(s) which elute near peridinin and 19'-butanoyloxyfucoxanthin in reversed-phase HPLC (Mantoura, unpublished), and therefore are unsuitable for use with HPLC and TLC. Bidigare *et al.* (1985, 1990) used the insoluble 0.4 µm polyester filter (Nuclepore). Although membrane filters are widely used for spectrophotometric and fluorometric determinations of chlorophyll *a*, almost all TLC and HPLC studies have used glass-fibre filters.

10.2.2 *Storage of filtered samples of phytoplankton*

Storage temperatures

A survey of filter storage conditions and temperatures used by marine scientists is shown in Figure 10.1. Although freezing at -18° to -20°C was the most popular storage method, due to the ready availability of deep freezers in the laboratory, the range of storage temperatures was great (+22°C to -196°C), but in no case was the choice supported by scientific evidence that it was 'the best'. The majority (56%) opted for

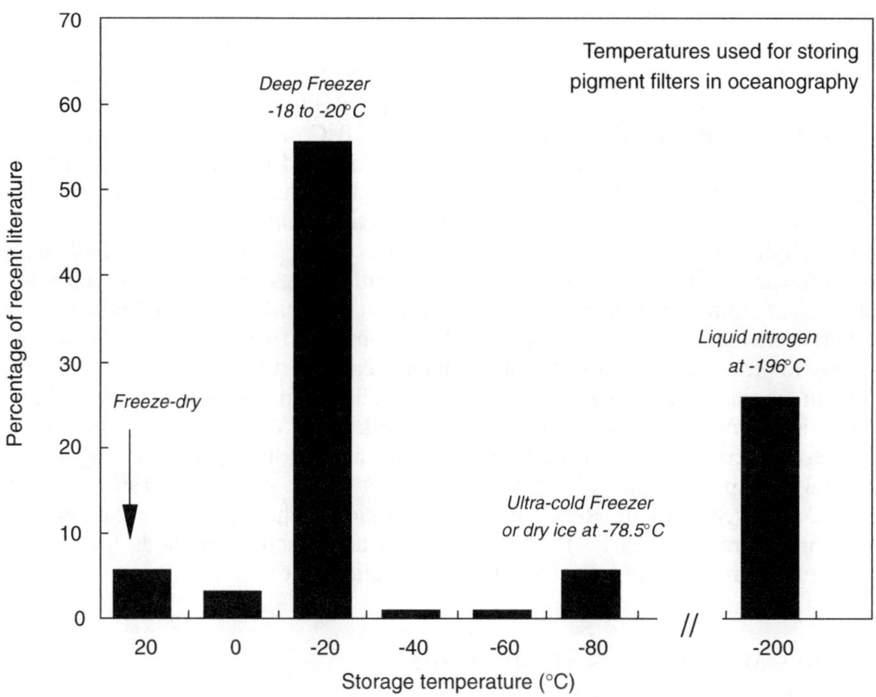

Figure 10.1
Temperatures used for storage of filtered phytoplankton pigment samples based on a survey of 104 research papers published between 1980 and 1993.

the convenience of deep-freezing at -18 to -20°C as recommended by UNESCO (1966), Strickland and Parsons (1968), and Parsons *et al.* (1984). However, an increasing proportion (30%) of marine scientists now store phytoplankton samples in liquid nitrogen at -196°C (Bidigare *et al.* 1985; Wright *et al.* 1991; Barlow *et al.* 1993; Goericke and Repeta, 1993; Strom, 1993), or in ultracold freezers (-30 to -80°C) or under solid CO_2 (-78°C) (e.g., Buma *et al.* 1990; Letelier *et al.* 1993; Olaizola *et al.* 1992). In only a few cases, room temperatures were used for dark storage of wet filters (Dandonneau, 1982) or for transporting freeze-dried microalgae harvested from cultures (Foss *et al.* 1984; Fiksdahl *et al.* 1984). As will be shown in the SCOR storage experiments (section 10.4) both room temperature storage and freeze-drying procedures are not recommended.

Storage time

In over 91% of recent (1980–1993) studies on marine algal pigments, filters were stored from a few days to several months, before analysis, in spite of surprisingly little knowledge of the effects of storage conditions on pigment stability. A few studies analysed the effects of storage on phytoplankton chlorophyll *a*, using simple spectrophotometric or fluorimetric assay methods. None used TLC or HPLC to study degradative processes, so any formation of chlorophyll allomers, epimers, chlorophyllides, carotenoid isomers or oxidation products would have been undetected.

Humphrey and Wooton (1966) monitored spectrophotometrically the concentrations of chlorophyll *a, b* and *c* in replicate filters of phytoplankton stored at various times and temperatures. At +10°C, chlorophyll *a* declined by 42% over 84 days, but at 1°C and -10°C chlorophyll *a* did not decline for up to 8 weeks. Jeffrey and Hallegraeff (1980) found that 5–10% of chlorophyll *a* from East Australian Current phytoplankton degraded within six weeks at -20°C. Tolstoy (1977) found that chlorophyll *a* on filters of freshwater phytoplankton did not degrade in a vacuum dessicator at -20°C for six weeks, and that only a 6% decrease was observed after six months of storage. Sand-Jensen (1976) observed significant losses (13–49%) of chlorophyll *a* during three days storage of filters of the chlorophyte *Scenedesmus quadricauda* and the diatom *Nitzschia palea* in either a dessicator at 4°C or a freezer at -15°C. Summarizing the results of a freshwater pigment workshop, Marker *et al.* (1980) concluded that '... a recommendation on deep freezing, drying or freeze-drying to preserve the filters cannot be given because results are contradictory ...'.

In summary, there were no systematic studies of pigment storage for HPLC analysis until the SCOR Working Group 78 Workshops. We now present the results of these experiments. Replicate filters of four cultured species of microalgae and natural phytoplankton were stored by freezing at -20°C, -90°C and -196°C; and at +22°C after freeze-drying. Recovery and degradation of chlorophylls and carotenoids were monitored by HPLC. The most efficient and practical methods for storing phytoplankton on filters were identified (see section 10.6).

10.3 Materials and methods

10.3.1 *Microalgae and phytoplankton*

The species of cultured microalgae selected for the storage experiments had key signature carotenoids well separated in the HPLC systems used. Some contained pigment-degrading enzymes. *Dunaliella tertiolecta* (CS-175) and *Phaeodactylum tricornutum* (CS-29) contained hydrolytic (e.g., chlorophyllase) and oxidative enzymes and *Phaeocystis pouchetii* (CS-165) produced acidic exudates. *Emiliania huxleyi* (CS-57) was used together with *D. tertiolecta* and *P. tricornutum* in the replicated experiment. Recoveries of species-specific chlorophylls and carotenoids were tracked during all storage conditions. The three SCOR cultures had the pigment compositions described in Chapter 13. The pigments of *Phaeocystis pouchetii* matched to those of *E. huxleyi* (Wright and Jeffrey, 1987), except in the fucoxanthin : 19'-hexanoyloxyfucoxanthin ratios.

The microalgae were cultured using standard SCOR protocols (see Chapter 6). *Phaeocystis pouchetii* (not a standard SCOR culture) was grown in G medium containing selenium, which promoted excellent growth and the formation of colonies (Harrison *et al.* 1988). Cultures were harvested at mid- to late-log phase and carefully decanted to produce uniform cell suspensions (free from clumps). Aliquots of the decanted suspension (10 ml) were filtered on Whatman GF/F (25 mm) filters, rinsed with sterile medium, folded, blotted with absorbent paper and immediately transferred into capped cryotubes for storage. Filters destined for liquid nitrogen storage were 'snap-frozen' for a few seconds in liquid nitrogen before being placed in a cryotube. The time from start of filtration to storage was less than 30 seconds.

Phytoplankton were obtained from 1 litre volumes of a well-mixed 120 l sample of seawater from the English Channel, which had been first screened through 100 μm mesh to exclude zooplankton and large debris. Replicate filters (25 mm GF/F) were immediately stored at the four temperatures, and extracted at the set times described below.

10.3.2 *Extraction protocol*

At pre-set storage times (Table 10.2), filters were withdrawn, cut into small pieces (10 s), put into the extraction solvent (3 ml buffered methanol) and sonicated for 30 s (see Chapters 9 and 12). After centrifugation, washing and re-sonicating, the pigment extracts were transferred to 5.0 ml amber volumetric flasks and made up to volume ready for quantitative HPLC. Pigment recoveries are expressed as weight (μg) of pigment per filter or as a percentage of the starting concentration.

10.3.3 *HPLC, pigment standards and chemicals*

The recovery and degradation of chlorophylls and carotenoids in frozen filters of microalgae and phytoplankton were followed in a long-term storage experiment (328 days), and a medium-term replicated experiment (60 days). Pigments were analysed by the reversed-phase gradient HPLC method of SCOR WG 78 (see Chapter 12, Wright *et al.* 1991) in the long term experiment and a similar binary gradient method (Barlow *et al.* 1993) in the medium-term replicated experiment. All solvents were HPLC-grade (Fisons or BDH) and were filtered (<1 μm) and degassed (He) before use.

The HPLC systems were calibrated with chlorophyll *a* and *b* standards (Sigma, Poole, UK) and fucoxanthin, 19'-hexanoyloxyfucoxanthin, diadinoxanthin, lutein, neoxanthin, violaxanthin, diatoxanthin, β,β-carotene, and chlorophylls c_1, c_2 and c_3, which were isolated from the SCOR reference microalgae (see Chapters 7, 8 and 13).

TABLE 10.2 Pigment storage experiments with filtered microalgae

Samples		Storage treatments[a]					Storage times[b] (days)
	Pretreatment:	Freeze-dry	-196°C	none	none	none	
	Temperature:	+22°C	-20°C	-20°C	-90°C	-196°C	
Long-Term Screening Experiment							
Emiliania huxleyi		✓	✓	✓		✓	0,1,4,17,21,40,328
Dunaliella tertiolecta		✓	✓	✓		✓	
Phaeodactylum tricornutum		✓	✓	✓		✓	
Phaeocystis pouchetii		✓	✓	✓		✓	
Medium-Term Replicated Experiment							
Emiliania huxleyi+ Dunaliella tertiolecta + *Phaeodactylum tricornutum*				✓✓✓	✓✓✓	✓✓✓	*,0,1,3,7,14,28,42,60
Phytoplankton, English Channel				✓✓✓	✓✓✓	✓✓✓	

a. HPLC analysis of pigments from single (✓) and from triplicate (✓✓✓) filters.
b. Pigment extractions were carried out immediately after filtering the microalgae or the phytoplankton sample, and after up to 4 hours' freezing. These are shown as * and 0 here and in Figures 10.5 and 10.6.

10.3.4 *Storage experiments*

The most frequently used temperatures and conditions for storing phytoplankton filters for pigment analysis (Figure 10.1) were evaluated in two experiments (Table 10.2).

Experiment 1 screened the long-term (328 days) recovery and degradation of pigments from single non-replicated filters of *Emiliania huxleyi, Dunaliella tertiolecta, Phaeodactylum tricornutum* and *Phaeocystis pouchetii*, all stored in the dark under the following conditions:

- freeze-drying followed by storage at room temperature (+22°C)
- snap-freezing in liquid nitrogen (-196°C) followed by storage at -20°C (deep freezer)
- immediate storage at -20°C (deep freezer)
- immediate storage at -196°C (liquid nitrogen)

Experiment 2 evaluated a more precise medium-term (60 days) study of pigment recoveries from triplicate filters of mixed cultured microalgae (*Emiliania* + *Dunaliella* + *Phaeodactylum*), and phytoplankton field samples, all stored in the dark under the following conditions:

- immediate storage at -20°C (deep freezer)
- immediate storage at -90°C (ultra-cold freezer)
- immediate storage at -196°C (liquid nitrogen)
- freshly filtered (no storage) for initial pigment concentrations

Pigment reproducibility in Experiment 1 was tested on three replicate filters available at the time of harvest. The variability of the chlorophyll *a* analysis from *Emiliania, Dunaliella* and *Phaeodactylum* was ±5%, ±3% and ±10% respectively. *Phaeocystis*, which was difficult to sample quantitatively because it forms colonies, had the greatest variability at ±18%, and was not used in the second experiment. In the medium-term storage experiment, the combined reproducibility of filtration, storage, extraction and HPLC analysis of chlorophyll *a* and of fucoxanthin from the triplicate filters were ±3.5% and ±1.9% respectively.

10.4 RESULTS

10.4.1 *Long-term screening experiment*

The HPLC chromatograms of pigment extracts from freshly filtered (non-stored) cultures of *Emiliania huxleyi, Dunaliella tertiolecta, Phaeodactylum tricornutum* and *Phaeocystis pouchetii* are shown in Figures 10.2 A–D, respectively. The chromatograms of the first three match those described in detail in Chapter 13. *Phaeocystis pouchetii*, which is not a SCOR reference alga, is a prymnesiophyte that forms massive blooms in the Southern Ocean and the North Sea. It contains chlorophylls *a*, c_2 and c_3, fucoxanthin, 19'-hexanoyloxyfucoxanthin, diadinoxanthin, diatoxanthin and β,β-carotene (Wright and Jeffrey, 1987).

The only traces of alteration products in these freshly filtered (non-stored) microalgae were pheophytin *a* (2%) and artefactual *cis* carotenoids (8%). Different storage treatments of *Phaeodactylum*, an alga rich in the hydrolytic enzyme chlorophyllase (Jeffrey and Hallegraeff, 1987), are shown on the same scale in Fig. 10.2 D–F. Compared to freshly extracted *Phaeodactylum* (Fig. 10.2 D), storage for 328 days at -196°C (Fig. 10.2 E) still gave high pigment recoveries (76–93%) with only traces (3%) of chlorophyll degradation products. However, major losses of all pigments (59–98%) and the appearance of several degradation products, including chlorophyllide *a* and chlorophyll *a* allomers and epimers, occurred in long-term storage of freeze-dried samples (Fig. 10.2 F).

The recoveries and degradation data for ten major pigments from filters of *Emiliania, Dunaliella, Phaeodactylum* and *Phaeocystis* after 328 days' storage at -196°C, -20°C (with and without initial snap freezing at -196°C) and at +22°C (after freeze-drying) are shown in Table 10.3 and Fig. 10.3 A–C.

The best pigment recoveries (averaging 96%) were obtained with liquid nitrogen storage (-196°C) (see Fig. 10.3 A). This treatment also had the lowest degradation of chlorophylls (3 mole %) compared to storage at -20°C (15 mol %) and +22°C after freeze-drying (15 mole %). Snap-freezing the filters in liquid nitrogen before long-term storage at -20°C did not significantly improve pigment recoveries (69% *versus* 75%; Table 10.3; Fig. 10.3 A).

In the four storage treatments the recoveries of chlorophyll *a* were consistently lower than those for chlorophyll *b* or chlorophylls $c_1 + c_2$ (see Fig. 10.3 B; Table 10.3). The recoveries of carotenoids, particularly of lutein and 19'-hexanoyloxyfucoxanthin (Fig. 10.3 C), were generally superior to those of chlorophyll *a* (Fig. 10.3 B). The lowest recoveries of both chlorophylls and carotenoids occurred with freeze-drying, specially with diadinoxanthin (16%), violaxanthin (13%) and β,β-carotene (1%).

Of the four microalgae tested, *Phaeocystis* consistently exhibited the lowest recoveries of individual and total pigments, irrespective of the storage method (see Fig. 10.4 A).

The recoveries of chlorophyll *a* averaged for the four species and the concurrent build-up of degradation products during 328 days' storage are shown in Figure 10.4 A and B, respectively. Most of the chlorophyll losses and degradation

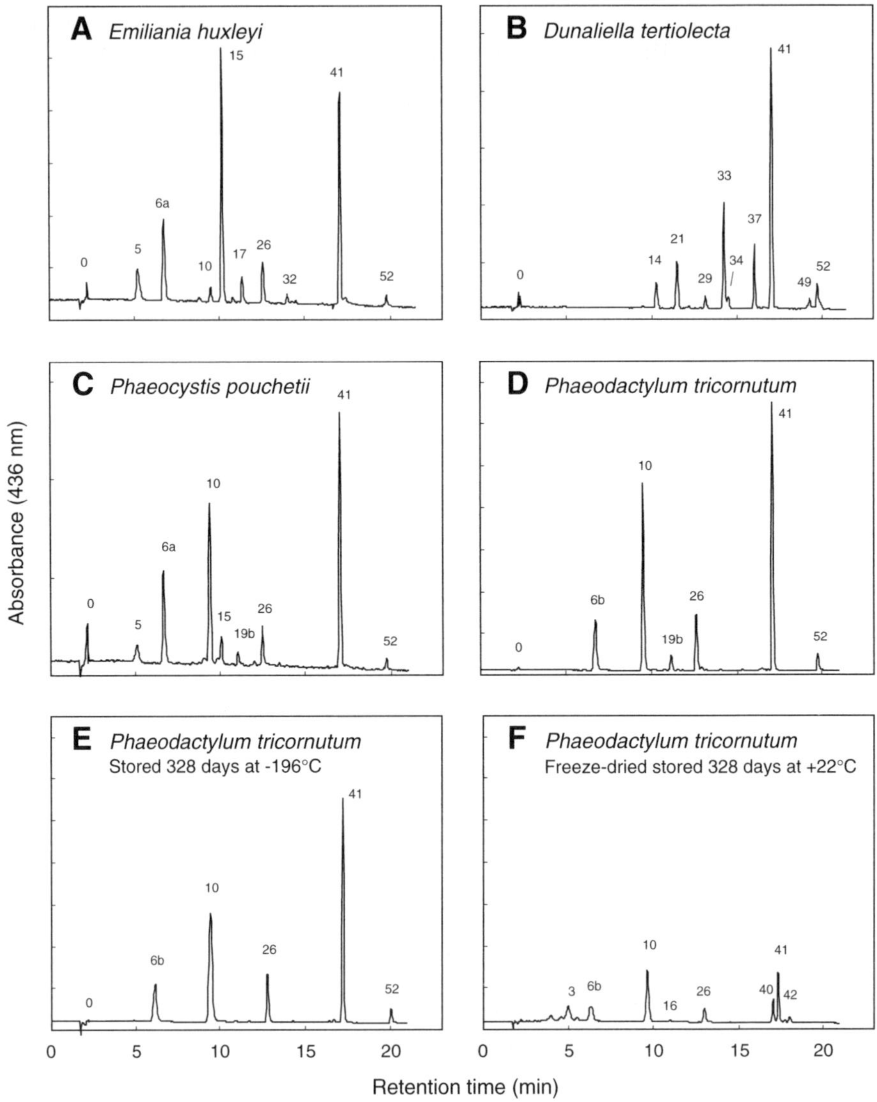

Figure 10.2
HPLC absorbance chromatograms (436 nm) of pigments extracted from fresh (A–D) and stored (E–F) filters of microalgae. A, *Emiliania huxleyi* (CS-57); B, *Dunaliella tertiolecta* (CS-175); C, *Phaeocystis pouchetii* (CS-165); D, *Phaeodactylum tricornutum* (CS-29); E, *P. tricornutum* after 328 days storage at -196°C; F, *P. tricornutum* after freeze-drying and storage at +22°C. Chromatograms in panels D, E and F are on identical scales.
HPLC peaks are as listed in Chapter 12: 3, chlorophyllide *a*; 5, chlorophyll c_3; 6a, chlorophyll c_2; 6b, chlorophylls c_{1+2}; 10, fucoxanthin; 14, neoxanthin; 15, 19'-hexanoyloxyfucoxanthin; 16 and 19b, *cis*-fucoxanthin; 17, *cis*-19'-hexanoyloxyfucoxanthin; 21, violaxanthin; 26, diadinoxanthin; 29, antheraxanthin; 32, diatoxanthin; 33, lutein; 34, zeaxanthin; 37, chlorophyll *b*; 40, chlorophyll *a* allomer; 41, chlorophyll *a*; 42, chlorophyll *a* epimer; 49, β,ψ-carotene; 52, β,β-carotene.

TABLE 10.3 Pigment recoveries (%) from filtered microalgae stored for 328 days

Storage (Pretreatment)	–196°C (None)					–20°C (None)				
Microalga	*Emiliania*	*Dunaliella*	*Phaeodactylum*	*Phaeocystis*	Average[a]	*Emiliania*	*Dunaliella*	*Phaeodactylum*	*Phaeocystis*	Average[a]
Pigment										
Chlorophyll *a*	80	85	86	81	83	83	58	41	46	57
Chlorophyll *b*		104			104		83			83
Chlorophylls c_{1+2}	104		90	89	94	104		87	53	81
Chlorophyll c_3	118			91	105	104			87	96
Fucoxanthin			93	95	94			72	32	52
Diadinoxanthin	104		91	83	93	67		25	18	37
19'-hexanoyloxyfucoxanthin	111			99	105	114			44	79
Lutein		99			99		104			104
Violaxanthin		103			103		107			107
ß,ß–carotene	92	84	76	67	80		76	41	33	50
					96					75
Average[b]	102	95	87	86	93[c]	94	86	53	45	69[c]

Storage (Pretreatment)	–20°C (Snap–Freeze in Liquid Nitrogen)					+22°C (Freeze–Dry)				
Microalga	*Emiliania*	*Dunaliella*	*Phaeodactylum*	*Phaeocystis*	Average[a]	*Emiliania*	*Dunaliella*	*Phaeodactylum*	*Phaeocystis*	Average[a]
Pigment										
Chlorophyll *a*	67	59	50	38	54	25	43	19	12	25
Chlorophyll *b*		77			77		70			70
Chlorophylls c_{1+2}	81		85	46	71	33		41	11	28
Chlorophyll c_3	90			85	88	54			93	74
Fucoxanthin			71	34	53			39	17	28
Diadinoxanthin	56		29	18	34	6		30	12	16
19'-hexanoyloxyfucoxanthin	90			61	76	28			26	27
Lutein		95			95		35			35
Violaxanthin		95			95		13			13
ß,ß–carotene	50	70	41	17	45	0	1	2	0	1
					69					32
Average[b]	72	79	55	43	62[c]	24	32	26	24	27[c]

a. Average recovery of individual pigments from all algae.
b. Average recovery of all pigments from individual algae.
c. Global average of all pigments from all algae.

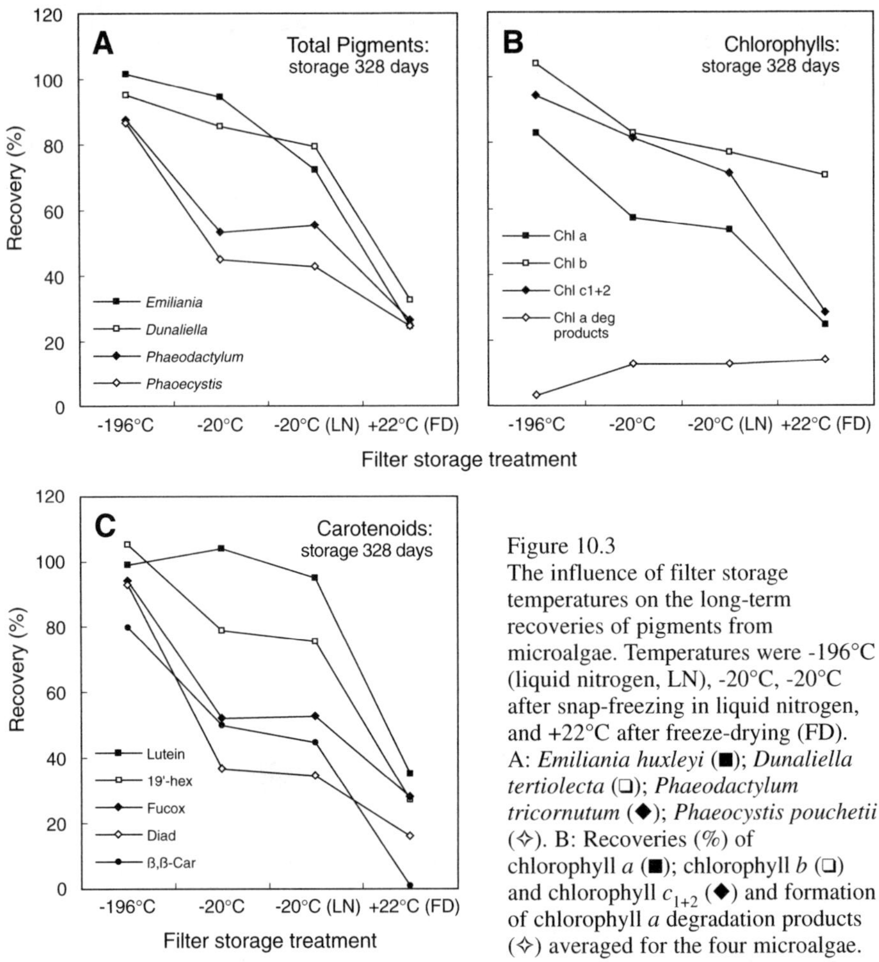

Figure 10.3
The influence of filter storage temperatures on the long-term recoveries of pigments from microalgae. Temperatures were -196°C (liquid nitrogen, LN), -20°C, -20°C after snap-freezing in liquid nitrogen, and +22°C after freeze-drying (FD). A: *Emiliania huxleyi* (■); *Dunaliella tertiolecta* (❑); *Phaeodactylum tricornutum* (◆); *Phaeocystis pouchetii* (✧). B: Recoveries (%) of chlorophyll *a* (■); chlorophyll *b* (❑) and chlorophyll c_{1+2} (◆) and formation of chlorophyll *a* degradation products (✧) averaged for the four microalgae. C: Recoveries (%) of lutein (■), 19'-hexanoyloxyfucoxanthin (❑); fucoxanthin (◆); diadinoxanthin (✧) and β,β-carotene (●) averaged for the microalgae.

occurred during the first two weeks of storage. Chlorophyll *a* degradation products were always dominated (65–90%) by chlorophyllide *a* and chlorophyll *a*-allomers and epimers, but their average concentrations accounted for only 24 mole % of the decline in chlorophyll *a*. With freeze-drying there was a dramatic but transient rise (32 mole %) in chlorophyll degradation products after 4–7 days storage (Fig. 10.4 B). They were subsequently further degraded to undetectable products. Carotenoid recoveries at 328 days (Fig. 10.4 C) were close to 100% for storage at -196°C, approximated to 80% for storage at -20°C and fell to 25% for freeze-dried samples.

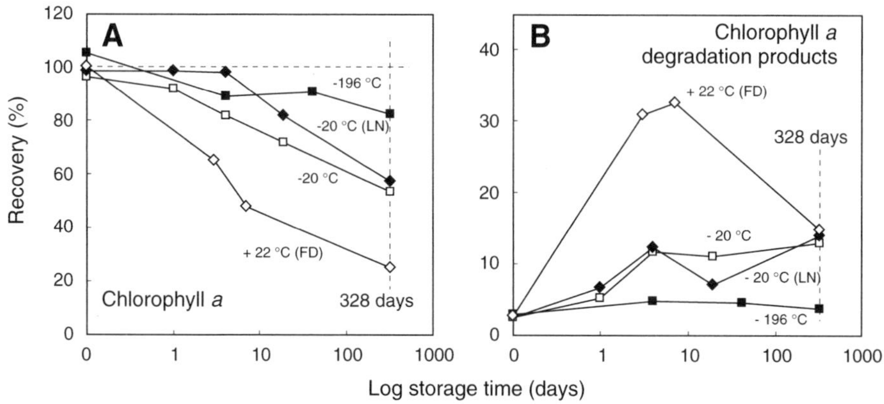

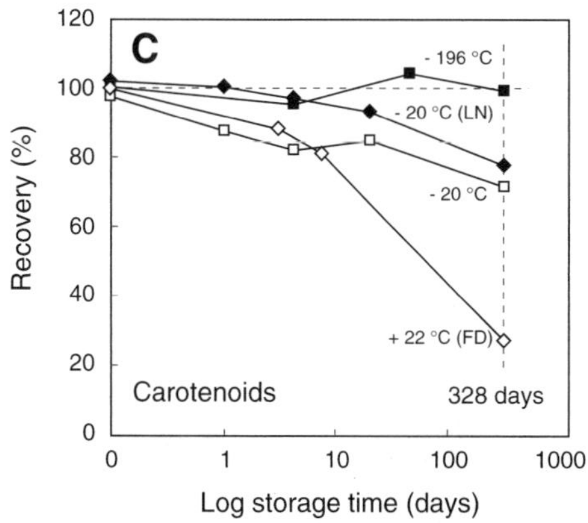

Figure 10.4
The influence of storage time and temperature on the recovery of (A) chlorophyll *a*, (B) production of chlorophyll *a* degradation products (in chlorophyll *a* equivalents), and (C) total carotenoids from filtered microalgae. Note logarithmic scale for storage time. LN =snap-freezing in liquid nitrogen; FD =freeze-drying.

10.4.2 *Medium-term replicated experiment*

This experiment focussed on frequent and precise tracking of pigment recoveries from triplicate filters of mixed microalgae and phytoplankton stored for up to 60 days at -196°C, -90°C and -20°C (Table 10.2). Because freeze-drying resulted in severe loss of pigments in the first screening experiment, this condition of storage was not tested in the medium-term replicated experiment. Pigment recoveries and degradation after 14, 28 and 60 days storage are summarized in Table 10.4. Detailed recovery trends over 60 days for seven of the most important pigments are shown in Figs. 10.5 and 10.6.

Storage of filters for 60 days at -196°C in liquid nitrogen produced excellent recoveries of pigments from both mixed cultured microalgae and phytoplankton field samples (111%, 116%) with only a slight increase in chlorophyll degradation products (9 to 14 mole %) in the case of the microalgal filters. Storage at -90°C also gave very high pigment recoveries (113%, 116%) and low chlorophyll degradation (<12 mole %). However, storage at -20°C over 60 days caused significant losses of pigment in both microalgae (81%) and phytoplankton (76%). The results are consistent with those of the long-term storage experiment. Pigment recoveries from both the mixed cultured microalgae and the phytoplankton field samples were in

TABLE 10.4 Pigment recoveries from frozen filters of cultured microalgae and phytoplankton stored for 14, 28 and 60 days

	Pigment concentration	Pigment recovery (%)								
Storage temperature		-20°C			-90°C			-196°C		
Storage times (days)	0	14	28	60	14	28[a]	60	14	28	60
Mixed microalgae	(µg/filter ± s.d.)									
Chlorophyll *a*	10.7 ± 0.3	72	75	82	101	103	105	105	99	98
Chlorophyll *b*	1.40 ± 0.06	83	82	95	108	111	115	119	112	110
Chlorophylls c_{1+2}	0.46 ± 0.06	112	104	105	117	116	115	124	116	116
Fucoxanthin	1.25 ± 0.01	107	87	85	118	118	118	127	116	117
19'-hexanoyloxyfucoxanthin	1.09 ± 0.01	105	87	88	116	116	116	128	118	117
Lutein	0.70 ± 0.05	86	66	68	114	114	115	120	111	110
ß,ß-carotene	0.42 ± 0.01	70	50	46	112	111	110	122	112	109
Average Pigment Recovery (%)		91	74	81	112	112	113	121	112	111
Chlorophyll *a* mole mass balance (%)										
Chlorophyll *a*	100	72	75	82	101	103	105	105	99	98
Chlide *a* + Chl *a*-allom	3	18	16	11	8	7	6	13	11	10
Phytin *a* + Phide	6	4	4	5	6	6	6	4	4	4
Phytoplankton	(ng/filter ± s.d.)									
Chlorophyll *a*	2585 ± 250	100	109	85	99		120	121	110	125
Chlorophyll *b*	243 ± 15	109	–	113	121		133	118	121	127
Chlorophylls c_{1+2}	190 ± 12	92	98	77	89		109	95	96	109
Peridinin	154 ± 11	100	90	76	107		122	109	101	110
Fucoxanthin	294 ± 21	104	97	96	120		134	126	118	138
19'-hexanoyloxyfucoxanthin	93 ± 11	93	101	83	115		123	111	107	118
Alloxanthin	323 ± 20	73	84	55	84		108	98	109	120
Diadinoxanthin	113 ± 12	70	78	46	87		101	86	87	96
Lutein	30 ± 7	74	71	67	109		97	102	98	100
Average Pigment Recovery (%)		91	91	76	103		116	107	96	116
Chlorophyll *a* mole mass balance (%)										
Chlorophyll *a*	100	100	109	85	99		120	121	110	125
Chlide *a* + Chl *a*-allom	2	6	5	4	3		5	3	4	3
Phytin *a* + Phide	3	5	5	4	3		1	2	3	2

a. Interpolated values between 21 and 42 days.

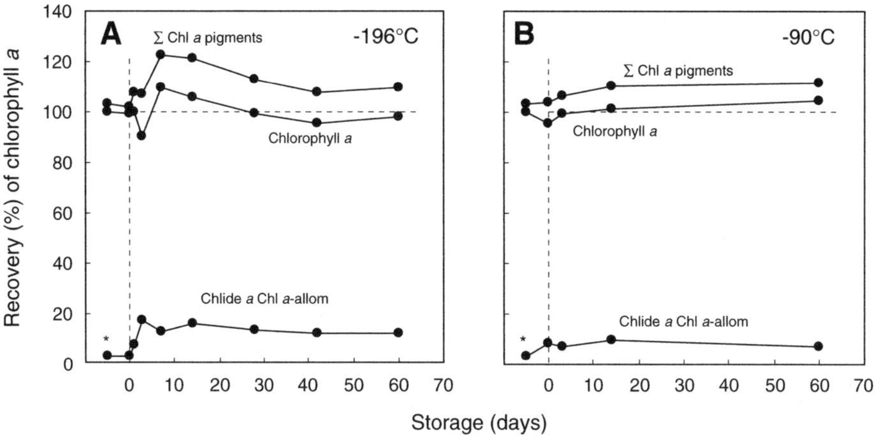

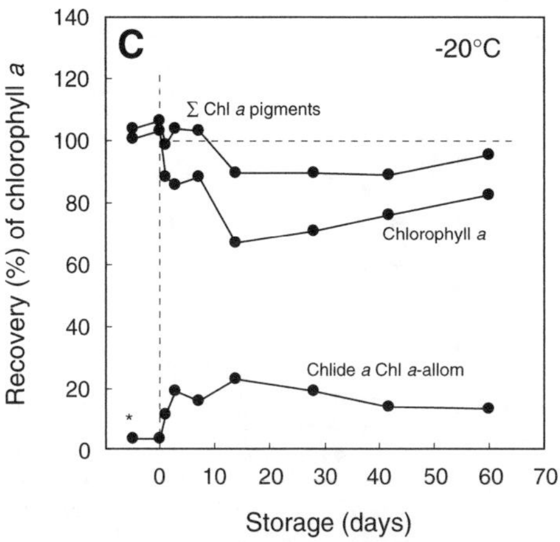

Figure 10.5
The influence of storage time and temperature on the recovery and degradation of chlorophyll *a* from frozen filters of mixed microalgae (*E. huxleyi* + *D. tertiolecta* + *P. tricornutum*). Each point is the mean of three filters with a typical standard deviation of 3.5%, *i.e.* twice the size of the data points. Σ Chl *a* corresponds to the molar sum of chlorophyll *a* and its degradation products (mainly chlorophyllide *a* and chlorophyll *a* allomer). Chlorophyll *a* recoveries obtained from the extraction of freshly filtered (*i.e.* non-stored) microalgae are shown beneath the symbol * on the time axis. Filters extracted less than 4 hours after freezing are shown as t = 0 on the time axis.

general agreement, indicating that the former could be regarded as representative of the latter.

The stability and recovery of chlorophyll *a* from microalgae stored on filters at -196°C, -90°C and -20°C are shown in Fig. 10.5 A–C. Comparisons of pigment recoveries at point t = * (freshly filtered samples, no storage) and t = 0 days (frozen and stored <4 hours before extraction) in Figs. 10.5 and 10.6 indicate that short-term freezing (<4 hours) did not affect pigment recoveries or cause degradation of chlorophyll *a*. However, there were pronounced changes in chlorophyll *a* recovery and degradation during the subsequent two weeks of storage.

An unexpected finding of storage at -196°C and -90°C was that recoveries of chlorophyll *a increased* from that of t = * and t = 0, resulting in a yield of 122% relative to the original concentrations of extractable chlorophyll *a* (Fig. 10.5 A). In con-

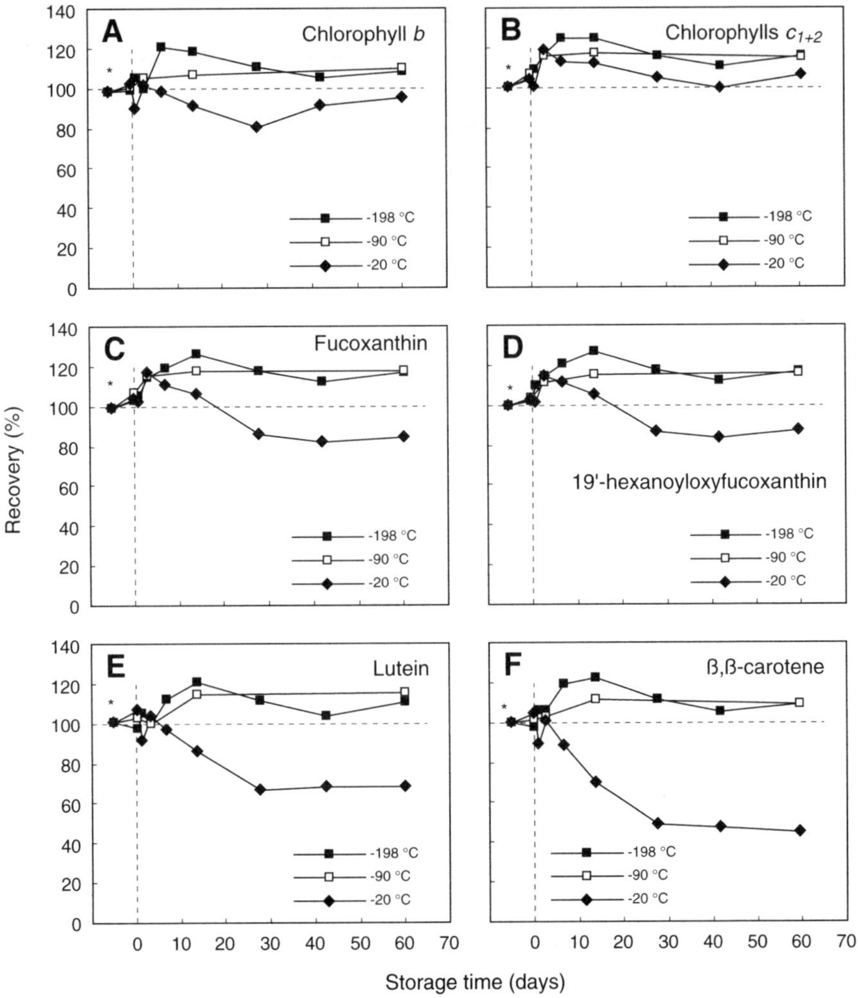

Figure 10.6
The influence of storage time and temperature on the recoveries of pigments from frozen filters of mixed microalgae (*E. huxleyi* + *D. tertiolecta* + *P. tricornutum*). A, chlorophyll *b*; B, chlorophylls c_{1+2}; C, fucoxanthin; D, 19'-hexanoyloxyfucoxanthin; E, lutein; F, β,β-carotene. Data marked * denotes non-stored freshly extracted filters as explained in Fig. 10.5.

trast, storage at -20°C initially resulted in a decline in chlorophyll *a* (72%) over the first two weeks, followed by a steady recovery to 82% after 60 days' storage (Fig. 10.5 C). Freezing of filters at -90°C yielded an intermediate trend: a small but continuous increase in the recoveries and degradation of chlorophyll *a* over the same period (Fig. 10.5 B). Any degradation of chlorophyll *a* during storage was accompanied by the formation of chlorophyllide *a*, chlorophyll *a* allomers and epimers (Table 10.4 and Fig. 10.5). The traces (6 mole %) of pheophytin *a* and pheophorbide *a* occurring in the freshly filtered microalgae did not change during the 60-day storage period.

Recoveries of chlorophylls b and $c_1 + c_2$ from microalgae stored on filters at -196°C, -90°C and -20°C are shown in Figures 10.6 A and B. In agreement with the long-term storage experiment, the recoveries of chlorophylls b and $c_1 + c_2$ over 60 days were consistently higher than those of chlorophyll a, even when filters were stored at -20°C.

The recoveries of four major carotenoids — fucoxanthin, 19'-hexanoyloxyfucoxanthin, lutein and β,β-carotene — are shown in Figs. 10.6 C–F. Similar to the chlorophylls, recoveries of all carotenoids from filtered microalgae stored at -196°C and -90°C increased by up to 28% during the first two weeks of storage, and thereafter showed either a small or no decline. Storage at-20°C, however, caused major losses of carotenoids, particularly lutein (32%) and β,β-carotene (54%) after 60 days' storage. The phytoplankton stored on filters at -20°C also had large losses of diadinoxanthin (46%) and alloxanthin (55%) after 60 days' storage (Table 10.4).

10.5 Discussion

This study tracked the recovery and degradation of 14 individual pigments from 274 filtered samples of microalgae and phytoplankton stored for up to eleven months under a range of storage conditions. HPLC was used for the first time to test storage conditions thoroughly in a large-scale study. Microalgal species known to contain high levels of degrading enzymes (e.g., chlorophyllase and oxidative enzymes in *Phaeodactylum tricornutum*, Barrett and Jeffrey, 1964, 1971; Mg^{2+}-dechelatase in *Dunaliella tertiolecta*, Owens and Falkowski, 1982) or acids (acrylic acid in *Phaeocystis pouchetii*, Liaaen-Jensen, 1990) were deliberately selected. Because *Phaeocystis pouchetii* consistently yielded the lowest pigment recoveries, especially in freeze-dried filters, it was excluded from the second storage experiment. Pigment recoveries and degradation rates with cultured *Emiliania, Dunaliella* and *Phaeodactylum* were consistent between both storage experiments and also consistent with the phytoplankton field samples.

10.5.1 *The fate of pigments during storage*

The recovery of pigments was not affected by the initial freezing treatments (<4 hours). The results were close to those from fresh filters (see t = * and t = 0 in Figs. 10.5 and 10.6). However, during the subsequent 2–3 weeks of storage, pigment recoveries increased significantly, particularly at -196°C, and then declined, particularly from filters stored at -20°C. The trends in recoveries of pigments may be explained in terms of two processes that operate concurrently but at different rates at different temperatures: release of bound pigment during storage and degradation of the free pigment.

Release of bound pigment

Liquid nitrogen released additional quantities of chlorophylls a, b and c_{1+2} (22%, 19% and 24%, respectively) and carotenoids (fucoxanthin 27%, 19'-hexanoyloxyfucoxanthin 28% and lutein 20%) during the first seven days of storage (Figs. 10.5 and 10.6). Storage at -90°C and -20°C also slightly increased the recovery of pigments. This suggests the existence of a 'bound' pool of pigments in microalgae, which is not normally extractable with methanol but is somehow 'freed' by extended freezing, after which it becomes methanol-extractable. Freeze-thawing at -70°C has previously been found to increase the acetone extractability of chlorophyll *a* from sediments by 80% (Sun *et al.* 1991).

The existence of a methanol-nonextractable pool was shown during the SCOR extraction experiments (Chapter 9). Methanol extracted 11–16% less of the pigments than dimethylformamide. This recalcitrant fraction may correspond to one or more of the different pigment-protein complexes within the photosynthetic apparatus (Young, 1993*a*; see also Chapter 3).

The data in Figures 10.5 and 10.6 clearly showed that the nett release of bound pigment was most pronounced with filters stored at -196°C and least for storage at -20°C. However, the rate of pigment degradation at -20°C (discussed below) may be sufficient to mask some of the release of bound pigment.

Degradation of free pigments

The decline in the concentrations of chlorophyll *a* and other pigments, most apparent when microalgal filters are stored at -20°C, probably arises from chemical and enzymic oxidation reactions. The storage temperature of -20°C is above the eutectic temperature (-22°C) for the crystallization of cell fluids (Morris, 1981). Consequently, a residual volume (5–10% of the original cell water) of hypersaline fluid rich in catabolites and oxygen will remain, and will induce biochemical damage and oxidation of pigments during storage. Formation of intracellular ice is known to damage membrane structures that could, on thawing, not only release pigment-protein complexes but also release membrane-bound enzymes such as chlorophyllase (Barrett and Jeffrey, 1971; Jeffrey and Hallegraeff, 1987). This would lead to the production of chlorophyllide *a* (see Figure 10.5) and esters during pigment extraction with solvents (e.g., acetone, methanol).

Samples stored below the eutectic temperature (-90°C and -196°C) had much better recoveries of pigments. Allomerization, epimerization and early oxidation of chlorophyll *a* occurs during senescence (Hendry *et al.* 1987; Chapter 3), and leads to the opening of the macrocycle and the formation of acyclic colourless products, which escape detection by HPLC. This explains why degradation products accounted for only a minor proportion (~24%) of the chlorophyll *a* losses during storage. It is noteworthy that pheophytin *a* and pheophorbide *a* were not produced during any of the storage experiments.

The selective destruction of β,β-carotene evident at -20°C in Figure 10.6 F was also found by Young (1993*b*) during storage studies on stressed and photobleached plants and algae. Degradation of β,β-carotene yielded 5,6 epoxy-β,β-carotene and peroxyl radicals, which fragmented the polyene chain into smaller apo-carotenals (which would not be visible in our HPLC system). The allenic carotenoids diadinoxanthin and alloxanthin (Table 10.4) were also degraded rapidly at -20°C. *Cis* isomers never accounted for more than 9% of the carotenoids. As with chlorophylls, loss of carotenoids during storage was not accompanied by a large build-up of detectable breakdown products.

10.5.2 *Optimizing the storage of filtered microalgae*

Storage of filters in liquid nitrogen (-196°C) frees up bound pigments and arrests oxidative degradation of the free pigments for up to 11 months. Filters should not be stored at -20°C for more than one week, since thereafter, oxidative degradation reactions lead to significant pigment losses, particularly of the non-polar pigments, chlorophylls *a* and *b*, β,β-carotene, and lutein. The quantitative reliability of the large body of literature reporting pigment HPLC data obtained from filtered phytoplankton stored at -20°C for more than one week, must now be viewed with some caution.

Storage at -90°C yields an intermediate trend between -20°C and -196°C, in which good (>96%) recoveries were obtained for all pigments throughout 60 days of storage. It follows that the solid CO_2 containers (-78°C) that are popular for the cryopreservation and carriage of perishable medical material could also be ideal for the medium-term (~1 month) storage and transport of phytoplankton filters between laboratories and research vessels.

Freeze-drying of microalgal cultures has been a common practice when shipping them between laboratories for semi-preparative isolation and identification of new marine carotenoids (Fiksdahl *et al.* 1984; Foss *et al.* 1984; Hooks *et al.* 1988). Our results extend the observations of Bjørnland *et al.* (1988) and Liaaen-Jensen (1990) on the formation of artefacts in freeze-dried samples of *Phaeocystis* to at least three other microalgae. Millie *et al.* (1993*a, b*) suggested that freeze-dried filters of reference cultures could be dispatched as a source of pigment standards for HPLC. Our results show that extensive degradation of pigments occurs on freeze-drying, and we cannot therefore support this recommendation.

10.6 Recommendations on filters and storage

10.6.1 *Filters*

Table 10.1 summarizes the properties of filters used for harvesting phytoplankton and particulate material from seawater samples. Whatman GF/F (0.7 µm) or equivalent filters are recommended for most sampling procedures in which pigments are analysed by TLC or HPLC. Solvent-resistant membrane filters (e.g., polycarbonate, polyethylene) allow for the size-fractionation of phytoplankton, but they should be carefully checked before use with HPLC. Cellulose ester filters dissolve in methanol or acetone, and so can only be used with spectrophotometric or fluorometric analyses. Careful attention must be given to the blanks.

10.6.2 *-196°C (Liquid nitrogen)*

Storage of filters of phytoplankton under liquid nitrogen (-196°C) is highly recommended for the preservation and recovery (96%) of pigments from oceanographic and microalgal samples for up to 328 days. Longer storage times were not examined. Liquid nitrogen also frees up a pool of 'bound' pigments, thereby improving pigment yields when extracted with methanol. This pool probably corresponds to the extra pigment extractable by DMF (see Chapter 9), but this was not tested directly.

10.6.3 *-90°C (Ultra-cold freezer)*

Storage of filters in an ultra-cold freezer (-90°C) achieves excellent pigment recoveries (~110%) with minimum degradation (<6 mole % chlorophyll *a*) for at least 60 days. Longer term storage at -90°C was not tested, but may be adequate. Solid-CO_2 containers (-78°C) popular for cryo-carriage of perishable medical material, could be ideal for medium-term (~1 month) storage and dispatch of phytoplankton filters between laboratories and research vessels.

10.6.4 *-20°C (Deep freezer)*

Long-term storage of phytoplankton filters at -20°C is not recommended, but if storage is short-term (not more than one week) good pigment recoveries (>95%) with minimum degradation of chlorophyll *a* may be obtained. Snap-freezing of filters in liquid nitrogen (-196°C) before storage at -20°C did not improve pigment recoveries and is therefore not necessary unless required for other purposes.

10.6.5 *Freeze-drying (+22°C)*

Freeze-drying causes rapid loss and extensive degradation of chlorophylls and carotenoids in filters of all microalgae tested: about 25% loss in 1 day and at least 80% loss in 328 days. Freeze-drying is therefore not recommended for storage or transport of filtered samples of microalgae or phytoplankton.

10.6.6 *Degradation during storage*

Whenever chlorophyll *a* was lost during storage, chlorophyllide *a*, chlorophyll *a* allomers and epimers were consistently the most prominent and diagnostic degradation products of inadequate treatment. They accounted for only 18–32 mole % of the chlorophyll *a* losses, probably because they are intermediates in the further degradation of chlorophyll *a* to colourless, acyclic products. Pheophytins and pheophorbides were not formed during storage of filtered samples, and if present they must have been in the original samples.

Acknowledgements

We would like to thank Ms J-M. LeRoi for assistance with the algal cultures, and Ms C. A. Llewellyn for help with preparing the filtered samples.

References

Barlow, R. G., Mantoura, R. F. C., Gough, M. A., Fileman, T. W. (1993). Pigment signatures of the phytoplankton composition in the northeastern Atlantic during the 1990 spring bloom. *Deep-Sea Res.* II 40: 459–478.

Barrett, J., Jeffrey, S. W. (1964). Chlorophyllase and formation of an atypical chlorophyllide in marine algae. *Plant Physiol.* 39: 44–47.

Barrett, J., Jeffrey, S. W. (1971). A note on the occurrence of chlorophyllase in marine algae. *J. Exp. Mar. Biol. Ecol.* 7: 255–262.

Bidigare, R. R., Kennicutt, M. C. II, Brooks, J. M. (1985). Rapid determination of chlorophylls and their degradation products by high performance liquid chromatography. *Limnol. Oceanogr.* 30: 432–435.

Bidigare, R. R., Marra, J., Dickey, T. D., Iturriaga, R., Baker, K. S., Smith, R. C., Pak, H. (1990). Evidence for phytoplankton succession and chromatic adaptation in the Sargasso Sea during springtime 1985. *Mar. Ecol. Progr. Ser.* 60: 113–121.

Bjørnland, T., Guillard, R. R. L., Liaaen-Jensen, S. (1988). *Phaeocystis* sp. clone 677-3, a tropical marine planktonic prymnesiophyte with fucoxanthin and 19'-acylfucoxanthins as chemosystematic carotenoid markers. *Biochem. System. Ecol.* 16: 445–452.

Buma, A. G. J., Treguer, P., Kraay, G. W., Morvan, T. (1990). Algal pigment patterns in different water masses of the Atlantic section of the Southern Ocean during fall 1987. *Polar Biol.* 11: 55–62.

Chisholm, S. W., Olson, R. J., Zettler, E. R., Goericke, R., Waterbury, J. B., Welschmeyer, N. A. (1988). A novel free-living prochlorophyte abundant in the oceanic euphotic zone. *Nature* 344: 340–343.

Daly, R. J., Gray, C. B. J., Brown, S. R. (1973). A quantitative, semi-routine method for determining algal and sedimentary chlorophyll derivatives. *J. Fish. Res. Bd. Can.* 30: 345–356.

Dandonneau, Y. (1982). A method for rapid determination of chlorophylls plus phaeopigments in samples collected by merchant ships. *Deep-Sea Res.* 29: 647–654.

Fiksdahl, A., Withers, N., Guillard, R. R. L., Liaaen-Jensen, S. (1984). Carotenoids of the Raphidophyceae — a chemosystematic contribution. *Comp. Biochem. Physiol.* 78B: 265–271.

Foss, P., Guillard, R. R. L., Liaaen-Jensen, S. (1984). Prasinoxanthin – a chemosystematic marker for algae. *Phytochem.* 23: 1629–1633.

Goericke, R., Repeta, D. J. (1992). The pigments of *Prochlorococcus marinus*: the presence of divinyl chlorophyll *a* and *b* in a marine prokaryote. *Limnol. Oceanogr.* 37: 425–433.

Goericke, R., Repeta, D. J. (1993). Chlorophylls *a* and *b* and divinyl chlorophylls *a* and *b* in the open subtropical North Atlantic Ocean. *Mar. Ecol. Progr. Ser.* 101: 307–313.

Guillard, R. R. L., Murphy, L. S., Foss, P., Liaaen-Jensen, S. (1985). *Synechococcus* spp. as likely zeaxanthin-dominant ultraphytoplankton in the North Atlantic. *Limnol. Oceanogr.* 30: 412–414.

Harrison, P. J., Yu, P. W., Thompson, P. A., Price, N. M., Phillips, D. J. (1988). Survey of selenium requirements in marine phytoplankton. *Mar. Ecol. Progr. Ser.* 47: 89–96.

Hendry, S. B., Houghton, J. D., Brown, S. B. (1987). The degradation of chlorophyll — a biological enigma. *New Phytol.* 107: 255–302.

Holm-Hansen, O. (1978). Chlorophyll *a* determination: improvement in methodology. *Oikos* 30: 438–447.

Hooks, C. E., Bidigare, R. R., Keller, M. D., Guillard, R. R. L. (1988). Coccoid eukaryotic marine ultraplankton with four different HPLC pigment signatures. *J. Phycol.* 24: 571–580.

Humphrey, G. F., Wooton, M. (1966). Comparison of the techniques used in the determination of phytoplankton pigments. In: *Determination of Photosynthetic Pigments in Seawater.* UNESCO Monograph on Oceanographic Methodology No. 1, UNESCO, Paris, 37–63.

Hurley, J. P., Armstrong, D. E. (1990). Fluxes and transformation of aquatic pigments in Lake Mendota, Wisconsin. *Limnol. Oceanogr.* 35: 384–398.

Hurley, J. P., Watras, C. J. (1991). Identification of bacteriochlorophylls in lakes *via* reverse-phase HPLC. *Limnol. Oceanogr.* 36: 307–315.

Jeffrey, S. W., Hallegraeff, G. M. (1980). Studies of phytoplankton species and photosynthetic pigments in a warm-core eddy of the East Australian Current. *Mar. Ecol. Progr. Ser.* 3: 285–294.

Jeffrey, S. W., Hallegraeff, G. M. (1987). Chlorophyllase distribution in ten classes of phytoplankton: a problem for chlorophyll analysis. *Mar. Ecol. Progr. Ser.* 35: 293–304.

Joint, I., Pomroy, A., Savidge, G., Boyd, P. (1993). Size-fractionated primary production in the northeast Atlantic in May–July 1989. *Deep-Sea Res.* II 40: 423–440.

Kana, T. M., Glibert, P. M., Goericke, R., Welschmeyer, N. A. (1988). Zeaxanthin and β-carotene in *Synechococcus* WH7803 respond differently to irradiance. *Limnol. Oceanogr.* 33: 1623–1627.

Letelier, R. M., Bidigare, R. R., Hebel, D. V., Ondrusek, M., Winn, C. D., Karl, D. M. (1993). Temporal variability of phytoplankton community structure based on pigment analysis. *Limnol. Oceanogr.* 38: 1420–1437.

Li, W. K. W. (1986). Experimental approaches to field measurement: methods and interpretation. *Can. Bull. Fish. Aquat. Sci.* 214: 251–286.

Li, W. K. W., Subba Rao, D. V., Harrison, W. G., Smith, J. C., Cullen, J. J., Irwin, B., Platt, T. (1983). Autotrophic picoplankton in the tropical ocean. *Science* 219: 292–295.

Liaaen-Jensen, S. (1990). Artifacts of natural carotenoids — unintended carotenoid synthesis, In: Krinsky, N. I., Mathews-Roth, M. M., Taylor, R. F. (eds), *Carotenoids – Chemistry and Biology*. Plenum Press, New York, p. 149–165.

Marker, A. F. H., Nutsch, E. A., Rai, H., Rieman, R. (1980). The measurement of photosynthetic pigments in freshwaters and standardization of methods: conclusions and recommendations. *Archiv. fur Hydrobiol.* 14: 91–106.

Millie, D. F., Baker, M. C., Tucker, C. S., Vinyard, B. T., Dionigi, C. P. (1992). High resolution remote sensing of bloom-forming phytoplankton. *J. Phycol.* 28: 281–290.

Millie, D. F., Paerl, H. W., Hurley, J. P. (1993*a*). Microalgal pigment assessments using high-performance liquid chromatography: a synopsis of organismal and ecological applications. *Can. J. Fish. Aquat. Sci.* 50: 2513–2527.

Millie, D. F., Paerl, H. W., Hurley, J. P., Kirkpatrick, G. J. (1993*b*). Algal pigment determination in aquatic ecoystems: analytical evaluations, applications and recommendations. In: Menon J., (ed.), *Current Topics in Botanical Research*. Council of Scientific Integration, Trivandrum, India, p. 1–13.

Morris, G. J. (1981). Cryopreservation: an introduction to cryopreservation in culture collections. Institute of Terrestrial Ecology, Cambridge. 27 pp.

Olaizola, M., Bienfang, P. K., Zieman, D. A. (1992). Pigment analysis of phytoplankton during a subarctic spring bloom: xanthophyll cycling. *J. Exp. Mar. Biol. Ecol.* 158: 59–74.

Owens, T. G., Falkowski, P. G. (1982). Enzymic degradation of chlorophyll *a* in phytoplankton *in vitro*. *Phytochem.* 21: 979–984.

Parsons, T. R., Maita, Y., Lalli, C. M. (1984). *A Manual of Chemical and Biological Methods for Seawater Analysis.* Pergamon Press, Oxford, 173 pp.

Partensky, F., Hoepffner, N., Li, W. K. W., Ulloa, O., Vaulot, D. (1993). Photoacclimation of *Prochlorococcus* sp. (Prochlorophyta) strains isolated from the North Atlantic and the Mediterranean Sea. *Plant Physiol.* 101: 285–296.

Phinney, D. A., Yentsch, C. S. (1985). A novel phytoplankton chlorophyll technique: toward automated analysis. *J. Plank. Res.* 7: 633–642.

Repeta, D. J., Simpson, D. J., Jorgensen, B. B., Jannasch, H. W. (1989). Evidence for anoxygenic photosynthesis from the distribution of bacteriochlorophylls in the Black Sea. *Nature* 342: 69–72.

Sand-Jensen, K. (1976). A comparison of chlorophyll *a* determination of unstored and stored plankton filters extracted by methanol and acetone. *Vatten* 4: 259–266.

SCOR-UNESCO Working Group 17 (1966). *Determination of Photosynthetic Pigments in Seawater*. UNESCO Monograph on Oceanographic Methodology 1. Paris, 69 pp.

Strickland, J. D. H., Parsons, T. R. (1968). *A Practical Handbook for Seawater Analysis*. Fish. Res. Bd. Can. Bull. 167. 2nd Edn. Ottawa, 311 pp.

Strom, S. L. (1993). Production of phaeopigments by marine protozoa: results of laboratory experiments analyzed by HPLC. *Deep-Sea Res.* 40: 57–80.

Sun, M., Aller, R. C., Lee, C. (1991). Early diagenesis of chlorophyll *a* in Long Island Sound sediments: a measure of carbon flux and particle reworking. *J. Mar. Res.* 49: 379–401.

Taguchi, S., Laws, E. D. (1988). On the microparticles which pass through glass-fibre filter type GF/F in coastal and open waters. *J. Plank. Res.* 10: 999–1008.

Tolstoy, A. (1977). Method for the determination of chlorophyll *a* in phytoplankton. *Statens naturvardsverk* PM 831, University of Uppsala, Sweden.

Vaulot, D., Partensky, F., Neveux, J., Mantoura, R. F. C., Llewellyn, C. (1990). Wintertime presence of prochlorophytes in surface waters of the north-western Mediterranean Sea. *Limnol. Oceanogr.* 35: 1156–1164.

Venrick, E. L. (1987). On fluorometric determination of filter-retained pigments. *Limnol. Oceanogr.* 32: 492–493.

Waterbury, J. B., Watson, S. W., Guillard, R. R. L., Brand, L. E. (1979). Widespread occurrence of a unicellular, marine, planktonic, cyanobacterium. *Nature* 277: 293–294.

Wright, S. W., Jeffrey, S. W. (1987). Fucoxanthin pigment markers of marine phytoplankton analyzed by HPLC and HPTLC. *Mar. Ecol. Prog. Ser.* 38: 259–266.

Wright, S. W., Jeffrey, S. W., Mantoura, R. F. C., Llewellyn, C. A., Bjørnland, T., Repeta, D. J., Welschmeyer, N. (1991). Improved HPLC method for the analysis of chlorophylls and carotenoids from marine phytoplankton. *Mar. Ecol. Prog. Ser.* 77: 183–196.

Yentsch, C. S. (1983). A note on the fluorescence characteristics of particles that pass through glass-fibre filters. *Limnol. Oceanogr.* 28: 597–599.

Young, A. J. (1993*a*). Carotenoids in pigment-protein complexes. In: Young, A., Britton, G. (eds), *Carotenoids in Photosynthesis*, Chapman and Hall, Andover, UK., p. 72–95.

Young, A. J. (1993*b*). Factors that affect the carotenoid composition of higher plants and algae. In: Young, A., Britton, G. (eds), *Carotenoids in Photosynthesis*, Chapman and Hall, Andover, UK., p. 160–205.

11

Simple isocratic HPLC methods for chlorophylls and their degradation products

R. F. C. Mantoura, R. G. Barlow and E. J. H. Head

11.1 Introduction

In marine ecology, an accurate knowledge of the concentration of chlorophyll *a* rather than of the full suite of chlorophylls, carotenoids and degradation products is often all that is required. Some instances are: when monitoring phytoplankton biomass in seasonal studies or in cultures, doing large scale oceanographic mapping or profiling of chlorophyll, and calibrating bio-optical instruments (such as *in situ* fluorometers and transmissometers).

If the determination of chlorophyll *a* is to be reliable, it must be separated from any interfering chlorophyll degradation products. One of the objectives of SCOR Working Group 78 (see Chapter 1) was to develop a simple, isocratic HPLC system capable of rapidly separating chlorophyll *a* from interfering pigments. The present chapter gives a critical summary of isocratic HPLC methods from the literature, and

TABLE 11.1. Isocratic HPLC systems for separation of chlorophyll and carotenoid pigments.

Number	Reference	Sample information			Isocratic elution			Column, length and particle size	Detection	RT (min) Chl *a*	DL (ng) Chl *a*	Comments on separations
		Type	Solvent	Inj (µl)	Isocratic steps	Eluant (% v/v)	Flow (ml/m)					
Reversed-phase HPLC of Chlorophylls												
1	Eskins *et al.* (1977)	Ph, P	MeOH/Ac	?	7	80, 90, 95, 97.5, 100% MeOH / 10, 50, 75% Ether/MeOH, 28°C	2.5	Bondapak C-18 122 cm, 37–75 µm	Abs, 440 nm	193	NI	Diatom pigments separated
2	Rebeiz *et al.* (1978)	P, Stds	90% Ac	10	0	75% MeOH/20% Ac/5% H_2O	1	Spherisorb C-18 25 cm, 20 µm	F: ex 425 nm em > ? nm	4.9	8	Chl *a*, *b*, Phytin *a*, *b*.
3	Shoaf (1978)	Ph, Std	Ac	?	0	95% MeOH/5% H_2O	4	Partisil PXS 1025 ODS-2, 25 cm	Abs: 654 nm	21	NI	Stds and lake pigments
4	Braumann and Grimme (1979)	Ph	Ether	10	4	85, 95, 100% MeOH/H_2O 50% EtOH/MeOH	2	Sorb-Sil 60D C-18 30 cm	Abs: 440 nm	32	NI	Pigments from *Chlorella* sp.
5	Liebezeit (1980)	Ph, P, Std	90% Ac	100	0	94.5% MeOH/5.5% H_2O	2	Spherisorb C-18 23 cm, 5 µm	F: ex 390 nm em > 580 nm	9.7	0.9	Misidentified Phytins
6	Falkowski and Sucher (1981)	Ph, Std	MeOH	25	2	A: 90% MeOH/10% H_2O B: 98% MeOH/2% H_2O	1.5	RP 8, 15 cm	F: ex 420 nm em > ?	7	0.1	Misidentified Chlides
7	Bessiere and Montiel (1982)	Ph, Std	MeOH	80	0	97.3% MeOH/2.7% H_2O	1	Spherisorb C-18 25 cm, 10 µm	F: ex 427 nm em > 470 nm	16	0.5	Fluoranthene used as Int Std
8	Goeyens *et al.* (1982)	Ph, Std, P	MeOH	50	0	75% MeOH/22% Ac/3% H_2O	1	Altex RP 18 25 cm, 5 µm	Abs: 428 nm F: ex 420-460 nm em 630-680 nm	29	0.08	Optim. chromatogr. resolution
9	Shioi *et al.* (1983)	P, Std, Bac	MeOH/Ac	50	0	MeOH, 40°C	1.5	Zorbax C-18 25 cm	F: ex433-448 nm em 650-655 nm	13	1.8	Separation and identification Chl esters. Not Phytins and chlides.
10	Bidigare *et al.* (1985)	Ph, Std	90% Ac+IP	100	2	A: 80% MeOH/20% IP B: 100% MeOH	9.9	Radial PAK C-18 ? cm, 10 µm	F: ex 434 nm em 670 nm	6.1	0.013	Phytoplankton pigments
11	Trees *et al.* (1985)	Ph, Std	90% Ac	?	0	93% MeOH/7% H_2O	2	LiChrosorb C8 20 cm, 10 µm	F: ex 435 nm em 670 nm	12	?	
12	Baweja (1986)	Std, Ph	THF	2	0	23% THF/38% ACN/28% MeOH	1	Utrosphere C-18 25 cm, 5 µm	Abs: 663 nm	19	NI	Separated Deutero- from normal Chl *a*

13	Otsuki *et al.* (1987)	Ph, Std	MeOH	5	0	MeOH	1	Unisil Q C-18 25 cm, 5 µm	F: ex 430 nm em 670 nm	16	NI	Separation of 10 OH- and 10 methoxy-Chl *a* and lactones
14	de las Rivas *et al.* (1989)	P, Std	Ac	20	3	A: 70% ACN/30% MeOH B: ACN/MeOH/H_2O/EthAc	2	Novapak C-18 10 cm, ? µm	Abs: 450 nm	8.3	NI	Separated Lut from Zea
15	Mantoura *et al.* (1996): this work	Ph, Std	90% Ac	50	0 2	90% MeOH/10% Ac A: 80% MeOH/20% 0.5 M Amm Acet pH = 7.1 B: 90% MeOH/10% Ac	1	Pecosphere C-18 3 cm, 3 µm	F: ex 408 nm em > 530 nm Abs: 440 nm	2.5	0.050	Phytoplankton + faecal pigments
Reversed-phase HPLC of Bacterial Chlorophylls												
16	Steiner *et al.* (1982)	Bact, Std	?	var	0	95% MeOH/5% H_2O	1.5	RP-8 Knauer 30 cm	Abs: 600 nm	11	NI	Separation of BChl ester homologues
Reversed-phase HPLC of Carotenoids only												
17	Nells & de Leenheer (1983)	P, Ph, Std	Eluant	50	0	70% ACN/20% DCM/10% MeOH	1	Zorbax ODS 25 cm, 7 µm	Abs: 450 nm	NI	NI	Biomedical + vegetable pigments
18	Craft *et al.* (1992)	Std	EtOH	20	0	95% MeOH/5% THF	1	Vydac 201 TP C-18, 25 cm, 5 µm	Abs: 445 nm	NI	NI	Polymeric C-18; Lut separated from Zea
Normal phase HPLC of Chlorophylls and Carotenoids												
19	Fiksdahl *et al.* (1978)	Std	Hex	?	0	0.1% MeOH/99.9% Hex	1.4	Spherisorb 25 cm, 5 µm	Abs: 445 nm	carotenes		Ac gradients required for elution of Chl and polar carotenoids
20	Iriyama *et al.* (1978)	P	10% Ether in dry Hex	1	4	1, 2, 5, 10% isopropanol in hexane	0.016	Silica 6.5 cm, 0.5 µm	Abs: 380 nm	15.6	1	Early example of microbore HPLC
21	Stransky (1978)	Ph, P, Std	Ac	100	0	90% iso-octane/10% EtOH	1	Nucleosil 50 25 cm, 5 µm	Abs: 445 nm	2.4	9	
22	Abaychi and Riley (1979)	Ph, Std	dry MeOH	20	2	A: Petr/Ac/DMSO/DEA 75/23.25/1.5/0.25% B: Petr/Ac/MeOH/DMSO 30/40/27/3%	2	Partisil 10 30 cm	Abs: 440 nm	4.9	80	Chlides, Phytins; Phides poorly resolved, dry samples only.
23	Gieskes and Kraay (1983)	Ph, Std	dry ether	50	2	A: Petr/Ac/DMSO/DEA 75/22.75/2/0.25% B: 85% A + 17% MeOH	1	µ-Porasil 15 cm, 5 µm	Abs: 436 nm F: ex 436 nm em 660 nm	7.5	NI	Separated Chl *a* from DV Chl *a*

Key: Ph = phytoplankton; P = plants; Std = standards; Int Std = internal standard; Bact = bacteria; Ac = acetone; ACN = acetonitrile; Petr. = petroleum ether; MeOH = methanol; ACN = acetonitrile; Hex = hexane; THF = tetrahydrofuran; EtOH = ethanol; DMSO = dimethylsulphoxide; DEA = diethylamine; DCM = dichloromethane; Abs = absorbance; F = fluorescence; IC = isocratic; RT = retention time; DL = detection limit; NI = not investigated; ? = not reported. Pigments abbreviated as shown in Appendix A.

describes two new 'simple' but efficient HPLC systems developed during the SCOR Workshops. The first is an isocratic HPLC for rapidly separating chlorophyll *a* from other pigments, and the second is a step-isocratic HPLC for separating and selectively detecting three principal chlorophylls *a*, *b* and c_{1+2} and ten additional chlorophyllides and pheopigments.

11.2 Review of isocratic HPLC methods for chlorophylls

Several simple isocratic (i.e. one solvent) or step-isocratic (two solvents) HPLC methods for separating chlorophylls from other plant pigments have been published. These methods, together with their analytical details, are summarized in Table 11.1. Some separate not only chlorophyll *a*, but also chlorophylls *b* and *c* and their degradation products, as well as bacterial chlorophylls and carotenoids.

The isocratic HPLC methods for pigments fall into two groups: normal-phase and reversed-phase systems. The normal-phase HPLC systems are all based on silica columns. These can only tolerate dry solvents and are best suited for the separation of limited pigment groups with similar polarities and structures (e.g., *cis*- and *trans*-carotenoid isomers, Fiksdahl *et al.* 1978; chlorophyll *a* and divinyl chlorophyll *a*, Gieskes and Kraay, 1983).

The reversed-phase HPLC systems for pigments are mostly based on octadecyl (C-18) bonded spherical silica particles. These stationary phases can separate pigments covering a wide range of polarities from the anionic chlorophyllides and chlorophyll *c* pigments to the non-polar β,β- and β,ε-carotenes and pyropheophytins. The chromatographic reproducibility of reversed-phase columns and their compatibility with aqueous-organic solvents make them ideal not only for the analytical laboratory but also for routine shipboard applications in marine science.

A critical factor in selecting a routine HPLC method is that the separation time of chlorophyll *a* and the volume of eluant required should both be kept to a minimum. The analysis times of chlorophyll *a* listed in Table 11.1 vary dramatically from more than 3 hours (Eskins *et al.* 1977) to 6–7 minutes (Falkowski and Sucher, 1981; Bidigare *et al.* 1985). Improvements in analysis times have been achieved mainly by using progressively smaller (75 μm to 10 μm) and more uniformly shaped reversed-phase particles. However, this improvement has sometimes been achieved at the expense of high eluant flow rates. In the 2 step-isocratic HPLC method of Bidigare *et al.* (1985), the rapid (6 min) elution of chlorophyll *a* requires about 60 ml of expensive solvent per analysis. Other systems require complex multistep changes in solvents (Braumann and Grimme, 1979; Eskins *et al.* 1977) or high temperatures (40°C, Shioi *et al.* 1983) to resolve chlorophyll *a*.

The analytical performances of all the reversed-phase HPLC methods listed in Table 11.1 are compared in Figure 11.1 in terms of analysis times and eluant volumes needed to separate chlorophyll *a*. The most rapid (5–7 min) and least wasteful (5–11 ml) analyses of chlorophyll *a* were obtained by Rebeiz *et al.* (1978) and Falkowski and Sucher (1981) who used 20 μm C-18 and 10 μm C-8 reversed-phase columns, respectively. Neither method used the more efficient 3 μm reversed-phase packing, which might have achieved more rapid separation of chlorophyll *a*.

The detection limits for HPLC-separated chlorophyll *a* are also listed in Table 11.1. These confirm the well known fact that fluorescence is not only more

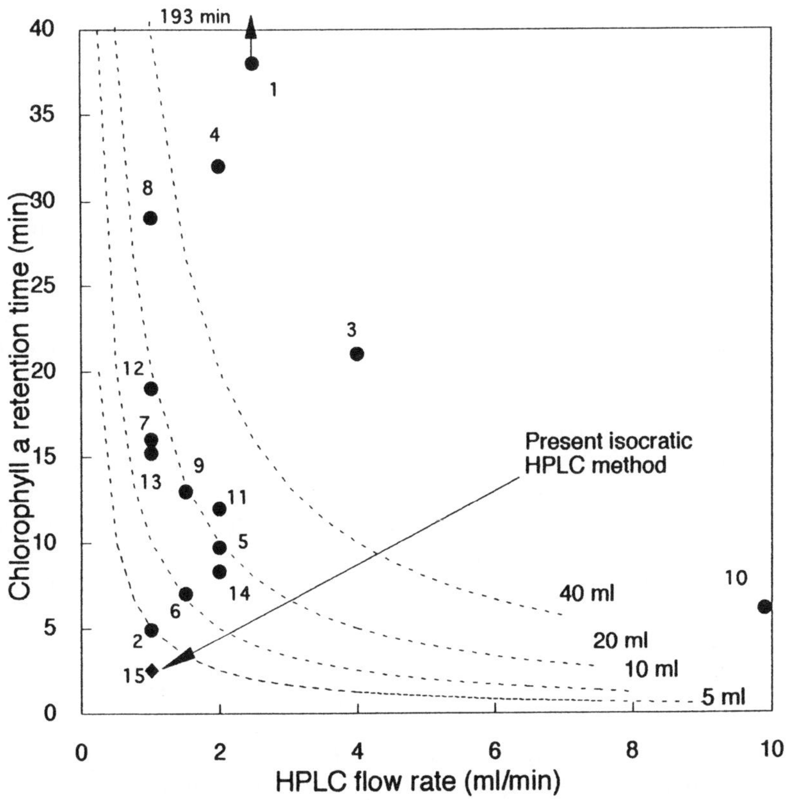

Figure 11.1
Comparative performances of published isocratic and step-isocratic HPLC methods for the reversed-phase separation of chlorophyll *a*. The numbers indicate methods listed in Table 11.1. The retention times of chlorophyll *a* are plotted against the HPLC flow rate. The dotted curves indicate the volume of eluant used to separate chlorophyll *a*, compared with the isocratic HPLC method (♦) developed here. HPLC separations of chlorophyll *a* with short retention times and low flow rates are the most efficient.

sensitive (detection limits = 0.1 ng at $\lambda_{em} > 600$ nm) than absorbance (detection limits = 10 ng at $\lambda = 440$ nm; Mantoura and Llewellyn, 1983), but it is also selective for chlorophylls in the presence of carotenoids. Consequently, for simple isocratic HPLC systems for chlorophylls only, fluorometry is clearly the preferred method of detection.

This chapter reports the results of SCOR WG 78 experiments to develop simple isocratic HPLC system(s) based on a 3 µm reversed-phase column for routine, accurate and sensitive analysis of chlorophyll *a* from marine samples. The use of autosamplers and optimization of fluorescence detectors are also reported. Recommendations on the operation and calibration of such simple HPLC systems for oceanographic applications are also made.

11.3 Materials and methods

11.3.1 *Standards, samples and reagents*

Chlorophyll *a* and *b* standards were purchased from Sigma (Poole, UK) or extracted from cultures of the chlorophyte *Dunaliella tertiolecta* (Plymouth Culture Collection PCC 83 ≡ CSIRO Algal Culture Collection, CS-175), and purified by HPLC, using the semi-preparative procedures detailed in Chapter 7. Stock standards of chlorophylls were made up in 1.0 ml amber volumetric flasks in acetone and their concentrations determined spectrophotometrically from the pigment extinction coefficients listed in the Part IV Data sheets and Appendix E. Other pigment standards were obtained from cultured microalgae including the diatom *Phaeodactylum tricornutum* (PCC 1052 ≡ CS-29) and the green alga *Dunaliella tertiolecta* (PCC 83 ≡ CS-175).

Phytoplankton samples were obtained from the English Channel, the Tamar Estuary (UK) and the northeast Atlantic Ocean by filtering 0.5–2 litres of seawater through 47 mm Whatman GF/F filters and extracting the pigments by ultrasonication in methanol or acetone (see Chapter 9). Samples rich in degraded pigments included: (1) sediment traps (Chesapeake Bay, Dr R. Dawson), (2) faecal pellets from the grazing copepod *Calanus helgolandicus* (Dr R. Harris) and (3) faecal debris from the mussel *Mytilus edulis*. These were analysed to evaluate the ability of the HPLC methods to also resolve complex mixtures of pheopigments. Pigments were identified by co-elution with authentic standards and by comparing the UV-visible and fluorescence spectra with the authentic spectra as documented in Part IV Data sheets.

All solvents and eluants were HPLC-grade (FISONS, UK) and were filtered (0.45 µm) and degassed with helium prior to HPLC. The ammonium acetate buffer (0.5 M, pH 7.1) was prepared in MilliQ-distilled/deionised water and filtered (0.7 µm Whatman, GF/F) before mixing with methanol for HPLC.

11.3.2 *HPLC hardware and elution systems*

The components of the isocratic and step-isocratic HPLC systems used in this study are schematically illustrated in Figure 11.2. The HPLC was based on the Perkin Elmer (PE) system, although other systems would do equally well. An LC-410 pump, which included a solvent selector valve, was used to deliver solvents at 1.0 ml min^{-1}. Pigment extracts were injected into a Rheodyne 7125 injector fitted with a 50 µl sample loop. The separated pigments were serially detected on a LC-75 spectrophotometer (λ = 440 nm) followed by a LS-40 fluorescence detector with variable excitation and emission options. An ISS-100 autosampler was used to automate the injection of extracts into the HPLC. The chromatograms were recorded and integrated *via* a Nelson 900 Series data interface on an Epson PC. The computer also controled the LC-410 pump and the LC-75 detector.

The analytical HPLC column selected for this study was Perkin Elmer 3 µm Pecosphere C-18, as this column-type has been reliably used at the Plymouth Marine Laboratory for over five years. The columns are available as replaceable cartridges (33 mm long and 4.5 mm diameter) packed with octadecyl-silane monomerically bonded onto 3 µm spherical silica particles (80 Å porosity). Residual silanol sites are reported as 'end capped' (with trimethyl silane), yielding a total car-

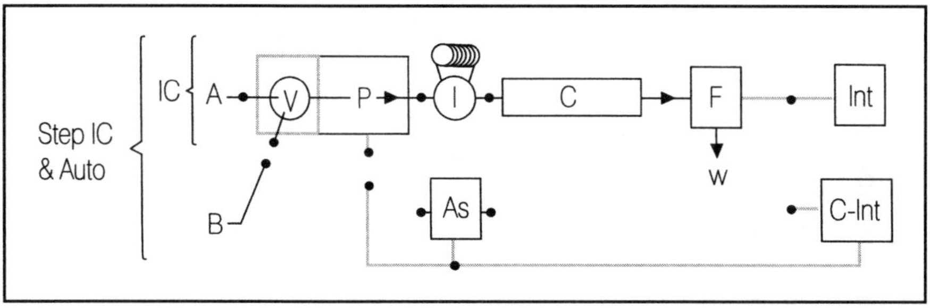

Figure 11.2
Schematic diagram of the isocratic and step-isocratic HPLC systems used for determining chlorophylls. The components include eluants A and B (described in Table 11.2), two-way valve (V), reciprocating pump (P), sample loop injector (I), analytical column (C), fluorometer (F) and integrator (Int). Automation options include a cooled autosampler (As) and a computing integrator (C-Int). The bold lines indicate the flow of eluant through the HPLC and out to waste (W) and shaded lines the flow of data.

bon content of 12% and a separation efficiency of 150,000 plates m^{-1}. Similar 3 μm reversed-phase C-18 columns are also available from other manufacturers (Beckman, Hewlett Packard, Shandon Hypersil) and should perform equally well.

11.3.3 *HPLC optimization experiments*

Various solvent combinations and fluorescence detection conditions were evaluated on a range of pigment standards and extracts of different complexities to develop the simplest eluant and detection system for HPLC-separation of chlorophyll *a*. Four experiments were undertaken to:

- develop an isocratic HPLC system for separating chlorophyll *a* from complex pigment mixtures
- develop a two-solvent step-isocratic HPLC system for separating three chlorophylls and up to ten degradation products
- optimize the excitation and emission wavelengths for fluorescence detection of chlorophylls and their breakdown products following separation by HPLC
- investigate the stability of pigments in buffered extracts held in an autosampler awaiting injection on the HPLC.

11.4 Results and discussion

11.4.1 *Isocratic HPLC analyses of chlorophyll a*

The polarity of methanol as an eluant for chlorophyll *a* was increased by the addition of 0.5 M ammonium acetate buffer and decreased by the addition of acetone. The effect of changing the polarity of methanol eluants on the isocratic retention time of chlorophyll *a* is shown in Figure 11.3. The results are plotted relative to the

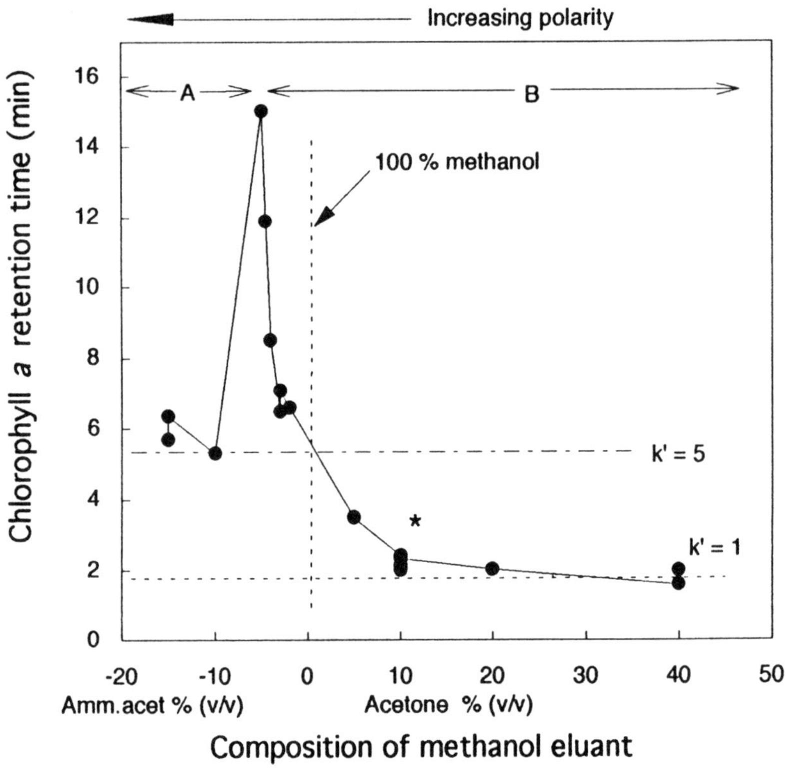

Figure 11.3
The effect of solvent composition and polarity on the retention time of chlorophyll *a* in the isocratic HPLC method. Region A corresponds to the polar methanol eluants that inhibit the retention and HPLC separation of chlorophyll *a*, and region B corresponds to the less polar methanol eluants that promote the reversed-phase separation of chlorophyll *a*. The region between k' = 1 and k' = 5 (where k' is the capacity ratio; Skoog, 1985) defines the optimum range of retention times for isocratic HPLC. * Denotes the optimal compromise between speed, solvent polarity and resolution which was adopted for the isocratic method.

column capacity ratios (k'; Skoog, 1985) of k' = 1 to 5, which is the optimal range of retention times to separate chlorophyll *a* efficiently in isocratic HPLC.

The retention time of chlorophyll *a* increased from 2 min to a maximum of 14.8 min with increasing polarity of the eluant (see region B, Figure 11.3). This is consistent with a reversed-phase partition mechanism (Ahuja, 1989; Skoog, 1985) controlling the retention of hydrophobic compounds such as chlorophyll *a* in HPLC. Beyond this maximum (95:5 (v/v) methanol:buffer), increasing the polarity of eluant produced poorly resolved and tailing peaks of chlorophyll *a* at ~6 min, probably because chlorophyll *a* tends to aggregate in polar solvents (e.g. aqueous ethanol, acetone; Scherz *et al.* 1991).

The optimum separation of chlorophyll *a* was obtained with a 90:10 (v/v) mixture of methanol and acetone. Elution of chlorophyll *a* was rapid (2.4 min, k' = 1.7) and still fully resolved from neighbouring allomer and chlorophyll *b*. This can be seen in the HPLC traces of chlorophyll pigments extracted from the English Channel (Fig. 11.4A) and of pheopigments from faecal debris produced by the filter-feeding mussel *Mytilus edulis* (Fig. 11.4B). Pheophytin *a* and even pyropheophytin *a* (not shown) could also be resolved, but their retention times of 6.5 min (k' = 6.3) and about 11 min (k' = 11.4) were impractically long. Pigments more polar than chlorophyll *b* (e.g., chlorophyll *c*, chlorophyllide *a* and pheophorbide *a*) co-eluted as a broad peak (1.1–1.3 min, k' = 0.9) just after the solvent front, but were well resolved from chlorophyll *a*. These optimimum conditions for separating chlorophyll *a* by this rapid isocratic HPLC method are listed in Table 11.2.

The analytical performance of our isocratic HPLC is compared to other systems in Figure 11.1. The isocratic HPLC system based on 3 μm C_{18} reversed-phase columns and elution with 90:10 (v/v) mixture of methanol:acetone significantly improves analysis time of chlorophyll *a* and solvent consumption. Adopting a generous 8-minute elution cycle, it follows that automating the isocratic HPLC system described above can achieve ~7 analyses per hour with a solvent consumption of only 8 ml per sample. To avoid potential co-elution errors between chlorophyll *a* and any pyropheophytins from a preceding sample, it is good practice to flush the HPLC column periodically with acetone to remove accumulated pyropheophytins.

11.4.2 *Step-isocratic HPLC of chlorophylls and degradation products*

The simple addition of a two-way valve (see V in Fig. 11.2) to the inlet of a basic HPLC pump is an inexpensive way of transforming an isocratic HPLC into a two-eluant step-isocratic HPLC. Many HPLC pumps have a manual or programmable solvent-switching valve usually fitted with a solvent-mixing column to dampen pressure fluctuations and solvent fronts. Consequently, a step change in eluants is in practice delivered as a sigmoid gradient to the column.

To profile the solvent change actually obtained on the LC-410, the two eluants were step-changed after 3 min of flow at 1.0 ml min^{-1}, and the UV absorbance (λ = 320 nm) of acetone in the secondary eluant was monitored during a blank run on the HPLC. The corresponding trace illustrated in Figure 11.5A shows an asymptotic rise in the second eluant starting at 6.5 min and ending at 12 min characteristic of a very damped system.

The choice of the first eluant (methanol:ammonium acetate, 80:20 (v/v)) was based on the success of methanol-based solvents in separating the more polar chlorophylls (chlorophyllides, chlorophyll *c*, pheophorbides) in many *gradient* HPLC systems (Mantoura and Llewellyn, 1983; Zapata *et al.* 1987; Wright *et al.* 1991). The second, less polar, eluant (methanol:acetone 90:10 (v/v)) corresponded to the isocratic HPLC system and served to elute the hydrophobic pigments (e.g., chlorophyll *a*, β,β-carotene).

The step-isocratic HPLC fluorescence traces of nine chlorophyll pigments are shown in Figure 11.5B–E. These were obtained from 90% acetone extracts of *Phaeodactylum tricornutum* (chlorophylls *a*, c_1, c_2 and chlorophyllide *a*), *Dunaliella tertiolecta* (chlorophylls *a* and *b*, their allomers and chlorophyllides) and phytoplankton from the deep chlorophyll maximum in the northeast Atlantic Ocean (pheophorbide *a*, pheophorbide *a*-like and chlorophyll *a*). The methanol or acetone

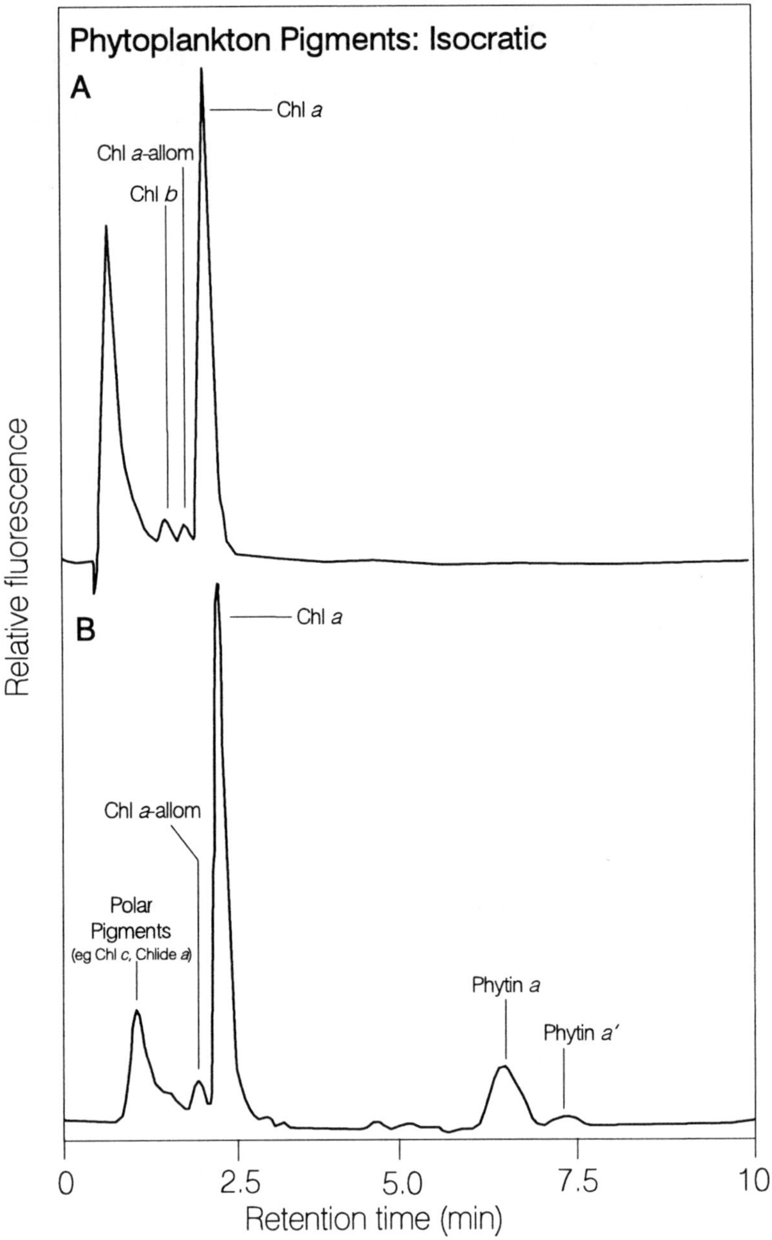

Figure 11.4
Fluorescence HPLC chromatograms (λ_{ex} = 440 nm, λ_{em} > 530 nm) for the isocratic (methanol:acetone, 90:10 (v/v)) separation of chlorophyll *a* from other chlorophylls and pheopigments in (A) a phytoplankton sample from the English Channel, (B) faecal detritus from the mussel *Mytilus edulis*.

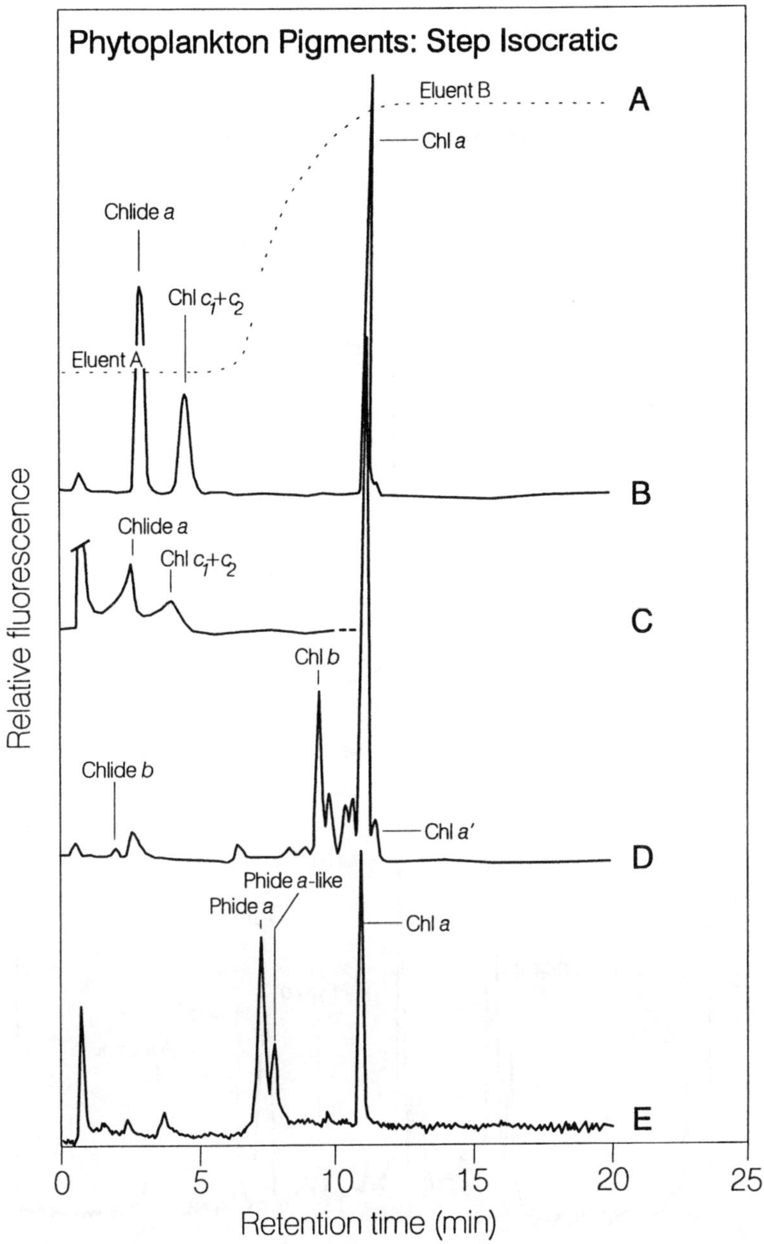

Figure 11.5
Step-isocratic HPLC separation of phytoplankton pigments: (A) UV absorbance ($\lambda = 320$ nm) profile of the solvent change obtained during a blank HPLC run in which solvent A (methanol:0.5 M ammonium acetate, 80:20 (v/v) pH = 7.1) is pumped for 3 minutes then step-changed to solvent B (methanol:acetone, 90:10 (v/v)) and run for 20 min. (B) Step-isocratic HPLC fluorescence ($\lambda_{ex} = 408$ nm, $\lambda_{em} >$ 530 nm) chromatograms of acetone extracts from the diatom *Phaeodactylum tricornutum* pre-mixed (3:1, v/v) with 0.5 M ammonium acetate pH 7.1. (C) As in B but omitting the premixing of extracts with buffer. (D) Step-isocratic HPLC fluorescence chromatogram of the chlorophyte *Dunaliella tertiolecta* with enhanced detection ($\lambda_{ex} = 440$ nm) of chlorophyll *b*. (E) Phytoplankton sample from the chlorophyll maximum (28 m) of the northeast Atlantic Ocean. Pigment abbreviations as in Appendix A.

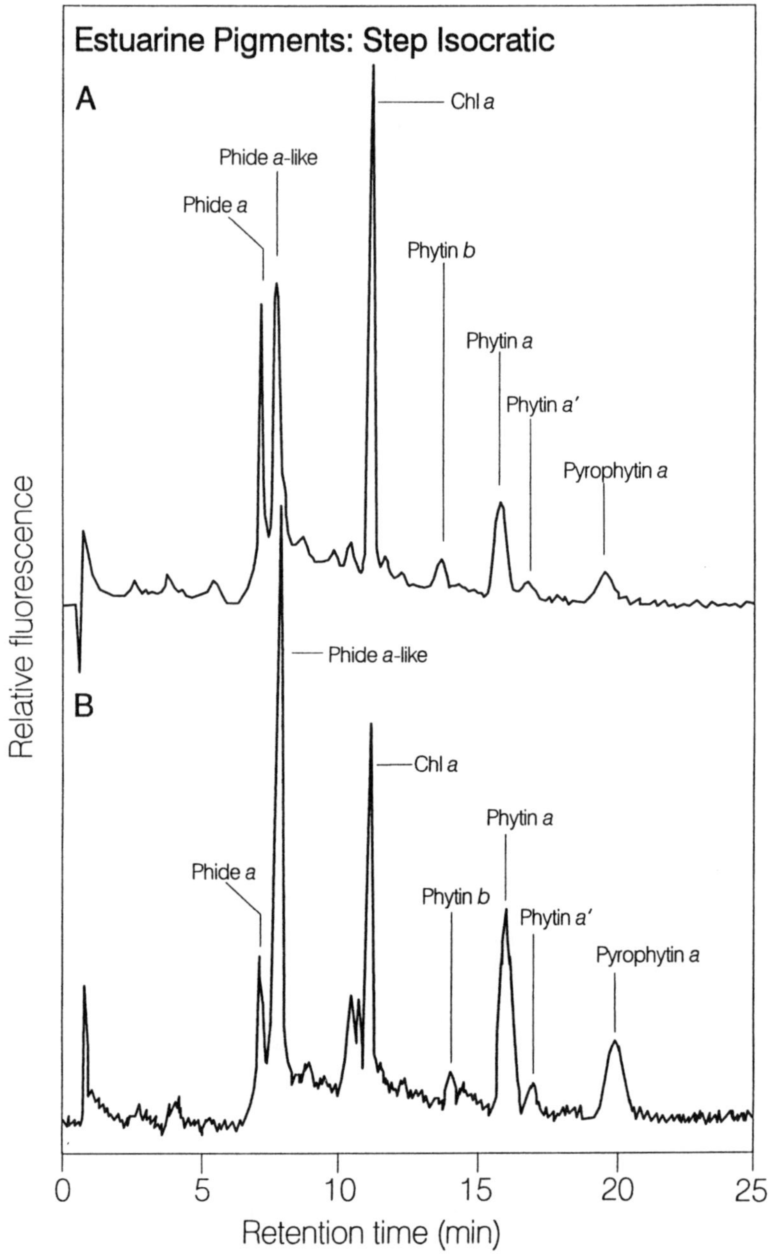

Figure 11.6
Step-isocratic HPLC separation and fluorescence (λ_{ex} = 408 nm, λ_{em} > 530 nm) detection of pheopigments from acetone extracts of (A) suspended particulate material from the turbid Tamar Estuary, UK, and from (B) a sediment trap sample collected in Chesapeake Bay, Maryland, USA.

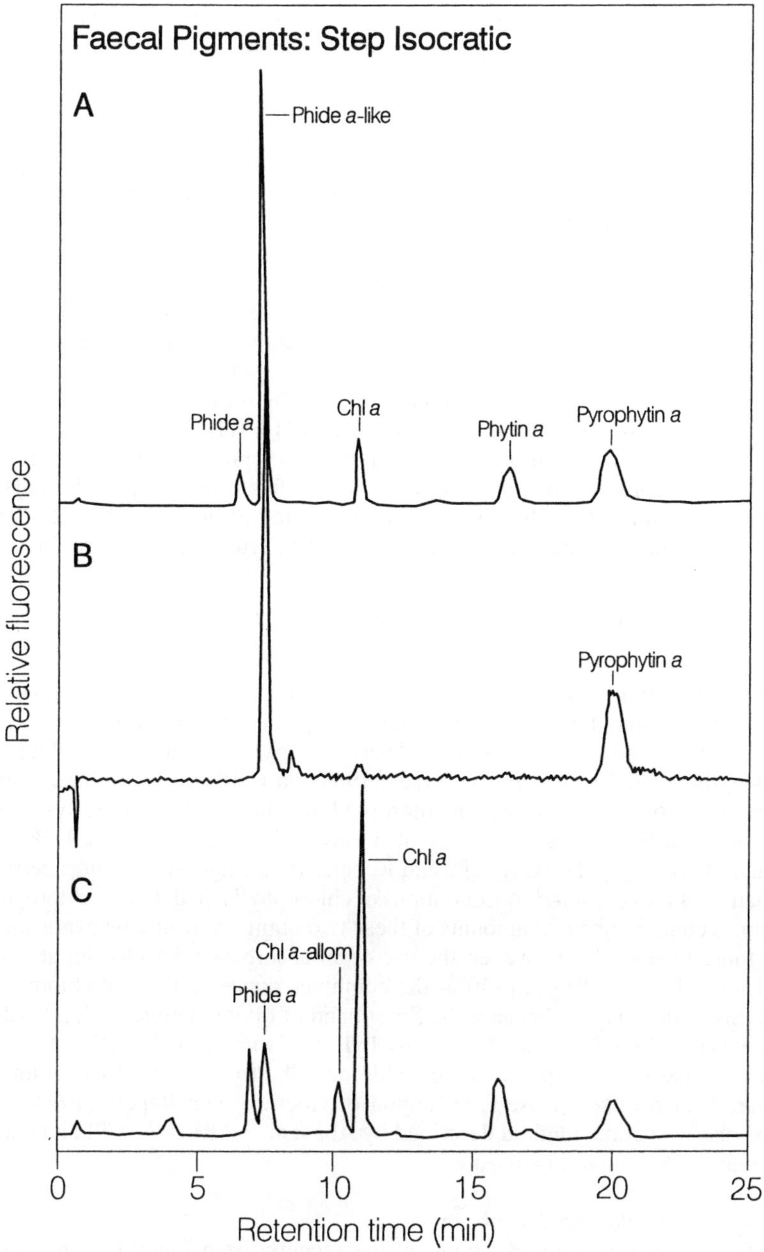

Figure 11.7
Step-isocratic HPLC separation and fluorescence (λ_{ex} = 408 nm, λ_{em} > 530 nm) detection of pheopigments from acetone extracts of faecal pellets from the copepod *Calanus helgolandicus* after (A) a period of heavy grazing on the chlorophyte microalga *Dunaliella tertiolecta*, and (B) after a period of starvation. (C) Chlorophyll pigments in the faecal debris produced by the mussel *Mytilus edulis* filter-feeding on the diatom *Phaeodactylum tricornutum*. Pigment abbreviations as in Appendix A.

extracts must be mixed with 0.5 M ammonium acetate buffer (3:1, v/v) immediately before injection, to ensure that even the most polar pigments are adsorbed as a narrow zone on the top of the HPLC column (Wright *et al.* 1991). The omission of this step results in poorly resolved peaks of chlorophyllide *a* and of chlorophylls c_1 and c_2 (Fig. 11.5C).

The step-isocratic HPLC chromatograms of acetone extracts rich in chlorophyll degradation products are shown in Figures 11.6 and 11.7. A complex mixture of at least seven pheopigments was separated from extracts of suspended particles in a sample from the Tamar Estuary (Fig. 11.6A) and from small quantities (<1 mg) of sediment-trap material collected in Chesapeake Bay (Fig. 11.6B). Similarly complex mixtures were obtained from extracts of faecal material produced by the copepod *Calanus helgolandicus* (Fig. 11.7A, B) and the common mussel *Mytilus edulis* (Fig. 11.7C). Variations in the food quality and ingestion rates appear to alter profoundly the composition of pheopigments released by these herbivores.

It is evident from the HPLC traces in Figures 11.5–11.7 that the three principal chlorophylls *a, b* and c_{1+2} and ten other chlorophyllides, pheopigments and pyropheophytin *a* were reproducibly separated in 20 min by this simple step-isocratic HPLC system. The optimized conditions, retention times and identities of the chlorophyll-derived pigments separated in these step-isocratic HPLC runs are detailed in Table 11.2.

11.4.3 *Pigments co-eluting with chlorophyll a*

Divinyl chlorophyll a

Divinyl chlorophylls *a* and *b* occur in prochlorophytes and may contribute up to 40% of the *total* chlorophyll *a* in the tropical and subtropical oceans (Goericke and Repeta, 1992; Veldhuis and Kraay, 1990). These divinyl derivatives co-elute with chlorophylls *a* and *b* (Gieskes and Kraay, 1983) on C-18 reversed-phase columns and may therefore give falsely high chlorophyll *a* values on our HPLC systems. As the broad-band fluorescence intensity of divinyl chlorophyll *a* is only 11% higher than that of chlorophyll *a* (Goericke and Repeta, 1993), our HPLC-fluorimetric determination of the *combined* concentrations of chlorophyll *a* and divinyl chlorophyll *a* in samples containing equal amounts of these two pigments would be overestimated by no more than 4–5%. However, the use of an absorbance detector set at 440 nm would severely overestimate (~40%) the combined concentrations of chlorophyll *a* and divinyl chlorophyll *a* because the Soret band of divinyl chlorophyll *a* (442 nm) in eluant is red-shifted compared to chlorophyll *a* (431 nm) (see Part IV Data sheets). When it is necessary to separate divinyl chlorophyll *a* from chlorophyll *a*, either the multi-gradient reversed-phase C-8 method of Goericke and Repeta (1993) or the normal-phase isocratic method described by Abaychi and Riley (1979) and Gieskes and Kraay (1983) should be used.

Phytylated chlorophyll c-like pigment

A phytylated chlorophyll *c*-like pigment has recently been found to elute close to chlorophyll *a* in a reversed-phase HPLC system (Nelson and Wakeham, 1989). It would therefore co-elute with chlorophyll *a* in our isocratic systems. However, in a study of 29 species (50 strains) of prymnesiophytes, Jeffrey and Wright (1994) found this pigment to account for less than 5% of the chlorophyll *a* concentration. It is therefore unlikely to interfere significantly in the isocratic HPLC determination of chlorophyll *a*.

TABLE 11.2. Recommended protocols for isocratic and step-isocratic HPLC of chlorophylls.

Parameter	Isocratic HPLC	Step-isocratic HPLC
Eluant	A: 90:10 (v/v) methanol:acetone B: none	A: 80:20 (v/v) MeOH:0.5 M AmAc B: 90:10 (v/v) MeOH:Acetone
Program	A: 0–8 min	A: 0–3 min B: 3–17 min
Flow Rates	1.0 ml min^{-1}	1.0 ml min^{-1}
Column	3 μm C-18 Pecosphere	3 μm C-18 Pecosphere
Detector	fluorescence Ex = 400–450 nm Em = >530 nm red-sensitive PMT	fluorescence Exc = 400–450 nm Em = >530 nm red-sensitive PMT
Sample Extracts	methanol or acetone	methanol or acetone
Pre-injection mix	3:1 (v/v) Sample:0.5 M AmAc	3.1 (v/v) Sample:0.5 M AmAc
Injection volume	50 μl	50 μl

Compound	Retention time (min) Figure 11.4	Retention time (min) Figure 11.5 and Figure 11.6	
Chlorophyllide *b*	1.1	2.04	
Chlorophyllide *a*	1.1	2.70	
Chlorophyll c_{1+2}	1.1	4.06	
Pheophorbide *a*	1.1	7.04	7.04
Pheophorbide *a*-like	1.1	7.50	7.50
Chlorophyll *b*	1.8	9.47	
Chlorophyll *a*-allomer	2.05		
Chlorophyll *a*	2.4	11.02	11.0
Chlorophyll *a*'	na	11.60	
Pheophytin *b*	na		13.42
Pheophytin *a*	6.5		15.68
Pheophytin *a*'	7.4		16.11
Pyropheophytin *a*	11.0		19.63

na = not analysed; AmAc = ammonium acetate; MeOH = methanol;
Exc = excitation wavelength; Em = emission wavelength; PMT = photomultiplier.

11.4.4 *Fluorescence detection of HPLC-separated chlorophylls*

The effect of varying the excitation wavelength on the fluorimetric detection of chlorophylls and their degradation products was determined by repeated HPLC analyses of pigments at different excitation wavelengths. The results, shown in Figure 11.8, clearly indicate that pheopigments are optimally excited at 400–415 nm, which is consistent with their known excitation maxima (pheophytin *a* = 409 nm in diethyl ether; pheophorbide *a* = 406 nm in acetone, see Part IV Data sheets). The fluorescence of both chlorophylls *a* and *c* are more sensitively excited at 430–440 nm. Narrow-band spectrofluorometers such as the LC 75 are selective and would be diagnostic of different groups of chlorophyll pigments. The less expensive filter fluorometers, when fitted with a broad-band (400–450 nm) excitation filter, would maximize sensitivity for all chlorophyll pigments. A detection limit of 0.05 ng chlorophyll *a* should be easily achievable with broad-band excitation, a long-pass (>600 nm) emission filter and a red-sensitive photomultiplier fitted to the fluorometer.

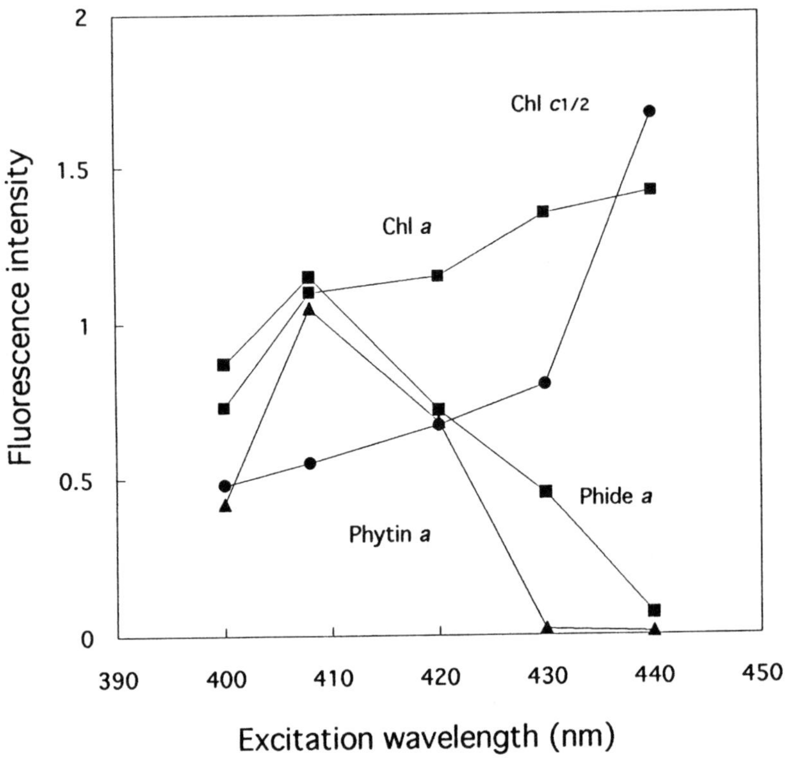

Figure 11.8
The effect of varying excitation wavelength from 400 to 440 nm on the fluorescence ($\lambda_{em} > 530$ nm) detection of chlorophylls *a* and $c_1 + c_2$, pheophorbide *a* and pheophytin *a* after their separation by step-isocratic HPLC.

11.4.5 *Pigment stability during autosampling*

We tested the stability of pigments of the diatom *Phaeodactylum tricornutum* in buffered methanol extracts kept for over 4 hours in a refrigerated (4°C) autosampler before injection. The concentration of chlorophyll *a* declined significantly over this period (Fig. 11.9). Increasing the buffer content of the acetone extracts from 10 to 40% (v/v) resulted in greater losses of chlorophyll *a* (1 to 16%) over the 4 hours (Fig. 11.9). There were no corresponding losses with fucoxanthin or chlorophyll *c*, and no breakdown products were formed in any of the samples. This selective decline in the recovery of non-polar pigments in the buffer-treated samples was probably due to adsorptive losses of these hydrophobic molecules onto the walls of the vials or to the formation of micelles (Scherz *et al.* 1991). These losses can be avoided

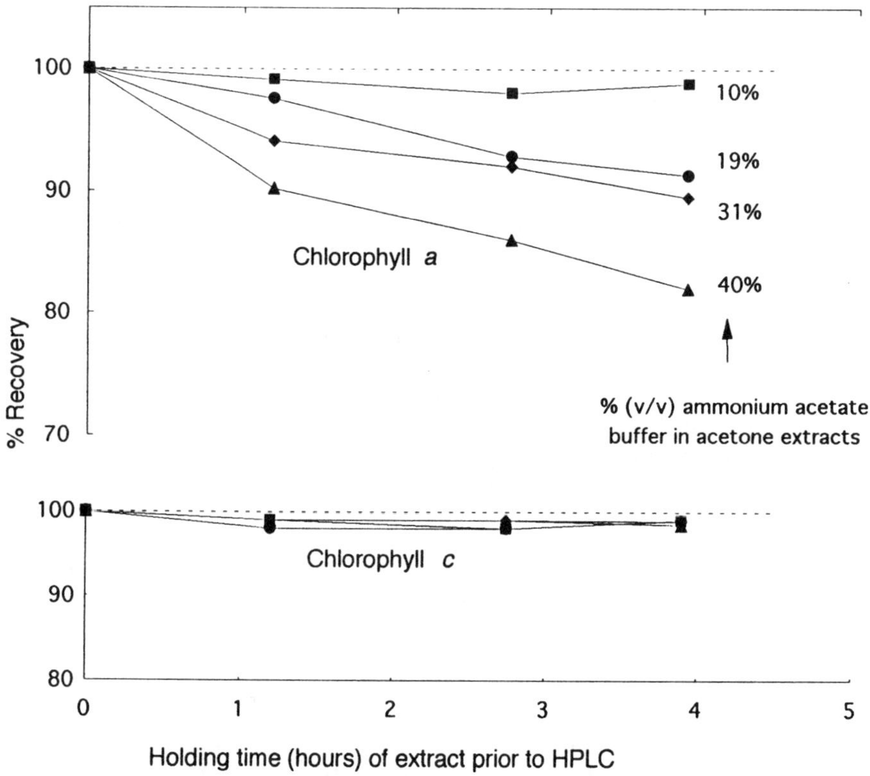

Figure 11.9
Recovery of chlorophyll a and $c_1 + c_2$ from acetone extracts of *P. tricornutum* premixed with increasing volumes (%, v/v) of 0.5 M ammonium acetate buffer during a four hour wait period in a refrigerated (4°C) autosampler, prior to HPLC.

if a 'derivatizing' autosampler is used to add and mix buffer to the extracts (1:3, v/v) immediately prior to injection. Excellent recoveries (103 ± 5%) of all pigments can be achieved even after 24 hours storage (Dr S. Wright, personal communication).

If the above precautions are observed, automation of step-isocratic HPLC analysis of extracts is practicable. Around 60 samples can be analyzed in 24 hours. A computer for data logging, integrating and reprocessing of chromatograms (see Fig. 11.1 and Chapter 15) is essential for large numbers of samples.

11.5 Conclusions and recommendations

The operational and chromatographic features of the isocratic HPLC and step-isocratic HPLC systems developed during the SCOR Workshops are summarized in Table 11.2. The following points should be considered when choosing which isocratic and detection system to use:

- The isocratic HPLC system with fluorescence detection is ideally suited for rapid (<5 min) and sensitive (0.1 ng) determination of chlorophyll *a* in marine phytoplankton samples. This is achieved with small (33 mm) cartridge columns packed with 3 µm C-18 bonded reversed-phase particles. Co-elution of any divinyl chlorophyll *a* or phytylated chlorophyll *c*-like pigments with chlorophyll *a* will bias the accuracy of *total* chlorophyll *a* (i.e. divinyl- and chlorophyll *a*) by a maximum of 5%. The modest cost of an isocratic HPLC is comparable to a bench-top scanning spectrophotometer, which might be used for spectrophotometric chlorophyll *a* measurement (Chapter 4; Appendix F, G). A later option to automate the analyses of large batches of samples by isocratic HPLC should also be possible.
- The step-isocratic HPLC system developed here features a damped asymptotic change in secondary eluant. This separates three chlorophylls (*a, b* and *c*) and ten additional chlorophyllides and pheopigments in under 21 min. The step-isocratic HPLC has the same column and fluorometer as the isocratic HPLC system. It is best suited for the quantitation of chlorophylls and breakdown products in samples rich in pheopigments (e.g., estuarine and faecal extracts). Methanol or acetone extracts should be mixed with buffer (3:1, v/v) immediately before injection to ensure polar pigments are adsorbed as a narrow band on the top of the column. Autosamplers must be capable of mixing preset volumes of buffer with refrigerated extracts just before injection on the HPLC.
- The fluorescence detector, when fitted with broad band-filters (λ_{ex} = 400–450 nm; λ_{em} >600 nm) and a red-sensitive photomultiplier, is an inexpensive, sensitive and selective detector of HPLC-separated chlorophylls and pheopigments. Since carotenoids are not detected, HPLC fluorescence chromatograms exhibit fewer peaks than absorbance chromatograms, which makes it easier to identify and quantify the chlorophylls. Calibration requires purified and spectrophotometrically-quantified standards of the chlorophylls and their degradation products to obtain relative fluorescence response factors (Chapters 7, 12). The linear detection range should be checked for quenching, as with all fluorometers. Subsequent calibration with chlorophyll *a* as external standard or with one of the promising new synthetic internal standards (Chapter 16) should be done regularly.

It should now be possible for any user to assemble an isocratic HPLC with the components specified here from many manufacturers (see Appendix K) and be assured of a reliable determination of chlorophyll *a*.

The practical limit on the polarity range of compounds that can realistically be separated is the ultimate limitation of step-isocratic HPLC. If a complete separation of the 50 or so marine phytoplankton chlorophylls and carotenoids is required, then the ternary gradient HPLC system developed at the SCOR Workshops (Chapter 12 and Wright *et al.* 1991) should be used.

ACKNOWLEDGEMENTS

We are grateful to Perkin Elmer (Beaconsfield) for the loan of HPLC apparatus during the SCOR WG 78 Workshop in Plymouth and to Tim Fileman for help with the figures.

REFERENCES

Abaychi, J. K., Riley, J. P. (1979). The determination of phytoplankton pigments by high-performance liquid chromatography. *Anal. Chim. Acta,* 107: 1–11.

Ahuja, S. (1989). *Selectivity and Detectability Optimizations in HPLC.* Wiley Interscience, New York, 610 pp.

Baweja, R. (1986). High-performance liquid chromatography separation of deuterated photosynthetic pigments from their protio analogues. *J. Chromatogr.* 369: 125–131.

Bessiere, J., Montiel, A. (1982). A selective and rapid method of chlorophylls *a* and *b* separative determination by high-pressure liquid chromatography. *Wat. Res.* 16: 987–993.

Bidigare, R. R., Kennicutt, M. C., Brooks, J. M. (1985). Rapid determination of chlorophylls and their degradation products by high-performance liquid chromatography. *Limnol. Oceanogr.* 30: 432–435.

Braumann, T., Grimme, H. L. (1979). Single step separation and identification of photosynthetic pigments by high-performance liquid chromatography. *J. Chromatogr.* 170: 264–268.

Craft, N. E., Wise, S. A., Soares, J. H. (1992). Optimization of an isocratic liquid chromatographic separation of carotenoids. *J. Chromatogr.* 589: 171–176.

de las Rivas, J., Abadia, A., Abadia, J. (1989). A new reversed-phase HPLC method resolving all major higher plant photosynthetic pigments. *Plant Physiol.* 91: 190–192.

Eskins, K., Schofield, C. R., Dutton, H. J. (1977). High-performance liquid chromatography of plant pigments. *J. Chromatogr.* 135: 217–220.

Falkowski, P. G., Sucher, J. (1981). Rapid, quantitative separation of chlorophylls and their degradation products by high-performance liquid chromatography. *J. Chromatogr.* 213: 349–351.

Fiksdahl, A., Mortensen, F. T., Liaaen-Jensen, S. (1978). High pressure liquid chromatography of carotenoids. *J. Chromatogr.* 157: 111–117.

Gieskes, W. W., Kraay, G. W. (1983). Unknown chlorophyll *a* derivatives in the North Sea and the tropical Atlantic Ocean revealed by HPLC analysis. *Limnol. Oceanogr.* 28: 757–766.

Goericke, R., Repeta, D. J. (1992). The pigments of *Prochlorococcus marinus*: the presence of divinyl chlorophyll *a* and *b* in a marine prokaryote. *Limnol. Oceanogr.* 37: 425–433.

Goericke, R., Repeta, D. J. (1993). Chlorophylls *a* and *b* and divinyl chlorophylls *a* and *b* in the open subtropical North Atlantic Ocean. *Mar. Ecol. Progr. Ser.* 101: 307–313.

Goeyens, L., Post, E., Dehairs, F., Vandenhoudt, A., Baeyens, W. (1982). The use of high pressure liquid chromatography with fluorimetric detection for chlorophyll *a* determination in natural extracts of chloropigments and their degradation products. *Inter. J. Environ. Anal. Chem.* 12: 51–63.

Iriyama, K., Yoshiura, M., Shiraki, M. (1978). Micro-method for the qualitative and quantitative analysis of photosynthetic pigments by high-performance liquid chromatography. *J. Chromatogr.* 154: 302–305.

Jeffrey, S. W., Wright, S. W. (1994). Photosynthetic pigments in the Haptophyta. In: Green, J. C., Leadbeater, B. S. (eds), *The Haptophyte Algae.* Oxford University Press, Oxford, p. 111–132.

Liebezeit, G. (1980). Chlorophyll *a* in marine phytoplankton: separation by HPLC and specific fluorimetric detection. *J. High Res. Chromatogr. Comm.* 3: 531–533.

Mantoura, R. F. C., Llewellyn, C. A. (1983). The rapid determination of algal chlorophyll and carotenoid pigments and their breakdown products in natural waters by reverse-phase high-performance liquid chromatography. *Anal. Chim. Acta,* 151: 297–314.

Nells, H. C. F., DeLeenheer, A. P. (1983). Isocratic non aqueous reverse-phase liquid chromatography of carotenoids. *Anal. Chem.* 55: 270–275.

Nelson, J. R., Wakeham, S. G. (1989). A phytol-substituted chlorophyll *c* from *Emiliania huxleyi*. *J. Phycol.* 25: 761–766.

Otsuki, A., Watanake, M. M., Sugahara, K. (1987). Chlorophyll pigments in methanol extracts from ten axenic cultured diatoms and three green algae as determined by reverse-phase HPLC with fluorometric detection. *J. Phycol.* 23: 406–414.

Rebeiz, C. A., Bazzaz, M. B. Belanger, F. (1978). The separation of chlorophyll and phaeophytins by reverse-phase HPLC. *Chromatogr. Rev.* 4: 8–9.

Scherz, A., Rosenbach-Belkin, V., Fisher, J. R. E. (1991). Chlorophyll aggregation in aqueous solutions. In: Scheer, H. (ed.), *Chlorophylls*. CRC Press, Boca Raton, p. 237–268.

Shioi, Y., Fukae, R., Sasa, T. (1983). Chlorophyll analysis by high performance liquid chromatography. *Biochim. Biophys. Acta*, 722: 72–79.

Shoaf, W. T. (1978). Rapid method for the separation of chlorophylls *a* and *b* by high-performance liquid chromatography. *J. Chromatogr.* 152: 247–249.

Skoog, D. A. (1985). *Principles of Instrumental Analysis*. Saunders Publishing, Philadelphia, 928 pp.

Steiner, R., Weischhoff, W., Scheer, H. (1982). High-performance liquid chromatography of bacteriochlorophyll *b* and its derivatives as an aid for structural analysis. *J. Chromatogr.* 242: 127–134.

Stransky, H. (1978). Die quantitative Bestimmung von Chloroplastenpigmenten im picomol-Bereich mit hilfe einer isochratischen HPLC-methode. *Z. Naturwissen.* 33: 836–840.

Trees, C. C., Kennicutt, M. C., Brooks, J. M. (1985). Errors associated with the standard fluorimetric determination of chlorophylls and phaeopigments. *Mar. Chem.* 17: 1–12.

Velduis, M. J. W., Kraay, G. W. (1990). Vertical distribution and pigment composition of a picoplanktonic prochlorophyte in the subtropical North Atlantic: a combined study of HPLC analysis of pigments and flow cytometry. *Mar. Ecol. Prog. Ser.* 68: 121–127.

Wright, S. W., Jeffrey, S. W., Mantoura, R. F. C., Llewellyn, C. A., Bjørnland, T., Repeta, D., Welschmeyer, N. (1991). Improved HPLC method for the analysis of chlorophylls and carotenoids from marine phytoplankton. *Mar. Ecol. Prog. Ser.* 77: 183–196.

Zapata, M., Ayala, A. M., Franco, J. M., Garrido, J. L. (1987). Separation of chlorophylls and their degradation products in marine phytoplankton by reversed-phase high-performance liquid chromatography. *Chromatographia*, 23: 26–30.

12

High-resolution HPLC system for chlorophylls and carotenoids of marine phytoplankton

S. W. Wright and S. W. Jeffrey

12.1 INTRODUCTION

Up until the SCOR Workshops in 1988, a number of HPLC methods for separating phytoplankton pigments had been published, but no one method was ideal for all pigments (see Chapter 4). Techniques for the separation of chlorophylls and their derivatives included those of Mantoura and Llewellyn (1983), Bidigare *et al.* (1985), Gieskes and Kraay (1986*a*) and Zapata *et al.* (1987), while the best methods for carotenoid resolution were those of Repeta and Gagosian (1982) and Wright and Shearer (1984). The first method had the disadvantage of using a silica column (which causes degradation of chlorophylls), while Wright and Shearer's did not resolve the polar chlorophylls c_1, c_2, c_3 and Mg-2,4 divinyl pheoporphyrin a_5 monomethyl ester (MgDVP), which coeluted as a single peak. Also the chlorophyllides *a* and *b* were not adequately resolved. A single method suitable for the full range of important chlorophylls, chlorophyll derivatives and taxonomically significant carotenoids found in phytoplankton was urgently needed, particularly for the international oceanographic climate-change studies (e.g., the Joint Global Ocean Flux Study).

At the SCOR Chlorophyll Workshop (Hobart, Australia, November 1988)

Dr N. Welschmeyer suggested that a more comprehensive separation might be achieved by combining his initial solvent system for separating polar chlorophylls with that of the Wright and Shearer (1984) system for carotenoids and nonpolar chlorophylls. The improved method that resulted separates more than 50 chlorophylls, carotenoids, their derivatives and isomers from marine phytoplankton in one ternary gradient system. This paper details the HPLC method, and shows clearly the resolution of pigments from the polar to non-polar ends of the chromatogram. While chlorophyll c_3 was now separated from the other polar chlorophyll *c* pigments (which still coeluted as a single peak following that of c_3) and chlorophyllides *a* and *b* were also resolved, some problems still remained in complex carotenoid mixtures. Two pigment pairs coeluted (19'-butanoyloxyfucoxanthin and siphonaxanthin, and 19'-hexanoyloxyfucoxanthin and 9'-*cis*-neoxanthin) but these could be distinguished by their on-line diode array spectra. This method has been published as Wright *et al.* (1991).

Since the SCOR Workshops, divinyl chlorophylls *a* and *b* have been identified in cultured samples of the prochlorophyte *Prochlorococcus marinus*. First located in field samples by normal-phase HPLC systems (Gieskes and Kraay, 1983*a*), they can now be separated from chlorophylls *a* and *b* on C_8-reversed-phase HPLC columns (Goericke and Repeta, 1992).

New advances have also been made in improved resolution of the chlorophyll *c* pigments by HPLC. Earlier, polyethylene columns had been used to separate the four chlorophyll *c* pigments chlorophyll c_1, c_2, c_3 and MgDVP (Jeffrey, 1989), but Kraay *et al.* (1992) and Garrido and Zapata (1993) have now claimed full separation of these chlorophyll *c* pigments on polymeric C_{18} reverse-phase columns.

These systems, excellent for the *c* pigments, do not resolve the carotenoids as well as our present method, so it appears that several systems will still be needed for those investigations that require full resolution of all pigments.

The present chapter describes the method of HPLC separation of about 50 algal pigments derived from mixed algal extracts, while the next chapter (Chapter 13) gives HPLC analyses of pigments from each standard algal culture.

12.2 Materials and methods

Algal cultures for HPLC tests were maintained in the CSIRO Algal Culture Collection (Jeffrey, 1980; and Chapter 6). Because of the commercial non-availability of pure pigment standards (except for a very small number, see Appendix J) SCOR Working Group 78 decided that the source of such standards should be reference algal cultures whose pigments had been fully characterized chemically. A small number of species were selected (Table 13.1, Chapter 13), which together contained the most significant pigments known in marine phytoplankton.

12.2.1 *Culture harvest*

Cultures for the HPLC trials were grown as outlined in Chapter 6. Cells were harvested before the end of log phase and 10 ml culture aliquots were taken for pigment analysis. These were filtered through 25 mm Whatman GF/F filters under low vacuum (250 mmHg) and rinsed with filtered medium or seawater . The filter was extracted immediately, or was folded, placed in a labelled cryotube and immediately frozen in liquid nitrogen. Several replicates of each culture were made. The time taken from the start of filtration to liquid nitrogen was not more than 30 seconds, and storage of the samples in liquid nitrogen before HPLC analysis was between 1-10 days.

12.2.2 *Field samples*

Seawater (1 to 5 l) was filtered through a 25 mm diameter Whatman GF/F filter. This was folded, blotted dry and stored in a cryotube under liquid nitrogen for laboratory analysis. The sample illustrated in Fig. 12.3 was collected on 9 June 1990 from 90 m depth at 51°54'S, 73°14'E in the Southern Ocean near Heard Island. The particular sample illustrated was stored for 3 months before analysis. The changes occuring over this time period are relatively small (see Chapter 10).

12.2.3 *Pigment extraction*

Cultures

The frozen filter containing the algal sample was cut into small slices (several mm × 1 cm) and ground for 30 seconds in a motorized grinder with a teflon pestle in 2.7 ml 100% acetone. (While Chapter 9 recommends buffered methanol as an optimal solvent for all phytoplankton, the 100% acetone used with the 'easy to extract' cultures gave excellent and quantitative results; see Chapter 13). Water (0.3 ml) was then added to make up to 90% acetone, and the sample reground. The filter plus acetone extract was transferred to a plastic centrifuge tube cut down to 5 ml, which had been "needle-pierced" at the bottom (Wright and Shearer 1984) and covered with a 5 mm circle of GF/F filter. The cut-down centrifuge tube containing the extract was then placed in the top of a 10 ml conical glass centrifuge tube, and both tubes were centrifuged in tandem for 3 minutes at 700 g. The clear green-yellow extract was collected in the lower glass centrifuge tube, while the filter debris remained in the upper tube. The homogenizer and pestle were rinsed with 1 ml 90% acetone, which was added to the filter debris in the upper tube, and recentrifuged. The clear combined extracts from the lower tube were then transferred to a 5 ml volumetric flask and made up exactly to 5 ml with 90% acetone to ensure volumetric accuracy. After filtering through a Millex-SR 0.5 µm filter (Millipore), an aliquot of the extract (about 25 µl) was injected immediately into the chromatography system for pigment analysis.

Field samples

The procedure for field samples was slightly different from that of the cultured samples because of the need to keep extraction volumes to a minimum to achieve a higher final concentration of pigments. Filters were cut into small pieces (several mm × 1 cm) and sonicated for 30 s in the cut-down centrifuge tubes, using 1.5 ml buffered methanol (or acetone). An anonymous reviewer of the Wright *et al.* (1991) paper suggested that adding 2% ammonium acetate buffer to the methanol extraction solvent would improve the height and sharpness of the chlorophyll *c* peaks. This was confirmed, and is now recommended. A Braun Labsonic 1510 sonicator equipped with a 4 mm diameter probe was operated at 50W. The bottom needle-hole in the centrifuge tube was securely covered with a single layer of Nescofilm or Parafilm to prevent solvent leakage during sonication. During centrifugation (as above), the film burst, allowing the extract to collect in the lower tube. The extract was filtered (Millex SR 0.5 µm) and then, immediately before injection, was diluted with water to 80% MeOH (or 66% acetone if acetone was used). An amount of 125 µl of the diluted extract was injected.

In general, as the field samples were much less concentrated than the culture samples, a larger injection volume was required. However, if a large aliquot of strong solvent (e.g., methanol or 90% acetone) was injected, then the pigment bands, particularly the early ones, were too broad and retention times decreased. Dilution of the extract increased the affinity of pigments for the column during the loading step, which resulted in sharper peaks and allowed greater overall loading than was possible with undiluted extract. It is important that the extract is not diluted until immediately before injection, since highly aqueous extracts are not stable and pigment is lost through adsorption and precipitation (particularly the more hydrophobic pigments) on standing (see Chapter 11).

12.2.4 *High-performance liquid chromatography*

Three HPLC instruments were used for comparing separation efficiencies: (a) a Waters Associates liquid chromatograph comprising M6000A, M45 and M501 pumps, Valco injector, Hewlett-Packard 8450A diode array spectrophotometer, and a Waters 440 absorbance detector connected via a Waters System Interface Module to a microcomputer running Waters Maxima software; (b) a Varian Model 5000 liquid chromatograph with Varian UV-50 variable wavelength detector, Varian Fluorichrome fluorescence detector, and Varian 'Vista' 402 data module; and (c) a Spectraphysics HPLC comprising an SP8800 ternary pump, SpectraFocus detector and a Gilson 231 autosampler (200 µl loop) with samples refrigerated to -10°C. A Hitachi F1000 fluorometric detector was used in some experiments. The dual monochrometers were both used in zero order mode and substituted with broad band filters (340-520 nm, excitation; >604 nm emission) from a Turner fluorometer. Chromatograms from the Varian 5000 were used in Fig. 12.1, and from the Spectraphysics instrument in Fig. 12.2 and 12.3. The reversed-phase columns used were Spherisorb ODS2, 25 cm × 4.6 mm ID, 5 µm particle size, packed by Australian Government Analytical Laboratories, Melbourne (90,000 to 140,000 plates m^{-1}). Similar resolution was obtained with commercial columns such as Waters Resolve C_{18} and Activon Ultratechsphere, although some C_{18} columns tested were not as efficient, particularly in the ability to resolve lutein and zeaxanthin.

Pigment detection was routinely at 436 nm for all chlorophylls and carotenoids and 405 nm for pheophytin *a* and pheophorbide *a*. With the Spectraphysics set-up, absorbance detection at 435, 470 and 665 nm was also used in conjunction with full spectrum collection to distinguish different pigment types (see below).

The solvent systems used were as follows:

- Solvent A: 80:20 methanol : 0.5 M ammonium acetate (aq.; pH 7.2 v/v)
- Solvent B: 90:10 acetonitrile (210 nm UV cut-off grade) : water (v/v)
- Solvent C: ethyl acetate (HPLC grade)

Flow rate was 1 ml min^{-1}. The gradient systems used are shown in Table 12.1. In the analytical system an initial four-minute gradient from 100% A to 100% B improved the separation of polar compounds. The program returned to initial conditions and re-equilibrated for 5 min before the next sample injection. Table 12.1a shows the gradient used on the Varian system to obtain the results presented in Fig. 12.1. Table 12.1b shows the modified procedure for the Spectraphysics system. In this case, the gradient for solvent C was reduced between 2.6 and 13.6 min. for the region of the chromatogram between fucoxanthin and chlorophyll *a* (increasing the resolution in the most complex region of the chromatogram) and increased elsewhere to reduce the overall time for analysis. Table 12.1c shows the shutdown protocol whereby the column is washed in the most hydrophobic eluant (ethyl acetate) to remove any retained material before being shutdown overnight or for storage. The start-up protocol is the reverse of this program.

Methanol, acetonitrile and ethyl acetate were HPLC grade reagents from Waters and BDH (Hypersolve), used without further purification other than filtration and degassing by sonication. Water was purified with a Millipore Milli-Q system. Ammonium acetate was A.R. grade.

TABLE 12.1. HPLC solvent system programs

Time (min)	Flow rate (ml min^{-1})	% A[1]	% B[2]	% C[3]	Conditions
a. Analytical gradient protocol (Waters and Varian systems)					
0	1.0	100	0	0	Injection
4	1.0	0	100	0	Linear gradient
18	1.0	0	20	80	Linear gradient
21	1.0	0	100	0	Linear gradient
24	1.0	100	0	0	Linear gradient
29	1.0	100	0	0	Equilibration
b. Modified protocol on Spectraphysics systems					
0	1.0	100	0	0	Injection
2	1.0	0	100	0	Linear gradient
2.6	1.0	0	90	10	Linear gradient
13.6	1.0	0	65	35	Linear gradient
20	1.0	0	31	69	Linear gradient
22	1.0	0	100	0	Linear gradient
25	1.0	100	0	0	Linear gradient
30	1.0	100	0	0	Equilibration
c. Shutdown protocol					
0	1.0	100	0	0	Analysis complete
3	1.0	0	100	0	Linear gradient
6	1.0	0	0	100	Linear gradient
16	1.0	0	0	100	Washing
17	0	0	0	100	Shut down; linear gradient

1. = 80:20 methanol : 0.5M ammonium acetate (pH 7.2, v/v)
2. = 90:10 acetonitrile : H_2O (v/v)
3. = ethyl acetate

12.2.5 *Pigment identification*

Pigments were identified by co-chromatography with authentic standards prepared at the SCOR Chlorophyll and Carotenoid Workshops (see Chapters 7 and 8), by diode array spectroscopy during elution, and by transferring HPLC fractions to standard solvents and comparing their visible absorption spectra with reference standards (see Part IV Data sheets, and references therein). A Shimadzu Model RPI recording spectrophotometer was used for UV-visible spectroscopy of purified pigments.

12.2.6 *Additional standard pigments*

Echinenone (synthetic) and lycopene (from tomatoes) were gifts from Hoffmann La Roche, Basel, to Dr S. Liaaen-Jensen, Trondheim, Norway. Canthaxanthin was a synthetic standard provided by Dr D. Repeta. Ethyl 8'-apo-β-carotenoate, purchased from Fluka Chemicals, USA, was suggested by Dr Repeta as a possible quantitative internal standard (see Chapter 16), and it was included in these tests to ensure that it did not co-chromatograph with known marker pigments. Another synthetic carotenoid, *trans*-β-apo-8'-carotenal, purchased from Sigma Chemical Co., was included as an internal standard in the field sample. All pigments were stored in the dark at -20°C or in vials submerged in liquid nitrogen until used. In our experience, if the solvent remains frozen the pigments are more stable, otherwise pigments degrade.

12.2.7 *Pigment recovery*

Recovery of pigments was measured by comparing peak areas for peridinin, lutein, chlorophyll *b*, chlorophyll *a*, and β,β-carotene, using the gradient and column described above with those obtained using isocratic buffered methanol (first phase of the gradient) and stainless steel tubing instead of the column (5 replicates). Figures were not corrected for differences in extinction coefficients in the different mobile phases.

12.3 RESULTS

Fig. 12.1 shows the resolution of a comprehensive pigment mixture prepared by pooling extracts from seven reference algae supplemented with additional carotenoid and chlorophyll derivatives. It gave a suite of over 40 pigments, of which 39 pigments were resolved and 37 securely identified. Table 12.2 lists the pigments separated in increasing elution order, together with a summary of their UV-visible spectral characteristics. Full visible spectra of each pigment in appropriate solvents (including HPLC eluant) may be found in the Part IV Data sheets.

At the polar end of the chromatogram, the phytol-free chlorophyll derivatives chlorophyllide *b* (peak 1), chlorophyllide *a* (peak 3) and chlorophyll c_3 (peak 5) were almost completely resolved from chlorophyll $c_1 + c_2$ + MgDVP, which coeluted as a single fraction (peak 6).

In the complex central region of the chromatogram, most of the known carotenoid bio-markers (peridinin, fucoxanthin, prasinoxanthin, violaxanthin, alloxanthin, lutein, zeaxanthin, canthaxanthin and siphonein) were adequately resolved. Only two pairs of carotenoids were not separated: 19'-butanoyloxyfucoxanthin from siphonaxanthin, and 19'-hexanoyloxyfucoxanthin from 9'-*cis*-neoxanthin, although the individual components of each pair could be distinguished spectrally by diode array detection. Many minor carotenoid components were also resolved and identified (Figs 12.1 and 12.3, Table 12.2 and Chapter 13).

At the non-polar end of the chromatogram, chlorophylls *a* and *b*, the pheophytins, and the less-polar carotenoids and carotenes were resolved. In fresh extracts, the

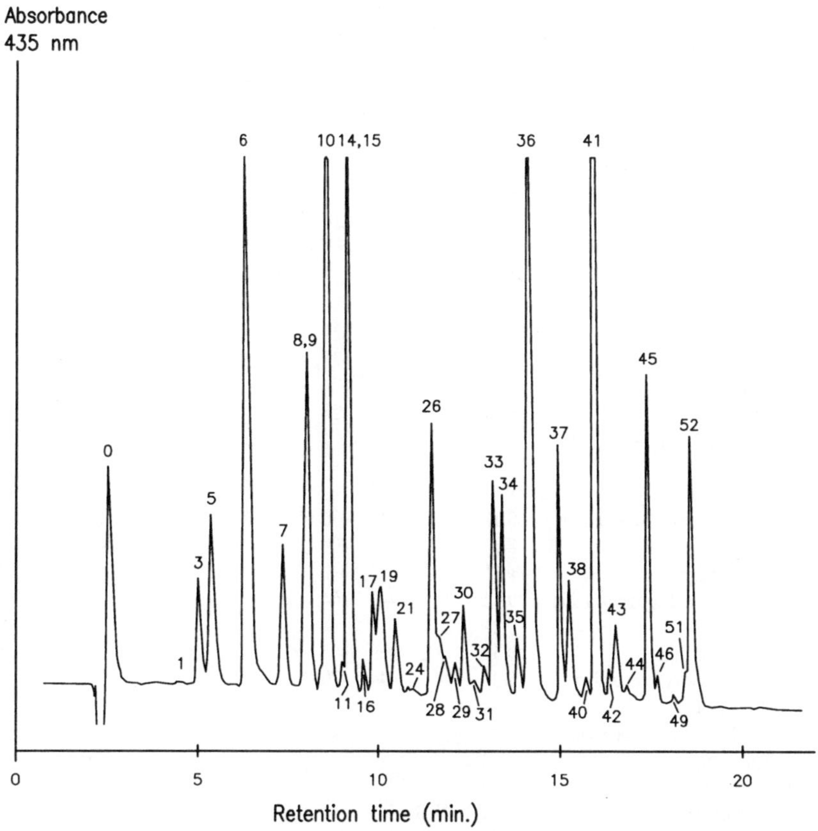

Figure 12.1
HPLC traces of a mixed pigment extract from algal cultures of *Dunaliella tertiolecta* (green alga), *Phaeodactylum tricornutum* (diatom), *Synechococcus* sp. (cyanophyte), *Pycnococcus provasolii* (prasinophyte), *Chroomonas salina* (cryptomonad), *Emiliania huxleyi* (prymnesiophyte), *Pelagococcus subviridis* (chrysophyte), and the authentic carotenoids diadinochrome, siphonein, siphonaxanthin, echinenone, canthaxanthin, lycopene, peridinin and the internal synthetic standard ethyl 8'-β-apo-carotenoate. For identity of pigment fractions see Table 12.2. Chromatogram obtained with the Varian system.

native chlorophylls predominated, with a very small proportion of chlorophyll *a* allomer, epimer and trace unknown chlorophyll *a* derivatives (less than 1%). We can not determine whether such trace pigments are present *in situ*, or are formed during the short (1 to 10 day) low-temperature storage period (see Chapter 10) or the short 'extraction to injection' period. The internal carotenoid standard, ethyl 8'-β-apo-carotenoate (peak 38), separated as a single peak between chlorophyll *b* and chlorophyll *a* (Fig. 12.1). The other internal standard *trans*-8'-β-apo-carotenal co-chromatographed with canthaxanthin, eluting before chlorophyll *b* (Fig. 12.1 and

TABLE 12.2. Elution order of pigments from standard cultures and their visible absorption characteristics

Peak no.	Pigment	Retention time (min)	Acetone		Eluant	
			Maxima (nm)	%III:II[a]	Maxima (nm)	%III:II[a]
0	(Solvent front)	2.56				
1	Chlorophyllide *b*	4.48	453 598 645			
2	Carotenoid P468[b]	4.50			470	
3	Chlorophyllide *a*	5.01	430 577 617 663			
4	Carotenoid P457[b]	5.11			458	
5	Chlorophyll c_3	5.38	452 584 627			
6a	Chlorophyll c_2	6.40	449 581 629		446 580 626	
6b	Chlorophyll $c_1 + c_2$	6.40				
6c	MgDVP	6.40	437 575 625			
6d	Chlorophyll $c_1 + c_2$ + MgDVP	6.40				
7	Peridinin	7.42	474		472	
8	Siphonaxanthin	8.11	445		442 (466)	
9	19'-Butanoyloxyfucoxanthin	8.11	445 470	44	(415) 444 469	42
10	Fucoxanthin	8.70	446		446 (466)	
11	*trans*-Neoxanthin	9.11			416 441 470	100
12	*cis*-19'-Butanoyloxyfucoxanthin	9.12	440 464	33		
13	Neochrome	9.21			400 422 450	90
14	9'-*cis*-Neoxanthin	9.31	415 439 467	65	412 436 466	88
15	19'-Hexanoyloxyfucoxanthin	9.31	444 470	48	418 445 469	38
16	*cis*-Fucoxanthin	9.68			444 (464)	
17	*cis*-19'-Hexanoyloxyfucoxanthin	9.97				
18	Unknown siphonaxanthin-like	10.12			453	
19	Prasinoxanthin	10.20	451		450 (470)	
20	Pheophorbide *a*	10.39	410 505 535 609 666			
21	Violaxanthin	10.59	419 443 472	96	416 440 470	101
22	Pheophorbide *a*-like[(c)]	10.62	410 505 535 609 666			
23	Dinoxanthin	10.76	418 442 471	86	415 440 470	87
24	*cis*-Prasinoxanthin	11.11				
25	Unknown siphonaxanthin-like	11.15			456	
26	Diadinoxanthin	11.61			422 446 476	64
27	Diadinochrome I	11.79	431 458	63	406 430 458	76
28	Diadinochrome II	11.96			406 428 458	76
29	Antheraxanthin	12.24			(423) 446 476	67
30	Alloxanthin	12.51	454 484	48	(428) 454 484	63
31	Monadoxanthin	12.78			(428) 448 480	40
32	Diatoxanthin	13.08	454 483	42	(424) 452 482	33
33	Lutein	13.36	425 446 476	68	(422) 446 476	64
34	Zeaxanthin	13.59	454 481	32	(428) 454 478	11
35	Canthaxanthin	14.00	472		472	
35b	*trans*-8'-β-apo-Carotenal	14.00	456			
36	Siphonein	14.36	452		455	
37	Chlorophyll *b*	15.15	453 598 645		456 596 646	
38	Ethyl 8'-apo-β-caroteneoate	15.43	444		446 (467)	
39	Crocoxanthin	15.87			(427) 449 480	58
40	Chlorophyll *a* allomer	15.87			425 616 664	
41	Chlorophyll *a*	16.15	412 431 581 616 663		431 618 665	
41a	Phytylated chlorophyll *c*	*	454 583 632		456 584 638	

Peak no.	Pigment	Retention time (min)	Acetone Maxima (nm)	Acetone %III:II[a]	Eluant Maxima (nm)	Eluant %III:II[a]
42	Chlorophyll *a* epimer	16.53			431 618 666	
43	Echinenone	16.74	461		464	
44	Unknown carotenoid	16.99			420 443 471	59
45	Lycopene	17.59	448 474 505	77	438 472 505	70
46	Pheophytin *b*	17.68	434 527 600 654		434 528 654	
47	*cis*-Lycopene	17.84			365, 444, 470 502	67
48	Pheophytin *a*	18.56	410 505 535 609 666		408, 504, 534, 608, 666	
49	β,ψ Carotene	18.26	464 495	54	438 462 496	44
50	ε,ε-Carotene	18.40			416 440 471	94
51	β,ε-Carotene	18.64	447 475	55	(425) 446 476	60
52	β,β-Carotene	18.76	453 480	13	427 462 480	18
53	*cis*-β,ε-Carotene	18.83			(419) 442 470	37
54	*cis*-β,β-Carotene	18.94			(422) 446 470	

a. See Ke *et al.* (1970)

b. Following the nomenclature of Johansen *et al.* (1974)

c. Probably pyropheophorbide *a* (unverified)

* Variable retention time before or after chlorophyll *a*; now stabilized to after chl *a* (Jeffrey and Wright, 1994).

Table 12.2; peak 35b). The hydrocarbons, lycopene (ψ,ψ-carotene) and β,ψ-carotene, were completely resolved. β,ε-Carotene eluted as the leading shoulder of the β,β-carotene peak. *Cis*-β,β-carotene eluted as the trailing shoulder of the β,β-carotene peak (see Fig. 13.1A, Chapter 13).

Recovery of pigments from the column was excellent. The recoveries measured were: peridinin, 102%; lutein, 107%; chlorophyll *b*, 100%; chlorophyll *a*, 107%; and β,β-carotene, 98%. Recovery figures greater than 100% may reflect the different extinction coefficients in the mobile phases used.

Two pheophorbide *a*-like peaks co-eluted with carotenoids but did not interfere with detection of the carotenoids because they did not absorb at 436 nm. The pheophorbides can be detected without interference from carotenoids by fluorescence (excitation 407 nm, emission 672 nm for pheophorbide *a*; excitation 432 and emission 659 nm for pheophorbide *b*) or absorbance at 405 or 665 nm.

Figures 12.2A-D show HPLC traces from a sample containing a mixture of algal pigments, that have been detected by their absorbance at 435 nm, 470 nm and 665 nm, as well as fluorescence. The differential responses of pigments in each channel reflect the different absorption maxima of the various pigment types and can be used both to aid identification and to resolve compounds that co-chromatograph.

Figure 12.2A shows the HPLC trace detected at 435 nm, a wavelength at which nearly all carotenoids and chlorophylls have a significant absorbance. Pheophytin *a* and pheophorbide *a* do not absorb in this region and cannot be detected if only this wavelength is used. Apart from these compounds, absorbance detection in the region 435–440 nm is the best compromise for all pigments and is recommended when only a single channel is available. Many single or multi-channel detectors have mercury-vapour lamps that produce a strong emission line at 436 nm, which is ideal for this purpose. Figure 12.2B shows the absorption at 470 nm. Nearly all carotenoids of

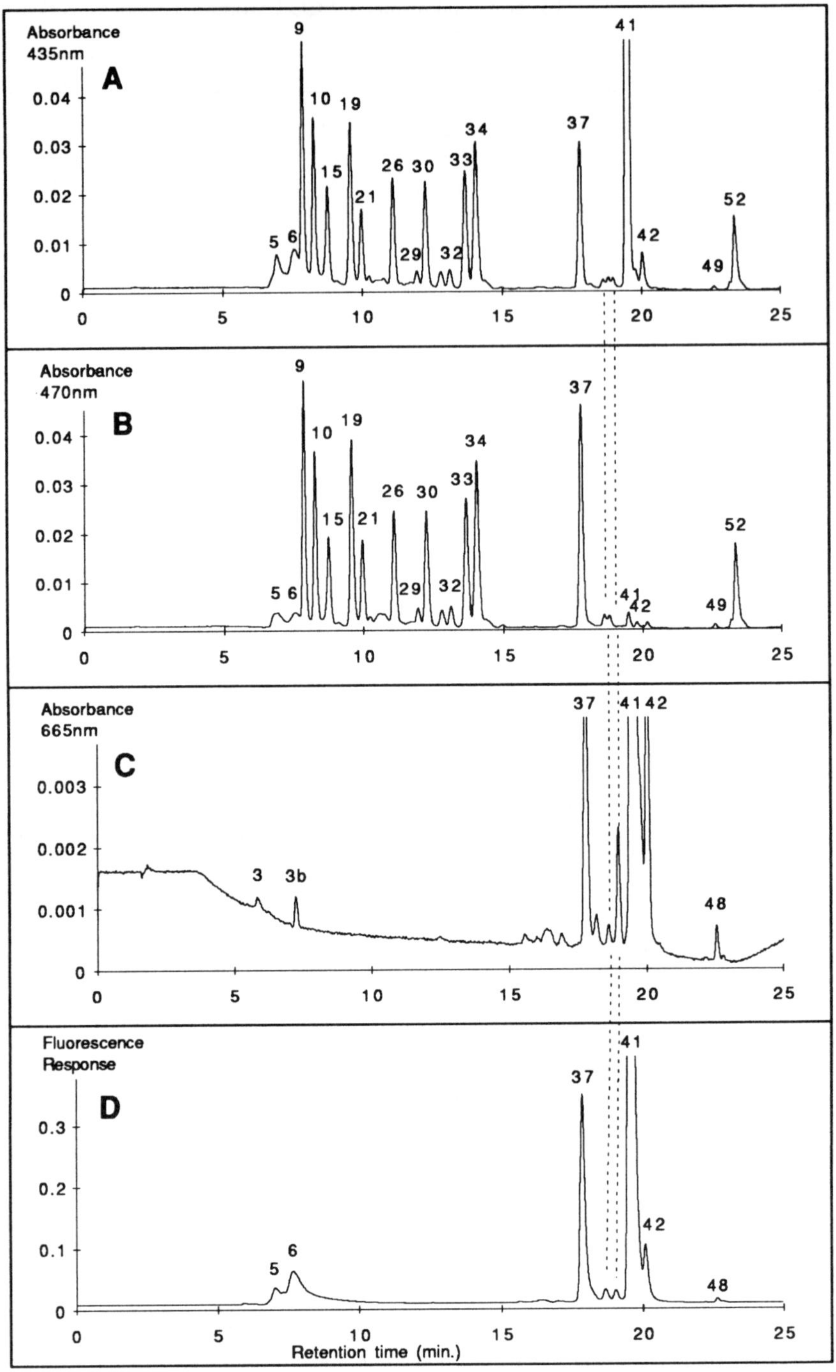

Figure 12.2
HPLC traces of a mixed pigment extract derived from *Pavlova lutheri*, *Synechococcus* sp., *Chroomonas salina*, *Pycnococcus provasolii*, *Dunaliella tertiolecta* and *Pelagococcus subviridis*, detected by absorbance at 435 (A), 470 (B) and 665 (C) nm and fluorescence (excitation 340–520 nm; emission >604 nm)(D). The dotted lines between the traces highlight the three peaks referred to in the text. For identity of pigment fractions see Table 12.2. Chromatograms obtained with the Spectraphysics system.

interest absorb strongly at this wavelength, as do chlorophylls c_1, c_2, c_3 and *b*, although chlorophyll *a* compounds are virtually absent from this trace. At 665 nm (Figure 12.2C) only chlorophyll *a* compounds are detected, apart from chlorophyll *b*, which has a small absorbance at this wavelength. Figure 12.2D shows the response from a fluorescence detector in which broad band-pass filters were used for excitation (340–520 nm) and emission (>604 nm), allowing detection of all chlorophylls and derivatives.

Although not significant in this sample, pheophytins and pheophorbides can be detected with great sensitivity by the fluorescence channel (or absorbance at 405 nm). This can be used to great advantage to detect small amounts of pheophorbide *a* in the presence of violaxanthin, with which it co-chromatographs. Note that in this trace, the peaks are broader due to a large volume flow cell in the fluorometric detector and peak-spreading occurring in the tubing and connectors between the absorbance and fluorometric detectors. There is also a slight time delay between peaks arriving at the two detectors.

Comparison of the channels can be very helpful in identifying peaks. For instance in the 435 nm trace, a small triplet of peaks was observed at about 18 min retention time between the peaks of chlorophylls *b* and *a*. The first peak also absorbs at 470 nm, not 665 nm, and is fluorescent, suggesting that it is a chlorophyll *b* derivative, probably the epimer (based on its retention time). The second peak also absorbs at 470 nm but is non-fluorescent and is hence probably a carotenoid. The third peak does not absorb at 470 nm but does at 665 nm, and is fluorescent, suggesting that it is a chlorophyll *a* derivative, probably the allomer (again based on retention time). Full-spectrum data would of course strengthen these assignments, but the example shows that much can be deduced from a combination of simpler (cheaper) detectors (see Mantoura and Llewellyn, 1983).

Figure 12.3 shows the resolution of chlorophylls and carotenoids from a field sample from the Southern Ocean near Heard Island. This field sample was analysed with the Spectraphysics HPLC system and the retention times do not exactly match those of the Varian HPLC system listed in Table 12.2. The absence of chlorophyll *a* epimer and chlorophyllides indicated that little degradation had occurred. The chromatogram shows a good resolution of the major pigments. Apart from chlorophyll *a*, the chromatogram is dominated by fucoxanthin, its 19'-hexanoyloxy- and 19'-butanoyloxy- derivatives, chlorophylls c_3 and c_1+c_2, and diadinoxanthin. Smaller amounts of peridinin, chlorophyll *b*, prasinoxanthin and antheraxanthin were present along with traces of alloxanthin and zeaxanthin.

Examination of the field sample by light microscopy showed it to be dominated by small diatoms (consistent with the abundance of fucoxanthin) and a numerically smaller dinoflagellate population (indicated by peridinin) with many unidentifiable flagellates. The pigment composition suggests that the flagellates were mainly prymnesiophytes (chlorophyll c_3, fucoxanthin, 19'-hexanoyloxyfucoxanthin and 19'-butanoyloxyfucoxanthin) together with smaller populations of prasinophytes (chlorophyll *b*, prasinoxanthin and antheraxanthin) and cryptomonads (alloxanthin).

12.4 Discussion

In the development of this HPLC method, 12 species from 10 algal classes were examined (see Table 13.1, Chapter 13). While this range is comprehensive, one algal class with representatives widely distributed in the world's oceans is the

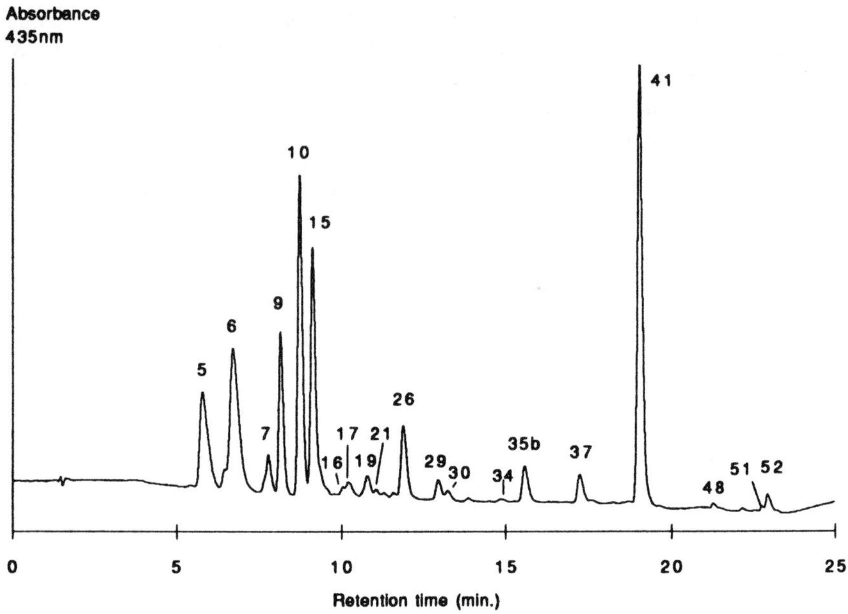

Figure 12.3
HPLC trace of phytoplankton sample from the Southern Ocean (51°54'S, 73°14'E, 9 June 1990). For identity of pigment fractions see Table 12.2. Chromatogram obtained with the Spectraphysics system.

Prochlorophyceae which was omitted because no culture of an isolated representative was available in Australia at the time of the Workshop (late 1988-89). Prochlorophytes contain chlorophyll *a*-like and chlorophyll *b*-like pigments instead of chlorophyll *a* and *b* (Chisholm *et al.* 1988, Veldhuis and Kraay, 1990). These are now identified as 8-desethyl-8-vinyl chlorophyll *a* and *b* (Goericke and Repeta, 1992). These two pigments normally known as divinyl chlorophylls *a* and *b*, were originally separated in normal-phase HPLC systems (Gieskes and Kraay, 1983*a*). In reverse phase systems they coeluted with normal chlorophylls *a* and *b* unless converted to their respective pheophytins, a technique subject to interference by an unknown carotenoid, according to an anonymous reviewer of Wright *et al.* (1991). New advances in HPLC methods now allow both divinyl chlorophylls *a* and *b* to be separated from chlorophylls *a* and *b* in C_8–reversed phase systems (Goericke and Repeta, 1992; Partensky *et al.* 1993; and Fig. 13.2, Chapter 13), and these are recommended for use when prochlorophytes are of significance in field samples.

The present method has a number of advantages over other published methods for phytoplankton pigments (Roy, 1987), in that separation of over 50 key carotenoids, chlorophylls and chlorophyll derivatives is achieved in one simple ternary gradient system.

In comparison with the technique of Wright and Shearer (1984), the present method gives better resolution of the polar chlorophylls and carotenoids while retaining the resolution of later-eluting pigments. Of particular importance is the pigment pair lutein and zeaxanthin, not separated by Mantoura and Llewellyn (1983) or Gieskes and Kraay (1986*b*). The new method retains the almost baseline resolution for this pigment pair of Wright and Shearer (see Fig. 12.1), similar to that of Ben-Amotz *et al.* (1982), although it does not match the excellent separation achieved by Zapata *et al.* (1987). However, Zapata's resolution of other carotenoids is inferior to our improved method, as shown by comparing the resolution of *Dunaliella tertiolecta* pigments (their Fig. 4B, and our Fig. 13.1A, Chapter 13).

Further advantages of the method are that no ion-pairing reagent is necessary as in the method of Mantoura and Llewellyn (1983) and Bidigare *et al.* (1985), and no expensive radial compression module is used as in the original method of Wright and Shearer (1984). Further, our early method trials, using two different liquid chromatographs (a Waters system and a Varian system), showed only slightly different retention times between the two instruments. Some differences in retention times but not retention sequence occurred in the Spectraphysics system (Fig. 12.3), which had a different dead volume between the pump and column.

Apart from the 39 pigments resolved in Fig. 12.1, two important fucoxanthin derivatives — 19'-hexanoyloxyfucoxanthin and 19'-butanoyloxyfucoxanthin — were not chromatographically separated from 9'-*cis* neoxanthin and siphonaxanthin respectively. Both 19'-acyloxyfucoxanthins are important indicators of particular prymnesiophytes, chrysophytes and dinoflagellates (see Chapter 2). A diode array detector is essential to track spectroscopic, if not chromatographic, resolution, and is an essential adjunct to pigment identification of field samples. To our knowledge, these pigment pairs have not been resolved in other reversed-phase HPLC systems.

While chlorophyll c_3 was clearly resolved from other chlorophyll *c* pigments, the important chlorophylls c_1, c_2 and the chlorophyll *c*-like prasinophyte pigment MgDVP could not be resolved from each other, which is a problem common to other HPLC systems using octadecylsilica columns (Mantoura and Llewellyn, 1983, Bidigare *et al.* 1985, Zapata *et al.* 1987; Gieskes *et al.* 1988). Until 1992–93 the only known HPLC method capable of resolving these four chlorophyll *c* pigments was that of Jeffrey (1989), which uses a special polyethylene column. Shioi and Beale (1987) had previously separated chlorophylls c_1 and c_2 on polyethylene HPLC. Such a method, or the newly described reversed-phase method of Garrido and Zapata (1993), could be used in addition to the present one to obtain more detail on the chlorophyll *c* pigments present. Confirmation that the prasinophyte chlorophyll *c*-like pigment is MgDVP and not chlorophyll c_1 as was suggested by Wilhelm (1987) has come from spectral and chemical data (Jeffrey, 1989; Fookes and Jeffrey, unpublished). A fifth chlorophyll *c*-like pigment, designated chl $c_{CS\text{-}170}$, was recently characterized from a prasinophyte, *Micromonas pusilla* (CS-170) (Jeffrey, 1989; Fookes and Jeffrey, unpubl.). A sixth non-polar esterified chlorophyll *c* derivative has also been described (Nelson and Wakeham, 1989; Jeffrey and Wright, 1994; and Chapter 13).

Our improved method presented here fulfils an objective of SCOR WG 78 to recommend an HPLC method for resolution of phytoplankton chlorophylls and carotenoids in seawater and cultures of algae. Due to the complexity of pigment composition in field samples, caution should be exercised when identifying pigments. Retention times, spectral analysis, chemical properties, and co-chromatography with known standards from reference cultures should all be included to confirm identifications, wherever possible.

Important pigments characteristic of algal classes, species or processes can be determined by HPLC techniques (see Chapters 2 and 4), but for quantitative assessment of the biomass of taxonomic groups, the natural variability of pigment ratios must be taken into account mathematically (Gieskes and Kraay, 1983*b*; Gieskes *et al.* 1988). If definitive taxonomic characterization of individual species is required, HPLC should be complemented by other techniques such as light and electron microscopy of non-destructively preserved samples.

ACKNOWLEDGEMENTS

We thank Dr M. Brown for running pigment samples on the Varian HPLC, Dr S. Blackburn for assistance with algal cultures, Professor S. Liaaen-Jensen for gifts of carotenoid standards, and Mr J. Cox for preparing the figures.

REFERENCES

Ben-Amotz, A., Katz, A., Avron, M. (1982). Accumulation of β-carotene in halotolerant algae: purification and characterisation of β-carotene-rich globules from *Dunaliella bardawil* (Chlorophyceae). *J. Phycol.* 18: 529–537.

Bidigare, R. R., Kennicutt, M. C., Brookes, J. M. (1985). Rapid determination of chlorophylls and their degradation products by high performance liquid chromatography. *Limnol. Oceanog.* 30: 432–435.

Chisholm, S. W., Olson, R. J., Zettler, E. R., Goericke, R., Waterbury, J. B., Welschmeyer, N. A. (1988). A novel, free-living prochlorophyte abundant in the oceanic euphotic zone. *Nature*, 334: 340–343.

Garrido, J. L., Zapata, M. (1993). High performance liquid chromatographic separation of polar and non-polar chlorophyll pigments from algae using a wide pore polymeric octadecylsilica column. *J. High Res. Chromatog.* 16: 229–233.

Gieskes, W. W. C., Kraay, G. W. (1983*a*). Unknown chlorophyll *a* derivatives in the North Sea and the tropical Atlantic Ocean revealed by HPLC analysis. *Limnol. Oceanogr.* 28: 757–766.

Gieskes, W. W. C., Kraay, G. W. (1983*b*). Dominance of Cryptophyceae during the phytoplankton spring bloom in the central North Sea detected by HPLC analysis of pigments. *Mar. Biol.* 75: 179–185.

Gieskes, W. W. C., Kraay, G. W. (1986*a*). Analysis of phytoplankton pigments by HPLC before, during and after mass occurrence of the microflagellate *Corymbellus aureus* during the spring bloom in the open northern North Sea in 1983. *Mar. Biol.* 92: 45–52.

Gieskes, W. W. C., Kraay, G. W. (1986*b*). Floristic and physiological differences between the shallow and the deep nanophytoplankton community in the euphotic zone of the open tropical Atlantic revealed by HPLC analysis of pigments. *Mar. Biol.* 91: 567–576.

Gieskes, W. W. C., Kraay, G. W., Nontji, A., Setiapermana, D., Sutomo. (1988). Monsoonal alternation of a mixed and a layered structure in the phytoplankton of the euphotic zone of the Banda Sea (Indonesia): a mathematical analysis of algal pigment fingerprints. *Neth. J. Sea Res.* 22: 123–137.

Goericke, R., Repeta, D. J. (1992). The pigments of *Prochlorococcus marinus*: the presence of divinyl chlorophyll *a* and *b* in a marine procaryote. *Limnol. Oceanogr.* 37: 425–433.

Jeffrey, S. W. (1980). Cultivating unicellular marine plants. CSIRO (Hobart, Tasmania) Fish. and Oceanogr. *Ann. Rep.* 1977–1979, p. 22–43.

Jeffrey, S. W. (1989). Chlorophyll *c* pigments and their distribution in the chromophyte algae. In: Green, J. C., Leadbeater, B. S. C., Diver, W. L. (eds), *The Chromophyte Algae: Problems and Perspectives*, Clarendon Press, Oxford, pp. 13–36.

Jeffrey, S. W., Wright, S. W. (1994). Photosynthetic pigments in the Haptophyta. In: Green, J.C., Leadbeater, B.S.C. (eds), *The Haptophyte Algae*, Clarendon Press, Oxford, p. 111–132.

Johansen, J. E., Svec, W. A., Liaaen-Jensen, S., Haxo, F. T. (1974). Carotenoids of the Dinophyceae. *Phytochem.* 13: 2261–2271.

Ke, B., Imsgard, F., Kjøsen, H., Liaaen-Jensen, S. (1970). Electronic spectra of carotenoids at 77°K. *Biochim. Biophys. Acta*, 210: 139–152.

Kraay, G. W., Zapata, M., Veldhuis, M. J. W. (1992). Separation of chlorophylls c_1, c_2 and c_3 of marine phytoplankton by reversed-phase C_{18} high performance liquid chromatography. *J. Phycol.* 28: 708–712.

Mantoura, R. F. C., Llewellyn, C. A. (1983). The rapid determination of algal chlorophyll and carotenoid pigments and their breakdown products in natural waters by reverse-phase high-performance liquid chromatography. *Analytica Chim. Acta*, 151: 297–314.

Nelson, J. R., Wakeham, S. G. (1989). A phytol-substituted chlorophyll *c* from *Emiliania huxleyi* (Prymnesiophyceae). *J. Phycol.* 25: 761–766.

Partensky, F., Hoepffner, N., Li, W. K. W., Ulloa, O., Vaulot, D. (1993). Photoacclimation of *Prochlorococcus* sp. (Prochlorophyta) strains isolated from the North Atlantic and the Mediterranean Sea. *Plant Physiol.* 101: 285–296.

Repeta, D. J., Gagosian, R. B. (1982). Carotenoid transformations in coastal marine waters. *Nature*, 295: 51–54.

Roy, S. (1987). High performance liquid chromatographic analysis of chloropigments. *J. Chromatog.* 391: 19–34.

Shioi, Y., Beale, S. I. (1987). Polyethylene-based high-performance liquid chromatography of chloroplast pigments: resolution of mono- and divinyl-chlorophyllides and other pigment mixtures. *Anal. Biochem.* 162: 493–499.

Veldhuis, M. J. W., Kraay, G. W. (1990). Vertical distribution and pigment composition of a picoplanktonic prochlorophyte in the subtropical North Atlantic: a combined study of HPLC analysis and flow cytometry. *Mar. Ecol. Prog. Ser.* 68: 121–127.

Wilhelm, C. (1987). Purification and identification of chlorophyll c_1 from the green alga *Mantoniella squamata*. *Biochim. Biophys. Acta*, 892: 23–29.

Wright, S. W., Shearer, J. D. (1984). Rapid extraction and high-performance liquid chromatography of chlorophylls and carotenoids from marine phytoplankton. *J. Chromatog.* 294: 281–295.

Wright S. W., Jeffrey S. W., Mantoura R. F. C., Llewellyn C. A., Bjørnland T., Repeta D., Welschmeyer N. A. (1991). Improved HPLC method for the analysis of chlorophylls and carotenoids from marine phytoplankton. *Mar. Ecol. Progr. Ser.* 77: 183–196.

Zapata, M., Ayala, A. M., Franco, J. M., Garrido, J. L. (1987). Separation of chlorophylls and their degradation products in marine phytoplankton by reversed-phase high-performance liquid chromatography. *Chromatographia*, 23: 26–30.

13

Qualitative and quantitative HPLC analysis of SCOR reference algal cultures

S. W. Jeffrey and S. W. Wright

13.1 INTRODUCTION

In order to set up an HPLC system for pigments, a source of authentic standards is essential to confirm pigment identities, determine retention times and calibrate detector responses. However, few algal pigment standards are available commercially.

Chlorophylls and carotenoids are notoriously labile to light, oxygen, acids, etc. They readily allomerize and epimerize (chlorophylls) and oxidize and isomerize (carotenoids), and researchers must have considerable expertise to make their own highly purified crystalline products. For this reason (and perhaps a perceived 'lack of market') only a handful of 'pure' pigments are available commercially (see Appendix J): the rest must either be prepared by the chromatographer or requested as private gifts from experts.

At the SCOR Working Group 78 meeting in 1986, a viable alternative to the problem of obtaining pure pigments for oceanographers was chosen: that of using standard algal cultures whose pigment complement had been identified by rigorous chemical analysis (Bjørnland and Liaaen-Jensen, 1986). Such cultures, grown under standard conditions, would provide researchers with valid pigment fractions for standardization and identification. Cultures on frozen filters (see Chapter 10), stored in liquid nitrogen, would also be suitable for on-board field work.

The cultures chosen (Table 13.1), which together contained all the significant pigments so far identified in marine phytoplankton (Table 13.2), included a representative from each of 10 algal classes: a diatom, a dinoflagellate, two prymnesiophytes, a chrysophyte, a cryptomonad, a red alga, a blue-green alga (cyanobacterium), and green algae (a chlorophyte, a prasinophyte and an euglenophyte). Where possible, the exact clone that had been analysed chemically was used. At the time of the SCOR Workshops (1988–89) no culture of a prochlorophyte, which contains the newly described pigments divinyl chlorophylls *a* and *b* (Chisholm *et al.* 1992), was available at CSIRO in Hobart, but a chromatogram of a field sample containing *Prochlorococcus marinus*, published by Goericke and Repeta (1993), is included here (see Fig. 13.2).

Algal cultures grown in the same conditions of light, temperature, day-length and culture medium and harvested at the same culture age and the same time of the light:dark cycle should produce reproducible pigment suites for each alga, and these could be used with authority for pigment identifications. Analyses of replicate cultures, grown under identical conditions, should be reliable at least as semi-quantitative culture standards.

This Chapter describes the HPLC pigment profile of each SCOR reference culture, and presents quantitative pigment analyses of cultures grown under rigorously defined standard conditions. Cultures suitable for preparative (larger scale) analyses of key pigments are also identified.

13.2 Materials and methods

13.2.1 *Algal cultures and culture harvest*

Qualitative analyses

The reference algal cultures, clone designations and publications relating to chemical identification of pigments are listed in Table 13.1. All cultures were maintained in the CSIRO Algal Culture Collection (Jeffrey, 1980) under the rigorously controlled culture conditions specified in Chapter 6. The 75 ml cultures were grown in 125 ml glass Erlenmeyer flasks at 20°C at a light intensity of 80 $\mu E\ m^{-2}\ s^{-1}$ on 12 : 12 h light : dark cycles (except *Synechocococcus* CS-197, which was grown at 17.5°C and 20 $\mu E\ m^{-2}\ s^{-1}$). Light was provided by Philips 'White Deluxe' (TL3694) fluorescent tubes, and was measured by a Biospherical Optics light meter QSL-100 inserted into the medium of a replicate flask without an algal inoculum (see Section 6.3.6, and Fig. 6.3E). Culture media used were: medium f (Guillard and Ryther, 1962) at half strength (f2) for *Dunaliella tertiolecta, Emiliania huxleyi, Pavlova lutheri, Pelagococcus subviridis, Phaeodactylum tricornutum, Porphyridium cruentum, Pycnococcus provasolii* and *Synechococcus* sp.; medium fE (medium f + EDTA, Jeffrey, 1980) for *Chroomonas salina*; MBL/NB_2 (Guillard's MBL in Nichols, 1973) for *Euglena gracilis* and medium G (Loeblich and Smith, 1968) for *Amphidinium carterae* and *Micromonas pusilla.* Cultures were acclimatized to the above conditions for 7–10 days, before a small volume (0.5 ml) of adapted culture was used to inoculate the experimental cultures.

TABLE 13.1. Microalgal species recommended for authentic sources of pigments and references to their chemical identification

Species	Class	Clonal designation and origin	CSIRO culture code	Reference to chemical analysis of pigments[a]
Amphidinium carterae	Dinophyceae	AMPH1, Plymouth 450	CS-212	Johansen *et al.* (1974)
Chroomonas salina	Cryptophyceae	CCMP 56/00/00	CS-174	Pennington *et al.* (1985)
Dunaliella tertiolecta	Chlorophyceae	CCMP	CS-175	Jeffrey (1968)[b]
Emiliania huxleyi	Prymnesiophyceae	BT6	CS-57	Arpin *et al.* (1976)
Euglena gracilis	Euglenophyceae	Univ. NSW, strain Z	CS-66	Bjørnland *et al.* (1987)
Micromonas pusilla	Prasinophyceae	UTEX LB991	CS-86	Ricketts (1966*a,b*; 1970)[b,c]
Pavlova lutheri	Prymnesiophyceae	CCMP	CS-182	Berger *et al.* (1977)
Pelagococcus subviridis	Chrysophyceae[d]	East Aust. Current	CS-99	Bjørnland *et al.* (1989)
Phaeodactylum tricornutum	Bacillariophyceae	Plymouth (1052/1)	CS-29	Pennington *et al.* (1988)
Porphyridium cruentum	Rhodophyceae	Halifax, Canada	CS-25	Stransky and Hager (1970)[b]
Pycnococcus provasolii[e]	Prasinophyceae	CCMP, 48-23AX	CS-185	Foss *et al.* (1984)
Synechococcus sp.	Cyanophyceae	DC2	CS-197	Guillard *et al.* (1985)

All cultures were grown at 20°C at 80 μE m^{-2} sec^{-1} light intensity on a 12 : 12 h light: dark cycle, with the exception of Synechococcus sp. (17.5°C and 20 μE m^{-2} sec^{-1}). For culture media used for growth see Table 6.1.

a. Analysis in the cited reference may have been done on a different clone or strain to that used in the present work.
b. Reinvestigation by modern methods required.
c. Recently examined chemically by Egeland and Liaaen-Jensen (1992, 1993).
d. Included in the new class Pelagophyceae by Andersen *et al.* (1993).
e. Previously known as Prasinophyte 48-23AX (Guillard *et al.* 1991).

Quantitative analyses

Three replicate flasks of each culture were set up and grown under the standard conditions outlined above. A fourth flask containing artificial seawater prepared from Sigma sea salts, to which f2 nutrients were added, was used for comparison with media prepared from natural CSIRO purified seawater (see Chapter 6). Cultures were not aerated. For both quantitative and qualitative analyses, cultures were harvested at mid- to late-log phase (7–10 days), usually 4–6 hours into the daily light cycle. We have not checked whether this degree of variability in the harvest conditions affected the pigment content of the cultures, but from our results (Table 13.3) we expect it to be no more than 20%.

The cultures were gently decanted into a fresh flask, leaving behind any clumps or cell residues. Ten ml aliquots were filtered through 25 mm Whatman GF/F filters under low vacuum (250 mmHg), and rinsed with 2 × 2 ml filtered medium or seawater. The filters were immediately frozen in liquid nitrogen, and stored in vials either in liquid N_2 or in a freezer at -135°C. The pigments were extracted from the filters with 3.5 ml of 98 : 2 = methanol : ammonium acetate buffer, 0.5 M, pH 7.1

(v/v) within 10 days (a time within which there is no significant degradation of pigments, see Chapter 10) and analysed by HPLC within 16 hours. During this time they were stored in an autosampler at -10°C. Replicate HPLC analyses of the same extract showed a standard error of no more than 2% for the pigment concentrations.

Cell counts were determined immediately after cell harvest, by fixing in dilute Lugol's solution (0.2% final concentration), using a Neubauer haemocytometer and phase-contrast illumination. The average standard deviation of the cell counts was 6.1% (range 3.6–14.5%), see Table 13.3.

13.2.2 *HPLC analysis*

The HPLC systems described in Chapter 12 were used. The solvent systems and reversed-phase columns were also as specified. The data shown in Fig. 13.1 were obtained with the Varian system. The reversed-phase system of Goericke and Repeta (1992) for the separation of the divinyl chlorophylls (Fig. 13.2) is described in the Part IV Data sheets preamble, Section IV.1.7.2. Extinction coefficients for the quantitative analyses of major pigments are listed individually in the Part IV Data sheets, and summarized for all pigments in Appendix E. Calibration of HPLC systems for quantitative pigment analyses are described in Chapter 16.

13.3 Results

Chromatograms of 12 unialgal pigment extracts taken from log-phase cultures are shown in Fig. 13.1. Excellent resolution of almost all individual pigments was achieved for each species. The elution order of all pigments and their visible absorption characteristics are given in Table 12.2. An additional HPLC chromatogram (Fig. 13.2) of a field sample containing *Prochlorococcus marinus* (Goericke and Repeta, 1993), shows the separation of the newly-described pigments, divinyl chlorophyll *a* and divinyl chlorophyll *b* from chlorophylls *a* and *b* (Goericke and Repeta, 1992; Partensky *et al.* 1993).

The pigment content of the SCOR reference cultures grown under the standard conditions described above, expressed as $\mu g/10^6$ cells, are listed in Table 13.3. For simplicity, alteration products, such as carotenoid *cis*-isomers and chlorophyll allomers, epimers and chlorophyllides, have been included in the concentrations of the parent compounds. Definitive pigment ratios are also given: chlorophyll *a* : *b*, chlorophyll *a* : *c*, and chlorophyll *a* : major carotenoid. Cell numbers per ml for log-phase cultures ranged from 1.1×10^6 cells per ml for *Chroomonas salina* to 9.4×10^7 cells per ml for the picoplanktonic cyanophyte, *Synechococcus* sp.

The standard deviations of the cell counts are listed separately and averaged 6.1%, with a range of 3.6–14.5%. The average standard deviation of the pigment analyses (pigment/cell) was 15.0%, with a range of 8.3–31.4%. The high standard deviation of *Synechococcus* sp. (31.4%) was due to the difficulty of counting bacterial-sized cells. In view of the variability of algal cultures, we consider our standard deviations (pigment/cell) of less than 20% to be quite acceptable.

Peak no.	Pigment	*Amphi-dinium carterae* CS-212	*Chroo-monas salina* CS-174	*Dunaliella tertiolecta* CS-175	*Emiliania huxleyi* CS-57	*Euglena gracilis* CS-66	*Micro-monas pusilla*[a] CS-86	*Pavlova lutheri* CS-182	*Pelago-coccus subviridis* CS-99	*Phaeodac-tylum tricornutum* CS-29	*Porphy-ridium cruentum* CS-25	*Pycno-coccus provasolii* CS-185	*Synecho-coccus* sp. CS-197
2	Carotenoid P 468[b]	+											
4	Carotenoid P 457[b]	+											
5	Chlorophyll c_3[c]				+				+				
6a	Chlorophyll c_2	+	+		+				+				
6b	Chlorophyll c_1+ c_2							+		+			
6c	MgDVP[d]						+					+	
7	Peridinin	+											
9	19'-Butanoyloxyfucoxanthin								+				
10	Fucoxanthin				+[e]			+	+	+			
14	9'-*cis*-Neoxanthin			+		+	+					+	
15	19'-Hexanoyloxyfucoxanthin				+								
19	Prasinoxanthin						+					+	
21	Violaxanthin			+			+					+	
23	Dinoxanthin	+											
26	Diadinoxanthin	+			+	+		+	+	+			
29	Antheraxanthin			+			+						
30	Alloxanthin		+										
31	Monadoxanthin		+										
32	Diatoxanthin	+			+			+	+	+			
33	Lutein			+								+	
34	Zeaxanthin			+			+				+	+	+
37	Chlorophyll *b*			+		+	+					+	
39	Crocoxanthin		+										
41	Chlorophyll *a*	+	+	+	+	+	+	+	+	+	+	+	+
49	β,ψ-carotene (= γ)			+								+	
50	ε,ε-carotene (= ε)								+				
51	β,ε-carotene (= α)		+	+							+		
52	β,β-carotene (= β)	+		+	+	+	+	+	+	+		+	+

a. For new data on *Micromonas pusilla*, refer to Egeland and Liaaen-Jensen (1992, 1993)
b. Following the nomenclature of Johansen *et al.* (1974)
c. Chemical confirmation of structure by Fookes and Jeffrey (1989)
d. Chemical confirmation of structure by Fookes and Jeffrey (unpubl.)
e. Fucoxanthin is a minor pigment in *E. huxleyi*
* Divinyl chlorophylls *a* and *b* are present in cultures of the prochlorophyte, *Prochlorococcus marinus*, but were not available to us at the time of the SCOR workshops (see Part IV Data sheets).

TABLE 13.3. Pigment content of 10 SCOR reference algal cultures. All cultures were grown under 12 : 12 h light : dark cycles at 80 µE m^{-2} s^{-1} and 20°C except for *Synechococcus* (20 µE m^{-2} s^{-1} and 17.5°C). Pigment concentrations are the means of triplicate cultures and are expressed as µg per 10^6 cells, and include identified alteration products.

Peak no.	Pigment	*Dunaliella tertiolecta* CS-175	*Phaeodactylum tricornutum* CS-29	*Amphidinium carterae* CS-212	*Pavlova lutheri* CS-182	*Emiliania huxleyi* CS-57	*Pelagococcus subviridis* CS-99	*Pycnococcus provasolii* CS-185	*Micromonas pusilla* CS-86	*Chroomonas salina* CS-174	*Synechococcus* sp. CS-197
2	Carotenoid P 468			0.21							
4	Carotenoid P 457			0.21							
5	Chlorophyll c_3					0.049	0.019				
6a	Chlorophyll c_2			1.68			0.034			0.25	
6b	Chlorophyll c_1+ c_2		0.051		0.055	0.051					
6c	MgDVP							0.0050	0.0012		
7	Peridinin			3.36							
	Unknown[a]								0.0016		
9	19'-Butanoyloxyfucoxanthin						0.028				
10	Fucoxanthin		0.21		0.21		0.074				
14	9'-*cis*-Neoxanthin	0.084						0.011	0.0029		
15	19'-Hexanoyloxyfucoxanthin					0.29					
19	Prasinoxanthin							0.023	0.0044[e]		
21	Violaxanthin	0.073						0.0045	0.0021		
23	Dinoxanthin			0.088							
	Unknown[b]								0.0017		
26	Diadinoxanthin		0.038	0.76	0.11	0.017	0.0068				
29	Antheraxanthin	0.021						0.00046			
30	Alloxanthin									0.33	
31	Monadoxanthin									0.032	

32	Diatoxanthin		0.0005	0.0083	0.0064	0.0027	0.0014				
33	Lutein	0.27						0.00064	0.00008		
34	Zeaxanthin	0.012									0.0016
	Unknown[c]							0.0035	0.00042		
37	Chlorophyll *b*	0.035						0.069	0.016		
39	Crocoxanthin									0.011	
41a	Phytylated Chlorophyll *c*					0.0054					
41	Chlorophyll *a*	1.33	0.278	3.16	0.49	0.17	0.076	0.073	0.020	1.44	0.0046
	Unknown[d]								0.00094		
49	β,ψ-Carotene	0.013									
50	ε,ε-Carotene						0.00011				
51	β,ε-Carotene	0.0078				0.0018		0.00078	0.00006	0.082	
52	β,β-Carotene	0.048	0.0076	0.074	0.014	0.0046	0.0012	0.0021	0.00048		0.00044
	Av. % Std Dev (pigment/cell)	16.3	10.2	17.8	9.9	9.6	8.3	10.9	14.7	20.8	31.4
	Av. cell count (cells/ml)	2.4×10^6	3.9×10^6	4.3×10^6	1.9×10^6	3.4×10^6	5.7×10^6	1.0×10^7	3.7×10^7	1.1×10^6	9.4×10^7
	Av. % Std Dev (cell count)	5.6	3.6	8.1	4.5	3.7	4.1	6.6	6.7	3.7	14.5
	Pigment Ratios:										
	Chl *a* : *b*	3.8						1.1	1.2		
	Chl *a* : *c*		5.5	1.9	8.9	1.8	1.4	14.7	16.0	5.9	
	Chl *a* : Major carotenoid	4.9	1.3	0.9	2.3	0.61	1.0	3.2	4.4	4.4	2.8

a. Unknown carotenoid, λ_{max} = 452 nm, %III/II = 72;
b. Unknown carotenoid, λ_{max} = 457 nm, %III/II = 0;
c. Unknown carotenoid, λ_{max} = 456 nm, %III/II = 200;
d. Unknown carotenoid, λ_{max} = 445 nm, %III/II = 56;
e. Contained about 30% of an unknown carotenoid, λ_{max} = 455 nm, %III/II = 104.

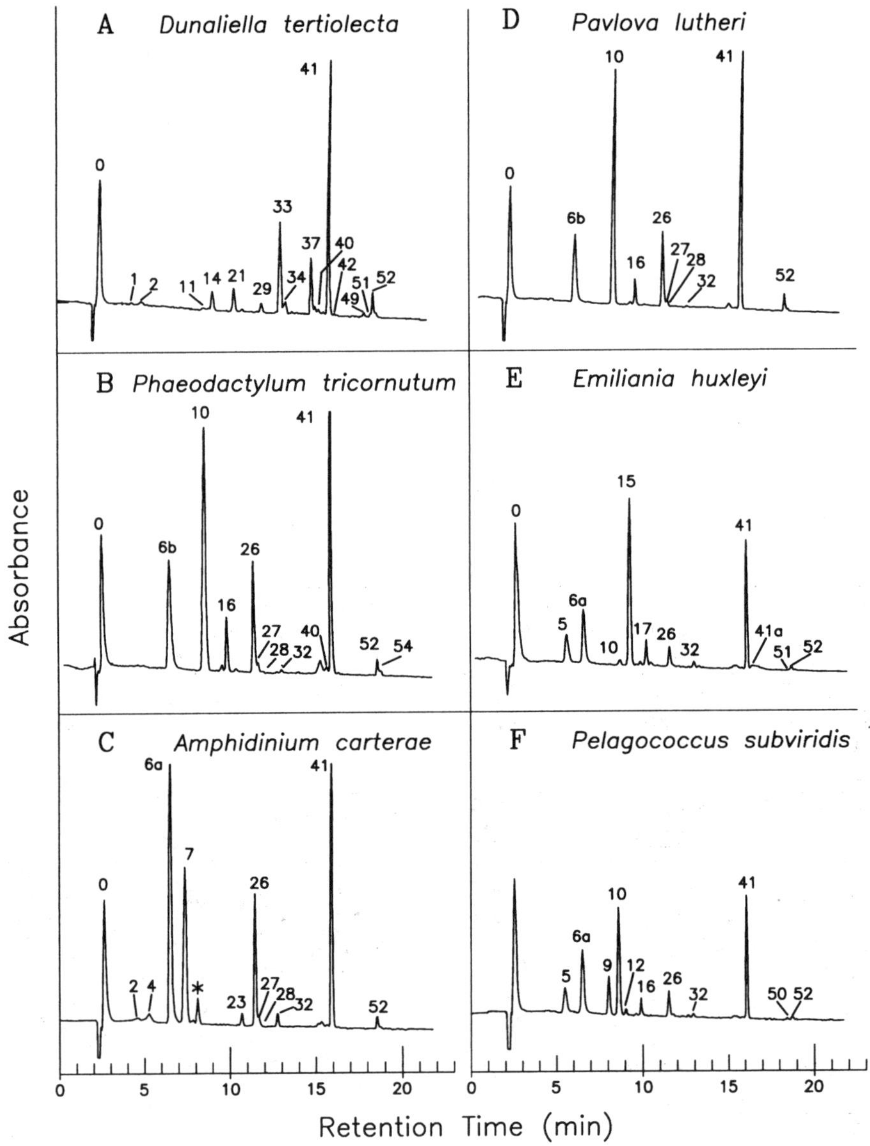
A Dunaliella tertiolecta
0
1 2 11 14 21 29 33 34 37 40 41 42 49 51 52
B Phaeodactylum tricornutum
0
6b 10 16 26 27 28 32 40 41 52 54
C Amphidinium carterae
0
2 4 6a 7 * 23 26 27 28 32 41 52
D Pavlova lutheri
0
6b 10 16 26 27 28 32 41 52
E Emiliania huxleyi
0
5 6a 10 15 17 26 32 41 41a 51 52
F Pelagococcus subviridis
5 6a 9 10 12 16 26 32 41 50 52
Absorbance
0 5 10 15 20
Retention Time (min)

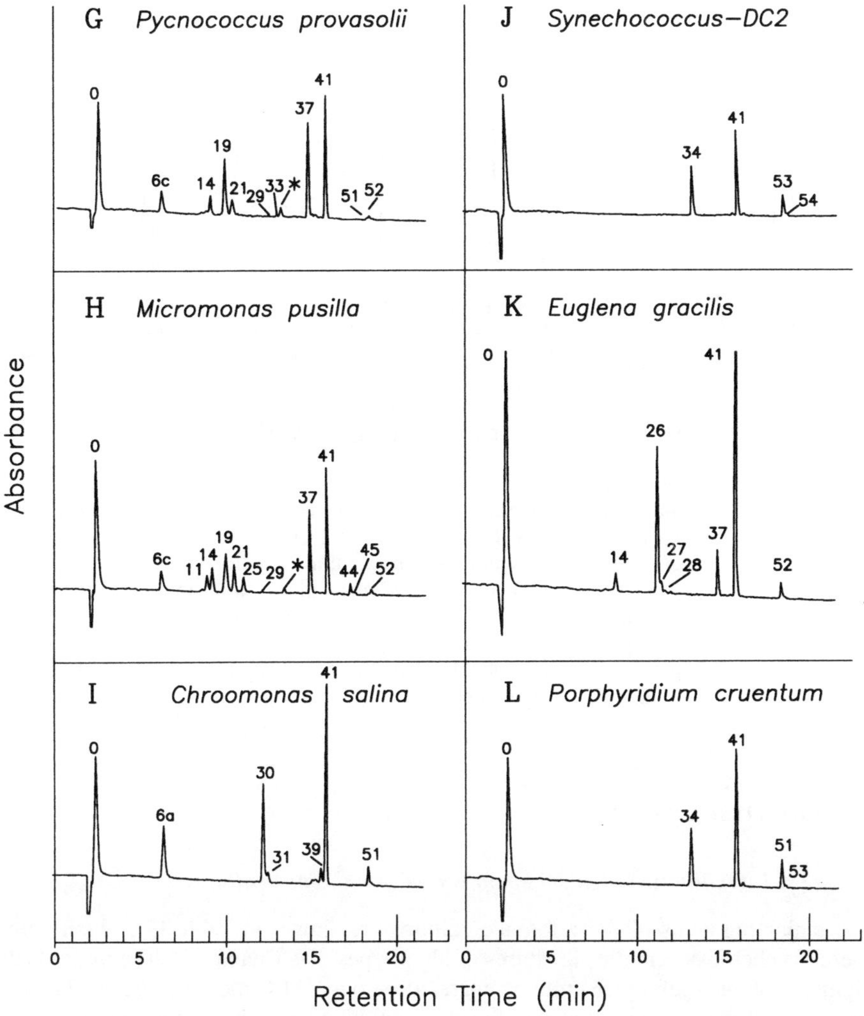

Figure 13.1
HPLC absorbance (λ = 436 nm) chromatograms of pigment separations of unialgal cultures of 12 reference microalgae: (A) *Dunaliella tertiolecta* (green alga); (B) *Phaeodactylum tricornutum* (diatom); (C) *Amphidinium carterae* (dinoflagellate); (D) *Pavlova lutheri* (prymnesiophyte); (E) *Emiliania huxleyi* (prymnesiophyte); (F) *Pelagococcus subviridis* (chrysophyte); (G) *Pycnococcus provasolii* (prasinophyte); (H) *Micromonas pusilla* (prasinophyte); (I) *Chroomonas salina* (cryptomonad); (J) *Synechococcus* sp. (cyanophyte); (K) *Euglena gracilis* (euglenoid); and (L) *Porphyridium cruentum* (red alga). For identity of pigment fractions see Tables 13.2, 13.3 and Table 12.2. The peaks labelled * in Fig 13.1C, G, H are unknowns. Chromatograms were obtained with the Varian system.

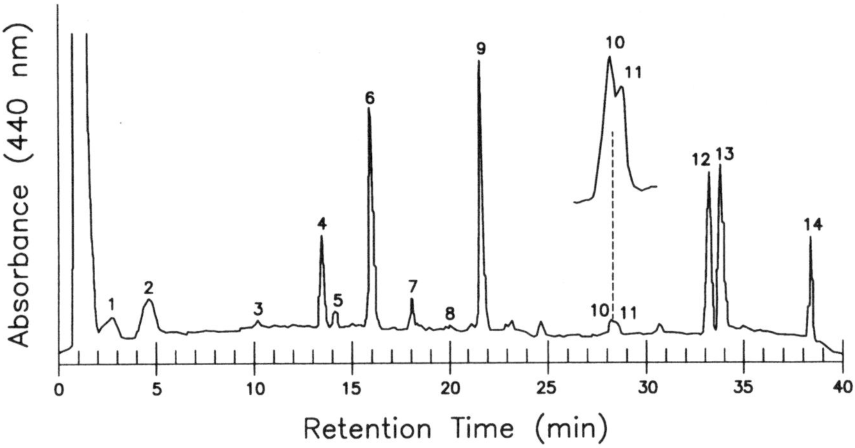

Figure 13.2
Chromatogram of Sargasso Sea sample containing *Prochlorococcus marinus*. Pigments: 1, chlorophyll c_3; 2, chlorophylls c_1 and c_2; 3, peridinin; 4, 19'-butanoyloxyfucoxanthin; 5, fucoxanthin; 6, 19'-hexanoyloxyfucoxanthin; 7, diadinoxanthin; 8, alloxanthin; 9, zeaxanthin + lutein; 10, DV chlorophyll *b*; 11, chlorophyll *b*; 12, DV chlorophyll *a*; 13, chlorophyll *a*; 14, β,ε- and β,β-carotene. Note baseline separation of chl *a* and DV chl *a*; partial separation of chl *b* and DV chl *b* (after Goericke and Repeta, 1993).

13.4 Discussion

13.4.1 *Qualitative pigment composition of reference cultures*

The major pigments of the reference cultures are shown in Table 13.2. Many are useful as chemosystematic signatures of algal types (see Chapter 2). The individual pigments of each reference culture separated by our HPLC method (Fig. 13.1) will now be discussed with reference to the chemical identifications from previous studies referred to in Table 13.1.

The chlorophyte *Dunaliella tertiolecta* (Fig. 13.1A) contained chlorophylls *a* (41) and *b* (37), lutein (33), violaxanthin (21) and 9'-*cis*-neoxanthin (14) as major pigments; zeaxanthin (34), antheraxanthin (29), β,ψ- and β,β-carotene (49, 52, respectively) as minor pigments; and a small amount of chlorophyllide *a* (3) as a degradation product (which probably resulted from chlorophyllase activity during filtration, storage or extraction). Trace amounts of other pigments were also present on the chromatogram: β,ε-carotene (51) and the alteration products, chlorophyllide *b* (1), all-*trans*-neoxanthin (11), chlorophyll *a* allomer (40), and chlorophyll *a* epimer (42). These qualitative data confirm and extend the TLC analyses of *D. tertiolecta* by Jeffrey (1968) and HPLC analyses by Mantoura and Llewellyn (1983) and Wright and Shearer (1984). The present HPLC methods bring improved resolution and additional sensitivity to the detection of minor and trace pigments.

The diatom *Phaeodactylum tricornutum* (Fig. 13.1B) contained chlorophyll *a*

(41), chlorophylls c_1 and c_2 (6b), fucoxanthin (10), diadinoxanthin (26) and β,β-carotene (52) as major pigments; diatoxanthin (32) as a trace pigment; *cis*-fucoxanthin (16), diadinochrome (27, 28), chlorophyll *a* allomer (40) and *cis*-β,β-carotene (54) as minor or trace alteration products. These results agree with those of Stauber and Jeffrey (1988) and Pennington *et al.* (1988) for *P. tricornutum* pigments. The absence of chlorophyllide *a* in Fig. 13.1B reflects the undegraded nature of the extract. *P. tricornutum* has a highly active chlorophyllase (Barrett and Jeffrey, 1964; Jeffrey and Hallegraeff, 1987) and, unless great care is taken, significant de-esterification of chlorophyll *a* to chlorophyllide *a* and the atypical chlorophyllide *a* (Barrett and Jeffrey, 1964) can occur during harvest and extraction.

The dinoflagellate *Amphidinium carterae* (Fig. 13.1C) contained chlorophyll *a* (41), chlorophyll c_2 (6a), peridinin (7) and diadinoxanthin (26) as major pigments; and dinoxanthin (23), diatoxanthin (32), β,β-carotene (52) and carotenoids P-468 (2) and P-457 (4) as minor pigments (Johansen *et al.* 1974). The structure of P-457 has now been established as the 3'-lactoside of 7', 8'-dihydro-neoxanthin-20'-al (Liaaen-Jensen, 1990). The alteration products of diadinoxanthin, the diadinochromes (27, 28) were found in trace quantities. This pigment composition supports other analyses of *A. carterae* (e.g. Jeffrey *et al.* 1975), but in the chemical study of freeze-dried *A. carterae* preparations (Johansen *et al.* 1974), minor quantities of the peridinin-related xanthophylls pyrrhoxanthin and peridininol were also identified. These trace pigments were not detected in our *A. carterae* strain.

The prymnesiophyte *Pavlova lutheri* (Fig. 13.1D) contained chlorophyll *a* (41), chlorophylls c_1 and c_2 (6b), fucoxanthin (10), diadinoxanthin (26) and β,β-carotene (52) as major pigments; and diatoxanthin (32) as a minor or trace pigment. The alteration products found were *cis*-fucoxanthin (16) and diadinochrome (27, 28). These results confirm other studies of *P. lutheri* by Berger *et al.* (1977), who found in addition traces of β-carotene-5,6-epoxide, and cryptoxanthin-5,6,5',6'-diepoxide. *P. lutheri* belongs in Type 1 of the four pigment types identified in 29 species (50 strains) of prymnesiophytes (Jeffrey and Wright, 1994). Chlorophylls c_1 and c_2 and fucoxanthin are the definitive pigments of this type.

The prymnesiophyte *Emiliania huxleyi* (Fig. 13.1E) contained chlorophyll *a* (41), chlorophylls c_2 (6a) and c_3 (5), 19'-hexanoyloxyfucoxanthin (15), diadinoxanthin (26), diatoxanthin (32) and β,β-carotene (52) as major pigments; fucoxanthin (10) and β,ε-carotene (51) as minor or trace pigments; and *cis*-19'-hexanoyloxyfucoxanthin (17) as a minor alteration product. Trace amounts of phytylated chlorophyll *c* (41a), with a Soret absorption band resembling chlorophyll c_3, were also found. This strain of *E. huxleyi* (BT6; see Table 13.1) was chemically characterized by Arpin *et al.* (1976). They found, in addition to the above pigments, traces of 19'-hexanoyloxyparacentrone. *E. huxleyi* is a Type 3 prymnesiophyte which is characterized by 19'-hexanoyloxyfucoxanthin and chlorophyll c_3 (Jeffrey and Wright, 1994; see comments above for *Pavlova lutheri*).

The chrysophyte *Pelagococcus subviridis* (Fig. 13.1F) contained chlorophyll *a* (41), chlorophylls c_2 (6a) and c_3 (5), 19'-butanoyloxyfucoxanthin (9), fucoxanthin (10) and diadinoxanthin (26) as major pigments; diatoxanthin (32), ε,ε- and β,β-carotene (50 and 52, respectively) as minor or trace pigments; and the *cis*-derivatives of 19'-butanoyloxyfucoxanthin (12) and fucoxanthin (16) as alteration products. This pigment suite, first identified chromatographically and spectrally in the Australian strain (Vesk and Jeffrey, 1987), was confirmed by spectrometric and chemical analysis in the northern hemisphere strain (Bjørnland *et al.* 1989).

The picoplanktonic prasinophyte *Pycnococcus provasolii* (Fig. 13.1G) contained chlorophyll *a* (41), chlorophyll *b* (37) and MgDVP (6c), prasinoxanthin (19), 9'-*cis*-neoxanthin (14) and violaxanthin (21) as major pigments; and antheraxanthin (29), lutein (33), an unknown pigment that co-chromatographed with zeaxanthin (34), and β,ε- and β,β-carotene (51 and 52, respectively) as minor or trace pigments. This analysis is in agreement with the chemical analysis of the same strain by Foss *et al.* (1984). They found in addition two minor unknown carotenoids.

The prasinophyte *Micromonas pusilla* (Fig. 13.1H) was a problem species at the time of the SCOR Workshops, since its pigments had not been examined chemically, but it was included as an 'authentic' source of MgDVP (Ricketts, 1966*a*, *b*; Jeffrey, 1989). Some of the carotenoids posed difficulties because the spectral and chromatographic data of the major salmon-pink carotenoids suggested a similarity to siphonaxanthin and siphonein, rather than prasinoxanthin (Wright *et al.* 1991). As noted in Table 1 of Wright *et al.* (1991), full chemical identification of the carotenoids of *M. pusilla* was urgently needed, and Professor Liaaen-Jensen's group has now completed this task (Egeland and Liaaen-Jensen, 1992, 1993). Some new carotenoids (see below) have been identified, although chromatographic systems and spectral data for these new pigments have not yet been published (but see Note added in proof, p. 358).

The major pigment fractions in Fig. 13.1H are: chlorophyll *a* (41), chlorophyll *b* (37), MgDVP (6c), prasinoxanthin (19), 9'-*cis*-neoxanthin (14) and violaxanthin (21). The minor or trace components are: unknowns (peaks 25 and 44), and unknowns co-chromatographing with peaks 34 and 45, antheraxanthin (29) and β,β-carotene (52). Egeland and Liaaen-Jensen (1992) identified a new carotenoid, micromonal, as a major fraction, with uriolide, micromonol and anhydrouriolide as minor or trace constituents. Insufficient data was given in these references to allow identification of the unknown peaks in this study. So many pigments are present as minor constituents in *M. pusilla* that it is not recommended as a source of carotenoids at our present state of knowledge. However, it is an authentic source of MgDVP.

The cryptomonad *Chroomonas salina* (Fig. 13.1I) contained chlorophyll *a* (41), chlorophyll c_2 (6a), and alloxanthin (30) as major pigments, with monadoxanthin (31), crocoxanthin (39) and β,ε-carotene (51) as minor or trace components. This analysis agrees with that of Pennington *et al.* (1985) for *C. salina*. In other cryptomonad species they also found traces of ε,ε-carotene, β,β-carotene and lycopene. The red water-soluble phycobiliproteins were not examined.

The picoplanktonic cyanophyte *Synechococcus* sp. (Fig. 13.1J), which also contained phycobiliproteins, had chlorophyll *a* (41) and zeaxanthin (34) as major pigments, and β,β-carotene (52) as a minor component. The *cis*-isomer of β,β-carotene (54) occurred as an alteration product. This analysis confirms data of Guillard *et al.* (1985) for the same species.

The freshwater euglenophyte *Euglena gracilis* (Fig. 13.1K) was included because no culture of a marine euglenophyte was available at CSIRO at the time of the SCOR Workshops. This strain contained chlorophyll *a* (41), chlorophyll *b* (37) and diadinoxanthin (26) as the major pigments, with 9'-*cis*-neoxanthin (14) and β,β-carotene (52) as minor pigments. Diadinochromes (27, 28) occurred as alteration products of diadinoxanthin. Bjørnland *et al.* (1987), who examined the same strain, found in addition, diatoxanthin and minor amounts of alloxanthin diesters of myristic and palmitic acid. Ketocarotenoids were present in trace amounts only: in addition to echinenone and canthaxanthin the monoacetylenic ketocarotenoids pectenolone and 7, 8-didehydro-astaxanthin were also identified.

The unicellular rhodophyte *Porphyridium cruentum* (Fig. 13.1L) had a simple pigment profile similar to that of the cyanophyte *Synechococcus* (Fig. 13.1J). The major pigments were chlorophyll *a* (41) and zeaxanthin (34) with β,ε-carotene (51) as a minor component. The *cis*-isomer of β,ε-carotene (53) occurred as an alteration product. The phycobiliproteins of this alga were not examined. Stransky and Hager (1970) examined this species by TLC, but a re-examination by modern chemical methods including mass spectrometry and NMR now seems warranted.

While neither *E. gracilis* nor *P. cruentum* add new pigment standards to the previously listed species, they are good sources of diadinoxanthin and zeaxanthin, respectively. The chromatograms are simple and 'clean', without many trace carotenoids to contaminate the major peaks.

In summary, the HPLC separation of pigments of the SCOR reference cultures (Fig 13.1) by the SCOR-developed HPLC method (Chapter 12 and Wright *et al.* 1991) shows an almost complete agreement with previous analyses of pigments identified chemically in these or similar algal strains (Table 13.1). The exception is *Micromonas pusilla*, in which new pigments identified chemically by Egeland and Liaaen-Jensen (1992,1993) have not yet been unequivocally separated and identified on our system.

Divinyl chlorophylls *a* and *b* can be separated from the prochlorophyte, *Prochlorococcus marinus* by the reverse-phase HPLC method of Goericke and Repeta (1993) (see Fig. 13.2). As noted in the Introduction to this chapter, no culture of *Prochlorococcus marinus* was available in Hobart at the time of the SCOR Workshops, and these difficult-to-grow strains must be obtained from researchers such as Partensky *et al.* (1993), who hold cultured strains of this species.

Researchers wishing to prepare pure pigments should refer to the summary Table 13.2 for guidance on which species provide the best source material for individual pigments. The Data and Graphics sheets in Part IV, and Chapters 7 (chlorophylls) and 8 (carotenoids) provide further detailed information. All algal cultures used here are easy to grow (see Chapter 6) and provide ample biomass in litre culture volumes for preparative scale purification of the major pigments, if needed.

13.4.2 *Quantitative pigment composition of 10 reference cultures grown under standard conditions*

The pigment content (μg per 10^6 cells) of 10 reference cultures listed in Table 13.3 showed excellent reproducibility between the three replicates, with standard deviations mostly less than 20%. About half of this variability was accounted for by the standard deviations of the cell counts.

Under our standard growth conditions the concentrations of chlorophyll *a* roughly followed the cell size, with the largest cells *Amphidinium carterae*, *Chroomonas salina*, and *Dunaliella tertiolecta* (range 10–20 μm, largest diameter) having 1 to 3 μg chlorophyll *a* per 10^6 cells, the smaller nanoplankton *Phaeodactylum tricornutum*, *Emiliania huxleyi*, and *Pavlova lutheri* (range 4–10 μm) having 0.2 to 0.5 μg chlorophyll *a* per 10^6 cells, and the picoplankton *Pycnococcus provasolii*, *Pelagococcus subviridis*, *Synechococcus*-DC2 and *Micromonas pusilla* (range 0.5–3 μm) having less than 0.1 μg chlorophyll *a* per 10^6 cells. These quantitative pigment compositions for each species are valid only for the growth conditions used here, the most important factors influencing pigment composition being light intensity, day-length, time of harvest, age of culture and algal strain.

To obtain an idea of the range of culture variation, we compared two sets of HPLC chlorophyll *a* analyses of identical strains carried out in 1988 (SCOR Chlorophyll Workshop) and in 1993 (present work). Although very similar culture conditions were employed, time of harvest and culture age may not have been as rigorously controlled in the 1988 as in the 1993 experiments. In spite of this, six strains gave very similar chlorophyll concentrations per 10^6 cells. Listing the 1988 values first, *C. salina* had chlorophyll *a* values of 1.8 and 1.4 μg per 10^6 cells; *E. huxleyi*, 0.17 and 0.21; *M. pusilla*, 0.020 and 0.016; *P. lutheri*, 0.49 and 0.46; *P. subviridis*, 0.076 and 0.085; *P. provasolii*, 0.07 and 0.11. In *A. carterae*, *D. tertiolecta* and *P. tricornutum* the variation in the two sets of analyses was greater than ±50%. While these analyses indicate the pigment constancy of many of the strains over a period of five years, more work needs to be done to assess the extent of culture variability in all strains grown under the specified standard conditions.

Very few published quantitative analyses of algal pigments have used the same algal strain and culture conditions and expressed the results on a per cell basis. Consequently, comparisons are difficult. Jeffrey *et al.* (1975) listed chlorophyll concentrations per 10^6 cells of 21 species of dinoflagellates and Stauber and Jeffrey (1988) of 51 species (71 strains) of diatoms respectively. The chlorophyll concentration of dinoflagellates ranged from 1.03 μg chlorophyll *a* + *c* per 10^6 cells in the smallest dinoflagellate *Gymnodinium punctatum* to 141 μg chlorophyll *a* + *c* per 10^6 cells in the largest cell *Gonyaulax polyedra*. A similar range was found for diatoms: 0.02 μg chlorophyll *a* + *c* per 10^6 cells in *Extubocellulus spinifer* to 174.4 μg chlorophyll *a* + *c* per 10^6 cells in the large centric diatom *Coscinodiscus* sp. Chlorophyll *a* in 29 species (50 strains) of nanoplanktonic prymnesiophytes, ranged from only 0.08 to 4.5 μg per 10^6 cells from the smallest to the largest cells (Jeffrey and Wright, 1994). The chlorophyll *a* concentrations per 10^6 cells we report here are close to those of similarly sized species grown under similar culture conditions and investigated in the above studies.

The pigment ratios were also consistent with previous literature values. In the green algae, the chlorophyte *D. tertiolecta* had a ratio of chlorophyll *a : b* of 3.8, whereas in the two prasinophytes *M. pusilla* and *P. provasolii*, the ratios were 1.1 and 1.2, reflecting the higher concentration of chlorophyll *b* in these species (Wood, 1979). In the six chromophyte algae, the ratios of chlorophylls *a : c* ranged from 1.4 in *P. subviridis* to 8.9 in *P. lutheri*, whereas the content of the chlorophyll *c*-like pigment (MgDVP) of the two prasinophytes *P. provasolii* and *M. pusilla* was much lower (*a : c* = 14.7 and 16.0 respectively). These chlorophyll *a:c* ratios were similar to previously published data from this laboratory for strains of the same species of *A. carterae* (1.9 this work versus 2.5), *P. tricornutum* (5.5 versus 4.6), *P. lutheri* (8.9 versus 8.5), *C. salina* (5.9 versus 4.9) and *E. huxleyi* (1.8 versus 3.9) (Jeffrey *et al.* 1975; Vesk and Jeffrey, 1987; Wright and Jeffrey, 1987; Jeffrey, 1989; Jeffrey and Wright, 1994). Where chlorophyll c_3 is a major component of the chlorophyll *c* fraction (*E. huxleyi* and *P. subviridis*) the *c* values of literature data are likely to be underestimated, due to the low 630 nm absorbance of this pigment and because we now recommend the blue Soret band be used for quantitative measurements (see Data Sheets, Part IV for chlorophyll c_3).

The relative content of the major carotenoid varied across the species with chlorophyll *a* : major carotenoid ratios of 0.6 to 4.9. This is the range for similar species grown under standard growth conditions and cited in the above references. On average, the chlorophyll *a* content exceeded that of the major carotenoid by approximately 2.5 times across the species. It is essential, however, to study this variability in more detail in future work if chlorophyll *a* : carotenoid ratios are to be used to calculate the relative biomass of different algal types in field populations.

The pigment apparatus of microalgae is very flexible to enable them to adapt to the fluctuating marine environment, particularly that of light quality and intensity. Blue-green light can increase chlorophyll concentrations up to 150% in some species (Vesk and Jeffrey, 1977). Also, in short-day and low-light regimes chlorophyll concentrations increase, while in long-day and high-light regimes they decrease (Sakshaug *et al.* 1987; Nielsen, 1992; Johnsen and Sakshaug, 1993). Not only the pigment concentration but also the relative proportions of pigments can change under different light regimes (Vesk and Jeffrey, 1977). Under a normal day : night cycle, the xanthophyll cycle carotenoids can also vary widely in concentration (diadinoxanthin → diatoxanthin in chromophyte algae; violaxanthin → zeaxanthin in chlorophyte algae). In the light, de-epoxidation occurs, with the reverse in darkness. This cycle is thought to exert a photoprotective function both in algae and in higher plants (Demers *et al.* 1991; Demmig-Adams and Adams, 1993). Even different culture media can cause qualitative pigment changes in prymnesiophytes (Jeffrey and Wright, 1994).

Finally, we emphasize that to obtain the pigment compositions described here for the SCOR cultures, rigorously controlled growth conditions, identical to those described in Chapter 6, must be used to achieve the same results. If these are closely matched, we are confident the results will be similar to those described in the present work.

In the quantitative experiments, we also made up a replicate culture in artificial seawater (Sigma Sea Salts) with f2 nutrients added. Although the cell density was often only half that of cultures grown in the natural CSIRO purified seawater used for standard media, the pigment content per 10^6 cells of most species was not significantly different from those listed in Table 13.2 The one exception was *P. tricornutum* which had 50% higher pigment content.

13.4.3 *Summary*

The present Chapter summarizes the pigment composition of each of the SCOR reference algal cultures. This suite of cultures contains almost all major pigments of known chemosystematic value that are regularly found in field phytoplankton populations, with the exception of divinyl chlorophylls *a* and *b* from the prochlorophyte *Prochlorococcus marinus*, which was not available to us at the time of the study. The cultures, when grown under the standard conditions used here, provide authentic pigment standards from which 'unknowns' in field samples and new algal cultures can be evaluated. Further, the choice of algal species for preparative pigment isolations can now readily be made (see Table 13.2 and the individual pigment Data sheets, Part IV). It is hoped that with the advice for growth and maintenance of algal cultures given in Chapter 6, the oceanographer may now have ready access to pigment standards, until such time as chemically stable crystalline preparations are commercially available for all key pigments.

Acknowledgements

We thank Ms Jeannie-Marie LeRoi and Ms Cathy Johnstone for assistance with algal cultures and cell counts, and Dr Malcolm Brown for the HPLC analyses on the Varian system (Fig. 13.1). Dr T. Bjørnland is specially thanked for his review of this chapter.

Note added in proof: A pigment analysis of the prasinophyte *Bathycoccus prasinos* has now been published by Egeland *et al.* (1995). *J. Phycol.* 31: 554–561.

References

Andersen, R. A., Saunders, G. W., Paskirt, H. P., Sexton, J. P. (1993). Ultrastructure and 18S rRNA gene sequence for *Pelagomonas calceolata* gen. et sp. nov. and the description of a new algal class, the Pelagophyceae *classis nov. J. Phycol.* 29: 701–715.

Arpin, N., Svec, W. A., Liaaen-Jensen, S. (1976). New fucoxanthin-related carotenoids from *Coccolithus huxleyi*. *Phytochem.* 15: 529–532.

Barrett, J., Jeffrey, S. W. (1964). Chlorophyllase and formation of an atypical chlorophyllide in marine algae. *Plant Physiol.* 39: 44–47.

Berger, R., Liaaen-Jensen, S., McAlister, V., Guillard, R.R.L. (1977). Carotenoids of the Prymnesiophyceae (Haptophyceae). *Biochem. Syst. Ecol.* 5: 71–75.

Bjørnland, T., Fiksdahl, A., Haxo, F. T., Liaaen-Jensen, S. (1987). New and unusual carotenoid features of *Euglena gracilis* (Euglenophyceae). Proc. 8th Int. Symp. Carotenoids. July 27–31, 1987, Boston.

Bjørnland, T., Liaaen-Jensen, S. (1986). Suggested algal cultures as pigment references for HPLC analyses of seawater samples. Unpublished report prepared for SCOR WG 78, 6 pp.

Bjørnland, T., Liaaen-Jensen, S., Throndsen, J. (1989). Carotenoids of the marine chrysophyte *Pelagococcus subviridis*. *Phytochem.* 28: 3347–3353.

Chisholm, S. W., Frankel, S., Goericke, R., Olsen, R., Palenik, B., Urbach, E., Waterbury, J., Zettler, E. (1992). *Prochlorococcus marinus* nov. gen. nov. sp.: an oxyphototrophic marine prokaryote containing divinyl chlorophyll *a* and *b*. *Arch. Microbiol.* 157: 297–300.

Demers, S., Roy, S., Gagnon, R., Vignault, C. (1991). Rapid light-induced changes in cell fluorescence and in xanthophyll cycle pigments in *Alexandrium excavatum* (Dinophyceae) and *Thalassiosira pseudonana* (Bacillariophyceae): a photo-protection mechanism. *Mar. Ecol. Progr. Ser.* 76: 185–193.

Demmig-Adams, B., Adams, W. W. (1993). The xanthophyll cycle. In: Young, A., Britton, G. (eds), *Carotenoids in Photosynthesis*. Chapman and Hall, London, p. 206–251.

Egeland, E. S., Liaaen-Jensen, S. (1992). Eight new carotenoids from a chemosystematic evaluation of Prasinophyceae. Proc. 7th Int. Symp. on Marine Natural Products, Capri, July 1992, p. 2.

Egeland, E. S., Liaaen-Jensen, S. (1993). New carotenoids and chemosystematics in the Prasinophyceae. Proc. 10th Int. Symp. Carotenoids, Trondheim, Norway, June 1993, pp. 4–6.

Fookes, C. J. R., Jeffrey, S. W. (1989). The structure of chlorophyll c_3, a novel marine photosynthetic pigment. *J. Chem. Soc. Chem. Comm.* 23: 1827–1828.

Foss, P., Guillard, R. R. L., Liaaen-Jensen, S. (1984). Prasinoxanthin – a chemosystematic marker for algae. *Phytochem.* 23: 1629–1633.

Goericke, R., Repeta, D. J. (1992). The pigments of *Prochlorococcus marinus*: the presence of divinyl chlorophyll *a* and *b* in a marine procaryote. *Limnol. Oceanogr.* 37: 425–433.

Goericke, R., Repeta, D. J. (1993). Chlorophylls *a* and *b* and divinyl chlorophylls *a* and *b* in the open subtropical North Atlantic Ocean. *Mar. Ecol. Progr. Ser.* 101: 307–313.

Guillard, R. R. L., Keller, M. D., O'Kelly, C. J., Floyd, G. L. (1991). *Pycnococcus provasolii* gen. et sp. nov., a coccoid prasinoxanthin-containing phytoplankter from the western North Atlantic and Gulf of Mexico. *J. Phycol.* 27: 39–47.

Guillard, R. R. L., Murphy, L. S., Foss, P., Liaaen-Jensen, S. (1985). *Synechococcus* spp. as likely zeaxanthin-dominant ultraphytoplankton in the North Atlantic. *Limnol. Oceanogr.* 30: 412–414.

Guillard, R. R. L., Ryther, J. H. (1962). Studies of marine plankton diatoms. I. *Cyclotella nana* Hustedt and *Detonula confervacea* (Cleve) Gran. *Can. J. Microbiol.* 8: 229–239.

Jeffrey, S. W. (1968). Quantitative thin-layer chromatography of chlorophylls and carotenoids from marine algae. *Biochim. Biophys. Acta,* 162: 271–285.

Jeffrey, S. W. (1980). Cultivating unicellular marine plants. CSIRO (Hobart, Tasmania) Fish. and Oceanogr. *Ann. Rep.* 1977–1979, pp. 22–43.

Jeffrey, S. W. (1989). Chlorophyll *c* pigments and their distribution in the chromophyte algae. In: Green, J. C., Leadbeater, B. S. C., Diver, W.L. (eds). *The Chromophyte Algae: Problems and Perspectives.* Clarendon Press, Oxford, p. 13–36.

Jeffrey, S. W., Hallegraeff, G. M. (1987). Chlorophyllase distribution in 10 classes of phytoplankton – a problem for chlorophyll analysis. *Mar. Ecol. Prog. Ser.* 35: 293–304.

Jeffrey, S. W., Sielicki, M., Haxo, F. T. (1975). Chloroplast pigment patterns in dinoflagellates. *J. Phycol.* 11: 374–384.

Jeffrey, S. W., Wright, S. W. (1994). Photosynthetic pigments in the Haptophyta. In: Green, J. C., Leadbeater, B. S. C. (eds) *The Haptophyte Algae*. Clarendon Press, Oxford, p. 111–132.

Johansen, J. E., Svec, W. A., Liaaen-Jensen, S., Haxo, F. T. (1974). Carotenoids of the Dinophyceae. *Phytochem.* 13: 2261–2271.

Johnsen, G., Sakshaug, E. (1993): Bio-optical characteristics and photoadaptive responses in the toxic and bloom-forming dinoflagellates *Gyrodinium aureolum*, *Gymnodinium galatheanum*, and two strains of *Prorocentrum minimum*. *J. Phycol.* 29: 627–642.

Liaaen-Jensen, S. (1990). Marine carotenoids – selected topics. *New J. Chem.* 14: 747–759.

Loeblich, A. R., Smith, V. E. (1968). Chloroplast pigments of the marine dinoflagellate *Gyrodinium resplendens*. *Lipids*, 3: 5–13.

Mantoura, R. F. C., Llewellyn, C. A. (1983). The rapid determination of algal chlorophyll and carotenoid pigments and their breakdown products in natural waters by reverse-phase high-performance liquid chromatography. *Anal. Chim. Acta*, 151: 297–314.

Nichols, H. W. (1973). Growth media – freshwater. In: Stein, J.R. (ed.), *Handbook of Phycological Methods.* Cambridge University Press, Cambridge, p. 7–24.

Nielsen, M. V. (1992). Irradiance and daylength effects on growth and chemical composition of *Gyrodinium aureolum* Hulburt in culture. *J. Plankt. Res.* 14: 811–820.

Partensky, F., Hoepffner, N., Li, W.K.W., Ulloa, O., Vaulot, D. (1993). Photoacclimation of *Prochlorococcus* sp. (Prochlorophyta) strains isolated from the North Atlantic and the Mediterranean Sea. *Plant Physiol.* 101: 285–296.

Pennington, F., Guillard, R. R. L., Liaaen-Jensen, S. (1988). Carotenoid distribution patterns in the Bacillariophyceae. *Biochem. Syst. Ecol.* 16: 589–592.

Pennington, F. C., Haxo, F. T., Borch, G., Liaaen-Jensen, S. (1985). Carotenoids of Cryptophyceae. *Biochem. Syst. Ecol.* 13: 215–219.

Ricketts, T. R. (1966*a*). Magnesium 2,4-divinylpheoporphyrin a_5 monomethyl ester, a protochlorophyll-like pigment present in some unicellular flagellates. *Phytochem.* 5: 223–229.

Ricketts, T. R. (1966*b*). The carotenoids of the phytoflagellate, *Micromonas pusilla*. *Phytochem.* 5: 571–580.

Ricketts, T. R. (1970). The pigments of the Prasinophyceae and related organisms. *Phytochem.* 9: 1835–1842.

Sakshaug, E., Demers, S., Yentsch, C. M. (1987). *Thalassiosira oceanica* and *T. pseudonana*: two different photoadaptational responses. *Mar. Ecol. Progr. Ser.* 41: 275–282.

Stauber, J. L., Jeffrey, S. W. (1988). Photosynthetic pigments in 51 species of marine diatoms. *J. Phycol.* 24: 158–172.

Stransky, H., Hager, A. (1970). Das Carotinoidmuster und die Verbreitung des lichtinduzierten Xanthophyllcyclus in verschiedenen Algenklassen. IV. Cyanophyceae und Rhodophyceae. *Arch. Mikrobiol.* 72: 84–96.

Vesk, M., Jeffrey, S. W. (1977). The effect of blue-green light on photosynthetic pigments and chloroplast structure in uni-cellular marine algae from six algal classes. *J. Phycol.* 13: 280–288.

Vesk, M., Jeffrey, S. W. (1987). Ultrastructure and pigments of two strains of the picoplanktonic algae, *Pelagococcus subviridis* (Chrysophyceae). *J. Phycol.* 23: 322–336.

Wood, A. M. (1979). Chlorophyll *a* : *b* ratios in marine planktonic algae. *J. Phycol.* 15: 330–332.

Wright, S. W., Jeffrey, S. W. (1987). Fucoxanthin pigment markers of marine phytoplankton analysed by HPLC and HPTLC. *Mar. Ecol. Progr. Ser.* 38: 259–266.

Wright, S. W., Jeffrey, S. W., Mantoura, R. F. C., Llewellyn, C. A., Bjørnland, T., Repeta, D., Welschmeyer, N. (1991). Improved HPLC method for the analysis of chlorophylls and carotenoids from marine phytoplankton. *Mar. Ecol. Progr. Ser.* 77: 183–196.

Wright, S. W., Shearer, J. D. (1984). Rapid extraction and high performance liquid chromatography of chlorophylls and carotenoids from marine phytoplankton. *J. Chromatog.* 294: 281–295.

14

Comparison between spectrophotometric, fluorometric and HPLC methods for chlorophyll analysis

R. F. C. Mantoura, S. W. Jeffrey, C. A. Llewellyn, H. Claustre and C. E. Morales

14.1 Introduction

Many methods of varying accuracy are available for the analysis of chlorophylls in oceanography (see Chapters 4 and 17). The SCOR-UNESCO sponsored intercalibration of Lorenzen and Jeffrey (1980) (see Section 1.2.3, Chapter 1; Appendix F), set a benchmark for objective comparisons by using analytically pure pigment standards and mixtures to test the accuracy of several spectrophotometric and fluorometric

TABLE 14.1. Literature comparisons of chlorophyll methods

Reference	Analytical methods compared							Standards and samples tested								
	HPLC	Spectrophotometry				Fluoro-metry	Spectro-fluoro-metry	Chlorophyll standards					No. and type of samples			
(Grouped by method)		JH	L	RT	PS	H-H	NP	Chl *a*	Chl *b*	Chl *c*	Phytins	Phide	SW	LW	MA	S
Murray *et al.* (1986)	•	•				m		+	+		+		22	5	2	
Mantoura and Llewellyn (1983)	•	•	•	•	•			+	+	+	+	+	1		4	
Neveux *et al.* (1990)	•	•	•			•	•	+	+	+	+	+	26		22	24
Barlow *et al.* (1990)	•		•					+		+	+	+				23
Brunet *et al.* (1992)	•		•					+					19			
Sartory (1985)	•		•		•			+		+	+			3	9	
Brown *et al.* (1981)	•		•			•		+	+	+	+					27
Webb *et al.* (1992)	•		m					+	+					81		
Jacobsen (1978)	•				•	•		+	+		+		8	6		
Schanz and Rai (1988)	•				•	m		+						3	1	
Lorenzen (1981)	•					•		+	+				84			
Trees *et al.* (1985)	•					•		+					400			
Porra *et al.* (1989)		•						+	+				*			
Lorenzen and Jeffrey (1980)		•	•			•		+	+	+	+	+	*			
Lutz and Carreto (1991)		•	•			•	m	+	+	+	+		600		11	
Marker and Jinks (1982)			•					+						26		
Welschmeyer (1994)						•,m		+	+		+		*			

Abbreviations

JH, Jeffrey and Humphrey (1975); L, Lorenzen (1967); RT, Richards and Thompson (1952); PS, Parsons and Strickland (1963); H-H, Holm-Hansen *et al.* (1965); NP, Neveux and Panouse (1987); SW, sea water; LW, lake water; MA, microalgae; S, sediments; •, method tested; m, method modified; +, standard used.

* no field samples tested

methods. In this Chapter, we summarize the conclusions of Lorenzen and Jeffrey (1980), review the most important comparisons between HPLC, spectrophotometric and fluorescence methods in the literature since 1980, and report the new results of SCOR WG78 comparisons between HPLC and one fluorometric and three spectrophotometric methods. Samples tested were mixtures of pure chlorophyll standards, extracts of log-phase microalgal cultures, and extracts of phytoplankton and sediments.

14.2 Comparison of methods

14.2.1 *Previous comparisons*

The 1979 SCOR-UNESCO-sponsored intercalibration tests (Lorenzen and Jeffrey, 1980) used one acid-fluorometric and five spectrophotometric methods with 20 mixtures of pure pigments to simulate a number of oceanographic situations. It was found that:

- Chlorophyll *a* was satisfactorily determined in mixtures free of degradation products by the methods of Jeffrey and Humphrey (1975), Parsons and Strickland (1963) and SCOR-UNESCO (1966) with only a 1–3% error, since all of the equations were based on reliable extinction coefficients for chlorophyll *a* (see Appendix F). None had terms to correct for degradation products.
- Chlorophyll *b* determination was satisfactory only with the Jeffrey and Humphrey (1975) equations when applied to pheopigment-free extracts. Chlorophyll *b* interfered with the acid-fluorometric method of Holm-Hansen *et al.* (1965) for the determination of chlorophyll *a* and pheopigments. Acidification has since been shown to cause a wavelength shift in chlorophyll *b* which reduces the fluorescence readings for chlorophyll *a* and increases the readings for pheopigments (Gibbs, 1979; Welschmeyer, 1994; see Chapter 4 and Appendix F).
- Chlorophyll *c* was measured satisfactorily (2–5% error) only with the Jeffrey and Humphrey (1975) method provided it occurred at >20% (w/w) relative to chlorophyll *a* (see Appendix G).
- Pheophytins and pheophorbides could only be measured by the acidification methods of Lorenzen (1967) and Holm-Hansen *et al.* (1965) and then only with varying degrees of precision. Even 10% (w/w) of chlorophyllide *a* or pheophorbide *a* caused significant errors in the Jeffrey and Humphrey (1975) estimates of chlorophyll *a*, because there were no terms to correct for them.

Since degradation products often dominate in deep waters, tidal estuaries, sediment traps and zooplankton faeces, Lorenzen and Jeffrey (1980) concluded that 'if precise knowledge of pigment composition is required, chromatography is the only method', and recommended TLC and HPLC for wider use in oceanography.

The development of pigment HPLC for oceanography (Abaychi and Riley, 1979; Mantoura and Llewellyn, 1983; Wright and Shearer, 1984) provided a powerful tool for chlorophyll and carotenoid separations and a new basis for evaluating the accuracy of spectrophotometric and fluorometric analyses of natural waters. Table 14.1 lists 12 reports published since 1980 comparing HPLC with spectrophotometric and fluorometric methods for chlorophylls.

The major conclusions are summarized below:

- *Jeffrey and Humphrey (1975).* Comparisons between HPLC and the spectrophotometric method of Jeffrey and Humphrey (1975) showed that chlorophyll *a* was overestimated in field samples by an average of 48% for 22 samples from Bass Strait in Australia (Murray *et al.* 1986) and 54% for 26 samples from the Charente Estuary, French Atlantic (Neveux *et al.* 1990). Most of the errors arose from spectroscopic interference from pheophorbides, chlorophyllide *a* and chlorophyll *a* epimers and allomers which were estimated as 'chlorophyll *a*'. In contrast, a high accuracy (2% error) was achieved for chlorophyll *a* analysis of 22 microalgal cultures, since no interfering degradation products were present (Neveux *et al.* 1990).

- *Lorenzen (1967).* Seven comparisons between HPLC and the acid spectrophotometric method of Lorenzen (1967) were inconsistent, possibly due to calibration inaccuracies. In extensive investigations of estuarine and sedimentary samples, Brown *et al.* (1981), Barlow *et al.* (1990) and Neveux *et al.* (1990) found the Lorenzen (1967) method slightly underestimated chlorophyll *a* compared to HPLC (8%, 5%, 12% average error, respectively). On the other hand, Brunet *et al.* (1992) found chlorophyll *a* was overestimated by Lorenzen's (1967) method in seawater samples from the English Channel (42% average error). Sartory (1985) and Webb *et al.* (1992) found errors of 23 and 61% for fresh water samples. All except Barlow *et al.* (1990) reported that pheopigment concentrations were severely overestimated (40–375% average error) by the Lorenzen (1967) method compared to HPLC, probably due to the presence of chlorophyll *b* (see Chapter 4).
- *Parsons and Strickland (1963).* Four comparisons between HPLC and the spectrophotometric method of Parsons and Strickland (1963) showed an overestimation of chlorophyll *a* in freshwater samples, ranging from 22% for microalgae (Sartory, 1985; Schanz and Rai, 1988) to 366% for rivers (Jacobsen, 1978). The latter were obviously rich in degradation products.
- *Holm-Hansen et al. (1965).* Four comparisons between HPLC and the acid-fluorometric method of Holm-Hansen *et al.* (1965) reported fluorometric underestimates of chlorophyll *a* in seawater samples ranging from 8 to 38% due to spectral interference from chlorophyll *b* (Lorenzen, 1981; Trees *et al.* 1985; Neveux *et al.* 1990). In estuarine and sedimentary samples, Jacobsen (1978) and Brown *et al.* (1981) found fluorometry to overestimate chlorophyll *a* by up to 70% compared to HPLC, probably due to fluorescent compounds abundant in sediments.

In most of the above comparisons with HPLC, the purity of calibration standards was not checked and mixtures containing degradation products were inadequately described. This makes it difficult to judge the conflicting results of these comparisons with HPLC, particularly in the case of the acid-spectrophotometric method of Lorenzen (1967).

14.2.2 *New developments*

Recent additions to spectroscopic methods since the 1979 SCOR-UNESCO intercalibration study (Lorenzen and Jeffrey, 1980) include

- new spectrophotometric equations for chlorophyll *a* and *b* in methanol and other solvents (Porra *et al.* 1989)
- the extension of Lorenzen's (1967) acid-spectrophotometric method to methanol and ethanol extracts (Marker and Jinks, 1982)
- an elaborate spectrofluorometric method for simultaneous determination of

chlorophylls *a, b* and *c* and their pheopigments (Neveux and Panouse, 1987) with a hexane back-extraction to reduce interferences from breakdown products (Lutz and Carreto, 1991)

- a new combination of lamps and optical filters for the standard fluorometer (Welschmeyer, 1994) to improve the selective detection of chlorophyll *a* in the presence of chlorophyll *b* and pheopigments.

14.2.3 *Present work*

Here we report the SCOR WG78 comparisons between HPLC and one fluorometric and three spectrophotometric methods (Table 14.2) tested on 11 mixtures of pure chlorophylls and derivatives, 29 microalgal extracts, and 14 seawater and sediment samples (Table 14.3). The HPLC method was first validated against the primary mixture of pure pigments, then used for the spectrophotometric and fluorometric method comparisons.

TABLE 14.2. Spectrophotometric, fluorometric and HPLC methods tested by SCOR WG 78

Method	Pigments analysed	Comments and recommended application by authors[b]
Spectrophotometric		
• Jeffrey and Humphrey (1975)	Chls *a, b*, $c_1 + c_2$	Designed for mixtures of pure chlorophyll pigments; no terms to correct for degradation products; recommended for use with algal cultures, 'natural phytoplankton' (Jeffrey and Humphrey, 1975) and for 'surface euphotic zone' (Lorenzen and Jeffrey, 1980).
• Lorenzen (1967)	Chl *a* and 'pheopigments'	Designed for chlorophyll *a* and the sum of pheophytin *a* and pheophorbide *a*; recommended for fresh, estuarine and coastal waters (Lorenzen, 1967).
• Porra *et al.* (1989)	Chls *a, b*	Designed for higher plants extracted with methanol; no terms for chl *c* or degradation products.
Fluorometric		
• Holm-Hansen *et al.* (1965)	Chl *a* and 'pheophytins'	Designed for chlorophyll *a* and pheophytin *a*; pheophorbide *a* is also detected. Chl *b* lowers chl *a* estimation and increases 'pheophytin' estimates. Recommended for surface to deep open ocean samples (Holm-Hansen *et al.* 1965).
HPLC		
• Mantoura and Llewellyn (1983)[a]	Chls *a, b*, chlides *a, b*, phytins *a, b, c*, phides *a, b*, pyro- and allomeric derivatives	Major and minor chlorophylls and derivatives separated (not chl c_1 from c_2, or DV Chl *a* from Chl *a*). Recommended for phytoplankton and detrital pigments in all natural waters.

a. The recommended Wright *et al.* (1991) HPLC system was not fully developed at the time of the SCOR WG 78 workshops.

b. Our recommendations may differ from those of the original authors as a result of the present tests (see section 14.5).

TABLE 14.3. Mixtures of pure pigments tested by SCOR WG 78: simulations of oceanographic conditions

Sample no. (see Table 14.4)	Pure pigments											Simulations
	Chl *a*	Chlide *a*	Phytin *a*	Phide *a*	Pyro-phytin *a*	Chl *b*	Chlide *b*	Phytin *b*	Chl c_1+c_2	Chl c_2	Chl c_3	
A Pure pigment mixtures												
1	+	+	+	+	+	+	+	+	+	+	+	Mixed pure pigment standards
3	+					+						Chl *a*, *b* (green algae)
4	+								+			Chl *a*, *c* (chromophyte algae)
5	+					+			+			Chl *a*, *b*, *c* (mixed algae)
6	+	+	+	+	+			+				Zooplankton faecal pellets
8	+	+	+	+	+	+	+	+	+			Sediment trap detritus
9	+	+	+	+	+				+			Deep phytoplankton (E Australian current)
10	+	+	+	+		+	+	+				Resuspended estuarine sediments
11	+	+	+	+	+							Salp faeces

B Algal cultures plus pure pigments

	Emiliania huxleyi extract (Chls *a*, c_2, c_3)	*Micromonas pusilla extract* (Chl *a*, *b*, Mg3,8 DVP)	Chlide *a*	Phytin *a*	Phide *a*	
26	++++					Surface euphotic zone (chromophyte algae)
27	+++	+				Euphotic zone
28	++	++				Euphotic zone
29	+	+++				Euphotic zone
30		++++				Deep chlorophyll maximum
12	+	+++	+	+	+	Deep chlorophyll maximum: some grazing
13	+	+++	++	++	++	Deep chlorophyll maximum: moderate grazing
14	+	+++	+++	+++	+++	Deep chlorophyll maximum: heavy grazing

14.3 Materials and methods

Comparisons between HPLC and spectrophotometric and fluorometric methods were carried out on mixtures of pure pigments (samples 1–11, Tables 14.3 and 14.4), on taxonomically diverse microalgal cultures (samples 15–40, Table 14.4) and on pigment extracts from seawater and sediments (samples 41–54, Table 14.4). The first two comparisons were carried out at the SCOR WG 78 Chlorophyll Workshop in Hobart, the third at the Field Workshop in Plymouth.

14.3.1 *Pigment standards*

Chlorophylls *a*, *b*, c_1+c_2, c_2 and c_3, pheophytins *a* and *b*, pyropheophytin *a*, chlorophyllides *a* and *b*, and pheophorbide *a* were isolated from microalgae and other sources and purified according to procedures detailed in Chapter 7. Each single pigment standard was prepared in acetone and, after checking the purity by HPLC, the concentration was determined spectrophotometrically using extinction coefficients listed in Appendix E and Chapter 7. Single pigment standards were used to calibrate the HPLC (Chapter 16) and the Turner fluorometer as well as to prepare the 11 pigment mixtures shown in Tables 14.3 and 14.4.

14.3.2 *Microalgae*

Twenty nine microalgal cultures (see Table 14.4), which included the twelve SCOR standard reference cultures (see Chapters 6 and 13), were grown and harvested according to standard protocols (Chapters 6 and 12). Replicate volumes (10–20 ml) were filtered (Whatman GF/F, 25 mm) under mild vacuum and pigments were immediately extracted by grinding the filters in cold 90% acetone (samples 15–30) or by sonicating in methanol (samples 31–40) and centrifuged. The concentrations of individual chlorophylls and their breakdown products were then analysed by spectrophotometry (Jeffrey and Humphrey, 1975; Lorenzen, 1967, Porra *et al.* 1989), fluorometry (Holm-Hansen *et al.* 1965) and HPLC (Mantoura and Llewellyn, 1983).

14.3.3 *Field samples*

Fourteen field samples (estuarine and oceanic seawater, zooplankton and mussel faeces and sediment traps; see samples 41–54, Table 14.4) were filtered and pigments extracted in methanol (see Chapter 9) and analyzed by spectrophotometry (Porra *et al.* 1989), fluorometry (after dilution in acetone; Holm-Hansen *et al.* 1965), and HPLC (Mantoura and Llewellyn, 1983).

14.3.4 *Spectrophotometry*

Pigment extracts were transferred to a 10 mm cuvette and scanned between 750 and 350 nm on a Shimadzu Model UV-240 or a Perkin Elmer Model Lambda II spectrophotometer. The concentrations of chlorophylls *a*, *b*, and *c* in 90% acetone extracts were determined using the equations of Jeffrey and Humphrey (1975) and the concentrations of chlorophyll *a* and of pheopigments by the acid-spectrophotometric method of Lorenzen (1967). The concentrations of chlorophylls *a* and *b* were determined in methanol extracts using the equations of Porra *et al.* (1989).

14.3.5 *Fluorometry*

Chlorophyll *a* and pheopigments were determined by the fluorometric method of Holm-Hansen *et al.* (1965). Turner fluorometers (Designs Model 10 and Sequoia Model 112) were used with the standard light source (F4T5-BL) and Corning filters (excitation CS-5-60; emission CS-2-64) and calibrated with HPLC-purified chlorophyll *a* standards.

14.3.6 *HPLC*

Pigments were separated using the reversed-phase binary gradient HPLC method of Mantoura and Llewellyn (1983). The SCOR-recommended HPLC method of Wright *et al.* (1991) with improved resolution for carotenoids (see Chapter 12) was not fully developed at the time of the SCOR Workshops.

The HPLC used at the Chlorophyll Workshop consisted of an ACS-model 300 HPLC binary pump, a Hewlett Packard 1040 Diode Array UV-visible detector (350–600 nm) and a Dupont 838 fluorescence detector (excitation 390–490 nm, emission >620 nm). Pigment standards and extracts were mixed 3:1 (v:v) with buffer (0.5 M ammonium acetate, pH 7) and 100 μl immediately injected *via* a Rheodyne 7125 injector (100 μl) onto an 8 cm C-18 Pecosphere (Perkin Elmer; 3 μm particle size) column and eluted at 1 ml/min. The HPLC used at the Field Workshop was a Shimadzu system (Dual LC6A pumps, SPD-6AV detector (440 nm) and SCL-6B controller) operated with the same injector and flow rates. The data was captured and processed using a Philips PU6000 2 channel system. Both HPLCs were calibrated with purified standards (five chlorophylls and six degradation products) using the external standardization procedure described in Chapter 16.

14.3.7 *Pigment definitions*

Since pheophytin *a* and pheophorbide *a* are spectrally similar, Lorenzen (1967) suggested the term 'pheopigment' to denote 'pheophytin and pheophorbide measured together' in his acid-spectrophotometric method. The Holm-Hansen *et al.* (1965) acid-fluorometric method also detects similar fractions. While we do not recommend the term 'pheopigments', it is used here to denote the concepts of the original authors. For comparisons with HPLC, the notations Σ chlorophyll *a* degradation products and Σ chlorophyll *c* are used to denote the summed weights of chlorophyll *a* degradation products and of chlorophyll *c* derivatives (chlorophylls c_1, c_2, c_3 and Mg3,8 DVP) separated by HPLC and listed in Table 14.5.

14.4 Results and discussion

Estimates of the concentrations of chlorophylls in the mixed standards, microalgal cultures and the field samples as determined by spectrophotometry, fluorometry and HPLC are shown in Tables 14.4 and 14.5 respectively.

The analytical accuracy of the HPLC method was determined by comparing HPLC chlorophyll concentrations against the actual concentrations of the pure pigments used to prepare the standard mixtures (see Table 14.3). Comparisons for five pigments: chlorophylls *a*, *b*, c_1+c_2, pheophytin *a* and pheophorbide *a* are shown in Fig. 14.1. Excellent regressions were obtained (overall correlation coefficient (r^2) =

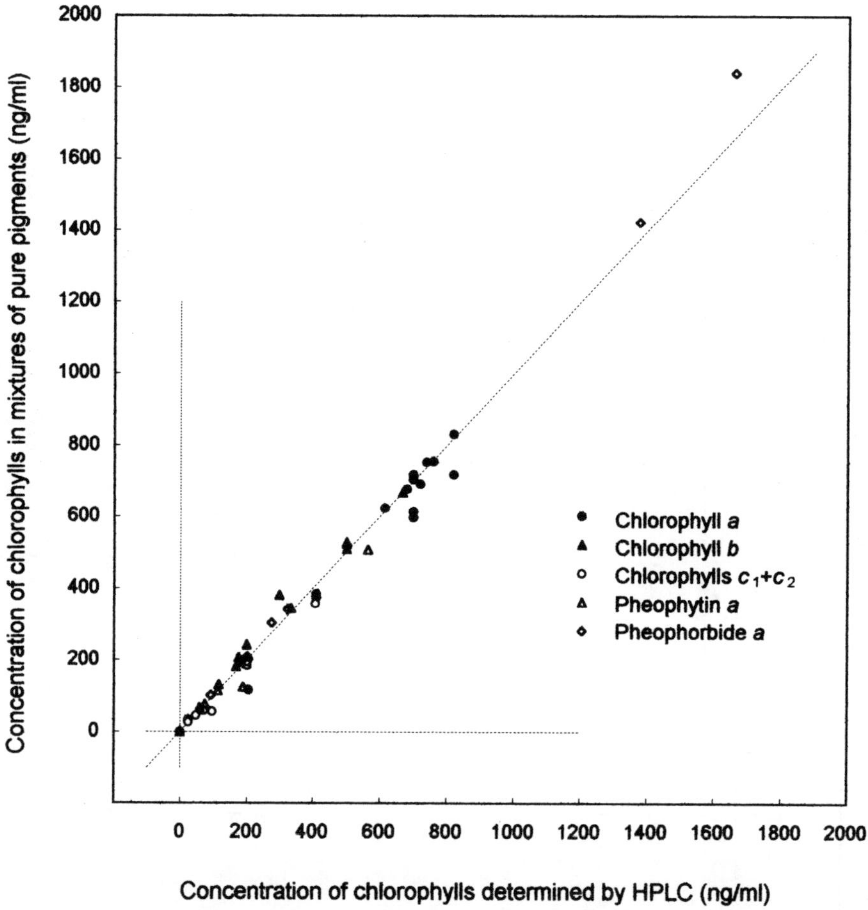

Figure 14.1
The concentrations of chlorophylls *a*, *b*, c_1+c_2, pheophytin *a* and pheophorbide *a* determined by HPLC compared to their actual concentration in the original mixture (r^2 = 0.998; slope = 1.019).

0.998; slope (s) = 1.019) indicating that the HPLC accurately determined the pigment concentrations of the standard mixtures. The overall pigment recovery was 101.9%. All comparisons with the three spectrophotometric and one fluorometric method were then made using the HPLC values as standard.

Chlorophyll *a* determined by HPLC and the methods of Jeffrey and Humphrey (1975), Lorenzen (1967), Porra *et al.* (1989) and Holm-Hansen *et al.* (1965) are shown in Figs. 14.2A, B, C and D, respectively. Chlorophylls *b* and *c* determined by HPLC and the method of Jeffrey and Humphrey (1975) are shown in Figs. 14.3A, B respectively. Pheopigments determined by HPLC and the methods of Lorenzen (1967) and Holm-Hansen *et al.* (1965) are shown in Figs. 14.3C and D, respectively. The chlorophyll data originating from the pigment standards (samples 1–14 in Table 14.4),

TABLE 14.4. Pigment concentrations in mixed standards, microalgal cultures and field samples determined by various spectrophotometric and fluorometric methods. For the composition of the mixed pigment standards see Table 14.3A and Table 14.5.

Sample No.	Sample description	Pigment concentration (ng/ml)						
		Spectrophotometry					Fluorometry	
		Jeffrey & Humphrey (1975)			Lorenzen (1967)		Holm-Hansen *et al.* (1965)	
		Chl *a*	Chl *b*	Chl *c*	Chl *a*	Pheopigment	Chl *a*	Pheopigment
Mixed pigment standards								
1	Pure pigment standards	952	499	210	508	708	674	1906
2	Pure pigment standards	–	–	–	–	–	554	1674
3	Pure pigment standards	615	156	3	535	83	532	273
4	Pure pigment standards	594	-20	101	535	27	654	54
5	Pure pigment standards	833	276	99	1042	-444	834	297
6	Pure pigment standards	2193	-8	-176	882	2112	698	3206
7	Pure pigment standards	1213	536	64	535	1112	454	1453
8	Pure pigment standards	–	–	–	–	–	342	5289
9	Pure pigment standards	–	–	–	–	–	358	2940
10	Pure pigment standards	2051	517	-263	535	2515	236	3984
11	Pure pigment standards	–	–	–	–	–	67	2134
12	*Emiliania/Micromonas* + pheopigment stds*	895	364	93	695	278	763	696
13	*Emiliania/Micromonas* + pheopigment stds*	1051	409	73	722	476	829	843
14	*Emiliania/Micromonas* + pheopigment stds*	1232	443	49	882	540	859	1037
Microalgal cultures								
15	*Chroomonas salina* CS-174	2089	93	240	1871	112	2121	154
16	*Phaeodactylum tricornutum* CS-29	2912	59	368	2700	88	2984	66
17	*Pycnococcus provasoli* CS-185	1792	1578	233	1497	430	1409	1704
18	*Synechococcus* sp. CS-197	837	92	-19	882	-115	830	71
19	*Dunaliella tertiolecta* CS-175	1518	577	71	1390	88	1449	698
20	*Amphidinium carterae* CS-212	2520	17	1025	2245	112	2876	165
21	*Pelagococcus subviridis* CS-99	687	73	270	642	-24	817	92
22	*Micromonas pusilla* CS-86	1670	1255	162	1417	361	1563	1039
23	*Emiliania huxleyi* CS-57	735	49	288	695	35	795	150
24	*Pavlova lutheri* CS-182	899	10	110	855	-32	938	-26
25	*Synechococcus* CS-94	835	-29	28	775	-8	810	35
26	*Emiliania/Micromonas* mix (8:0, v:v)	717	-9	175	668	24	250	134

No.	Sample							
27	*Emiliania/Micromonas* mix (6:2, v:v)	823	255	157	374	636	738	327
28	*Emiliania/Micromonas* mix (4:4, v:v)	708	278	91	668	61	613	447
29	*Emiliania/Micromonas* mix (2:6, v:v)	622	363	39	535	102	692	548
30	*Emiliania/Micromonas* mix (0:8, v:v)	723	529	37	561	225	634	697

		Porra *et al.* (1989)		Holm-Hansen *et al.* (1965)	
		Chl *a*	Chl *b*	Chl *a*	Pheopigment
31	*Phaeodactylum tricornutum* P-100	1700	-112	1365	85
32	*Phaeodactylum tricornutum* CS-29	2307	130	2110	68
33	*Pycnococcus provasoli* P-161	1641	1546	1405	2066
34	*Synechococcus* sp. CS-169	815	41	375	67
35	*Dunaliella tertiolecta* P-83	3281	2255	3142	1282
36	*Amphidinium carterae* P-450	415	97	399	80
37	*Micromonas pusilla* P-27	1417	1191	1287	1585
38	*Emiliania huxleyi* P-92d	961	110	835	162
39	*Pavlova lutheri* P-75	2216	-7	2031	106
40	*Gymnodinium galatheanum* P-357	714	-30	563	191
Field samples					
41	Seawater: N W Atlantic 40°N 47°E, 20 m, 1/5/89	543	207	370	306
42	Seawater: N W Atlantic 40°N 47°E, 10 m, 1/5/89	489	175	138	211
43	Seawater: Tamar Estuary, 1 m, 16/9/89	2509	686	675	2539
44	Seawater: N E Atlantic 52°N 20°W, 10 m, 28/6/89	393	53	101	229
45	Seawater: N E Atlantic 52°N 20°W, 70 m, 28/6/89	124	43	15	67
46	Seawater: N E Atlantic 47°N 17°W, 5 m, 28/6/89	184	10	115	72
47	Faecal pellets: *Calanus* sp. fed on diatoms	1618	-380	0	1212
48	Faecal pellets: mixed copepods	2434	308	102	2325
49	Faecal pellets: *Calanus* sp. starved	937	-230	66	1532
50	Faecal string: *Mytilus edulis*	4521	333	295	485
51	Sediment Trap: Chesapeake Bay, 10 m, 15/5/89	545	99	76	310
52	Sediment Trap: Chesapeake Bay, deep, 15/5/89	1156	58	500	1475
53	Sediment Trap: Chesapeake Bay, deep, 29/9/88	4364	-386	98	3068
54	Sediments: Lynher Estuary, 16/9/89	–	–	–	–

Microalgal clones from CSIRO and Plymouth Collections are designated CS and P, respectively; –, not analyzed
* See Table 14.3B

TABLE 14.5. Pigment concentrations in mixed standards, microalgal cultures and field samples determined by high performance liquid chromatography. For the composition of the mixed pigment standards see Table 14.3A.

Sample No.	Chlorophylls (ng/ml)							Chlorophyll degradation products (ng/ml)								
	Chl *a*	Chl *b*	Chl c_1+c_2	Chl c_2	Chl c_3	Mg3, 8D	Chl *c*	Chlide *a*	Phorbide *a*	Pyro-phorbide *a*	Chl *a*-allom	Phytin *a*	Pyro-phytin *a*	Chlide *b*	Phytin *b*	Chl *a* deg. prods.
Mixed pigment standards																
1	384	190	357		203		560	965	343			77	159	204	167	1544
2	116	206	197				197	343	209			124	90	213	167	1146
3	623	241					0									0
4	718		184				184									0
5	832	381	84				84									0
6	375						0	79	1503			115	361		182	2240
7	–	–	–	–	–	–	–	–	–	–	–	–	–	–	–	–
8	18	19	66				66	7	1842			1476	879		353	4557
9	25		479				479	17	747			738	271			1773
10	202	243					0		1981			429	552	243	552	2962
11							0		358			763	405			1526
12	718	528		33	1		34	35	102			67				204
13	596	509		27	2		29	124	197			131				452
14	613	523		27	1		28	134	304			207				645
Microalgal cultures																
15	2123			214			214									0
16	2916		327				327	76								76
17	1486	1638				120	120									0
18	901						0									0
19	1142	546					0	506								506
20	2520		1024				1024									0
21	688			135	68		203									0
22	1600	1358				79	79									0
23	862			119	84		203									0
24	868		55				55									0
25							0									0
26	755			56	27		83					1				1

27	753	181		58	15		73					2				2
28	692	345		45	10		55					1				1
29	704	525		27	3		30					1				1
30	678	668					0					1				1
31	1271		206				206	99								99
32	1597		268				268	168								168
33	1193	1304				96	96									0
34	688						0									0
35	3007	1112					0	586								586
36	367			168			168	28								28
37	1575	1440				114	114									0
38	1071		184		120		304									0
39	1911			139			139	51								51
40	656		94		35		129									0
Field samples																
41	325	93	61		28		89				6					6
42	300	81	50		24		74				6					6
43	494	155	44		0		44	63	231	430	25	1643	651		9842	2979
44	313	30	70		43		112				3	0	0			3
45	49	7	10		6		16		8	8	3	48	48			114
46	92	5	23		7		31			8	3	48	67			126
47	11						0			669		124	403			1196
48	257	98			1		1			1373		567	489		3249	2429
49	10						0			371		96	224			691
50	1723		331				331		15	31	136	313	320			815
51	160	28	33				33	167	8	15	74	70	48			215
52	155	0	23				23	43	18	224	26	126	612			1007
53	186		30				30	47	49	799	39	721	288			1895
54	2498		512				512	596	215	1491	102	1635	715		1214	4158

HPLC separations using the method of Mantoura and Llewellyn (1983).
Chl *c* and Chl *a* degradation products denote the summed weights of all chlorophyll *c* pigments and of all chlorophyll *a* degradation products, respectively.
Empty spaces denote the pigment was not detected on the HPLC, i.e. concentration = 0 ng/ml.
–, not analysed.

the microalgal cultures (samples 14–40) and the field samples (samples 41–54) are coded differently in these figures and are shown in relation to the 1:1 agreement line. The results for each pigment relative to HPLC are evaluated below in terms of their correlation coefficients (r^2) and slopes (s).

14.4.1 *Chlorophyll a*

Chlorophyll *a* was satisfactorily determined in standard mixtures of pure pigments and microalgal extracts not containing any degradation products by the methods of Jeffrey and Humphrey (1975), Lorenzen (1967), Porra *et al.* (1989) and Holm-Hansen *et al.* (1965) ($r^2 = 0.91–0.96$; s = 0.93–1.12).

Chlorophyll *a* was over-estimated in the presence of equal weights of pheopigments by approximately 60% when using methods that did not account for them e.g., Jeffrey and Humphrey (1975) and Porra *et al.* (1989). This is shown in Figs. 14.2A and C. The pheopigment-correcting acidification methods of Lorenzen (1967) and Holm-Hansen *et al.* (1965) significantly improved the overall chlorophyll *a* correlations with HPLC (from $r^2 = 0.48$ and 0.26 in Figs. 14.2A and C to $r^2 = 0.86$ and 0.85 in Figs. 14.2B and D respectively).

Chlorophyll *a* was under-estimated in the presence of equimolar concentrations of chlorophyll *b* (by 2–19%) using the acid-fluorometric method of Holm-Hansen *et al.* (1965). This confirms previous studies showing that chlorophyll *b* reduces the fluorescence readings for chlorophyll *a* and increases the readings for pheopigments (Gibbs, 1979; Trees *et al.* 1985; Welschmeyer, 1994; see Chapter 4 and Appendix F).

14.4.2 *Chlorophyll b*

Chlorophyll *b* was satisfactorily determined by the Jeffrey and Humphrey (1975) method with pure chlorophyll mixtures and in pheopigment-free microalgal extracts containing at least 5 % chlorophyll *b* relative to chlorophyll *a* (Fig. 14.3A: $r^2 = 0.98$, s = 0.90).

Chlorophyll *b* determined by the method of Porra *et al.* (1989) in green microalgal extracts also gave acceptable correlations with HPLC (not plotted; $r^2 = 0.81$, s = 1.19). However, poor correlations and 'negative' concentrations were obtained with field samples rich in pheopigments ($r^2 = 0.58$, s = 4.35). The Porra *et al.* (1989) equations cannot be used when chlorophyll *c* is present as they have no term for chlorophyll *c* and they do not correct for the spectral overlap of chlorophylls *a* and *c* as do the Jeffrey and Humphrey (1975) equations.

14.4.3 *Chlorophyll c*

Chlorophyll *c* was satisfactorily determined in pure chlorophyll mixtures and microalgal extracts by the method of Jeffrey and Humphrey (1975) (Fig. 14.3 B: $r^2 = 0.78$, s = 0.89). With the HPLC method, the different chlorophyll *c* compounds (c_1+c_2, c_3 and Mg3,8 DVP) must be summed to get the total chlorophyll *c* value.

14.4.4 *Pheopigments*

The concentrations of Σ chlorophyll *a* degradation products determined by HPLC are plotted against pheopigment values derived by the acid-spectrophotometric method

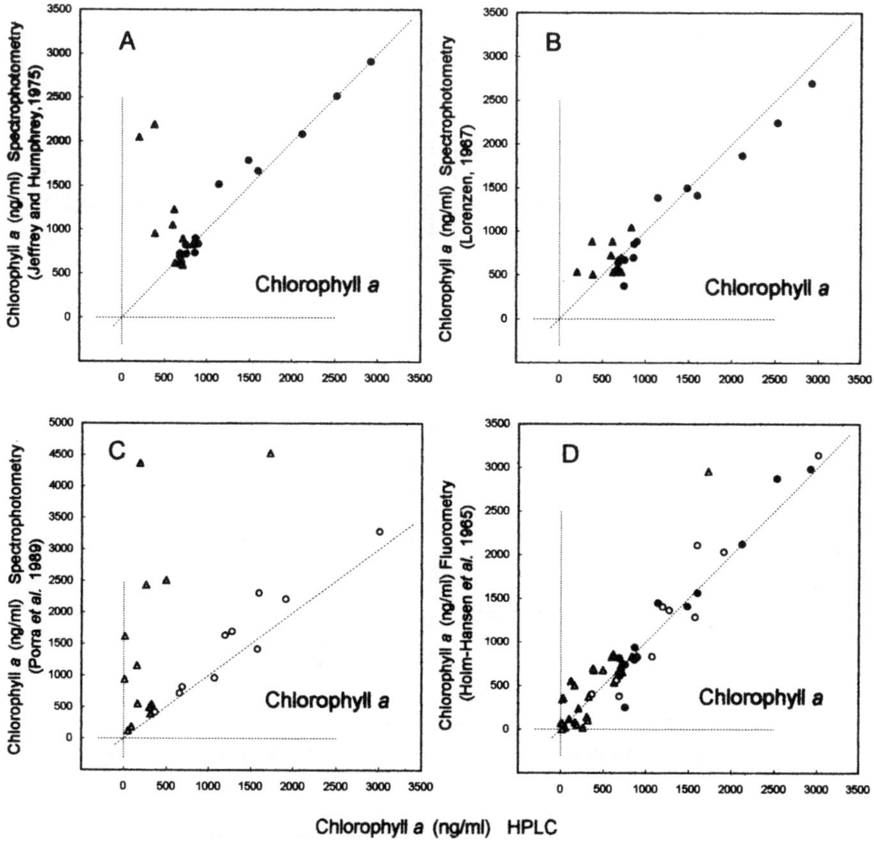

Figure 14.2
Comparisons of chlorophyll *a* concentrations determined by HPLC and the spectrophotometric methods of A, Jeffrey and Humphrey (1975); B, Lorenzen (1967); C, Porra *et al.* (1989); D, the fluorometric method of Holm-Hansen *et al.* (1965). Diagonal lines indicate a 1:1 agreement between methods. The data symbols correspond to: (▲) mixed standards of chlorophylls and degradation products prepared in acetone (sample number 1–14 see Table 14.4); (●) microalgal chlorophylls extracted in acetone (samples 15–30); (○) microalgal chlorophylls extracted in methanol (samples 31–40); (△) seawater, estuarine water, detrital and faecal samples extracted in methanol (samples 41–54).

of Lorenzen (1967) and the acid-fluorometric method of Holm-Hansen *et al.* (1965) in Figs. 14.3C and D respectively.

Pheopigments were satisfactorily determined in pure pigment mixtures and field samples rich in pheopigments using the Lorenzen (1967) and Holm-Hansen *et al.* (1965) methods (Fig. 14.3C: $r^2 = 0.88$, $s = 0.79$, Fig 14.3D: $r^2 = 0.89$, $s = 1.03$). However, in pheopigment-free samples, interference from chlorophyll *b* produced high spectrophotometric and fluorometric estimates of pheopigments (see Tables 14.4

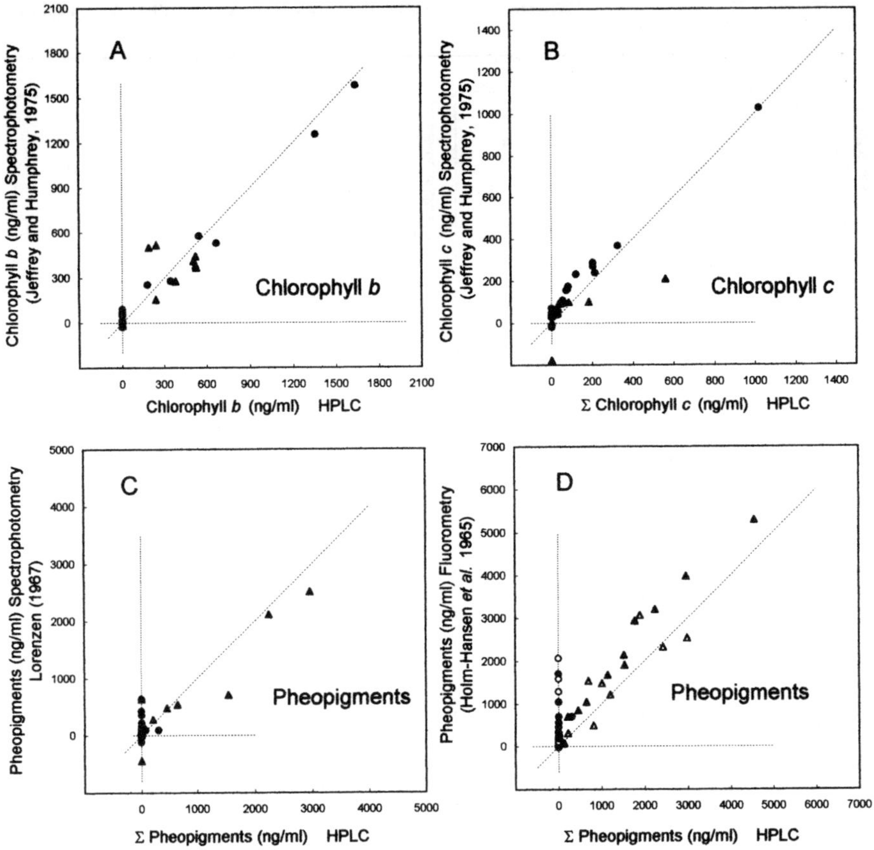

Figure 14.3
Comparisons are shown between chlorophyll *b* (A) and chlorophyll *c* (B) determined by HPLC and the spectrophotometric method of Jeffrey and Humphrey (1975) in mixed standards and microalgal extracts; and between pheopigments determined by HPLC and Lorenzen's (1967) spectrophotometric method (C) and Holm-Hansen's *et al.* (1965) fluorometric method (D). Data symbols correspond to those of Fig. 14.2.

and 14.5) as is clearly seen in the data cluster on the Y axes of Figs. 14.3C and D.

Very low concentrations of any chlorophyll in a mixture (less that 10% by weight relative to other chlorophylls) gave large spectrophotometric errors in recovery (up to 60%; see Appendix G).

14.5 Recommendations

Comparisons reported above between spectrophotometric, fluorometric and HPLC methods generally confirm and extend the results of the previous SCOR-UNESCO study (Lorenzen and Jeffrey, 1980).

- Reversed-phase HPLC (e.g., Mantoura and Llewellyn, 1983; Wright *et al.* 1991) can resolve and accurately determine most of the important chlorophyll pigments including chlorophylls *a*, *b*, c_1+c_2, c_3, their pheophytins, pheophorbides, allomers, epimers and pyro-derivatives in extracts of microalgal cultures, phytoplankton, and seawater samples. Coelution of divinyl chlorophyll *a* and chlorophyll *a*, and chlorophyll c_1 and c_2 may be overcome using other reversed-phase HPLC systems (e.g., Goericke and Repeta, 1993; Garrido *et al.* 1995; see section 4.4.2.2, Chapter 4).

- The spectrophotometric method of Jeffrey and Humphrey (1975) satisfactorily measures moderate to high concentrations (0.1–5 μg/ml) of chlorophylls *a*, *b* and *c* provided they occur as major (>20%) components and that chlorophyll degradation products are absent (see Appendix F and G). These conditions occur in log-phase microalgal cultures and early phytoplankton blooms before the onset of grazing or senescence. For satisfactory use of the equations, verification of the absence of degradation products should be done by TLC or HPLC.
- The pheopigment-correcting methods of Lorenzen (1967) and Holm-Hansen *et al.* (1965) for chlorophyll *a* significantly improve the reliability of spectrophotometric and fluorometric methods in samples containing >20% by weight of degradation products. These methods suffer from interference from chlorophyll *b* which is known to be widely distributed in coastal and oligotrophic oceans. Recently, Welschmeyer (1994) optimized a new set of lamps and filters making the fluorometric method of Holm-Hansen *et al.* (1965) much more selective for chlorophyll *a* in the presence of chlorophyll *b* and pheopigments (see Appendix F). The fifty-fold greater sensitivity of fluorometry over spectrophotometry makes this method particularly useful for oligotrophic waters containing <50 ng chlorophyll *a*/litre.
- The determination of 'pheopigments' by the Lorenzen (1967) and Holm-Hansen *et al.* (1965) methods should be regarded as approximate since both methods lump a wide molecular weight range (530–894) of pyro, meso, epimer and allomer derivatives of pheophytin *a* and pheophorbide *a* into a single 'pheopigment' category. Ultimately this term should disappear as detailed knowledge of the components is obtained.
- Chlorophyllides *a* and *b* cannot be distinguished from chlorophylls *a* and *b* by any of the spectrophotometric or fluorometric methods tested. A simple phase test was proposed to eliminate phytol-free chlorophylls (Whitney and Darley, 1979), but it has been little used in the literature.
- The Porra *et al.* (1989) equations for chlorophylls *a* and *b* in methanol are of particular significance since this solvent is the recommended extractant of pigments from microalgae (see Chapter 9). New equations need to be derived to include chlorophyll *c* so that accurate methanol equations are available for mixtures of the three major chlorophylls.
- HPLC has revealed over 20 new chlorophyll derivatives over the last decade including:
  - Divinyl chlorophyll *a* and *b* in prochlorophytes, abundant in oligotrophic waters (Gieskes and Kraay, 1983; Chisholm *et al.* 1988; Goericke and Repeta, 1993)
  - Bacteriochlorophylls in the hypoxic Black Sea (Repeta *et al.* 1989; Repeta and Simpson, 1991)
  - Homologous series of pyro-, meso- and steryl-pheophorbide esters in zooplankton faecal pellets, seawater and sediments (Head and Harris, 1992; Pearce *et al.* 1993)

- New chlorophyll *c* pigments including chlorophyll c_3 in prymnesiophytes (Jeffrey and Wright, 1987; 1994), the phytylated chlorophyll c_2-like pigment in *Emiliania huxleyi* (Nelson and Wakeham, 1989), chlorophyll $c_{CS\text{-}170}$ in *Micromonas pusilla* (Jeffrey, 1989), and two new chlorophyll *c* derivatives (Garrido *et al.* 1995).

This vindicates the previous recommendation of Lorenzen and Jeffrey (1980) that 'if precise knowledge of pigment composition is required, chromatography is the only method'.

14.6 Conclusions

The determination of chlorophyll *a* in simple pigment mixtures and extracts free of chlorophyll degradation products is accurately done using the Jeffrey and Humphrey (1975) spectrophotometric method (< 5% error) and the Holm-Hansen *et al.* (1965) fluorometric method if chlorophyll *b* is absent. The acid-spectrophotometric and fluorometric methods for chlorophyll *a* of Lorenzen (1967) and Holm-Hansen *et al.* (1965) reduce, but do not eliminate, interference from pheophytins and pheophorbides. Welschmeyer's (1994) recent optical improvements to the Holm-Hansen *et al.* (1965) fluorometric method were not tested here, but seem to significantly reduce interference from chlorophyll *b* and pheopigments and increase the fluorometric selectivity for chlorophyll *a*. Finally, since the spectra of chlorophyllide *a* and *b* are identical to chlorophylls *a* and *b*, they cannot be distinguished from their parent chlorophylls by either spectrophotometry or fluorometry, unless a phase separation step is included.

The correct use of the above spectroscopic methods in oceanography requires that the conditions listed in Table 14.2 are noted. In practice, a limited number of representative samples must first be screened by TLC or HPLC to ensure the absence of interfering degradation products. Thus the advantages of pigment spectroscopy in terms of simplicity, accuracy, cost and throughput can be fully realized. However, the only methods that can accurately assess all chlorophylls in the presence of degradation products are separation techniques, *viz* TLC, HPTLC, isocratic HPLC and reversed-phase HPLC (See Chapters 4, 7, 11 and 12).

Acknowledgments

Thanks are due to Drs. S.W. Wright, N. A. Welschmeyer, R. Dawson, T. Bianchi and D.L. Repeta for their contributions to the intercalibration tests at the SCOR Workshops in Hobart and Plymouth. Hewlett Packard are thanked for lending HPLC equipment for the Hobart and Plymouth Workshops.

References

Abaychi, J. K., Riley, J. P. (1979). The determination of phytoplankton pigments by high-performance liquid chromatography. *Anal. Chim. Acta,* 107: 1–11.

Barlow, R. G., Collos, Y., Maestrini, S. Y., Roy, S. (1990). Intercomparison of spectrophotometry and HPLC for estimating microphytobenthic pigment biomass in salt marsh ponds. *Mar. Microb. Food Webs,* 4: 117–128.

Brown, L. M., Hargrave, B. T., MacKinnon, M. D. (1981). Analysis of chlorophyll *a* in sediments by high-pressure liquid chromatography. *Can. J. Fish. Aquat. Sci.* 38: 205–214.

Brunet, C., Brylinski, J-M., Degros, N., Hilde, D. (1992). Etude de la biomasse phytoplanctonique en Manche Orientale: comparison de trois techniques de mesure des concentrations en chlorophylle *a*. *J. Rec. Oceanog.* 16: 95–99.

Chisholm, S. W., Olson, R. J., Zettler, E. R., Goericke, R., Waterbury, J. B., Welshmeyer, N. A. (1988). A novel free-living prochlorophyte abundant in the oceanic zone. *Nature,* 334: 340–343.

Garrido, J. L., Zapata, M., Muniz, S. (1995). Spectral characterization of new chlorophyll *c* pigments isolated from *Emiliania huxleyi* (Prymnesiophyceae) by high performance liquid chromatography. *J. Phycol.* 31: 761–768.

Gibbs, C. F. (1979). Chlorophyll *b* interference in the fluorometric determination of chlorophyll *a* and "phaeopigments". *Austr. J. Mar. Freshwater Res.* 30: 597–606.

Gieskes, W. W. C., Kraay, G. W. (1983). Unknown chlorophyll *a* derivatives in the North Sea and the tropical Atlantic Ocean revealed by HPLC analysis. *Limnol. Oceanogr.* 28: 757–766.

Goericke, R., Repeta D. (1993). Chlorophylls *a* and *b* and divinyl chlorophylls *a* and *b* in the open subtropical North Atlantic. *Mar. Ecol. Progr. Ser.* 101: 307–313.

Head, E. J. H., Harris, L. R. (1992). Chlorophyll and carotenoid transformation and destruction by *Calanus* spp. grazing on diatoms. *Mar. Ecol. Progr. Ser.* 86: 229–238.

Holm-Hansen, O., Lorenzen, C. J., Holmes, R. W., Strickland, J. D. H. (1965). Fluorometric determination of chlorophyll. *J. Cons. Perm. Int. Explor. Mer.* 30: 3–15.

Jacobsen, T. R. (1978). A quantitative method for the separation of chlorophylls *a* and *b* from phytoplankton pigments by high pressure liquid chromatography. *Mar. Sci. Commun.* 4: 33–47.

Jeffrey, S. W. (1989). Chlorophyll *c* pigments and their distribution in the chromophyte algae. In: Green, J. C., Leadbeater, B. S. C., Diver, W. L. (eds). *The Chromophyte Algae: Problems and Perspectives*, Clarendon Press, Oxford, p. 13–36.

Jeffrey, S. W., Humphrey, G. F. (1975). New spectrophotometric equations for determining chlorophylls *a, b.* c_1 and c_2 in higher plants, algae and natural phytoplankton. *Biochem. Physiol. Pflanzen,* 167: 191–194.

Jeffrey, S. W., Wright, S. W. (1987). A new spectrally distinct component in preparations of chlorophyll *c* from the microalga *Emiliania huxleyi* (Prymnesiophyceae). *Biochim. Biophys. Acta,* 894: 180–188.

Jeffrey, S. W., Wright, S. W. (1994). Photosynthetic pigments in the Haptophyta. In: Green, J. C., Leadbeater, B. S. C. (eds). *The Haptophyte Algae*, Oxford University Press, p. 111–132.

Lorenzen, C. J. (1967). Determination of chlorophyll and pheopigments: spectro-photometric equations. *Limnol. Oceanogr.* 12: 343–346.

Lorenzen, C. J. (1981). Chlorophyll *b* in the eastern North Pacific Ocean. *Deep-Sea Res.* 28A: 1049–1056.

Lorenzen, C. J., Jeffrey, S. W. (1980). Determination of chlorophyll in seawater. UNESCO, Technical papers in Marine Science No. 35, 21 pp.

Lutz, V. A., Carreto, J. I. (1991). A new spectrofluorometric method for the determination of chlorophylls and degradation products and its application in two frontal areas of the Argentine Sea. *Cont. Shelf Res.* 11: 433–451.

Mantoura, R. F. C., Llewellyn, C. A. (1983). The rapid determination of algal chlorophyll and carotenoid pigments and their breakdown products in natural waters by reverse-phase high-performance liquid chromatography. *Anal. Chim. Acta,* 151: 297–314.

Marker, A. F. H., Jinks, S. (1982). The spectrophotometric analysis of chlorophyll *a* and phaeopigments in acetone, ethanol and methanol. *Arch. Hydrobiol. Beih. Ergebn. Limnol.* 16: 3–17.

Murray, A. P., Gibbs, C. F., Longmore, A. R. (1986). Determination of chlorophyll in marine waters: intercomparison of a rapid HPLC method with full HPLC, spectrophotometric and fluorometric methods. *Mar. Chem.* 19: 211–227.

Nelson, J. R., Wakeham, S. G. (1989). A phytol-substituted chlorophyll *c* from *Emiliania huxleyi*. *J. Phycol.* 25: 761–766.

Neveux, J., Delmas, D., Romano, J. C., Algarra, P., Ignatiades, L., Herbland, A., Morand, P., Neori, A., Bonin, D., Barbe, J., Sukenik, A., Berman, T. (1990). Comparison of chlorophyll and phaeopigment determinations by spectrophotometric, fluorometric, spectrofluorometric and HPLC methods. *Mar. Microb. Food Webs*, 4: 217–238.

Neveux, J., Panouse, M. (1987). Spectrofluorometric determination of chlorophylls and pheophytins. *Arch. Hydrobiol.* 109: 567–581.

Parsons, T. R., Strickland, J. D. H. (1963). Discussion of spectrophotometric determination of marine-plant pigments, with revised equations for ascertaining chlorophylls and carotenoids. *J. Mar. Res.* 21: 155–163.

Pearce, G. E. S., Keely, B. J., Harradine, P. J., Eckhardt, C. B., Maxwell, J. R. (1993). Characterisation of naturally occuring steryl esters derived from chlorophyll *a*. *Tetrah. Lett.* 34: 2989–2992.

Porra, R. J., Thompson, W. A., Kriedermann, P. E. (1989). Determination of accurate extinction coefficients and simultaneous equations for assaying chlorophyll *a* and *b* extracted with four different solvents: verification of the concentration of chlorophyll standards by atomic absorption spectroscopy. *Biochim. Biophys. Acta*, 975: 384–394.

Repeta, D. J., Simpson, D. J. (1991). The distribution and recycling of chlorophyll, bacteriochlorophyll and carotenoids in the Black Sea. *Deep-Sea Res.* 39: 969–984.

Repeta, D. J., Simpson, D. J., Jorgensen, B. B., Jannasch, H. W. (1989). Evidence for anoxygenic photosynthesis from the distribution of bacteriochlorophylls in the Black Sea. *Nature*, 342: 69–72.

Richards, F. A., Thompson, T. G. (1952). The estimation and characterisation of plankton populations by pigment analysis. II. A spectrophotometric method for the estimation of plankton pigments. *J. Mar. Res.* 11: 156–172.

Sartory, D. P. (1985). The determination of algal chlorophyllous pigments by high performance liquid chromatography and spectrophotometry. *Water Res.* 19: 605–610.

Schanz, F., Rai, H. (1988). Extract preparation and comparison of fluorometric, chromatographic (HPLC) and spectrophotometric determinations of chlorophyll *a*. *Arch. Hydrobiol.* 112: 533–539.

SCOR-UNESCO (1966). *Determination of Photosynthetic Pigments in Sea water.* UNESCO Monographs on Oceanographic Methodology. I, Paris, 69 pp.

Trees, C. C., Kennicutt, M. C., Brooks, J. M. (1985). Errors associated with the standard fluorometric determination of chlorophylls and phaeopigments. *Mar. Chem.*, 17: 1–12.

Webb, D. J., Burnison, B. K., Trimbee, A. M., Prepas, E. E. (1992). Comparisons of chlorophyll *a* extractions with ethanol and dimethylsulfoxide/acetone, and a concern about spectrophotometric phaeopigment correction. *Can. J. Fish. Aquat. Sci.* 49: 2331–2336.

Welschmeyer, N. A. (1994). Fluorometric analysis of chlorophyll *a* in the presence of chlorophyll *b* and pheopigments. *Limnol. Oceanogr.* 39: 1985–1992.

Whitney, D. E., Darley, W. M. (1979). A method for the determination of chlorophyll *a* in samples containing degradation products. *Limnol. Oceanogr.* 24: 183–186.

Wright S. W., Shearer, J. D. (1984). Rapid extraction and high-performance liquid chromatography of chlorophylls and carotenoids. *J. Chromatogr.* 294: 281–295.

Wright, S. W., Jeffrey, S. W., Mantoura, R. F. C., Llewellyn, C. A., Bjørnland, T., Repeta, D., Welschmeyer, N. (1991). Improved HPLC method for the analysis of chlorophylls and carotenoids from marine phytoplankton. *Mar. Ecol. Prog. Ser.* 77: 183–196.

Part III

Guidelines for HPLC systems, sampling and selecting methods for pigments

15

Guidelines for selecting and setting up an HPLC system and laboratory

S. W. Wright and R. F. C. Mantoura

15.1 INTRODUCTION

Chapters 11 and 12 described new isocratic and gradient methods, respectively, for HPLC analysis of pigments. This chapter describes the HPLC hardware and software required to implement those methods, with particular emphasis on optimizing resolution and reproducibility. These are key aspects of the system's performance, since pigments from field samples often produce very complex chromatograms that test the limits of HPLCs and their software. Peaks must be adequately resolved and retention times must be kept as reproducible as possible for reliable peak identification, particularly with large batches of samples.

For more general information on HPLC hardware and operation, Snyder and Kirkland (1979), Smith (1988), Meyer (1988), Szepesi (1992) or Scott (1994) are recommended.

15.2 HPLC HARDWARE

The HPLC system consists of a solvent delivery system (one or more high-pressure pumps), a sample injector, the HPLC column, one or more detectors, and a computer/integrator (Fig. 15.1A). An autosampler and a fraction collector are optional, although the former is strongly recommended for automation of the system. A column-bypass valve diverts solvent from the column while priming the pump or flushing solvents.

15.2.1 *Pumps and solvent delivery*

Modern HPLC pumps accurately deliver a pulseless supply of one or more solvents at flow rates of 0.1–10 ml min^{-1} and pressures of 3000–7000 psi. Most modern high-quality pumps are suitable for pigment chromatography. In a test of 4 major commercial brands, no significant difference was found in reproducibility of retention times (Wright, unpublished). Other pump features are ease and flexibility of programming, ability to self-prime, programmable shutdown procedures, and automatic shutdown in the event of solvent depletion.

HPLC pumps normally have (virtually incompressible) sapphire pistons operating in a stainless steel block with ruby or stainless steel valves. Two pistons are often used with alternating strokes to minimize pressure pulsing or an 'accumulator' piston provides solvent during the intake stroke of the main piston. Some modern pumps have microprocessor-controlled feedback from the pressure sensor and adjust the piston speed continuously to produce extremely stable flows.

A gradient in solvent strength is produced either by mixing the output from two

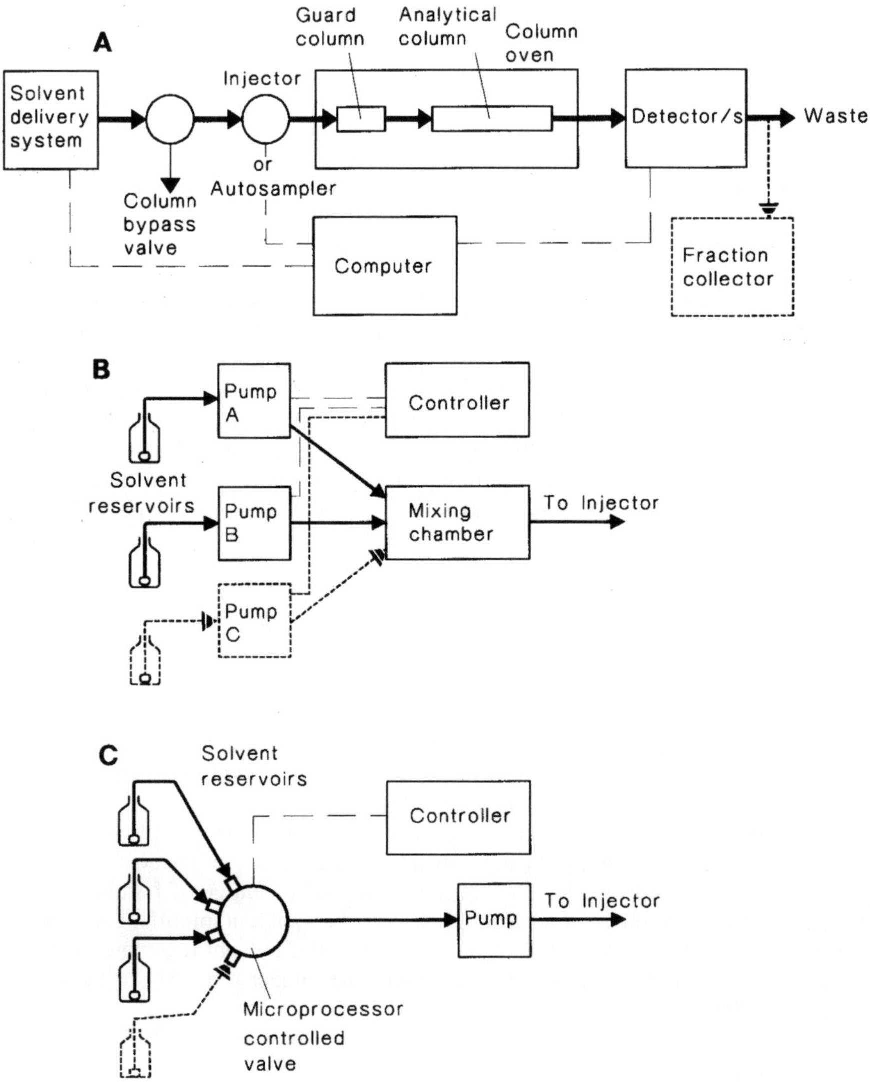

Figure 15.1
A, Diagram of an HPLC system, showing the major components. Solid lines with arrow heads denote solvent flow paths, dashed lines indicate electrical connections, dotted lines indicate optional components. B and C show two types of solvent delivery systems: B, High-pressure mixing, with control of two or more pumps by an external controller; C, Low-pressure mixing, with control of solvent inlet valves by a controller, which is normally incorporated into the pump.

or more externally controlled pumps ('high-pressure mixing', Fig. 15.1B), or by using a single pump that has 3 or 4 high-speed valves on solvent inlet lines. The valves open sequentially during the intake stroke of the piston to produce any desired solvent mix to within about 0.5% ('low pressure mixing', Fig. 15.1C). Most modern pumps use low pressure mixing, as it is cheaper and more compact (requiring only one pump), more reliable (fewer valves and seals), and generally provides a more stable baseline. With high-pressure mixing, slight variations in the output of the two pumps can be compounded at points in the gradient where the pumps come into phase, which produces baseline irregularities. Low-pressure mixing pumps use a short mixing column to ensure uniform composition. Manufacturers claim that the accuracy and precision of low-pressure mixing is such that, even for isocratic work, it is more accurate and reproducible to use the pump to mix solvents than to pre-mix them manually, although not with extreme mixes (e.g., 99:1).

Particulate matter in unfiltered solvents or air locks can disable pump valves and particles can wear out seals. System reliability is greatly improved by always filtering and degassing solvents. Buffered solutions must be flushed from the system with a compatible solvent before shutting down, as they promote corrosion and can precipitate. Pump spares must always be stocked, especially during shipboard work.

15.2.2 *Solvents*

Solvents should be HPLC grade. Water should be double-distilled or produced by a commercial purifier such as Millipore's Milli-Q system, and should be used before it becomes contaminated by bacteria.

Solvents can be mixed more reproducibly by weight than volume (see Chapter 17). They should always be filtered through a solvent-resistant <1 µm filter before being transferred to the solvent reservoirs. A 2 µm filter (stainless steel or plastic frit) should be fitted on each solvent intake line.

Solvents must be degassed continuously with helium (sparging) to prevent air bubbles forming in the pump during solvent mixing or in the detector where the pressure drops to atmospheric. The latter is a notorious source of baseline noise. After an initial 5 min purge, the helium flow must be kept to a minimum so that the solvent does not evaporate and change its composition. Solvent reservoirs that can be sealed after a short sparging and then maintained under a blanket of helium at 1–2 psig are best.

15.2.3 *The injector*

Most modern injectors have a 6-port rotary valve (e.g., Rheodyne 7125) so samples can be injected manually or automatically without stopping the solvent flow (Fig. 15.2 A, B). In the 'loading' position, the solvent from the pump is conducted directly to the column, and the sample loop is unpressurised, although full of eluant. Samples are injected into the loop, displacing the eluant through the waste port. When the valve is rotated to the 'inject' position, solvent from the pump flows though the sample loop, carrying the sample on to the column.

Sample loops are interchangeable. They vary in volume from 5 µl to 5 ml, but for best peak shape with a 4.6 mm I.D. column, injection volumes of < 20 µl are preferred (Smith 1988). However dilute pigment extracts typical of field samples require injections of 100–150 µl.

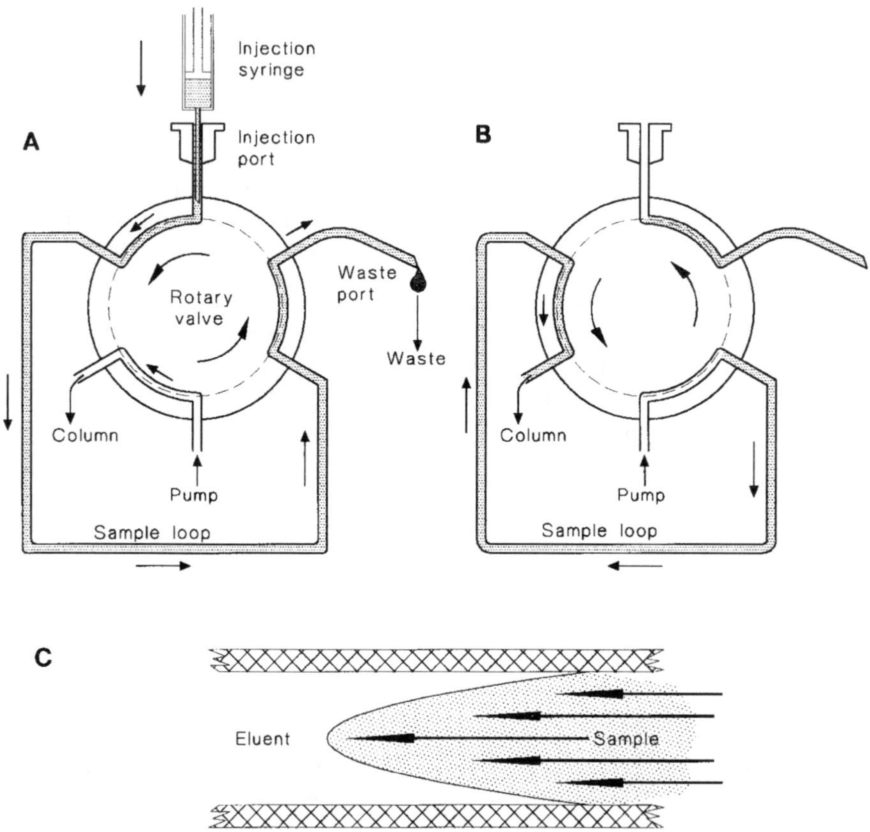

Figure 15.2
Operation of a six-port injection valve showing three internal solvent paths in the rotary valve that connect different adjacent ports, depending on their position. A, 'Loading' position, showing injection of sample (dotted) into the sample loop, while eluent from the pump flows directly to the column; B, 'Inject' position, after rotation of the valve, eluent from the pump flows through the sample loop, carrying the sample onto the column; C, Flow of sample solution within the sample loop.

Loops can be used in two ways: partial or complete filling. With partial filling, the injection volume is controlled with the sample syringe. It is more versatile, uses much less sample, but is less reproducible than complete filling, where the injection volume is determined by the loop volume. Both types of injection are subject to wall effects in the loop. When a sample is injected into the loop, drag from the walls of the tube retards the displacement of the eluant, so that the sample travels preferentially down the centre (Fig. 15.2 C). In complete filling mode, 3 to 5 volumes of sample must be flushed through the loop to ensure that the eluant along the walls is displaced and the loop is completely full of sample. Thus, for a 100 µl loop, 500 µl of sample may be required, 400 µl of which goes to waste. For partial filling, only

50–80% of the loop volume can be injected before sample emerges from the other end. The proportion of the loop volume useable for partial-filling injections depends on the internal diameter of the loop tubing and the injection speed and should be calibrated for each instrument.

A further consideration for partial loop injections is that a six-port valve may have a small dead volume of about 5μl. This can be calibrated by injecting replicates of 5, 10, 15, 20 μl of a standard solution and plotting the resultant peak areas *vs.* injection volume (Fig. 15.3). The value for the dead volume is found where the line of best fit crosses the volume axis. This can also serve as a basis for calibration with an external standard (Chapter 16).

Injection can be inaccurate if the injector waste port outlet is connected directly by a tube to a waste reservoir, since the tube can set up a siphon to draw sample from the loop (or waste back into it). The waste should be allowed to drip into an open tube or funnel connected to a glass waste bottle. Also, the syringe must be left inserted until after the injector valve is rotated to prevent leaks through the injection port seal.

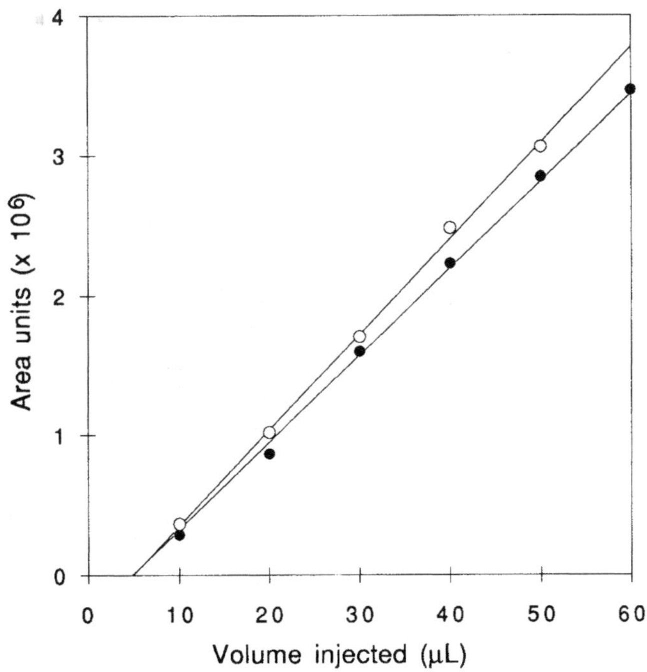

Figure 15.3
Determination of the dead volume of an injector. Areas of peaks derived from injections of various volumes of solutions of either 19'-hexanoyloxyfucoxanthin (○) or chlorophyll *a* (●). Equations to the lines of best fit were determined by least-squares regression and used to calculate the dead volume (4.8 μl, Hex-fuco data; 4.9 μl, Chl *a* data) where the lines intersect the volume axis.

15.2.4 *Automatic injection*

Automatic injection by an autosampler is highly desirable for routine analysis of pigment samples. Autosamplers normally use a standard 6-port injector with an automatic syringe and a robotic arm sampling from a rack. An autosampler allows unattended, accurate, reproducible injections yielding precise retention times from large batches of samples. Not all autosamplers can effectively add buffer to samples and mix (recommended procedure, Chapters 11, 12); this must be checked before purchase. Autosamplers without this capability should be avoided for pigment HPLC, since losses of hydrophobic pigments occur within 1 hour if the extracts are pre-buffered (see Chapter 11). The autosampler should have a refrigerated sample rack so that pigment samples can be kept cold (preferably -10°C) while awaiting injection.

Tubing in manual or automatic injectors may clog up with particles from poorly filtered extracts or from erosion of septa by the autosampler needle. Monitoring for blockages is important, particularly with autosamplers, to prevent compromising a large batch of samples.

15.2.5 *Columns*

Types of columns

Virtually all recent methods for analysing pigments use reversed-phase octadecyl silica (C-18) columns (see Chapter 11). Normal-phase silica columns have been used for pigment analyses and to separate chlorophyll *a* from divinylchlorophyll *a* (Gieskes and Kraay 1983). Similar separations of DVChl *a* from Chl *a* were achieved with multi-gradient C-8 columns (Goericke and Repeta, 1993). They are useful for carotenoid analysis (Chapter 8), but are now seldom used for field samples because of the reactivity of the stationary phase and the incompatibility of the extraction solvents with the technique (Roy, 1987), and the sensitivity to the presence of water in the sample (Smith, 1988). Conventional stainless steel columns are cheap, reliable, and can be repacked commercially. They are normally 4.6 mm internal diameter, and either 25 cm long with 5 µm particle size, or 5–10 cm long with 3 µm particle size. The latter give greater plate counts per metre (Appendix H) and faster analyses, but are more prone to clogging and injection volume overload (Appendix H). Radial compression columns (cf. Wright and Shearer, 1984) are no longer recommended since they no longer offer the differential resolution over the improved packings now available in stainless steel and are more expensive. Axial compression columns, where adjustable end-fittings compress the column packing, are useful in that headspaces caused by settling can be eliminated, yet the column can still be repacked.

Column packings

An octadecyl silica stationary phase consists of spherical or irregular particles of silica to which octadecyl (C-18) chains are covalently bound (Fig. 15.4). Most pigment HPLC methods (including the new methods recommended in Chapters 11 and 12) use monomeric coatings of C-18. These are produced by reacting the silanol (Si-OH) groups on the silica surface (Fig. 15.4A) with octadecylmonochlorosilane. A single layer of C-18 is bound to the silica surface (Fig. 15.4B). Steric hindrance prevents total coverage of the silica surface with C-18 groups. The free silanol groups that remain are usually deactivated during manufacture with the sterically smaller trimethyl silane, a process called 'end-capping' (Fig. 15.4 C). Recently, polymeric

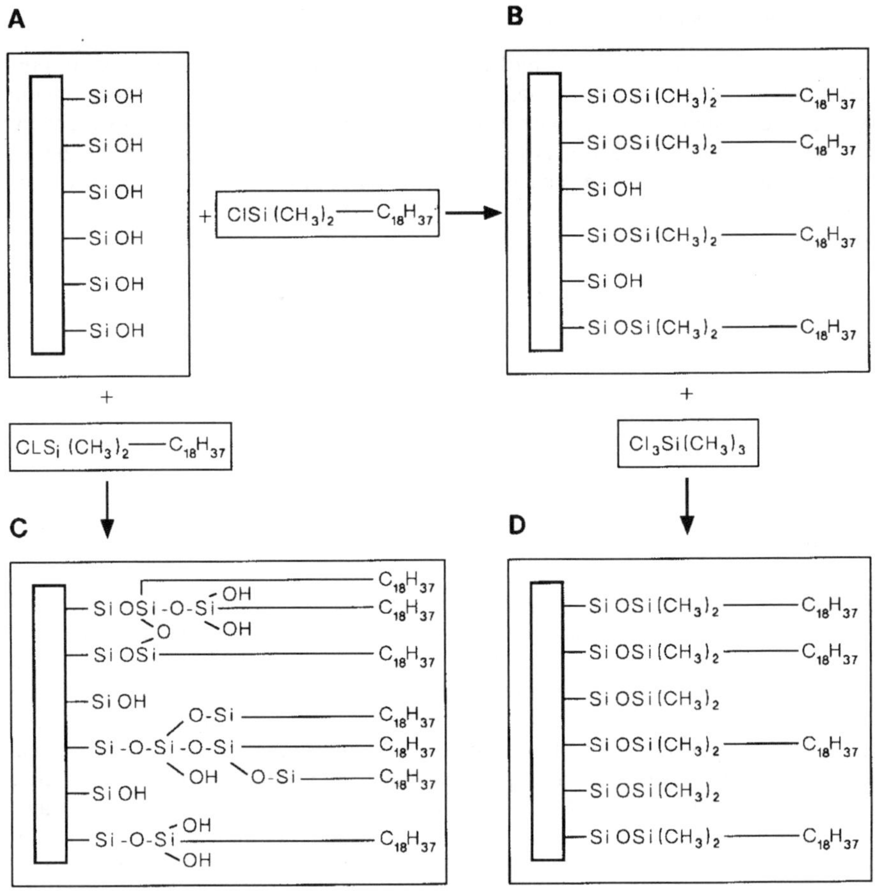

Figure 15.4
Production of octadecyl silica. A, untreated silica surface covered with free silanol groups; reacted with octadecylmonochlorosilane ($ClSi(CH_3)_2$——$C_{18}H_{37}$) to produce, B, a monomeric coating. Remaining free silanols on the monomeric coating may be reacted with trimethylchlorosilane ($ClSi(CH_3)_3$) to produce, D, an end-capped monomeric coating. Alternatively, the silica may be reacted with octadecyltrichlorosilane (Cl_3Si——$C_{18}H_{37}$) in the presence of water to produce, C, a polymeric coating. (Adapted from Sander and Wise, 1990).

coatings, which have multiple levels of C-18 bound to the silica (Fig. 15.4 D), have been used for pigment analysis (van Heukalem *et al.* 1992, Garrido and Zapata, 1993*a*, *b*). They have much greater shape selectivity than monomeric coatings (Sander and Wise, 1990) but fewer theoretical plates, due to the thicker coating (Appendix H). Polymeric columns can resolve chlorophylls c_1, c_2, c_3, and MgDVP (Garrido and Zapata, 1993*a*, *b*), which has not been achieved with a monomeric C-18 coating. The selectivity of polymeric coatings is much more sensitive to temperature than monomeric types, and hence the column must be thermostatted.

Column packings from different companies differ in their carbon load (amount of C-18 bound) and the type and extent of 'end-capping', resulting in different chromatographic properties. There may also be batch-to-batch differences between columns from a single manufacturer. An example of column differences can be seen in the resolution of chlorophyll *c* pigments: when the SCOR solvent system is used with a Spherisorb ODS2 column, chlorophylls c_1 and c_2 co-chromatograph but are resolved from chlorophyll c_3, but with a Phenomenex Selectosil C-18 column, chlorophyll c_1 co-chromatographs with chlorophyll c_3 but is resolved from chlorophyll c_2. Furthermore, as a column ages, the properties change as octadecyl residues are lost and contaminants bind to the packing, with a resultant change in retention times and resolution, and possibly even elution order.

A standard test mixture is not yet formulated for characterizing all properties of reversed-phase HPLC columns as does the 'Grob mixture' for capillary GC. However Chretien *et al.* (1986) ranked 14 columns for 'retention power' and similarity, but did not test columns for pigment selectivity. Epler *et al.* (1992) compared the properties of 60 commercially available columns for separation of 7 carotenoids, while Garrido and Zapata (1993*a*) compared 3 columns for separation of chlorophylls c_1, c_2, and c_3. Both studies found that polymeric columns gave better selectivity than monomeric columns.

Column maintenance

Columns require gentle treatment. Mechanical or chemical shocks can disturb the packing. The composition of eluting solvents should always be changed gradually and the column must never be allowed to dry out. A guard column can be used to protect the analytical column from clogging and contamination but will introduce some peak-spreading. A guard column is strongly recommended when sample loading is heavy e.g., from cultures. Some laboratories do not use a guard column but regularly remove the inlet fittings on the analytical column, scrape out the first few millimetres of packing, and replace it with new packing (as a solvent-wetted paste). Axial compression columns would aid this process by compressing the repacked region.

A semi-empirical but pragmatic way to check the performance of columns is to daily log the column back-pressure and the chromatographic resolution (*R,* see Appendix H) for pairs of closely eluting pigments—e.g., chlorophyllide *a* and chlorophyll c_1+c_2; fucoxanthin and 19'-hexanoyloxyfucoxanthin; lutein and zeaxanthin; and ß,ß- and ß,ε–carotene.

15.2.6 *Column oven*

A column oven is normally used to keep the temperature of the column constant in order to stabilize retention times, rather than to heat it. Temperature strongly affects the elution of compounds—a diurnal change in ambient temperature of 5°C (typical of most laboratories, and often exceeded at sea) can produce a change of up to 10% in retention times. It is strongly recommended that the HPLC column should be thermostatted to at least 5–8°C above ambient temperature (30°C is common) to stabilize retention times and facilitate subsequent peak identification.

Although temperature affects retention times, it does not strongly affect the selectivity of monomeric C-18 columns (Appendix H). On the other hand, polymeric coatings—such as used in the pigment methods of van Heukelem *et al.* (1992)

and Garrido and Zapata (1993*a*, *b*)—do respond to temperature and the resolution of particular peaks can be enhanced by operating the column at a different temperature.

15.2.7 *Detectors*

General considerations

Four types of detectors are commonly used in HPLC analysis of pigments:

- fixed wavelength absorbance detector
- variable wavelength absorbance detector
- full spectrum absorbance detector (photodiode array or rapid scanning monochromator)
- fluorescence detector.

For all of them, the important specifications are the wavelength range available (ideally 350 to 800 nm), the bandwidth (the spread of wavelengths around a nominal wavelength, usually 2–10 nm), the sensitivity, linear range, and reproducibility. Other important criteria are sensitivity to shipboard vibrations and roll as well as to variations in refractive index of the eluent (some older detectors produce large baseline irregularities when monitoring gradients). Detectors with a digital output (RS-232 or IEEE) of the signal to the computer/integrator yield less noisy chromatograms (see Section 15.3) than those with analog output.

The sensitivity of the detector can be increased by increasing the bandwidth (*i.e.* extending the range of wavelengths above and below the nominal wavelength) but this may be at the expense of linearity of response. The Beer-Lambert law relating absorbance and concentration applies only to monochromatic light, and does not hold for a range of wavelengths if the absorbing solute does not absorb equally for each. Unfortunately, this applies to chlorophyll *a* in the region most commonly used for detection (430 to 440 nm) since the absorbance at 440 nm is only half that at 430 nm.

All detectors have a limited linear range within which detector output is directly related to concentration of the peak. At higher concentrations, the response is non-linear and should be avoided in quantitative HPLC. Note that fluorescence detectors inherently depart from linearity at a lower concentration than absorbance detectors, due to self-absorption of the emitted photons by other molecules of the sample peak.

Fixed-wavelength detectors

Fixed-wavelength absorbance detectors are the simplest, cheapest and most robust of the detectors commonly used (Fig. 15.5A). They use narrow-bandpass filters to select emission lines from mercury, zinc or cadmium discharge lamps to produce a light beam that is essentially monochromatic with a closely defined wavelength. Mercury lamps produce discrete emission lines at 435.8 nm and 404.7 nm, which are ideal for chlorophylls + carotenoids and pheopigments respectively and can be used simultaneously on a dual-cell detector (e.g., Wright and Shearer 1984).

Variable-wavelength detectors

Variable-wavelength absorbance detectors use a continuous spectrum lamp (deuterium, xenon arc or tungsten) with wavelength dispersal from a diffraction grating monochromator to produce light of a narrow bandwidth (2–10 nm, Fig. 15.5B). They

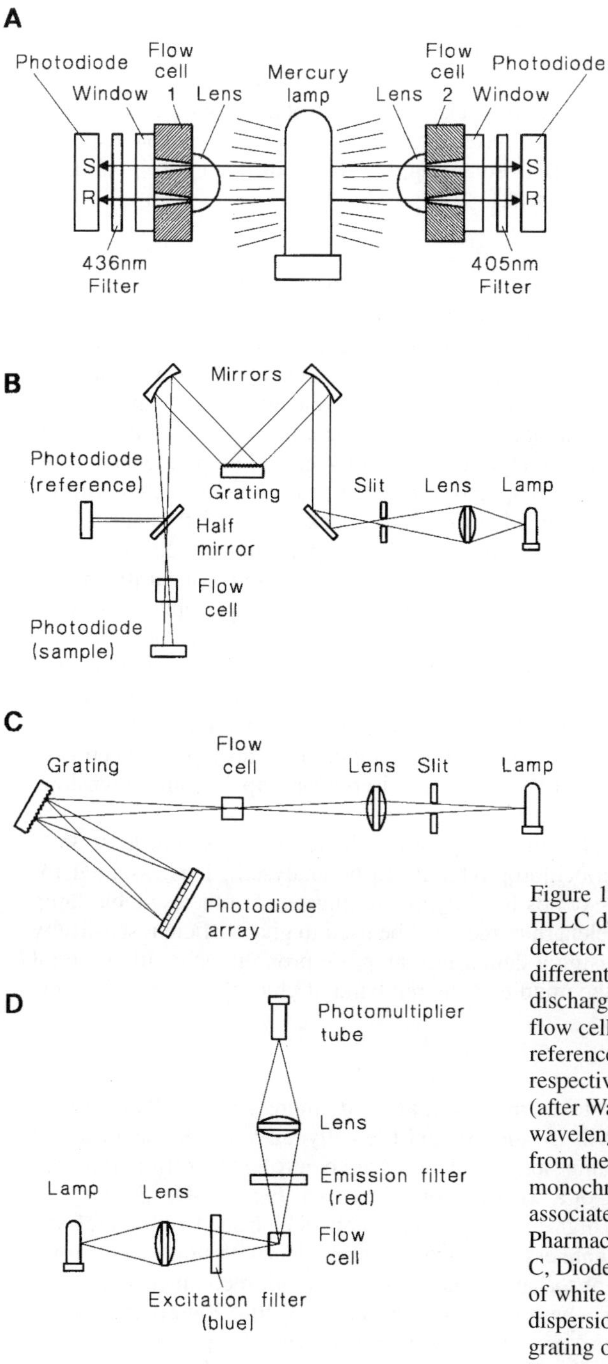

Figure 15.5
HPLC detectors. A, Fixed-wavelength detector showing dual flow cells using different wavelengths from a mercury discharge lamp, selected by filters. Each flow cell incorporates sample and reference lightpaths, labelled S and R respectively, at the photodiode detector (after Waters Associates); B, Variable-wavelength detector showing light path from the lamp, through the monochromator (diffraction grating and associated optics) and the flow cell (after Pharmacia Biotech AB, Uppsala, Sweden); C, Diode-array detector showing passage of white light through the flow cell and dispersion of the spectrum by a diffraction grating onto a photodiode array (after Hewlett-Packard); D, Fluorescence detector showing wavelength selection by filters. For chlorophyll *a*, the excitation filter is 400-450 nm band-pass (blue) and the emission filter is >600 nm long-pass (red).

are more versatile but can be 20 times less sensitive than fixed-wavelength detectors (Meyer 1988) due to lower light intensity from the monochromator. Many can be programmed to change wavelength during a run, but this is not usually practical for pigments, due to the close proximity of peaks with different absorption wavelengths in the chromatogram. Such detectors have been used to collect spectra on peaks by stopping the eluant flow as a peak is emerging, but this is difficult and invalidates the retention times of subsequent peaks.

Full-spectrum detectors

Full-spectrum detectors are of two types. The most common are the photodiode array detectors in which polychromatic light is directed through the sample flow cell (which absorbs light at some wavelengths). The absorbance spectrum is then separated down-beam from the flow cell on a diffraction grating and projected onto an array of up to 400 photodiodes, each of which monitors a discrete wavelength with 1–5 nm resolution (Fig. 15.5C). Alternatively, a rapid scanning monochromator, which scans 2 full spectra per second, can be used. Both types are extremely useful for pigment chromatography in that they allow the full spectrum of each peak to be collected without stopping the flow, greatly facilitating identification of peaks, although they are usually 5 times less sensitive than fixed wavelength detectors. Approximately 20 spectra can be collected across a peak, allowing checks of spectral identity and peak purity on both upslope and downslope of the peak (Fig. 15.6). This would be impossible by conventional fraction collection and spectrophotometry. A full-spectrum detector is strongly recommended for detailed analysis of complex pigment mixtures. Three-dimensional projections or contour maps of full-spectrum data, often produced by sales staff, are in practice very difficult to interpret and take considerable computer and plotter time to produce. A simple chromatogram coupled with an automatic display of spectra for each peak and a software 'peak purity check' (which compares upslope and downslope spectra) is much more useful.

The preferred wavelength range for photosynthetic pigments is at least 350–700 nm (or 800 nm if bacteriochlorophylls are to be analysed). It is essential for chlorophylls that the range extends to 700 nm—a tungsten or other suitable lamp with high output in this wavelength range must be used to give sufficient sensitivity in this region. Instruments using a deuterium lamp are prone to annoying spectral irregularities at 656 nm (in the middle of the red band of chlorophylls *a* and *b*) due to an emission line.

Fluorescence detectors

Fluorescence detectors are 1 to 2 orders of magnitude more sensitive than absorbance detectors for chlorophylls (Mantoura and Llewellyn, 1983) but cannot detect carotenoids. Excitation and emission wavelengths are controlled by diffraction grating monochromators, filters, or a combination of the two (Fig. 15.4D). For optimal detection of chlorophylls, a high-intensity lamp (such as a strobed xenon arc) and red-sensitive photomultiplier are essential. Older instruments with fluorescent lamps are often less sensitive than absorbance detectors and are not recommended.

Filter fluorometers with a band-pass excitation filter (340–520 nm) and a long-pass emission filter (>600 nm) are generally the most useful for pigments, as they can detect all chlorophylls and pheopigments in a sample. A dual-monochromator instrument is more sophisticated and allows the excitation and emission wavelengths to be set for particular compounds, but there is no single combination of wavelengths

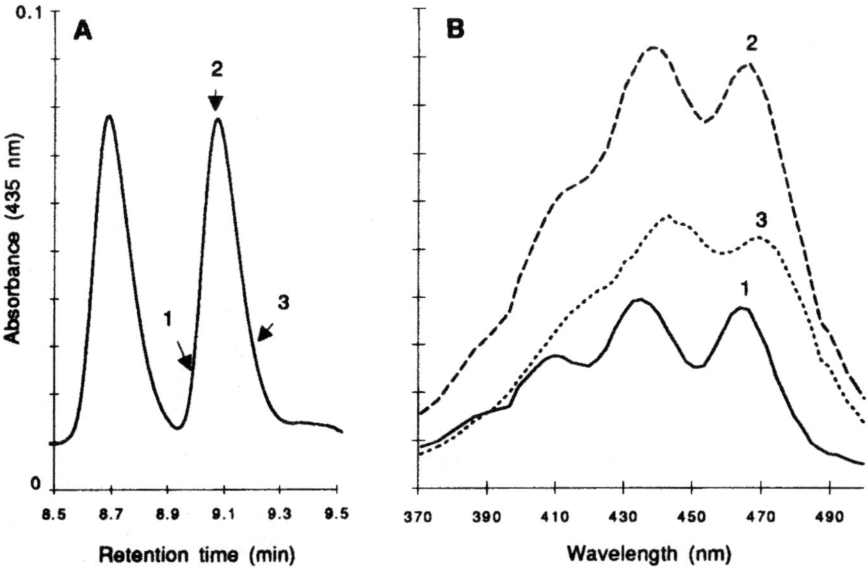

Figure 15.6
A. Segment of an HPLC chromatogram showing the elution of fucoxanthin and neoxanthin at 8.7 and 9.1 min., respectively. Positions 1, 2 and 3 on the neoxanthin peak denote the upslope, maximum and downslope events at which UV-visible spectra were collected using the diode-array detector.
B. UV-visible spectra of neoxanthin are evident in the upslope (1) and maximum (2) portion of the chromatographic peak. The downslope (3) portion shows the spectrum affected by co-elution with traces of 19'-hexanoyloxyfucoxanthin.

that will detect all chlorophylls and pheopigments (see Chapter 11). A dual-monochromator instrument may be used in 'zero-order' mode (reflecting white light) with wavelength discrimination provided by filters placed across the appropriate windows of the fluorometer. This arrangement enables all chlorophylls and pheopigments to be detected and provides 12-fold greater sensitivity for chlorophyll *a* than in normal mode (Wright, unpublished).

Fluorescence detectors are often used in sequence with an absorbance detector, as the data produced are complementary (for measuring chloropigments by fluorescence and all pigments by absorbance). The fluorescence detector should be positioned after the absorbance detector, since the larger fluorescence flow cell (usually 30 μl) causes more peak spreading than the absorbance cell (usually 8 μl) and the fluorescence chromatogram, having fewer peaks, is less sensitive to the small amount of peak spreading introduced by the absorbance detector.

15.2.8 *Tubing and fittings*

HPLC tubing has traditionally been made from 1/16 inch (outside diameter) stainless steel, connected by stainless steel nuts, ferrules and 'zero-dead-volume' connectors. Unfortunately, manufacturers (Waters, Valco, Rheodyne etc.) have produced end-fittings that are not interchangeable.

It is crucial that dead volumes are avoided when tubes are connected to equipment or joined to each other. Tubing ends must be cut cleanly and squarely. Stainless steel tubing is cut by scratching with a sharp-edged file, fracturing at the scratch by flexing the tube between two pairs of pliers, and squaring and polishing the ends with a fine file or grindstone before flushing the tube with solvent to remove debris. Alternatively, pre-cut and polished stainless tubing is commercially available in various lengths. The stainless steel ferrules are permanently fixed to the tubing ends when first compressed into a union. They should not then be used for a different junction because they do not adjust and may leave a dead volume.

Recently PEEK™ tubing has been introduced; it is more flexible and easier to cut than stainless steel. Interchangeable plastic fingertight fittings can be used either on steel or PEEK tubing and are much more convenient than stainless steel nuts and ferrules. Those rated to 6000 psi or more are preferable to the standard fittings as they are much less likely to slip off. Stainless steel tubing and fittings are recommended for the permanently fixed connections (before the column) because of their reliability. Plastic fittings are recommended on the column and detectors (to facilitate interchange and prevent wear of the sockets). PEEK tubing is recommended after the column (for flexibility and fitting columns of different lengths).

To reduce peak spreading, tubes should be as short as practical and with a narrow internal diameter (0.010 inch), especially between the injector, column and detectors. Wider-bore tubing (0.030 inch) can be used before the injector to reduce back-pressure, but it should be kept short to avoid introducing extra dead volume between the pump and injector, which delays the start of a gradient from reaching the column and affects the chromatography of early-eluting peaks.

15.2.9 *Fraction collector*

A fraction collector is not necessary for analytical work, but is very useful for preparation of standards by HPLC. Most collectors can be set to collect peaks when the output from the detector passes a threshold, with the most sophisticated being able to detect and fractionate shoulders on peaks as well. Some autosamplers (e.g., Gilson 232) can be configured to collect fractions as well as perform injections.

15.2.10 *Signal cables*

Most detectors have an analog output. The detector signal is transmitted as a voltage (0–1 V), which is used directly by a recorder/integrator or converted to digital for use by a computer. Many modern detectors also have serial (RS-232) or the faster parallel (IEEE, GPIB) interfaces for direct digital communication with a computer. This is the preferred arrangement, as digital signal cables are relatively insensitive to electrical noise and can be run much further if necessary (6 m, IEEE; 500 m, RS-232).

Analog signal cables, on the other hand, are very sensitive to electrical noise from power cables, pumps and other components in the HPLC (especially those with electric motors), and must be positioned away from them. Simple two-wire or coaxial cables are not adequate for transmitting low-level analog signals. The cable's positive and negative conductors should be surrounded by a separate shield (which should be earthed at one end only to prevent ground loops between instruments). For connection of an analog detector to a computer, the simplest arrangement is to run the analog cable to an analog/digital board in the computer, but this arrangement is prone to electrical noise. A better method is to use an analog/digital box, located near the detector/s, which translates the signal and transmits it to the computer via digital cables.

Other signal cables are used if the HPLC is not under direct computer control. The autoinjector, pump and integrator are connected so that the autoinjector (normally the controller) checks that the pump is signalling 'ready' (has not run out of solvent etc.) before injecting the sample. After injection it signals the pump to start the gradient and the integrator to collect data, synchronizing their start times. This system is not as flexible as direct computer control, but it is often necessary if components from different manufacturers are not software-compatible.

15.2.11 *Data collection system*

In early HPLC systems, output from the detectors was displayed on a chart recorder and peaks had to be identified and integrated manually. For those researchers who still have to do this, Dyson's (1990) critical review of methods is invaluable.

In the 1970s and 80s, dedicated integrators were standard for HPLC systems, but in the 1990s a personal computer running chromatography software provides much more flexibility and power, and either the software or computer can be upgraded when necessary.

The demands on the computer have steadily increased as software has become more complex and diode-array detectors, which produce about 1 megabyte of data per sample, have been introduced. The HPLC computer needs to be as fast as one can afford (and it should be upgradable) because much of the operator's time is spent waiting for the computer. For shipboard operations, the portability and small benchspace requirement of a laptop computer is an advantage. Storage capacity is important. It is easy to fill a 200 megabyte hard disk with data from a single cruise if using a diode-array detector, and consideration must be given to archiving such data until analysis is complete. Fortunately, mass storage on DAT tape, WORM or magneto-optical drives is relatively cheap, so that archiving gigabytes of data is not unreasonable.

15.3 HPLC SOFTWARE

HPLC software is an extremely important part of the system, as it is the operator's main link to the system and is responsible for the accuracy and reliability of the results. However, it is often under-rated. As software is developing rapidly, this section gives only an overview of software capabilities and performance.

The functions of the software system are: system control, sample queue entry, data collection and presentation, peak detection, integration, identification, quantification, and report generation. It is extremely advantageous if the system is multitasking (e.g., one can process chromatograms while new data are being acquired).

15.3.1 *System control*

In integrated HPLC systems from a single manufacturer, the computer normally controls the pump, autosampler and other ancillary equipment while collecting detector data. Unfortunately, mixed-brand equipment does not normally have this very desirable capability, so we recommend using equipment from a single manufacturer. Centralised sample and method logging facilitates good laboratory practice. Flexibility is important. For example, the pump control should allow the flow rate and composition to be set directly rather than by writing and executing a pump control table, and the autosampler control should allow samples to be run out of sequence (such as periodically running a standard sample during a queue).

15.3.2 *Sample queue entry*

Software should allow automatic filling and editing of the sample queue and preferably be able to import data from a spreadsheet. This is normally not a problem if working on a Macintosh or with 'Windows™' software, but many programs require each sample in a queue to be entered manually, which can be very frustrating, particularly if they exist elsewhere in computer files. For shipboard use, sample queues should be linked to station number, latitude, longitude, time and depth information imported from the CTD computer.

15.3.3 *Data collection*

The data sampling rate determines the accuracy of the peak integration. A sampling rate of 100 data points per peak gives near-maximum accuracy (Chesler and Cram, 1971), but manufacturer's manuals suggest that 20 data points per peak will give reasonably accurate quantitation. For most systems, this means a sampling rate of 2 or more points per second. The software should display the chromatogram on the screen as it is collected, allow re-scaling, and display spectral data if available.

15.3.4 *Presentation of data*

It must be possible to display, compare and overlay chromatograms, and zoom in easily on particular segments. Printing 'publication-quality' chromatograms (and UV-visible spectra, if available) with ability to format the output is essential, and the ability to transfer such data into standard word-processing or graphics packages is very useful. The program must be able to locate maxima and minima within spectra from a diode-array detector (if used) and list absorbances to enable calculation of %III/II ratios (see Section IV). Background subtraction of spectra (so that baseline drift can be subtracted from small peaks) is essential.

15.3.5 *Detection and integration of peaks*

The detection and integration of peaks by computer integrators is discussed in detail by Dyson (1990). Briefly, peaks are detected by a sequential search of the baseline for an upward slope that is greater than random noise or baseline drift (controlled by the slope-sensitivity parameter). Having confirmed a peak upslope and marked the peak start, it then searches for a zero slope (peak top) and a downslope. It must then locate the end of the peak, where the peak tail meets the baseline. A straight baseline is interpolated between the start and end of the peak, and the area above it is computed. The software will normally check that the resultant peak fulfils criteria for allowable peak widths and minimum area set by the operator. Integration parameters are now normally set by the software when the operator indicates an area of baseline and an example of the smallest expected peak in a chromatogram.

Integration software can produce excellent accuracy and precision for an ideal chromatogram, but if peaks are severely asymmetric (see Appendix H), or incompletely resolved, or if negative peaks are present, automatic integration can make lamentable mistakes. Many field samples are at or beyond the limit of most integration algorithms, and have large numbers of incompletely resolved peaks with often a poor signal-to-noise ratio. Misplacement of the starts and ends of peaks can produce a significant error in area calculations since they affect the base, the widest part of the peak.

It is essential therefore that the analyst can view and edit the integrated chromatogram—where necessary improving accuracy by adjusting baselines and the starts and ends of peaks. The overall efficiency and ease-of-use of the program depends largely on the ease of doing this. However, editing chromatograms in this way is directly contrary to the U.S. Environmental Protection Agency's requirements for 'Good Laboratory Practice' (since the analyst can 'tamper' with results) and is not allowed at all by some software. Unfortunately, results from such software are quite unreliable for complex chromatograms. Other integration software allows peak editing but indicates in the report that this has been done.

15.3.6 *Identification of peaks*

Identification of peaks is the most difficult part of the software's task, especially with complex field samples (see Appendix I).

If the chromatogram has been monitored at a single wavelength, the program has only retention time as a criterion for identification. The retention time of a peak is compared with a table (created by the operator for that system) which lists peak identities, retention times and tolerances on retention times. Normally, software can update the table to follow the drift of retention times during the day. The more reproducible the retention times, the more successful is peak identification—hence the desirability of thermostatting the HPLC column (see section 15.2.5). Identification is aided if the software can utilize reference peaks. Field samples almost invariably contain chlorophylls c_1 and c_2, fucoxanthin, diadinoxanthin, chlorophyll *a* and β,β–carotene, which software can locate and then use to adjust the expected retention times of other pigments.

The most recent software for diode-array detectors can compare the spectrum of an unknown peak with library spectra for compounds within the retention time window, and give a goodness-of-fit estimate. This is extremely desirable, greatly improving the reliability of the results and saving hours of operator time.

15.3.7 *Quantitation of peaks*

The concentration of each pigment in the sample is calculated by the integration program from peak areas, using tabulated response factors (ng injected per area unit) determined by the operator from external calibration (Chapter 16). It is important that the program will allow external calibration response factors to be entered manually. This is again contrary to the U.S. E.P.A. 'Good Laboratory Practice', which requires a standard solution to be injected with known concentrations of each pigment, for each batch of analyses. This is not feasible for pigment studies, because of the large number of pigments and their instability.

15.3.8 *Preparation of reports*

All programs will produce a report with sample names, retention times, peak areas and peak types etc., but it is very helpful if the program writes the data to a disk file compatible with the spreadsheet or database program used in the laboratory. The report should list all operational parameters used in the analysis, such as analyst, date, column, solvents, injection volume, to ensure coherency of the analysis conditions with the results.

15.3.9 *Ergonomics*

The ergonomics of the software interface is extremely important to the operation of the HPLC. If hundreds of field samples are to be analysed, everything must be optimised to prevent frustrating repetitive actions. User-definable default settings and keyboard short-cuts greatly facilitate analysis. Operations should involve a minimum of key strokes and/or mouse movements. Integrated programs that have all functions available on one screen are preferred to those in which one must constantly switch between different subprograms.

Purchasers of a new system are strongly advised to thoroughly test the software systems on offer with real samples so as to find the strengths and limitations of each.

15.3.10 *File maintenance*

It is important to note how and where sample data are stored by the software and if there is provision for archiving data.

As soon as one has a working version of the software running in the laboratory, it is essential to backup the computer's startup files ('autoexec.bat' and 'config.sys') since these can be modified by non-chromatography programs. Also, any software-specific method and configuration files should be backed up regularly.

15.4 Laboratory layout

A laboratory for pigment HPLC should have the following attributes for best efficiency and safety.

- Sample preparation area: about 3 m length of bench space for a small working area with running water, sink, a sonicator and centrifuge.
- Fume cupboard: for handling solvents.
- HPLC area: 2–3 m of free-standing bench below an extraction hood with access to power cables.
- Computer desk: close to the HPLC for convenience and monitoring, yet situated so that the operator is not exposed to solvent fumes while processing data. A mobile desk or portable computer is an advantage.

Part of the laboratory should be capable of being darkened, so that samples can be handled without photodegradation. Adjustable lighting is an advantage.

15.5 Ancillary equipment

In addition to normal HPLC equipment, the following ancillary equipment is required for preparation and analysis of pigment samples.

- Spectrophotometer: for preparing calibration standards and identifying peaks; 350–800 nm range; spectral band width, 2 nm or less with double-beam optics; capable of using micro (0.8 ml) or semi-micro (1–2 ml) cuvettes.
- Probe-type sonicator: for extracting pigments from filtered samples: 50W power or more, with a needle probe of 4–8 mm diameter; mounted in a sound-proof cabinet.

- Refrigerated centrifuge (-10°C) with rotor for 8 × 15 ml tubes (large multi-adaptor heads are slow to accelerate and to stop).
- Filtration equipment: for filtration of algae, a range of filter holders 25 mm and 47 mm in diameter and vacuum pump and gauge; a filtration manifold with water trap for simultaneous filtration. For filtration of solvents, a 47 mm diameter all-glass filter holder and flask (without rubber stoppers).
- Freezer: a small domestic freezer (-20°C) for short-term storage of samples (not more than one week, see Chapter 10); a liquid nitrogen Dewar or low-temperature freezer (-90°C or colder, with over-temperature alarm and racks for sample boxes) for longer-term storage. Low-temperature freezers may require special refrigerants, to be held by the maintenance company.
- Freezer containers: cryotubes for sample storage; small styrofoam containers for 'snap-freezing' samples in liquid nitrogen (glass vacuum Dewars can implode with thermal shock).
- Solid-phase extraction (SPE) equipment: for preparing and manipulating pigment solutions; SPE cartridges or columns; SPE vacuum manifold—can be used for filtering extracts as well as SPE.
- Access to culture facilities (described in Chapter 6)
- Checklist: for safety and reliability of data. Following a checklist (see Table 15.1) is essential during the HPLC startup and shutdown procedures.

15.6 Operations at sea

Scientific operations at sea push the performance of analytical instrumentation and the scientist to their limits. Automation and the implementation of good laboratory practice can significantly improve the quality of pigment data obtained at sea, as will planning and simplifying all operations. Everything possible, such as mixing and filtering solvents, should be done ashore before departure.

15.6.1 *Mounting the HPLC*

The HPLC system should be mounted in portable (50 kg max.) wooden units to minimize the commissioning time when on board. The boxes should be constructed of strong marine ply, have heavy-duty recessed handles, ventilation ports, and metallic plates to protect corners during transport. The front and upper faces should form a single detachable lid to allow access to the front, side and upper parts of the HPLC, as shown in Figure 15.7. Where possible, components should be stacked vertically and secured firmly to a base mounted on vibration-dampening pads, leaving a ventilation space of 10-15 cm from the back of the box. Solvent bottles should be mounted in a high-lipped metal tray near the HPLC pump to contain any spillage. Waste eluent should be ducted with wide-bore stainless steel or teflon tubing to a waste bottle, held in a metal drum or similar container in case of overflow, below the bench. A power-stabiliser unit should be used to filter spikes and fluctuations in voltage and frequency from the ship's power supply. The computer should be protected from solvents but close enough to allow HPLC monitoring.

The HPLC should be installed away from wet operations (such as filtration of seawater), or drafts and ventilation grills (to prevent contamination from dust and minimise temperature fluctuations). The deep freezer (-18°C) for temporary storage

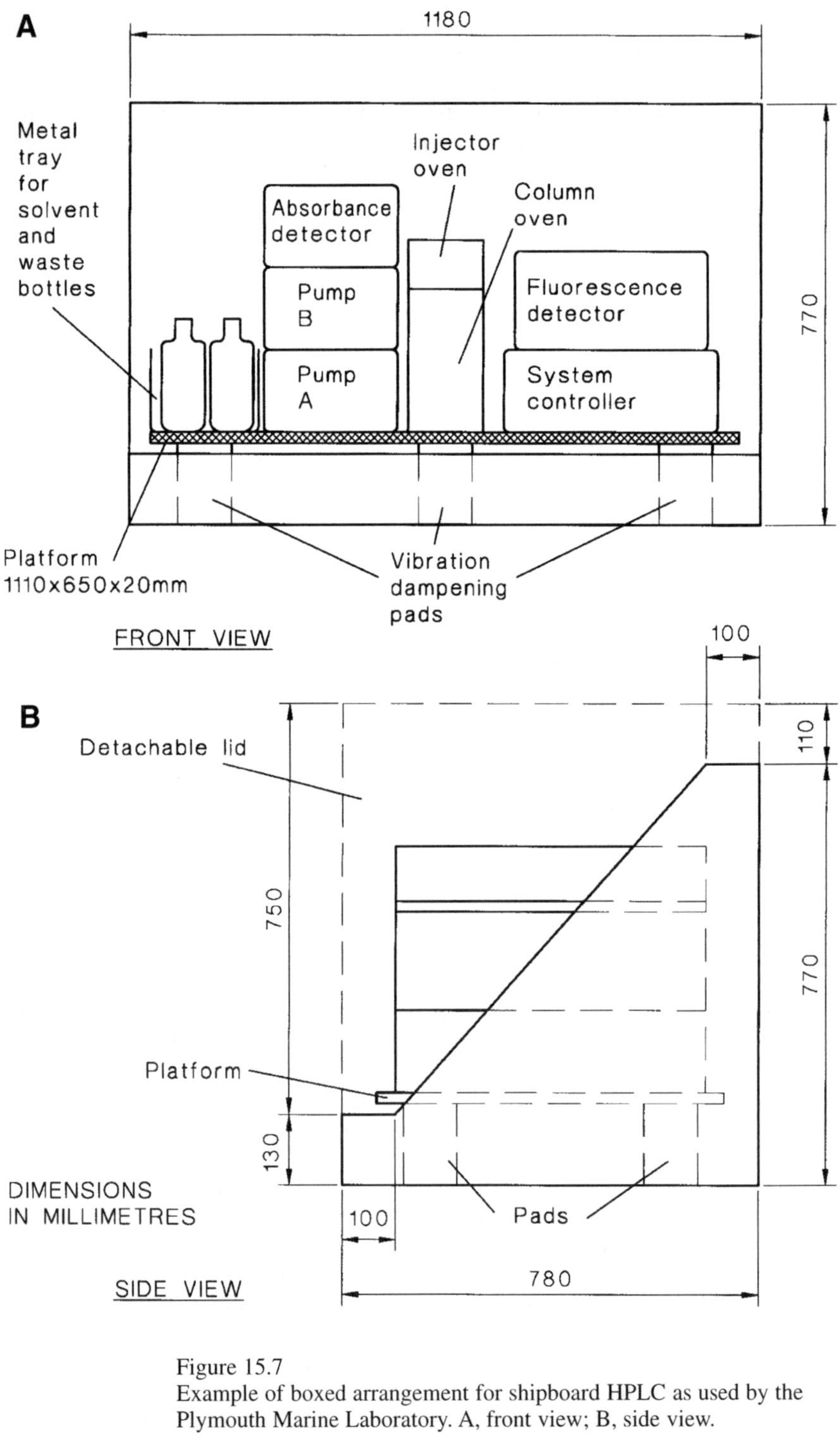

Figure 15.7
Example of boxed arrangement for shipboard HPLC as used by the Plymouth Marine Laboratory. A, front view; B, side view.

TABLE 15.1. Example of a checklist for use during HPLC startup procedure.

Component	Check
Pump	Solvent reservoirs full
	Helium sparging rate correct
	Startup programme completed
	Flow rate correct
	Analytical programme ready
	Pressure checked and logged
Autoinjector	Sample vials ready
	Mixing vials ready
	Buffer vials full
	Needle straight
	Solvent reservoir full
	Correct injector programme ready
Tubing and column	No leaks
Absorbance detector	Lamps on
	Remote control mode ready
	Correct wavelength set
Fluorescence detector	Lamp on and warmed-up to constant intensity
	Correct wavelengths set
Eluent waste bottle	Previous eluent emptied
	Eluent being received
Computer	Disk space available
	Correct detector file
	Correct method file
	Ready to receive data
AFTER START	
Autosampler	Sample drawn up—no bubbles
	Injection proceeding
	No blockage
Pump	Gradient started
Computer	Data collection started
Log book	Completed

and the ultrasonicator cabinet for extracting filtered samples should be located close to the HPLC.

15.6.2 *Filtration apparatus*

A filtration manifold system (Fig. 15.8) is strongly recommended for filtering multiple samples from a hydrographic cast. The shipboard setup should facilitate filling the sample bottles, handling filters and accurately measuring the volume filtered. The system must allow the filtration of a particular sample to be stopped (if the filter is starting to clog) without having to pour the unfiltered water from the filter (which risks disturbing the filtered material). A stainless-steel vacuum flask should be used to collect the filtrate — glass vessels should not be used due to the risk of implosion. A 20 l Millipore dispensing pressure vessel fitted with a vacuum closure (or similar) is recommended. These have a dip-tube that can be used to pump filtrate from the vessel if the filtrate tube is disconnected and pressure applied to the vacuum port (Fig. 15.8). Thus the operator need not lift the heavy vessel to empty it, which can be difficult on a rolling ship. The filtration pump must be heavy-duty, designed for continuous use, with a vacuum control and gauge and tube connection on the air outlet so that the filtrate can be pumped out as described above.

15.6.3 *Pigment standards*

Pigment standards and perishable chemicals and samples can be transported under solid CO_2 (-78°C) in an insulated box, or in a liquid nitrogen Dewar.

15.6.4 *Spare parts*

As with all operations at sea, it is essential to carry spare parts: HPLC columns, frits, unions, connectors, tubing and cutting gear; pump seals, pistons, valves; detector lamps; injection syringes, cleaning wires; fuses, voltmeter, tool box, spare filtration pump/s, sample log books, instrument operation manuals, service manuals with full circuit diagrams; pigment reference books (e.g., this UNESCO Monograph); back-up copies of all computer software files including system files (MS-DOS, Windows, or Macintosh System etc.), control and integration software, method and data files. Also very useful at sea: cordless drill/screwdriver, hot glue gun, 'gaffer' or 'ducting' tape, string, silicone sealant, sticky plastic matting to stop things sliding (sold by ship's chandlers). Manufacturers or their agents should be consulted for recommended spares for particular instruments.

Finally, there should be a fall-back strategy for storing samples in case of total failure of the HPLC. A suitable 43 l liquid nitrogen Dewar will freeze and maintain 600 samples in 2.5 ml cryotubes for over 3 months before the liquid nitrogen evaporates.

15.7 Safety precautions at sea

International Maritime Organisation (IMO) regulations should be followed for the safe carriage and use of all chemicals, solvents and gases at sea; and the research vessel's Safety Policy and Guidelines should be consulted.

Use of flammable, toxic solvents at sea presents a significant safety hazard. Stocks of solvent bottles should be stored in moulded styrofoam containers (in which they are normally supplied) and stowed in a fire-proof cabinet when not in use. Absorbent material should be on hand to contain any large spills. To reduce concentrations of solvent vapours in the air, open containers should not be used.

Half-size helium cylinders are much safer to install than full-size ones. They should be securely strapped externally to the bulkhead and fitted with a pressure regulator, toggle valve with 1/8 inch copper piping (preferably) ducted into the HPLC laboratory.

Appropriate protective gear (e.g., cryo-gloves, ear pads, face-mask) should always be worn when using liquid nitrogen, solvents or ultrasonicator. All solvent operations should be done in a fume cupboard, which should be checked to see that it complies with standards for handling flammable solvents. Manipulation of column packings must also be done in a fume cupboard, as inhalation of particles of silica < 5 μm dia. is a potential long-term hazard (Meyer, 1988).

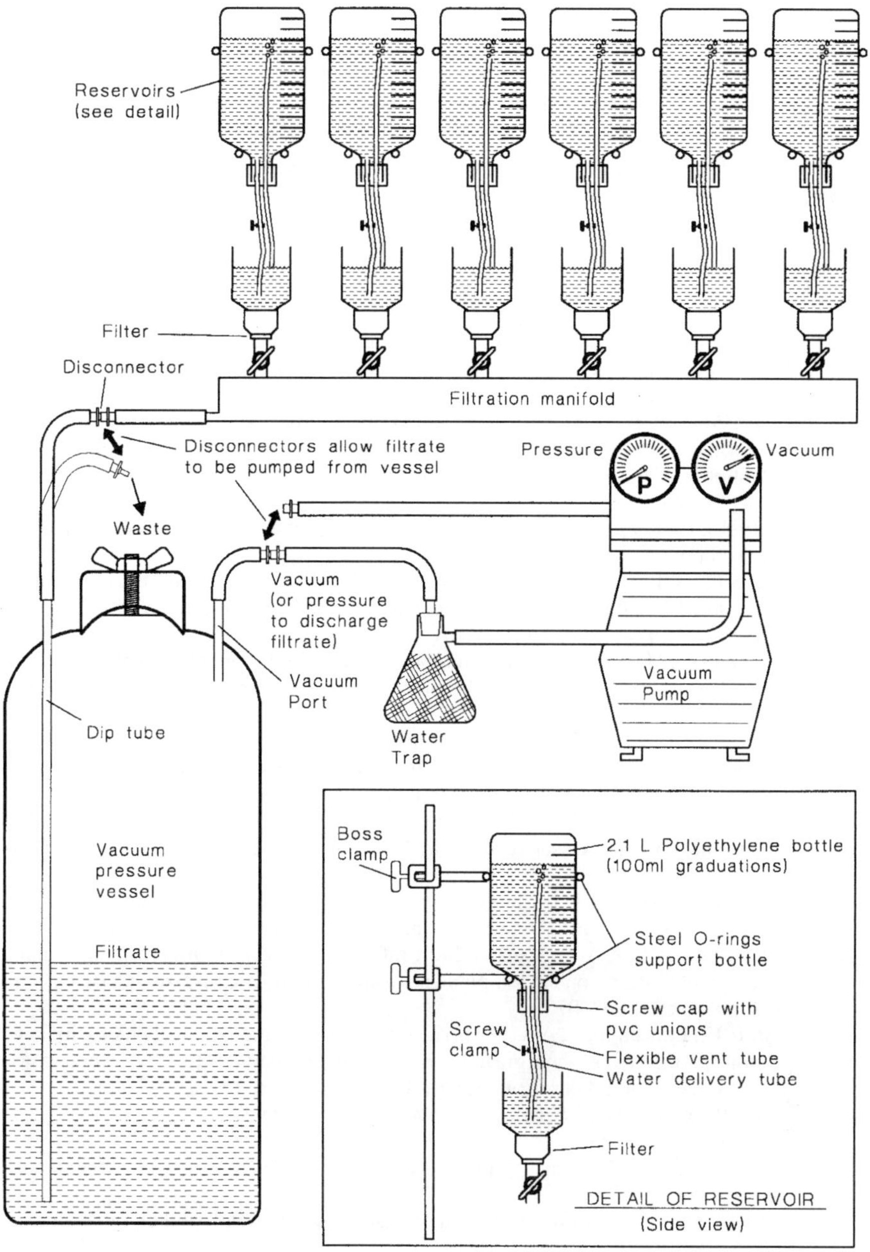

Figure 15.8
Example of apparatus for filtering six seawater samples simultaneously. Six plastic graduated bottles containing seawater are connected to a stainless steel filtration manifold equipped with 6 × 25 mm dia. filter holders. Water is drawn by vacuum through the filters and into a vacuum/pressure vessel, which can be mounted below the bench. After filtration, the filtrate can be pumped from the vessel by disconnecting the filtrate tube from the filtration manifold (dotted lines) and applying air pressure at the vacuum port, using the air outlet from the vacuum pump. Inset: Detail of the sample reservoir that delivers water into the filtration cup. Seawater flows from the sealed bottle through the water delivery tube, while air enters through the vent tube. When the water in the filtration cup reaches the level of the vent tube, no further air can enter the reservoir and flow from the reservoir stops until the level in the filtration cup falls. This arrangement can be operated unattended without the risk of flooding. A screw clamp allows the filtration to be stopped if the filter starts to clog.

Acknowledgments

We thank Mr J. Cox for preparing the diagrams and Dr R. Barlow (Plymouth Marine Laboratories) for supplying Figure 15.7.

References

Chesler, S. N., Cram, S. P. (1971). Effect of peak sensing and random noise on the precision and accuracy of statistical moment analyses from digital chromatographic data. *Anal. Chem.* 43: 1922–1933.

Chretien, J. R., Walczac, B., Morin-Allory, L., Dreux, M., Lafosse, M. J. (1986). Factor analysis and experiment design in high-performance liquid chromatography. VI. Comparison of the retention mechanism for fourteen octadecylsilica packings. *J. Chromatogr.* 371: 253–267.

Dyson, N. (1990). *Chromatographic Integration Methods*. Royal Society of Chemistry, Cambridge, 160 pp.

Epler, K. S., Sander, L. C., Ziegler, R. G., Wise, S. A., Craft, N. E. (1992). Evaluation of reversed-phase liquid chromatographic columns for recovery and selectivity of certain carotenoids. *J. Chromatogr.* 595: 89–101.

Garrido, J. L., Zapata, M. (1993*a*). High-performance liquid chromatography of chlorophylls c_3, c_1, c_2, and *a* and of carotenoids of chromophyte algae on a polymeric octadecyl silica column. *Chromatographia,* 35: 543–547.

Garrido, J. L., Zapata, M. (1993*b*). High-performance liquid chromatographic separation of polar and non-polar chlorophyll pigments from algae using a wide-pore polymeric octadecylsilica column. *J. High Res. Chromatogr.* 16: 229–233.

Gieskes, W. W. C., Kraay, G. W. (1983). Unknown chlorophyll *a* derivatives in the North Sea and the tropical Atlantic Ocean revealed by HPLC analysis. *Limnol. Oceanogr.* 28: 757–766.

Goericke, R., Repeta, D. J. (1993) Chlorophylls *a* and *b* and divinyl chlorophylls *a* and *b* in the open sub-tropical North Atlantic Ocean. *Mar. Ecol. Progr. Ser.* 101: 307–313.

van Heukelem, L., Lewitus, A. J., Kana, T. M., Craft, N. E. (1992). High-performance liquid chromatography of phytoplankton pigments using a polymeric reversed-phase C_{18} column. *J. Phycol.* 28: 867–872.

Mantoura R. F. C., Llewellyn, C. A. (1983). The rapid determination of algal chlorophyll and carotenoid pigments and their breakdown products in natural waters by reverse-phase high-performance liquid chromatography. *Anal. Chim. Acta,* 151: 297–314.

Meyer, V. R. (1988). *Practical High-performance Liquid Chromatography*. John Wiley and Sons, Chichester, 310 pp.

Roy, S. (1987). High-performance liquid chromatographic analysis of chloropigments. *J. Chromatogr.* 391: 19–34.

Sander, L. C., Wise, S. A. (1990). Evaluation of shape selectivity in liquid chromatography. *LC.GC* 8: 378–390.

Scott, R. P. W. (1994). *Liquid Chromatography for the Analyst*. Chromatographic Science series, vol. 67. Marcel Dekker, New York, 328 pp.

Smith, R. M. (1988). *Gas and Liquid Chromatography in Analytical Chemistry*. John Wiley and Sons, Chichester, 402 pp.

Snyder, L. R., Kirkland, J. J. (1979). *Introduction to Modern Liquid Chromatography*, 2nd edn., John Wiley and Sons, New York., 534 pp.

Szepesi, G. (1992). *How to Use Reverse-phase HPLC*. VCH Publishers, New York, 356 pp.

Wright, S. W., Shearer, J. D. (1984). Rapid extraction and high-performance liquid chromatography of chlorophylls and carotenoids from marine phytoplankton. *J. Chromatogr.* 294: 281–295.

16

Calibration methods for HPLC

R. F. C. Mantoura and D. J. Repeta

16.1 INTRODUCTION

Separation by HPLC of complex mixtures of pigments present in seawater permits the quantitative analysis of individual chlorophylls and carotenoids in the sample. Although the analytical accuracy and precision of a carefully-calibrated HPLC may reach 0.5%, such precision is not often required in oceanography since pigment concentrations vary by two or three orders of magnitude.

The calibration of the HPLC should be carried out only after the HPLC elution conditions have been optimized for resolution (see Chapters 11, 12 and 15) and the UV-visible and fluorescence detectors have been set for maximum sensitivity (see Chapters 11 and 15; Appendix H). The factors controlling detector sensitivity, linearity and noise are discussed in Chapter 15.

The HPLC may be calibrated by either external or internal standard methods:

- With external standards, the HPLC is first calibrated with a standard that contains the pigment(s) actually found in the sample. The chromatographic peak areas are correlated with the mass of pigment injected to yield response factors for each pigment. These response factors are then used to calculate the mass of pigments injected into the HPLC from peak areas in chromatograms of samples. The pigment concentration in sea water can then be calculated from the filtration, extraction and injection volumes. All liquid transfers must be volumetric and precisely controlled to ensure accurate calculation of pigment concentrations. A precision of ~1% can be obtained during calibration, but the overall precision for samples is typically 3% because of inaccuracies in the volume measurements as well as dilution or evaporation of the sample during handling (see Chapters 9, 10 and 17).
- With internal standards, a *relative* response factor is obtained for each pigment in relation to a calibration pigment (the internal standard, which must not occur in the samples). A known mass of the internal standard is added to each sample *before* extraction and the recoveries of sample pigments are subsequently calculated relative to this internal standard. The effects of sample handling, dilution, evaporation etc. are cancelled out, since they affect the sample pigments and internal standard equally. The overall precision is about 1%.

Both methods require a set of pure pigment standards. The preparation of these standards is discussed in the following section. The rest of the chapter describes the general principles of calibrating the HPLC with external and internal pigment standards. The merits of different internal standards, including a newly synthesized standard, are reviewed.

16.2 Preparation, purity and quantitation of pigment standards

The first steps in calibrating an HPLC are to prepare, characterize and standardize solutions of pigment standards. While there are commercial suppliers for chlorophylls *a* and *b*, β,β-carotene, and several other carotenoids (see Appendix J), most pigments are not commercially available, but may be obtained from the SCOR-recommended algal cultures (see Chapters 6 and 13). Methods of isolating and purifying chlorophylls and carotenoids are described in Chapters 7 and 8 respectively. All standard solutions must be carefully prepared and characterized before use. The equipment and the steps for doing this are summarized in Fig. 16.1 and detailed below.

16.2.1 *Hardware, reagents and instrumentation*

All volumetric flasks (10.0, 25.0 and 50.0 ml) and pipettes (1.0, 2.0, 5.0 and 10.0 ml) used for preparing standards must be calibrated to a known accuracy, be leak-proof and preferably made of amber glass. If amber glass is not available, work must be carried

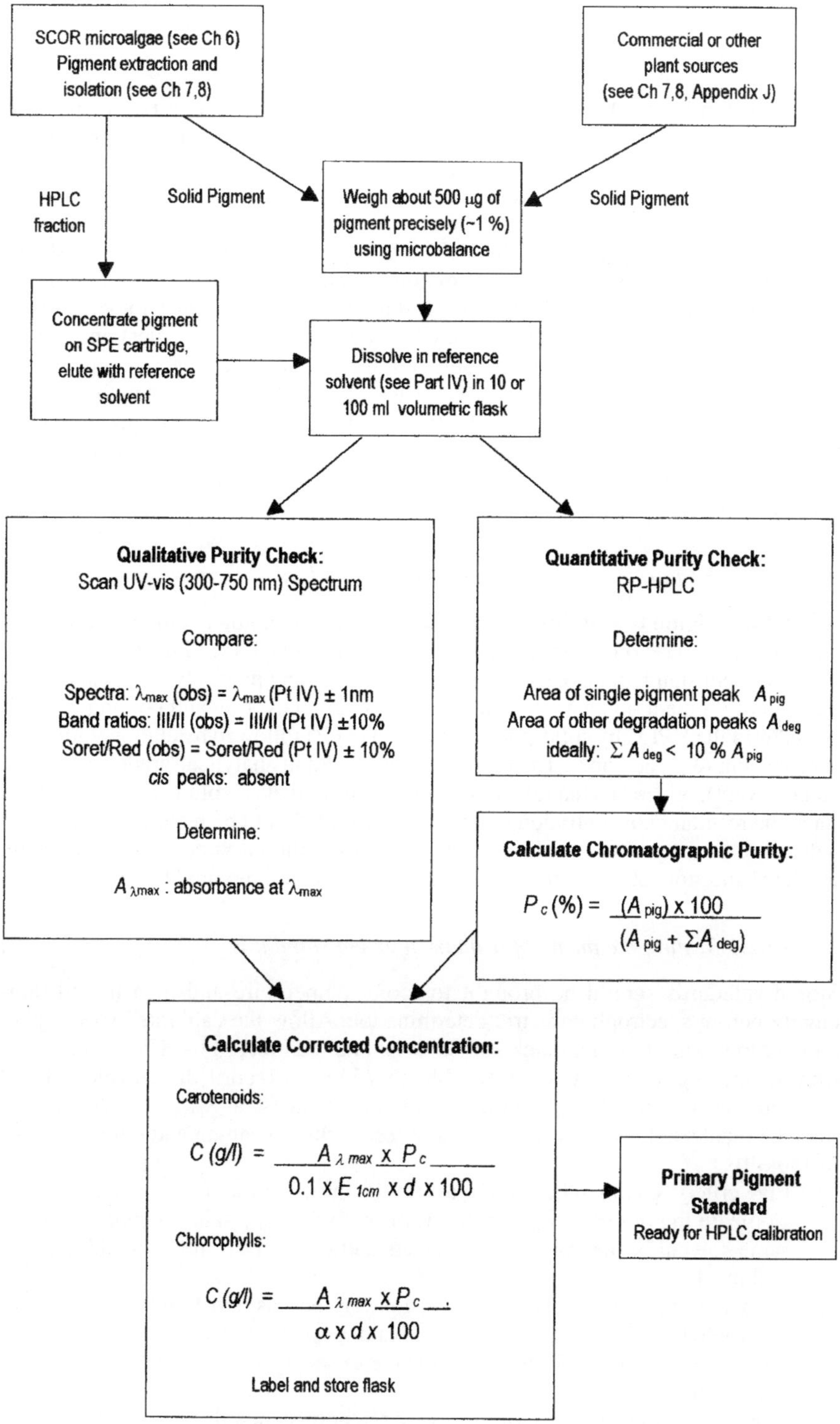

Figure 16.1
Schematic steps for the preparation, purity checks and quantitation of Primary Pigment Standards. Relevant terms and equations are described in the text and in Table 16.1 respectively.

out under dim light. Micropipettes (e.g., Eppendorfs, Gilsons) and HPLC glass syringes can only be used after checking their accuracy and precision. Solvents (methanol, acetone, benzene) and buffers (NH_4-acetate) must be spectroscopic and analytical grade; they must be handled with due regard to Health and Safety recommendations.

A calibrated analytical microbalance with ±1 µg precision should be used for weighing crystalline pigments. The UV-visible spectra of pigment solutions should be obtained with a scanning spectrophotometer, preferably with double-beam monochromator optics and adjustable slitwidth. The spectrophotometer must be regularly calibrated for wavelength accuracy (with a holmium oxide filter) and for absorbance (with a potassium chromate solution; see manufacturer's recommendations). Since the highest spectrophotometric precision (0.8–1.5%) is attained with solutions with an absorbance between 0.2 and 1.0 (Fifield and Kealey, 1990), pigment solutions should be scanned at concentrations that absorb in this range. A pair of optically matched 1 cm precision (±1%) cuvettes should be used for both blank (solvent) and sample spectra scanned from 340 to 750 nm.

16.2.2 *Preparation of a primary pigment standard*

A **Primary Pigment Standard** is a stock standard containing a known mass of one pigment dissolved volumetrically in an appropriate solvent (see Fig. 16.1). The primary pigment standard is made up by either (1) weighing precisely (±1%) about 500 µg (W_g) of pigment on an analytical balance, or (2) recovering and pooling pigment peaks from the HPLC by solid-phase extraction (SPE) followed by elution with reference solvent (e.g., acetone). The pigment is dissolved in analytical-grade acetone (or other solvent), mixed and made up to volume in a 100 ml glass volumetric flask. At this stage the resultant concentration of about 50 $\mu g.ml^{-1}$ should be regarded as **nominal** only, since any impurities would affect the accuracy of this gravimetric estimate. Stopper, label and store at -20°C for no more than 8 hours until required for HPLC.

16.2.3 *Testing the purity of primary pigment standards*

Stored standards should be brought to room temperature and then mixed thoroughly before spectrophotometric determination. Allow the calibrated spectrophotometer to warm up, run a blank solvent scan and then record the UV-visible spectrum of the pigment standard from 340 to 750 nm. Record the wavelengths of maximum absorbance (λ_{max}) and the absorbance value(s) at λ_{max} ($A_{\lambda max}$) and subtract the blanks at these wavelengths. (The terms, their meaning and units are listed in Table 16.1.)

Five criteria can then be used to assess the purity of the pigment standard:

1. Compare the observed λ_{max} values (nm) with the λ_{max} values tabulated for the same solvent in the Part IV Data sheets and Chapter 7. These should agree to within ±1 nm.
2. Compare the band ratios (III:II (%) for carotenoids and Soret: red ratios for chlorophylls) with the corresponding band ratios tabulated in the Part IV Data sheets and Chapter 7. These should agree to within 10% (carotenoids) and 5% (chlorophylls).
3. Check the presence of *cis*-isomers in carotenoid standards by their characteristic absorbance bands in the near-UV region of the spectra (Ke *et al.* 1970; Appendix D). Re-purify if *cis*-isomer contamination is significant.

TABLE 16.1. Explanation of terms used in equations

Parameter	Explanation	Units
$A_{\lambda max}$	Absorbance of a pigment solution at λ_{max} in a 1 cm cuvette	ND* ($0<A<2$)
A_{deg}	HPLC peak area of a degraded pigment	μV.s
A_{pig}	HPLC peak area of the pigment standard	μV.s
A_p	HPLC peak area of a pigment (*p*) from seawater	μV.s
A_{is}	HPLC peak area of the internal standard (*is*)	μV.s
α	Specific extinction coefficient: absorbance of a one $g.l^{-1}$ solution of pigment in a 1 cm cuvette at λ_{max} in a specific solvent	$l.g^{-1}cm^{-1}$
B	Buffer-extract dilution factor before HPLC injection	ND, ($B<1.0$)
C	Concentration of a primary pigment standard	$g.l^{-1}$
C_p	Concentration of pigment (*p*) in seawater	$ng.l^{-1}$
d	Cuvette path length	cm
$E^{1\%}_{1cm}$	The 1% extinction coefficient: absorbance of 1% solution of pigment in a 1 cm cuvette at λ_{max} in a specific solvent	100 $ml.g^{-1}cm^{-1}$
f_p	Response factor for pigment (*p*) derived from the external standardization of the HPLC	$ng.(\mu V.s)^{-1}$
f^{is}_p	Response factor of pigment (*p*) with respect to the internal standard (*is*)	ND, typically ($0.2<f^{is}_p<3.0$)
λ_{max}	Wavelength of maximum absorbance	nm
P_c	Chromatographic purity of the primary pigment standard	%
P_g	Gravimetric purity of the primary pigment standard	%
v_{ext}	Volume of pigment extract	ml
v_{filt}	Volume of seawater filtered	l
v_{inj}	Volume injected into the HPLC	μl
W_p	Weight or mass of pigment (*p*) injected during the internal standardization of the HPLC	ng
W_{is}	Weight or mass of internal standard (*is*) injected during the internal standardization of the HPLC	ng
W_g	Weight or mass of pigment used for standards	g
W_{sp}	Weight or mass of pigment derived spectrophotometrically	g
w_{is}	Weight or mass of internal standard (*is*) added to the sample extract	ng

* ND: non-dimensional, with typical values shown

4. Check the chromatographic purity of the standard by injecting it into a reversed-phase HPLC (see Chapters 11 and 12). Pure pigments should exhibit only one chromatographic peak. Any additional minor peaks indicate the presence of degraded pigments (e.g., *cis*-isomers of carotenoids, allomers or epimers of chlorophylls or other contaminants). In such cases the chromatographic purity (P_c, %) of the pigment standard can be determined approximately from the ratio of HPLC peak areas by Equation 16.1:

$$P_c \approx \left(\frac{A_{pig}}{A_{pig} + \sum A_{deg}}\right) \times 100\%, \qquad 16.1$$

where A_{pig} and ΣA_{deg} are the peak areas of the pigment standard and all the degradation products, respectively. If the purity of a pigment standard is less than 90%, it should be re-purified before using it for spectrophotometric or HPLC calibrations (see Chapters 7 and 8).

5. A gravimetric estimate of pigment purity (P_g, %) can also be derived by comparing the measured weight of the pigment standard (W_g, g) with the weight determined spectrophotometrically (W_{sp}; see Section 16.2.4). In this case:

$$P_g = \left(\frac{W_{sp}}{W_g}\right) \times 100\% = \left(\frac{C \times V}{W_g}\right) \times 100\%, \qquad 16.2$$

where C is the pigment concentration (g/l) derived spectrophotometrically from Equations 16.3 or 16.4, and V is volume (l) of the volumetric flask. This purity estimate includes any contribution of non-pigmented substances such as residual lipids or salts that would escape detection in the HPLC. Commercial sources of chlorophyll *a* may contain 5–10% (w/w) colourless contaminants and up to 20% of chlorophyll *a* epimers (Mantoura, Jeffrey, unpublished observations; see Chapter 7).

16.2.4 Quantitative spectrophotometry

Once the purity of the pigment standard has been determined, the purity-corrected pigment concentration (C, g/l) can be accurately determined (see Fig. 16.1) by using an extended version of the Beer-Lambert equation (see Preamble to Part IV).

For carotenoids:
$$C = \frac{A_{\lambda\max}}{0.1 \times E^{1\%}_{1cm} \times d} \times \frac{P_c}{100} \qquad 16.3$$

For chlorophylls:
$$C = \frac{A_{\lambda\max}}{\alpha \times d} \times \frac{P_c}{100}, \qquad 16.4$$

where $A_{\lambda max}$ is the absorbance (absorbance units) at $\lambda_{\max}$, d is the cuvette path length (cm), and $E^{1\%}_{1\text{cm}}$ and α are the 1% extinction coefficient (used for carotenoids) and the specific extinction coefficient (used for chlorophylls), respectively (see Table 16.1 and the Part IV Data sheets). This value of C should now be adopted as the **actual concentration** of the pigment standard, and the flask labelled accordingly.

16.3 HPLC CALIBRATION WITH EXTERNAL STANDARDS

In HPLC calibration by external standards, a series of known masses of pigment standards are injected. The resultant chromatographic peak areas are related to pigment masses to yield a response factor (or calibration curve) for each pigment.

16.3.1 *Initial calibration with multi-pigment working standards*

Multi-pigment calibration of the HPLC is the first step for both the external and internal standards methods. This is shown schematically in Fig. 16.2 and detailed below:

1. Prepare a 10 ml volumetric solution of **Primary Pigment Standard** containing 25 μg.ml^{-1} in acetone for each pigment. Determine the purity and concentration of the standard by the HPLC and spectrophotometric procedures detailed

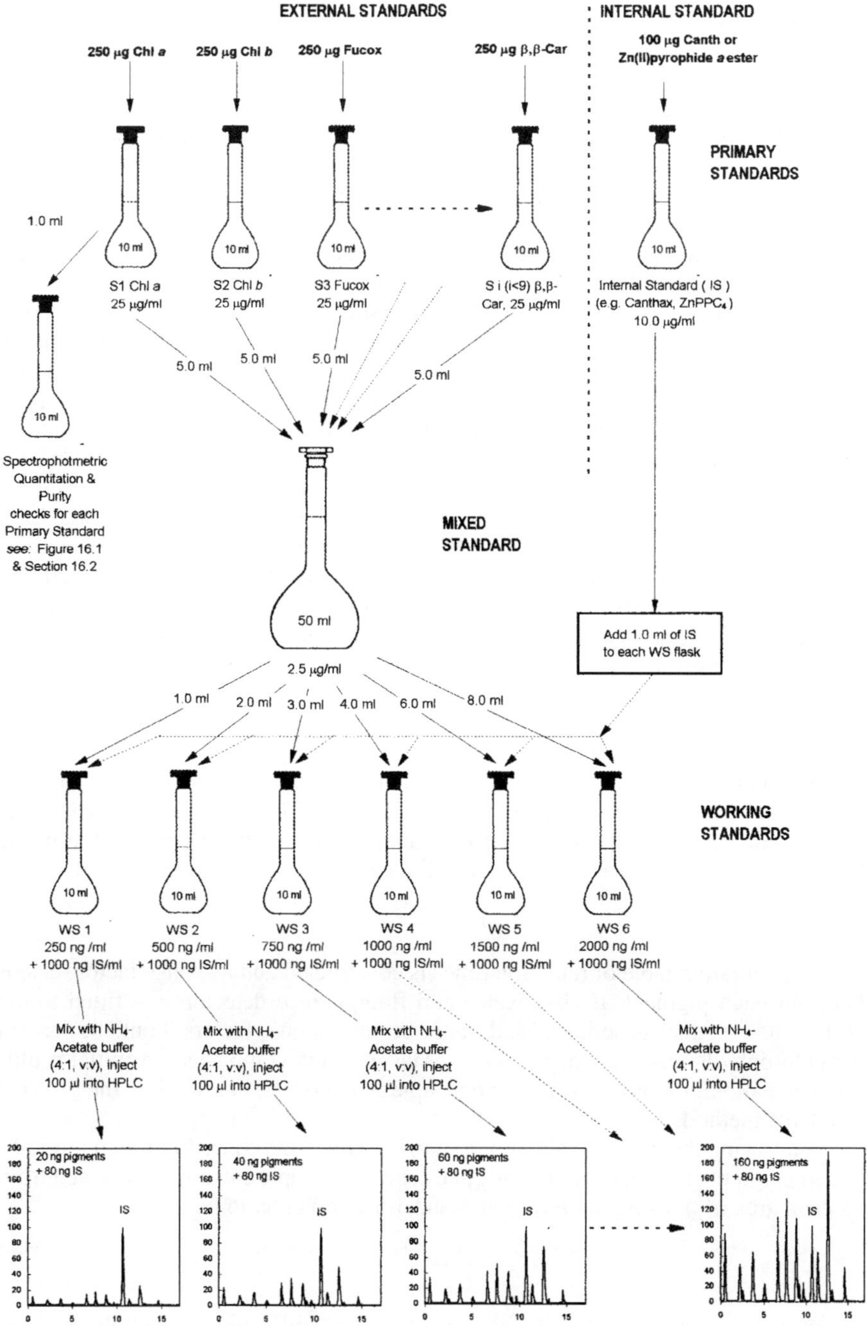

Figure 16.2
Schematic steps for the preparation of external and internal pigment standards, and their volumetric dilution into Mixed and Working Standards required for HPLC calibration. The weights and concentrations of pigments are suggested for guidance. The abbreviations S, MS, WS and IS denote the Primary, Mixed, Working and Internal Standards respectively.

in Section 16.2 and illustrated in Fig. 16.1. Additional standards are needed for each major pigment occuring in oceanographic samples. These primary pigment standard solutions are labelled: S1, S2 etc.

2. Prepare a **Mixed Standard** by volumetrically transferring 5.0 ml of primary pigment standard solutions S1, S2, etc. into a 50.0 ml volumetric flask and making up to volume with acetone. This mixed standard, which is labelled MS, contains about 2.5 μg ml^{-1} of each pigment.
3. Finally, prepare a series of six **Working Standards** by pipetting 1.0, 2.0, 3.0, 4.0, 6.0 and 8.0 ml of Mixed Standard (MS) into six 10.0 ml volumetric flasks and making up to volume with acetone (see Fig. 16.2). These Working Standards, labelled WS1, WS2–WS6, contain 250, 500, 750, 1000, 1500 and 2000 ng ml^{-1} of individual pigments, respectively. A higher range of concentrations may be needed to check calibration linearity of pigment-rich samples (>2 μg l^{-1} Chl *a*).
4. The HPLC should be calibrated as soon as possible after the working standards have been prepared. To 1.0 ml of Working Standard, add 300 μl of ammonium acetate buffer, mix and immediately inject 100 ml of the mixture into the HPLC. Run the HPLC and, after ensuring the peaks have been correctly integrated, record the peak areas (A_p, μV.s) for each pigment (p). Inject and analyse each WS in triplicate.

16.3.2 *Calculation of pigment response factors*

Calibration factors for each pigment are obtained by plotting peak area (A_p) against the weight of pigment injected (W_p; see Fig. 16.3.A). The regressions for each pigment should be linear with correlation coefficients of $r^2 > 0.95$. The pigment response factor (f_p , ng.(μV.s)$^{-1}$) is derived from the calibration slope as expressed by the equation:

$$f_p = \frac{\Delta A_p}{\Delta W_p}. \qquad 16.5$$

A calibration table of retention times is constructed and response factors calculated for each pigment. If absorbance and fluorescence detectors are fitted to the HPLC, then separate absorbance and fluorescence response factors should be derived for chlorophylls and their degradation products. This completes the initial multi-pigment calibration of the HPLC, and samples can now be analysed by the external standards method.

Following the HPLC analysis of a sample extract, the concentration of any calibrated pigment (C_p; ng.1^{-1}) in the original seawater sample can now be derived from its peak area (A_p), using the External Standardization equation:

$$C_p = \frac{A_p . f_p . v_{ext} . 10^3}{v_{inj} . v_{filt} . B} \qquad 16.6$$

where v_{ext}, v_{inj}, and v_{filt} are the volumes of extract (ml), injected sample (μl) and filtered seawater sample (l), respectively. B is the buffer dilution factor (<1.0); it equals the ratio of extract volume to the sum of extract and buffer volumes mixed before injection (see Sections 11.4.2 and 17.5.4). The External Standardization equation 16.6, together with pigment retention times, can be included in the HPLC integration program so that peak areas are automatically converted to pigment concentrations in seawater (see Section 17.5.5).

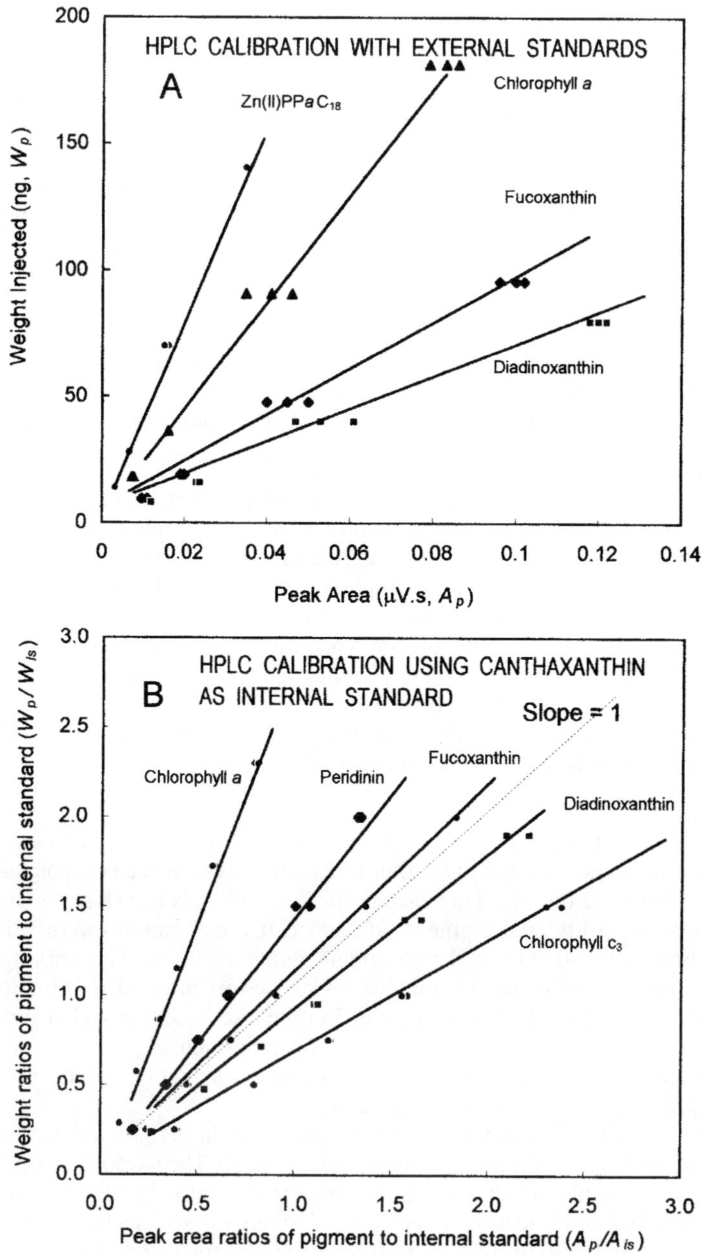

Figure 16.3
HPLC pigment calibration plots derived from A, the external and B, the internal standardization procedures detailed in Table 16.2 and Fig. 16.2, respectively. The corresponding pigment response factors f_p and f_p^{is} can be calculated from the slopes of Figs. 16.3A and B, respectively.

16.3.3 *Routine re-calibration*

For day-to-day operation of the HPLC, recalibration is more practical with a few readily available pigment standards such as chlorophylls *a, b* and lutein. Provided the injection and elution conditions (injector loop, columns, flow rates, gradient programme, etc.) and the detector settings (lamps, absorbance and fluorescence wavelengths, etc.) have not been altered, then variations in pigment response factors should not exceed 5%. A full multi-pigment recalibration of the HPLC is recommended every 3–4 months, or after a major service, or if f_p variability exceeds 5%.

16.3.4 *Shipboard re-calibration*

Two approaches for calibrating the HPLC at sea are now described. In each case the concentration of a primary standard is determined with a spectrophotometer ashore. Replicate working standards are prepared from this for shipboard calibration.

Method 1
Before going to sea, replicate amounts (500 µg) of pigment are carefully weighed from the same stock and stored under N_2 in stoppered glass vials in a solid-CO_2 (dry ice) container. The contents of one vial are then made up to volume and the purity and concentration are determined spectrophotometrically and by HPLC as described in Sections 16.2.3 and 16.2.4. At sea, another replicate pigment vial is taken from the dry-ice container and its contents are made up to volume. The concentration is then derived from the known weight of pigment and the shore-based calibration. If a spectrophotometer is available on board, the concentration can be verified. The stability and purity of working standards made up at sea should be monitored daily by HPLC. They should be replaced when necessary.

Method 2
A primary standard is prepared in a solvent and the pigment purity and concentration are measured (as described in Section 16.2). The solvent is then evaporated under N_2 and the pigment is dissolved in benzene and quantitatively transferred to a 100 ml volumetric flask. Additional pigments are added if a calibration **mixture** is to be prepared. Benzene is added to make up a final volume of 100 ml. The working standards are prepared by pipetting 1.0 ml aliquots of the mixture into 10 ml volumetric flasks, and after purging the headspace with helium, the flasks are sealed and frozen at -20°C. Note that benzene is a potent carcinogen and must be handled in a fume hood. It is recommended here because it solidifies at conventional freezer temperatures and prevents isomerization of pigments.

At sea, the working standards in benzene are brought to room temperature, and methanol or methanol/water added (final volume 10 ml). The working standard, dissolved in 10% benzene in methanol, is then injected to calibrate the HPLC or added to the sample before extraction as an internal standard (see Section 16.4 below). Replicate working standards (e.g., Chl *a*) have shown little degradation (<3%, *n*=3) during 6 months' storage under these conditions.

16.3.5 *Example*

An example of the calibration of a Waters HPLC system with external primary standards of fucoxanthin, diadinoxanthin, chlorophyll *a* and Zn(II) pyropheophorbide *a*

TABLE 16.2. External standard calibration of a Waters HPLC for fucoxanthin, diadinoxanthin, chlorophyll *a* and Zn(II) pyropheophorbide *a* octadecyl ester. Data used in the preparation of Figure 16.3A.

Pigments	UV-visible spectra[a]		Primary standard			HPLC absorbance (440 nm) peak area			
	$E^{1\%}_{1cm}$ (λ_{max}) (100 ml.g^{-1}cm^{-1})	(λ_{max}) (nm)	Purity[b] P_c (%)	Absorbance[c] ($A_{\lambda max}$)	Concentration[d] C (µg.ml^{-1})	Dilutions of Mixed Standard[e] 0	1:1	1:4	1:9
Fucoxanthin	1660	443	95	0.167	9.56	0.100	0.050	0.019	0.011
						0.096	0.045	0.020	0.0097
						0.102	0.040	0.020	0.0097
Diadinoxanthin	2230	448	90	0.198	7.99	0.120	0.061	0.023	0.012
						0.118	0.053	0.023	0.012
						0.122	0.047	0.024	0.012
Chlorophyll *a*	1084	430	95	0.207	18.14	0.079	0.041	0.016	0.0076
						0.083	0.046	0.016	0.0074
						0.086	0.035	0.016	0.0077
Zn(II) pyropheophorbide *a* octadecyl ester	1175	427	100	0.165	14.04	0.035	0.016	0.0067	0.0032
						0.035	0.016	0.0070	0.0032
						0.035	0.015	0.0067	0.0032

a. Values of $E^{1\%}_{1cm}$ and λ_{max} for fucoxanthin, diadinoxanthin and chlorophyll *a* are in acetone (see Part IV, Data sheets). The $E^{1\%}_{1cm}$ value for Zn(II) pyropheophorbide *a* octadecyl ester is approximate and was derived from a nominal ε value = 100,000 and a molecular weight of 850.

b. Determined by HPLC and Equation 16.1: $P_c = \frac{A_{pig}}{A_{pig} + \sum A_{deg}} \times 100\%$.

c. Absorbance of a 10% dilution of the pigment standards.

d. Determined by Equation 16.3: $C = \frac{A_{\lambda max}}{0.1 \times E^{1\%}_{1cm} \times d} \times \frac{P_c}{100}$.

e. The Mixed Standard was prepared with a 10-fold dilution of each Primary Standard. The Mixed Standard was further diluted 1:1, 1:4 and 1:9 before 100 µl was injected into the HPLC in triplicate.

octadecyl ester is shown in Table 16.2. The table also includes the spectrophotometric and purity data for deriving the purity-corrected concentrations. The HPLC was calibrated over three days (not recommended; immediate calibration is better) by triplicate injections of a mixed standard and 1:1, 1:4 and 1:9 dilutions of the mixed standard. The resultant peak areas are listed in Table 16.2 and plotted in the calibration graph in Fig. 16.3A.

The response factors (f_p) for diadinoxanthin, fucoxanthin, chlorophyll *a* and the Zn(II) pyropheophorbide *a* octadecyl ester standard, calculated from the slopes, were 666, 954, 2144 and 4003 ng.(μV.s)$^{-1}$, respectively. Diadinoxanthin, with the lowest slope or response factor, corresponded to the most sensitively detected pigment.

16.4 HPLC CALIBRATION WITH INTERNAL STANDARDS

With the internal standards method, a known mass of pigment (the internal standard) is added before sample extraction (to the phytoplankton sample on the filter), extracted, and resolved by HPLC with the other pigments in that sample. Use of an internal standard minimizes errors arising from sample handling (evaporation of solvent, incomplete transfer during extraction and concentration etc.) and HPLC analysis (errors in injection volume), since sample and internal standard are treated identically.

16.4.1 *Evaluation of internal standards*

The ideal internal standard (IS) should be a readily available and stable compound with chemical and spectral properties similar to the other pigments. It should absorb strongly at 430–440 nm and fluoresce above 600 nm. In addition, it should elute in a portion of the HPLC chromatogram that is free from other pigments (e.g., between zeaxanthin and chlorophyll *b* or between chlorophyll *a* and β,β-carotene).

Several internal standards are potentially suitable: three porphyrins, three phorbins and four carotenoids. Their structures and properties are shown in Fig. 16.4 and Table 16.3, respectively. All porphyrins tested had fluorescence peaks but absorbed weakly at 440 nm consistent with their λ_{max} = 395–400 nm. Deuteroporphyrin IX dimethyl ester (II) and etioporphyrin (III) co-eluted with pheophorbide *a* and chlorophyll *b*, and so were rejected. Mesoporphyrin IX dimethyl ester (I) was more useful, since it eluted between zeaxanthin and chlorophyll *b*. Hurley and Armstrong (1990) successfully used this pigment as a fluorescent internal standard in an HPLC study of pigment fluxes in Lake Mendota.

Like the porphyrins, *n*-butyl-pheophorbide *a* (IV) fluoresced but absorbed weakly (λ_{max} = 410 nm) and eluted optimally between zeaxanthin and chlorophyll *b* (C. Llewellyn, pers. comm.). However, as this compound degraded in acetone after a few hours, it was rejected.

The evaluation of the recently synthesized Zn(II) pyropheophorbide *a* alkyl esters (V) (Repeta *et al.* in preparation) as a potential internal standard for pigment HPLC is described in Section 16.4.2.

Recently, Moore *et al.* (1994) synthesized a family of caroteno-porphyrin and caroteno-pyropheophorbide 'dyads' to investigate the nature of photochemical coupling of chlorophylls and carotenoids. The pyropheophorbide *a*-β,χ-carotene 'dyad',

TABLE 16.3. Properties of internal standards for pigment HPLC

Internal standard	Structure[a]	Source[b]	Cost ($/g)	λ_{max} (nm) (solvent)	$E^{1\%}_{1cm}$	Fluorescent	HPLC elution[c]	Comments	Ref	Recommendation[d]
Porphyrins and Phorbins										
Mesoporphyrin IX dimethyl ester	I	A, S, P	800	399 (acetone)	2930	*	Just after Lut and Zea	Weak absorbance at 440 nm, useful with fluorescence detection	e, f	✓
Deuteroporphyrin IX dimethyl ester	II	A, F, P	800	400 ($CHCl_3$)	3430	*	Coelutes with Phide *a*	Weak absorbance at 440 nm, useful with fluorescence detection	g	X
Etioporphyrin I	III	A	400	400 ($CHCl_3$)	?	*	Coelutes with Chl *b*	Weak absorbance at 440 nm, useful with fluorescence detection	h, f	X
n-Butyl-pheophorbide *a*	IV	Syn		410 (acetone)	1490	*	Between Lut and Chl *b*	Moderate absorbance at 440 nm; moderately stable in solution	i, f	✓
Zn(II) pyropheophorbide *a* C_4–C_{20} esters	V	Syn, EPA		427 (acetone)	1175[o]	*	See Fig. 16.6. C_8-ester between Chl *a* and Chl *b*	Excellent absorbance at 440 nm and fluorescence detectability; stable	j, f	✓✓✓
Pyropheophorbide *a* caroteno 'dyads'	VI	Syn		430–460 (various)	2000[p]	*	Unknown, close to Chl *a*?	Untested, potentially good absorbance and fluorescence properties and stable	k	✓✓
Carotenoids										
Canthaxanthin	VII	F	20	472 (acetone)	2200		Between Lut and Chl *b*	Inexpensive, sensitive at 440 nm, suitable for phytoplankton but not sediments; traces may occur in green algae	l, f	✓✓✓
Ethyl-8'-apo-β-carotenoate	VIII	F	18	448 (pet ether)	2360		Between Chl *b* and Chl *a*	Sensitive at 440 nm, may isomerize after 1 day	m, f	✓✓
8'-apo-β-carotenal	IX	F	18	452 (ethanol)	2540		Between Lut and Chl *b*	Sensitive at 440 nm, stability not tested	m, f	✓
Nonapreno-β,β-carotene	X	Syn		480 (acetone)	2780		4 min after β,β-carotene	Late-eluting; used with food carotenoids	n	X

a. See Fig. 16.4. b. Commercial suppliers include A: Aldrich; S: Sigma; P: Porphyrin Products; F: Fluka; Syn: synthesized by authors; EPA: Environmental Protection Agency. c. SCOR pigment abbreviations (see Appendix A). d. ✓✓✓: highly recommended; ✓✓: recommended; ✓: useful; X: not recommended. e. Hurley and Armstrong (1990). f. Tested by SCOR WG 78. g. Evaluated by D. J. Repeta. h. Evaluated by R. Bidigare. i. Synthesized by D. J. Repeta. j. Synthesized by Repeta *et al.* (in prep.). k. Synthesized by Moore *et al.* (1994). l. Letelier *et al.* (1993). m. Wright *et al.* (1991). n. Khachik and Beecher (1986). o. Based on $\varepsilon = 10^5$ $l.mole^{-1}.cm^{-1}$ assumed for Zn(II) pyropheophorbide *a* octadecyl ester. p. Estimated between carotenoid and pheophorbide.

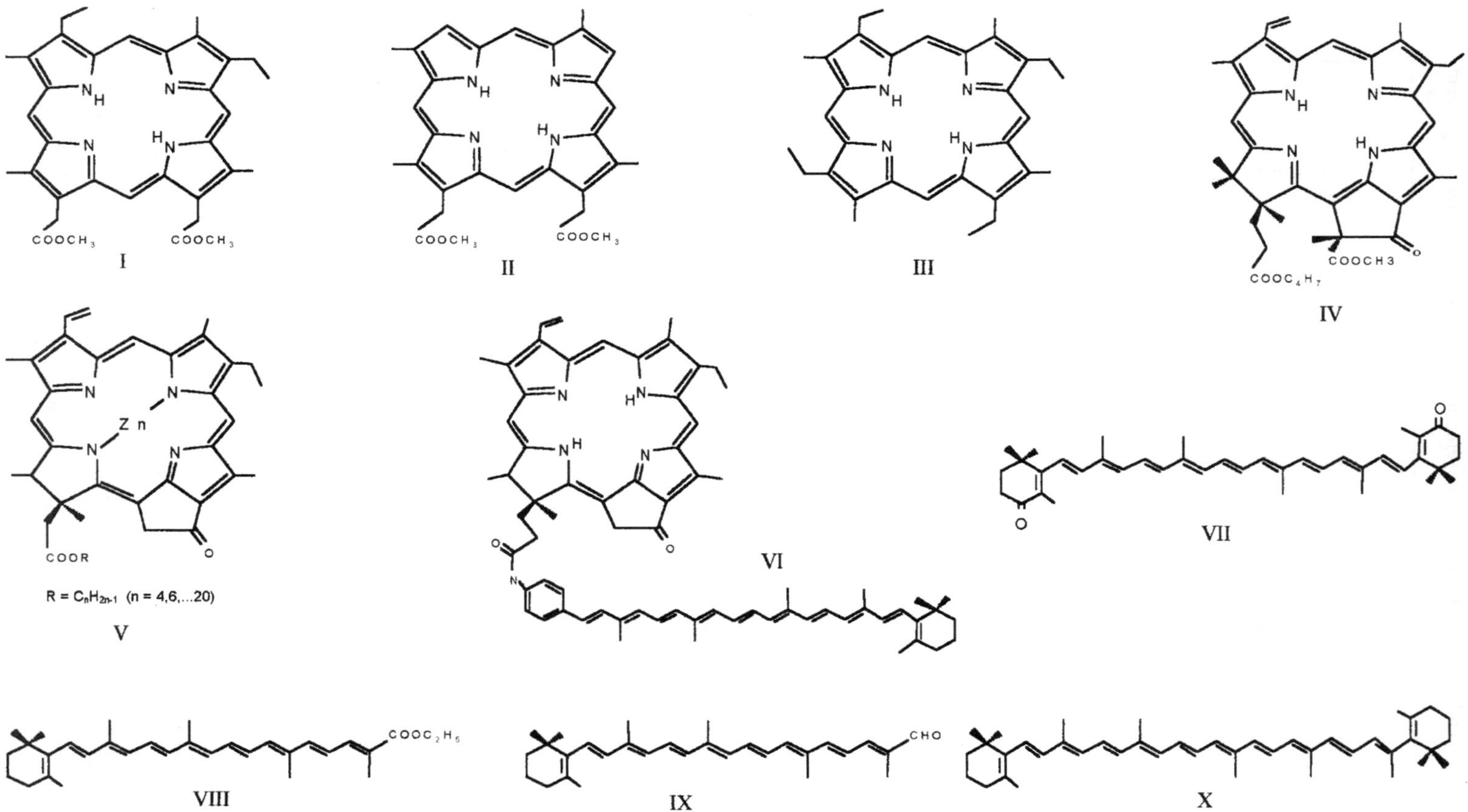

Figure 16.4

Structures of various internal standards evaluated in Table 16.3. I: Mesoporphyrin IX dimethyl ester; II: Deuteroporphyrin IX dimethyl ester; III: Etioporphyrin I; IV: Pheophorbide *a* *n*-butyl ester; V: Zn(II) pyropheophorbide *a* alkyl (C_4-C_{20}) esters; VI: Pyropheophorbide *a* -,β,χ-carotene 'dyad'; VII: Canthaxanthin; VIII: Ethyl-8'-apo-β-carotenoate; IX: 8'-apo-β-carotenal; X: Nonapreno-β,β-carotene.

shown in Fig. 16.4 (VI), is an example of a 'dyad' with a saturated bond linkage. This compound exhibits spectral properties similar to its two precursors, pyropheophorbide *a* and β,χ-carotene, but with an intermediate polarity and retention time. Although not tested on our HPLC, these 'dyad' pigments may prove to be useful internal standards.

All four carotenoids listed in Table 16.3 absorbed strongly at 440 nm, but as with most carotenoids, they did not fluoresce. Only canthaxanthin (VII) and 8'-apo-β-carotenal (IX) eluted between zeaxanthin and chlorophyll *b*. Both carotenoids are stable at -20°C for up to 1–2 years in the crystalline state and for 1 or 2 days in solution, after which isomerization and oxidation become significant. Although canthaxanthin may occur in trace quantities in a few green algae and zooplankton (Antia and Cheng, 1983), it is usually absent from marine phytoplankton. Letelier *et al.* (1993) used canthaxanthin as an internal standard in an HPLC study of phytoplankton successions off Hawaii. However it should not be used in the HPLC analyses of sediments, since early diagenesis operates to selectively preserve 4-oxo-carotenoids such as canthaxanthin (Repeta, 1990).

Ethyl 8'-apo-β-carotenoate (VIII), which is more stable than 8'-apo-β-carotenal (IX), elutes between chlorophyll *a* and *b*. It is not known to occur in the marine environment, and is therefore a suitable internal standard.

Nonapreno-β,β-carotene (X), synthesised by Khachick and Beecher (1986), has been advocated as an internal standard for the HPLC of carotenoids in food. Although more stable than canthaxanthin, it is not generally available and elutes several minutes after β,β-carotene.

The use of structurally unrelated compounds such as fluoranthene (Bessiere and Montiel, 1982) or Sudan II (Pearl and Millie, 1993) as internal standards for HPLC of phytoplankton pigments is not recommended, since their recovery from the column may differ from the sample pigments.

16.4.2 *Synthesis of a new internal standard for oceanography*

An alternative to using commercially available pigments as internal standards is to synthesize a pigment with the desired properties from readily available starting material. For HPLC analyses of pigments in seawater, the most promising internal standards are synthetic derivatives of chlorophyll *a*. The retention time, chemical stability, and spectroscopic properties of chlorophyll *a* can be readily manipulated by using such well-known reactions as trans-esterification of the phytol ester, pyrolysis of the 10-carboxymethyl ester, transmetallation, and reduction of the 9-oxo and 2-vinyl groups.

Repeta *et al.* (in preparation) synthesized Cu, Ni, and Zn derivatives of pyropheophorbide *a* and investigated their stabilities, spectra and HPLC properties. The Zn(II) pyropheophorbide *a* C_4–C_{20} alkyl esters were chosen because of their stability and the similarity of their UV-visible absorbance and fluorescence to the chlorophylls. Their synthesis and spectral properties are summarized in Fig. 16.5. Pyropheophorbide *a* is prepared by refluxing pheophorbide *a* in collidine. It is then esterified with an alcohol in the presence of N,N-dicyclohexylcarbodiimide. The product is refluxed in a saturated methanol solution of zinc acetate to yield Zn(II) pyropheophorbide *a* alkyl ester. By using different alkyl esters (e.g., C_4 to C_{20}), the retention times of the internal standard can be optimized as shown in Fig. 16.6.

Zn(II) pyropheophorbide *a* alkyl esters are very stable, since they cannot

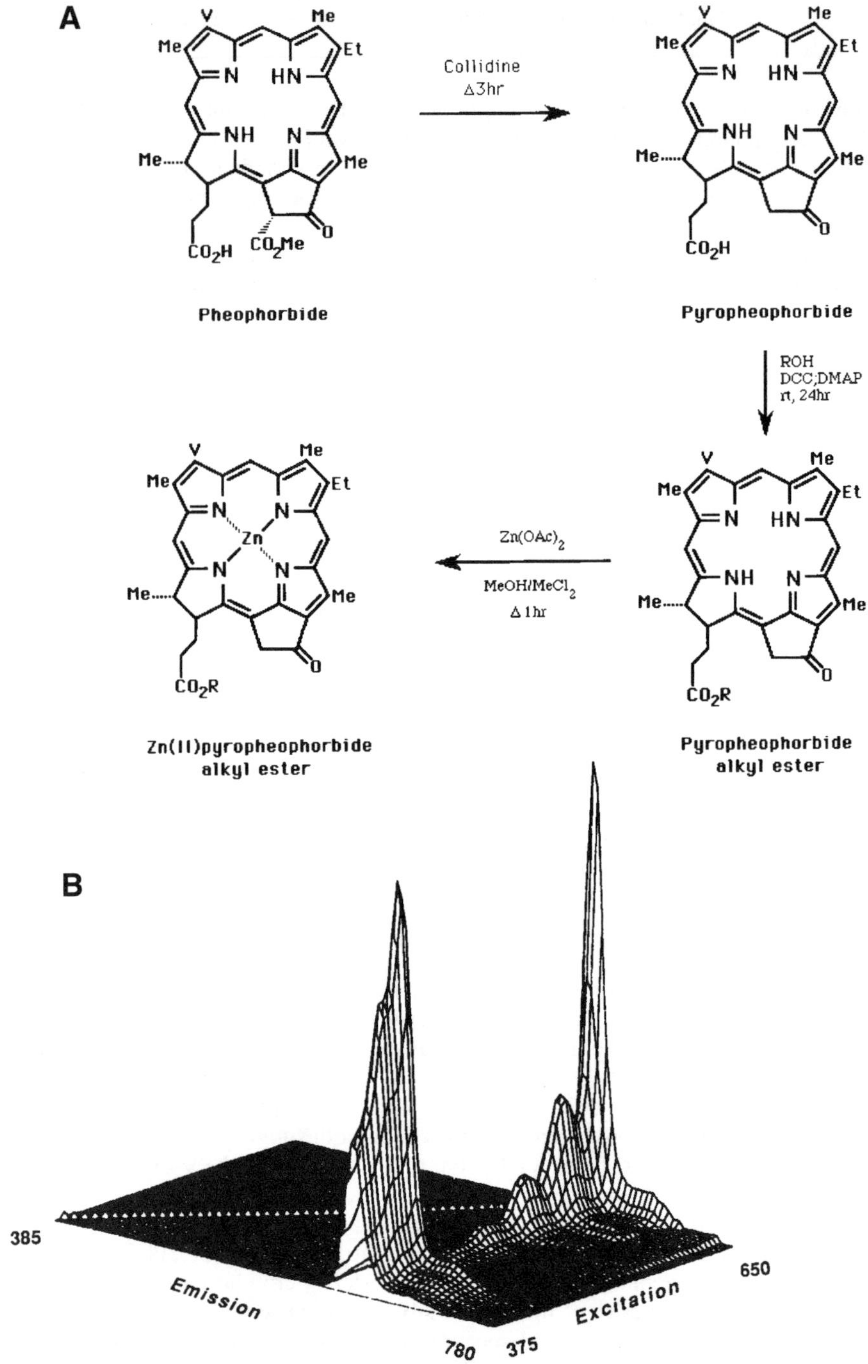

Figure 16.5
A. Synthesis of Zn(II) pyropheophorbide *a* alkyl (C_4-C_{20}) esters from pheophorbide *a* (Repeta *et al.* in prep.). B. Excitation and emission spectra in acetone.

epimerize or allomerize, and the Zn^{2+} complex is more resistant than Mg^{2+} to acid-catalyzed demetallation (Hynninen, 1991). As the rates of photochemical oxidation of the Zn complex are similar to chlorophyll *a*, the standard must, like chlorophyll *a*, be protected from light.

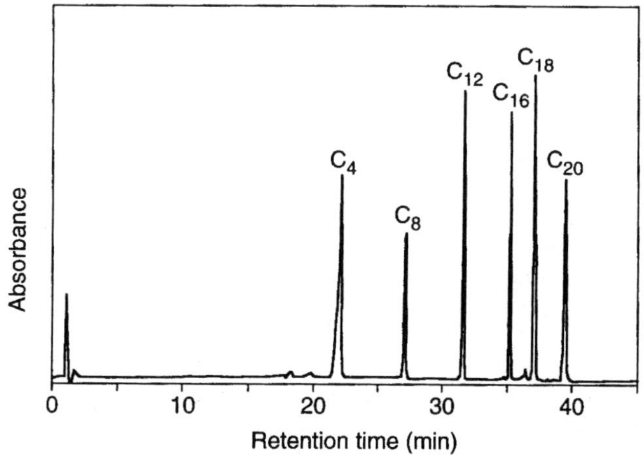

Figure 16.6
HPLC separation of Zn(II) pyropheophorbide *a* alkyl (C_4-C_{20}) esters (Repeta *et al.* in prep.). By using esterified alcohols of different molecular weight, the retention time of the internal standard can be adjusted to elute in a region that is free from major phytoplankton pigments. HPLC conditions: 15 cm ODS column (3 μm adsorbosphere) eluted with a linear gradient of 100% A to 100% B in 30 minutes, where A = 20/80 0.5 M ammonium acetate/methanol and B = 20/80 acetone/methanol. Absorbance detection was at 659 nm. Under these conditions, chlorophyll *a* elutes at 31 min.

16.4.3 *Initial calibration procedure*

The procedure for determining the relative response factors of the internal standard with respect to other marine pigments is illustrated on the right side of Fig. 16.2 and described below:

1. Prepare a 10 ml volumetric solution containing 10 μg.ml^{-1} of **Primary Internal Standard** in acetone (e.g., canthaxanthin or Zn(II) pyropheophorbide *a* octadecyl ester; see Section 16.4.1 and 16.4.2). Determine the purity and concentration of the internal standard by HPLC and spectrophotometric steps detailed in Section 16.2 and in Fig. 16.1. The flask is labelled IS and, if necessary, stored in a freezer.
2. Prepare the Mixed Standard as described previously for the external standards (Section 16.3.1). After the serial addition of 1.0, 2.0–8.0 ml of Mixed Standard into the 10 ml flasks for Working Standards (Fig. 16.2), pipette 1.0 ml of internal standard into each flask and make up to volume with acetone. These six Working Standards (WS1, WS2, ...) now contain 250, 500, 750, 1000, 1500

and 2000 ng ml^{-1} of pigment together with a fixed concentration of 1000 ng ml^{-1} of internal standard.

3. To 1.0 ml of Working Standard, add 300 ml of ammonium acetate buffer, mix and immediately inject 100 µl of the mixture into the HPLC. Analyse each Working Standard in triplicate and ensure the HPLC peaks have been correctly integrated. Tabulate the peak areas of pigments (A_p) and of internal standard (A_{is}) and the corresponding weights (W_p, W_{is}) injected into the HPLC.

16.4.4 *Calculation of relative response factors*

For each pigment injected into the HPLC, plot the ratio of its weight (W_p) to that of the internal standard (W_{is}) on the Y axis, against the corresponding ratio of peak areas (A_p/A_{is}) on the X axis (see Fig. 16.3.B). In this ratio plot, any volumetric errors made during injection are cancelled out. The relative response factor (f_p^{is}) for each pigment with respect to the internal standard can now be derived from the slope and is equal to:

$$f_p^{is} = \frac{A_{is}.W_p}{A_p.W_{is}}. \qquad 16.7$$

When f_p^{is} <1.0, the pigment is more sensitively detected than the internal standard, and vice versa when f_p^{is} >1.0.

The relative response factors for chlorophyll *a*, peridinin, fucoxanthin, diadinoxanthin and chlorophyll c_3 with respect to the internal standard (canthaxanthin) are shown in Table 16.4. Provided the HPLC elution and detector settings are not

TABLE 16.4 Pigment concentrations from the HPLC analysis of the seawater sample shown in Figure 16.7[a] using canthaxanthin as internal standard.

HPLC peak no. (see Fig. 16.7)	Pigment	Peak area A_p (µV.s)	Relative response factor f_p^{is} (ND)[b]	Pigment concentration in seawater[c] C_p (ng.l^{-1})
5	Chl c_3	70.06	0.55	152
6b	Chl c_{1+2}	55.00	0.63	139
7	Perid	24.12	1.49	143
9	But-fuco	20.73	1.24	102
10	Fuco	17.94	1.09	78
15	Hex-fuco	94.94	1.26	478
26	Diadino	43.44	0.86	148
33	Lut	2.41	0.72	7
33	Zea	3.70	0.86	13
IS	Cantha	50.76	1.00	202
37	Chl *b*	5.86	2.51	59
41	Chl *a*	85.78	2.89	989
52	β,β-Car	4.13	0.77	13

a. Volume of seawater filtered (v_{filt}) = 2.05 l
Weight of internal standard added to sample (w_{is}) = 415 ng
b. ND: non-dimensional
c. Calculated from Equation 16.8: $C_p = \frac{A_p.w_{is}.f_p^{is}}{A_{is}.v_{filt}}$

changed, the f_p^{is} values obtained from this initial calibration should be constant. A table of pigment retention times and relative response factors is then constructed for analysing the seawater samples.

16.4.5 *Using internal standards with samples*

Before introducing the internal standard into sample extracts, it is essential (1) to verify that no pigment peaks co-elute with it (see Section 16.4.1 and 16.4.2), and (2) to ensure the mass of internal standard added is comparable to the mass of pigment in the extract. For example, filtering 2 l of seawater containing 1 μg.l^{-1} chlorophyll *a* requires about 2 μg of internal standard to be carefully micropipetted (typically 100 μl) onto the filter just before pigment extraction (see Chapter 17). The pigment concentration (C_p; ng.l^{-1}) in a seawater sample can now be derived from its peak area in the HPLC chromatogram by the Internal Standards equation:

$$C_p = \frac{A_p . w_{is} . f_p^{is}}{A_{is} . v_{filt}}, \qquad 16.8$$

where A_{is}, is the peak area (μV.s), w_{is} the added mass (ng) of internal standard and v_{filt} the volume (l) of seawater filtered.

Note that the volumes of extract, buffer and sample injected are not required in Equation 16.8. Any handling errors during extraction, centrifugation, evaporation etc. are eliminated when using an internal standard, because these errors affect the extracted pigments and the standard equally. Much higher precision (~1%) can be obtained with internal than with external standards (3%).

16.4.6 *Example*

A fully worked example with canthaxanthin as the internal standard in a seawater sample is shown in Fig. 16.7 and Table 16.4. A volume of 2.05 l seawater was filtered through a 25 mm GF/F, and 100 μl of canthaxanthin internal standard (4.15 μg.ml^{-1}) was added to the filter immediately before extraction. The resultant HPLC chromatogram is shown in Fig. 16.7. The calculations of pigment concentrations in the seawater sample using peak areas and relative response factors are shown in Table 16.4.

16.5 Analytical quality control

HPLC analyses of large batches of field samples over extended periods in the same laboratory may involve several analysts using different batches of solvents, standards, HPLC columns etc. It is essential that analytical quality be maintained. Excellent discussions of quality control are given by Juran (1974) and Keith *et al.* (1983).

Analytical protocols must be fully documented, intercalibrated and monitored to ensure a high reproducibility between analysts. The analyst must log a full description of the sampling, filtration, extraction, preparation and use of internal standards, chromatographic separation, detector calibration and data processing (peak integration, coefficients and formulae for quantitative calculations).

It is essential to regularly monitor the recoveries and reproducibility of replicate samples from the field and from the standard microalgal cultures (see Chap-

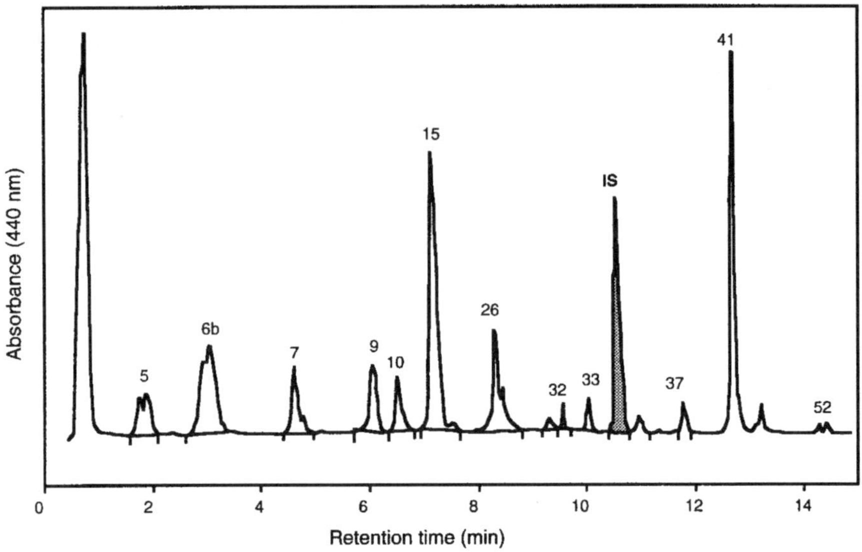

Figure 16.7
HPLC separation of phytoplankton pigments from the northeast Atlantic Ocean (52°N, 20°W, 10 m depth, 28 June 1989) sampled during JGOFS North Atlantic Bloom Experiment. Canthaxanthin was used as internal standard. HPLC conditions: 3 cm column ODS 3 μm Pecosphere, eluted at 1 ml/min with a linear gradient of 100% A to 100% B in 12 min. A = 80/20 methanol/ammonium acetate, and B = 80/20 methanol/acetone. Pigments detected by absorbance at 440 nm. Pigment identities: 5, chlorophyll c_3; 6b, chlorophyll c_{1+2}; 7, peridinin; 9, 19'-butanoyloxyfucoxanthin; 10, fucoxanthin; 15, 19'-hexanoyloxyfucoxanthin; 26, diadinoxanthin; 32, diatoxanthin; 33, lutein + zeaxanthin; IS, internal standard canthaxanthin; 37, chlorophyll *b*; 41, chlorophyll *a*; 52, β,β-carotene.

ter 6). A plot of recoveries and standard deviations through time would indicate any degradation or drift over the course of the analyses (see Chapter 17).

Finally, for quantitative analyses, the limits of both detection and quantitation need to be defined by the analyst. The limit of detection is the lowest concentration of pigment that can be detected as statistically different from a blank (Keith *et al.* 1983; Long and Windfordner, 1983). For a pigment peak to be 'detected' it must be significantly greater than 0 at the 99% confidence limit. This corresponds to when [sample] – [blank] exceeds 3 standard deviations (s.d.) of the baseline noise from a blank HPLC run. Signals below 3 s.d. should be reported as 'not detected'. The limit of quantitation is the minimum concentration above which quantitative data can be obtained within a specified confidence limit. For environmental samples, Keith *et al.* (1983) recommend that the limit of quantitation of 10 s.d. be adopted, which corresponds to an uncertainty of 30% in the concentration.

16.6 Conclusions

The methods of calibration and the calculations of pigment response factors described in this chapter will result in accurate and precise quantitation of pigment analyses by HPLC. The most precise method uses internal standards, but the external standard method is also highly recommended. At present, canthaxanthin and ethyl-8'-*apo*-β-carotenoate are the only widely available compounds recommended as internal standards for oceanographic samples. The recently synthesised Zn(II) pyropheophorbide *a* alkyl esters and possibly the caroteno-chlorin 'dyads' may, in the future, provide a more useful internal standard. We understand the US Environmental Protection Agency may soon distribute the Zn (II) pyropheophorbide *a* alkyl esters, which would become available to the oceanographic community.

Acknowledgements

We thank Drs J. I. Irvine and D. J. Simpson for synthesizing the Zn(II) pyropheophorbide *a* alkyl esters, and Ms C. A. Llewellyn for HPLC screening of selected internal standards and helping with the graphics.

References

Antia, N. J., Cheng, J. Y. (1983). The keto-carotenoids of two marine coccoid members of the Eustigmatophyceae. *Br. Phycol. J.* 17: 39–50.

Bessiere, J., Montiel, A. (1982). A selective and rapid method of chlorophylls *a* and *b* separative determination by high-performance liquid chromatography. *Water Res.* 16: 987–993.

Fifield, F. W., Kealy, D. (1990). *Principles and Practice of Analytical Chemistry.* Blackie, Glasgow, 3rd ed., 521 pp.

Hurley, J. P., Armstrong, D. E. (1990). Fluxes and transformations of aquatic pigments in Lake Mendota, Wisconsin. *Limnol. Oceanogr.* 35: 384–398.

Hynninen, P. H. (1991). Chemistry of chlorophylls: modifications. In: Scheer, H. (ed.),*Chlorophylls.* CRC Press, Ann Arbor, Michigan, p. 145–209.

Juran, J. M.(1974). *Quality Control Handbook.* McGraw-Hill, New York, 3rd edn.

Ke, B., Imsgard, F., Kjosen, H., Liaaen-Jensen, S. (1970). Electronic spectra of carotenoids at 77°K. *Biochim. Biophys. Acta*, 210: 139–152.

Keith, L. H., Crummett, W., Deegan, J., Libby, R. A., Taylor, J. K., Wentler, G. (1983). Principles of environmental analysis. *Anal. Chem.* 55: 2210–2218.

Khachick, F., Beecher, B.R. (1986). Synthesis of C_{45}-β-carotene, a potentially useful internal standard for quantification of hydrocarbon carotenoids by high-performance liquid chromatgraphy. *J. Ind. Eng. Chem. Prod. Res. Dev.* 25: 671–675.

Letelier, R. M., Bidigare, R. R., Hebel, D. V., Ondrusek, M., Winn, C. D., Karl, D. M. (1993). Temporal variability of phytoplankton community structure based on pigment analysis. *Limnol. Oceanogr.* 38: 1420–1437.

Long G. L., Windforner, J. D. (1983). Limit of detection. A closer look at the IUPAC definition. *Anal. Chem.* 55: 712A–724A.

Moore, T. A., Gust, D., Moore, A. L. (1994). Carotenoids: nature's unique pigments for light and energy processing. *Pure Appl. Chem.* 66: 1033–1040.

Paerl, H. W., Millie, D. F. (1993). Evaluations of spectrophotometric, fluorometric and high performance liquid chromatographic methods for algal pigment determinations in aquatic ecosystems. *Workshop Report to US Environmental Protection Agency*, 43 pp.

Repeta, D.J. (1990). Carotenoid diagenesis in pleistocene to miocene sediments from the Peru Margin. In: Suess, E., von Huene, R. (eds), *Proceedings of the Ocean Drilling Program, Scientific Results*, vol. 112, US Government Printing Office, Washington, p. 657–572.

Repeta, D. J., Irvine, J. I., Simpson, D. J. (1996). Preparation of Zn(II) pyropheophorbide *a* alkyl esters as internal standards for the HPLC analyses of pigments in seawater (in preparation).

Wright, S. W., Jeffrey, S. W., Mantoura, R. F. C., Llewellyn, C. A., Bjørnland, T., Repeta, D., Welschmeyer, N. (1991). Improved HPLC method for the analysis of chlorophylls and carotenoids from marine phytoplankton. *Mar. Ecol. Prog. Ser.* 77: 183–196.

17

Guidelines for collection and pigment analysis of field samples

S. W. Wright and R. F. C. Mantoura

17.1 MATCHING THE METHOD TO THE QUESTION

The concentration of pigments in phytoplankton provides a measure of the phytoplankton biomass (see Chapter 1). This information is valuable for oceanographers, fisheries and water management authorities, and environmental agencies. The type of data required varies from an 'approximate' measurement of chlorophyll *a* to a full analysis of chlorophylls, carotenoids and degradation products. The aim may be to produce surface maps, vertical profiles, time series, grazing or sedimentation profiles, pigment mass balances and so on. The choice of method depends on the type of data required, the type and number of samples, and the equipment and time available (see Chapter 4).

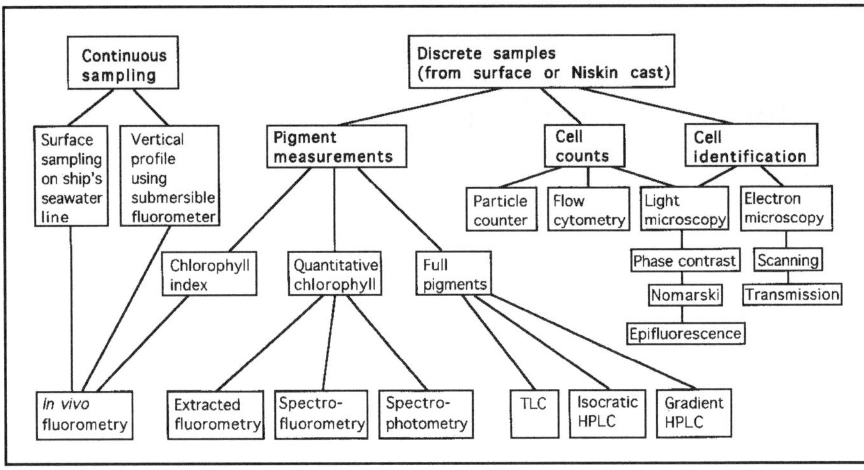

Figure 17.1
Techniques commonly used for phytoplankton and pigment analyses.

17.1.1 *Methods of measuring pigments*

Methods available for analysing phytoplankton pigments are shown in Fig. 17.1. Measurement of chlorophyll *a* with or without other pigments is most common, either alone or in conjunction with microscopy or particle counting. Methods of analysing pigments are reviewed in Chapters 1, 4 and Appendices F, G, and their suitability for various measurements in phytoplankton ecology is summarized in Table 17.1. Approximate analysis times and volumes of seawater required are also given.

Fluorometric methods — in vivo fluorometry

The most sensitive but least accurate method of measuring chlorophyll *a* is fluorometry. *In vivo* fluorometry (the measurement of chlorophyll within living cells) is used over wide geographic ranges to detect phytoplankton in seawater, e.g.:

- for continuous measurements of water pumped from the ship's seawater supply to produce an index of surface chlorophyll abundance along the ship's track,
- to measure vertical profiles of chlorophyll by lowering a submersible profiling fluorometer through the water column,
- for fast measurements on discrete water samples such as those from a Niskin cast, and
- to measure phytoplankton chlorophyll *a* fluorescence by flow cytometry.

In vivo fluorometry is generally only semi-quantitative — the fluorescence of chlorophyll depends on the species present, the time of day, accessory pigments and the physiological condition of the cells (see Chapter 4; Falkowski and Kiefer, 1985). Nevertheless, the ability to collect continuous data without sample preparation makes this an excellent technique for mapping chlorophyll distribution, monitoring phytoplankton abundance and obtaining rapid order-of-magnitude chlorophyll analyses before using other, more accurate, chlorophyll techniques.

Table 17.1. Suitability of various techniques for pigment analysis in phytoplankton ecology (see also Chapter 4).

Technique	Data obtained	Analysis time[a] (min)	Volume required[b] (ml)	Advantages	Disadvantages
In vivo fluorometry	Approximate Chl *a* only	<1	15	Immediate data; sensitive.	Not quantitative; fluorescence signal depends not only on chlorophyll concentration but also on species composition, time of day, accessory pigments, physiological status, fluorescence quenching[c].
Extracted fluorometry	Accurate Chl *a*[d,e]	10	20	Very sensitive; accurate if no Chl *b* present; inexpensive	Interference from chlorophyll derivatives[d,e]
Spectrofluorometry	Accurate Chl *a, b, c*[d]	15	50	Very sensitive	Interference from chlorophyll derivatives; needs continual calibration[d]
Spectrophotometry	Accurate Chl *a, b, c*[d]	10	500	Accurate if no Chl derivatives present	Interference from chlorophyll derivatives[d]
Thin-layer chromatography	Chl *a, b, c*, carotenoids, degradation products	60	30000	2-dimensional TLC gives very good resolution; inexpensive; excellent for pigment purification.	Not suitable for routine analysis of oceanographic samples.
Isocratic HPLC	Accurate Chl *a, b, c*, some carotenoids, degradation products	25	1000[f]	Good for major chlorophylls[g]; simpler and faster than gradient HPLC; suitable for shipboard use.	Medium resolution permits analysis of simple samples only (e.g. cultures or chlorophylls only in field samples)
Gradient HPLC	Accurate Chl *a, b, c*, carotenoids, degradation products	40	1000[f]	Excellent resolution[h] and quantitation; very sensitive for chlorophylls and derivatives with fluorescence detection; suitable for shipboard use.	Expensive to set up; time-consuming for sample preparation, analysis, and data workup (not shown in column 3).

a. Time for analysis, including extraction time (excluding filtration time).
b. Approximate volume of oligotrophic seawater (0.1 µg Chl *a* l^{-1}) required for analysis.
c. Reviewed by Falkowski and Kiefer (1985).
d. Not accurate if samples contain Chl degradation products (see Lorenzen and Jeffrey, 1980; Appendix F and G; Chapter 14).
e. Not accurate if Chl *b* present (see Lorenzen and Jeffrey, 1980).
f. If using fluorescence detection of chlorophylls only, 200 ml can be used.
g. See Chapter 11.
h. Over 50 pigments can be resolved in a 30 min run (see Chapter 12).

The laser-saturated fluorescence of cultured microalgae in an analytical flow cytometer appears to be less susceptible to cell-dependent factors, since it shows a linear relationship with chlorophyll *a* (as determined by HPLC) over a range of three orders of magnitude (Burkill and Mantoura, 1990).

A new technique, fast repetition-rate fluorometry, measures the gradual saturation of photosystem 2 reaction centres (see Chapter 3) by a series of light pulses (Kolber and Falkowski, 1993). The change in fluorescence yield reflects the physiological status of cells and can be used to infer nutrient limitation (Greene *et al.* 1994). This technique shows great promise for biological oceanography (see Chapter 4).

Fluorometric methods — extracted fluorometry

If the chlorophyll is extracted into methanol or acetone before measurement ('extracted fluorometry', Holm-Hansen *et al.* 1965, Table 17.1), the response is quantitative only if chlorophyll *a* is the dominant chlorophyll and there are no chlorophyll *b*, pheophytins, pheophorbides or chlorophyllides present (see Chapter 14). The fluorometer must also be regularly calibrated. Non-linearity may occur at high chlorophyll concentrations due to absorption of emitted photons by other chlorophyll molecules and scattering from non-fluorescent particles.

Some problems of interference from other chlorophylls and degradation products may be reduced by spectrofluorometry (Neveux and Lantoine, 1993), where the excitation and emission wavelengths are varied to detect chlorophylls *a*, *b* and *c* and their degradation products separately. This technique has not been adopted so far by the oceanographic community, since it still needs extensive testing, and pure standards must be provided for frequent calibration of the spectrofluorometer.

Spectrophotometry

Spectrophotometric analysis of pigment extracts can accurately measure chlorophylls *a*, *b* and *c* (if these are the only pigments present) through the use of trichromatic equations (e.g., Jeffrey and Humphrey, 1975; Lorenzen and Jeffrey, 1980). Any pheopigments and chlorophyllides in the sample are not measured and will interfere with the measurement of the chlorophylls (see Chapter 14). Carotenoids cannot be differentiated in mixtures.

Chromatographic techniques

Thin-layer chromatography (TLC) allows analysis of all major phytoplankton pigments, and historically this technique was the basis of pigment chromatographic studies in oceanography (Jeffrey, 1974; see Chapters 1, 4). The volumes of water and analyst's time required now preclude the use of this technique for routine oceanographic samples, except for very small sample numbers. TLC is still invaluable in preparing pure pigments for standards or structure determinations (see Chapters 7 and 8).

Isocratic HPLC can rapidly analyse chlorophylls and derivatives from relatively small volumes of seawater (see Chapter 11). It is especially sensitive when coupled with fluorescence detection. Its limited resolution, however, usually prevents its use for full pigment analysis.

Gradient HPLC is the method of choice for full analysis of chlorophylls and carotenoids in oceanography (see Chapter 12 and Wright *et al.* 1991). The equipment is, however, expensive and the analyst requires more time and technical expertise than for non-chromatographic techniques.

17.2 Field-sampling stategy

The main considerations in field sampling are the sites, frequency and spatial distribution of the samples. To a large extent these depend on the objectives of the cruise, knowledge of the area, the equipment available and the requirements of other programs.

Sampling sites are normally on a regular transect or grid, usually with greater frequency in areas of known discontinuities, such as shelf breaks or fronts. Sampling sites can also be targetted by 'real-time' satellite images of the area. Continuous measurement of water conditions with a thermosalinograph and of chlorophyll concentrations with a flow-through fluorometer on the ship's seawater line can alert the researcher to changes in conditions and the need to sample more frequently. Similarly, a CTD can identify the position of a pycnocline and a profiling fluorometer can locate layers of phytoplankton in the water column that would otherwise be easily missed if sampling 'blind' at standard depths. The continuous data provided by a flow-through or a profiling fluorometer can greatly facilitate interpretation of the discrete data produced by HPLC or other techniques.

When sampling for HPLC, samples should also be collected for microscopy of living cells, preparation of preserved material for 'settled' cell counts and identification (e.g., with Lugol's iodine or buffered glutaraldehyde), flow cytometry (Lepesteur *et al.* 1993) and scanning and transmission electron microscopy. Each of these techniques provides valuable species data for interpretation of HPLC pigment profiles. Sometimes the HPLC data can be used to determine which samples should be studied in detail by microscopy, thus reducing this time-consuming but now not too difficult task (see Chapter 2).

If samples are collected with a CTD/rosette sampler, there is always the possibility of a misfire and the sample set not corresponding to the correct depths. The rosette must be checked after each cast to ensure that the firing mechanism is aligned to the correct position. Also, the sampling depths can be checked if reversing thermometers are attached to the Niskin bottles or if salinity samples are taken for comparison with the CTD profile. On a multidisciplinary cruise, these checks will normally be made routinely by physical oceanographers.

17.3 Field-sampling methods

The general protocol for collecting field samples is shown in Fig. 17.2, and described in more detail in the following sections.

17.3.1 *Collection of samples*

Seawater samples are collected from the surface by bucket, from the ship's seawater line (which may require frequent decontamination with biocide to prevent fouling with algae), from depth by a Niskin bottle or similar sampler or from up to 50 m depth by a deck-mounted pump. Samples must be collected in opaque bottles, and filtered immediately (not more than one hour's delay). They must be protected from light and warmth at all times during collection and filtration (particularly deep ocean samples). The sample should be gently mixed before subsampling to resuspend any

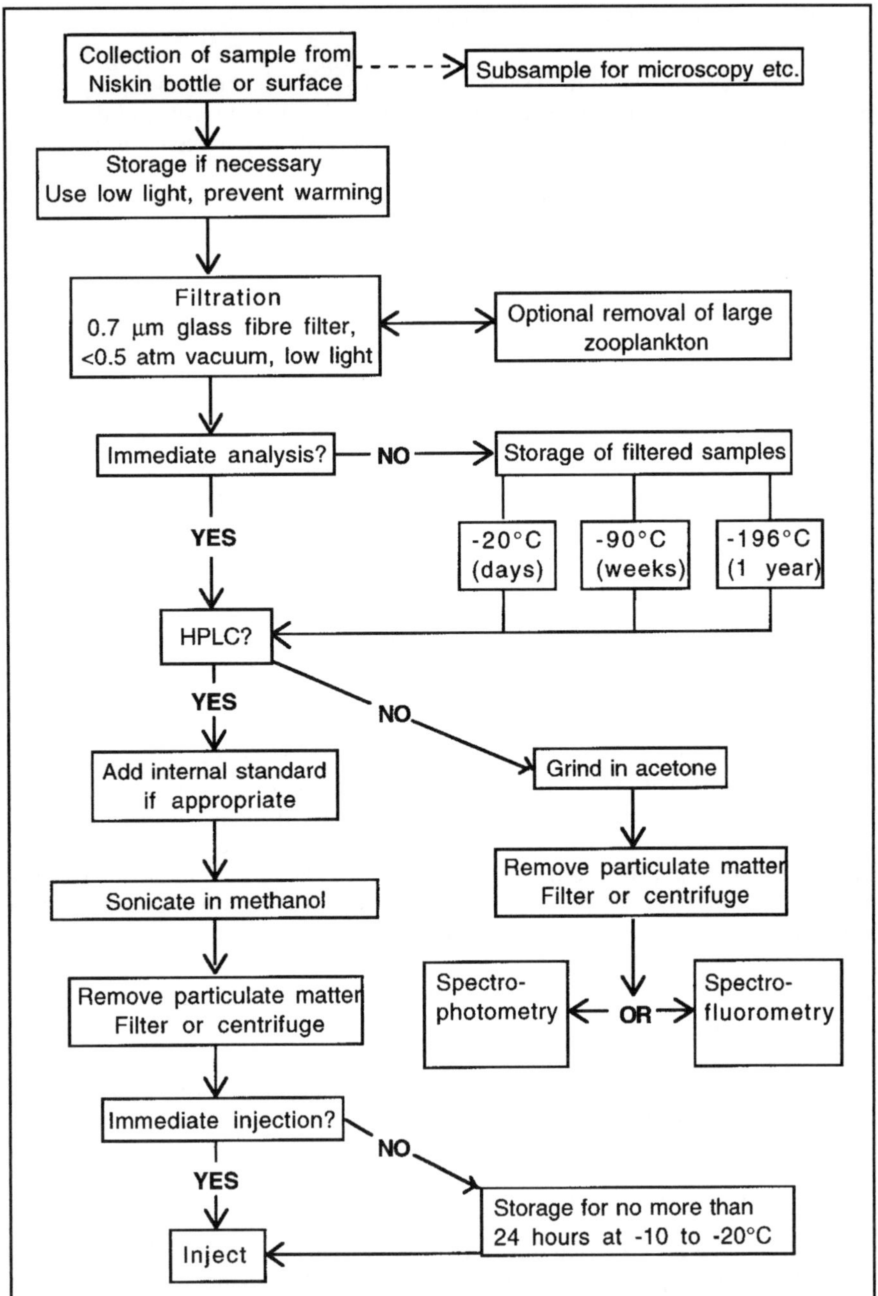

Figure 17.2
Summary of protocols for collection and analysis of phytoplankton pigments.

large cells that may have settled. A subsample should be preserved for microscopy as described above.

17.3.2 *Removal of zooplankton*

Large zooplankton present significant problems for pigment analysis. For example:
- their lipids interfere with chromatography,
- a single organism may contain sufficient chlorophyll in its gut to overwhelm the phytoplankton chlorophyll,
- they are common enough to be encountered in samples, yet very rare compared to phytoplankton — typically ranging from 0.1–2 copepods per litre. Many hundreds of litres of water would be required to accurately sample total particulate chlorophyll including the zooplankton contribution.

There are three strategies:
- ignore the problem and accept the additional variability in the results,
- remove zooplankton by pre-filtering the sample through 100 µm nylon mesh. This will remove copepods, salps and other small zooplankton animals but could also remove much of the phytoplankton if large diatoms, such as *Thalassiothrix*, or colonial or chain-forming species such as *Phaeocystis*, *Trichodesmium*, *Chaetoceros* or *Rhizosolenia* are present. The material retained on the prefilter should be washed off with filtered seawater and re-filtered to ensure that phytoplankton has not been removed.
- examine the filtered material before freezing and remove any large zooplankton with fine forceps. They are easily visible and quickly removed.

17.3.3 *Filtration*

A fine glass-fibre filter, such as a Whatman GF/F (0.7 µm, nominal pore size) should be used for HPLC sampling (see Chapter 10). If necessary, a larger pore-size filter, such as a Whatman GF/D, can be used on top of the GF/F to delay clogging with turbid samples. Normally, a 25 mm diameter filter is used — a larger filter is not necessary unless large amounts of pigments are required. On most multidisciplinary cruises, only a small volume of water is available from a Niskin bottle for pigment analysis (e.g., 1 l). A small filter will collect more sample per unit area of the filter and so achieve greater concentrations in the final extract. For size-fractionation, a controlled pore-size membrane filter such as Nuclepore, Microporetics or Millipore Isopore can be used, although the Nuclepore filters release a yellow pigment during subsequent extraction. (The other filters were not tested by the SCOR Working Group.) Cellulose acetate/nitrate membrane filters such as Millipore MF will partly dissolve during extraction in methanol or acetone and cannot be used for HPLC, although they can be used for spectrophotometry or fluorometry. They also have slower filtration rates than glass-fibre filters with corresponding pore-sizes.

Samples should be filtered with a vacuum of not more than 0.5 atm to avoid damaging cells, and the sample should be mixed occasionally during filtration. Magnesium carbonate should not be used as a filter aid (UNESCO, 1966; Daley *et al.* 1973) since it preferentially adsorbs chlorophyllides (Jeffrey and Hallegraeff, 1987). The filtration process should be monitored continuously and the water stopped before the filter clogs. Usually there will be sufficient pigment for analysis if any colour is visible on the filter to the naked eye, unless sediments are present. The last of the water should be gently sucked through the filter and any visible zooplankton removed from its surface. Then the filter should be folded once (algae inside), blotted between several layers of absorbent paper, and placed in a labelled container for immediate freezing.

17.3.4 *Storage of filtered samples*

Ideally, samples should be frozen in liquid nitrogen, but they may be kept for several days at -20°C, or for weeks at -90°C, without substantial loss of pigment (see Chapter 10). Samples for liquid nitrogen freezing must be stored in proper cryotubes or wrapped in aluminium foil. Other sorts of sealed plastic or glass tubes are very dangerous, as they can explode on warming if liquid nitrogen has leaked into them. Conventional plastic bags disintegrate in liquid nitrogen and must not be used. Cryotubes can be labelled with a vibrating engraver to ensure permanent marking — ink labels may be washed off by drips of solvent during handling.

For short-term storage of samples in a conventional freezer, thin plastic bags can be used. They allow rapid freezing and are cheap and convenient.

17.3.5 *Logging sample data*

At the time of sample collection, the following data must be logged onto the cruise data sheets: station number, date, Greenwich Mean Time, latitude, longitude, depth, bottle number, volume filtered, sample tube number, comments about the sample or location. A hard copy of the filtration details must be kept for data security, rather than just a computer file.

17.4 Fluorometry and spectroscopy of sample extracts

If extracted samples are to be analysed by fluorometry or spectroscopy, the sampling procedure is identical to that for HPLC, except that less sample is required (Table 17.1). Although sonication in methanol has been recommended for extraction for HPLC, spectrophotometric equations to correct for absorption by chlorophylls *b* and *c* have not yet been developed for this solvent as they have been for 90% acetone (Jeffrey and Humphrey, 1975). Extinction coefficients for chlorophylls *a, b* and *c* in methanol are available (Jeffrey, unpublished: see Appendix F and G), and equations should soon be derived for this solvent. Until then, we recommend 90% acetone as an extraction solvent for spectrophotometry, with grinding rather than sonication (see Chapter 9). Soaking samples on filters overnight in acetone, although commonly done, is likely to produce serious underestimates of chlorophyll concentrations in some phytoplankton samples (see Chapter 9).

Extinction coefficients and binary simultaneous equations have recently been redetermined for chlorophyll *a* and *b* in methanol for higher plants and green algae (Porra *et al.* 1989). As methanol was superior to acetone for extracting pigments from chlorophytes (Chapter 9) it is recommended for samples dominated by chlorophytes. Methanol can not be used for extracted fluorometry, as no fluorometric equations in methanol are available. Since chlorophylls form allomers in methanol (see Chapter 3), analyses must be completed quickly if this solvent is used.

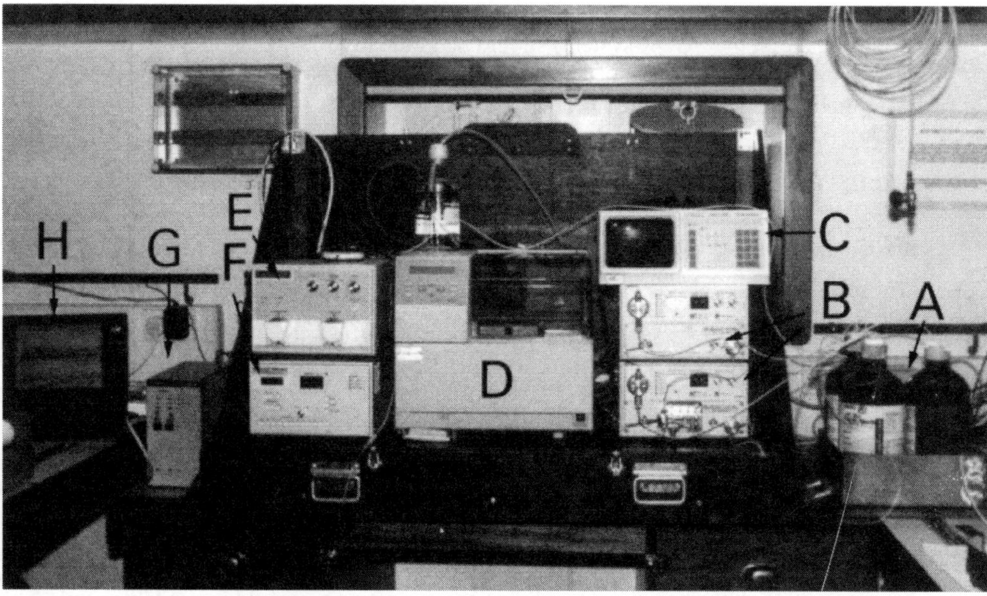

Figure 17.3
HPLC system for use at sea. A: Stainless-steel spill tray supporting two amber solvent degassing bottles and a waste bottle. B: Twin high pressure HPLC pumps (Shimadzu LC-6A). C: Controller (Shimadzu SCL-6B) for HPLC pumps, gradient program, and detectors.
D: Refrigerated (2°C) derivatising autosampler, injector and column oven (30°C) (Thermo-Separations AS-3000). E: UV-visible detector (Shimadzu SPD-6AV set at 440 nm). F: Fluorescence detector (Shimadzu RF-535 fitted with red-sensitive photomultiplier set at λ_{ex} = 405 nm, λ_{em} = 670 nm). G: Two-channel analog-digital converter (Phillips PU-6000). H: Laptop 486 microcomputer for recording and processing chromatographic data. System used by the Plymouth Marine Laboratory on a recent RRS *Discovery* cruise, Arabian Sea, 1994.

17.5 HPLC ANALYSIS OF FIELD SAMPLES

17.5.1 *Preparing the HPLC system*

After the HPLC has been set up (see Chapter 15 and Fig. 17.3) and calibrated with external standards of each major pigment (see Chapter 16), the system can be prepared for analysis.

The elution solvents should be sparged with helium with a reasonable stream of bubbles (about 100 ml min^{-1}) for about 5 min before starting the HPLC system, after which the stream should be reduced to a trickle to avoid changing the composition of the solvents by evaporation.

Air-saturated solvent in the inlet tubes to the pump should be purged by operating the pump at about 10 ml min^{-1} with the column bypass valve open. Then the valve should be closed and solvent in which the column has been stored overnight (ethyl acetate in the SCOR system) should be gradually flushed and replaced by the first eluent (buffered methanol), using the pump startup program (Chapter 12). The system should be carefully checked against a check list (Chapter 15) before proceeding.

17.5.2 *Extraction — general*

Immediately before extraction, an aliquot of internal standard should be added to the filter with a precision pipette, as described in Chapter 16. Methanol should be added (3.5 ml for a 25 mm GF/F filter, less for a smaller one) and the pigments extracted with a probe sonicator (50W, 30 s). Buffered methanol is not as effective an extractant as ethanol. Dimethyl formamide (DMF) is the best solvent for extracting pigments from recalcitrant algae (Chapter 9), but it is not recommended for routine use as it is more hazardous to handle. Representative duplicate samples should occasionally be extracted in DMF to ensure the pigment content has not been seriously underestimated.

Before an aliquot can be injected into the HPLC, particulates must be removed. Filter debris can be removed by centrifugal filtration (Fig. 17.4A,B) which gives almost complete recovery of pigment-containing solvent in a single run (Wright and Shearer, 1984). Alternatively, if a solid-phase extraction manifold is available, multiple samples can be conveniently filtered through methanol-resistant Luer-lock filters (<0.5 μm pore-size) instead of the SPE cartridges (Fig 17.4C,D). The clarified extract should be transferred to an amber autosampler vial and stored in a freezer or refrigerated autosampler until injected (no more than a few hours).

17.5.3 *Sample-injection sequence (i.e. sample queue)*

Before a group of samples is analysed, the HPLC column should be equilibrated by running the analytical solvent-gradient program. Next, a reference chromatogram should be run (from a standard SCOR reference algal mixture, see Chapter 13), to determine the retention times and resolution of key pigments for the day's analyses. Filters of SCOR cultures can be conveniently carried to sea in liquid nitrogen (Chapter 10) for this purpose.

Next, the field samples should be injected. The time between injections should be kept as constant as possible, so that differences in column re-equilibration do not affect retention times. Finally, the reference sample should be re-injected to confirm the correct functioning of the system, to check that retention times have not drifted so far that the HPLC software misidentifies peaks, and to check the injection volume if an internal standard is not used.

17.5.4 *Injection protocol*

Immediately before injection, the sample is diluted by taking 120 μl of sample, adding 30 μl of buffer (ammonium acetate, 0.5 M, pH 7.1) and mixing well. Normally, 125 μl of the mixture is injected, but if the pigment concentration is sufficient, a smaller amount can be used, which will produce sharper peaks (Appendix H). Any dead volume in the injector must be taken into account during the injection and in the subsequent calculations if using external standard calibration.

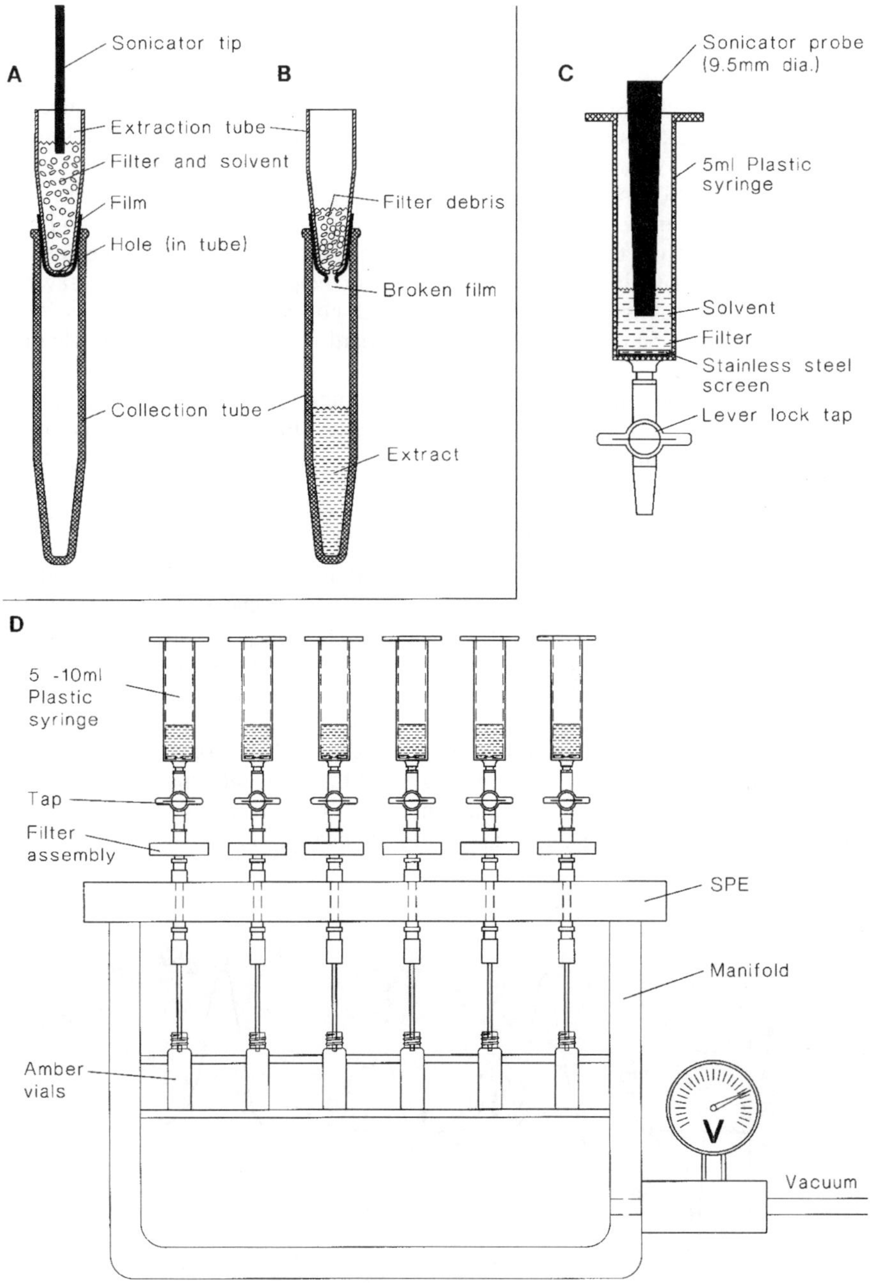

Figure 17.4
Sonication and centrifugal extraction. A. Filter pieces are sonicated in solvent in an extraction tube made by cutting a 15 ml plastic centrifuge tube at the 7 ml mark and making a tiny hole at the bottom. The hole is covered by a single layer of Nescofilm or Parafilm. When the extraction tube is centrifuged (700 g, 3 min) on top of a collection tube, the solvent bursts through the film and collects in the bottom tube. B. After centrifugation, the filter debris remains in the upper tube. C. The filter is sonicated in the barrel of a plastic syringe (13 mm internal diameter, with stainless steel screen from 13 mm Swinney filter and Luer-lock tap attached). D. Filtration using an SPE manifold into an autosampler vial with a solvent-resistent Luer-lock filter instead of an SPE cartridge.

Particular care must be taken to avoid injecting bubbles, since they reduce the volume injected and cause baseline irregularities. For manual injection, the sample can be drawn slowly into the syringe and any bubbles removed by tapping. Persistent bubbles can be dislodged by holding the syringe needle vertically against a rubber stopper and compressing the syringe contents (eye protection!) — the bubbles compress, detach from the syringe wall and float to the top, from where they can be ejected. For automatic injection, the syringe should be programmed to withdraw the sample from the vial as slowly as possible, which normally prevents bubbles forming. A reduction in the recovery of internal standard will indicate that a bubble may have been injected.

The injection syringe should be left inserted until the sample has been injected onto the column (see Chapter 15), as this will lessen the chances of sample or air leaking through the injection port.

17.5.5 *Checking chromatogram data*

Most chromatography software will automatically integrate the peaks after collecting the data. The chromatograms should immediately be checked for any obvious problems that may require re-injection of sample. Then the peak integration must be checked (see Chapter 15). Manuals for the particular software used describe how to set integration variables (slope sensitivity, noise rejection parameters etc.) to optimize integration, but it is still very likely that the program will make mistakes, such as those illustrated in Fig. 17.5 and described in Table 17.2. The operator must check each chromatogram to ensure that baselines and the starts and ends of peaks have been correctly located by the program, and adjust them if not. Dyson (1990) discusses integration problems and strategies for their solution.

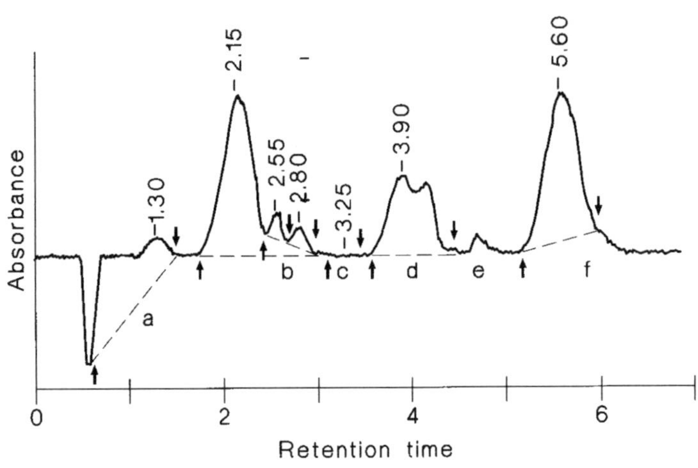

Figure 17.5
Examples of problems encountered during integration of an HPLC chromatogram. Six peaks have been recognized and integrated by the software. For each peak, the start (↑) and end(↓) are marked, along with the retention time (top) and projected baseline (dotted lines). Integration problems **a** to **f** are described in Table 17.2. Retention time is in min.

Table 17.2. Problems with integration of a chromatogram, and their solution (see Fig. 17.4).

Peak area	Problem	Solution
a	Upslope of negative peak has been interpreted as the start of a peak and not resolved until downslope of peak 1.30.	Inhibit integration until after negative peak has reached the baseline.
b	Two trailing peaks have been interpreted as riding on a long tail of peak 2.15, and 'skimmed', and their areas underestimated.	Reduce skim height ratio.*
c	Baseline noise has been interpreted as a peak.	Increase noise rejection and/or slope sensitivity parameters
d	Partially resolved peaks not separated by software.	Decrease slope-sensitivity parameter.
e	Peak has been missed.	Decrease noise rejection and/or slope sensitivity parameters.
f	Peak end has been misplaced.	Decrease slope-sensitivity parameter.

* The ratio at which the software defines the baseline of a small peak as the tailing edge of a big peak.

17.5.6 *Identifying pigment peaks*

Peaks are routinely identified by their retention time, together with their spectra. Injection of a known sample at the start and end of each sample queue (Section 17.5.3) greatly simplifies identification, but where there is any doubt, the sample should be co-injected with a known standard (Appendix I). An aliquot of the sample containing the unknown pigment should be mixed with an appropriate volume of an extract from a SCOR culture (or other standard) containing the authentic pigment so that the peak height in the standard equals that in the unknown. When the mixture is chromatographed, there should be a single peak without shoulders. The identification should be confirmed by co-elution in a different chromatography system (Appendix I).

Spectral data can be obtained automatically if a diode-array or other full-spectrum detector is used. The software may allow a direct comparison of the spectra with a library of known spectra (see Part IV Graphics sheets) as well as matching the 'goodness-of-fit'. If there is a poor signal-to-noise ratio (Appendix H), the spectra may be inadequate for identification. In this case, or if there is no diode-array detector available, spectra must be obtained by collecting the emerging peaks and examining them with a spectrophotometer. It may be necessary to pool pigments from several samples and concentrate them by solid-phase extraction to get sufficient pigment for fractionation of minor peaks in a chromatogram. Criteria for spectral identification are given in the Part IV Data and Graphics sheets.

If still unsure of the identity of a pigment, it should be temporarily named by reference to a similar compound e.g., 'prasinoxanthin-like', but its spectral information (λ_{max} and band ratios) and position on the chromatogram (retention time) must also be given to enable subsequent identification (see Appendix I).

After completion of a batch of samples, it is useful to plot the retention times of

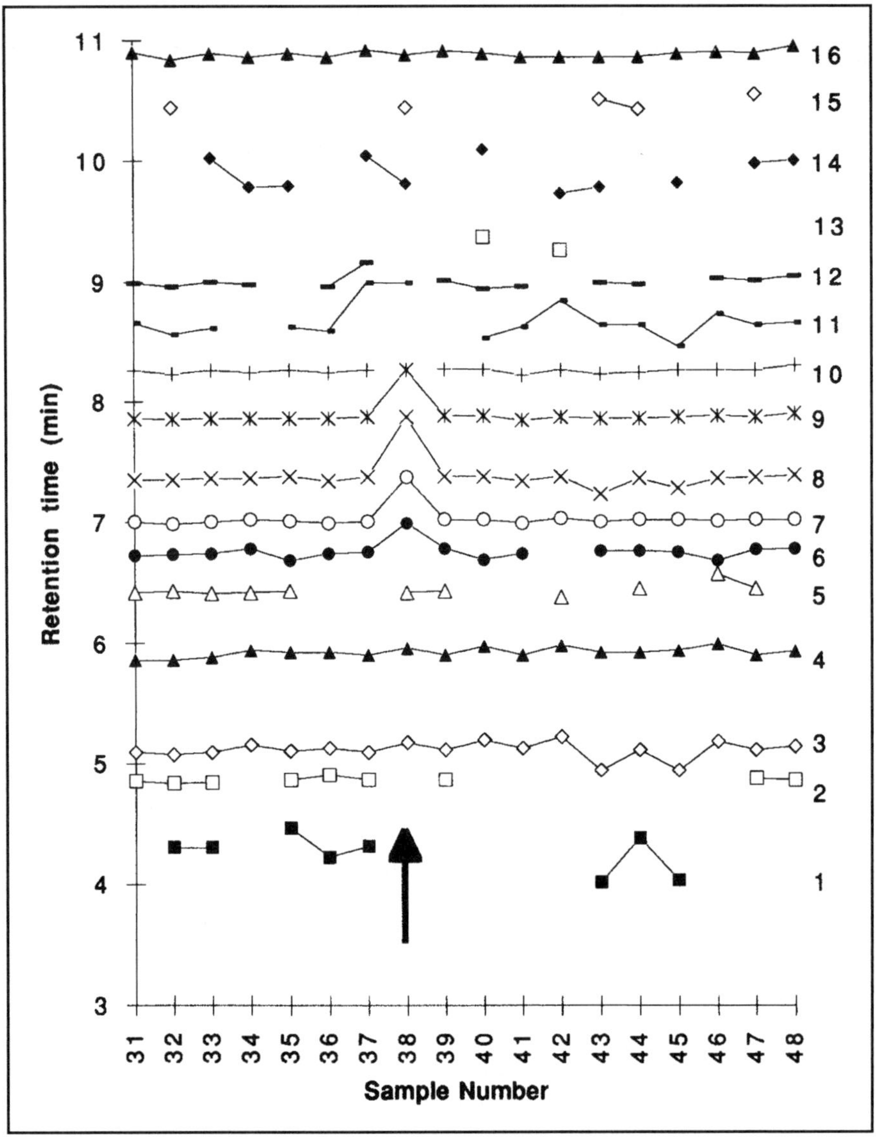

Figure 17.6
Plot of retention times (RT) of 16 pigments (numbered at right) from each of 18 consecutive HPLC samples for checking peak identification. The discontinuity for sample 38 (arrowed at RT 6.7–8.3 min.) indicates probable misidentification of pigments. The range of RT for the peak identified as violaxanthin (pigment 14) suggests that there may be two components. In some samples, peaks 2 and 11 appear to have been misidentified as peaks 3 and 12 respectively.
Peak identification: 1, unknown; 2, chlorophyllide *a*; 3, chlorophyll c_3; 4, chlorophylls c_1 and c_2; 5, methyl chlorophyllide *a*; 6, *cis*-peridinin; 7, peridinin; 8, 19'-butanoyloxyfucoxanthin; 9, fucoxanthin; 10, 19'-hexanoyloxyfucoxanthin; 11, *cis*-fucoxanthin; 12, *cis*-19'-hexanoyloxyfucoxanthin; 13, prasinoxanthin; 14, violaxanthin; 15, diadinoxanthin derivative; 16, diadinoxanthin.

each pigment for each sample in a queue in order of analysis (Fig. 17.6). This allows the amount of drift and scatter in retention times through the day to be seen, and to check that a peak has not migrated into the retention time window of another pigment or otherwise been misidentified.

17.5.7 *Quantitating pigments on HPLC*

Pigment quantitation by HPLC is described in detail in Chapter 16, to which the reader should refer.

17.5.8 *Quantitating unknown or uncalibrated pigments*

When an unknown peak is seen on a chromatogram, it can be approximately quantified if its spectrum is obtained in eluent. The spectra and molar extinction coefficients of carotenoids are very similar for compounds with the same chromophore, irrespective of the side-group constituents (see Chapter 8, Part IV Data sheets and Appendix D). Spectra within a family of chlorophylls or chlorophyll derivatives are often identical (e.g., Chl *a* and Chlide *a*; Chl *b* and Chlide *b*). If the unknown spectrum matches that of a known pigment, the response factor can be transferred on a molar basis. If not, an estimate of the response factor can be calculated as follows:

Figs 17.7 A, B show the stylized spectrum of a compound (chlorophyll *a*) for which the HPLC has been calibrated and an HPLC-uncalibrated pigment, respectively, in eluent. For the calibrated compound, A_{max_c} is the maximum absorbance (at wavelength λ_{max_c}, for which the specific extinction coefficient, α_{max_c}, is known) and

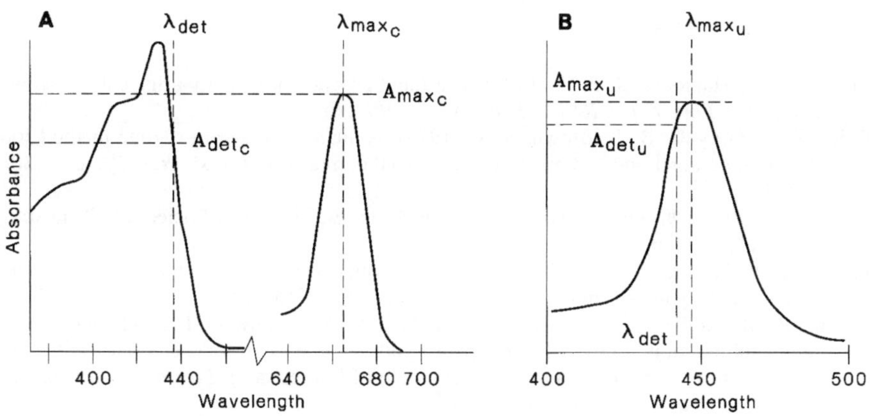

Figure 17.7
Measurements for calculating an approximate response factor for an unknown or uncalibrated pigment (B) from the spectra of a calibrated pigment (A). Parameters: λ_{det}, detection wavelength (436 nm); A_{det_c}, absorbance at detection wavelength for calibrated (c) compound; A_{max_c}, maximum absorbance of calibrated compound (at wavelength λ_{max_c}); similarly A_{det_u}, A_{max_u}, λ_{max_u} refer to parameters for the uncalibrated (u) compound. Note the discontinuous ordinate axis in A.

A_{det_c} is the absorbance at the detection wavelength, λ_{det}. The absorbance ratio, R_C, is the ratio of absorbances at the detection and maximum wavelengths:

$$R_c = \frac{A_{det_c}}{A_{max_c}}$$

A_{max_u}, λ_{max_u}, A_{det_u} and R_u are the corresponding parameters for the uncalibrated compound. For the calibrated compound, the response factor at the detection wavelength (R_{det_c}) is proportional to the specific extinction coefficient (α_{max_c}) at λ_{max_c} (see Part IV, section 1.6) and the ratio of absorbances at λ_{det} and λ_{max_c}:

$$R_{det_c} = k \,.\, \alpha_{max_c} \,.\, R_c \qquad \text{(equation 17.1)}$$

where k is a constant. Similarly, for the uncalibrated compound:

$$R_{det_u} = k \,.\, \alpha_{max_u} \,.\, R_u \qquad \text{(equation 17.2)}$$

Dividing equation 17.1 by equation 17.2 and multiplying by R_{det_c} produces:

$$R_{det_u} = R_{det_c} \cdot \frac{\alpha_{max_u} \cdot R_u}{\alpha_{max_c} \cdot R_c} \qquad \text{(equation 17.3)}$$

This can be used to approximately calibrate a pigment whose identity and specific extinction coefficient (α_{max_u}) are known but for which an appropriate calibration standard is not available. If the pigment elutes in a different part of the chromatogram from the reference compound, solvent differences will reduce the accuracy of this technique. If the peak is unknown, an appropriate α_{max} must be chosen from a similar compound. For carotenoids, an $E^{1\%}_{1\,cm}$ of 2500 (ie. $\alpha_{max_u} = 250$) is normally used. The method and bases of such calculations must be reported for correction later if necessary.

REFERENCES

Burkill, P. H., Mantoura, R. F. C. (1990). The rapid analysis of single marine cells by flow cytometry. *Phil. Trans. R. Soc. Lond. A*, 333: 99–112.

Daley, R. J., Gray, C. B. J., Brown, S. R. (1973). A quantitative, semi-routine method for determining algal and sedimentary chlorophyll derivatives. *J. Fish. Res. Bd. Can.* 30: 345–356.

Dyson, N. (1990). *Chromatographic Integration Methods*, The Royal Society of Chemistry, Cambridge, 160 pp.

Falkowski, P., Kiefer, D. A. (1985). Chlorophyll *a* fluorescence in phytoplankton: relationship to photosynthesis and biomass. *J. Plankt. Res.* 7: 715–731.

Greene, R. M., Kolber, Z. S., Swift, D. G., Tindale, N. W., Falkowski, P. G. (1994). Physiological limitation of phytoplankton photosynthesis in the eastern equatorial Pacific determined from variability in the quantum yield of fluorescence. *Limnol. Oceanogr.* 39: 1061–1074.

Holm-Hansen, O., Lorenzen, C. J., Holmes, R. W., Strickland, J. D. H. (1965). Fluorometric determination of chlorophyll. *J. Cons. Perm. Int. Explor. Mer.* 30: 3–15.

Jeffrey, S. W. (1974). Profiles of photosynthetic pigments in the ocean using thin-layer chromatography. *Mar. Biol.* 26: 101–110.

Jeffrey, S. W., Hallegraeff, G. M. (1987). Chlorophyllase distribution in ten classes of phytoplankton: a problem for chlorophyll analysis. *Mar. Ecol. Progr. Ser.* 35: 293–304.

Jeffrey, S. W., Humphrey, G. F. (1975). New spectrophotometric equations for determining chlorophyll *a*, *b*, c_1 and c_2 in higher plants, algae and natural phytoplankton. *Biochem. Physiol. Pflanz.* 167: 191–194.

Kolber, Z., Falkowski, P. G. (1993). Use of active fluorescence to estimate phytoplankton photosynthesis *in situ*. *Limnol. Oceanogr.* 38: 1646–1665.

Lepesteur, M., Martin, J. M., Fleury, A. (1993). A comparative study of different preservation methods for phytoplankton cell analysis by flow cytometry. *Mar. Ecol. Progr. Ser.* 93: 55–63.

Lorenzen, C. J., Jeffrey, S. W. (1980). *Determination of Chlorophyll in Seawater: Report of Intercalibration Tests.* UNESCO Technical Papers in Marine Science, 35, 21 pp.

Neveux, J., Lantoine, F. (1993). Spectrofluorometric assay of chlorophylls and phaeopigments using least squares approximation technique. *Deep-Sea Res.* 40: 1747–1765.

Porra, R. J., Thomson, W. A., Kriedemann, P. E. (1989). Determination of accurate extinction coefficients and simultaneous equations for assaying chlorophylls *a* and *b* extracted with four different solvents: verification of the concentration of chlorophyll standards by atomic absorption spectroscopy. *Biochim. Biophys. Acta*, 975: 384–394.

UNESCO (1966). *Determination of Photosynthetic Pigments in Seawater.* UNESCO Monographs on Oceanographic Methodology, Paris, 69 pp.

Wright, S. W., Shearer, J. D. (1984). Rapid extraction and high-performance liquid chromatography of chlorophylls and carotenoids from marine phytoplankton. *J. Chromatogr.* 294: 281–295.

Wright, S. W., Jeffrey, S. W., Mantoura, R. F. C., Llewellyn, C. A., Bjørnland, T., Repeta, D. J., Welschmeyer, N. (1991). Improved HPLC method for the analysis of chlorophylls and carotenoids from marine phytoplankton. *Mar. Ecol. Progr. Ser.* 77: 183–196.

Part IV

Data for the identification of 47 key phytoplankton pigments

Part IV

Data for the identification of 47 key phytoplankton pigments

Compiled by S. W. Jeffrey, R. F. C. Mantoura and T. Bjørnland

IV.1 PREAMBLE: NOTES ON THE USE OF THE DATA AND GRAPHICS SHEETS

IV.1.1 *Introduction*

The correct identification of pigments from marine samples is dependent on the availability of reliable spectral, chromatographic and chemical data for each pigment. This information is widely scattered in the literature and, although excellent reviews and listings such as those of Vernon and Seely (1966) and Scheer (1991) for chlorophylls, and Davies (1965, 1976), Isler (1971), Goodwin (1980) and Rowan (1989) for carotenoids are available, these deal with pigments of the entire plant and animal kingdoms. We believe that a compilation of pigment data relating only to phytoplankton would greatly assist oceanographers and phycologists. We have therefore listed the most accurate data for 17 chlorophylls (including derivatives) and

30 carotenoids that are most frequently encountered in HPLC analyses of phytoplankton and particulate material from the marine environment. The primary literature has been supplemented with results obtained during the SCOR WG 78 Workshops. The phycobiliproteins, and pigments of photosynthetic bacteria and marine animals, are not included.

The data on each pigment are arranged on two pages. The right-hand page (Data sheet) lists the name, occurrence, physico-chemical, spectroscopic and chromatographic properties of each pigment. The left-hand page (Graphic sheet) shows the UV-visible spectrum in reference solvent, the diode-array spectrum in HPLC eluant, the algal source, the HPLC chromatogram from which the pigment was isolated, and the molecular structure of the pigment.

IV.1.2 *Pigment name*

The common or trivial name is given on the top of both sheets, together with the full IUPAC name, including stereochemistry where appropriate. The chlorophylls are taken from Scheer (1991) and the carotenoids from IUPAC (1972), Straub (1987) and Köst (1988).

IV.1.3 *Abbreviation and peak number*

As with many biochemical compounds, such as amino acids and nucleotides, some of the lengthy names of pigments must be abbreviated for tables and figures (e.g., Mg-3,8-divinyl-pheoporphyrin a_5 mono-methyl ester and 19'-hexanoyloxy-fucoxanthin). A useful set of abbreviations (e.g., MgDVP and Hex-fuco, respectively), which retain their phonetic origins, is included for each pigment. These are also listed for all pigments in Appendix A.

The pigment peak number comes from the SCOR HPLC reversed-phase method of Wright *et al.* (1991), which we describe in Chapter 12. In the Graphics sheet, the relevant peak is highlighted in the HPLC chromatogram of the SCOR culture from which the pigment was obtained.

IV.1.4 *Occurrence and colour*

The distribution of each pigment in the different algal classes is given. In addition, its relative abundance is classified as follows:

Notation	Abundance relative to total pigment
Trace	< 1 %
Minor	1–10 %
Major	> 10 %

The colour of each pigment is that observed on TLC plates (e.g., cellulose, sucrose, silica gel, Merck RP8 HPTLC). The colour is determined by the absorbance characteristics of the chromophore, the pigment concentration, the adsorptive characteristics of the TLC solid phase and the solvent polarity. Because pigments bleach rapidly on air-dried TLC plates, the colours should be observed in dim light immedi-

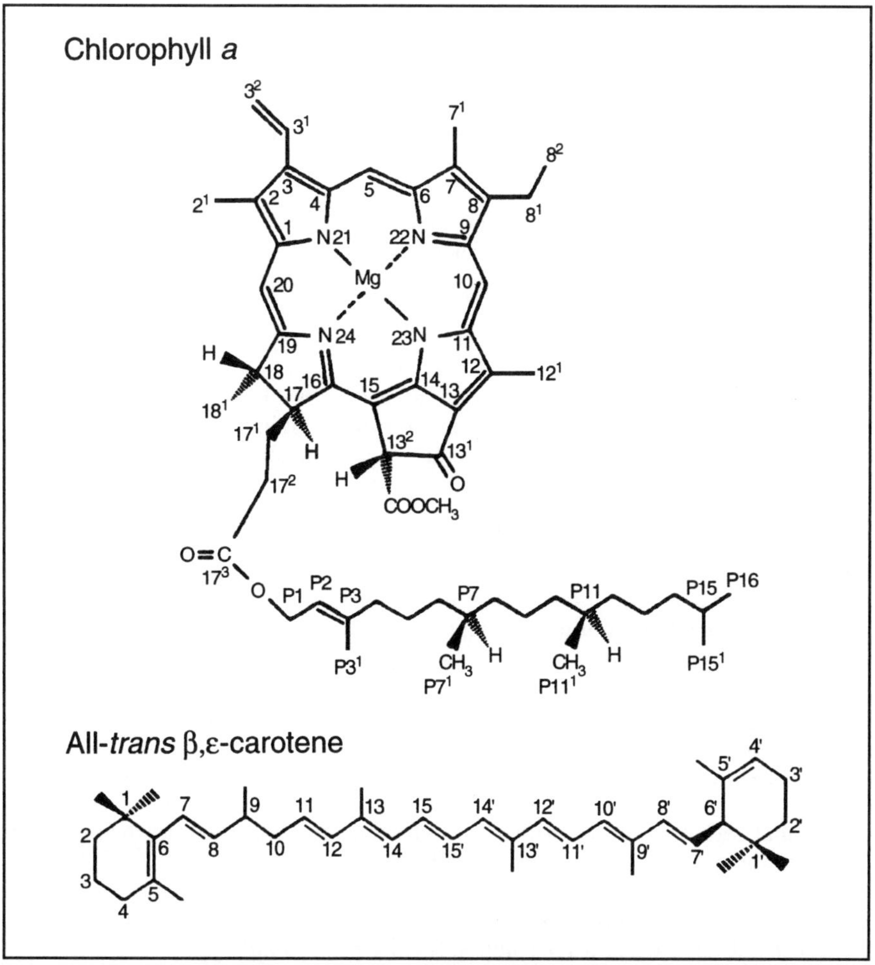

Figure IV.1
The IUPAC structure, numbering system and stereochemistry of chlorophyll *a* and all-*trans* ß,ε-carotene.

ately after completion of the chromatogram, while the plate is still wet and before the solvent has completely evaporated. Wet plates can be wrapped in aluminium foil to exclude light and immediately frozen (-20 °C), which will preserve their colour for several hours.

IV.1.5 *Molecular structure, formula and weight*

All molecular structures and names are based on the IUPAC (1972) numbering and notation system detailed by Scheer (1991) for chlorophylls and by Isler (1971), Davies (1976) and Straub (1987) for the carotenoids. The full stereochemical structures for chlorophyll *a* and β,ε-carotene are shown in Figure IV.1, together with the position numbering system used in the IUPAC names. The stereochemistry of each pigment is listed in the Data sheets, but several may differ slightly from those shown in the Graphics sheets due to recent advances in chiral technologies.

The IUPAC system of atomic weights (C = 12.011; H = 1.0079; O = 15.9994; N = 14.0067; Köst, 1988) was used together with the molecular formula to calculate the molecular weight. The molecular formulae were derived from the primary literature and from Davies (1976), Köst (1988) and Straub (1987). If structures are only tentative (e.g., phytylated chlorophyll c_2-like pigment; Nelson and Wakeham, 1989) then any necessary assumptions in the calculations of molecular structures and weights are also indicated.

IV.1.6 *Extinction coefficients*

Extinction coefficients express the quantitative relationship between the absorption at a specific wavelength of a known amount of pigment in a known volume of solvent. They are used for the spectrophotometric determination of pigment concentrations. The relationships are defined as:

$$A = \log \frac{I_o}{I} = \varepsilon \cdot C_m \cdot d = \alpha \cdot c \cdot d = E^{1\%}_{1\,cm} \cdot C \cdot d$$

where:

d = the inside length of the cuvette in centimetres,
c = the concentration of the pigment in g l^{-1},
C_m = molar concentration
C = concentration in g/100 ml
I_o, I = the light intensities transmitted by pure solvent and solution respectively,
log = $\log_{10}$, and
A = absorbance (optical density).

The formal definitions of the extinction coefficients α, $E^{1\%}_{1\,cm}$ and ε and their relationships are summarized in the equations below.

α = the absorbance of 1 g per litre solution at λ_{max} in a 1 cm cuvette.
$E^{1\%}_{1\,cm}$ = the absorbance of a 1% solution (i.e. 1 g per 100 ml) at λ_{max} in a 1 cm cuvette.
ε = the absorbance of a molar solution at λ_{max} in a 1 cm cuvette.
$\varepsilon = 0.1\, E^{1\%}_{1\,cm} \cdot MW = \alpha \cdot MW$

In the Data sheets the specific extinction coefficients of chlorophylls (α, l g^{-1} cm^{-1}) at the absorption maxima (λ_{max}) in reference solvent(s) are listed. For carotenoids it is more conventional to use the extinction of a 1% (w/v) solution ($E^{1\%}_{1\,cm}$; 100 ml g^{-1} cm^{-1}) at λ_{max} in reference solvent(s); Davies, 1976). We also give the molar extinction coefficients (ε; l mol^{-1} cm^{-1}) of all pigments, using accurately derived molecular weights (MW); see formula below. Both $E^{1\%}_{1\,cm}$ and α at λ_{max} in various solvents are given in the data sheets. These are also tabulated for all pigments in Appendix E.

There are a few marine pigments (e.g., 19'-butanoyloxyfucoxanthin) for which extinction coefficients have not been determined in the literature. In most instances, these may be estimated from the $E^{1\%}_{1\,cm}$ value of carotenoids with identical or similar chromophores (e.g., fucoxanthin) by assuming that their molar extinction coefficients (ε) are similar. Thus if:

$$\varepsilon_{\text{But-fuco}} = \varepsilon_{\text{Fuco}}$$

it follows that:

$$E^{1\%}_{\text{But-fuco}} = E^{1\%}_{\text{Fuco}} \cdot MW_{\text{Fuco}} / MW_{\text{But-fuco}}$$

Whenever such a calculation of $\varepsilon^{1\%}_{1\,cm}$ or e is carried out, it is highlighted in the Data sheets. Where extinction coefficients have not been reported or where no structurally or spectrally similar pigments exist, then we report this as 'not determined', and use the value of = 2500 for β,β-carotene, as suggested by Davies (1976).

IV.1.7 *UV-visible spectra*

Members and participants in the SCOR WG 78 Workshops isolated and purified 17 chlorophylls and derivatives and 30 carotenoids. UV-visible absorption spectra of all pigments were recorded in one or more reference solvents using a Shimadzu Model PRI double-beam scanning spectrophotometer, and in HPLC eluant using a diode-array spectrophotometer. In the case of three carotenoids, antheraxanthin, crocoxanthin and monadoxanthin, no spectra in standard solvents were available, but spectra in HPLC eluant are given.

IV.1.7.1 *Spectra in reference solvents*

The reference solvents used by SCOR WG 78 for the chlorophylls included peroxide-free diethyl ether and HPLC-grade acetone, since these solvents enhance spectral fine structure. Acetone and *n*-hexane were also used for carotenoids for the same reasons (see Appendix D). Ethanol and methanol were used because they resembled HPLC eluant. Although pigment purification and crystallization were often of very high quality in the early work on pigment characterization, the wavelength accuracy of the absorption spectra was often limited by the bandwidth (2–5 nm) of the manual spectrophotometers available at the time. In contrast, the spectral resolution of the modern recording spectrophotometers used during SCOR workshops was typically 0.2 nm. The reference spectra listed in the Data sheets should be consulted for the correct identification of pigments. It should be noted that ligroine, light petroleum and petroleum ether are North American and European terms for related solvents, which come in a range of boiling point fractions (e.g., 30–40, 40–60, 60–80, 80–110 °C). Readers must consult the original literature for the precise fraction used by the authors. In this monograph, we are standardizing on the term 'petroleum ether'.

The values of λ_{max} in various solvents are vertically aligned in the Data sheets and tabulated by band (e.g., band I, II, III for carotenoids, 'Soret' (blue maximum) and 'red' for chlorophylls). Bracketed values of λ denote an inflection in the absorption spectrum. In addition, the ratios of the UV-visible absorption bands are characteristic for each pigment and reflect the purity of the pigment fraction. For chlorophylls, the Soret (blue maximum) band : red band ratios are used (Smith and Benitez, 1955) whereas for the carotenoids, Band III : II ratios (expressed in %) are given (Ke *et al.* 1970). These ratios are illustrated in Figure IV.2.

IV.1.7.2 *Eluant spectra from diode-array detectors*

The UV-visible absorption spectra of HPLC-separated pigments were also recorded in HPLC eluant using a variety of diode-array detectors. The results are tabulated in the Data sheets as 'HPLC eluant' in terms of λ_{max} and band ratios. The sensitivity, spectral resolution and accuracy of diode-array detectors (typically 0.05 Absorbance Units, 2–5 nm and 1–2 nm respectively) are significantly less than the double-beam scanning spectrophotometers used to obtain the reference spectra (0.005 Absorbance Units, 1 nm and 0.2 nm respectively). It should be noted that the band ratios (III : II) for carotenoids are only valid when data is obtained with a high resolution spectro-

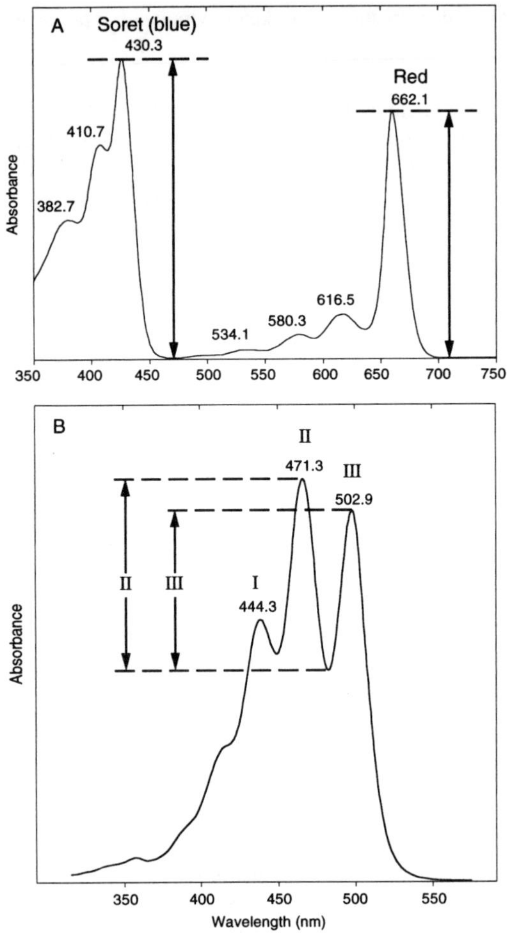

Figure IV.2
UV-visible spectra showing A: the location of the Soret (blue maximum) and red bands of a representative chlorophyll, chlorophyll *a* (after Smith and Benitez, 1955) and B: Bands I, II and III of a carotenoid, lycopene (after Ke *et al.* 1970). Vertical arrows denote the absorbance baselines and maxima used to derive band ratios.

photometer. They are *not* valid for data obtained with diode-array detectors, since their lower spectral resolution affects the resolution and intensity of the minima.

During the SCOR workshops, diode-array spectra were obtained with the two HPLC separation methods of Mantoura and Llewellyn (1983) and Wright *et al.* (1991). In the former system, pigments were eluted by a linear binary gradient consisting of methanol buffered with ammonium-acetate (pH 7.2) (80:20 = v/v) and methanol/acetone (70:30 = v/v), with the PIC-reagent (now known to be unnecessary) omitted. With the Wright *et al.* (1991) method, a ternary gradient of solvents based on ammonium acetate/methanol, acetonitrile and ethyl acetate was used. The composition of the SCOR eluant changed dramatically during the course of a chromatogram. Table IV.1 shows the computed composition of the eluant. Note that in many cases the eluant composition significantly alters the pigment spectrum from that obtained in reference solvent (e.g., chlorophyllide *b* and fucoxanthin). Thus, although the diode-array spectra may vary slightly between instruments and with different elution systems, they should nevertheless be reproducible for any one HPLC system. Diode-array spectra are also extremely useful in the spectral characterization of nanogram quantities of pigment obtained during HPLC separations.

A third standard reversed-phase system from Dr D. Repeta's laboratory was also

TABLE IV.1. Composition of SCOR eluant during a standard run, using the Wright *et al.* (1991) method.
Solvent A: 80:20 = methanol: 0.5M ammonium acetate (pH 7.2 v/v);
Solvent B: 90:10 = acetonitrile: H_2O (v/v);
Solvent C: ethyl acetate.
These data were computed for the Spectraphysics system allowing for the dead volume between the pump and column, and the void volume of the column itself. Eluant compositions will be slightly different for instruments with different configurations and column properties. For the identity of pigment abbreviations see Appendix A.

Peak no.	Pigment	Retention time	Eluant composition (v/v/v)		
			% A	% B	% C
0	solvent front	2.56			
1	Chlide *b*	4.48	29.0	71.0	0.0
2	P-468	4.5	28.0	72.0	0.0
3	Chlide *a*	5.01	2.5	97.5	0.0
4	P-457	5.11	0.0	99.2	0.8
5	Chl c_3	5.38	0.0	94.7	5.3
6	Chl c_1 and c_2	6.4	0.0	88.3	11.7
7	Perid	7.42	0.0	86.0	14.0
8	Siphx	8.11	0.0	84.4	15.6
9	But-fuco	8.11	0.0	84.4	15.6
10	Fuco	8.7	0.0	83.1	16.9
11	*t*-Neo	9.11	0.0	82.2	17.8
12	*c*-But-fuco	9.12	0.0	82.1	17.9
13	Neochrome	9.21	0.0	81.9	18.1
14	*c*-Neo	9.31	0.0	81.7	18.3
15	Hex-fuco	9.31	0.0	81.7	18.3
16	*c*-Fuco	9.68	0.0	80.9	19.1
17	*c*-Hex-fuco	9.97	0.0	80.2	19.8
18	Siphx-like	10.12	0.0	79.9	20.1
19	Pras	10.2	0.0	79.7	20.3
20	Phide *a*	10.39	0.0	79.3	20.7
21	Viola	10.59	0.0	78.8	21.2
22	Phide *a*-like	10.62	0.0	78.7	21.3
23	Dino	10.76	0.0	78.4	21.6
24	*c*-Pras	11.11	0.0	77.6	22.4
25	unk-Siphx	11.15	0.0	77.5	22.5
26	Diadino	11.61	0.0	76.5	23.5
27	Diadchr 1	11.79	0.0	76.1	23.9
28	Diadchr 2	11.96	0.0	75.7	24.3
29	Anth	12.24	0.0	75.0	25.0
30	Allo	12.51	0.0	74.4	25.6
31	Monado	12.78	0.0	73.8	26.2
32	Diato	13.08	0.0	73.1	26.9
33	Lut	13.36	0.0	72.5	27.5
34	Zea	13.59	0.0	72.0	28.0
35	Cantha	14.0	0.0	71.0	29.0
35b	Apo-8-al	14.0	0.0	71.0	29.0
36	Siphn	14.36	0.0	70.2	29.8
37	Chl *b*	15.15	0.0	68.4	31.6
38	Apo-8-ate	15.43	0.0	67.8	32.2

TABLE IV.1. *(contd)*

Peak no.	Pigment	Retention time	Eluant composition (v/v/v)		
			% A	% B	% C
39	Croco	15.87	0.0	66.8	33.2
40	Chl *a* allom	15.87	0.0	66.8	33.2
41	Chl *a*	16.15	0.0	66.2	33.8
42	Chl *a*'	16.53	0.0	65.3	34.7
43	Echin	16.74	0.0	64.6	35.4
44	unk-carot	16.99	0.0	63.2	36.8
45	Lyco	17.59	0.0	60.1	39.9
46	Phytin *b*	17.68	0.0	59.6	40.4
47	*c*-Lyco	17.84	0.0	58.7	41.3
48	Phytin *a*	18.56	0.0	54.9	45.1
49	βψ-Car	18.26	0.0	56.5	43.5
50	εε-Car	18.4	0.0	55.8	44.2
51	βε-Car	18.64	0.0	54.5	45.5
52	ββ-Car	18.76	0.0	53.8	46.2
53	*c* -βε-Car	18.83	0.0	53.5	46.5
54	*c* -ββ-Car	18.94	0.0	52.9	47.1

used to provide data not obtained at the SCOR WG 78 Workshops. This data is referred to as 'Repeta HPLC system' and 'Repeta eluant' in the Graphics sheets. The instruments used were a Waters HPLC system (Waters 660 solvent delivery system and Waters 990 diode-array detector). For carotenoids, the 15 cm column was an Adsorbosphere 3 μm ODS with a two solvent gradient system.

Solvent A: 0.5 M ammonium acetate:methanol = 1:4 (v/v)

Solvent B: acetone:methanol = 1:4 (v/v)

The eluant composition in which astaxanthin was eluted was 25% A and 75% B; and that of ethyl-8'-apo-carotenoate was 100% B (see Data and Graphics sheets for these pigments, Part IV).

For the divinyl chlorophylls, a Rainin 10 cm 3 μm C-8 column was used with a two solvent gradient system.

A: 0.5 M ammonium acetate:methanol = 1:3 (v/v)

B: 100% methanol

The eluant composition in which divinyl chlorophyll *a* was eluted was 4% A and 96% B; and that of divinyl chlorophyll *b* was 27% A and 73% B (see Data and Graphics sheets for these pigments, Part IV).

IV.1.8 *Fluorescence*

The fluorescence excitation and emission spectra of chlorophyll pigments were determined in acetone during the SCOR workshops. These are listed, together with literature data, in the Data sheets.

IV.1.9 *Alteration products*

The principal alteration products of chlorophylls (Hynninen, 1991) and carotenoids (Zechmeister, 1944; Liaaen-Jensen, 1990*a*), particularly those that occur during handling, are listed here.

IV.1.10 *Algal sources of pure pigments*

The algal species from which pure pigments were isolated during the SCOR Workshops are listed. Synthetic and other sources are also given (see Chapters 7, 8).

IV.1.11 *HPLC chromatogram*

The HPLC absorbance chromatogram obtained by the Wright *et al.* (1991) method is shown highlighting the separation of the pigment (with peak number) from the SCOR algal extract. The Wright *et al.* (1991) method was not always used for the isolation of the pure pigment; sometimes TLC, HPTLC or other HPLC systems were more convenient (see Chapters 7 and 8).

IV.1.12 *References*

Citations are provided for further information on the structure, chemistry and occurrence of each pigment.

IV.1.13 *Acknowledgements*

We thank Professor P. H. Hynninen for critically reviewing the chlorophyll Data and Graphics sheets. Drs T. Bjørnland, S. Wright, D. Repeta, S. Jeffrey and Ms C. Llewellyn provided most of the spectra or diode-array spectra of the pure pigments prepared at the SCOR Workshops. We thank Drs R. Goericke, D. Simpson, R. Barlow, H. Claustre and L. King for additional spectra and HPLC traces.

IV.2. Chlorophyll Data and Graphics sheets

Chlorophyll *a* related pigments
- Chlorophyll *a*
- Chlorophyllide *a*
- Pheophytin *a*
- Pheophorbide *a*
- Pyropheophytin *a*
- Divinyl chlorophyll *a*

Chlorophyll *b* related pigments
- Chlorophyll *b*
- Chlorophyllide *b*
- Pheophytin *b*
- Pheophorbide *b*
- Divinyl chlorophyll *b*

Chlorophyll *c* related pigments
- Chlorophyll c_1
- Chlorophyll c_2
- Chlorophyll c_3
- Chlorophyll $c_{CS\text{-}170}$
- Mg-2,4-divinyl pheoporphyrin a_5 monomethyl ester
- Phytylated chlorophyll *c*-like pigment

Chlorophyll *a* — HPLC peak 41

Standard spectrum in reference solvent: acetone (100%)

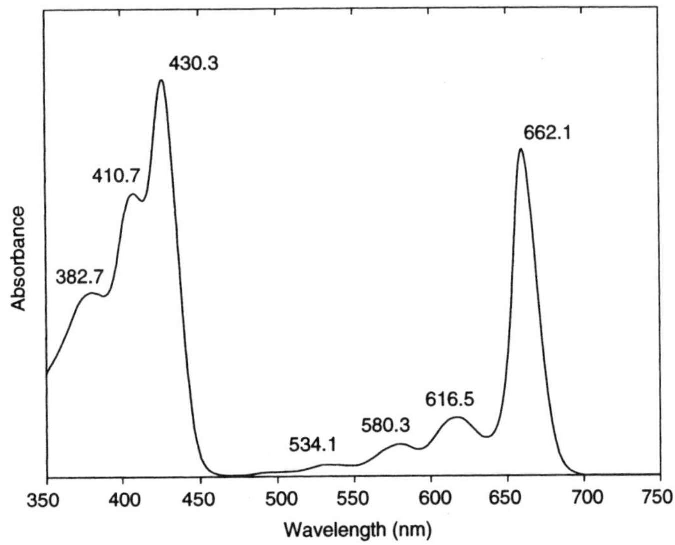

Molecular structure

Structure of chlorophyll *a*; left, the IUPAC numbering system; right, the older but still frequently used Fischer numbering system

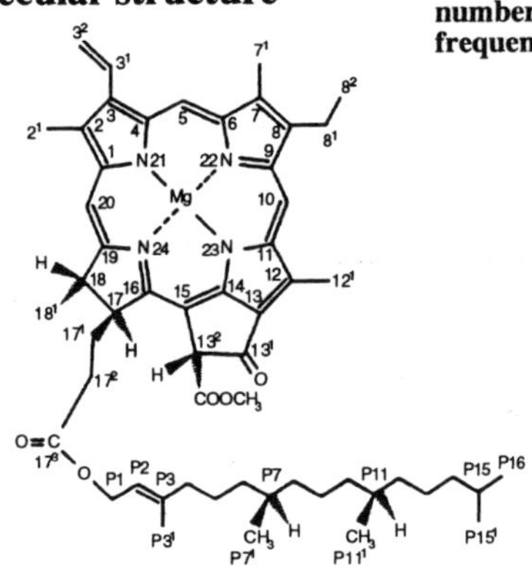

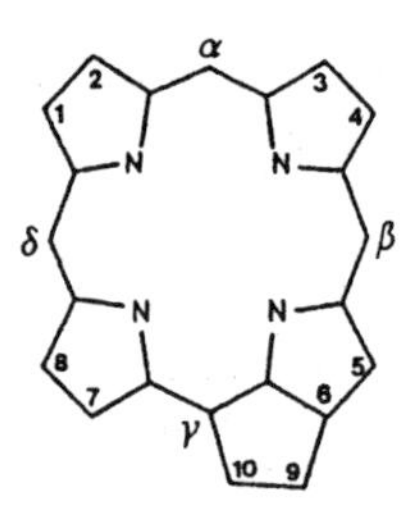

Diode array spectrum in SCOR eluant

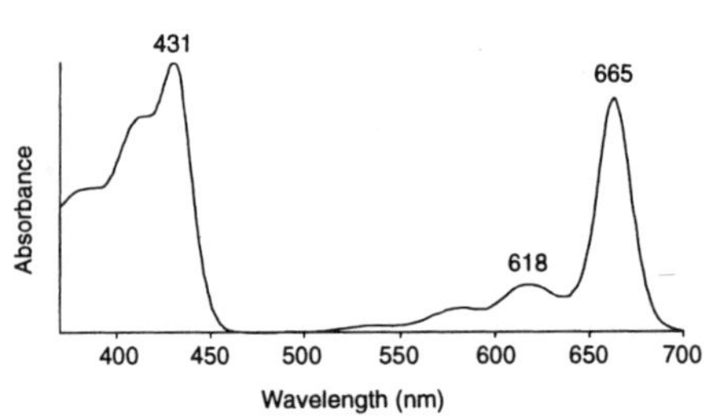

HPLC: Chl *a*, peak 41

Chroomonas salina

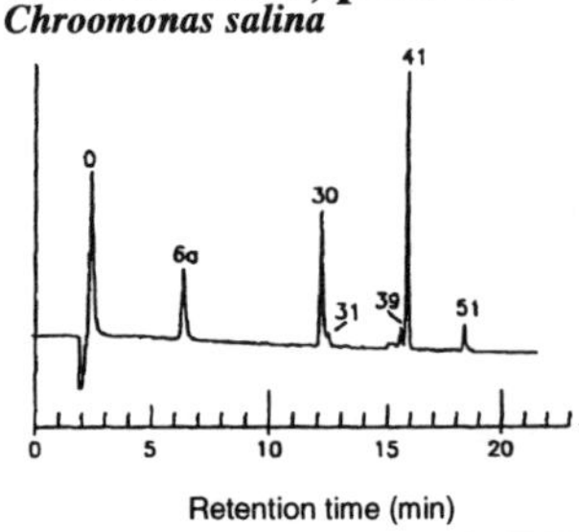

Chlorophyll *a*

Property		**Data**
Name:	(Trivial)	**Chlorophyll *a***
	(IUPAC)	Trivial name sufficient, see Hynninen (1991)
SCOR abbreviation:		Chl *a*
Occurrence:		All photosynthetic algae and higher plants
Colour:		Blue-green
Molecular formula:		$C_{55}H_{72}N_4O_5Mg$
Molecular weight:		893.50
Specific extinction coefficient: α (l g^{-1} cm^{-1})		88.15 (at 662.7 nm in 100% acetone) 87.67 (at 664.3 nm in 90% acetone) Jeffrey & Humphrey (1975)
Molar extinction coefficient: ε (l mol^{-1} cm^{-1})		78.75 x 10^3 (at 662.7 nm in 100% acetone) Calculated from α

UV-vis spectra:

Solvent	Absorbance maxima (nm)						Band ratio*	Reference
100% Acetone	410.7	430.3	534.1	580.3	616.5	662.1	1.23	SCOR WG 78 data
Diethyl ether	410	430.0	535.5	578.0	615.0	662.0	1.30	Smith & Benitez (1955)
Diethyl ether	409	428	530	575	614	660.0	1.28	Hynninen & Lötjönen (1983)
Methanol	417.6	431.8			618.2	665.2	0.96	Lichtenthaler (1987)
HPLC Eluant		431			618	665	1.17	SCOR WG 78: Wright *et al.* (1991) method

*Soret (blue maximum): red ratio

Fluorescence spectra:

Solvent	Excitation (nm)	Emission (nm)	Reference
Diethyl ether	428	666	Boardman & Thorne (1971)
Diethyl ether	427	666	SCOR WG 78 data
Acetone	430	668, 713	Jeffrey (1972)

Alteration products:	Chlorophyllide *a*, pheophytin *a*, pheophorbide *a*, epimers, allomers, pyro-derivatives
Culture from which SCOR data were obtained:	*Chroomonas salina* (cryptomonad)
Additional reference(s):	Scheer (1991)

Chlorophyllide *a* — HPLC peak 3

Standard spectrum in reference solvent: acetone (100%)

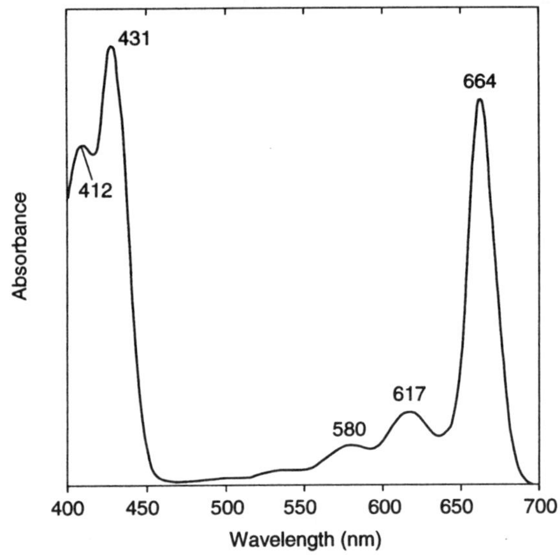

Molecular structure

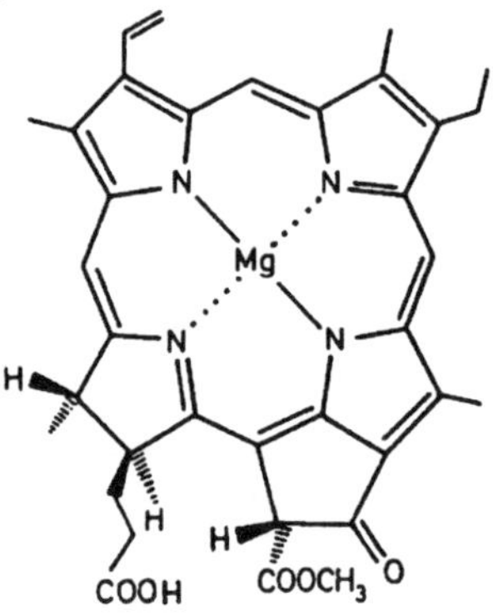

Diode array spectrum in SCOR eluant

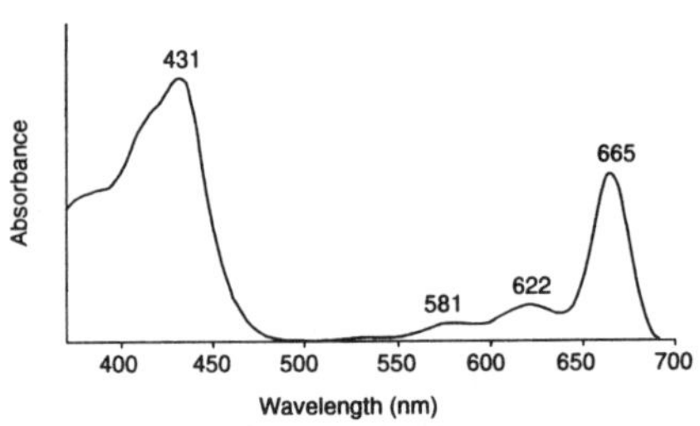

HPLC: Chlorophyllide *a*, peak 3

***Dunaliella tertiolecta*, hydrolysed extract**

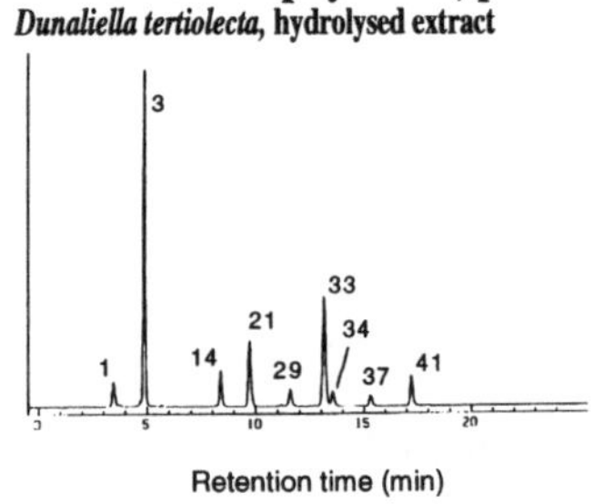

Chlorophyllide *a*

Property		**Data**
Name:	(Trivial)	**Chlorophyllide *a***
	(IUPAC)	Trivial name sufficient; see Hynninen (1991)
SCOR abbreviation:		Chlide *a*
Occurrence:		Senescent tissue, damaged centric diatoms, zooplankton faecal pellets
Colour:		Blue-green
Molecular formula:		$C_{35}H_{34}N_4O_5Mg$
Molecular weight:		614.97
Specific extinction coefficient: α ($l\ g^{-1}\ cm^{-1}$)		127 (at 664 nm in 90% acetone) Calculated from Lorenzen & Jeffrey (1980)
Molar extinction coefficient: ε ($l\ mol^{-1}\ cm^{-1}$)		78.3 x 10^3 (at 664 nm in 90% acetone) Calculated from Lorenzen & Jeffrey (1980)

UV-vis spectra:

Solvent	Absorbance maxima (nm)					Band ratio*	Reference
100% Acetone	412	431	580	617	664	1.14	Jeffrey & Lorenzen (unpublished)
HPLC eluant		431	581	622	665	1.11	SCOR WG 78: Wright *et al.* (1991) method

*Soret (blue maximum): red ratio

Fluorescence spectra:

Solvent	Excitation (nm)	Emission (nm)	Reference
100% Acetone	426	667	SCOR WG 78 data

Alteration products:	Pheophorbide *a* (Mg-free derivative)
Culture from which SCOR data were obtained:	*Phaeodactylum tricornutum* (diatom); *Dunaliella tertiolecta* (green flagellate)
Additional reference(s):	Barrett & Jeffrey (1964); Jeffrey & Hallegraeff (1987)

Pheophytin *a* — HPLC peak 48

Standard spectrum in reference solvent: acetone (100%)

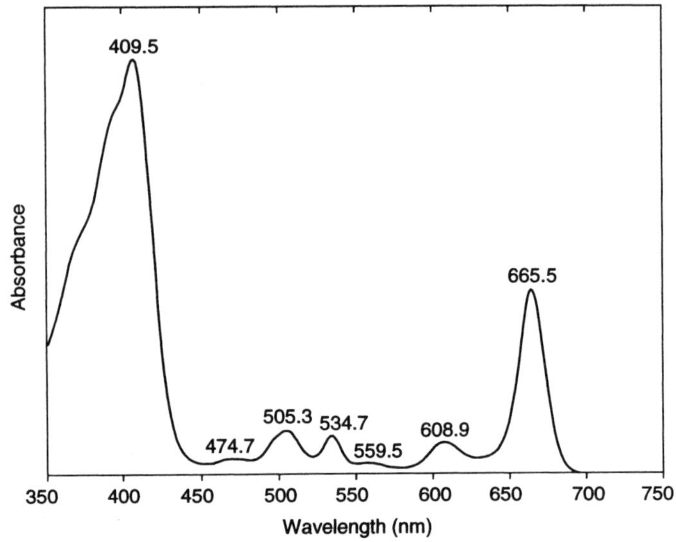

Molecular structure

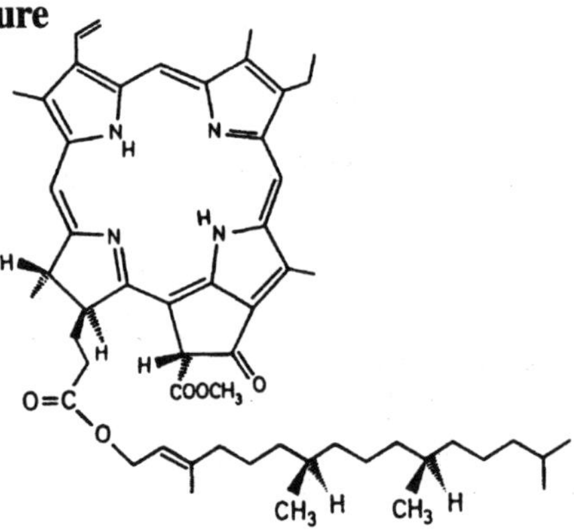

Diode array spectrum in SCOR eluant

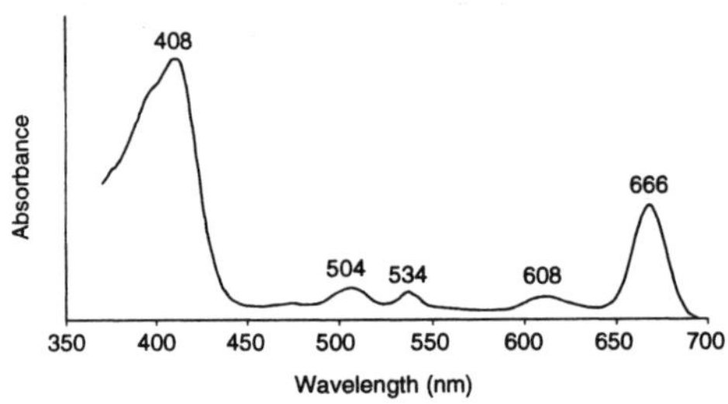

HPLC: Pheophytin *a*, peak 48

Acidified chlorophylls from *Dunaliella*

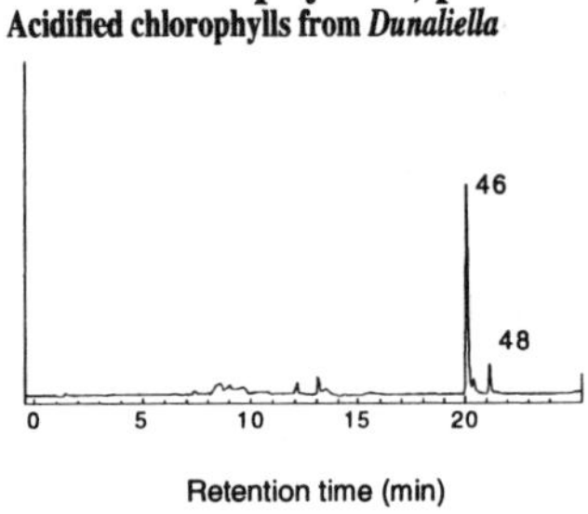

Pheophytin *a*

Property		**Data**
Name:	(Trivial)	**Pheophytin *a***
	(IUPAC)	Trivial name sufficient, see Hynninen (1991)
SCOR abbreviation:		Phytin *a*
Occurrence:		Reaction centres of photosynthetic apparatus of higher plants; plant and algal detritus
Colour:		Grey (red fluorescence) on TLC; yellow-grey (concentrated solution)
Molecular formula:		$C_{55}H_{74}N_4O_5$
Molecular weight:		871.21
Specific extinction coefficient: α (l g^{-1} cm^{-1})		51.2 (at 667 nm in 90% acetone) Lorenzen & Jeffrey (1980)
Molar extinction coefficient: ε (l mol^{-1} cm^{-1})		44.6 x 10^3 (at 667 nm in 90% acetone) Calculated from α above

UV-vis spectra:

Solvent	Absorbance maxima (nm)						Band ratio*	Reference
100% Acetone	409.5	505.3	534.7	559.5	608.9	665.5	2.26	SCOR WG 78 data
Diethyl ether	408.5					667.0	2.07	Smith & Benitez (1955)
HPLC Eluant	408	504	534		608	666	2.34	SCOR WG 78: Wright *et al.* (1991) method
Tetrahydro-furan	411.4	505	534	560	609	668.0	2.15	Lötjönen & Hynninen (1983)

*Soret (blue maximum): red ratio

Fluorescence spectra:

Solvent	Excitation (nm)	Emission (nm)	Reference
Diethyl ether	409	673	Smith & Benitez (1955)
Diethyl ether	408	672	Boardman & Thorne (1971)

Alteration products:	Epimers and allomers
Origin:	Purified from Sigma chlorophyll *a* that had been acidified
Additional reference(s):	Scheer (1991)

Pheophorbide *a* — HPLC peak 20

Standard spectrum in reference solvent: acetone (100%)

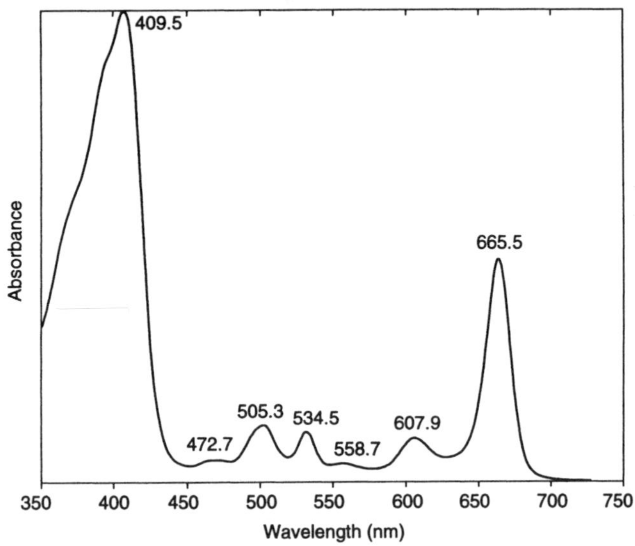

Molecular structure

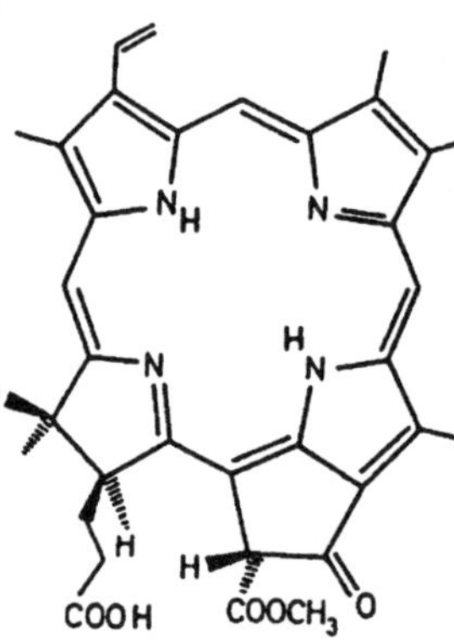

Diode array spectrum in SCOR eluant

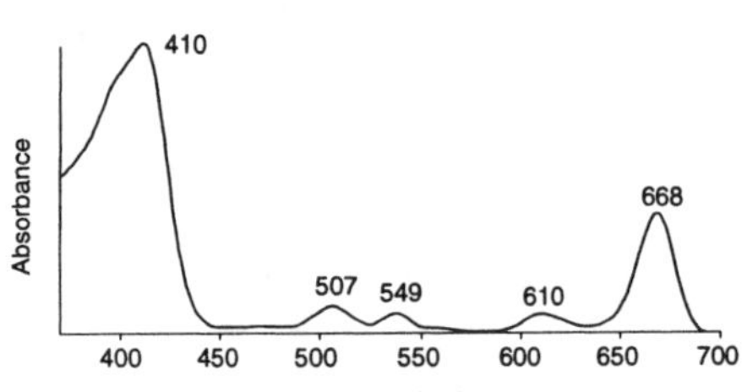

HPLC*: Pheophorbide *a*, peak 20

Mixture of standards

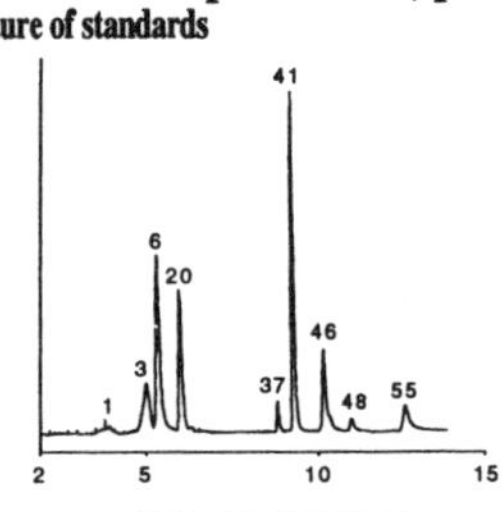

*** Mantoura and Llewellyn (1983) system**

Pheophorbide *a*

Property		**Data**
Name:	(Trivial)	**Pheophorbide *a***
	(IUPAC)	Trivial name sufficient; see Hynninen (1991)
SCOR abbreviation:		Phide *a*
Occurrence:		Marine detritus, zooplankton faecal pellets
Colour:		Grey (red fluorescence) on TLC; yellow-grey (concentrated solution)
Molecular formula:		$C_{35}H_{36}N_4O_5$
Molecular weight:		592.67
Specific extinction coefficient: α (l g^{-1} cm^{-1})		74.2 (at 667 nm in 90% acetone) Lorenzen & Jeffrey (1980)
Molar extinction coefficient: ε (l mol^{-1} cm^{-1})		43.98 x 10^3 (at 667 nm in 90% acetone) Calculated from α above

UV-vis spectra:

Solvent	**Absorbance maxima (nm)**						**Band ratio***	**Reference**
100% Acetone	409.5	505.3	534.5	558.7	607.9	665.5	2.16	SCOR WG 78 data
HPLC Eluant	410	507	549		610	668	2.35	SCOR WG 78: Wright *et al.* (1991) method

*Soret (blue maximum): red ratio

Fluorescence spectra:

Solvent	**Excitation (nm)**	**Emission (nm)**	**Reference**
Acetone	406	672	SCOR WG 78 data

Alteration products:	Epimers, allomers, pyro- and meso-derivatives
Origin:	Acidification of chlorophyllide *a*
Additional reference(s):	Scheer (1991)

Pyropheophytin *a* — HPLC peak 55

Standard spectrum in reference solvent: acetone (100%)

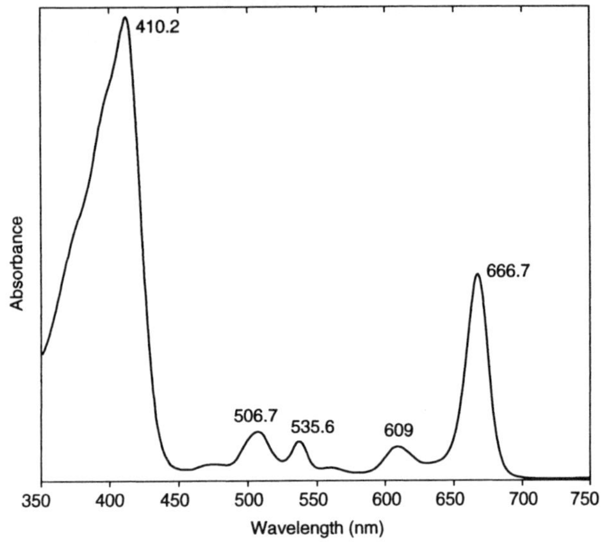

Molecular structure

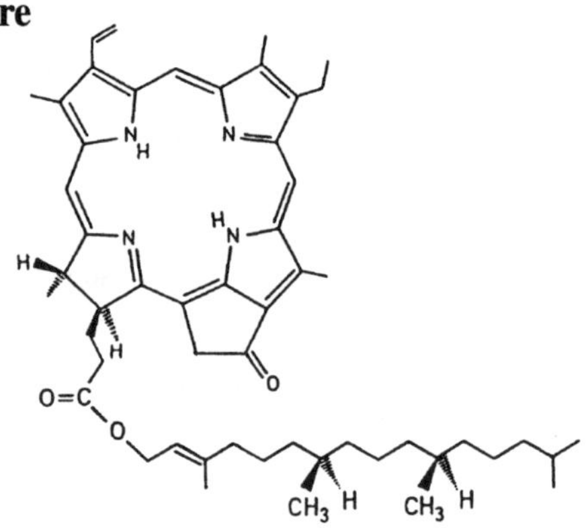

Diode array spectrum in SCOR eluant

No data available

HPLC*: Pyropheophytin *a*, peak 55

Mixture of standards

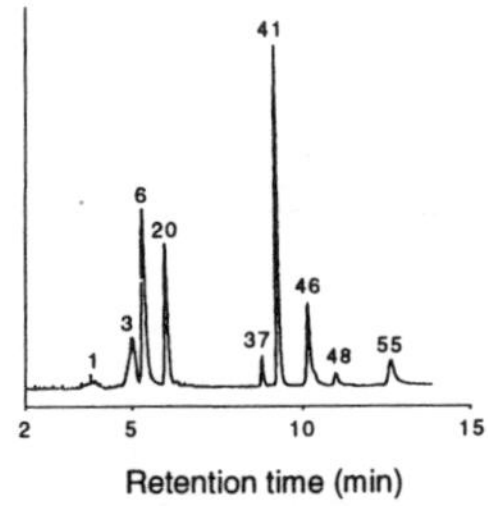

*** Mantoura and Llewellyn (1983) system**

Pyropheophytin *a*

Property		**Data**
Name:	(Trivial)	**Pyropheophytin *a***
	(IUPAC)	13^2-demethoxycarbonyl-pheophytin *a*
SCOR abbreviation:		Pyrophytin *a*
Occurrence:		Zooplankton faecal pellets, terrestrial detritus
Colour:		Grey
Molecular formula:		$C_{53}H_{72}N_4O_3$
Molecular weight:		813.17
Specific extinction coefficient: α ($l\ g^{-1}\ cm^{-1}$)		60.29 (at 667 nm in diethyl ether) Pennington *et al.* (1964)
Molar extinction coefficient: ε ($l\ mol^{-1}\ cm^{-1}$)		49.0×10^3 (at 667 nm in diethyl ether) Pennington *et al.* (1964)

UV-vis spectra:

Solvent	**Absorbance maxima (nm)**					**Band ratio***	**Reference**
Diethyl ether	409.0				667.0	2.09	Pennington *et al.* (1964)
Acetone	410.2	506.7	535.6	609	666.7	2.28	SCOR WG 78 data

*Soret (blue maximum): red ratio

Fluorescence spectra:

Solvent	**Excitation (nm)**	**Emission (nm)**	**Reference**
100% Acetone	407	672	SCOR WG 78 data

Alteration products:	Unknown
Origin:	Sample donated (B. Keely)
Additional reference(s):	Pennington *et al.* (1964); Hynninen (1991)

Divinyl chlorophyll *a* — HPLC peak 41*

Standard spectrum in reference solvent: acetone (100%)

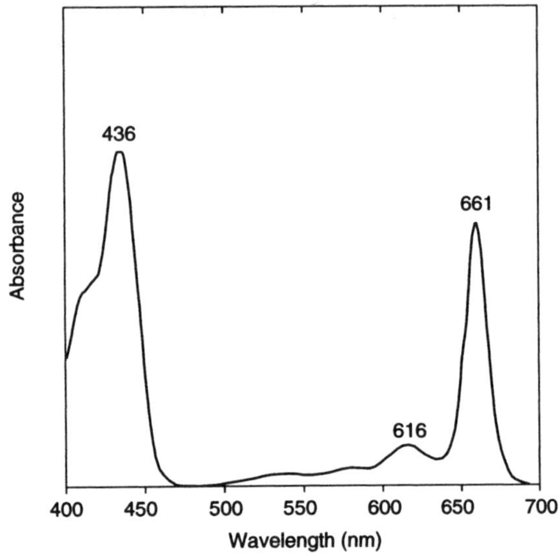

Data from Gieskes and Kraay (1983)

Molecular structure

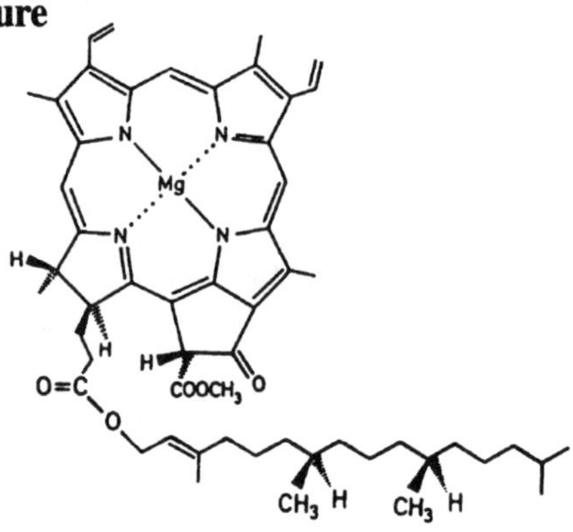

Diode array spectrum in Repeta eluant

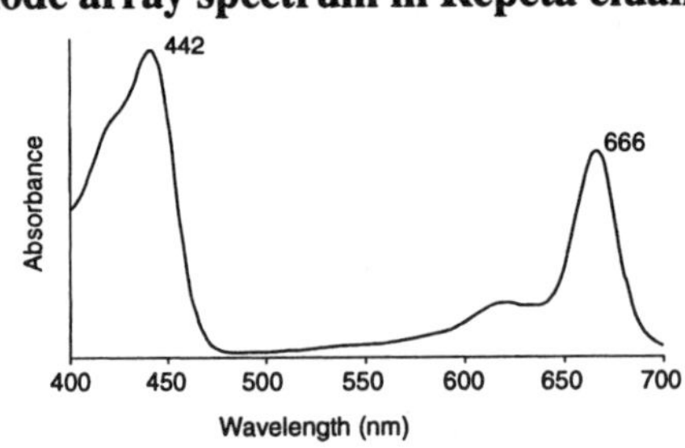

HPLC: Divinyl chlorophyll *a*, peak 41*†

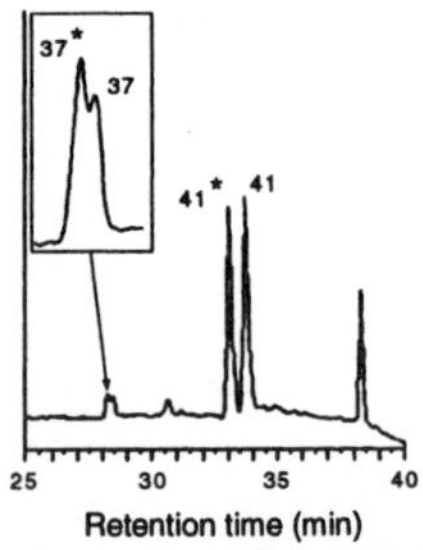

† Goericke and Repeta (1992) HPLC RP-8 system

Divinyl chlorophyll *a*

Property		**Data**
Name:	(Trivial)	**Divinyl chlorophyll *a***
	(IUPAC)	8-deethyl-8-vinyl-chlorophyll *a* = $8^1,8^2$-didehydro-chlorophyll *a*
SCOR abbreviation:		DV chl *a*
Occurrence:		*Prochlorococcus marinus* (class Prochlorophyceae)
Colour:		Green
Molecular formula:		$C_{55}H_{70}N_4O_5Mg$
Molecular weight:		891.48
Specific extinction coefficient: α ($l\ g^{-1}\ cm^{-1}$)		Not determined
Molar extinction coefficient: ε ($l\ mol^{-1}\ cm^{-1}$)		Assume identical to chlorophyll *a*

UV-vis spectra:

Solvent	**Absorbance maxima (nm)**			**Band ratio***	**Reference**
Acetone	436	616	661	1.28	Gieskes & Kraay (1983)
Diethyl ether	436		661		Bazzaz (1981)
Diethyl ether	435	615	660		Goericke & Repeta (1992)
HPLC Eluant	442		666	1.51	Repeta (unpub. data)

*Soret (blue maximum): red ratio

Fluorescence spectra:

Solvent	**Excitation (nm)**	**Emission (nm)**	**Reference**
No data available			

Alteration products:	8-deethyl-8-vinyl pheophytin *a*; easily bleached (Gieskes, 1991)
Culture from which SCOR data were obtained:	Culture of *Prochlorococcus* not available to SCOR WG 78
Additional reference(s):	Gieskes & Kraay (1983); Chisholm *et al.* (1988, 1992); Goericke & Repeta (1992)

Chlorophyll *b* — HPLC peak 37

Standard spectrum in reference solvent: acetone (100%)

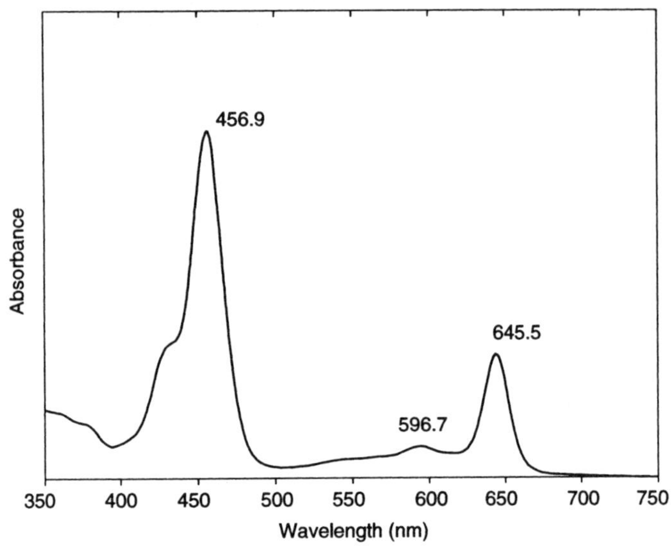

Molecular structure

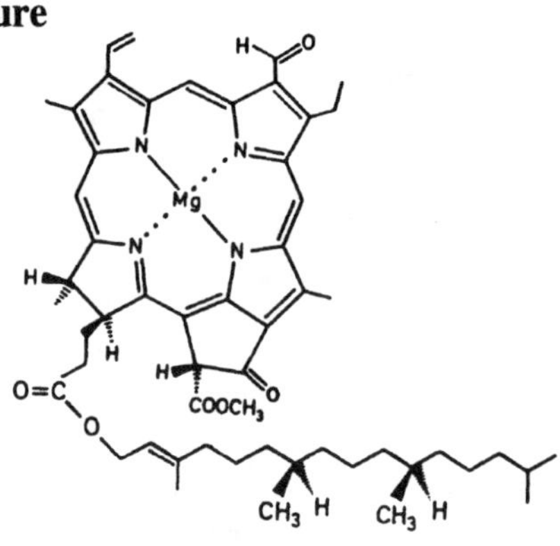

Diode array spectrum in SCOR eluant

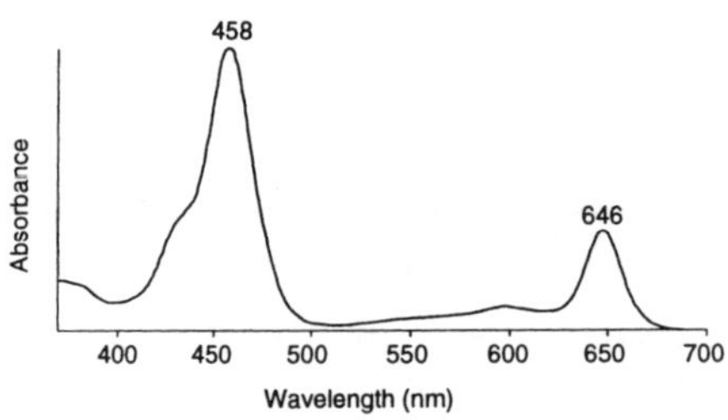

HPLC: Chl *b*, peak 37

Micromonas pusilla

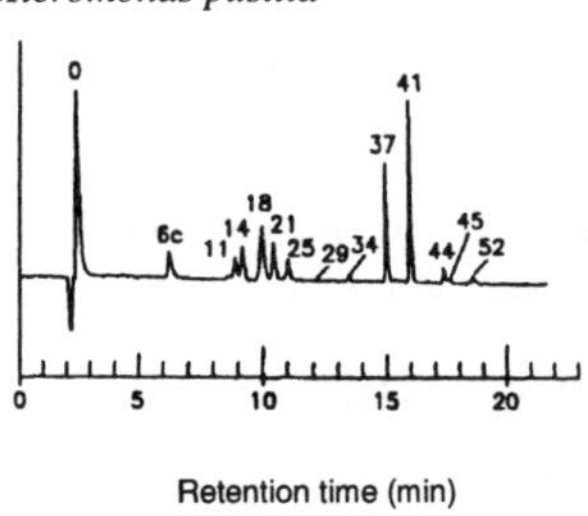

Chlorophyll *b*

Property		**Data**
Name:	(Trivial)	**Chlorophyll *b***
	(IUPAC)	Trivial name sufficient, see Hynninen (1991)
SCOR abbreviation:		Chl *b*
Occurrence:		Higher plants, green algae, symbiotic prochlorophytes
Colour:		Olive green (TLC); emerald green (concentrated solution)
Molecular formula:		$C_{55}H_{70}N_4O_6Mg$
Molecular weight:		907.49
Specific extinction coefficient: α (l g^{-1} cm^{-1})		51.36 (at 646.8 nm in 90% acetone) 62.00 (at 643.3 nm in diethyl ether) Jeffrey & Humphrey (1975)
Molar extinction coefficient: ε (l mol^{-1} cm^{-1})		46.61 x 10^3 (at 646.8 nm in 90% acetone) 56.26 x 10^3 (at 643.3 nm in diethyl ether) Calculated from α above

UV-vis spectra:

Solvent	**Absorbance maxima (nm)**			**Band ratio***	**Reference**
100% Acetone	456.9	596.7	645.5	2.82	SCOR WG 78 data
Diethyl ether	455	595	644	2.81	Smith & Benitez (1955)
Diethyl ether	453	593	642	2.77	Hynninen & Lötjönen (1983)
Ethanol	464		649		Griffiths (1991)
HPLC Eluant	458	596	646	2.83	SCOR WG 78: Wright *et al.* (1991) method

*Soret (blue maximum): red ratio

Fluorescence spectra:

Solvent	**Excitation (nm)**	**Emission (nm)**	**Reference**
Diethyl ether	453	646	Boardman & Thorne (1971)
100% Acetone	459	652, 710	Jeffrey (1972)
100% Acetone	453	652	SCOR WG 78 data

Alteration products:	Chlorophyllide *b*, pheophytin *b*, pheophorbide *b*, epimers, allomers, pyro-pheophytin *b*
Culture from which SCOR data were obtained:	*Dunaliella tertiolecta* (chlorophyte), *Micromonas pusilla* (prasinophyte)
Additional reference(s):	Scheer (1991)

Chlorophyllide *b*

HPLC peak 1

Standard spectrum in reference solvent: acetone (100%)

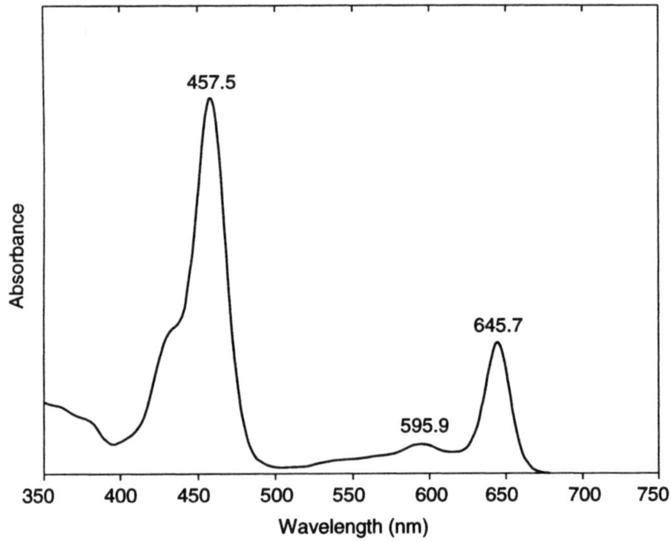

Molecular structure

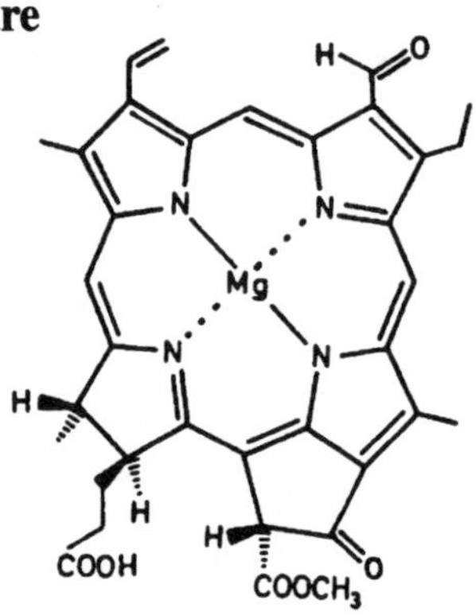

Diode array spectrum in SCOR eluant

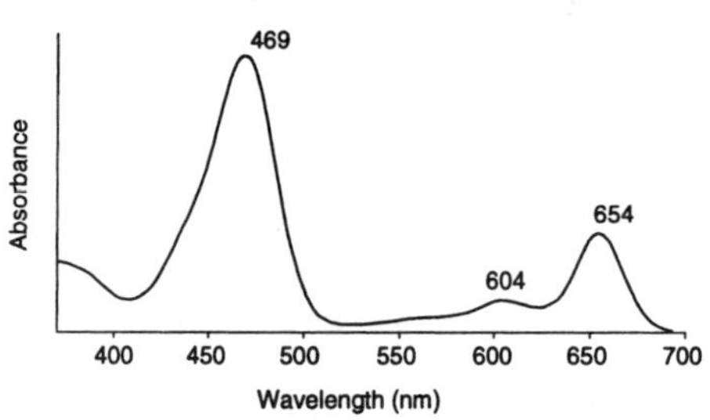

HPLC: Chlorophyllide *b*, peak 1

***Dunaliella tertiolecta*, hydrolysed extract**

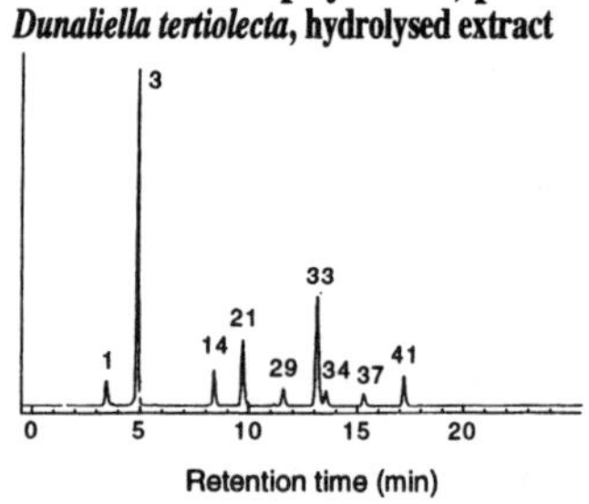

Chlorophyllide *b*

Property		**Data**
Name:	(Trivial)	**Chlorophyllide *b***
	(IUPAC)	Trivial name sufficient; see Hynninen (1991)
SCOR abbreviation:		Chlide *b*
Occurrence:		Senescent tissue, zooplankton faecal pellets
Colour:		Olive green
Molecular formula:		$C_{35}H_{32}N_4O_6Mg$
Molecular weight:		628.97
Specific extinction coefficient: α (l g^{-1} cm^{-1})		74.07 (at 645 nm in 90% acetone) Lorenzen & Jeffrey (1980)
Molar extinction coefficient: ε (l mol^{-1} cm^{-1})		46.6 x 10^3 (at 645 nm in 90% acetone) Calculated from α above

UV-vis spectra:

Solvent	Absorbance maxima (nm)			Band ratio*	Reference
100% Acetone	457.5	595.9	645.7	2.88	SCOR WG 78 data
HPLC eluant	469	604	654	2.83	SCOR WG 78: Wright *et al.* (1991) method

*Soret (blue maximum): red ratio

Fluorescence spectra:

Solvent	Excitation (nm)	Emission (nm)	Reference
100% Acetone	454	652	SCOR WG 78 data

Alteration products:	Pheophorbide *b*
Culture from which SCOR data were obtained:	*Dunaliella tertiolecta* (green flagellate)
Additional reference(s):	Jeffrey & Hallegraeff (1987)

Pheophytin *b* — HPLC peak 46

Standard spectrum in reference solvent: acetone (100%)

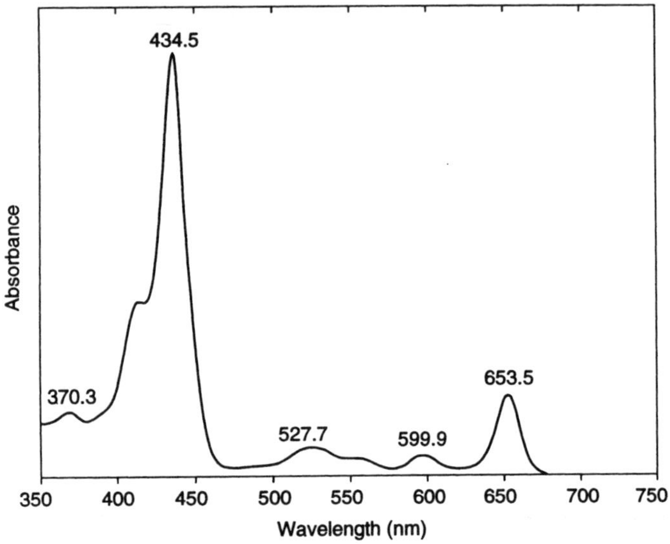

Molecular structure

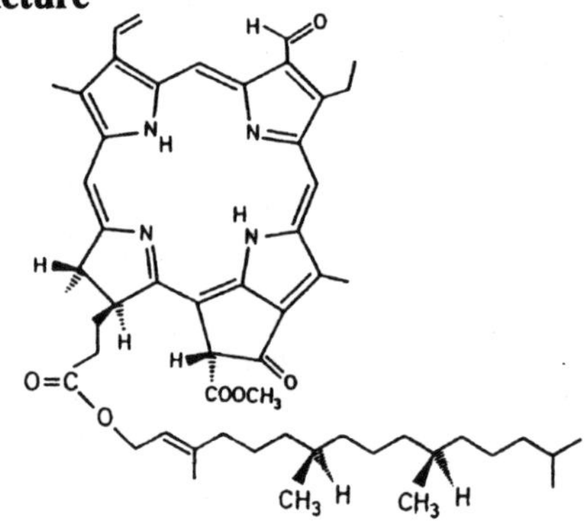

Diode array spectrum in SCOR eluant

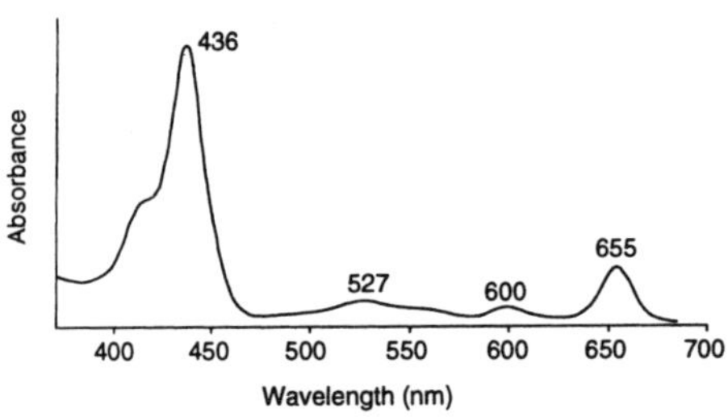

HPLC: Pheophytin *b*, Peak 46

Acidified chlorophylls from *Dunaliella*

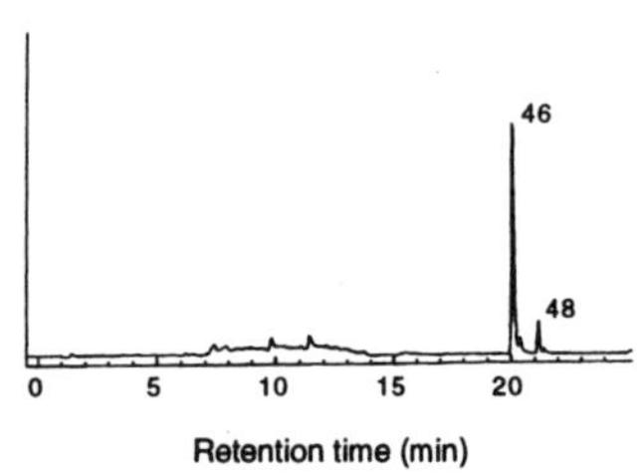

Pheophytin *b*

Property		**Data**
Name:	(Trivial)	**Pheophytin *b***
	(IUPAC)	Trivial name sufficient; see Hynninen (1991)
SCOR abbreviation:		Phytin *b*
Occurrence:		Terrestrial plant detritus
Colour:		Grey (red fluorescence) on TLC; yellow-grey (concentrated solution)
Molecular formula:		$C_{55}H_{72}N_4O_6$
Molecular weight:		885.20
Specific extinction coefficient: α (l g^{-1} cm^{-1})		42.1 (at 655 nm in diethyl ether) Smith & Benitez (1955) 31.8 (at 657 nm in 90% acetone) N. Welschmeyer, pers. comm.
Molar extinction coefficient: ε (l mol^{-1} cm^{-1})		37.3×10^3 (at 655 nm in diethyl ether) 28.1×10^3 (at 657 nm in 90% acetone) Calculated from α above

UV-vis spectra:

Solvent	Absorbance maxima (nm)				Band ratio*	Reference
100% Acetone	434.5	527.7	599.9	653.5	5.30	SCOR WG 78 data
Diethyl ether	434	525	599	655	5.13	Smith & Benitez (1955)
HPLC Eluant	436	527	600	655	5.23	SCOR WG 78: Wright *et al.* (1991) method
Tetrahydro-furan	435	525	599	654	4.92	Hynninen & Lötjönen (1983)

*Soret (blue maximum): red ratio

Fluorescence spectra:

Solvent	Excitation (nm)	Emission (nm)	Reference
Diethyl ether	434	661	Smith & Benitez (1955)
Diethyl ether	434	658	Boardman & Thorne (1971)
100% Acetone	432	659	SCOR WG 78 data

Alteration products:	Epimers and allomers
Origin:	Acidification of chlorophyll *b* preparation
Additional reference(s):	Scheer (1991)

Pheophorbide *b*

Standard spectrum in reference solvent: acetone (100%)

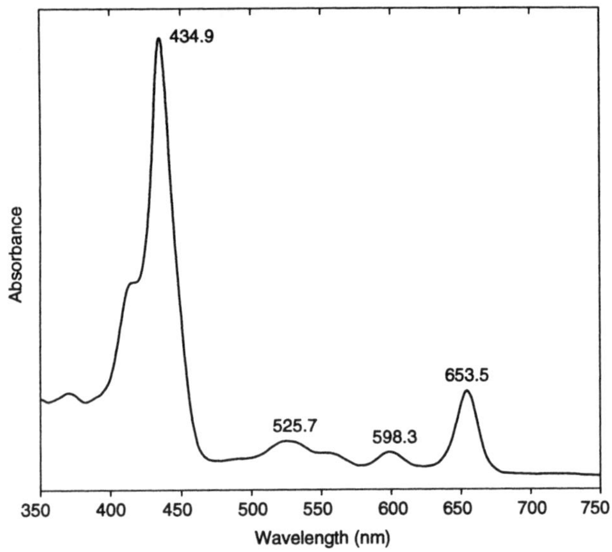

Molecular structure

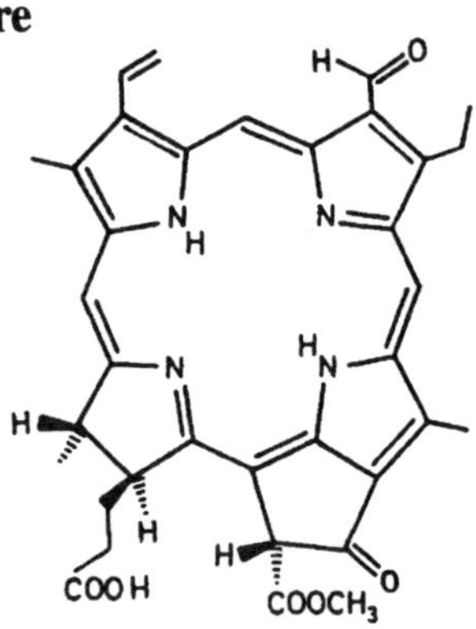

Diode array spectrum in SCOR eluant

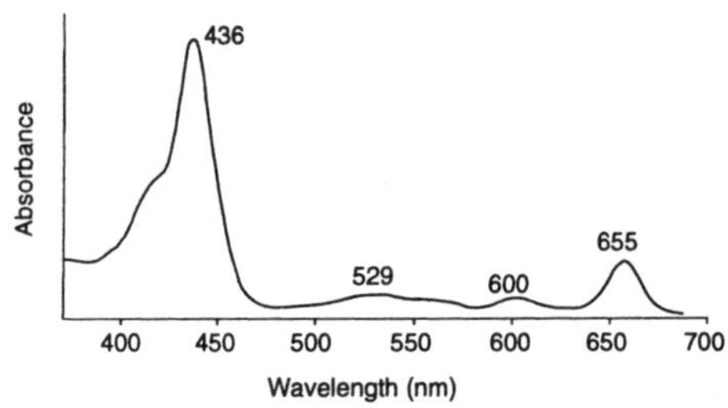

HPLC: Pheophorbide *b*

Prepared from acidified chlorophyllide *b*

Pheophorbide *b*

Property		**Data**
Name:	(Trivial)	**Pheophorbide *b***
	(IUPAC)	Trivial name sufficient, see Hynninen (1991)
SCOR abbreviation:		Phide *b*
Occurrence:		Terrestrial plant detritus, sediments
Colour:		Grey (red fluorescence) on TLC; yellow-grey (concentrated solution)
Molecular formula:		$C_{35}H_{34}N_4O_6$
Molecular weight:		606.67
Specific extinction coefficient: α (l g^{-1} cm^{-1})		46.37 (at 657 nm in 90% acetone) Calculated from Lorenzen & Jeffrey (1980)
Molar extinction coefficient: ε (l mol^{-1} cm^{-1})		28.1 x 10^3 (at 657 nm in 90% acetone) Calculated from α above

UV-vis spectra:

Solvent	Absorbance maxima (nm)				Band ratio*	Reference
100% Acetone	434.9	525.7	598.3	653.5	4.64	Repeta (unpub. data)
Diethyl ether	432			654	4.81	Brown (1968)
HPLC Eluant	436	529	600	655	5.23	SCOR WG 78: Wright *et al.* (1991) method

*Soret (blue maximum): red ratio

Fluorescence spectra:

Solvent	Excitation (nm)	Emission (nm)	Reference
No data available			

Alteration products:

Origin: Acidification of chlorophyllide *b*

Additional reference(s): Scheer (1991)

Divinyl chlorophyll *b*

HPLC peak 37*

Standard spectrum in reference solvent: diethyl ether†

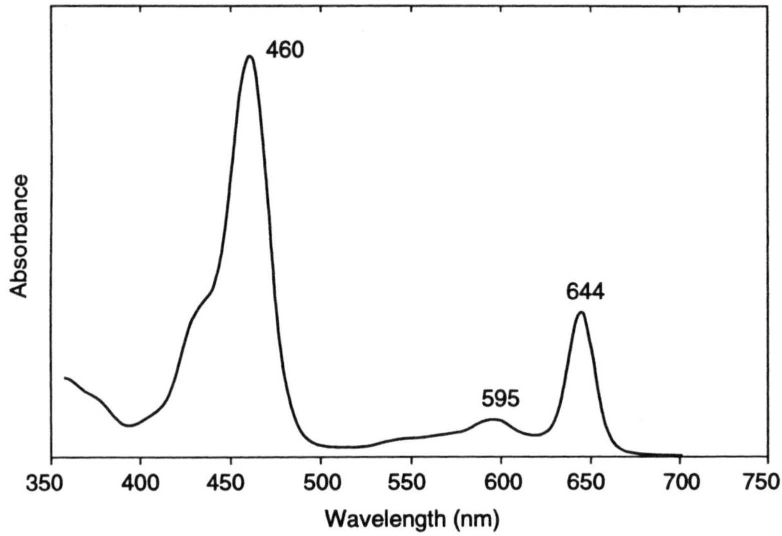

Molecular structure

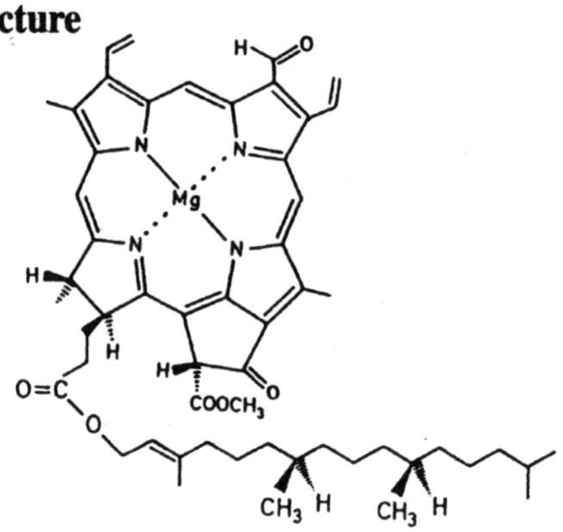

Diode array spectrum in Repeta eluant

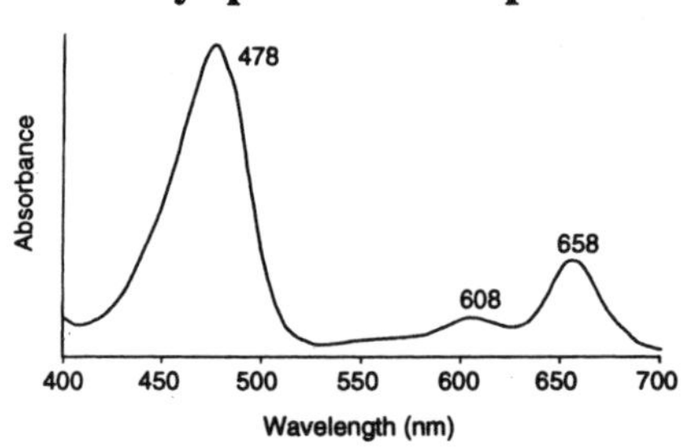

HPLC: Divinyl chlorophyll *b*, peak 37*†

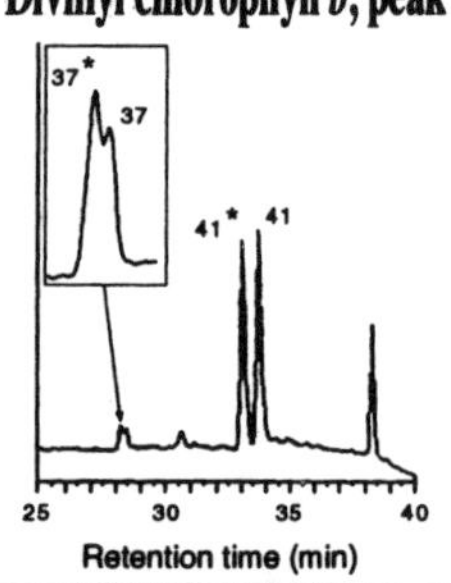

† Goericke and Repeta (1992) HPLC RP-8 system

Divinyl chlorophyll *b*

Property		**Data**
Name:	(Trivial)	**Divinyl chlorophyll *b***
	(IUPAC)	8-deethyl-8-vinyl chlorophyll *b* = $8^1,8^2$-didehydro-chlorophyll *b*
SCOR abbreviation:		DV chl *b*
Occurrence:		*Prochlorococcus marinus* (Class Prochlorophyceae)
Colour:		Brown-green
Molecular formula:		$C_{55}H_{68}N_4O_6Mg$
Molecular weight:		905.48
Specific extinction coefficient: α ($l\ g^{-1}\ cm^{-1}$)		Not determined
Molar extinction coefficient: ε ($l\ mol^{-1}\ cm^{-1}$)		Assume identical to chlorophyll *b*

UV-vis spectra:

Solvent	**Absorbance maxima (nm)**			**Band ratio***	**Reference**
Diethyl ether	460	595	644	2.78	Goerike & Repeta (1992)
Diethyl ether	462		648		Bazzaz (1981)
HPLC eluant	478	608	658	3.43	Repeta (unpub. data)

*Soret (blue maximum): red ratio

Fluorescence spectra:

Solvent	**Excitation (nm)**	**Emission (nm)**	**Reference**
No data available			

Alteration products:	8-deethyl-8-vinyl pheophytin *b;* easily bleached (Gieskes, 1991)
Culture from which SCOR data were obtained:	Culture of *Prochlorococcus* not available to SCOR WG 78
Additional reference(s):	Gieskes & Kraay (1987); Shioi & Beale (1987); Chisholm *et al.* (1988, 1992)

Chlorophyll c_1

Polyethylene HPLC*

Standard spectrum in reference solvent: diethyl ether

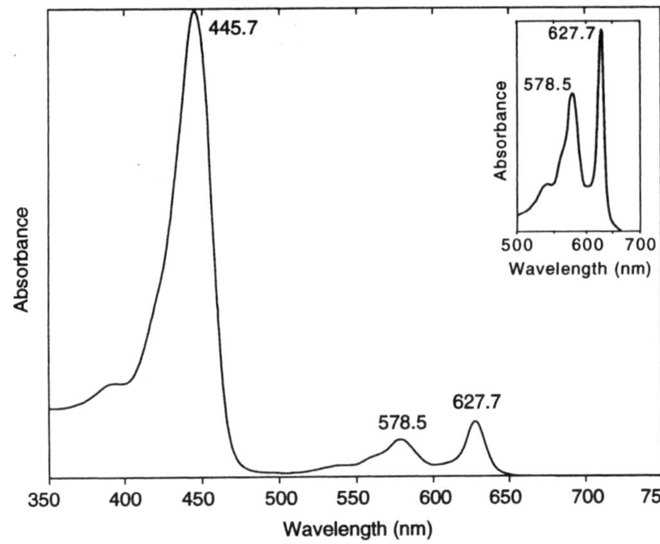

Molecular structure

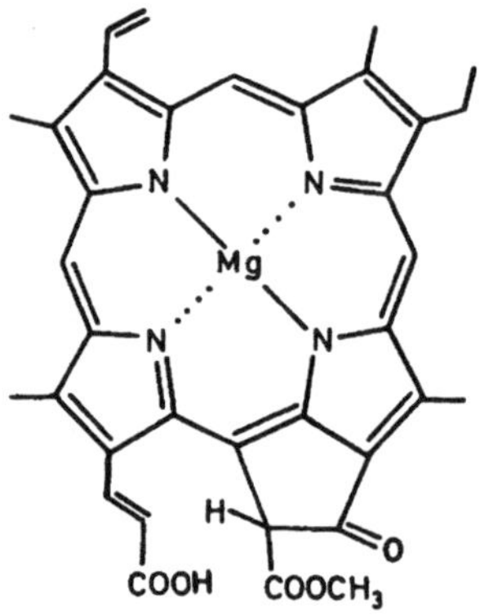

Diode array spectrum in SCOR eluant

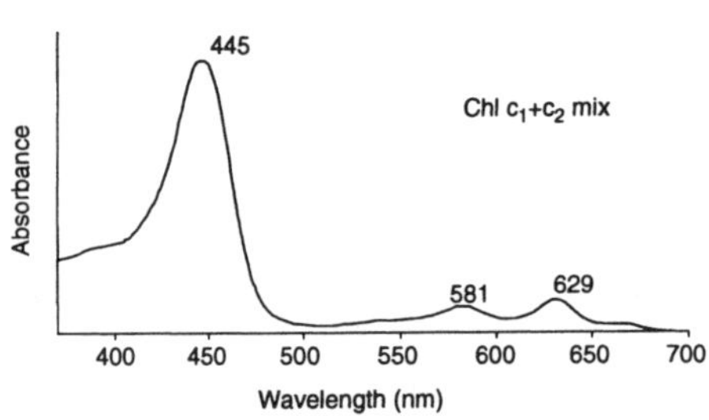

Polyethylene HPLC: Chl c_1*

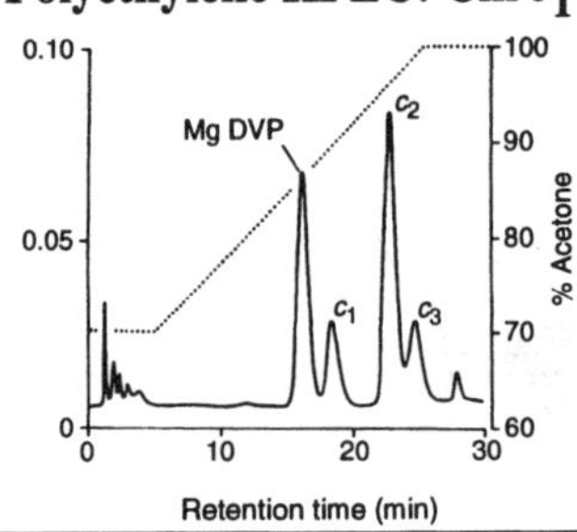

* **No resolution by the Wright *et al.* (1991) method; separated by polyethylene HPLC, using aqueous acetone eluant (Jeffrey, 1989)**

Chlorophyll c_1

Property		Data
Name:	(Trivial)	**Chlorophyll c_1**
	(IUPAC)	$3^1,3^2,17^1,17^2$-tetradehydro-13^2-methoxycarbonyl-phytoporphyrinato-Mg(II); see Hynninen (1991)
SCOR abbreviation:		Chl c_1
Occurrence:		Chromophyte algae, brown seaweeds (Jeffrey, 1976, 1989)
Colour:		Light green on TLC; emerald green (concentrated solution)
Molecular formula:		$C_{35}H_{30}N_4O_5Mg$
Molecular weight:		610.95
Specific extinction coefficient: † α (l g^{-1} cm^{-1})		318 (at 443.2 nm in 90% acetone + 1% pyridine) 44.8 (at 630.6 nm in 90% acetone + 1% pyridine) 39.2 (at 629.1 nm in 100% acetone + 1% pyridine) Jeffrey (1972)

† Where chlorophylls c_1 & c_2 are not resolved, the mean of the extinction coefficients was used. In 90% acetone, $\alpha = (44.8 + 42.4)/2 = 42.6$ l g^{-1} cm^{-1}

Molar extinction coefficient: ε (l mol^{-1} cm^{-1})	194.3×10^3 (at 443.2 nm in 90% acetone + 1% pyridine) 27.4×10^3 (at 630.6 nm in 90% acetone + 1% pyridine) 23.9×10^3 (at 629.1 nm in 100% acetone + 1% pyridine) Calculated from α above

UV-vis spectra:

Solvent	Absorbance maxima (nm)			Band ratio*	Reference
100% Acetone	446.1	577.9	629.1	8.88	Jeffrey (1972)
Diethyl ether	445.7	578.5	627.7	8.62	Jeffrey (1969)
HPLC Eluant (c_1+ c_2 mix)	445	581	629	7.50	SCOR WG 78: Wright *et al.* (1991) method

*Soret (blue maximum): red ratio

Fluorescence spectra:

Solvent	Excitation (nm)	Emission (nm)	Reference
100% Acetone	450	633, 694	Jeffrey (1972)

Alteration products:	Corresponding phytoporphyrin
Culture from which SCOR data were obtained:	*Lauderia annulata* (diatom)
Additional reference(s):	Jeffrey (1989); Scheer (1991)

Chlorophyll c_2

HPLC peak 6a

Standard spectrum in reference solvent: diethyl ether

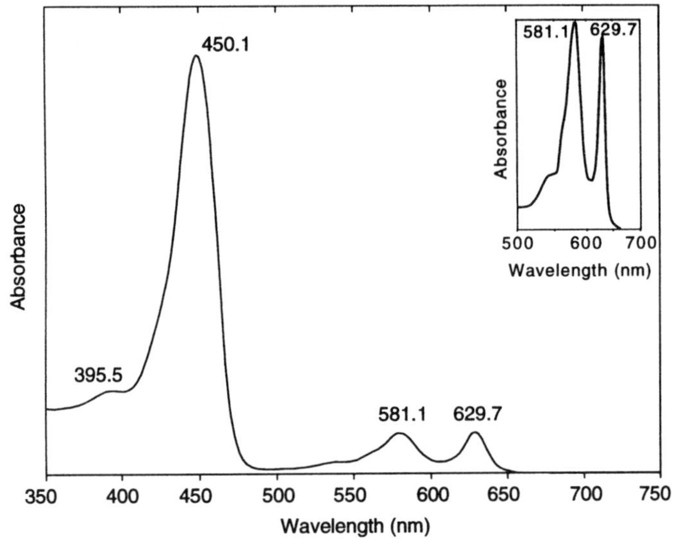

Molecular structure

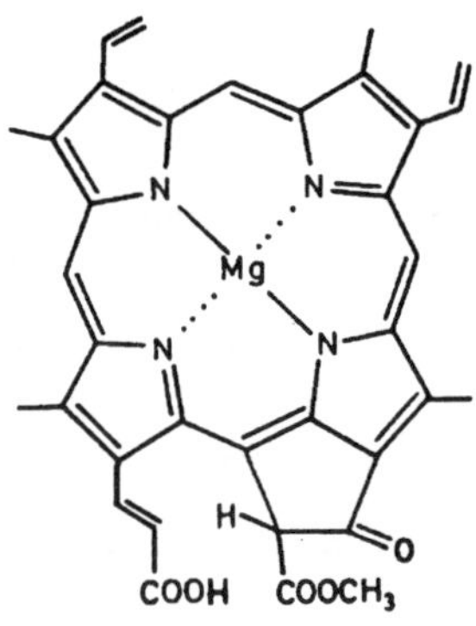

Diode array spectrum in SCOR eluant

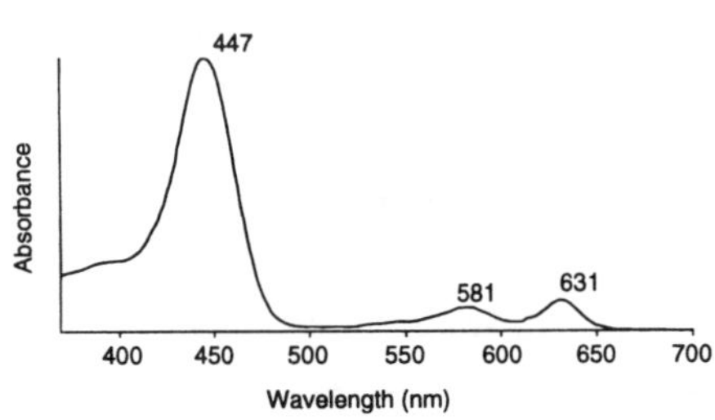

HPLC: Chl c_2, peak 6a

Amphidinium carterae

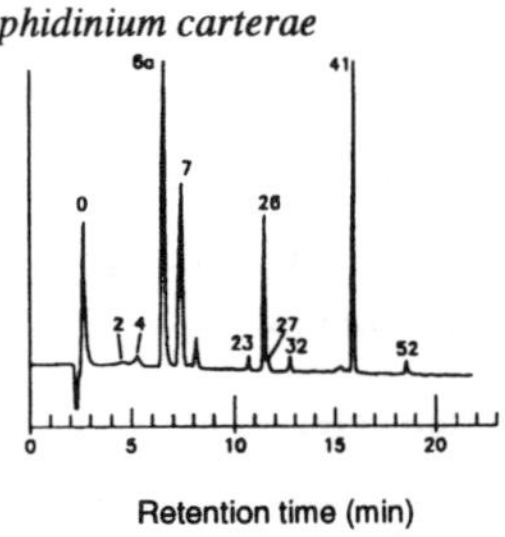

Chlorophyll c_2

Property		Data
Name:	(Trivial)	**Chlorophyll c_2**
	(IUPAC)	$3^1,3^2,8^1,8^2,17^1,17^2$-hexadehydro-$13^2$-methoxycarbonyl-phytoporphyrinato-Mg(II); see Hynninen (1991)
SCOR abbreviation:		Chl c_2
Occurrence:		Chromophyte algae, brown seaweeds (Jeffrey, 1976, 1989)
Colour:		Light green on TLC; emerald green (concentrated solution)
Molecular formula:		$C_{35}H_{28}N_4O_5Mg$
Molecular weight:		608.94
Specific extinction coefficient: † α ($l\ g^{-1}\ cm^{-1}$)		374 (at 443.8 nm in 90% acetone + 1% pyridine) 40.4 (at 630.9 nm in 90% acetone + 1% pyridine) 37.2 (at 629.6 nm in 100% acetone + 1% pyridine) Jeffrey (1972)

† Where chlorophylls c_1 & c_2 are not resolved, the mean of the extinction coefficients was used. In 90% acetone, $\alpha = (44.8 + 40.4)/2 = 42.6\ l\ gm^{-1}\ cm^{-1}$.

Molar extinction coefficient: ε ($l\ mol^{-1}\ cm^{-1}$)	227.7×10^3 (at 443.8 nm in 90% acetone + 1% pyridine) 24.6×10^3 (at 630.9 nm in 90% acetone + 1% pyridine) 22.7×10^3 (at 629.6 nm in 100% acetone + 1% pyridine) Calculated from α above

UV-vis spectra:

Solvent	Absorbance maxima (nm)			Band ratio*	Reference
100% Acetone	444.6	581.3	629.6	8.62	Jeffrey (1972)
Diethyl ether	450.1	581.1	629.7	10.35	SCOR WG 78 data
HPLC Eluant	447	581	631	8.48	SCOR WG 78: Wright *et al.* (1991) method

*Soret (blue maximum): red ratio

Fluorescence spectra:

Solvent	Excitation (nm)	Emission (nm)	Reference
100% Acetone	453	635, 696	Jeffrey (1972)
100% Ethanol	450	638, 694	Jeffrey (1972)

Alteration products:	Corresponding phytoporphyrin
Culture from which SCOR data were obtained:	*Amphidinium carterae* (dinoflagellate)
Additional reference(s):	Jeffrey (1989); Scheer (1991)

Chlorophyll c_3 — HPLC peak 5

Standard spectrum in reference solvent: acetone (100%)

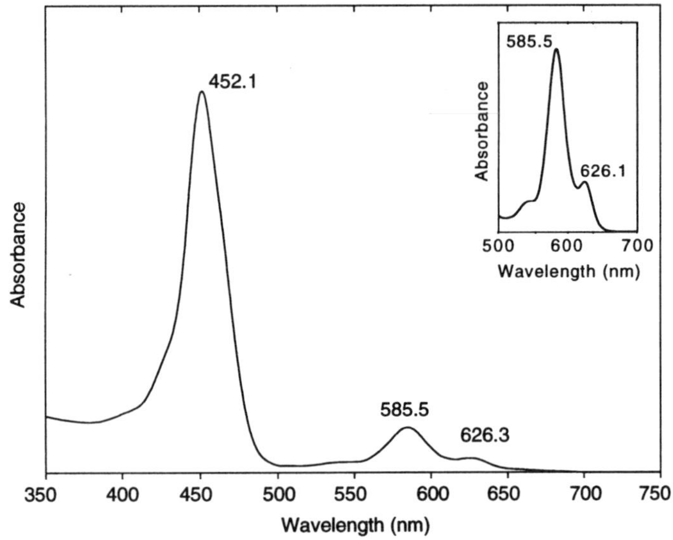

Molecular structure

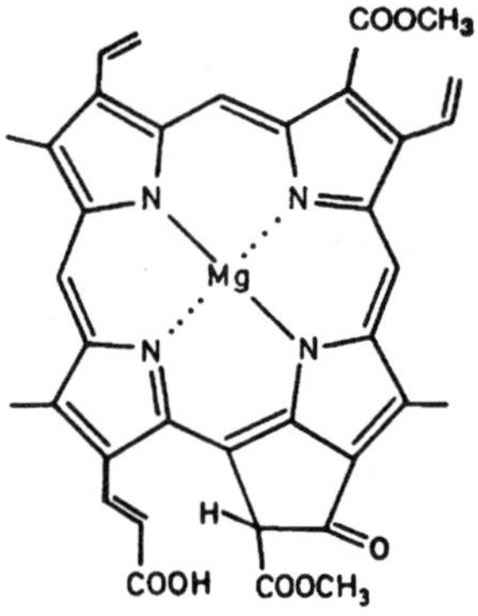

Diode array spectrum in SCOR eluant

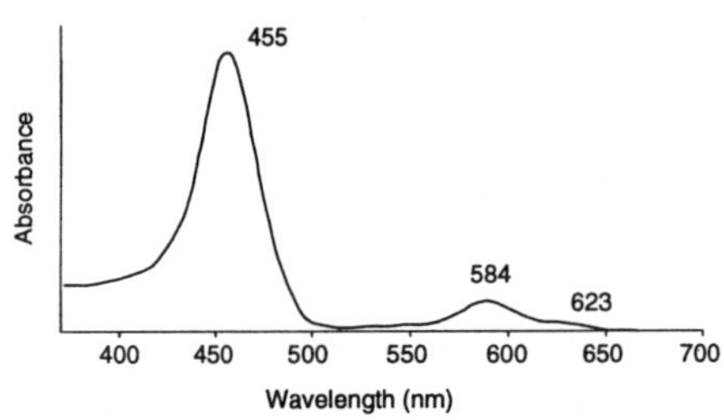

HPLC: Chl c_3, peak 5

Emiliania huxleyi

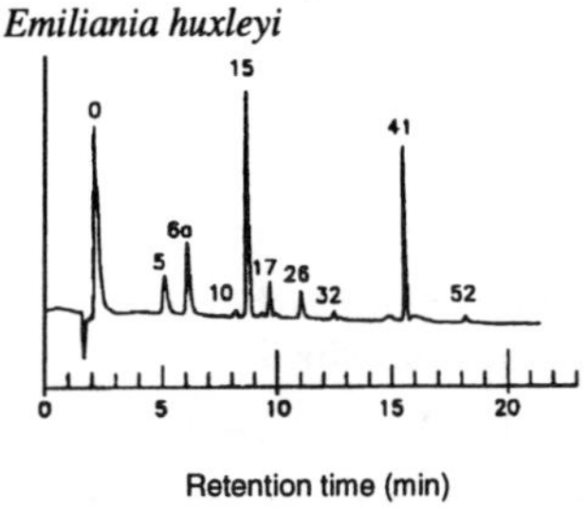

Chlorophyll c_3

Property		Data
Name:	(Trivial)	**Chlorophyll c_3**
	(IUPAC)	7-demethyl-7-methoxycarbonyl chlorophyll c_2
SCOR abbreviation:		Chl c_3
Occurrence:		Some prymnesiophytes, some diatoms, chrysophytes (Vesk & Jeffrey, 1987; Jeffrey & Wright, 1994)
Colour:		Light green on TLC; emerald green (concentrated solution)
Molecular formula:		$C_{36}H_{28}N_4O_7Mg$ Fookes & Jeffrey (1989)
Molecular weight:		652.95
Specific extinction coefficient: α (l g^{-1} cm^{-1})		Not known; use mean of chl c_1 + c_2 at Soret band. 334.5 (at 452.9 nm in 100% acetone + 1% pyridine) 346 (at 452.9 nm in 90% acetone + 1% pyridine) Jeffrey (1972)
Molar extinction coefficient: ε (l mol^{-1} cm^{-1})		Not known; 218.4 x 10^3 (at 452.9 nm in 100% acetone + 1% pyridine) Calculated from α above

UV-vis spectra:

Solvent	Absorbance maxima (nm)	Band ratio*	Reference
100% Acetone	452.1 585.5 626.3	28.3	SCOR WG 78 data
Diethyl ether	451.3 584.5 625.9 32.1		Jeffrey & Wright (1987)
HPLC Eluant	455 584 623	33.5	SCOR WG 78: Wright *et al.* (1991) method

*Soret (blue maximum): red ratio

Fluorescence spectra:

Solvent	Excitation (nm)	Emission (nm)	Reference
100% Acetone	452	635, 690	SCOR WG 78 data

Alteration products:	None known
Culture from which SCOR data were obtained:	*Emiliania huxleyi* (prymnesiophyte)
Additional reference(s):	Jeffrey & Wright (1987); Fookes & Jeffrey (1989); Jeffrey (1989)

Chlorophyll $c_{CS\text{-}170}$

Standard spectrum in reference solvent: diethyl ether

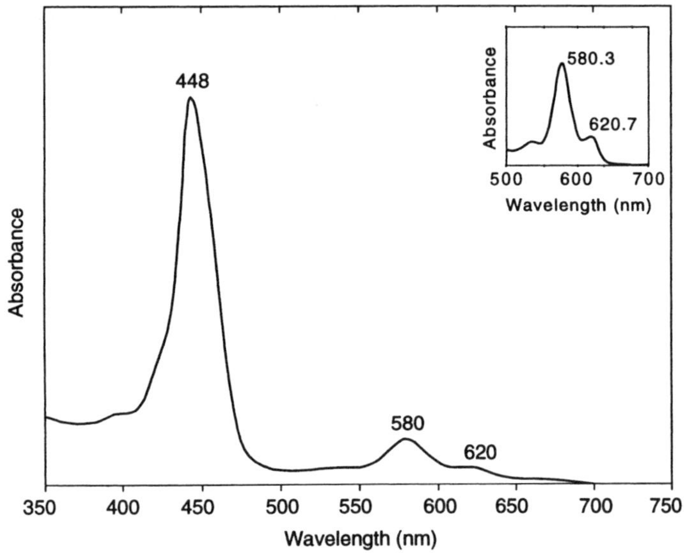

Molecular structure*

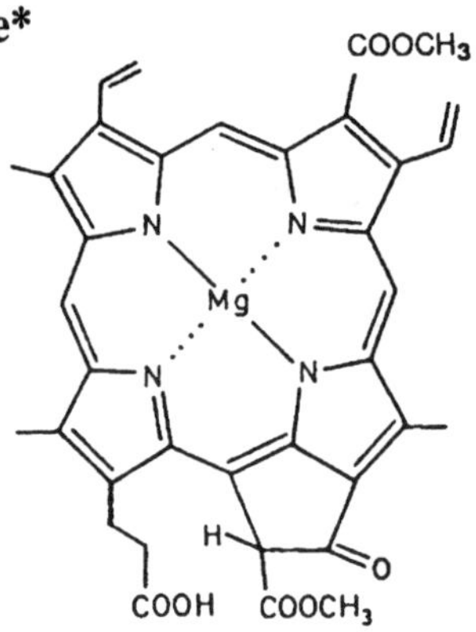

***Tentative evidence, Fookes and Jeffrey (unpublished)**

Spectrum in SCOR eluant

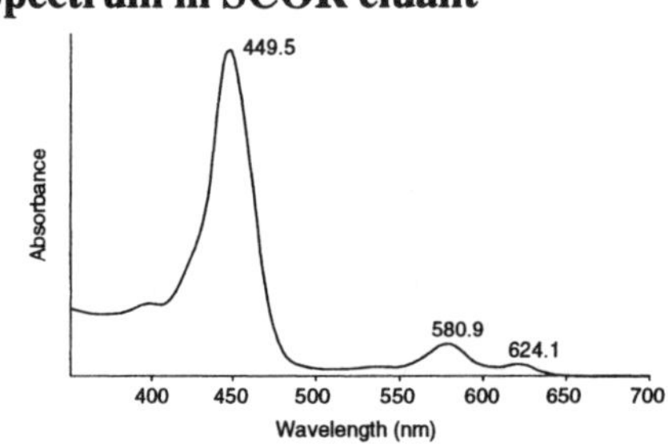

HPLC: Chl $c_{CS\text{-}170}$

***Micromonas pusilla*(CS-170)**

Does not separate by the Wright *et al.* (1991) HPLC method. Separated by HPTLC method of Jeffrey (1989); see Chapter 7

Chlorophyll $c_{CS\text{-}170}$

Property		Data
Name:	(Trivial)	**Chlorophyll $c_{CS\text{-}170}$**
	(IUPAC)	Structure not confirmed; evidence suggests a propionate derivative of chl c_3 (Fookes & Jeffrey, unpublished)
SCOR abbreviation:		Chl c_{CS-170}
Occurrence:		*Micromonas pusilla* (CS–170)
Colour		Light green on TLC; emerald green (concentrated solution)
Molecular formula:		$C_{36}H_{30}N_4O_7Mg$ (to be confirmed)
Molecular weight:		654.96
Specific extinction coefficient: α ($l\ g^{-1}\ cm^{-1}$)		Not known; use α for chl c_1 + c_2 mixture
Molar extinction coefficient: ε ($l\ mol^{-1}\ cm^{-1}$)		Not known; use ε for chl c_1 + c_2 mixture

UV-vis spectra:

Solvent	Absorbance maxima (nm)			Band ratio*	Reference
Diethyl ether	448	580	620	22.1	Jeffrey (1989)
HPLC eluant	449.5	580.9	624.1	26.6	Jeffrey (unpublished)

*Soret (blue maximum): red ratio

Fluorescence spectra:

Solvent	Excitation (nm)	Emission (nm)	Reference
No data available			

Alteration products:	Not known
Culture from which SCOR data were obtained:	*Micromonas pusilla* (CS-170) (prasinophyte); also contains Mg divinyl pheorphyrin a_5 monomethyl ester
Additional reference(s):	Jeffrey (1989)

Mg-2,4-divinyl pheoporphyrin a_5 monomethyl ester HPLC peak 6c

Standard spectrum in reference solvent: acetone (100%)

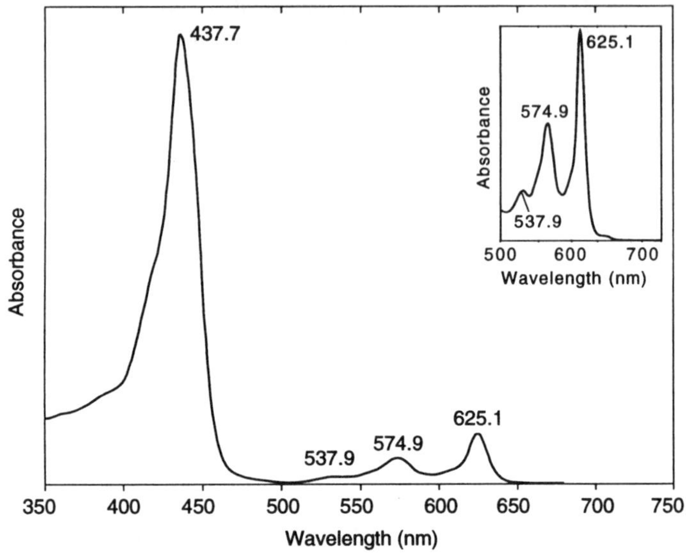

Molecular structure

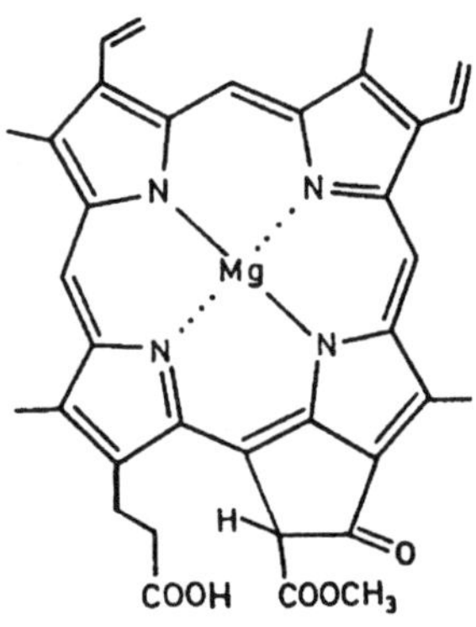

Diode array spectrum in SCOR eluant

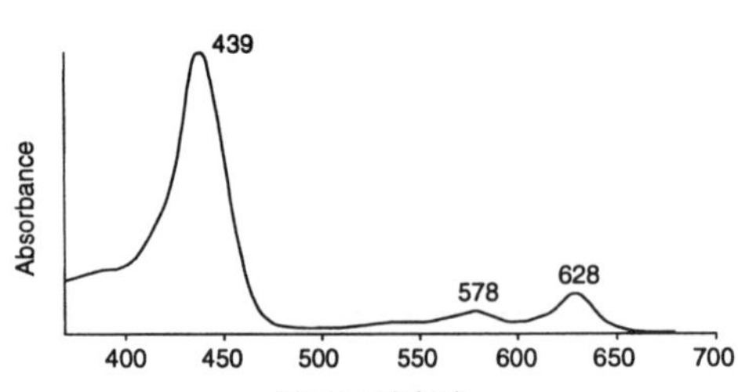

HPLC: Mg-2,4-divinyl pheoporphyrin a_5 monomethyl ester

Micromonas pusilla **peak 6c**

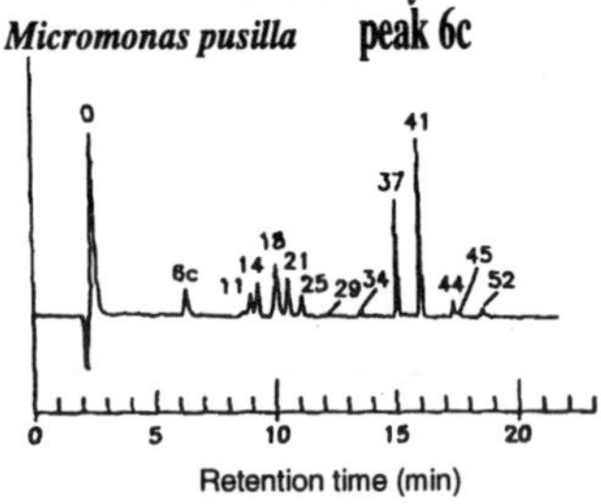

Mg-2,4-divinyl pheoporphyrin a_5 monomethyl ester

Property		Data
Name:	(Trivial: Fischer nomenclature)	**Mg-2,4-divinyl pheoporphyrin a_5 monomethyl ester**
	(Trivial: IUPAC)	Mg-3,8-divinyl phytoporphyrin-13^2-methyl carboxylate
	(IUPAC)	$8^1,8^2$,17,18-tetradehydro-chlorophyllide *a* = $3^1,3^2,8^1,8^2$-tetradehydro-13^2-methoxycarbonyl-phytoporphyrinato-Mg(II); see Hynninen (1991)
SCOR abbreviation:		Mg DVP
Occurrence:		Some prasinophytes; Ricketts (1966); Jeffrey (1989)
Colour:		Light green on TLC; emerald green (concentrated solution)
Molecular formula:		$C_{35}H_{30}N_4O_5Mg$
Molecular weight:		610.94
Specific extinction coefficient: α ($l\ g^{-1}\ cm^{-1}$)		Unknown; use 58.9 (at 623 nm in methanol) for the monovinyl derivative (Griffiths, 1991)
Molar extinction coefficient: ε ($l\ mol^{-1}\ cm^{-1}$)		Unknown; use $36x10^3$ (at 623 nm in methanol) for the monovinyl derivative (Griffiths, 1991)

UV-vis spectra:

Solvent	Absorbance maxima (nm)			Band ratio*	Reference
100% Acetone	437.7	574.9	625.1	9.33	SCOR WG 78 data
Diethyl ether	437.1	573.9	623.5	9.72	Jeffrey & Wright (1987)
HPLC Eluant	439	578	628	7.55	SCOR WG 78: Wright *et al.* (1991) method

*Soret (blue maximum): red ratio

Fluorescence spectra:

Solvent	Excitation (nm)	Emission (nm)	Reference
Diethyl ether	440	627	Chereskin *et al.*(1983)

Alteration products:	None known, but rapid bleaching can occur when handling
Culture from which SCOR data were obtained:	*Micromonas pusilla* (CS-86) (prasinophyte)
Additional reference(s):	Jeffrey (1989); Griffiths (1991)

Phytylated Chlorophyll *c*-like

HPLC peak 41a

Standard spectrum in reference solvent: acetone (100%)

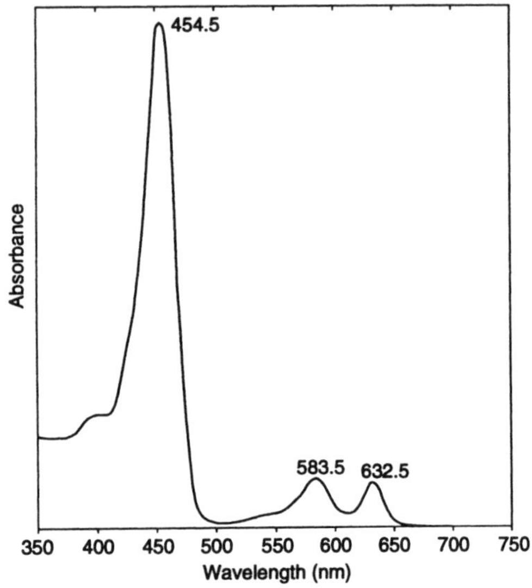

Data from *Emiliania huxleyi* (Nelson and Wakeham, 1989)

Molecular structure*

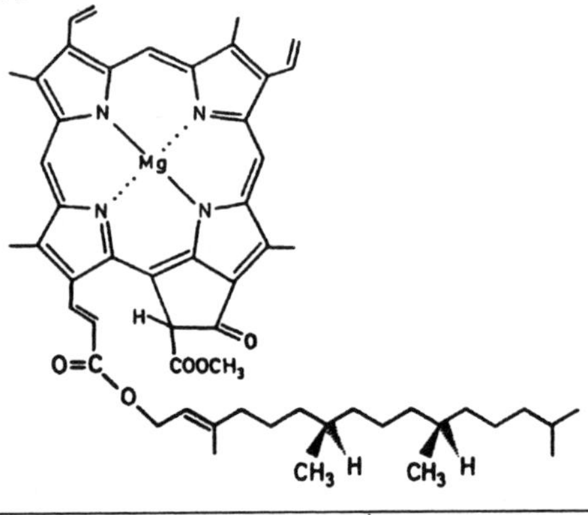

***Tentative structure only (Nelson and Wakeham, 1989)**

Diode array spectrum in SCOR eluant

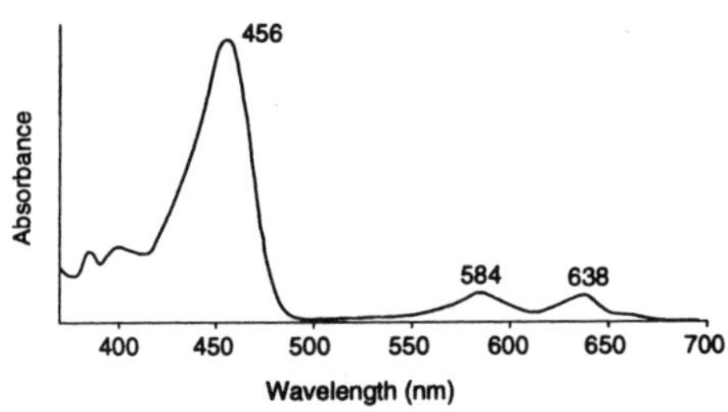

HPLC: Phytylated Chlorophyll *c*, peak 41a

Chrysochromulina minor

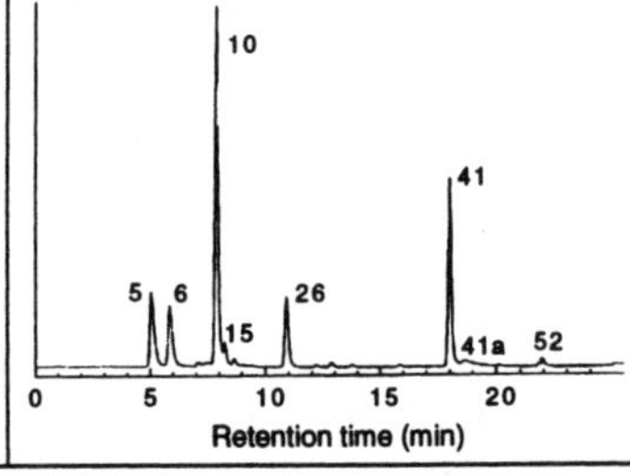

Property		**Phytylated chlorophyll *c*-like** **Data**
Name:	(Trivial)	**Phytylated chlorophyll *c*-like pigment***
	(IUPAC)	Not confirmed
		*All information provisional until chemical structure confirmed
SCOR abbreviation:		Phytyl-chl *c*
Occurrence:		*Emiliania huxleyi* (Nelson & Wakeham, 1989) Many prymnesiophytes (28 out of 50 strains tested; Jeffrey & Wright, 1994)
Colour:		Lightgreen
Molecular formula:		$C_{55}H_{66}N_4O_5Mg$
Molecular weight:		887.46 Molecular formula, molecular weight, and coefficient calculations assume phytylated chl c_2 ester
Specific extinction coefficient: α ($l\ g^{-1}\ cm^{-1}$)		25.6 (at 629 nm in 100% acetone +1% pyridine) Calculated from ε chlorophyll c_2
Molar extinction coefficient: ε ($l\ mol^{-1}\ cm^{-1}$)		22.7×10^3 (at 629 nm in 100% acetone + 1% pyridine); assumed equal to ε for chlorophyll c_2; Jeffrey (1972)

UV-vis spectra:

Solvent	**Absorbance maxima (nm)**			**Band ratio***	**Reference**
100% Acetone	454.5	583.5	632.5	10.30	Nelson & Wakeham (1989)
HPLC Eluant	456	584	638		Jeffrey & Wright (1994)

*Soret (blue maximum): red ratio

Fluorescence spectra:

Solvent	**Excitation (nm)**	**Emission (nm)**	**Reference**
No data available			

Alteration products:	None known
Culture from which SCOR data were obtained:	*Emiliania huxleyi, Chrysochromulina minor* (prymnesiophytes)
Additional reference(s):	Nelson & Wakeham (1989)

IV.3. Carotenoid Data and Graphics Sheets

Alloxanthin
Antheraxanthin
Astaxanthin
19'-Butanoyloxyfucoxanthin
Canthaxanthin
β,ε-Carotene
β,β-Carotene
β,ψ-Carotene
ε,ε-Carotene
Crocoxanthin
Diadinochrome
Diadinoxanthin
Diatoxanthin
Dinoxanthin
Echinenone
Ethyl-apo-8'-β-carotenoate
Fucoxanthin
19'-Hexanoyloxyfucoxanthin
Lutein
Lycopene
Monadoxanthin
Neoxanthin
P-457
Peridinin
Prasinoxanthin
Siphonaxanthin
Siphonein
Vaucheriaxanthin (ester)
Violaxanthin
Zeaxanthin

Alloxanthin

HPLC peak 30

Standard spectrum in reference solvent: acetone

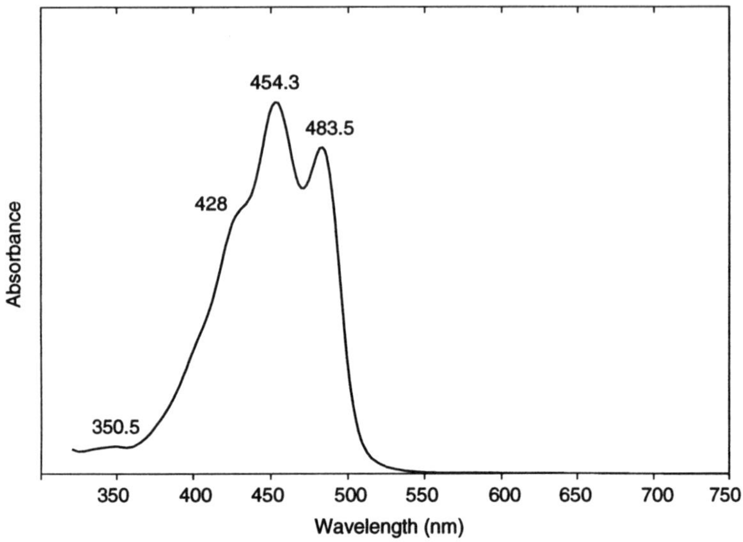

Molecular structure

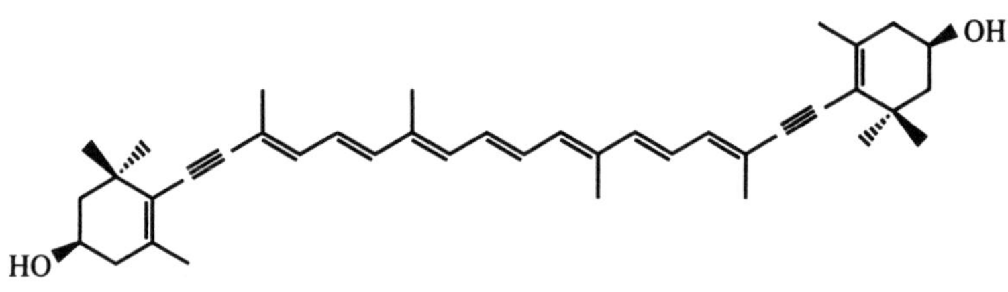

Diode array spectrum in SCOR eluant

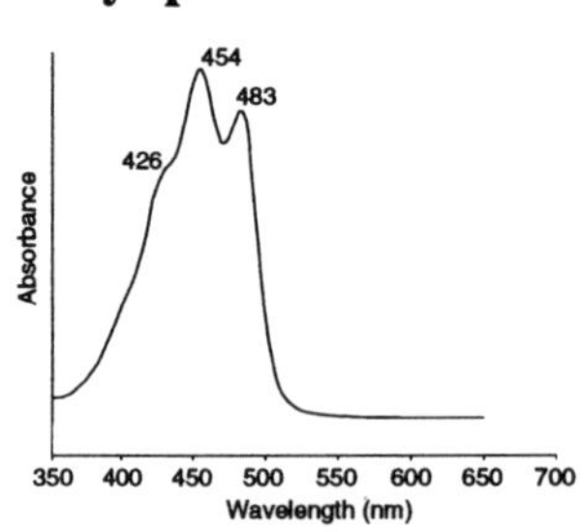

HPLC: Alloxanthin, peak 30

Chroomonas salina

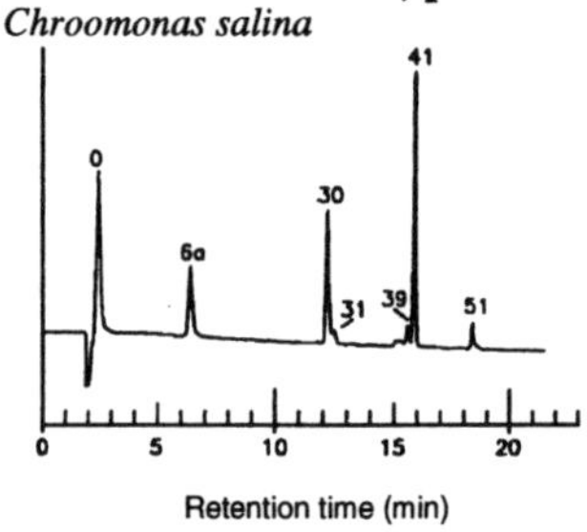

Alloxanthin

Property		**Data**
Name:	(Trivial)	**Alloxanthin**
	(IUPAC)	(3*R*,3'*R*)-7,8,7',8'-Tetradehydro-β,β-carotene-3, 3'-diol
SCOR abbreviation:		Allo
Occurrence:		Cryptomonads (major carotenoid)
Colour:		Yellow-orange
Molecular formula:		$C_{40} H_{52} O_2$
Molecular weight:		564.85
Specific extinction coefficient: $E^{1\%}_{1\,cm}$ (100 ml g^{-1} cm^{-1})		2500 (at 454 nm in acetone) Not determined; use $E^{1\%}_{1\,cm}$ for β,β-carotene; Davies (1976)
Molar extinction coefficient: ε (l mol^{-1} cm^{-1})		141 x 10^3 (at 454 nm in acetone) Calculated from $E^{1\%}_{1\,cm}$ above

UV-vis spectra:

Solvent	Maxima (nm) I	II	III	Band ratio %III:II	Reference
100% Acetone	(428)	454.3	483.5	50	SCOR WG 78 data
Ethanol	(427)	450	478	29	Hager & Stransky (1970c)
Diethyl ether	(430)	451	481	44	Pennington *et al.* (1985)
Hexane	(427)	451	480	50	Cheng *et al.* (1974)
HPLC Eluant	(427)	454	482	47	SCOR WG 78: Mantoura & Llewellyn (1983) method
HPLC Eluant	(428)	454	483	43	SCOR WG 78: Wright *et al.* (1991) method

Alteration products:	*Cis*-isomers
Culture from which SCOR data were obtained:	*Chroomonas salina* (cryptomonad)
Additional reference(s):	Chapman (1966); Pennington *et al.* (1985)

Antheraxanthin

HPLC peak 29

Standard spectrum in reference solvent:

No data available

Molecular structure

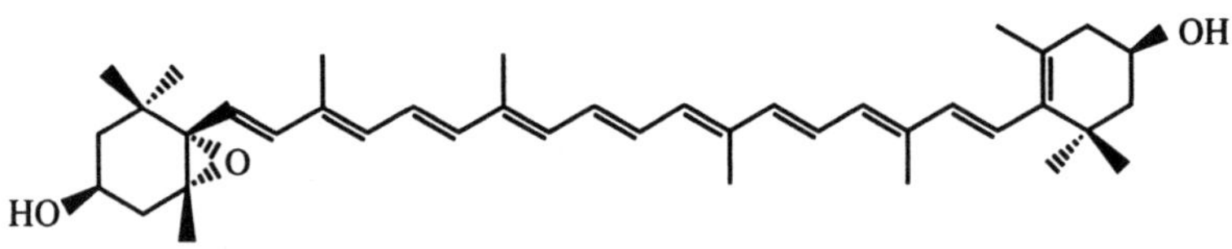

Diode array spectrum in Mantoura-Llewellyn eluant*

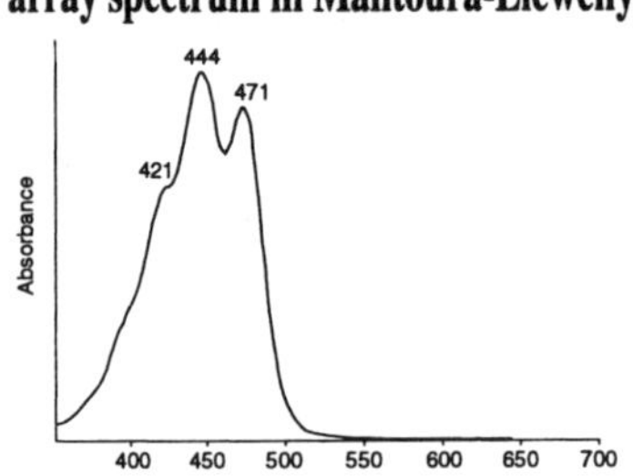

HPLC: Antheraxanthin, peak 29

Dunaliella tertiolecta

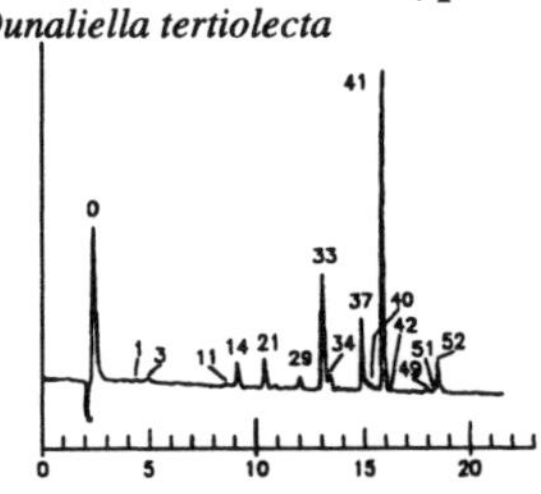

*Mantoura and Llewellyn (1983)

Antheraxanthin

Property		Data
Name:	(Trivial)	**Antheraxanthin**
	(IUPAC)	(3*S*,5*R*,6*S*,3'*R*)-5,6-Epoxy-5,6-dihydro-β, β-carotene-3,3'-diol
SCOR abbreviation:		Anth
Occurrence:		Green algae, higher plants (minor pigment)
Colour:		Yellow
Molecular formula:		$C_{40}H_{56}O_3$
Molecular weight:		584.88
Specific extinction coefficient: $E^{1\%}_{1\,cm}$ (100 ml g^{-1} cm^{-1})		2350 (at 446 nm in ethanol) Hager & Meyer-Bertenrath (1966)
Molar extinction coefficient: ε (l mol^{-1} cm^{-1})		137 x 10^3 (at 446 nm in ethanol) Calculated from $E^{1\%}_{1\,cm}$ above

UV-vis spectra:

Solvent	Maxima (nm) I	II	III	Band ratio %III:II	Reference
Ethanol	422	444	472	54	Stransky & Hager (1970a)
Ethanol	421	443	473		Jungalwala & Cama (1962)
Hexane	(421)	445	470		Valadon & Mummery (1967)
Petroleum ether	424	445	475	75	Bjørnland *et al.* (1984)
HPLC Eluant	(421)	444	471	57	SCOR WG 78: Mantoura & Llewellyn (1983) method

Alteration products:	*Cis*-isomers; furanoids (mutatoxanthins)
Culture from which SCOR data were obtained:	*Dunaliella tertiolecta* (green flagellate)
Additional reference(s):	Hager & Meyer-Bertenrath (1966); Stransky & Hager (1970c)

Astaxanthin

Standard spectrum in reference solvent: methanol

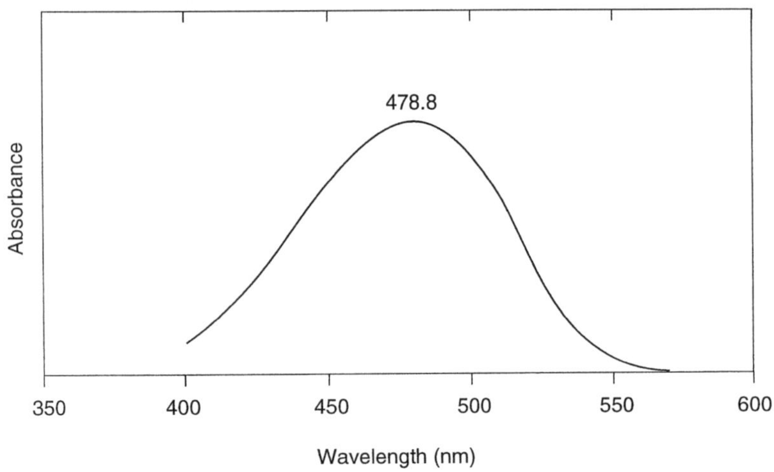

Molecular structure

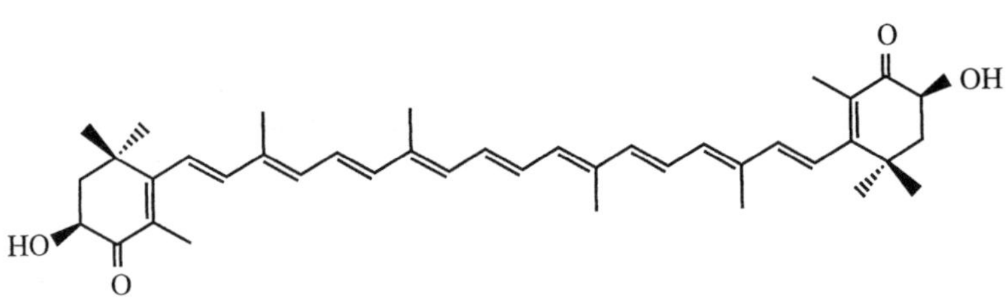

Diode array spectrum in Repeta eluant*

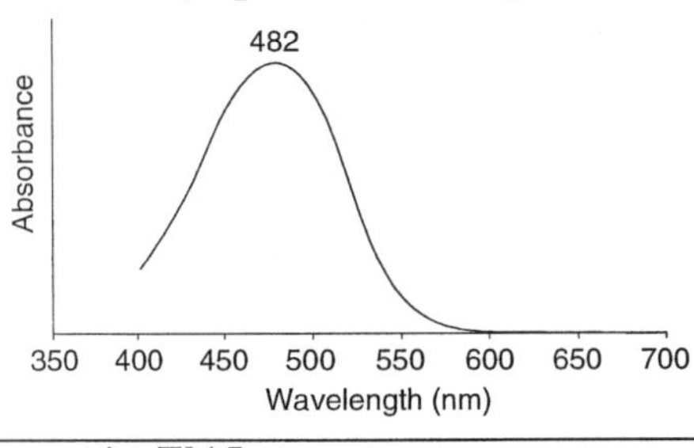

HPLC: Astaxanthin

No data available

* see section IV.1.7

Astaxanthin

Property		Data
Name:	(Trivial)	**Astaxanthin**
	(IUPAC)	(3*S*,3'*S*)-3,3'-Dihydroxy-β,β-carotene-4,4'-dione
SCOR abbreviation:		Asta
Occurrence:		Some green algae (minor pigment; becomes major during nutrient limitation in some species). Transformation product in marine animals. Frequently present together with its mixed fatty acid mono- and diesters.
Colour:		Salmon pink
Molecular formula:		$C_{40} H_{52} O_4$
Molecular weight:		596.85
Specific extinction coefficient: $E^{1\%}_{1\,cm}$ (100 ml g^{-1} cm^{-1})		2100 (at 468 nm in hexane) Withers *et al.* (1977)
Molar extinction coefficient: ε (l mol^{-1} cm^{-1})		125 x 10^3 (at 468 nm in hexane) Calculated from $E^{1\%}_{1\,cm}$ above

UV-vis spectra:

Solvent	Maxima (nm) I	II	III	Band ratio %III:II	Reference
Hexane		468			Withers *et al.* (1977)
Ethanol		472			Withers *et al.* (1977)
Petroleum ether		468			Katayama *et al.* (1970)
Acetone		480			Francis *et al.* (1972)
Methanol		478.8			Repeta (unpublished)
HPLC Eluant		482			SCOR WG 78 data; Repeta eluant (see Section IV.1.7)

Alteration products:	Astacene (oxidation product)
Origin:	Synthetic (Hoffman La Roche)
Additional reference(s):	Katayama *et al.* (1970)

19'-Butanoyloxyfucoxanthin

HPLC peak 9

Standard spectrum in reference solvent: acetone

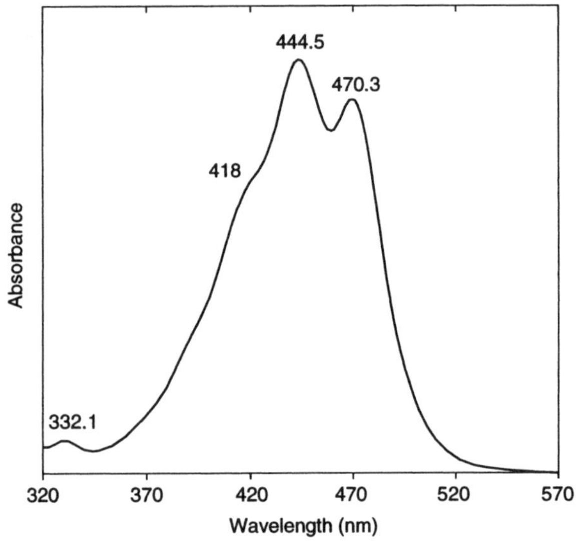

Molecular structure

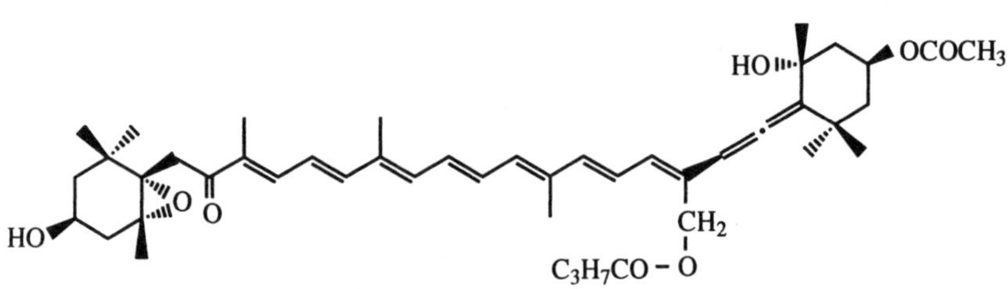

Diode array spectrum in SCOR eluant

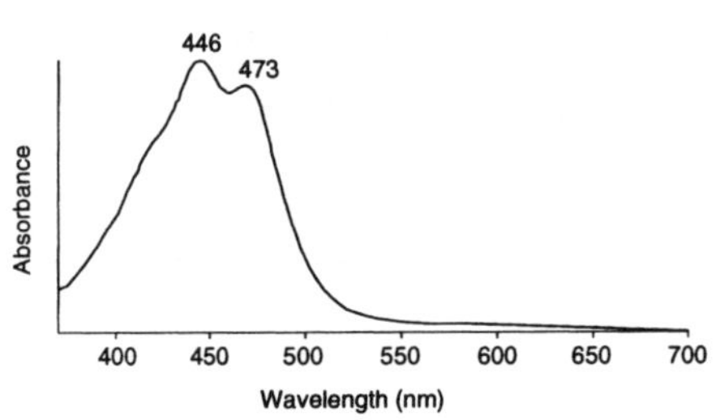

HPLC: 19'-Butanoyloxyfucoxanthin
Pelagococcus subviridis, **peak 9**

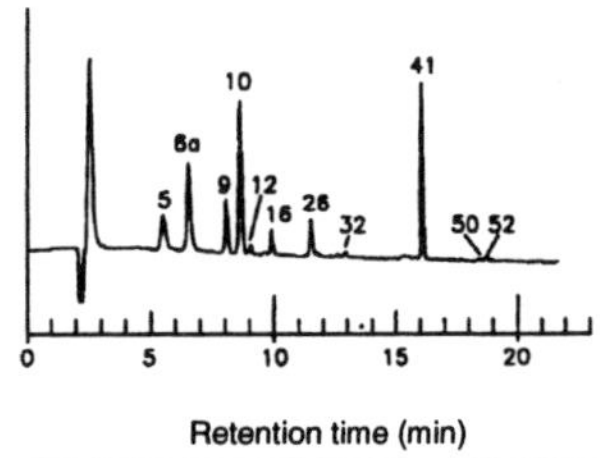

19'-Butanoyloxyfucoxanthin

Property		Data
Name:	(Trivial)	**19'-Butanoyloxyfucoxanthin**
	(IUPAC)	(3*S*,5*R*,6*S*,3'*S*,5'*R*,6'*S*)-5,6-Epoxy-3,3',5',19'-tetrahydroxy-6',7'-didehydro-5,6,7,8,5',6'-hexahydro-β,β-caroten-8-one 3'-acetate 19'-butanoate
SCOR abbreviation:		But-fuco
Occurrence:		Major pigment in some prymnesiophytes (eg *Phaeocystis*), marine chrysophytes (eg *Pelagococcus*) and 3 dinoflagellates. Trace pigment in other prymnesiophytes (e.g. *Emiliania*)
Colour:		Yellow-orange
Molecular formula:		$C_{46} H_{64} O_8$
Molecular weight:		745.01
Specific extinction coefficient: $E^{1\%}_{1\ cm}$ (100 ml g^{-1} cm^{-1})		1470 (at 445 nm in acetone) Calculated from ε for fucoxanthin, Haugan & Liaaen-Jensen (1989)
Molar extinction coefficient: ε (l mol^{-1} cm^{-1})		109 x 10^3 (at 445 nm in acetone) Assume ε But-fuco = ε for fucoxanthin (Haugan & Liaaen-Jensen, 1989); see Preamble

UV-vis spectra:

Solvent	Maxima (nm) I	II	III	Band ratio %III:II	Reference
Acetone	(418)	444.5	470.3	44	SCOR WG 78 data
Acetone		445	471	40	Vesk & Jeffrey (1987)
Ethanol		446	470	18	Wright & Jeffrey (1987)
n-Hexane	(426)	446	473	57	Bjørnland *et al.* (1989)
HPLC Eluant		448	469	21	SCOR WG 78: Mantoura & Llewellyn (1983) method
HPLC Eluant		446	473	25	SCOR WG 78: Wright *et al.* (1991) method

Alteration products:	*Cis*-isomers
Culture from which SCOR data were obtained:	*Pelagococcus subviridis* (chrysophyte)
Additional reference(s):	Vesk & Jeffrey (1987); Bjørnland *et al.* (1989); Bjørnland & Liaaen-Jensen (1989); Jeffrey & Wright (1994)

Canthaxanthin

HPLC peak 35

Standard spectrum in reference solvents

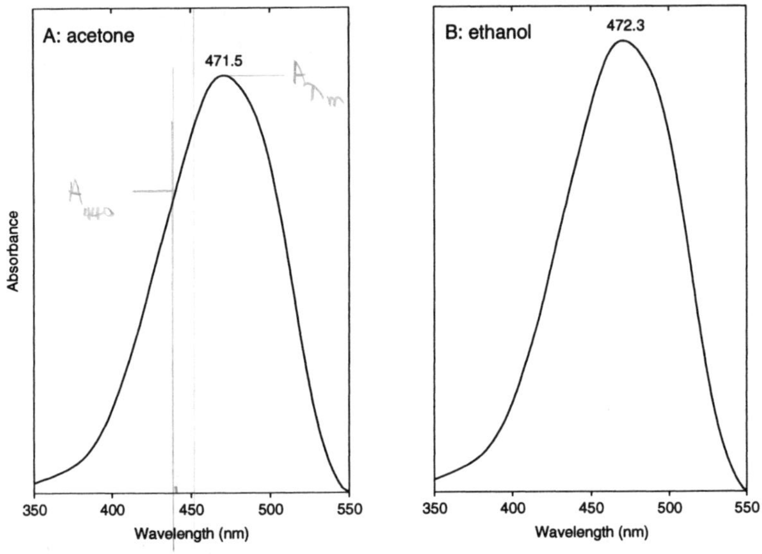

Molecular structure

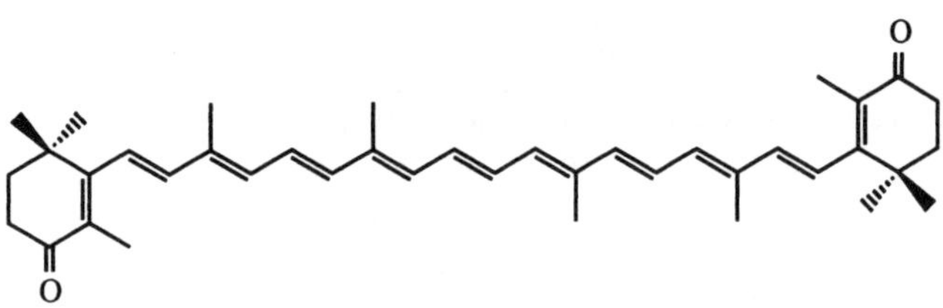

Diode array spectrum in SCOR eluant

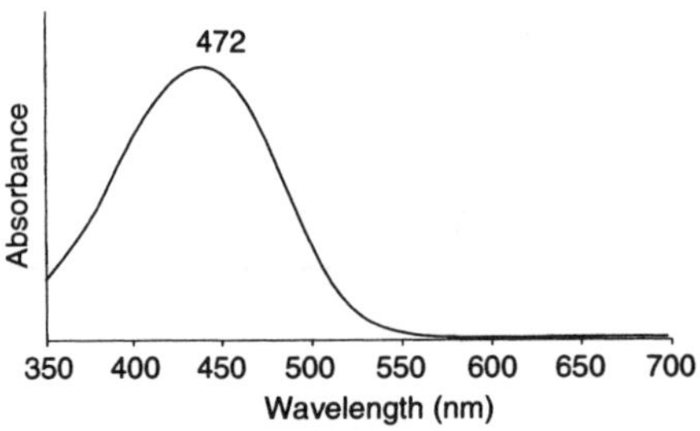

HPLC: Synthetic Canthaxanthin, peak 35

Pigment mixture

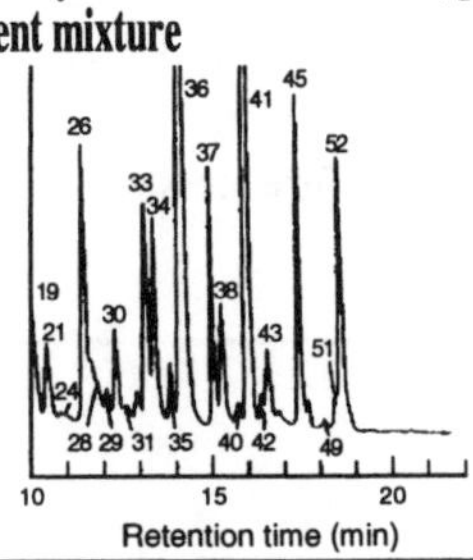

Canthaxanthin

Property		**Data**
Name:	(Trivial)	**Canthaxanthin**
	(IUPAC)	β,β-Carotene-4,4'-dione
SCOR abbreviation:		Cantha
Occurrence:		Minor or trace pigment in some green and blue-green algae, diatoms, prymnesiophytes, eustigmatophytes
Colour:		Red
Molecular formula:		$C_{40}H_{52}O_2$
Molecular weight:		564.85
Specific extinction coefficient: $E^{1\%}_{1\,cm}$ (100 ml g^{-1} cm^{-1})		2200 (at 466 nm in petroleum ether) Davies (1965) 2092 (at 480 nm in benzene) Warren & Weedon (1958)
Molar extinction coefficient: ε (l mol^{-1} cm^{-1})		124 x 10^3 (at 466 nm in petroleum ether) 118 x 10^3 (at 480 nm in benzene) Calculated from $E^{1\%}_{1\,cm}$ above

UV-vis spectra:

Solvent	**Maxima (nm) I**	**II**	**III**	**Band ratio %III:II**	**Reference**
Acetone		472			SCOR WG 78 data
Ethanol		472			SCOR WG 78 data
Ethanol		474			Hager & Stransky (1970b)
n-Hexane		467			Davies (1976)
HPLC Eluant		472			SCOR WG 78: Wright *et al.* (1991) method

Alteration products:

Culture from which SCOR data were obtained:	Synthetic
Additional reference(s):	Goodwin (1980); Antia & Cheng (1982)

β, ε-carotene **HPLC peak 51**

Standard spectrum in reference solvents

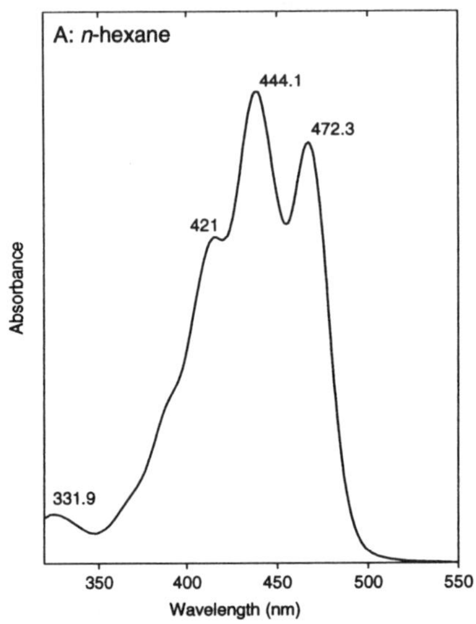

Molecular structure

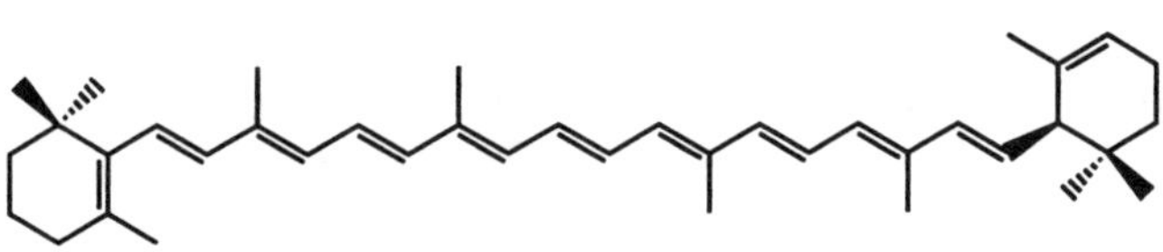

Diode array spectrum in SCOR eluant

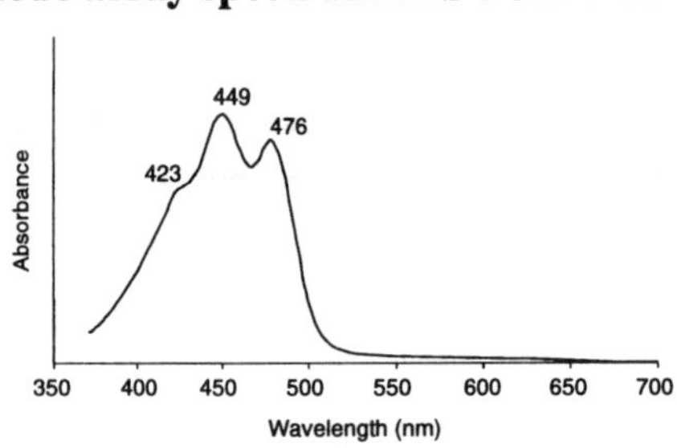

HPLC: β, ε-carotene, peak 51

Porphyridium purpureum

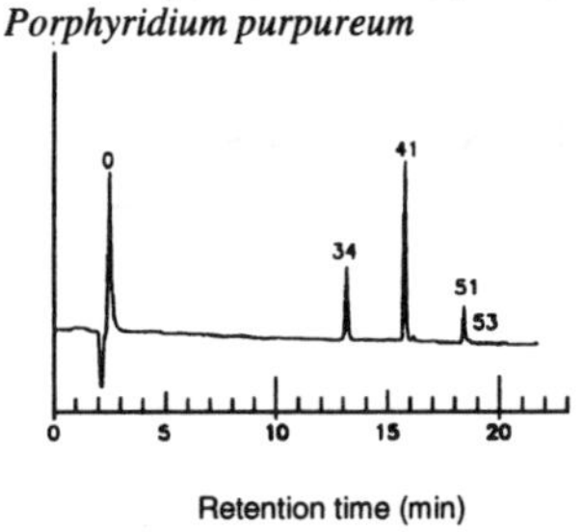

β,ε-Carotene

Property		Data
Name:	(Trivial)	**α-Carotene**
	(IUPAC)	(6'*R*)-β,ε-carotene
SCOR abbreviation:		βε-car
Occurrence:		Minor or trace pigment in green algae, cryptomonads, red algae
Colour:		Yellow
Molecular formula:		$C_{40}H_{56}$
Molecular weight:		536.88
Specific extinction coefficient: $E^{1\%}_{1\,cm}$ (100 ml g^{-1} cm^{-1})		2700 (at 448 nm in acetone) Hiyama *et al.* (1969) 2800 (at 444 nm in petroleum ether) Davies (1976)
Molar extinction coefficient: ε (l mol^{-1} cm^{-1})		145 x 10^3 (at 448 nm in acetone) 150 x 10^3 (at 444 nm in petroleum ether) Calculated from $E^{1\%}_{1\,cm}$ above

UV-vis spectra:

Solvent	Maxima (nm) I	II	III	Band ratio %III:II	Reference
Acetone	424	448	476		Hiyama *et al.* (1969)
Acetone	(422)	447	475	53	SCOR WG 78 data
Hexane	421	444	472	62	SCOR WG 78 data
Hexane	420	442	472		Valadon & Mummery (1967)
HPLC Eluant	(423)	447	474	58	SCOR WG 78: Mantoura & Llewellyn (1983) method
HPLC Eluant	(423)	449	476	47	SCOR WG 78: Wright *et al.* (1991) method

Alteration products:	*Cis*-isomers
Culture from which SCOR data were obtained:	*Porphyridium purpureum* (red alga), *Chroomonas salina* (cryptomonad)
Additional reference(s):	Goodwin (1980)

β, β-carotene **HPLC peak 52**

Standard spectrum in reference solvents

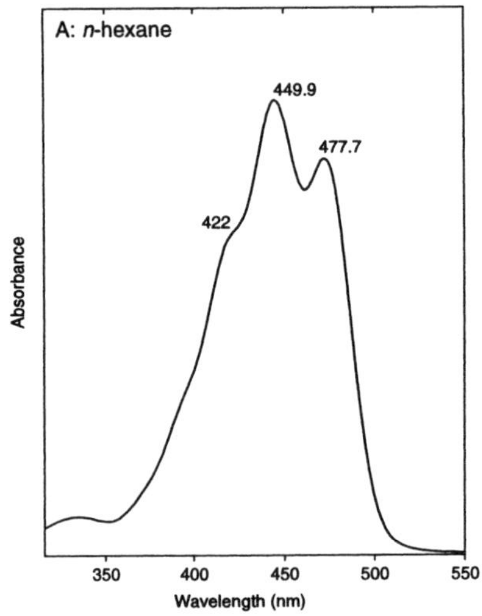

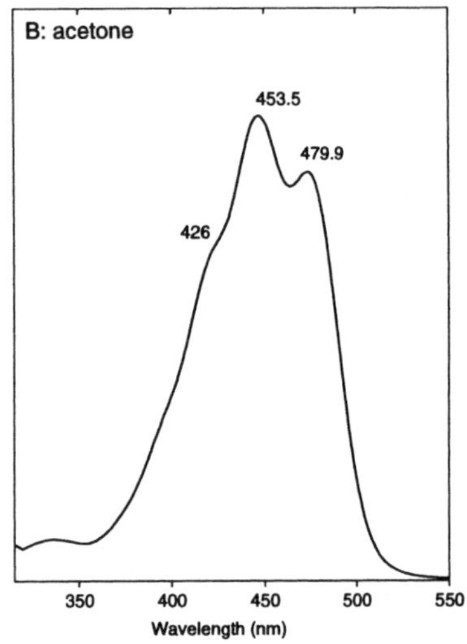

Molecular structure

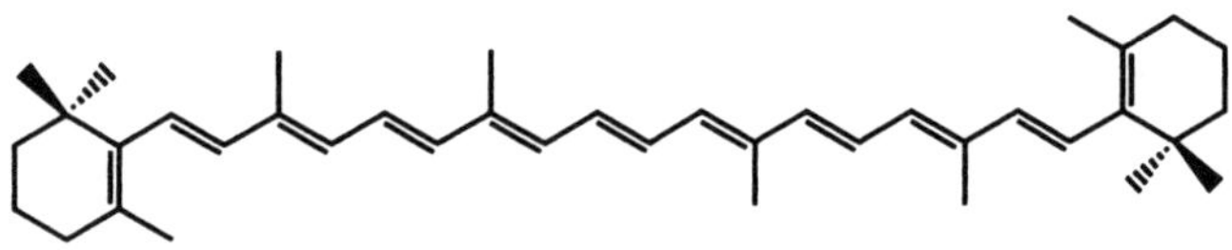

Diode array spectrum in SCOR eluant

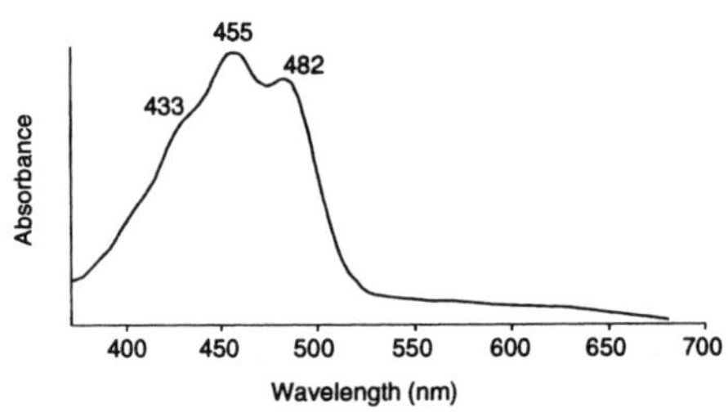

HPLC: β, β-carotene, peak 52

Pavlova lutheri

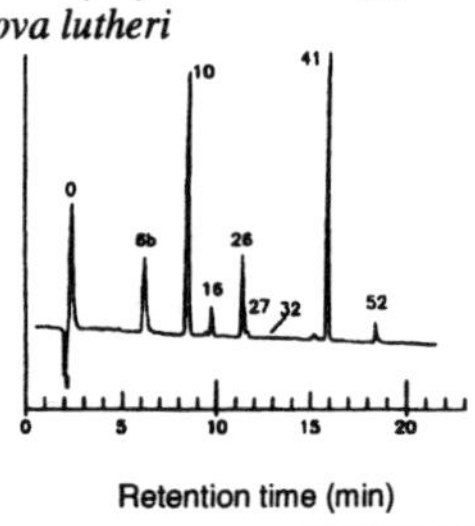

β,β-Carotene

Property		Data
Name:	(Trivial)	**β-Carotene**
	(IUPAC)	β,β-Carotene
SCOR abbreviation:		ββ-car
Occurrence:		Major pigment in higher plants, green algae; minor or trace pigment in chromophyte algae
Colour:		Yellow-orange
Molecular formula:		$C_{40} H_{56}$
Molecular weight:		536.88
Specific extinction coefficient: $E_{1\ cm}^{1\%}$ (100 ml g^{-1} cm^{-1})		2500 (at 454 nm in acetone) Hiyama *et al.* (1969) 2592 (at 453 nm in hexane) Isler *et al.* (1956) 2620 (at 453 nm in ethanol) Isler *et al.* (1956)
Molar extinction coefficient: ε (l mol^{-1} cm^{-1})		134 x 10^3 (at 454 nm in acetone) 139 x 10^3 (at 453 nm in hexane) 141 x 10^3 (at 453 nm in ethanol) Calculated from $E_{1\ cm}^{1\%}$ above

UV-vis spectra:

Solvent	Maxima (nm) I	II	III	Band ratio %III:II	Reference
Acetone	(426)	453.5	479.9	21	SCOR WG 78 data
Hexane	(422)	449.9	477.7	36	SCOR WG 78 data
Ethanol	(427)	449	475		Katayama *et al.* (1972)
HPLC Eluant	(425)	453	476	22	SCOR WG 78: Mantoura & Llewellyn (1983) method
HPLC Eluant	(433)	455	482	22	SCOR WG 78: Wright *et al.* (1991) method

Alteration products:	*Cis*-isomers
Culture from which SCOR data were obtained:	*Pavlova lutheri* (prymnesiophyte)
Additional reference(s):	Goodwin (1980)

β, ψ-carotene **HPLC peak 49**

Standard spectrum in reference solvent

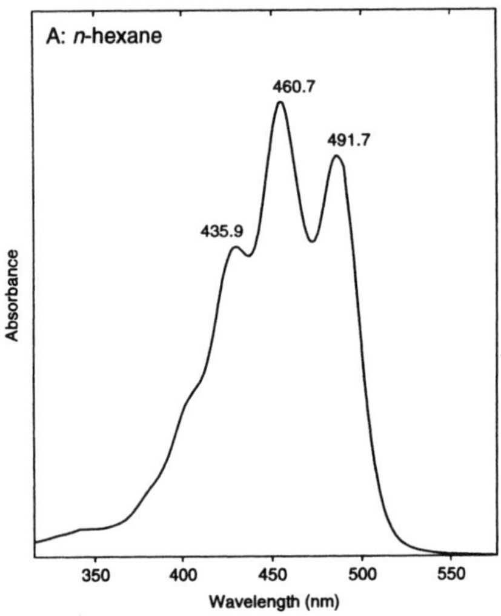

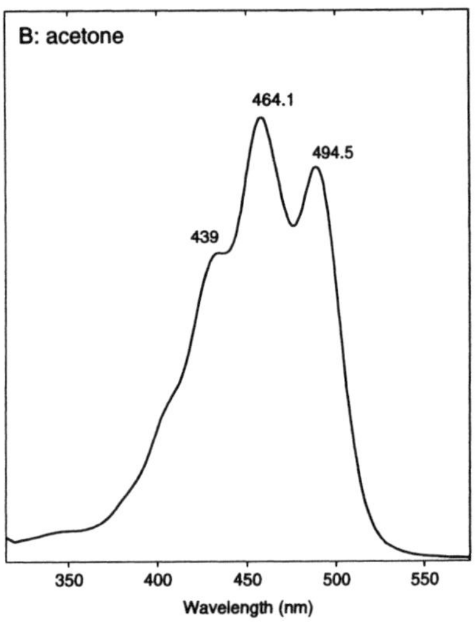

Molecular structure

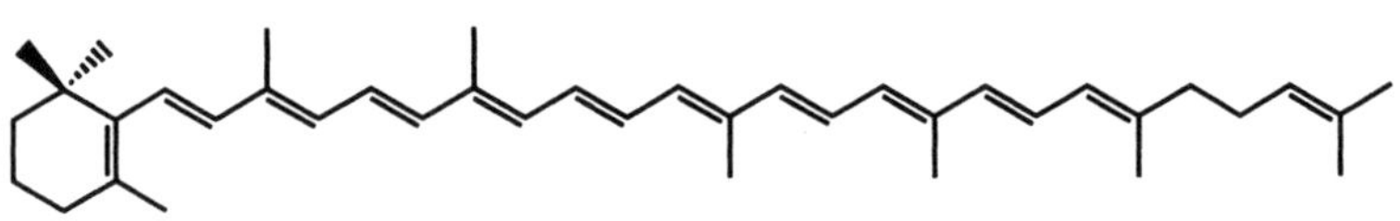

Diode array spectrum in SCOR eluant

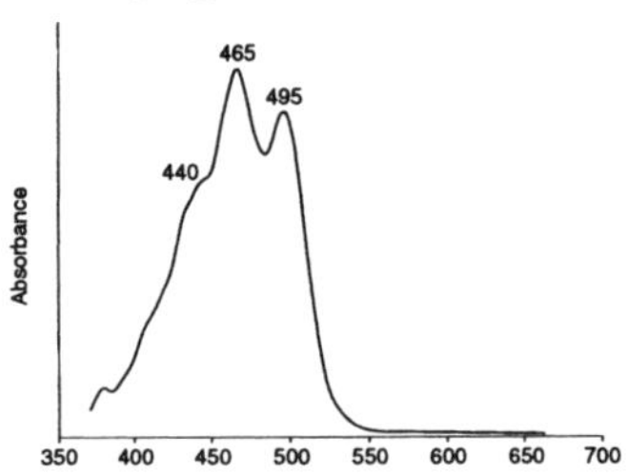

HPLC: β, ψ-carotene, peak 49

Dunaliella tertiolecta

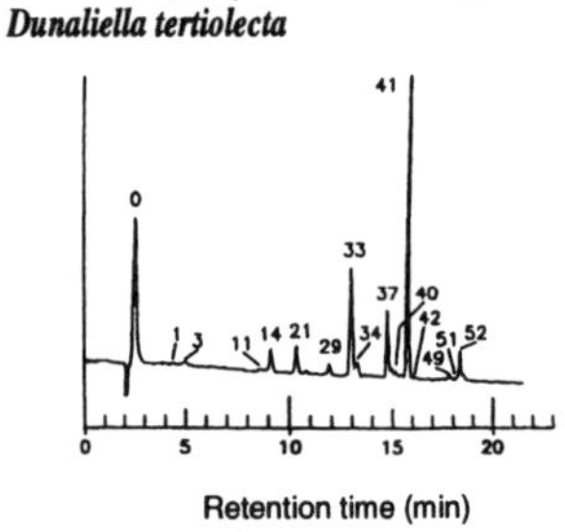

β,ψ-Carotene

Property		**Data**
Name:	(Trivial)	**γ-Carotene**
	(IUPAC)	β,ψ-Carotene
SCOR abbreviation:		βψ-car
Occurrence:		Minor or trace pigment in green algae, chromophyte algae
Colour:		Orange-red
Molecular formula:		$C_{40} H_{56}$
Molecular weight:		536.88
Specific extinction coefficient: $E^{1\%}_{1\ cm}$ (100 ml g^{-1} cm^{-1})		3185 (at 459 nm in petroleum ether) Manchand *et al.* (1965) 2760 (at 462 nm in hexane) Bindl *et al.* (1970)
Molar extinction coefficient: ε (l mol^{-1} cm^{-1})		171 x 10^3 (at 459 nm in petroleum ether) 148 x 10^3 (at 462 nm in hexane) Calculated from $E^{1\%}_{1\ cm}$ above

UV-vis spectra:

Solvent	**Maxima (nm) I**	**II**	**III**	**Band ratio %III:II**	**Reference**
Acetone	439	464	494	55	SCOR WG 78 data
Acetone	(439)	461	491		Francis & Halfen (1972)
Hexane	436	461	492	59	SCOR WG 78 data
HPLC Eluant	(436)	462	490	32	SCOR WG 78: Mantoura & Llewellyn (1983) method
HPLC Eluant	(440)	465	495	50	SCOR WG 78: Wright *et al.* (1991) method

Alteration products:	*Cis*-isomers
Culture from which SCOR data were obtained:	*Dunaliella tertiolecta* (green flagellate)
Additional reference(s):	Goodwin (1980)

ε, ε-carotene

HPLC peak 50

Standard spectrum in reference solvent: hexane

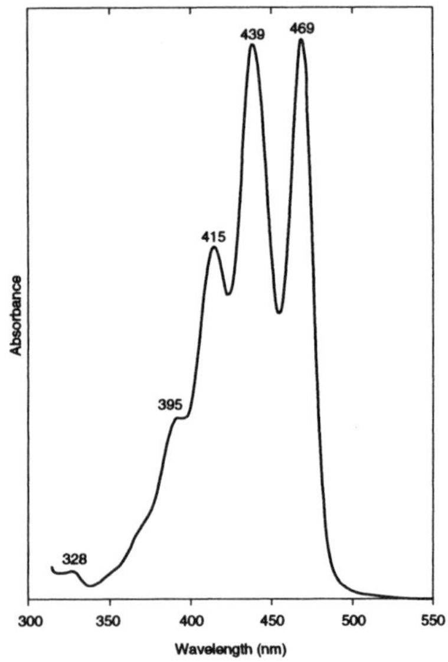

Molecular structure

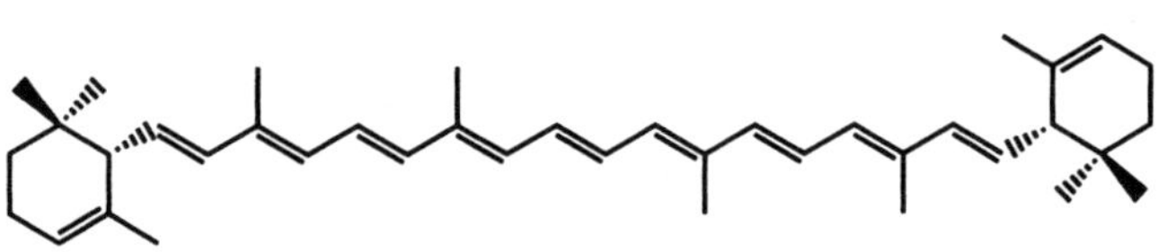

Diode array spectrum in SCOR eluant

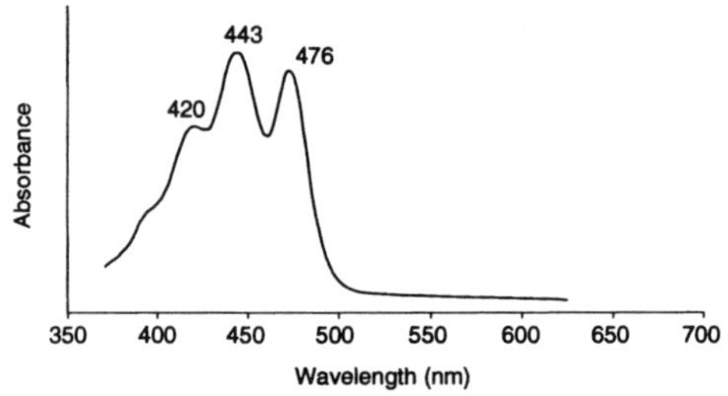

HPLC: ε, ε-carotene, peak 50

Pelagococcus subviridis

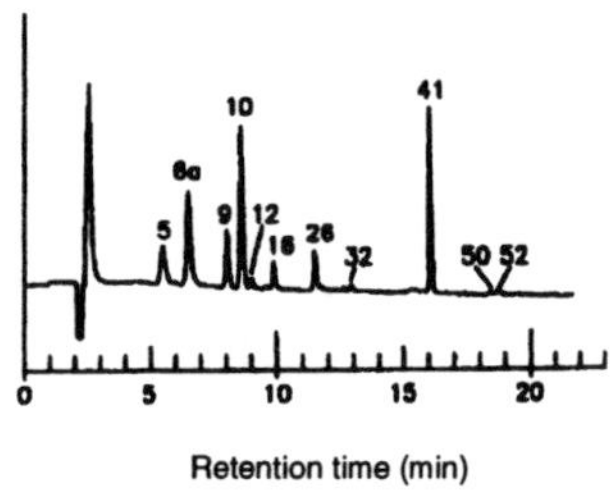

ε,ε-Carotene

Property		Data
Name:	(Trivial)	**ε-Carotene**
	(IUPAC)	(6*S*,6'*S*)-ε,ε-Carotene
SCOR abbreviation:		εε-car
Occurrence:		Minor or trace pigment in marine chrysophytes (e.g. *Pelagococcus subviridis*), cryptomonads
Colour:		Yellow
Molecular formula:		$C_{40}H_{56}$
Molecular weight:		536.88
Specific extinction coefficient: $E^{1\%}_{1\,cm}$ (100 ml g^{-1} cm^{-1})		2900 (at 439 nm in hexane) Chapman & Haxo (1963) 3010 (at 440 nm in petroleum ether) Schwieter *et al.* (1965)
Molar extinction coefficient: ε (l mol^{-1} cm^{-1})		156 x 10^3 (at 439 nm in hexane) 162 x 10^3 (at 440 nm in petroleum ether) Calculated from $E^{1\%}_{1\,cm}$ above

UV-vis spectra:

Solvent	Maxima (nm) I	II	III	Band ratio %III:II	Reference
Ethanol	417	440	470		Chapman & Haxo (1963)
Petroleum ether	416	440	470		Schwieter *et al.* (1965)
Benzene	425	451	482		Chapman & Haxo (1963)
Hexane	415	439	469	101	Bjørnland *et al.* (1989)
HPLC Eluant	420	443	476	79	SCOR WG 78: Wright *et al.* (1991) method

Alteration products:	*Cis*-isomers
Culture from which SCOR data were obtained:	*Pelagococcus subviridis* (chrysophyte)
Additional reference(s):	Goodwin (1980); Bjørnland *et al.* (1989)

Crocoxanthin

HPLC peak 39

Standard spectrum in reference solvent

No data available

Molecular structure

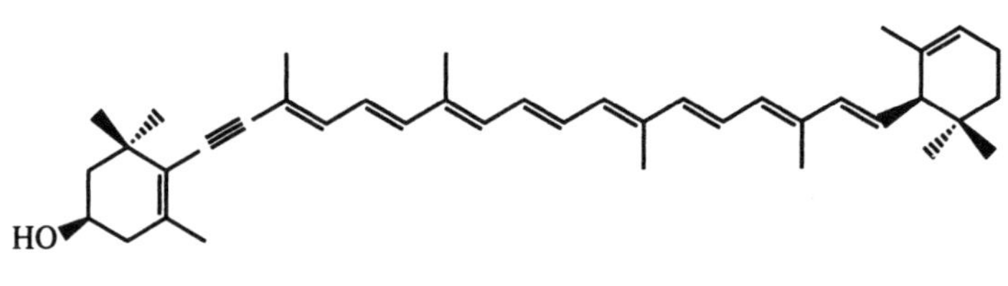

Diode array spectrum in SCOR eluant

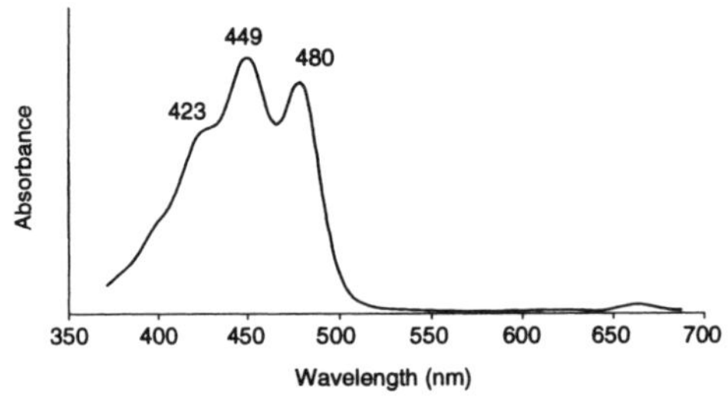

HPLC: Crocoxanthin, peak 39

Chroomonas salina

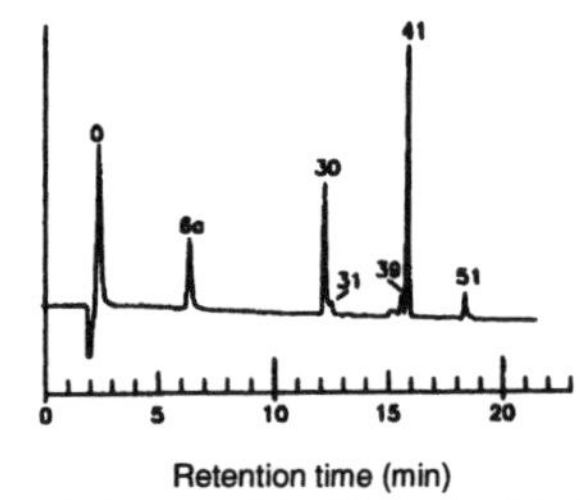

Crocoxanthin

Property		Data
Name:	(Trivial)	**Crocoxanthin**
	(IUPAC)	(3*R*,6'*R*)-7,8-Didehydro-β,ε-caroten-3-ol
SCOR abbreviation:		Croco
Occurrence:		Cryptomonads (minor carotenoid)
Colour:		Yellow
Molecular formula:		$C_{40} H_{54} O$
Molecular weight:		550.87
Specific extinction coefficient: $E^{1\%}_{1\,cm}$ (100 ml g^{-1} cm^{-1})		2500 (at 443 nm in ethanol) Not determined; use $E^{1\%}_{1\,cm}$ for β,β-carotene, Davies (1976)
Molecular extinction coefficient: ε (1 mol^{-1} cm^{-1})		138 x 10^3 (at 443 nm in ethanol) Calculated from $E^{1\%}_{1\,cm}$ above

UV-vis spectra:

Solvent	Maxima (nm) I	II	III	Band ratio %III:II	Reference
Ethanol	(421)	443	472	62	Hager & Stransky (1970c)
Hexane	422	445	475	69	Chapman (1966)
Diethyl ether	428	445	475	58	Pennington *et al.* (1985)
HPLC Eluant	423	449	480	61	SCOR WG 78: Wright *et al.* (1991) method

Alteration products:

Culture from which SCOR data were obtained: *Chroomonas salina* (cryptomonad)

Additional reference(s): Pennington *et al.* (1985)

Diadinochrome

HPLC peak 27

Standard spectrum in reference solvent: acetone

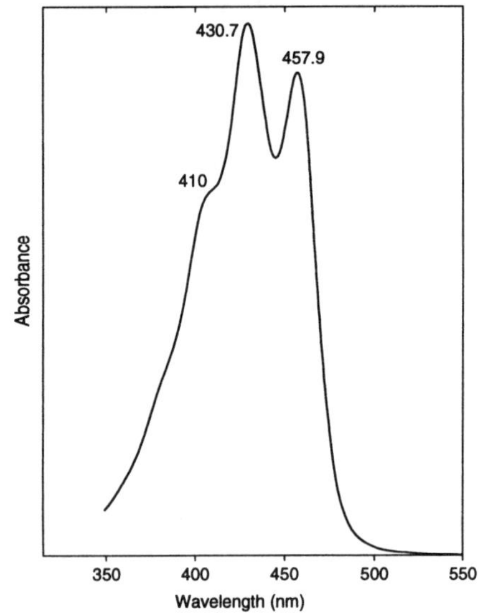

Molecular structure

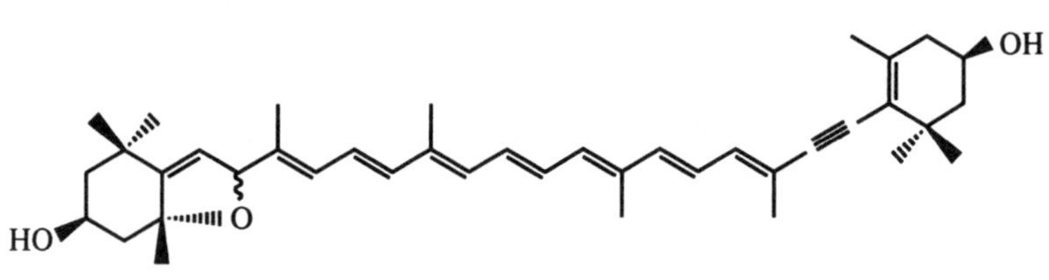

Diode array spectrum in SCOR eluant

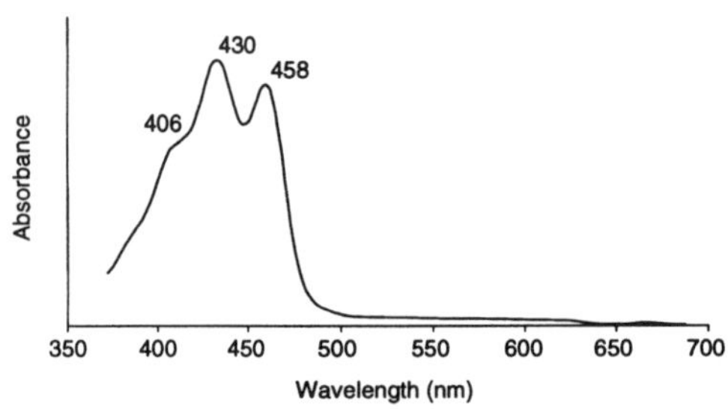

HPLC: Diadinochrome, peak 27

Amphidinium carterae

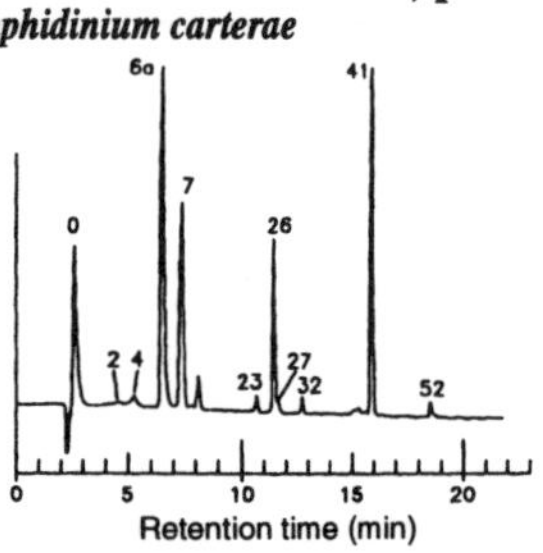

Diadinochrome

Property		Data
Name:	(Trivial)	**Diadinochrome I and II**
	(IUPAC)	5,8-Epoxy-7',8'-didehydro-5,8-dihydro-β, β-carotene-3,3'-diol
SCOR abbreviation:		Diadchr
Occurrence:		Rearrangement product of diadinoxanthin
Colour:		Yellow
Molecular formula:		$C_{40}H_{54}O_3$
Molecular weight:		582.86
Specific extinction coefficient: $E^{1\%}_{1\,cm}$ (100 ml g^{-1} cm^{-1})		Not known; use 2500 (at 428 nm in acetone), see Davies (1976)
Molar extinction coefficient: ε (l mol^{-1} cm^{-1})		146 x 10^3 (at 428 nm in acetone) Calculated from $E^{1\%}_{1\,cm}$ above

UV-vis spectra:

Solvent	Maxima (nm) I	II	III	Band ratio %III:II	Reference
Acetone	(410)	431	458	63	SCOR WG 78 data
Diethyl ether	405	429	456		Fiksdahl *et al.* (1984b)
HPLC Eluant	406	430	458	76	SCOR WG 78: Wright *et al.* (1991) method

Alteration products:	Readily formed by acid catalysed conversion of diadinoxanthin epoxide groups to furanoid groups. Both the 8*R* and 8*S* diadinochrome epimers are formed.
Culture from which SCOR data were obtained:	*Amphidinium carterae* (dinoflagellate)
Additional reference(s):	Liaaen-Jensen (1971,1990a)

Diadinoxanthin

HPLC peak 26

Standard spectrum in reference solvent: acetone

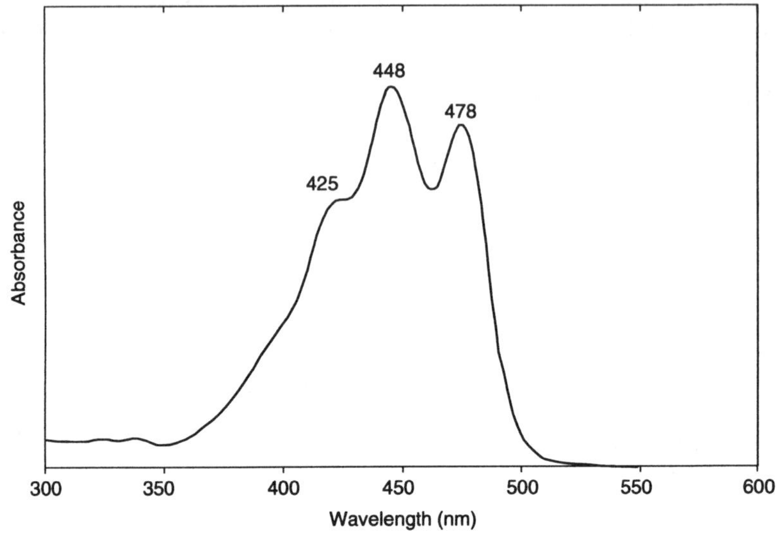

Molecular structure

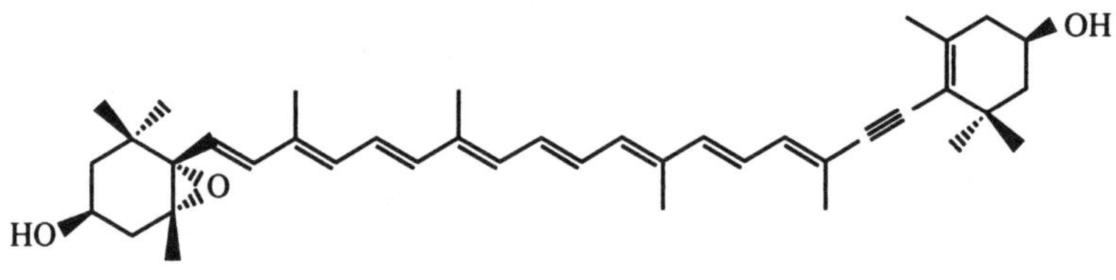

Diode array spectrum in SCOR eluant

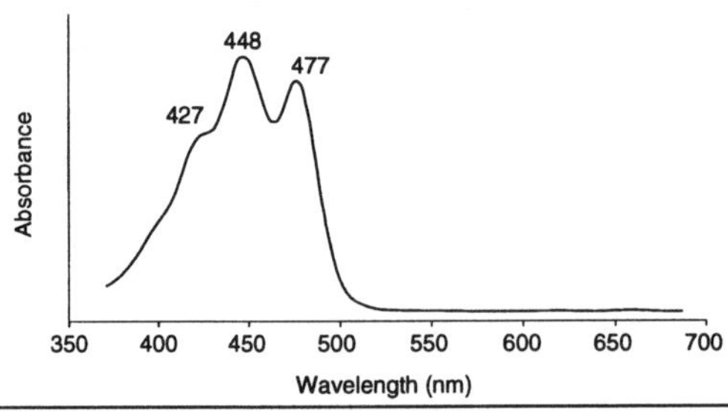

HPLC: Diadinoxanthin, peak 26

Amphidinium carterae

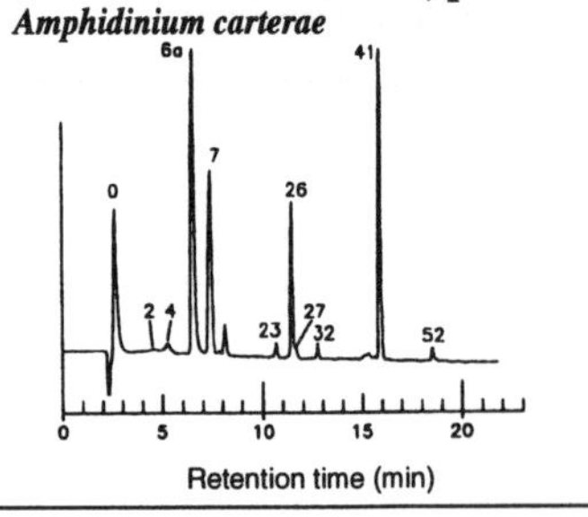

Diadinoxanthin

Property		Data
Name:	(Trivial)	**Diadinoxanthin**
	(IUPAC)	(3*S*,5*R*,6*S*,3'*R*)-5,6-Epoxy-7',8'-didehydro-5,6-dihydro-β,β-carotene-3,3'-diol
SCOR abbreviation:		Diadino
Occurrence:		Major pigment in diatoms, prymnesiophytes, some chrysophytes, dinoflagellates
Colour:		Yellow
Molecular formula:		$C_{40}H_{54}O_3$
Molecular weight:		582.86
Specific extinction coefficient: $E^{1\%}_{1\,cm}$ (100 ml g^{-1} cm^{-1})		2230 (at 447.5 nm in acetone) 2250 (at 444.5 nm in methanol) 2110 (at 445.5 nm in hexane) Johansen *et al.* (1974)
Molar extinction coefficient: ε (l mol^{-1} cm^{-1})		130 x 10^3 (at 447.5 nm in acetone) 131 x 10^3 (at 444.5 nm in methanol) 123 x 10^3 (at 445.5 nm in hexane) Calculated from $E^{1\%}_{1\,cm}$ above

UV-vis spectra:

Solvent	Maxima (nm) I	II	III	Band ratio %III:II	Reference
Acetone	426	447.5	478	61	Johansen *et al.* (1974)
Acetone	(425)	448	478	63	Bjørnland (1990b)
Methanol		444.5	474	54	Johansen *et al.* (1974)
Ethanol	(424)	446	476	67	Loeblich & Smith (1968)
Ethanol	425	446	476		Stauber & Jeffrey (1988)
Hexane	(424)	445.5	474.5	57	Johansen *et al.* (1974)
HPLC Eluant	(421)	446	475	63	SCOR WG 78: Mantoura & Llewellyn (1983) method
HPLC Eluant	(427)	448	477	63	SCOR WG 78: Wright *et al.* (1991) method

Alteration products:	*Cis*-isomers; furanoids (diadinochromes)
Culture from which SCOR data were obtained:	*Amphidinium carterae* (dinoflagellate), *Phaeodactylum tricornutum* (diatom)
Additional reference(s):	Stransky & Hager (1970c); Stauber & Jeffrey (1988)

Diatoxanthin

HPLC peak 32

Standard spectrum in reference solvents

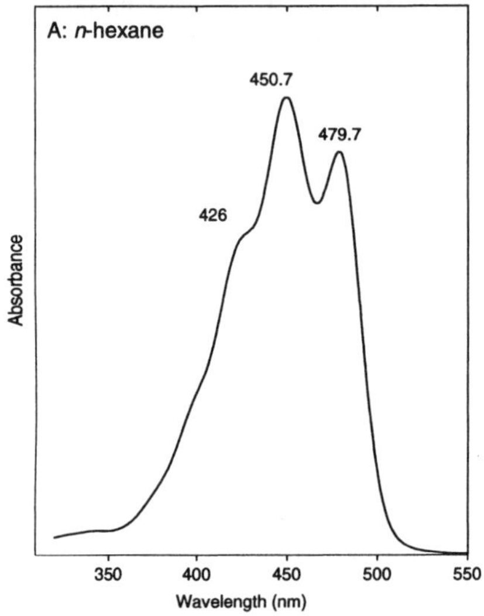

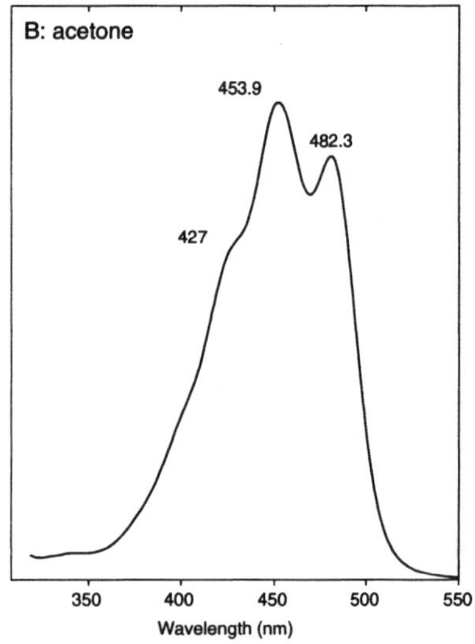

Molecular structure

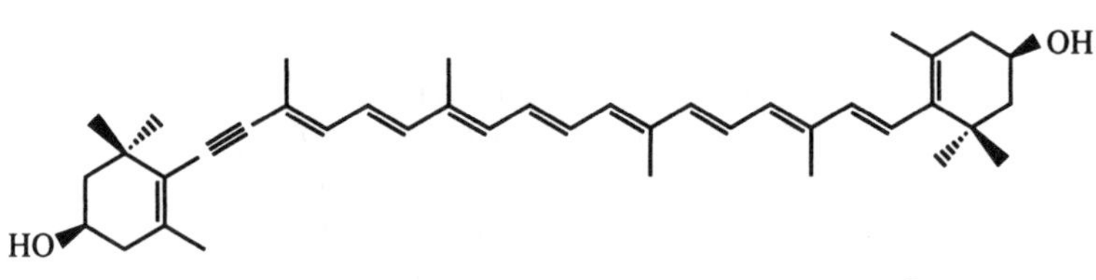

Diode array spectrum in SCOR eluant

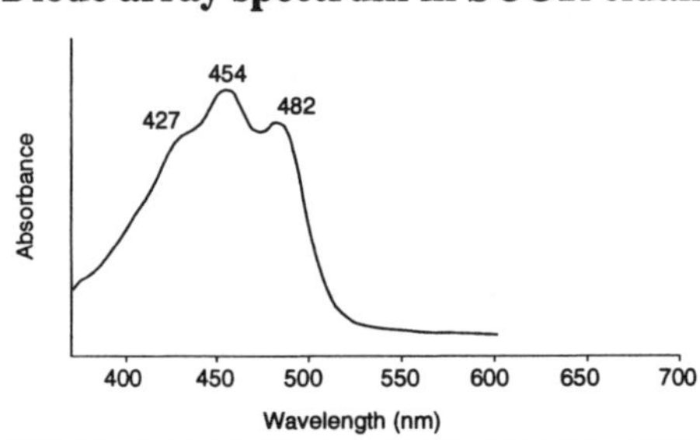

HPLC: Diatoxanthin, peak 32

Amphidinium carterae

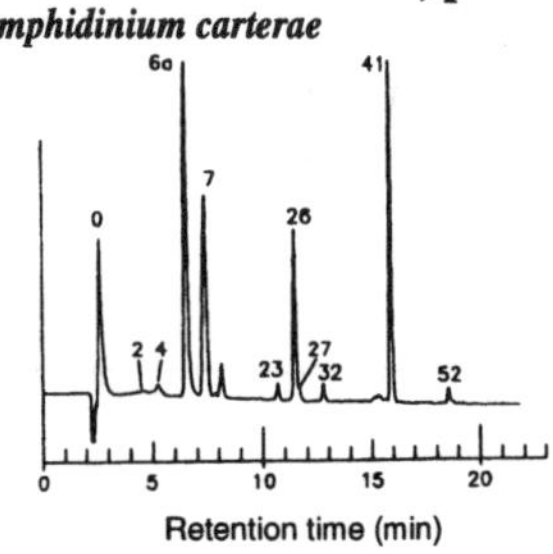

Diatoxanthin

Property		Data
Name:	(Trivial)	**Diatoxanthin**
	(IUPAC)	(3*R*,3'*R*)-7,8-Didehydro-β,β-carotene-3,3'-diol
SCOR abbreviation:		Diato
Occurrence:		Minor or trace pigment in diatoms, prymnesiophytes, some chrysophytes, dinoflagellates
Colour:		Pale orange
Molecular formula:		$C_{40}H_{54}O_2$
Molecular weight:		566.87
Specific extinction coefficient: $E^{1\%}_{1\,cm}$ (100 ml g^{-1} cm^{-1})		2100 (at 452 nm in acetone) Not determined; recommended by Johansen *et al.* (1974)
Molar extinction coefficient: ε (l mol^{-1} cm^{-1})		119 x 10^3 (at 452 nm in acetone) Calculated from $E^{1\%}_{1\,cm}$ above

UV-vis spectra:

Solvent	Maxima (nm) I	II	III	Band ratio %III:II	Reference
Acetone	(427)	454	482	42	SCOR WG 78 data
Acetone	(434)	454	482	35	Bjørnland (1982)
Acetone	429	454	482		Berger *et al.* (1977)
Ethanol		451	480		Stauber & Jeffrey (1988)
Hexane	(426)	451	480	31	SCOR WG 78 data
HPLC Eluant	(427)	454	482	26	SCOR WG 78: Wright *et al.* (1991) method

Alteration products:	*Cis*-isomers
Culture from which SCOR data were obtained:	*Amphidinium carterae* (dinoflagellate)
Additional reference(s):	Stauber & Jeffrey (1988); Stransky & Hager (1970b)

Dinoxanthin

HPLC peak 23

Standard spectrum in reference solvents

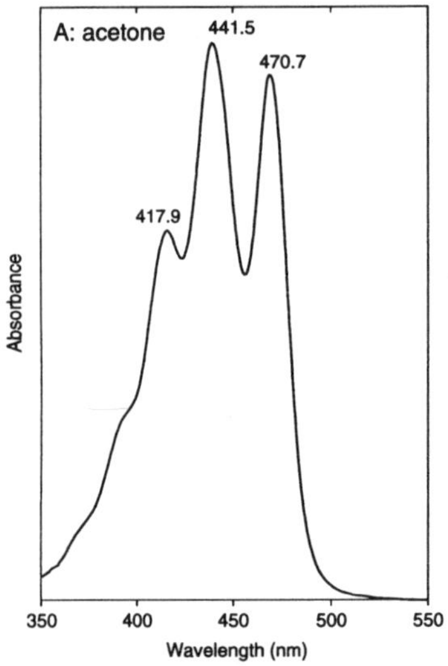

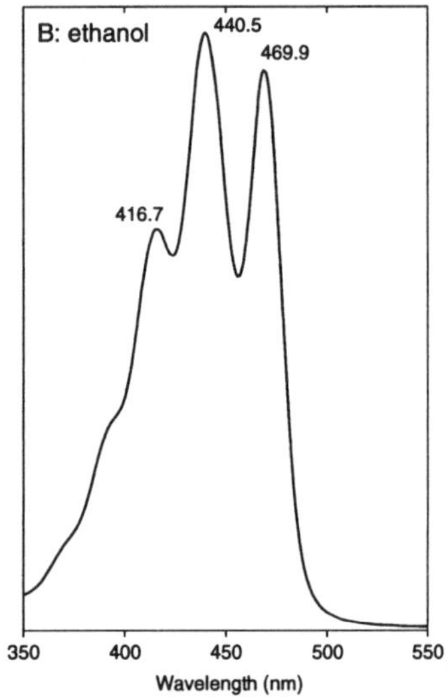

Molecular structure

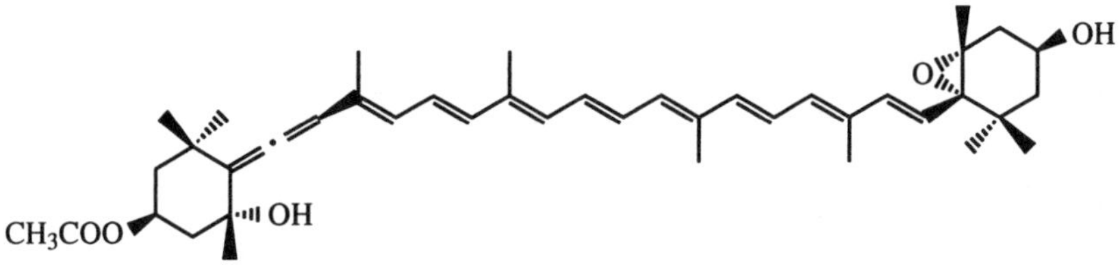

Diode array spectrum in SCOR eluant

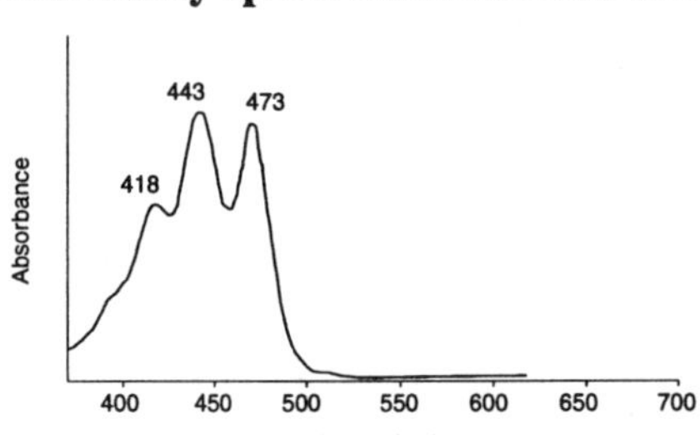

HPLC: Dinoxanthin, peak 23

Amphidinium carterae

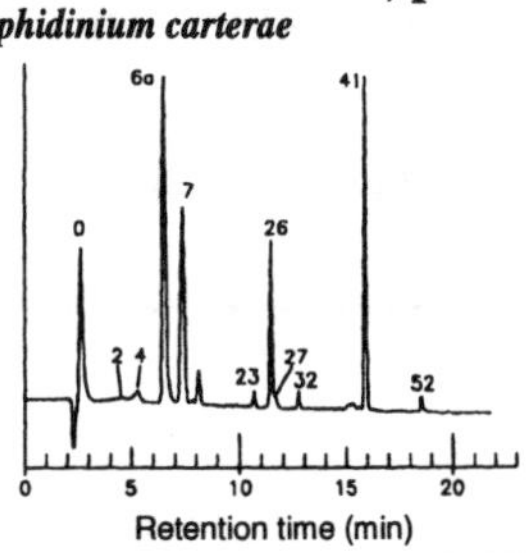

Dinoxanthin

Property		Data
Name:	(Trivial)	**Dinoxanthin**
	(IUPAC)	(3*S*,5*R*,6*R*,3'*S*,5'*R*,6'*S*)-5',6'-Epoxy-6,7-didehydro-5,6,5',6'-tetrahydro-β,β-carotene-3,5,3'-triol 3-acetate
SCOR abbreviation:		Dino
Occurrence:		Minor pigment in dinoflagellates, except those containing endosymbionts of other algal classes
Colour:		Bright yellow
Molecular formula:		$C_{42}H_{58}O_5$
Molecular weight:		642.92
Specific extinction coefficient: $E^{1\%}_{1\,cm}$ (100 ml g^{-1} cm^{-1})		2100 (at 442 nm in acetone) Not determined; recommended by Johansen *et al.* (1974)
Molar extinction coefficient: ε (l mol^{-1} cm^{-1})		135 x 10^3 (at 442 nm in acetone) Calculated from $E^{1\%}_{1\,cm}$ above

UV-vis spectra:

Solvent	Maxima (nm) I	II	III	Band ratio %III:II	Reference
Ethanol	417	441	470	85	SCOR WG 78 data
Ethanol	(416)	440	470		Jeffrey *et al.* (1975)
Ethanol	(419)	441	470		Loeblich & Smith (1968)
Acetone	418	442	471	86	SCOR WG 78 data
Hexane	416	439	469	83	Loeblich & Smith (1968)
Methanol	416	438	467	76	Loeblich & Smith (1968)
HPLC Eluant	418	443	473	87	SCOR WG 78: Wright *et al.* (1991) method

Alteration products:	*Cis*-isomers; (probably) furanoids (dinochromes)
Culture from which SCOR data were obtained:	*Amphidinium carterae* (dinoflagellate)
Additional reference(s):	Johansen *et al.* (1974); Jeffrey *et al.* (1975)

Echinenone

HPLC peak 43

Standard spectrum in reference solvent: acetone

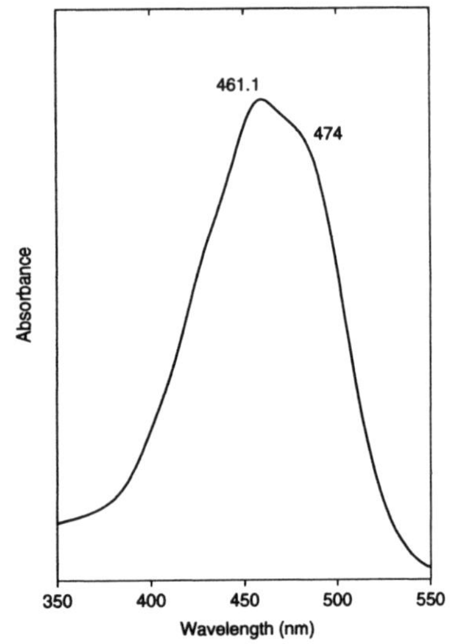

Molecular structure

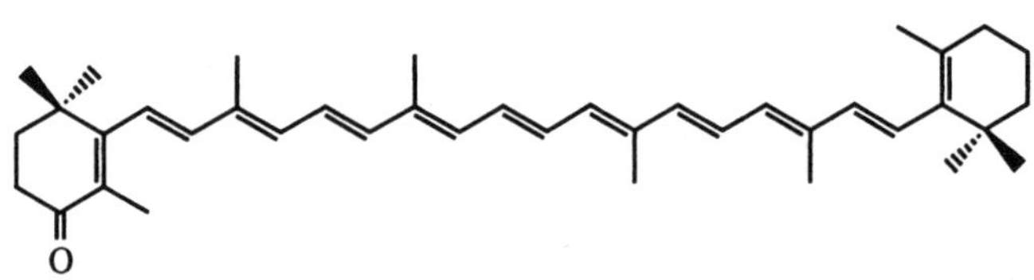

Diode array spectrum in SCOR eluant

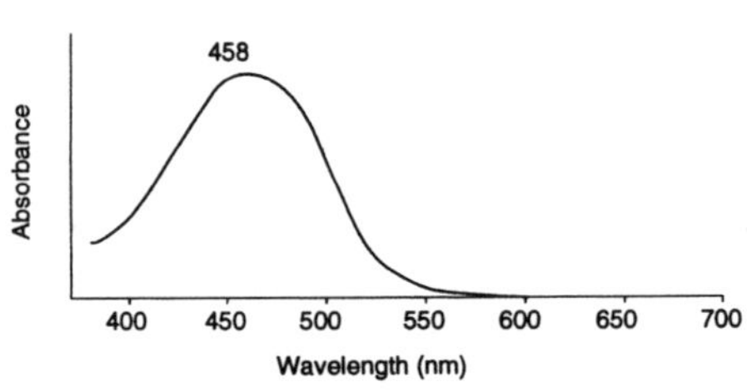

HPLC: Echinenone, peak 43

Pigment mixture

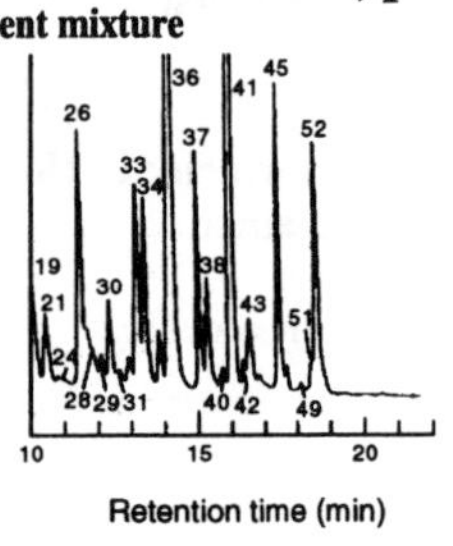

Echinenone

Property		**Data**
Name:	(Trivial)	**Echinenone**
	(IUPAC)	β,β-Caroten-4-one
SCOR abbreviation:		Echin
Occurrence:		Major carotenoid of freshwater & estuarine cyanobacteria (blue-green algae); minor or trace carotenoid in some green algae and euglenophytes.
Colour:		Red-orange
Molecular formula:		$C_{40}H_{54}O$
Molecular weight:		550.87
Specific extinction coefficient: $E^{1\%}_{1\,cm}$ (100 ml g^{-1} cm^{-1})		2158 (at 458 nm in petroleum ether) Davies (1965)
Molar extinction coefficient: ε (l mol^{-1} cm^{-1})		119 x 10^3 (at 458 nm in petroleum ether) Calculated from $E^{1\%}_{1\,cm}$ above

UV-vis spectra:

Solvent	**Maxima (nm) I**	**II**	**III**	**Band ratio %III:II**	**Reference**
Acetone		461	(474)	0	SCOR WG 78 data
Acetone		460			Francis & Halfen (1972)
Ethanol		461		0	Stransky & Hager (1970b)
Hexane	(432)	459	(483)		Davies (1976)
HPLC Eluant		458			SCOR WG 78: Wright *et al.* (1991) method

Alteration products:

Culture from which SCOR data were obtained:	*Oscillatoria agardhii* (*rubescens*) (blue-green alga)
Additional reference(s):	Nichols (1973)

Ethyl-apo-8'-carotenoate — HPLC peak 38

Standard spectrum in reference solvent: acetone

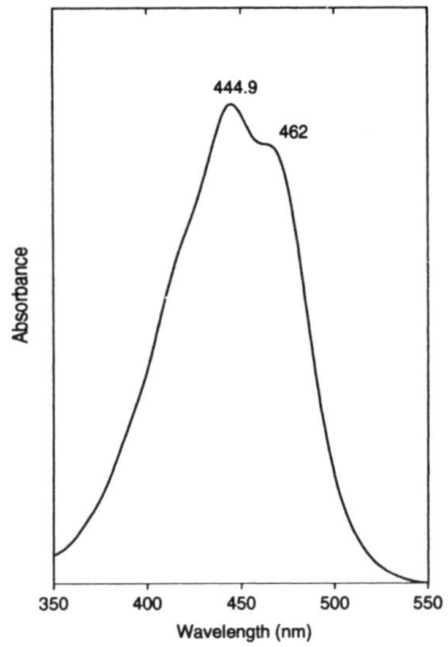

Molecular structure

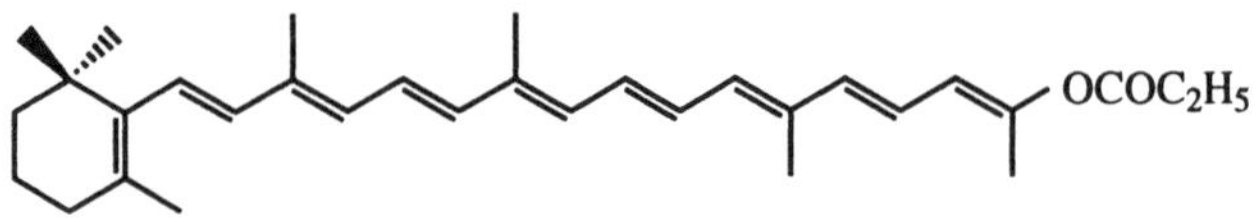

Diode array spectrum in Repeta eluant

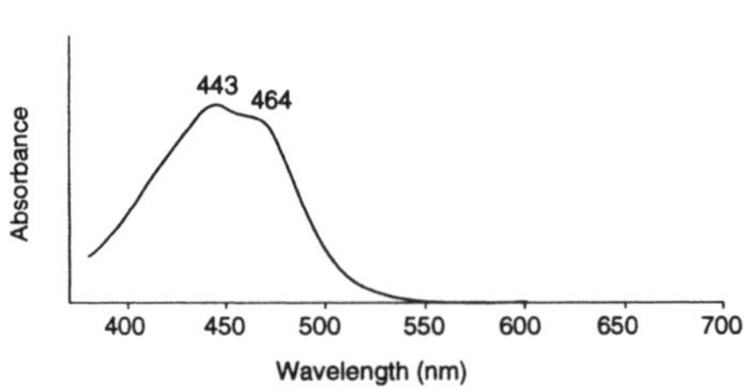

HPLC: Ethyl-apo-8'-carotenoate, peak 38

Pigment mixture

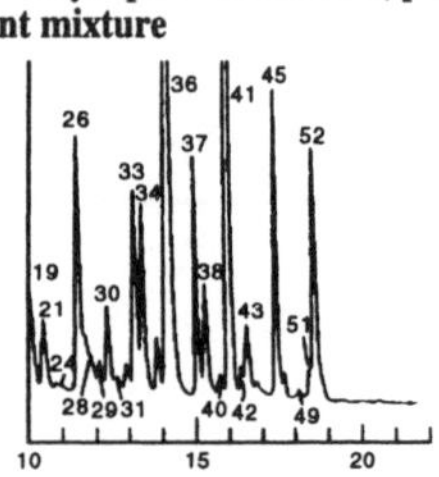

Ethyl-apo-8'-carotenoate

Property		Data
Name:	(Trivial)	**Ethyl-apo-8'-carotenoate**
	(IUPAC)	Ethyl 8'-apo-β-caroten-8'-oate
SCOR abbreviation:		Et-8-carot
Occurrence:		Synthetic (possible internal standard)
Colour:		Orange
Molecular formula:		$C_{32}H_{44}O_2$
Molecular weight:		460.71
Specific extinction coefficient: $E^{1\%}_{1\,cm}$ (100 ml g^{-1} cm^{-1})		2362 (at 448 nm in petroleum ether) Calculated from ε below
Molar extinction coefficient: ε (l mol^{-1} cm^{-1})		108.8 x 10^3 (at 448 nm in petroleum ether) Isler & Schudel (1963)

UV-vis spectra:

Solvent	Maxima (nm) I	II	III	Band ratio %III:II	Reference
Acetone		445	(462)	0	SCOR WG 78 data
HPLC Eluant		443	(464)	0	Repeta (unpublished data)

Alteration products:	
Origin:	Synthetic (Fluka Chemie)
Additional reference(s):	Vetter *et al.* (1971)

Fucoxanthin

HPLC peak 10

Standard spectrum in reference solvent: acetone

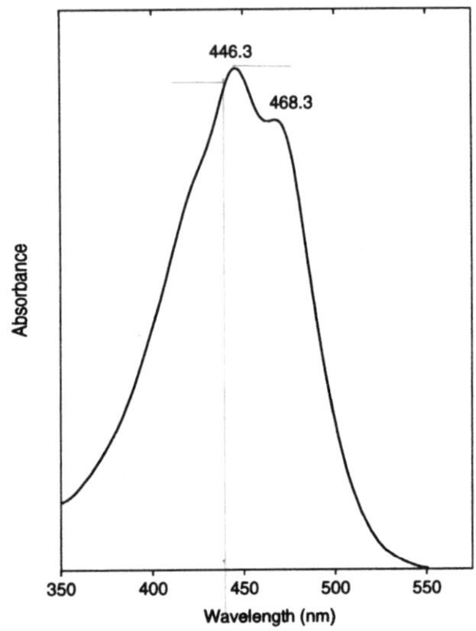

Molecular structure

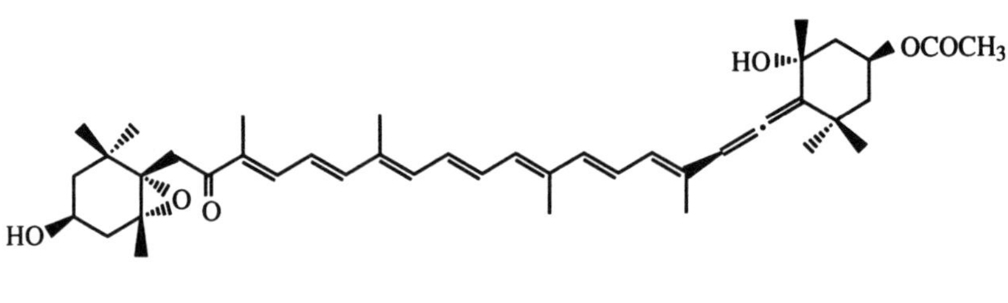

Diode array spectrum in SCOR eluant

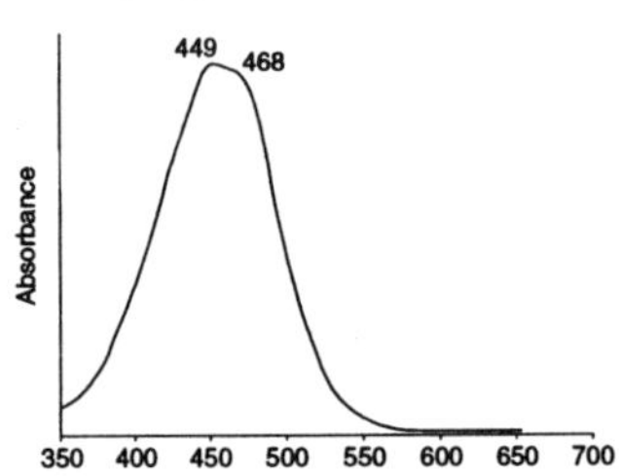

HPLC: Fucoxanthin, peak 10

Phaeodactylum tricornutum

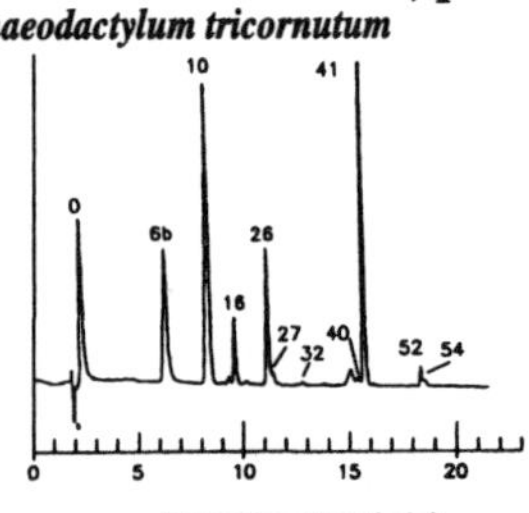

Fucoxanthin

Property		Data
Name:	(Trivial)	**Fucoxanthin**
	(IUPAC)	(3*S*,5*R*,6*S*,3'*S*,5'*R*,6'*R*)-5,6-Epoxy-3,3',5'-trihydroxy-6',7'-didehydro-5,6,7,8,5',6'-hexahydro-β,β-caroten-8-one 3'-acetate
SCOR abbreviation:		Fuco
Occurrence:		Major pigment in diatoms, prymnesiophytes, brown seaweeds, raphidophytes, some dinoflagellates with endosymbionts.
Colour:		Orange
Molecular formula:		$C_{42}H_{58}O_6$
Molecular weight:		658.92
Specific extinction coefficient: $E^{1\%}_{1\,cm}$ (100 ml g^{-1} cm^{-1})		1660 (at 443 nm in acetone) Haugen & Liaaen-Jensen (1989) 1660 (at 453 nm in petroleum ether) Jensen (1966a)
Molar extinction coefficient: ε (l mol^{-1} cm^{-1})		109 x 10^3 (at 443 nm in acetone) 109 x 10^3 (at 453 nm in petroleum ether) Calculated from $E^{1\%}_{1\,cm}$ above

UV-vis spectra:

Solvent	Maxima (nm) I	II	III	Band ratio %III:II	Reference
Acetone		446	468	4	SCOR WG 78 data
Acetone		447	468		Berger *et al.* (1977)
Acetone	(420)	443	467		Haugan & Liaaen-Jensen (1989)
Acetone		446	470	3	Wright & Jeffrey (1987)
Ethanol		448	470	0	Wright & Jeffrey (1987)
Hexane	427	450	476		Bonnett *et al.* (1969)
HPLC Eluant		450	(467)	0	SCOR WG 78: Mantoura & Llewellyn (1983) method
HPLC Eluant		449	(468)	0	SCOR WG 78: Wright *et al.* (1991) method

Alteration products:	*Cis*-isomers
Culture from which SCOR data were obtained:	*Phaeodactylum tricornutum* (diatom)
Additional reference(s):	Berger *et al.* (1977); Wright & Jeffrey (1987); Haugan & Liaaen-Jensen (1989); Bjørnland & Liaaen-Jensen (1989); Jeffrey & Wright (1994)

19'-Hexanoyloxyfucoxanthin — HPLC peak 15

Standard spectrum in reference solvent: acetone

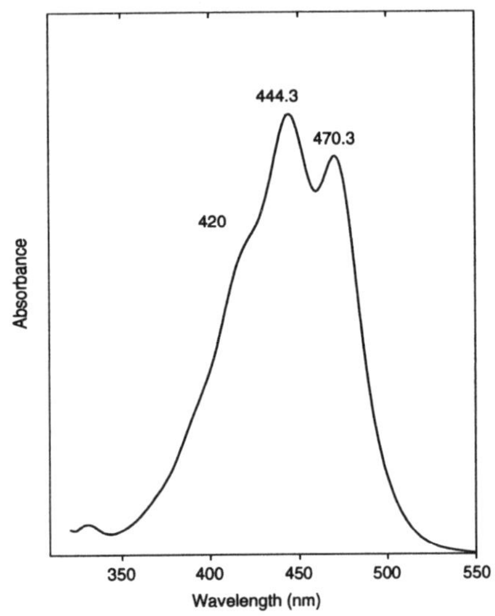

Molecular structure

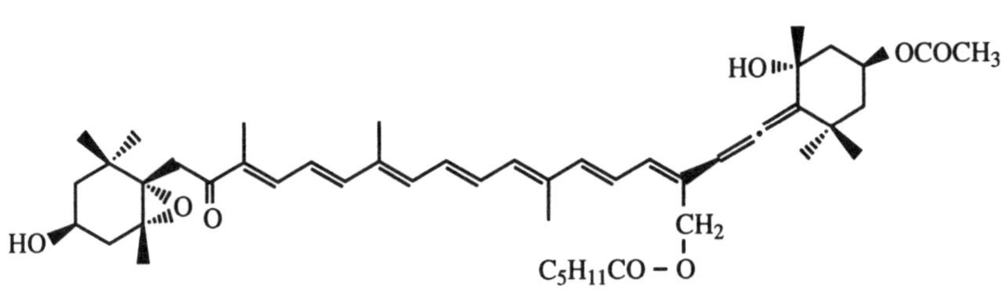

Diode array spectrum in SCOR eluant

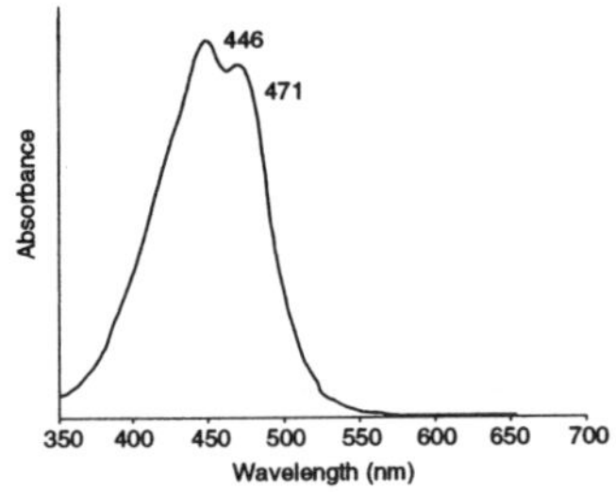

HPLC: 19'-Hexanoyloxyfucoxanthin, peak 15

Emiliania huxleyi

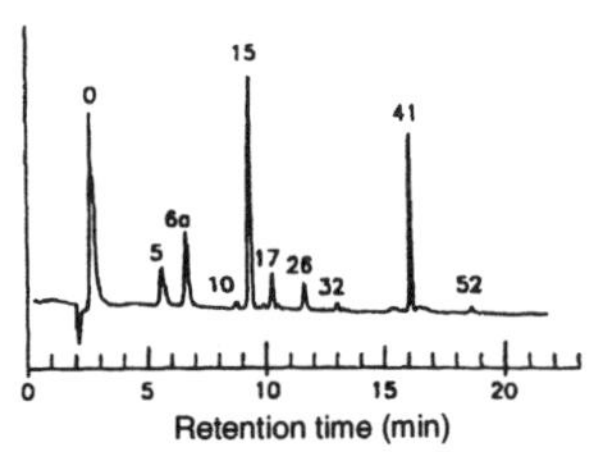

19'-Hexanoyloxyfucoxanthin

Property		**Data**
Name:	(Trivial)	**19'-Hexanoyloxyfucoxanthin**
	(IUPAC)	(3*S*,5*R*,6*S*,3'*S*,5'*R*,6'*S*)-5,6-Epoxy-3,3',5',19'-tetrahydroxy-6',7'-didehydro-5,6,7,8,5',6'-hexahydro-β,β-caroten-8-one 3'-acetate 19'-hexanoate
SCOR abbreviation:		Hex-fuco
Occurrence:		Major or minor pigment in several prymnesiophytes, some dinoflagellates
Colour:		Orange
Molecular formula:		$C_{48}H_{68}O_8$
Molecular weight:		773.06
Specific extinction coefficient: $E^{1\%}_{1\,cm}$ (100 ml g^{-1} cm^{-1})		1300 (at 445 nm in ethanol) Haxo (1985); Calculated from Jensen (1966a) 1420 (at 445 nm in acetone) Calculated from ε for fucoxanthin; Haugan & Liaaen-Jensen (1989), see Preamble
Molar extinction coefficient: ε (l mol^{-1} cm^{-1})		109 x 10^3 (at 445 nm in acetone) Assumed to be equal to ε for fucoxanthin, Haugan & Liaaen-Jensen (1989)

UV-vis spectra:

Solvent	**Maxima (nm) I**	**II**	**III**	**Band ratio %III:II**	**Reference**
Acetone	(420)	444.3	470.3	47	SCOR WG 78 data
Acetone		445	471	44	Wright & Jeffrey (1987)
n-Hexane	(423)	445	474	64	Bjørnland *et al.* (1988)
Ethanol		447	471	25	Wright & Jeffrey (1987)
Diethyl ether		444	470	45	Wright & Jeffrey (1987)
HPLC Eluant		447	470	22	SCOR W G 78: Mantoura & Llewellyn (1983) method
HPLC Eluant	418	446	471	34	SCOR WG 78: Wright *et al.* (1991) method

Alteration products:	*Cis*-isomers
Culture from which SCOR data were obtained:	*Emiliania huxleyi* (prymnesiophyte)
Additional reference(s):	Arpin *et al.* (1976); Hertzberg *et al.* (1977); Bjørnland & Tangen (1979); Haxo (1985); Wright & Jeffrey (1987); Bjørnland & Liaaen-Jensen (1989); Jeffrey & Wright (1994)

Lutein

HPLC peak 33

Standard spectrum in reference solvents

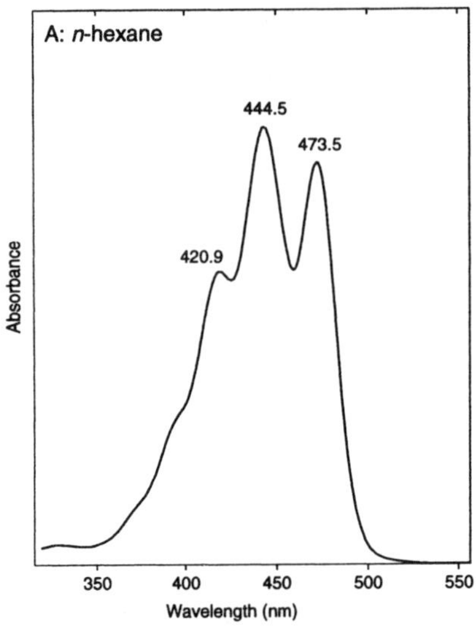

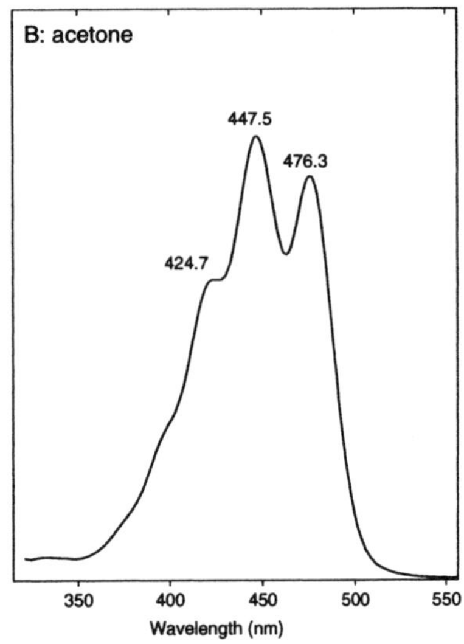

Molecular structure

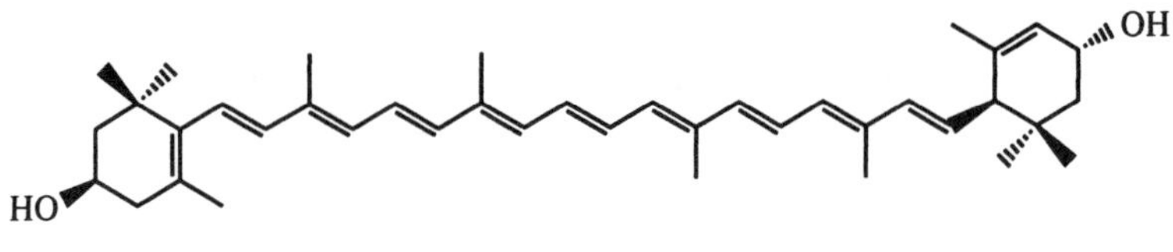

Diode array spectrum in SCOR eluant

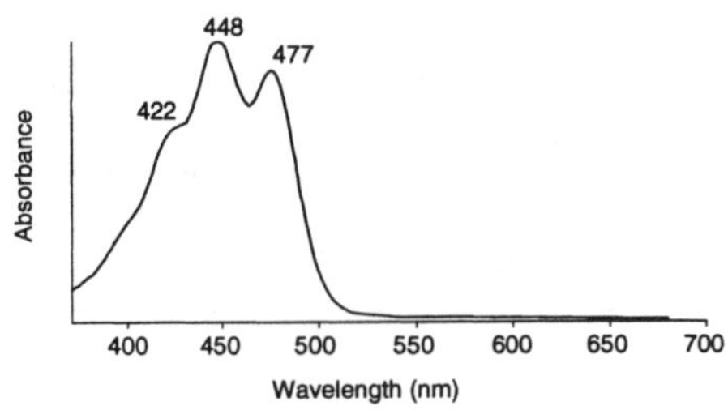

HPLC: Lutein, peak 33

Dunaliella tertiolecta

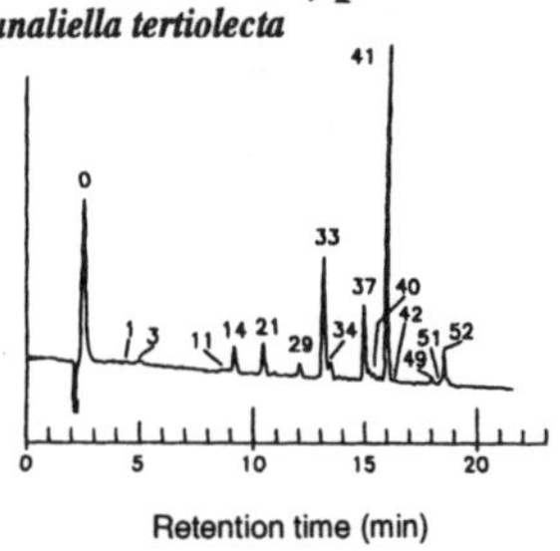

Lutein

Property		Data
Name:	(Trivial)	**Lutein**
	(IUPAC)	(3*R*,3'*R*,6'*R*)-β,ε-Carotene-3,3'-diol
SCOR abbreviation:		Lut
Occurrence:		Major pigment in red seaweeds, green algae, higher plants
Colour:		Yellow
Molecular formula:		$C_{40}H_{56}O_2$
Molecular weight:		568.88
Specific extinction coefficient: $E^{1\%}_{1\,cm}$ (100 ml g^{-1} cm^{-1})		2550 (at 445 nm in ethanol) Strain (1938) 2480 (at 445 nm in diethyl ether) Goodwin (1955)
Molar extinction coefficient: ε (l mol^{-1} cm^{-1})		145 x 10^3 (at 445 nm in ethanol) 141 x 10^3 (at 445 nm in diethyl ether) Calculated from $E^{1\%}_{1\,cm}$ above

UV-vis spectra:

Solvent	Maxima (nm) I	II	III	Band ratio %III:II	Reference
Acetone	425	447.5	476	67	SCOR WG 78 data
Ethanol	422	445	474	62	Hager & Stransky (1970*b*)
Ethanol		447	475		Jeffrey (1968a)
Methanol	418	444	474		Karrer & Jucker (1948)
n-Hexane	421	444.5	473.5	76	SCOR WG 78 data
HPLC Eluant	(421)	446	474	59	SCOR WG 78: Mantoura & Llewellyn (1983) method
HPLC Eluant	(422)	448	477	54	SCOR WG 78:Wright *et al.* (1991) method

Alteration products:	*Cis*-isomers
Culture from which SCOR data were obtained:	*Dunaliella tertiolecta* (chlorophyte), *Tetraselmis suecica* (prasinophyte)
Additional reference(s):	Hager & Stransky (1970b); Isler (1971); Krinsky *et al.* (1989)

Lycopene

HPLC peak 45

Standard spectrum in reference solvents

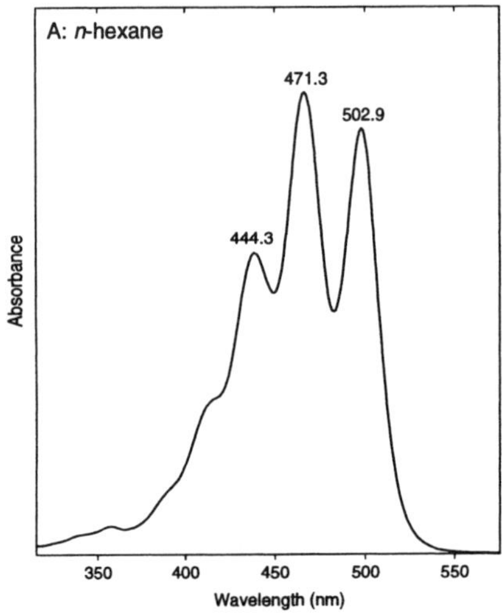

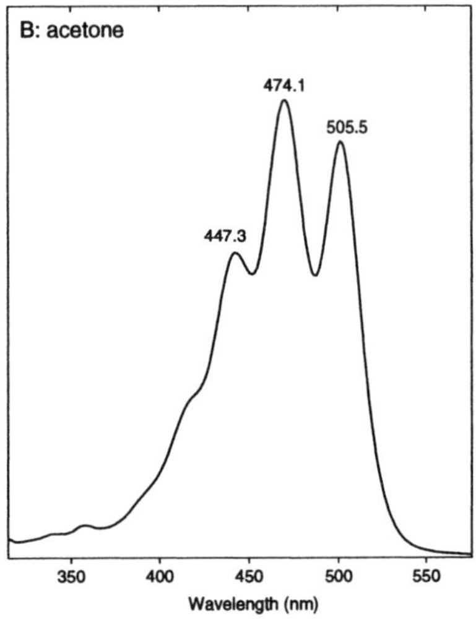

Molecular structure

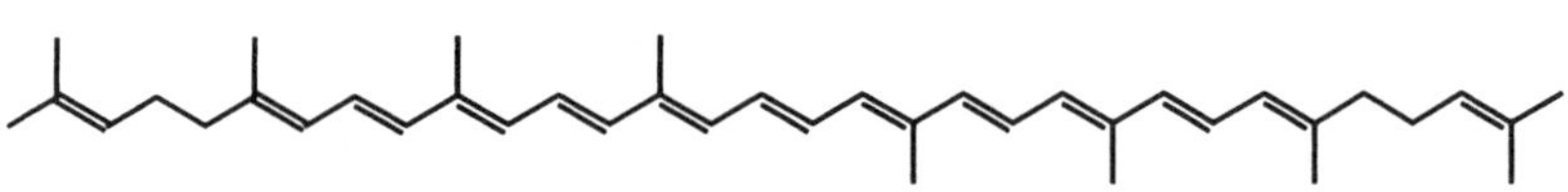

Diode array spectrum in SCOR eluant

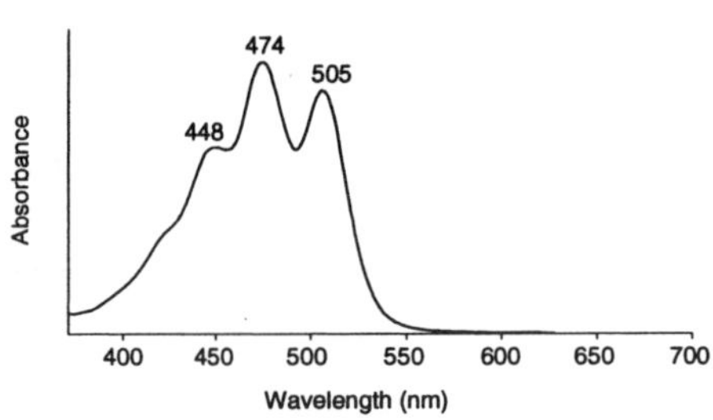

HPLC: Lycopene, peak 45

Micromonas pusilla

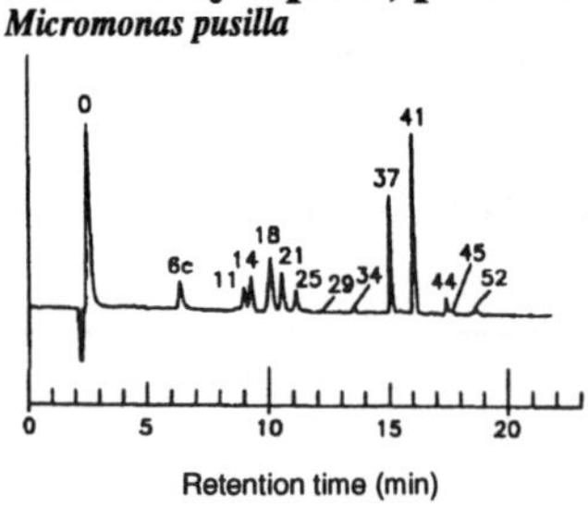

Lycopene

Property		Data
Name:	(Trivial)	**Lycopene**
	(IUPAC)	ψ,ψ-Carotene
SCOR abbreviation:		Lyco
Occurrence:		Minor pigment in prasinophytes, trace pigment in cyanophytes
Colour:		Red
Molecular formula:		$C_{40}H_{56}$
Molecular weight:		536.88
Specific extinction coefficient: $E^{1\%}_{1\ cm}$ (100 ml g^{-1} cm^{-1})		3446 (at 474 nm in acetone) Aasen & Liaaen-Jensen (1966a) 3470 (at 472 nm in *n*-hexane) Zechmeister (1944)
Molar extinction coefficient: ε (l mol^{-1} cm^{-1})		185 x 10^3 (at 474 nm in acetone) 186 x 10^3 (at 472 nm in *n*-hexane) Calculated from $E^{1\%}_{1\ cm}$ above

UV-vis spectra:

Solvent	Maxima (nm) I	II	III	Band ratio %III:II	Reference
Acetone	447	474	506	77	SCOR WG 78 data
Acetone	448	474	505		Hertzberg & Liaaen-Jensen (1967)
n-Hexane	444	471	503	83	SCOR WG 78 data
n-Hexane	448	473	504		Bindl *et al.* (1970)
Ethanol	446	472	503		Davies (1976)
HPLC Eluant	448	474	505	67	SCOR WG 78: Wright *et al.* (1991) method

Alteration products:	
Culture from which SCOR data were obtained:	*Micromonas pusilla* (prasinophyte)
Additional reference(s):	Goodwin (1980); Krinsky *et al.* (1989)

Monadoxanthin

HPLC peak 31

Standard spectrum in reference solvent

No data available

Molecular structure

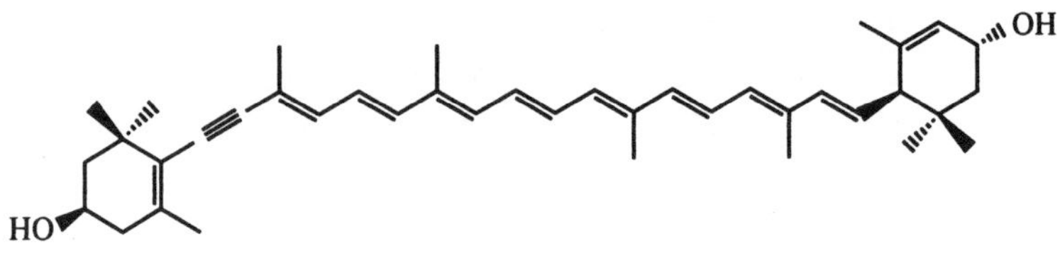

Diode array spectrum in SCOR eluant

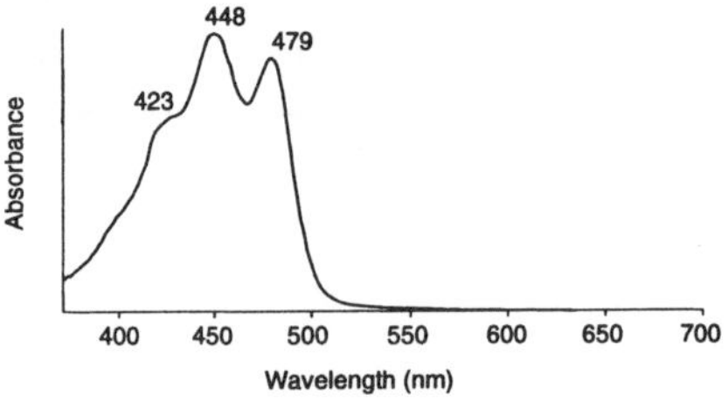

HPLC: Monadoxanthin, peak 31

Chroomonas salina

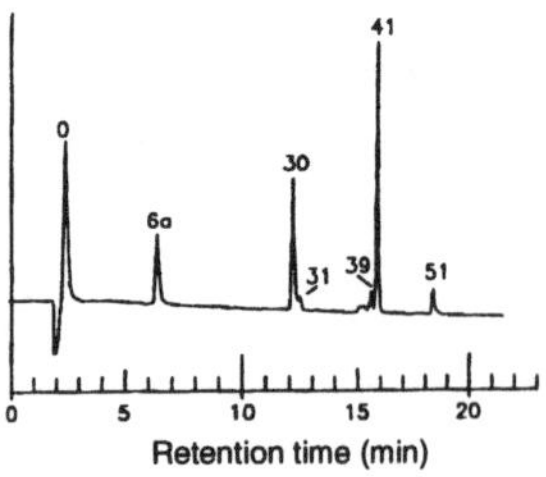

Monadoxanthin

Property		Data
Name:	(Trivial)	**Monadoxanthin**
	(IUPAC)	7,8-Didehydro- β,ε-carotene-3,3'-diol
SCOR abbreviation:		Monado
Occurrence:		Minor pigment in cryptomonads
Colour:		Yellow
Molecular formula:		$C_{40}H_{54}O_2$
Molecular weight:		566.87
Specific extinction coefficient: $E^{1\%}_{1\,cm}$ (100 ml g^{-1} cm^{-1})		2500 (at 446 nm in diethyl ether) Not determined, use $E^{1\%}_{1\,cm}$ for β,β-carotene; Davies (1976)
Molar extinction coefficient: ε (l mol^{-1} cm^{-1})		142 x 10^3 (at 446 nm in diethyl ether) Calculated from $E^{1\%}_{1\,cm}$ above

UV-vis spectra:

Solvent	Maxima (nm) I	II	III	Band ratio %III:II	Reference
Diethyl ether	428	446	476	60	Pennington *et al.* (1985)
n-Hexane	422	445	475	72	Chapman (1966)
Benzene	430	456	487		Cheng *et al.* (1974)
HPLC Eluant	(423)	448	479	65	SCOR WG 78: Wright *et al.* (1991) method

Alteration products:	*Cis*-isomers
Culture from which SCOR data were obtained:	*Chroomonas salina* (cryptomonad)
Additional reference(s):	Cheng *et al.* (1974); Pennington *et al.* (1985)

Neoxanthin

HPLC peak 14

Standard spectrum in reference solvent: acetone

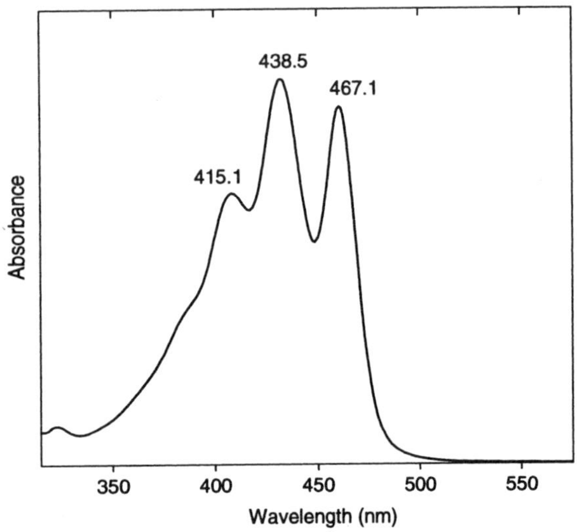

Molecular structure

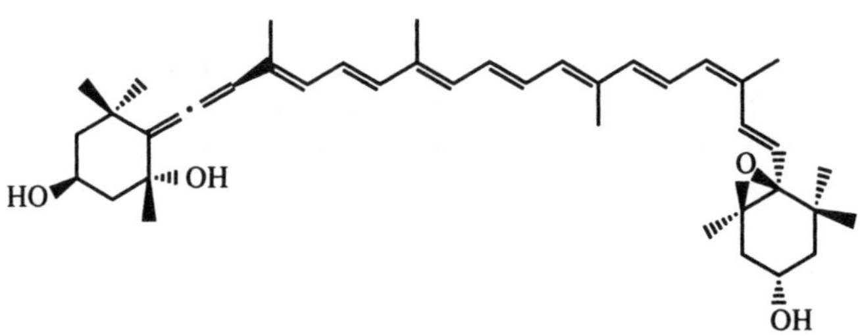

Diode array spectrum in SCOR eluant

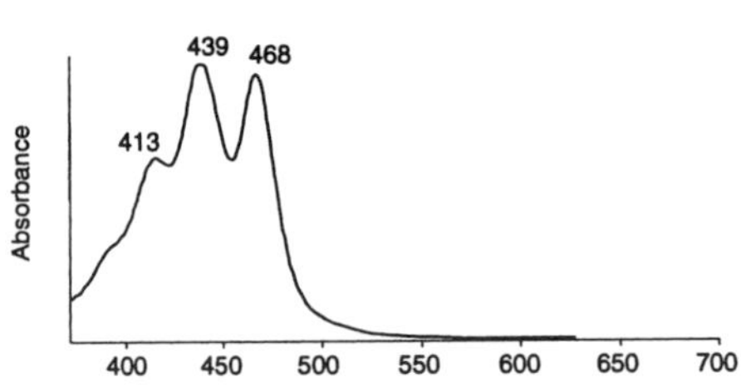

HPLC: Neoxanthin, peak 14

Dunaliella tertiolecta

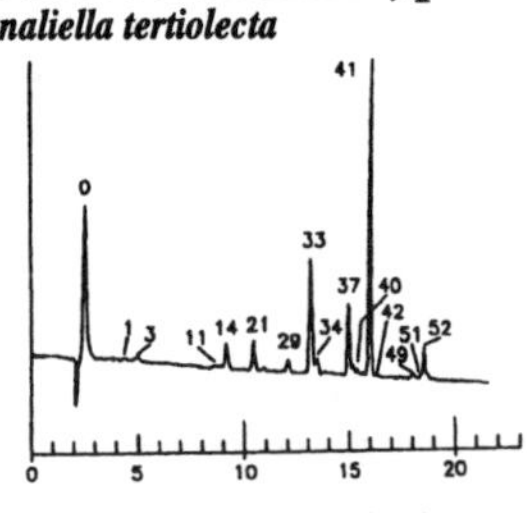

Neoxanthin

Property		Data
Name:	(Trivial)	**Neoxanthin**
	(IUPAC)	(9'-*cis*)(3*S*,5*R*,6*R*,3'*S*,5'*R*,6'*S*)-5',6'-Epoxy-6,7-didehydro-5,6,5',6'-tetrahydro-β,β-carotene-3,5,3'-triol
SCOR abbreviation:		Neo
Occurrence:		9'-*cis* isomer is minor to major pigment in green algae, euglenophytes, higher plants; all-*trans* form in brown seaweeds, chrysophytes
Colour:		Yellow
Molecular formula:		$C_{40}H_{56}O_4$
Molecular weight:		600.88
Specific extinction coefficient: $E^{1\%}_{1\,cm}$ (100 ml g^{-1} cm^{-1})		2270 (at 438 nm in ethanol) Banthorpe *et al.* (1972) 2243 (at 439 nm in ethanol) Cholnoky *et al.* (1966)
Molar extinction coefficient: ε (l mol^{-1} cm^{-1})		136 x 10^3 (at 438 nm in ethanol) Calculated from mean of $E^{1\%}_{1\,cm}$ above

UV-vis spectra:

Solvent	Maxima (nm) I	II	III	Band ratio %III:II	Reference
Acetone	415	439	467	83	SCOR WG 78 data
Acetone	413	438	466	90	Bjørnland *et al.* (1987)
Ethanol	415	438	467		Mallams *et al.* (1967)
n-Hexane	411	435	463	97	Bjørnland (pers. comm.)
n-Hexane	416	440	469	94	Bjørnland (pers. comm.); all *trans* isomer
HPLC Eluant	414	437	466	90	SCOR WG 78: Mantoura & Llewellyn (1983) method
HPLC Eluant	413	439	468	87	SCOR WG 78: Wright *et al.* (1991) method

Alteration products:	*Cis-trans* isomers; furanoids (neochromes)
Culture from which SCOR data were obtained:	*Dunaliella tertiolecta* (green flagellate) (9'-*cis* isomer)
Additional reference(s):	Bjørnland *et al.* (1987); Bjørnland (1990a)

P-457

HPLC peak 4

Standard spectrum in reference solvent: methanol

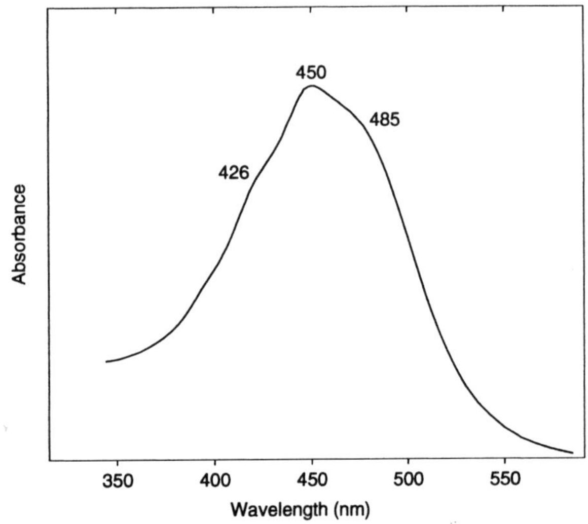

Data from Johansen *et al.* (1974)

Molecular structure

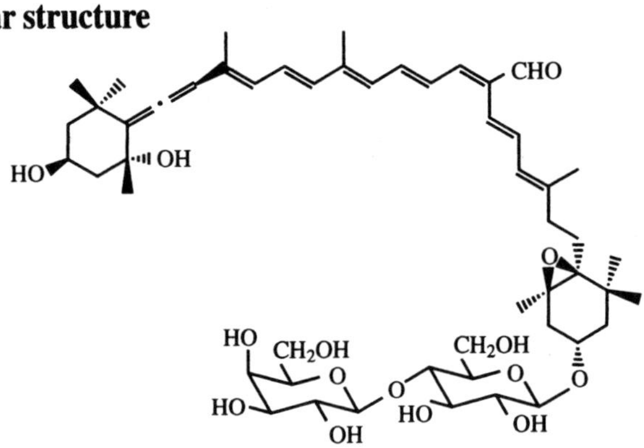

Diode array spectrum in SCOR eluant

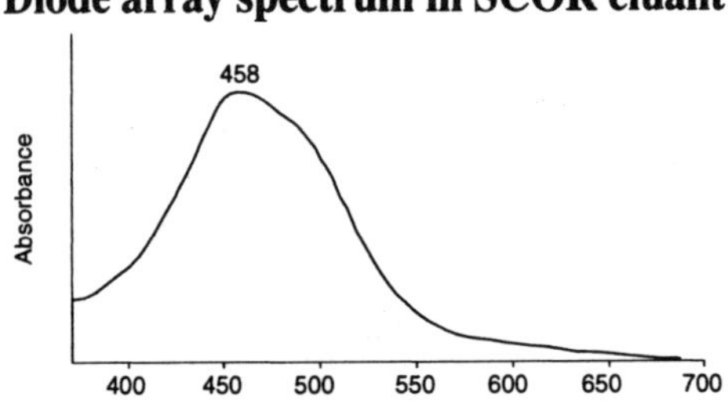

HPLC: P-457, peak 4

Amphidinium carterae

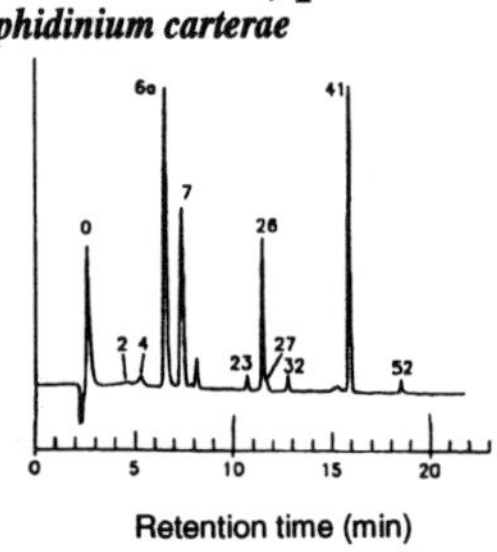

P-457

Property	Data
Name: (Trivial)	**P-457**
Semi-systematic name	7',8'-dihydroneoxanthin-20'-al 3'-β-lactoside
SCOR abbreviation:	P-457
Occurrence:	Minor pigment in photosynthetic peridinin-containing dinoflagellates; P457 = fraction 9, present in 23 dinoflagellate species (Jeffrey *et al.*, 1975)
Colour:	Pink
Molecular formula:	$C_{52}H_{76}O_{15}$
Molecular weight:	941.16
Specific extinction coefficient: $E^{1\%}_{1\,cm}$ (100 ml g^{-1} cm^{-1})	1640 (at 457 nm in acetone) Not determined; use $E^{1\%}_{1\,cm}$ = 2500 for the free xanthophyll $[\equiv 2500 \times \frac{616}{940} = 1638 \equiv 1640]$
Molar extinction coefficient: ε (l mol^{-1} cm^{-1})	154 x 10^3 (at 457 nm in acetone) Calculated from $E^{1\%}_{1\,cm}$ above

UV-vis spectra:

Solvent	Maxima (nm) I	II	III	Band ratio %III:II	Reference
Acetone		457			Johansen *et al.* (1974)
Methanol		454.5			Johansen *et al.* (1974)
HPLC Eluant		458			SCOR WG 78: Wright *et al.* (1991) method

Alteration products:	
Culture from which SCOR data were obtained:	*Amphidinium carterae* (dinoflagellate)
Additional reference(s):	Johansen *et al.* (1974); Jeffrey *et al.* (1975); Liaaen-Jensen (1990b)

Peridinin

HPLC peak 7

Standard spectrum in reference solvents

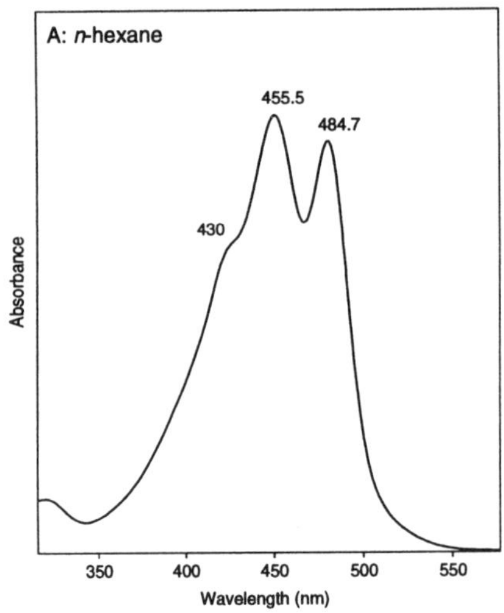

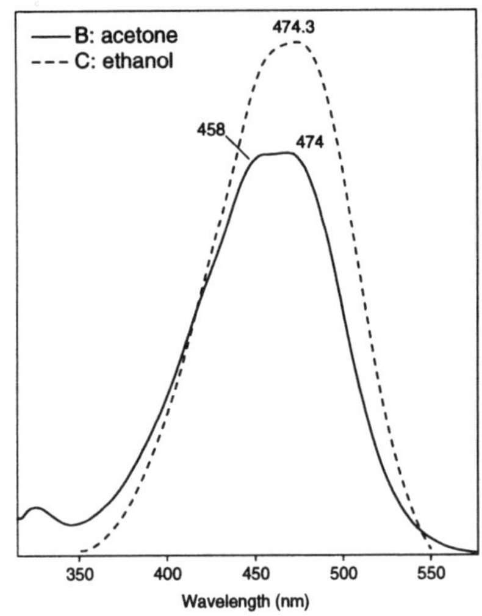

Molecular structure

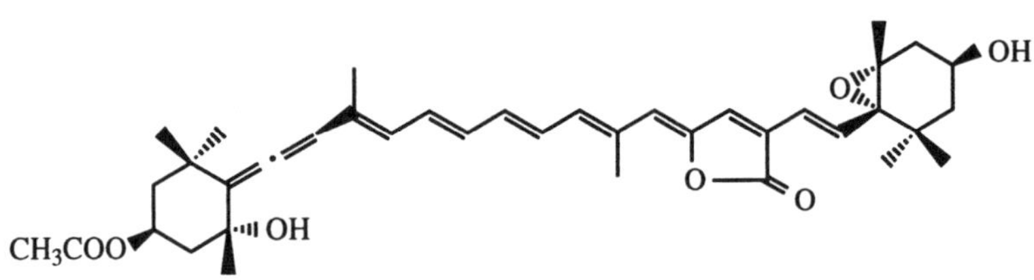

Diode array spectrum in SCOR eluant

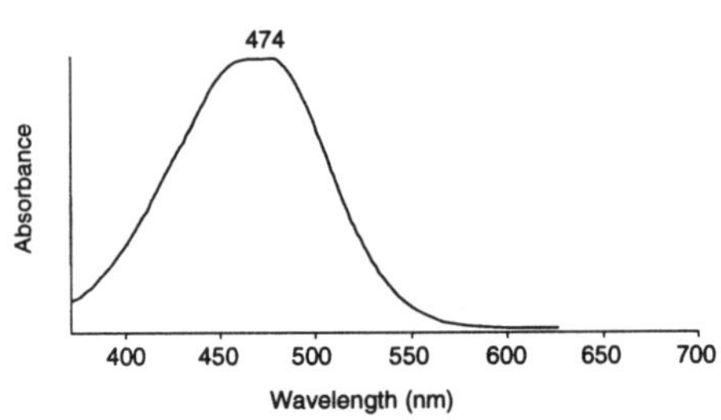

HPLC: Peridinin, peak 7

Amphidinium carterae

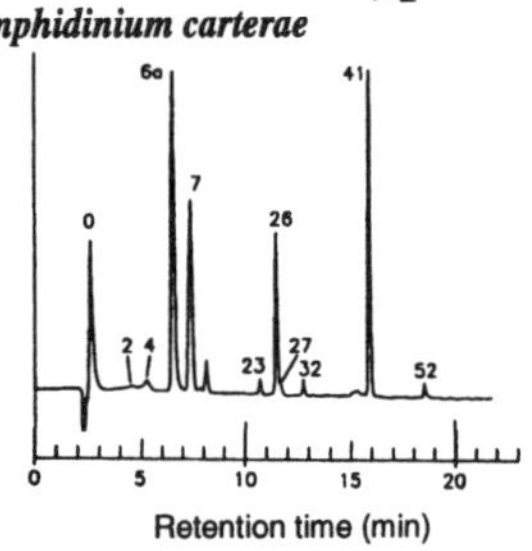

Peridinin

Property		Data
Name:	(Trivial)	**Peridinin**
	(IUPAC)	(3*S*,5*R*,6*S*,3'*S*,5'*R*,6'*R*)-5,6-Epoxy-3,3',5'-trihydroxy-6',7'-didehydro-5,6,5',6'-tetrahydro-10',11',20'-trinor-β,β-caroten-19,11-olide 3'-acetate
SCOR abbreviation:		Perid
Occurrence:		Photosynthetic dinoflagellates (major pigment), (except those containing endosymbionts of other algal classes)
Colour:		Brick red
Molecular formula:		$C_{39} H_{50} O_7$
Molecular weight:		630.82
Specific extinction coefficient: $E^{1\%}_{1\,cm}$ (100 ml g^{-1} cm^{-1})		1340 (at 466 nm in acetone) 1325 (at 472 nm in ethanol) 1360 (at 469 nm in methanol) Jeffrey & Haxo (1968)
Molar extinction coefficient: ε (l mol^{-1} cm^{-1})		84.5 x 10^3 (at 466 nm in acetone) 83.6 x 10^3 (at 472 nm in ethanol) 85.8 x 10^3 (at 469 nm in methanol) Calculated from $E^{1\%}_{1\,cm}$ above

UV-vis spectra:

Solvent	Maxima (nm) I	II	III	Band ratio %III:II	Reference
Acetone		458	474	0	SCOR WG 78 data
Ethanol			474	0	SCOR WG 78 data
Ethanol			472	0	Jeffrey & Haxo (1968)
n-Hexane	(430)	455.5	484.7	75	SCOR WG 78 data
n-Hexane	(428)	454	484	85	Jeffrey & Haxo (1968)
Methanol			469	0	Jeffrey & Haxo (1968)
HPLC Eluant			474	0	SCOR WG 78: Wright *et al.* (1991) method

Alteration products:	*Cis*-isomers
Culture from which SCOR data were obtained:	*Amphidinium carterae* (dinoflagellate)
Additional reference(s):	Jeffrey & Haxo (1968); Strain *et al.* (1971); Johansen *et al.* (1974); Jeffrey *et al.* (1975)

Prasinoxanthin

HPLC peak 19

Standard spectrum in reference solvents

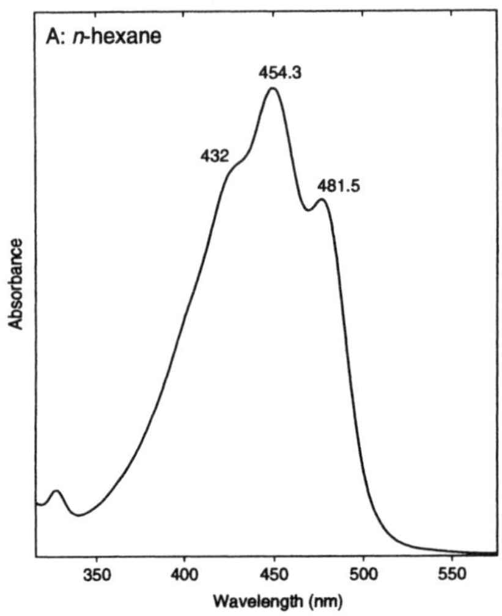

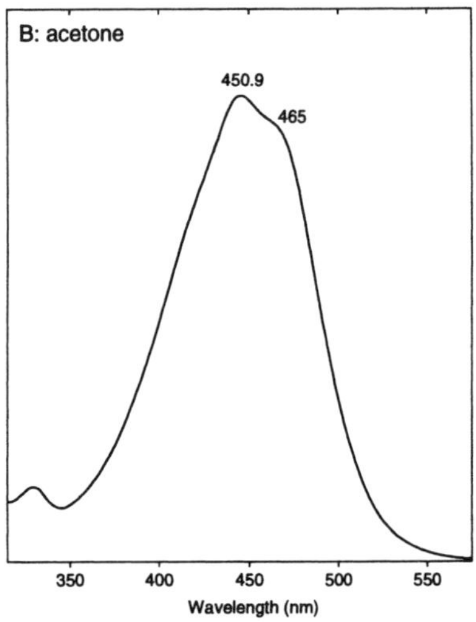

Molecular structure

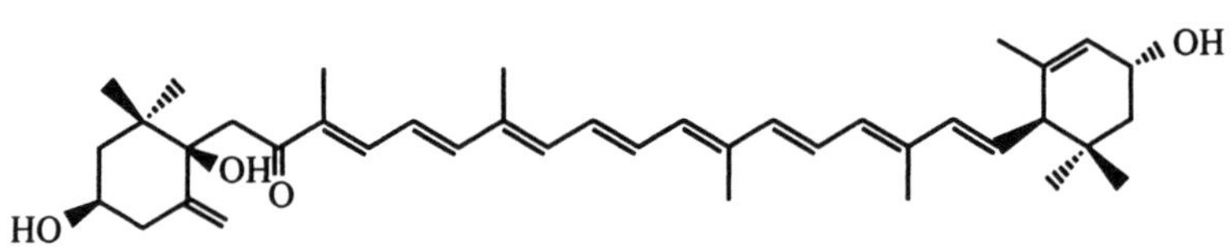

The chiralities at C-3 and C-6 are tentative and are based on biochemical arguments.

Diode array spectrum in Mantoura & Llewellyn* eluant

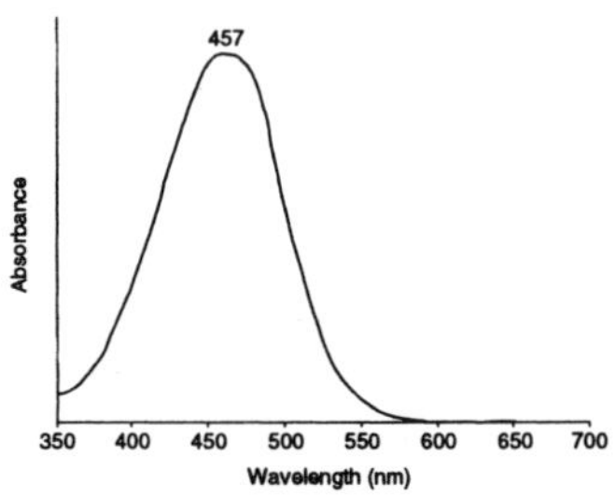

HPLC: Prasinoxanthin, peak 19

Pycnococcus provasolii

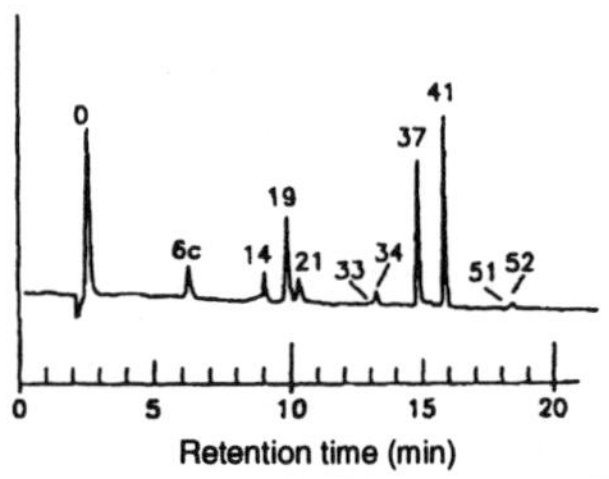

***Mantoura and Llewellyn (1983)**

Prasinoxanthin

Property		Data
Name:	(Trivial)	**Prasinoxanthin** Formerly Xanthophyll K, Ricketts (1970)
	(IUPAC)	(3'*R*,6'*R*)-3,6,3'-Trihydroxy-7,8-dihydro-γ, ε-caroten-8-one
SCOR abbreviation:		Pras
Occurrence:		Major pigment in some prasinophytes (e.g. some Micromonadophyceae)
Colour:		Deep pink
Molecular formula:		$C_{40}H_{56}O_4$
Molecular weight:		600.88
Specific extinction coefficient: $E^{1\%}_{1\,cm}$ (100 ml g^{-1} cm^{-1})		2500 (at 446 nm in diethyl ether) Not determined; recommended by Foss *et al.* (1984)
Molar extinction coefficient: ε (l mol^{-1} cm^{-1})		150 x 10^3 (at 446 nm in diethyl ether) Calculated from $E^{1\%}_{1\,cm}$ above

UV-vis spectra:

Solvent	Maxima (nm) I	II	III	Band ratio %III:II	Reference
Acetone		450.9	(465)	0	SCOR WG 78 data
Diethyl ether		446	(466)	0	Foss *et al.* (1984)
Hexane	(432)	454.3	481.5	9	SCOR WG 78 data
HPLC Eluant		457		0	SCOR WG 78: Mantoura & Llewellyn (1983) method
HPLC Eluant		450	(470)		SCOR WG 78 ; Wright *et al.* (1991) method

Alteration products:	*Cis*-isomers
Culture from which SCOR data were obtained:	*Pycnococcus provasolii* (prasinophyte)
Additional reference(s):	Ricketts (1970); Foss *et al.* (1984); Foss *et al.* (1986); Hooks *et al.* (1988)

Siphonaxanthin

HPLC peak 8

Standard spectrum in reference solvents

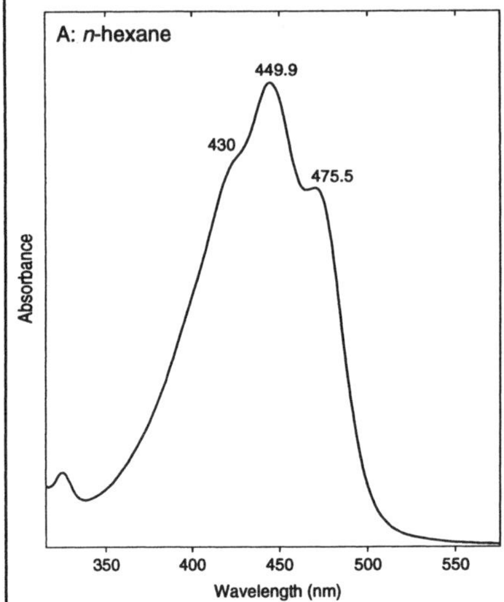

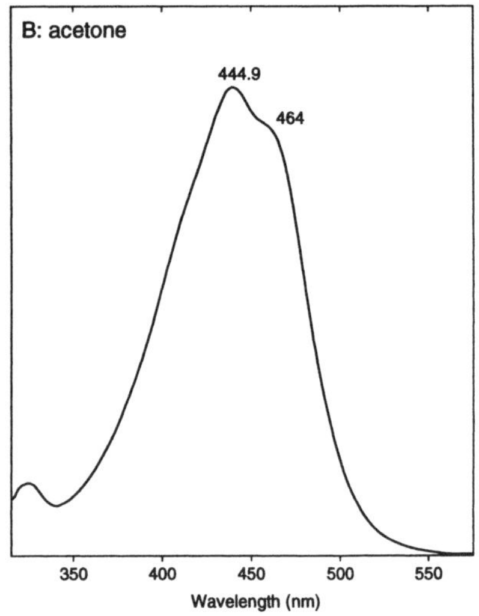

Molecular structure

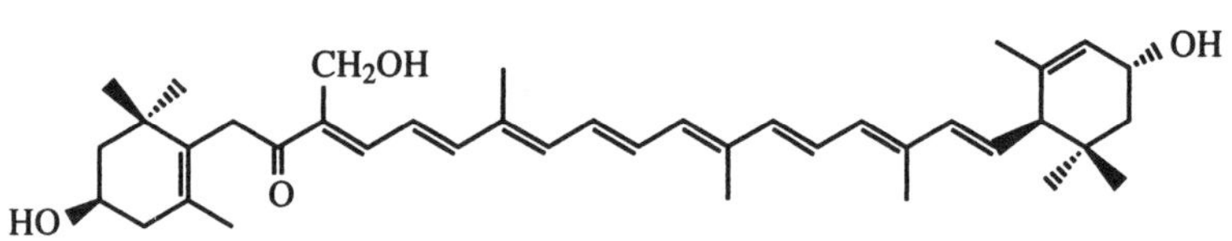

The chirality of siphonaxanthin is assumed to be identical to that of siphonein.

Diode array spectrum in SCOR eluant

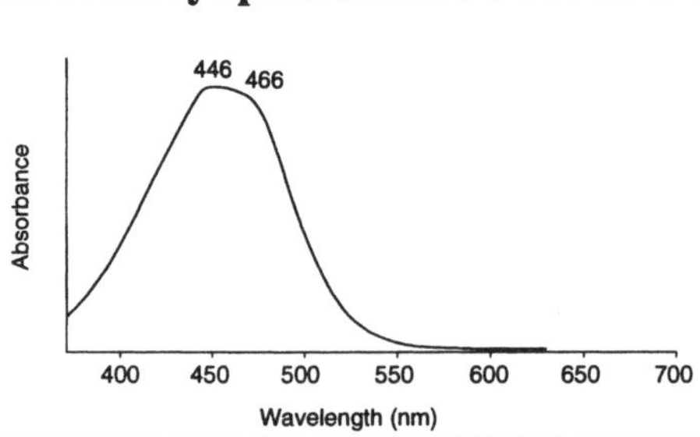

HPLC: Siphonaxanthin, peak 8

Codium fragile

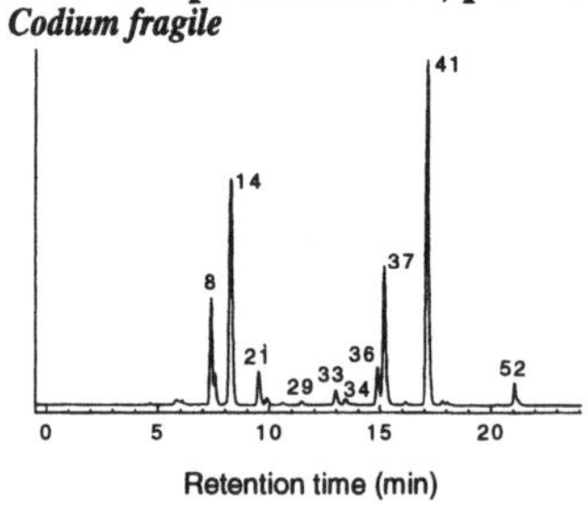

Siphonaxanthin

Property	**Data**
Name: (Trivial)	**Siphonaxanthin**
(IUPAC)	3,19,3'-Trihydroxy-7,8-dihydro-β, ε-caroten-8-one
SCOR abbreviation:	Siphx
Occurrence:	Major pigment in siphonous green seaweeds (e.g. *Caulerpa*), and *Ostreobium*. Suspected but not confirmed in some members of the Micromonadophyceae, Wright *et al.* (1991)
Colour:	Salmon pink
Molecular formula:	$C_{40}H_{56}O_4$
Molecular weight:	600.88
Specific extinction coefficient: $E^{1\%}_{1\,cm}$ (100 ml g^{-1} cm^{-1})	2500 (at 445 nm in acetone) Not determined; use $E^{1\%}_{1\,cm}$ for β,β-carotene, Davies (1976)
Molar extinction coefficient: ε (l mol^{-1} cm^{-1})	150 x 10^3 (at 445 nm in acetone) Calculated from $E^{1\%}_{1\,cm}$ above

UV-vis spectra:

Solvent	Maxima (nm) I	II	III	Band ratio %III:II	Reference
Acetone		445	(464)	0	SCOR WG 78 data
Ethanol		449			Jeffrey (1968b)
Ethanol		449			Yokohama *et al.* (1977)
n-Hexane	(430)	450	476	3	SCOR WG 78 data
HPLC Eluant		446	(466)		SCOR WG 78: Wright *et al.* (1991) method

Alteration products:	*Cis*-isomers
Origin:	*Codium fragile* fronds (siphonous green seaweed from a natural habitat))
Additional reference(s):	Jeffrey (1968b); Yokohama *et al.* (1977); Goodwin (1980); Anderson (1985); Fawley & Lee (1990)

Siphonein
HPLC peak 36

Standard spectrum in reference solvents

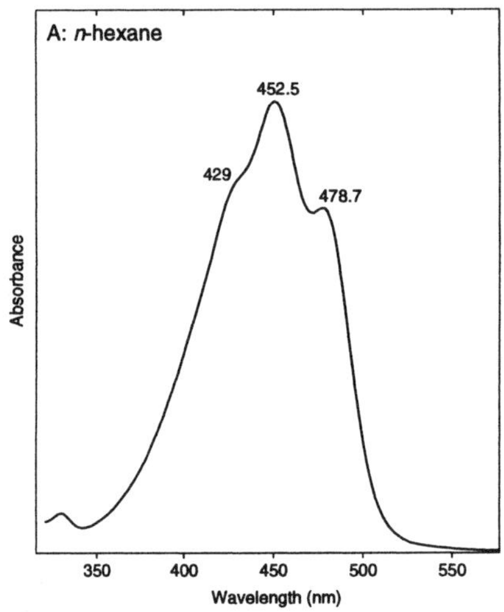

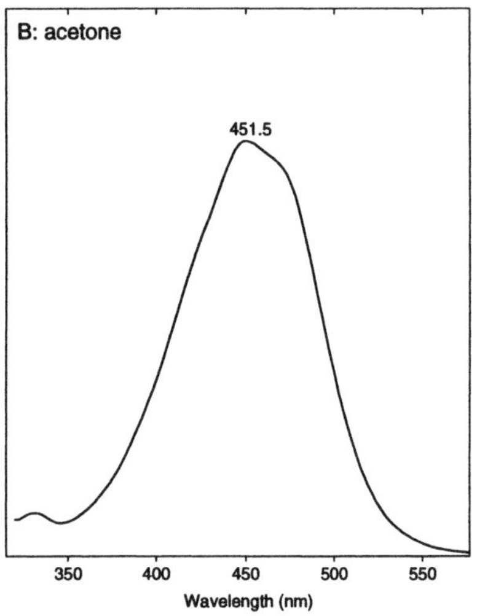

Molecular structure

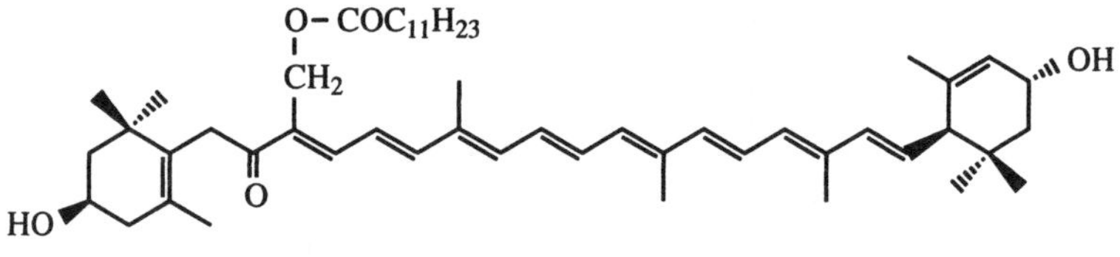

Diode array spectrum in SCOR eluant

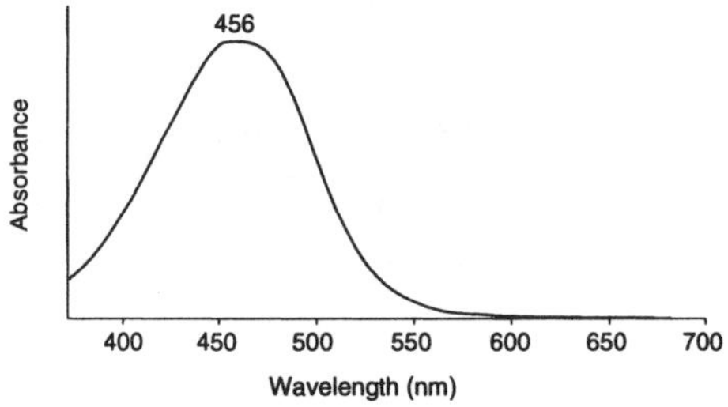

HPLC: Siphonein, peak 36
Codium fragile

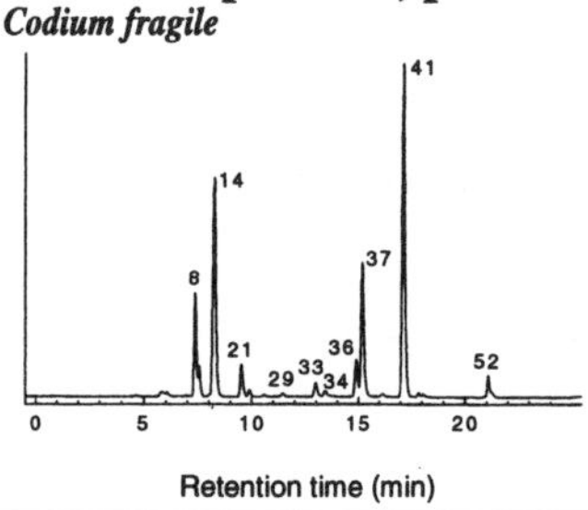

Siphonein

Property		Data
Name:	(Trivial)	**Siphonein**
	(IUPAC)	(3*R*,3'*R*,6*R*)-3,19,3'-Trihydroxy-7,8-dihydro- β,ε-caroten-8-one 19-laurate
SCOR abbreviation:		Siphn
Occurrence:		Major pigment in siphonous green seaweeds, green algae, some euglenophytes
Colour:		Salmon pink
Molecular formula:		$C_{52}H_{78}O_5$
Molecular weight:		783.19
Specific extinction coefficient: $E^{1\%}_{1\,cm}$ (100 ml g^{-1} cm^{-1})		1920 (at 462 nm in ethanol) Calculated from ε below
Molar extinction coefficient: ε (l mol^{-1} cm^{-1})		150 x 10^3 (at 462 nm in ethanol) Derived from ε for siphonaxanthin

UV-vis spectra:

Solvent	Maxima (nm) I	II	III	Band ratio %III:II	Reference
Acetone		452			SCOR WG78 data
Hexane	(429)	452.5	478.7	4	SCOR WG78 data
Petroleum ether		457	481		Jeffrey (1968b)
HPLC Eluant		456			SCOR WG 78: Wright *et al.* (1991) method

Alteration products:	*Cis*- isomers
Origin:	*Codium fragile* fronds (siphonous green seaweed from a natural habitat); *Micromonas pusilla* (prasinophyte); Wright *et al.* (1991)
Additional reference(s):	Fiksdahl *et al.* (1984a); Goodwin (1980)

Vaucheriaxanthin (ester)

HPLC peak V

Standard spectrum in reference solvents

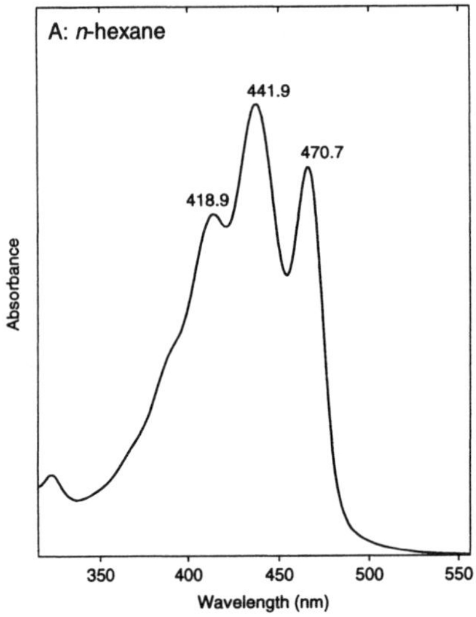

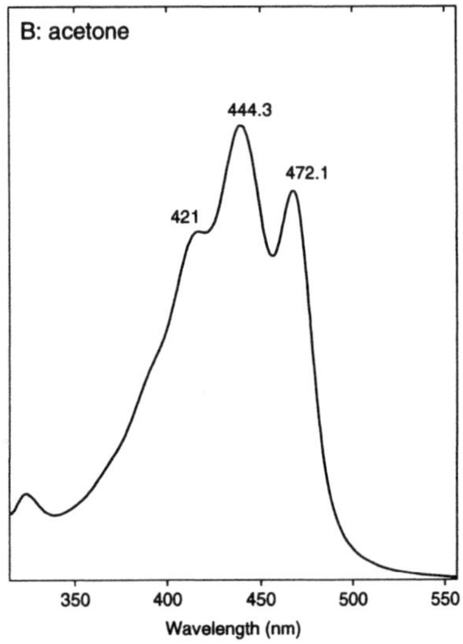

Molecular structure

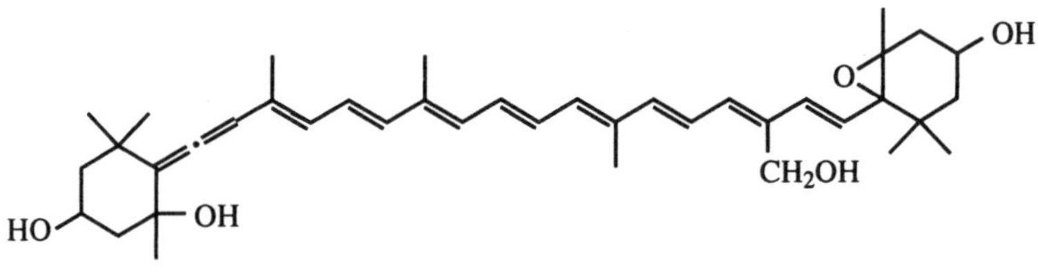

Diode array spectrum in SCOR eluant

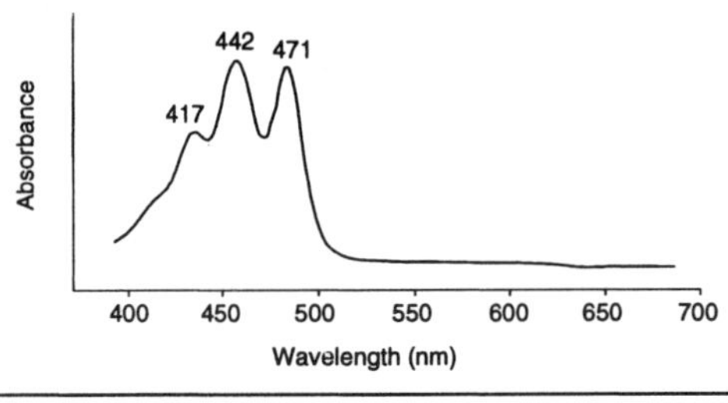

HPLC: Vaucheriaxanthin (ester), peak V

Nannochloropsis oculata

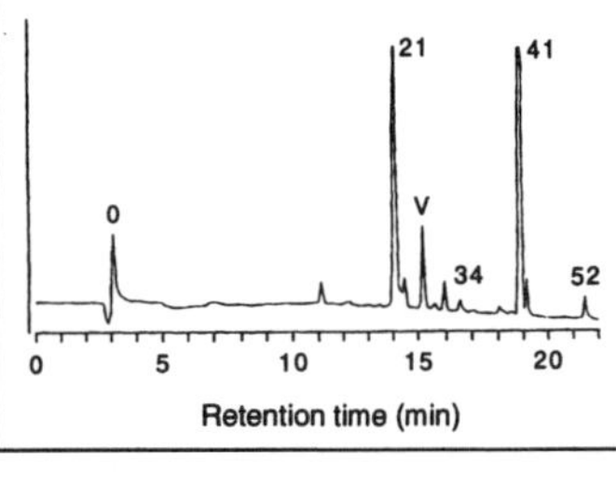

Vaucheriaxanthin (ester)

Property		**Data**
Name:	(Trivial)	**Vaucheriaxanthin (ester)**
	(IUPAC)	5',6'-Epoxy-6,7-didehydro-5,6,5',6'-tetrahydro-β,β-carotene-3,5,3',19'-tetrol ester
SCOR abbreviation:		Vauch
Occurrence:		Major pigment in xanthophytes and eustigmatophytes
Colour:		Yellow
Molecular formula:		$C_{40}H_{56}O_5$
Molecular weight:		616.88
Specific extinction coefficient: $E^{1\%}_{1\ cm}$ (100 ml g^{-1} cm^{-1})		2500 (at 444 nm in acetone) Not determined; use $E^{1\%}_{1\ cm}$ for β,β-carotene (Davies, 1976)
Molar extinction coefficient: ε (l mol^{-1} cm^{-1})		154 x 10^3 (at 444 nm in acetone) Calculated from $E^{1\%}_{1\ cm}$ above

UV-vis spectra:

Solvent	Maxima (nm) I	II	III	Band ratio %III:II	Reference
Acetone	421	444	472	50	SCOR WG 78 data
Acetone	420	441	467		Norgård *et al.* (1974)
n-Hexane	419	442	471	64	SCOR WG 78 data
Ethanol	418	441	470		Whittle & Casselton (1975) – non-esterified
Ethanol	419	442	471		Whittle & Casselton (1975) – esterified
Ethanol	419	443	470		Antia & Cheng (1982)
HPLC Eluant	417	442	471	92	SCOR WG 78: Wright *et al.* (1991) method

Alteration products:	*Cis*-isomers; (probably) furanoids
Culture from which SCOR data were obtained:	*Nannochloropsis oculata* (eustigmatophyte)
Additional reference(s):	Norgård *et al.* (1974); Whittle & Casselton (1975); Volkman *et al.* (1993)

Violaxanthin

HPLC peak 21

Standard spectrum in reference solvents

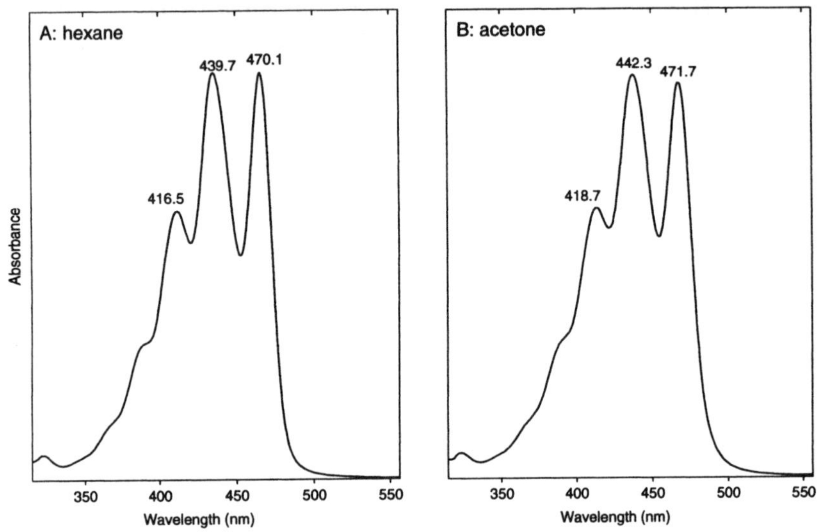

Molecular structure

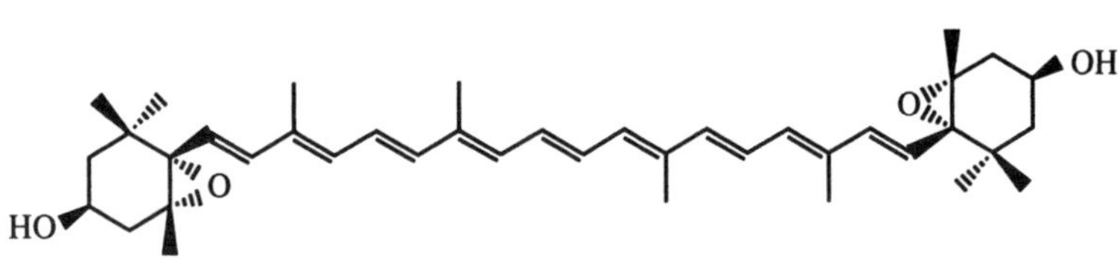

Diode array spectrum in SCOR eluant

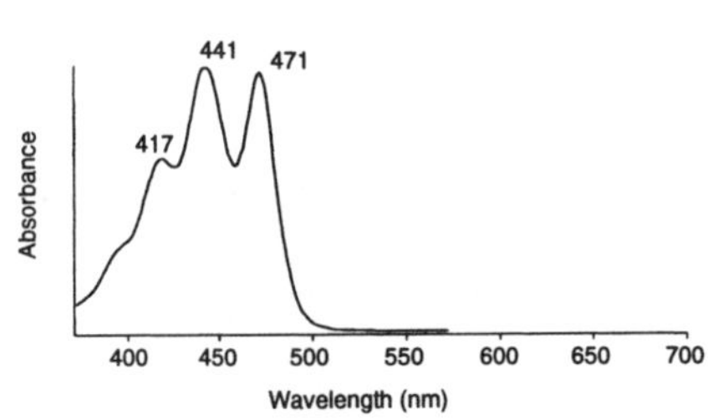

HPLC: Violaxanthin, peak 21

Dunaliella tertiolecta

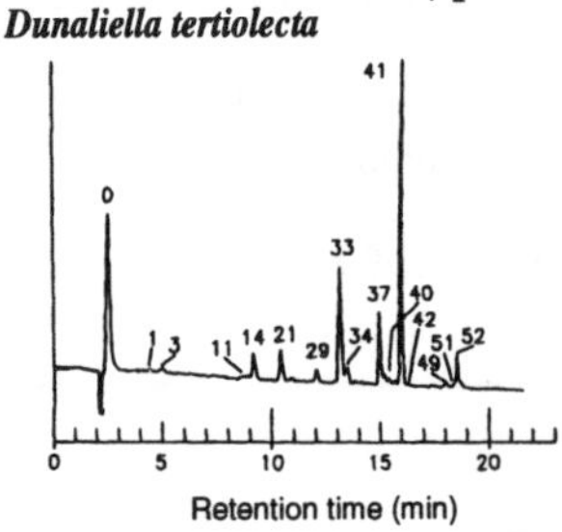

Violaxanthin

Property		Data
Name:	(Trivial)	**Violaxanthin**
	(IUPAC)	(3*S*,5*R*,6*S*,3'*S*,5'*R*,6'*S*)-5,6,5',6'-Diepoxy-5,6,5',6'-tetrahydro-β,β-carotene-3,3'-diol
SCOR abbreviation:		Viola
Occurrence:		Major pigment in higher plants, green algae, eustigmatophytes, brown seaweeds
Colour:		Yellow
Molecular formula:		$C_{40}H_{56}O_4$
Molecular weight:		600.88
Specific extinction coefficient: $E^{1\%}_{1\,cm}$ (100 ml g^{-1} cm^{-1})		2550 (at 443 nm in ethanol) Davies (1965) 2400 (at 442 nm in acetone) Jensen (1966b)
Molar extinction coefficient: ε (l mol^{-1} cm^{-1})		153×10^3 (at 443 nm in ethanol) 144×10^3 (at 442 nm in acetone) Calculated from $E^{1\%}_{1\,cm}$ above

UV-vis spectra:

Solvent	Maxima (nm) I	II	III	Band ratio %III:II	Reference
Acetone	419	442	472	95	SCOR WG 78 data
Acetone	417	440	470	76	Renstrom *et al.* (1981)
Ethanol	417	440	470	96	SCOR WG 78 data
Ethanol	417	440	469	93	Stransky & Hager (1970a)
n–Hexane	417	440	470	100	SCOR WG 78 data
Methanol	415	440	469		Karrer & Jucker (1948)
HPLC Eluant	417	440	469	96	SCOR WG 78: Mantoura & Llewellyn (1983) method
HPLC Eluant	417	441	471	93	SCOR WG 78: Wright *et al.* (1991) method

Alteration products:	*Cis*-isomers; furanoids (luteoxanthins and auroxanthins)
Culture from which SCOR data were obtained:	*Dunaliella tertiolecta* (green flagellate)
Additional reference(s):	Davies (1965); Isler (1971); Foppen (1971)

Zeaxanthin

HPLC peak 34

Standard spectrum in reference solvents

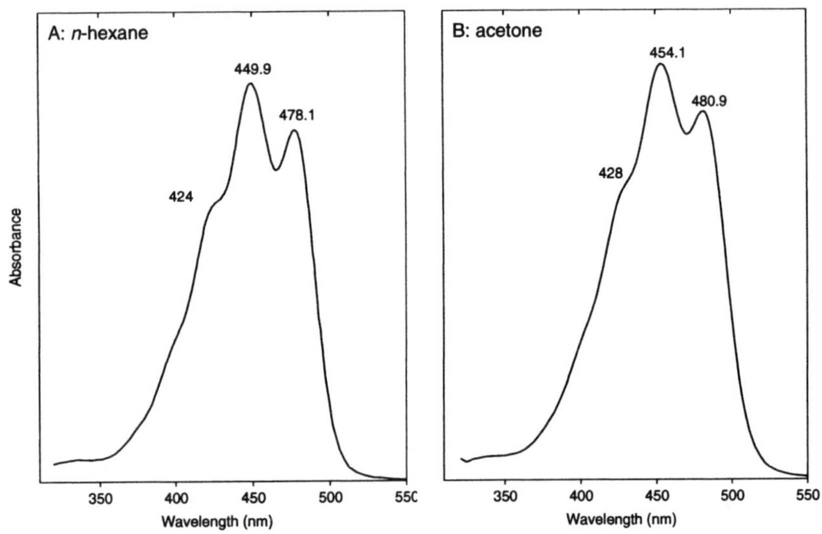

Molecular structure

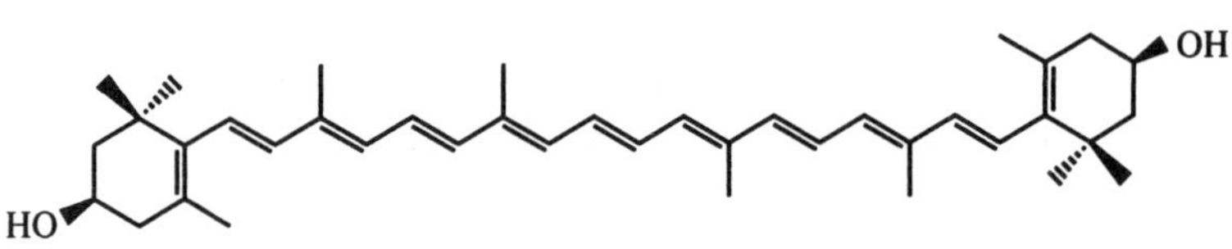

Diode array spectrum in SCOR eluant

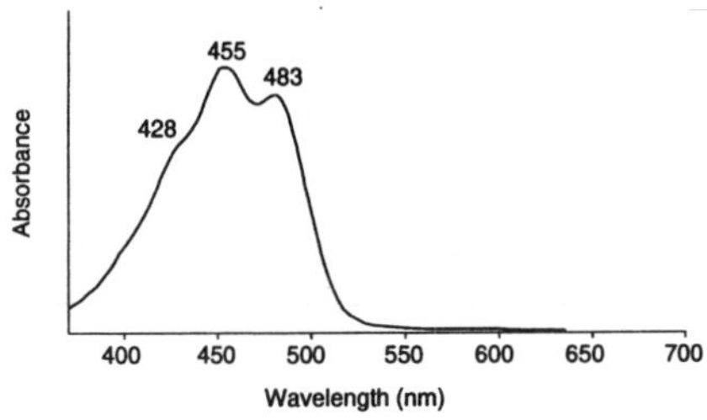

HPLC: Zeaxanthin, peak 34

***Synechococcus* sp.**

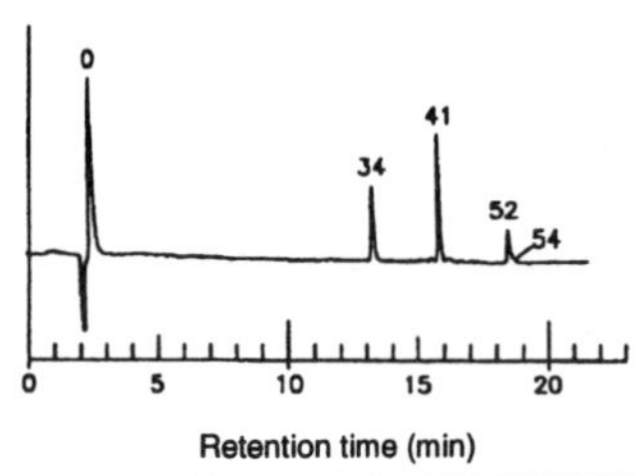

Zeaxanthin

Property		Data
Name:	(Trivial)	**Zeaxanthin**
	(IUPAC)	(3*R*,3'*R*)-β,β-Carotene-3,3'-diol
SCOR abbreviation:		Zea
Occurrence:		Major or minor pigment in prochlorophytes, cyanobacteria (coccoid), green algae, most chrysophytes, raphidophytes
Colour:		Yellow-orange
Molecular formula:		$C_{40}H_{56}O_2$
Molecular weight:		568.88
Specific extinction coefficient: $E^{1\%}_{1\,cm}$ (100 ml g^{-1} cm^{-1})		2340 (at 452 nm in acetone) Aasen & Liaaen-Jensen (1966b) 2540 (at 450 nm in ethanol) Strain (1938) 2350 (at 452 nm in petroleum ether) Davies (1976)
Molar extinction coefficient: ε (l mol^{-1} cm^{-1})		133 x 10^3 (at 452 nm in acetone) 145 x 10^3 (at 450 nm in ethanol) 134 x 10^3 (at 452 nm in petroleum ether) Calculated from $E^{1\%}_{1\,cm}$ above

UV-vis spectra:

Solvent	Maxima (nm) I	II	III	Band ratio %III:II	Reference
Acetone	(428)	454	481	33	SCOR WG 78 data
Acetone	(425)	450	478		Withers *et al.* (1981)
Ethanol	(428)	450	478	26	Stransky & Hager (1970a)
n–Hexane	(424)	450	478	46	SCOR WG 78 data
n–Hexane	(426)	450	480		Valadon & Mummery (1967)
Methanol	422	450	481		Karrer & Jucker (1948)
HPLC Eluant	(426)	453	479	33	SCOR WG 78: Mantoura & Llewellyn (1983) method
HPLC Eluant	(428)	455	483	23	SCOR WG 78: Wright *et al.* (1991) method

Alteration products:	*Cis*-isomers
Culture from which SCOR data were obtained:	*Synechococcus* sp. (DC-2)(blue-green alga)
Additional reference(s):	Guillard *et al.* (1985); Krinsky *et al.* (1989)

References

Aasen, A. J., Liaaen-Jensen, S. (1966*a*). The carotenoids of the flexibacteria. II. A new xanthophyll from *Saprospira grandis*. *Acta Chem. Scand.* 20: 811–819.

Aasen, A. J., Liaaen-Jensen, S. (1966*b*). Carotenoids of flexibacteria. IV. The carotenoids of two further pigment types. *Acta Chem. Scand.* 20: 2322–2324.

Anderson, J. M. (1985). Chlorophyll-protein complexes of a marine alga, *Codium* species (Siphonales). *Biochim. Biophys. Acta*, 806: 145–153.

Antia, N. J., Cheng, J. Y. (1982). The keto-carotenoids of two marine coccoid members of the Eustigmatophyceae. *Br. Phycol.* J. 17: 39–50.

Arpin, N., Svec, W. A., Liaaen-Jensen, S. (1976). New fucoxanthin-related carotenoids from *Coccolithus huxleyi*. *Phytochem.* 15: 529–532.

Banthorpe, D. V., Doonan, H. J., Wirz-Justice, A. (1972). Terpene biosynthesis. Part V. Interconversions of some monoterpenes in higher plants and their possible role as precursors of carotenoids. *J. Chem. Soc. Perkin Trans.* I: 1764–1769.

Barrett, J., Jeffrey, S. W. (1964). Chlorophyllase and formation of an atypical chlorophyllide in marine algae. *Plant. Physiol.* 39: 44–47.

Bazzaz, M. B. (1981). New chlorophyll chromophores isolated from a chlorophyll-deficient mutant of maize. *Photobiochem. Photobiophys.* 2: 199–207.

Berger, R., Liaaen-Jensen, S., McAlister, V., Guillard, R. R. L. (1977). Carotenoids of the Prymnesiophyceae (Haptophyceae). *Biochem. Syst. Ecol.* 5: 71–75.

Bindl, E., Lang, W., Rau, W. (1970). Light dependent carotenoid synthesis. VI. Time course of synthesis of various carotenoids in *Fusarium aquaeductuum* after various inductive treatments. *Planta*, 94: 156–174.

Bjørnland, T. (1982). Chlorophylls and carotenoids of the marine alga *Eutreptiella gymnastica*. *Phytochem.* 21: 1715–1719.

Bjørnland, T. (1990*a*). Carotenoid structures and lower plant phylogeny. In: Krinsky, N. I., Mathews-Roth, M. M., Taylor, R. F. (eds), *Carotenoids. Chemistry and Biology*. Plenum Press, New York, pp. 21–37.

Bjørnland, T. (1990*b*). Chromatographic separation and spectrometric characterization of native carotenoids from the marine dinoflagellate *Thoracosphaera heimii*. *Biochem. Syst. Ecol.* 18: 307–316.

Bjørnland, T., Borch, G., Liaaen-Jensen, S. (1984). Configurational studies on red algal carotenoids. *Phytochem.* 23: 1711–1715.

Bjørnland, T., Guillard, R. R. L., Liaaen-Jensen, S. (1988). *Phaeocystis* sp. clone 677-3 — a tropical marine planktonic prymnesiophyte with fucoxanthin and 19'-acyloxyfucoxanthins as chemosystematic carotenoid markers. *Biochem. Syst. Ecol.* 16: 445–452.

Bjørnland, T., Liaaen-Jensen, S. (1989). Distribution patterns of carotenoids in relation to chromophyte phylogeny and systematics. In: Green, J. C., Leadbeater, B. S. C., Diver, W. L. (eds), *The Chromophyte Algae: Problems and Perspectives*. Clarendon Press, Oxford, pp. 37–61.

Bjørnland, T., Liaaen-Jensen, S., Throndsen, J. (1989). Carotenoids of the marine chrysophyte *Pelagococcus subviridis*. *Phytochem.* 28: 3347–3353.

Bjørnland, T., Skjetne, T., Liaaen-Jensen, S. (1987). Studies on the stereoisomeric set of neoxanthin — non-existence of *R*/*S* allene isomerism in carotenoids. *Abst. 8th Int. Symp. Carotenoids*, Boston, p. 3.

Bjørnland, T., Tangen, K. (1979). Pigmentation and morphology of a marine *Gyrodinium* (Dinophyceae) with a major carotenoid different from peridinin and fucoxanthin. *J. Phycol.* 15: 457–463.

Boardman, N. K., Thorne, S. W. (1971). Sensitive fluorescence method for the determination of chlorophyll *a*/chlorophyll *b* ratios. *Biochim. Biophys. Acta*, 253: 222–231.

Bonnett, R., Mallams, A.K., Spark, A. A., Tee, J. L., Weedon, B. C. L., McCormick, A. (1969). Carotenoids and related compounds. Part XX. Structure and reactions of fucoxanthin. *J. Chem. Soc.* C: 429–454.

Brown, S. R. (1968). Absorption coefficients of chlorophyll derivatives. *J. Fish. Res. Bd. Can.* 25: 523–540.

Chapman, D. J. (1966). Three new carotenoids isolated from algae. *Phytochem.* 5: 1331–1333.

Chapman, D. J., Haxo, F. T. (1963). Identity of ε-carotene and ε_1-carotene. *Plant Cell Physiol.* 4: 57–63.

Cheng, J. Y., Don-Paul, M., Antia, N. J. (1974). Isolation of an unusually stable *cis*-isomer of alloxanthin from a bleached autolysed culture of *Chroomonas salina* grown photoheterotrophically on glycerol. Observations of *cis-trans* isomerization of alloxanthin. *J. Protozool.* 21: 761–768.

Chereskin, B. M., Castelfranco, P. A., Dallas, J. L., Straub, K. M. (1983). Mg-2,4–divinyl pheoporphyrin a_5: The product of a reaction catalysed *in vitro* by developing chloroplasts. *Arch. Biochem. Biophys.* 266: 10–18.

Chisholm, S. W., Frankel, S. L., Goericke, R., Olson, R. J., Palenik, B., Waterbury, J. B., West-Johnsrud, L., Zettler, E. R. (1992). *Prochlorococcus marinus* nov. gen. nov. sp: An oxytrophic procaryote containing divinyl-chlorophyll *b*. *Arch. Microbiol.* 157: 297–300.

Chisholm, S. W., Olson, R. J., Zettler, E. R., Goericke, R., Waterbury, J. B, Welschmeyer, N. A. (1988). A novel free-living prochlorophyte abundant in the oceanic euphotic zone. *Nature*, 334: 340–343.

Cholnoky, L., Györgyfy, K., Szabolcs, J., Weedon, B. C. L., Waight, E. S. (1966). Foliaxanthin. *Chem. Comm.* 404–405.

Davies, B. H. (1965). Analysis of carotenoid pigments. In: Goodwin, T. W. (ed.), *Chemistry and Biochemistry of Plant Pigments*. Academic Press, London, pp. 489–532.

Davies, B. H. (1976). Carotenoids. In: Goodwin, T.W., (ed.), *Chemistry and Biochemistry of Plant Pigments,* 2nd edn., vol. 2 Academic Press, London, p. 38–165.

Fawley, M. W., Lee, C. M. (1990). Pigment composition of the scaly green flagellate, *Mesostigma viride* (Micromonadophyceae) is similar to that of the siphonous green alga, *Bryopsis plumosa* (Ulvophyceae). *J. Phycol.* 26: 666-670.

Fiksdahl, A., Bjørnland, T., Liaaen-Jensen, S. (1984*a*). Algal carotenoids with novel end groups. *Phytochem.* 23: 649–655.

Fiksdahl, A., Withers, N., Guillard, R. R. L., Liaaen-Jensen, S. (1984*b*). Carotenoids of the Raphidophyceae — a chemosystematic contribution. *Comp. Biochem. Physiol.* 78B: 265–271.

Fookes, C. J. R., Jeffrey, S. W. (1989). The structure of chlorophyll c_3, a novel marine photosynthetic pigment. *J. Chem. Soc. Chem. Comm.* 23: 1827–1828.

Foppen, F. H. (1971). Tables for the identification of carotenoid pigments. *Chromatogr. Rev.* 14: 133–298.

Foss, P., Guillard, R. R. L., Liaaen-Jensen, S. (1984). Prasinoxanthin — a chemosystematic marker for algae. *Phytochem.* 23: 1629–1633.

Foss, P., Guillard, R. R. L., Liaaen-Jensen, S. (1986). Carotenoids from eucaryotic ultraplankton clones (Prasinophyceae). *Phytochem.* 25: 119–124.

Francis, G. W., Halfen, L. N. (1972). γ-Carotene and lycopene in *Oscillatoria princeps. Phytochem.* 11: 2347–2348.

Francis, G. W., Hertzberg, S., Upadhyay, R. R., Liaaen-Jensen, S. (1972). The structures of animal carotenoids VI. *Acta. Chem. Scand.* 26: 1097–1104.

Gieskes, W. W. C. (1991). Algal pigment fingerprints: clue to taxon-specific abundance, productivity and degradation of phytoplankton in seas and oceans. In: Demers, S. (ed.), *Particle Analysis in Oceanography*, NATO ASI Series, Vol. G27 Springer-Verlag, Berlin, p. 61–99.

Gieskes, W. W. C., Kraay, G. W. (1983). Unknown chlorophyll *a* derivatives in the North Sea and the tropical Atlantic Ocean revealed by HPLC analysis. *Limnol. Oceanogr.* 28: 757–766.

Goericke, R., Repeta, D. J. (1992). The pigments of *Prochlorococcus marinus*: the presence of divinyl chlorophyll *a* and *b* in a marine procaryote. *Limnol. Oceanogr.* 37: 425–433.

Goodwin, T. W. (1955). Carotenoids. In: Paech K., Tracey, M. V. (eds), *Modern Methods of Plant Analysis,* vol. 3 Springer, Berlin, pp. 272–311.

Goodwin, T. W. (1980). *The Biochemistry of the Carotenoids*. vol. 1, Plants, 2nd edn. Chapman and Hall, London, 377 pp.

Griffiths, W. T. (1991). Protochlorophyllide photoreduction. In: Scheer, H. (ed.), *Chlorophylls*. CRC Press, Boca Raton, Florida, pp. 433–449.

Guillard, R. R. L., Murphy, L. S., Foss, P., Liaaen-Jensen, S. (1985). *Synechococcus* spp. as likely zeaxanthin-dominant ultraphytoplankton in the North Atlantic. *Limnol. Oceanogr.* 30: 412–414.

Hager, A., Meyer-Bertenrath, T. (1966). Extraction and quantitative determination of carotenoids and chlorophylls of leaves, algae and isolated chloroplasts with the aid of thin-layer chromatography. *Planta*, 69: 198–217.

Hager, A., Stransky, H. (1970*a*). The carotenoid pattern and the occurrence of the light-induced xanthophyll cycle in various classes of algae. I. Methods for the identification of pigments. *Arch. Mikrobiol.* 71: 132–163.

Hager, A., Stransky, H. (1970*b*). The carotenoid pattern and the occurrence of the light-induced xanthophyll cycle in various classes of algae. III Green algae. *Arch. Mikrobiol.* 72: 68–83.

Hager, A., Stransky, H. (1970*c*). The carotenoid pattern and the occurrence of the light-induced xanthophyll cycle in various classes of algae. V. A few members of Cryptophyceae, Euglenophyceae, Bacillariophyceae, Chrysophyceae and Phaeophyceae. *Arch. Mikrobiol.* 73: 77–89.

Haugan, J. A., Liaaen-Jensen, S. (1989). Improved isolation procedure for fucoxanthin. *Phytochem.* 28: 2797–2798.

Haxo, F. T. (1985). Photosynthetic action spectrum of the coccolithophorid, *Emiliania huxleyi* (Haptophyceae): 19'-hexanoyloxyfucoxanthin as antenna pigment. *J. Phycol.* 21: 282–287.

Hertzberg, S., Liaaen-Jensen, S. (1967). Bacterial carotenoids. XX. The carotenoids of *Mycobacterium phlei* strain Vera. 2. The structures of the Phlei–xanthophylls — two novel tertiary glucosides. *Acta Chem. Scand.* 21: 15–41.

Hertzberg, S., Mortensen, T., Borch, G., Siegelman, H.W., Liaaen-Jensen, S. (1977). On the absolute configuration of 19'-hexanoyloxyfucoxanthin. *Phytochem.* 16: 587–590.

Hiyama, T., Nishimura, M., Chance, B. (1969). Determination of carotenes by thin-layer chromatography. *Anal. Biochem.* 29: 339–350.

Hooks, C. E., Bidigare, R. R., Keller, M. D., Guillard, R. R. L. (1988). Coccoid eucaryotic ultraplankters with four different HPLC pigment signatures. *J. Phycol.* 24: 571–580.

Hynninen, P. H. (1973). Chlorophylls IV. Preparation and purification of some derivatives of chlorophylls *a* and *b*. *Acta Chem. Scand.* 27: 1771–1780.

Hynninen, P. H. (1991). Chemistry of chlorophylls: modifications. In: Scheer, H. (ed.), *Chlorophylls.* CRC Press, Boca Raton, Florida, pp. 145–209.

Hynninen, P. H., Lötjönen, S. (1980). Preparation of phorbin derivatives from chlorophyll mixture utilizing the principle of selective hydrolysis. *Synthesis*, 7: 539–541.

Hynninen, P. H., Lötjönen, S. (1983). Large-scale preparation of crystalline (10*S*) – chlorophylls *a* and *b*. *Synthesis*, 9: 705–708.

Isler, O. (1971). *Carotenoids*, Birkhäuser, Basle, 932 pp.

Isler, O., Lindlar, H., Montavon, M., Rüegg, R., Zeller, P. (1956). Synthesen in der Carotinoid-Reihe. 1. Die technische Synthese von ß-Carotin. *Helv. Chim. Acta*, 39: 249–259.

Isler, O., Schudel, P. (1963). Synthese und Markierung von Carotinen und Carotinoiden. *Wiss. Veröff. DT. Ges. Ernähr.* 9: 54–104.

IUPAC Commission (1972). Tentative rules for the nomenclature of carotenoids. *Biochim. Biophys. Acta*, 286: 217–242.

Jeffrey, S. W. (1968*a*). Quantitative thin-layer chromatography of chlorophylls and carotenoids from marine algae. *Biochim. Biophys. Acta*, 162: 271–285.

Jeffrey, S. W. (1968*b*). Pigment composition of siphonales algae in the brain coral *Favia*. *Biol. Bull.* 135: 141–148.

Jeffrey, S. W. (1969). Properties of two spectrally different components in chlorophyll *c* preparations. *Biochim. Biophys. Acta*, 177: 456–467.

Jeffrey, S. W. (1972). Preparation and some properties of crystalline chlorophyll c_1 and c_2 from marine algae. *Biochim. Biophys. Acta*, 279: 15–33.

Jeffrey, S. W. (1976). The occurrence of chlorophylls c_1 and c_2 in algae. *J. Phycol.* 12: 349–354.

Jeffrey, S. W. (1989). Chlorophyll *c* pigments and their distribution in the chromophyte algae. In: Green, J. C., Leadbeater, B. S. C., Diver, W. L. (eds), *The Chromophyte Algae: Problems and Perspectives*. Clarendon Press, Oxford, pp. 13–36.

Jeffrey, S. W., Hallegraeff, G. M. (1987). Chlorophyllase distribution in 10 classes of phytoplankton — a problem for chlorophyll analysis. *Mar. Ecol. Prog. Ser.* 35: 293–304.

Jeffrey, S. W., Haxo, F. T. (1968). Photosynthetic pigments of dinoflagellates (zooxanthellae) from corals and clams. *Biol. Bull.* 135: 149–165.

Jeffrey, S. W., Humphrey, G. F. (1975). New spectrophotometric equations for determining chlorophylls *a, b,* c_1 and c_2 in higher plants, algae and natural phytoplankton. *Biochem. Physiol. Pflanzen.* 167: 191–194.

Jeffrey, S. W., Sielicki, M., Haxo, F. T. (1975). Chloroplast pigment patterns in dinoflagellates. *J. Phycol.* 11: 374–384.

Jeffrey, S. W., Wright, S. W. (1987). A new spectrally distinct component in preparations of chlorophyll *c* from the microalga *Emiliania huxleyi* (Prymnesiophyceae). *Biochim. Biophys. Acta*, 894: 180–188.

Jeffrey, S. W., Wright, S. W. (1994). Photosynthetic pigments in the Haptophyta. In: Green, J. C., Leadbeater, B. S. C. (eds), *The Haptophyte Algae*. Clarendon Press, Oxford, p. 111–132.

Jensen, A. (1966*a*). Algal carotenoids V. Iso-fucoxanthin — a rearrangement product of fucoxanthin. *Acta Chem. Scand.* 6: 1728–1730.

Jensen, A. (1966*b*). Carotenoids of Norwegian brown seaweeds and of seaweed meals. Norwegian Inst. of Seaweed Research, Report No. 31, 138 pp.

Johansen, J. E., Svec, W. A., Liaaen-Jensen, S., Haxo, F. T. (1974). Carotenoids of the Dinophyceae. *Phytochem.* 13: 2261–2271.

Jungalwala., F. B., Cama, H. R. (1962). Carotenoids in *Delonix regia* (Gul Mohr) flower. *Biochem.* J. 85: 1–8.

Karrer, P., Jucker, E. (1948). *Carotinoide*, Birkhäuser, Basle, 388 pp.

Katayama, T., Hirata, K., Yokoyama, H., Chichester, C. O. (1970). The biosynthesis of astaxanthin. III. The carotenoids in sea breams. *Bull. Jpn. Soc. Scient. Fish.* 36: 709–714.

Katayama, T., Katama, T., Chichester, C. O. (1972). The biosynthesis of astaxanthin. VI. The carotenoids in the prawn, *Penaeus japonicus* Bate (Part II). *Int. J. Biochem.* 3: 363–368.

Ke, B., Imsgard, F., Kjøsen, H., Liaaen-Jensen, S. (1970). Electronic spectra of carotenoids at 77°K. *Biochim. Biophys. Acta*, 210: 139–152.

Keely, B. (1989). *Early diagenesis of chlorophyll and chlorin pigments*. PhD thesis, Organic Geochemistry Unit, School of Chemistry, University of Bristol, UK, 212 pp.

Köst, H.-P. (ed.) 1988. *CRC Handbook of Chromatography: Plant Pigments*. vol. 1. Fat-soluble Pigments, CRC Press, Boca Raton, Florida, 328 pp.

Krinsky, N. I., Mathews-Roth, M. M., Taylor, R. F.(1989). *Carotenoids: Chemistry and Biology*. Plenum Press, New York, 382 pp.

Liaaen-Jensen, S. (1971). Isolation, Reactions. In: Isler, O. (ed.), *Carotenoids*. Birkhäuser, Basle, pp. 61–188.

Liaaen-Jensen, S. (1990*a*). Artifacts of natural carotenoids — unintended carotenoid synthesis. In: Krinsky, N. I., Mathews-Roth, M. M., Taylor, R. F. (eds), *Carotenoids: Chemistry and Biology*. Plenum Press, New York, pp. 149–165.

Liaaen-Jensen, S. (1990*b*). Marine carotenoids — selected topics. *New J. Chem.* 14: 747–759.

Lichtenthaler, H. K. (1987). Chlorophylls and carotenoids: pigments of photosynthetic biomembranes. *Methods Enzymol.* 148: 350–382.

Loeblich, A. R., Smith, V. E. (1968). Chloroplast pigments of the marine dinoflagellate *Gyrodinium resplendens*. *Lipids*, 3: 5–13.

Lorenzen, C. J., Jeffrey, S. W. (1980). Determination of chlorophyll in seawater. UNESCO *Technical Papers in Marine Science*, No. 35, 20 pp.

Lötjönen, S., Hynninen, P. H. (1983). An improved method for the preparation of (10*S*) – and (10*S*) – pheophytins *a* and *b*. *Synthesis*, 708–710.

Mallams, A. K., Waight, E. S., Weedon, B. C. L., Cholnoky, L., Györgyfy, K., Szabolcs, J., Krinsky, N. I., Schimmer, B. P., Chichester, C. O., Katayama, T., Lowry, L., Yokoyama, H. (1967). The identity of neoxanthin and foliaxanthin. *Chem. Comm.* p. 484–485.

Manchand, P. S., Rüegg, R., Schwieter, U., Siddons, P. T., Weedon, B. C. L. (1965). Carotenoids and related compounds. Part XI. Synthesis of δ-carotene and ε-carotene. *J. Chem. Soc.*: 2019–2026.

Mantoura, R. F. C., Llewellyn, C. A. (1983). The rapid determination of algal chlorophyll and carotenoid pigments and their breakdown products in natural waters by reverse-phase high-performance liquid chromatography. *Anal. Chim. Acta,* 151: 297–314.

Nelson, J. R., Wakeham, S. G. (1989). A phytol-substituted chlorophyll *c* from *Emiliania huxleyi* (Prymnesiophyceae). *J. Phycol.* 25: 761–766.

Nichols, B. W. (1973). Lipid composition and metabolism. In: Carr, N. G., Whitton, B. A. (eds), *The Biology of Blue-green Algae*. Blackwells, Oxford, pp. 144–161.

Norgård, S., Svec, W. A., Liaaen-Jensen, S., Jensen, A., Guillard, R. R. L. (1974). Chloroplast pigments and algal systematics. *Biochem. Syst. Ecol.* 2: 3–6.

Pennington, F. C., Haxo, F. T., Borch, G., Liaaen-Jensen, S. (1985). Carotenoids of Cryptophyceae. *Biochem. Syst. Ecol.* 13: 215–219.

Pennington, F. C., Strain, H. H., Svec, W. A., Katz, J. J. (1964). Preparation and properties of pyrochlorophyll *a*, methyl pyrochlorophyllide *a*, pyropheophytin *a* and methyl pyropheophorbide *a* derived from chlorophyll by decarbomethoxylation. *J. Am. Chem. Soc.* 86: 1418–1426.

Pennington, F., Guillard, R. R. L., Liaaen-Jensen, S. (1988). Carotenoid distribution patterns in the Bacillariophyceae. *Biochem. Syst. Ecol.* 16: 589–592.

Rebeiz, C. A., Wu, S. M., Kuhadja, M., Daniell, H., Perkins, E. J. (1983). Chlorophyll *a* biosynthetic routes and chlorophyll *a* chemical heterogeneity in plants. *Mol. Cell. Biochem.* 57: 97–125.

Renstrøm, B., Borch, G., Skulberg, O. M., Liaaen-Jensen (1981). Optical purity of (3*S*, 3'*S*) - astaxanthin from *Haematococcus fluvialis*. *Phytochem.* 20: 2561–2564.

Ricketts, T. R. (1966). Magnesium 2,4-divinyl pheoporphyrin a_5 monomethyl ester, a protochlorophyll-like pigment present in some unicellular flagellates. *Phytochem.* 5: 223–229.

Ricketts, T. R. (1970). The pigments of the Prasinophyceae and related organisms. P*hytochem.* 9: 1835–1842.

Rowan, K. S. (1989). *Photosynthetic Pigments of Algae*. Cambridge Univ. Press, Cambridge, U.K., 334 pp.

Scheer, H. (ed.) (1991). *Chlorophylls*. CRC Press, Boca Raton, Florida, 1257 pp.

Schwieter, U., Bolliger, H. R., Chopard-dit-Jean, L. H., Englert, G., Kofler, M., Koenig, A., von Planta, C., Rüegg, R., Vetter, W., Isler, O. (1965). Synthesen in der Carotinoid-Reihe. 19. Mitteilung: Physikalische Eigenschaften der Carotine. *Chimia (Aarau)*, 19: 294–302.

Shioi, Y., Beale, S. I. (1987). Polyethylene-based high performance liquid chromatography of chloroplast pigments: resolution of mono- and divinyl- chlorophyllides and other pigment mixtures. *Anal. Biochem.* 162: 493–499.

Smith, J. H. C., Benitez, A. (1955). Chlorophylls: analysis in plant materials. In: Paech, K., Tracey, M. (eds), *Modern Methods of Plant Analysis*, vol. 4 Springer, Heidelberg, pp. 142–196.

Stauber, J. L., Jeffrey, S. W. (1988). Photosynthetic pigments in fifty-one species of marine diatoms. *J. Phycol.* 24: 158–172.

Strain, H. H. (1938). *Leaf xanthophylls.* Carnegie Institution of Washington, Publication No. 490, Washington, 147 pp.

Strain, H. H., Svec, W. A., Aitzetmüller, K., Grandolfo, M. C., Katz, J. J., Kjøsen, H., Norgård, S., Liaaen-Jensen, S., Haxo, F.T., Wegfahrt, P., Rapoport, H. (1971). The structure of peridinin, the characteristic dinoflagellate carotenoid. *J. Am. Chem. Soc.* 93: 1823–1825.

Stransky, H., Hager, A. (1970*a*). The carotenoid pattern and the occurrence of the light-induced xanthophyll cycle in various classes of algae. II. Xanthophyceae. *Arch. Mikrobiol.* 71: 164–190.

Stransky, H., Hager, A. (1970*b*). The carotenoid pattern and the occurrence of the light-induced xanthophyll cycle in various classes of algae. IV. Cyanophyceae and Rhodophyceae. *Arch. Mikrobiol.* 72: 84–96.

Stransky, H., Hager, A. (1970*c*). The carotenoid pattern and the occurrence of the light-induced xanthophyll cycle in various classes of algae. VI. Chemosystematic study. *Arch. Mikrobiol.* 73: 315–323.

Straub, O. (1987). 2nd edn. *Key to Carotenoids.* Pfander, H., Gerspacher, M., Rychener, M., Schwabe, R. (eds), Birkhäuser, Basel, 296 pp.

Valadon, L. R. G., Mummery, R. S. (1967). Carotenoids of certain Compositae flowers. *Phytochem.* 6: 983–988.

Vernon, L. P., Seely, G. R. (eds) (1966). *The Chlorophylls: Physical, Chemical and Biological Properties.* Academic Press, New York, 679 pp.

Vesk, M., Jeffrey, S. W. (1987). Ultrastructure and pigments of two strains of the picoplanktonic alga, *Pelagococcus subviridis* (Chrysophyceae). *J. Phycol.* 23: 322–336.

Vetter, W., Englert, G., Rigassi, N., Schwieter, U., (1971). Spectroscopic methods. In: Isler, O. (ed.), *Carotenoids.* Birkhäuser, Basle, pp. 189–266.

Volkman, J. K., Brown, M. R., Dunstan, G. A., Jeffrey, S. W. (1993). The biochemical composition of marine microalgae from the class Eustigmatophyceae. *J. Phycol.* 29: 69–78.

Warren, C. K., Weedon, B. C. L. (1958). Carotenoids and related compounds. Part VII. Synthesis of canthaxanthin and echinenone. *J. Chem. Soc.* 3986–3993.

Whittaker, R. H., Likens, G. E. (1975). The biosphere and man. In: Leith, H., Whittaker, R. H. (eds), *Primary Productivity of the Biosphere.* Springer, Berlin, pp. 305–328.

Whittle, S.J., Casselton, P.J. (1975). The chloroplast pigments of the algal classes Eustigmatophyceae and Xanthophyceae. I. Eustigmatophyceae. *Br. Phycol. J.* 10: 179–191.

Withers, N. W., Cox, E. R., Tomas, R., Haxo, F. T. (1977). Pigments of the dinoflagellate *Peridinium balticum* and its photosynthetic endosymbiont. *J. Phycol.* 13: 354–358.

Withers, N. W., Fiksdahl, A., Tuttle, R. C., Liaaen-Jensen, S. (1981). Carotenoids of the Chrysophyceae. *Comp. Biochem. Physiol.* 68: 345–349.

Wright, S. W., Jeffrey, S. W. (1987). Fucoxanthin pigment markers of marine phytoplankton analysed by HPLC and HPTLC. *Mar. Ecol. Progr. Ser.* 38: 259–266.

Wright, S. W., Jeffrey, S. W., Mantoura, R. F. C., Llewellyn, C. A., Bjørnland, T., Repeta, D., Welschmeyer, N. (1991). Improved HPLC method for the analysis of chlorophylls and carotenoids from marine phytoplankton. *Mar. Ecol. Progr. Ser.* 77, 183–196.

Yokohama, Y., Kageyama, A., Ikawa, T., Shimura, S. (1977). A carotenoid characteristic of Chlorophycean seaweeds living in deep coastal waters. *Botanica Marina*, 20: 433–436.

Zechmeister, L. (1944). *Cis-trans* isomerization and stereochemistry of carotenoids and diphenylpolyenes.*Chem. Rev.* 34: 267–344.

Part V

Appendices

Part V

Appendices

Appendix A

Pigment abbreviations used by SCOR WG 78

Compiled by S. W. Jeffrey and R. F. C. Mantoura

Pigment	Abbreviation
Chlorophylls	
Chlorophyll *a, b,* c_1, c_2 . . .	Chl *a, b,* c_1, c_2 . . .
Chlorophyll *a* allomers	Chl *a* allom
Chlorophyll *a* epimer	Chl *a*'
Chlorophyllide *a, b*	Chlide *a, b*
Methyl chlorophyllide *a, b*	Me-chlide *a, b* . . .
Ethyl chlorophyllide *a, b*	Et-chlide *a, b* . . .
Pheophytin *a, b, c* . . .	Phytin *a, b, c* . . .
Pheophorbide *a, b, c* . . .	Phide *a, b* . . .
Pyropheophytin *a, b*	Pyrophytin *a, b* . . .
Divinyl chlorophyll *a, b*	DV chl *a, b*
Phytylated chlorophyll *c*-like	Phytyl-chl *c*
Mg 2,4 divinyl pheoporphyrin a_5 monomethyl ester	Mg DVP
Carotenoids	
Alloxanthin	Allo
Antheraxanthin	Anth
Astaxanthin	Asta
19'-Butanoyloxyfucoxanthin	But-fuco
Canthaxanthin	Cantha
β,ε-Carotene	βε-Car
β,β-Carotene	ββ-Car
β,ψ-Carotene	βψ-Car
ε,ε-Carotene	εε-Car
Crocoxanthin	Croco
Diadinochrome	Diadchr
Diadinoxanthin	Diadino
Diatoxanthin	Diato
Dinoxanthin	Dino
Echinenone	Echin
Ethyl-apo-8'-carotenoate	Et-8-carot
Fucoxanthin	Fuco
19'-Hexanoyloxyfucoxanthin	Hex-fuco
Lutein	Lut

Lycopene	Lyco
Monadoxanthin	Monado
Neoxanthin	Neo
P-457	P-457
Peridinin	Perid
Prasinoxanthin	Pras
Siphonaxanthin	Siphx
Siphonein	Siphn
Vaucheriaxanthin-ester	Vauch
Violaxanthin	Viola
Zeaxanthin	Zea

Note: For *cis*- and *trans*- prefixes use *c*- and *t*- before abbreviation, e.g. *cis*-Fucoxanthin = *c*-Fuco; *cis*-β,β-Carotene = *c*-ββ-Car.

Appendix B

Structural relationships between algal chlorophylls

S. W. Jeffrey

B.1 INTRODUCTION

Algal and bacterial chlorophylls are magnesium coordination complexes of cyclic tetrapyrroles containing a fifth isocyclic ring, ring E (see Fig. B.2), and they possess the oxidation state of a porphyrin, chlorin or bacteriochlorin (Scheer, 1991; Fig. B.1; and Chapter 3). Chlorophylls *a* and *b* are chlorins with ring D reduced (Fig. B.1B), the chlorophyll *c* family are porphyrins with ring D fully unsaturated (Fig. B.1A), and the bacteriochlorophylls are either chlorins (Fig. B.1B) with ring D reduced (bacteriochlorophylls *c, d* and *e*) or are bacteriochlorins (Fig. B.1C) with both rings B and D reduced (bacteriochlorophylls *a, b* and *g*). For a review of the bacteriochlorophylls see Scheer (1991).

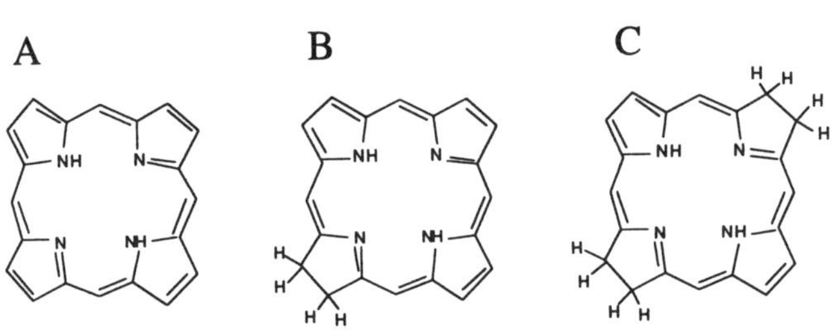

Figure B.1
Oxidation states of the tetrapyrrole macrocycle. A, porphyrin; B, 17,18-dihydroporphyrin (= chlorin); C, 7,8, 17',18'-tetrahydroporphyrin (= bacteriochlorin) (after Scheer, 1991).

The metabolism and function of chlorophylls and derivatives are discussed in Chapter 3, and chemical and spectroscopic data of 17 important chlorophylls and derivatives are listed in the Part IV Data and Graphics sheets. In this Appendix the structures of important algal chlorophylls and derivatives are presented side by side, so that the small structural differences between them are clearly visible.

B.2 Comparison of structures

Chlorophylls and derivatives important in phytoplankton ecology are shown in Figs. B.2 and B.3. The chlorophyll *a*, *b* and *c* series are shown in Fig. B.2. The structures of chlorophylls *a* and *b* were confirmed by total synthesis (Woodward, 1960), and those of chlorophyll c_1 and c_2 (originally separated by Jeffrey, 1969) were established by NMR and mass spectroscopy (Dougherty *et al.* 1970; Budzikiewicz and Taraz, 1971). The structure of chlorophyll c_3 was proposed from data obtained by nuclear magnetic resonance (NMR), nuclear Overhauser enhancement (NOE) and mass spectroscopy of crystalline preparations (Fookes and Jeffrey, 1989). The structures of two other chlorophyll *c* pigments — chlorophyll $c_{\mathrm{CS\text{-}170}}$ and the phytylated chlorophyll *c*-like pigment from prymnesiophytes — are tentative only. The nature of the ring substituents of chlorophyll $c_{\mathrm{CS\text{-}170}}$ has been established from studies of crystalline preparations (Fookes and Jeffrey, unpublished), but their location, suggested by analogy with other pigments, needs confirmation. Similarly, the stereochemistry at the only asymmetric carbon atom, $C13^2$, is probably analogous to that observed in chlorophylls *a* and *b* but has not been confirmed experimentally. The structure of the phytylated chlorophyll *c*-like pigment (Nelson and Wakeham, 1989) has not yet been determined by rigorous chemical analysis. These molecular structures of the chlorophyll *c* family are presented as the best evidence available in late 1995.

The structure of epimers, pyro-derivatives and some allomers are illustrated for the chlorophyll *a* series in Fig. B.3. A complete description of the origin of these derivatives may be found in Chapter 3 and Hynninen (1991).

The gross structural features of important algal chlorophylls and their derivatives are compared in Table B.1, and a list of some macrocycle substituents is given in Table B.2. Again, it should be noted that the structures given for chlorophyll $c_{\mathrm{CS\text{-}170}}$ and the phytylated chlorophyll *c*-like pigment are tentative only.

B.3 Distribution

The phylogenetic distribution of the algal chlorophylls as currently known is discussed in Chapter 2, and summarized in Table 2.3.

References

Budzikiewicz, H., Taraz, K. (1971). Chlorophyll *c*. *Tetrahedron Lett.* 27: 1447–1460.

Dougherty, R. C., Strain, H. H., Svec, W. A., Uphaus, R. A., Katz, J. J. (1970). The structure, properties and distribution of chlorophyll *c*. *J. Am. Chem. Soc.* 92: 2826–2833.

Fookes, C. J. R., Jeffrey, S. W. (1989). The structure of chlorophyll c_3, a novel marine photosynthetic pigment. *J. Chem. Soc. Chem. Comm.* 23: 1827–1828.

Chlorophyll *a* and derivatives

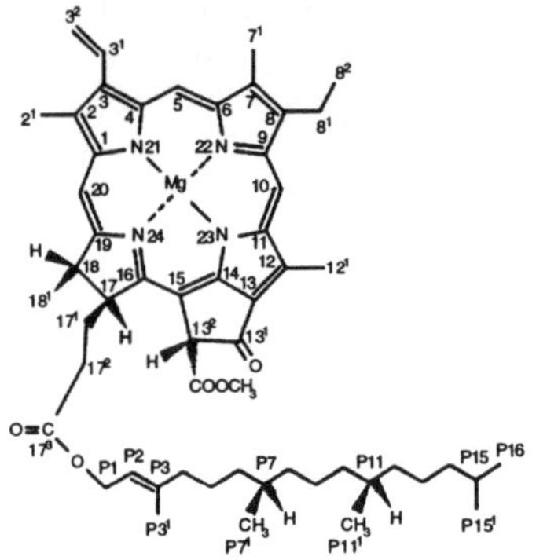

Chlorophyll *a*

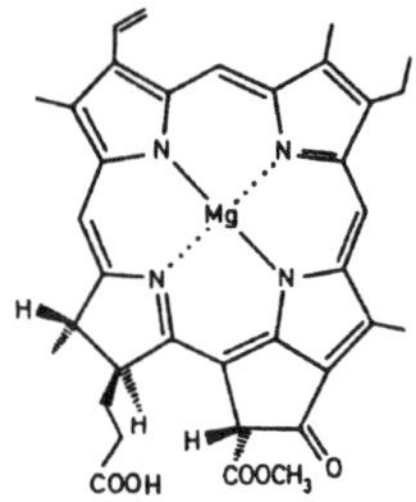

Chlorophyllide *a*

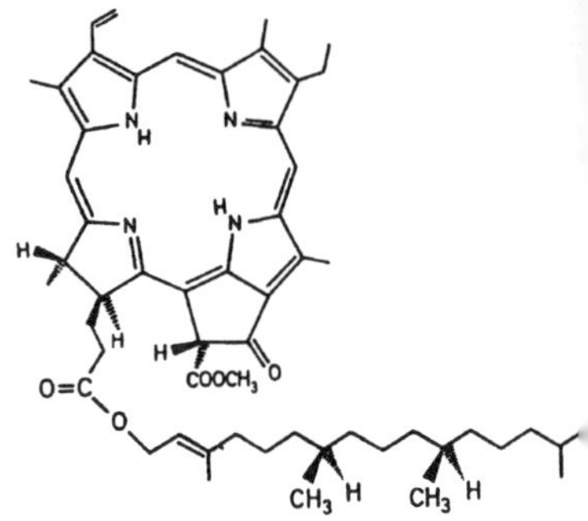

Pheophytin *a*

Chlorophyll *b* and derivatives

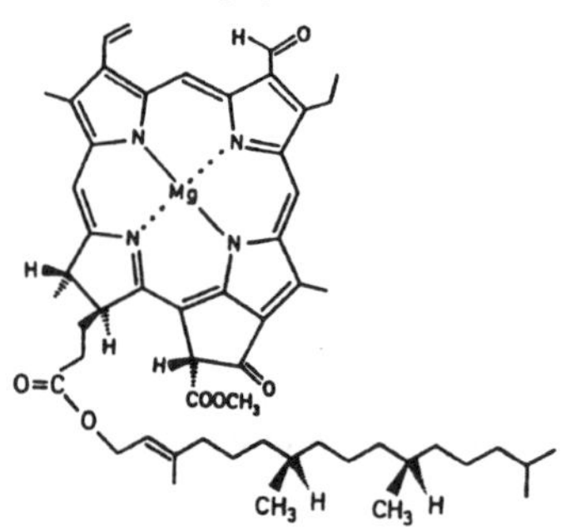

Chlorophyll *b*

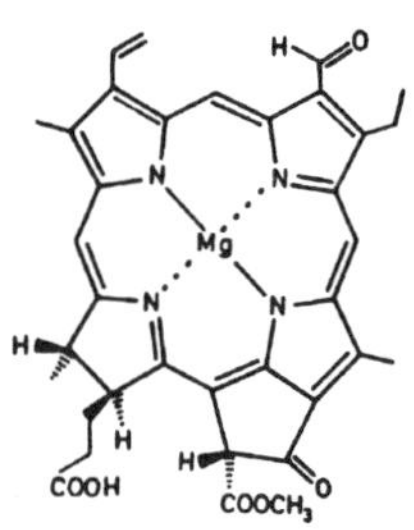

Chlorophyllide *b*

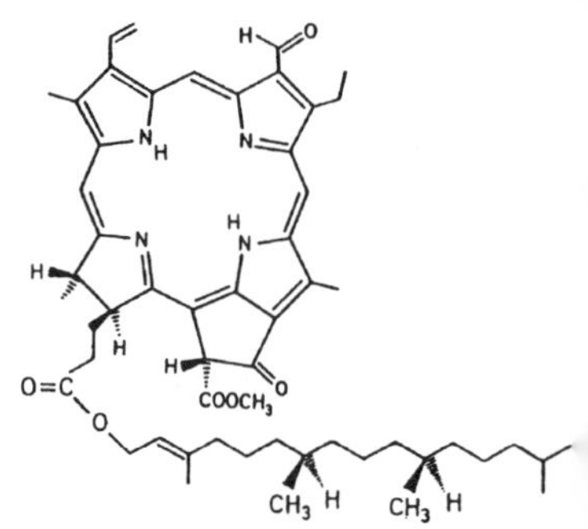

Pheophytin *b*

Chlorophyll *c* and derivatives

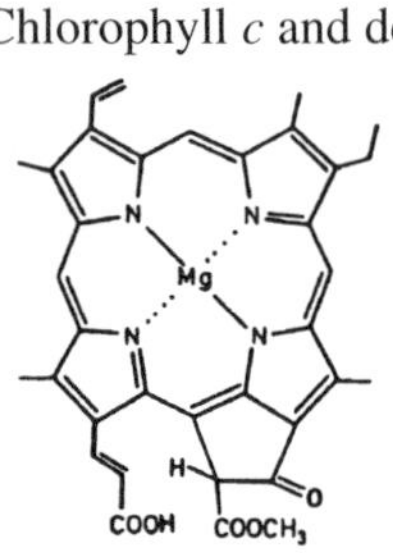

Chlorophyll c_1

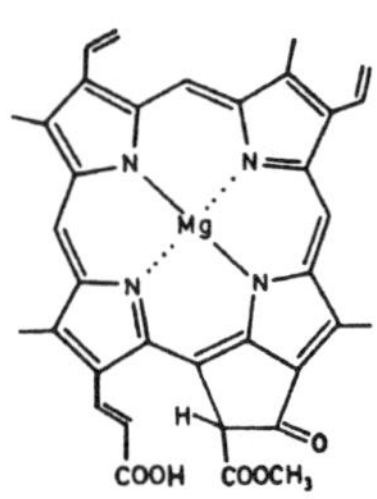

Chlorophyll c_2

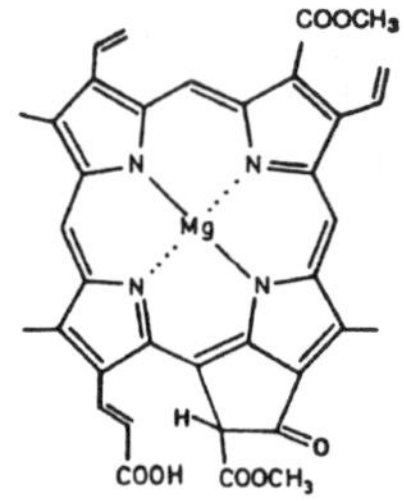

Chlorophyll c_3

Figure B.2
Structure of chlorophylls *a*, *b* and *c* and derivatives important in marine phytoplankton ecology. The structure of chlorophyll $c_{CS\text{-}170}$ is from unpublished data of Fookes and Jeffrey; that of the phytylated chlorophyll *c*-like pigment is tentative only (Nelson and Wakeham, 1989). The stereochemistry of the chlorophyll *c* pigments is unknown.

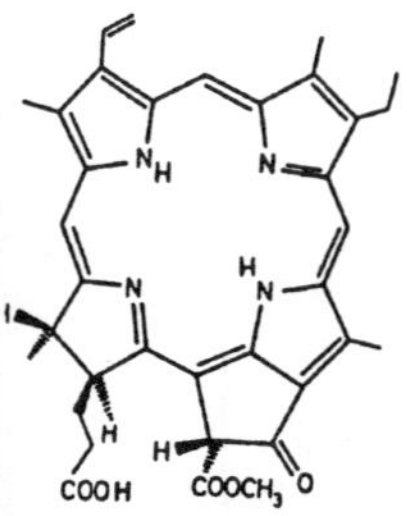

Pheophorbide *a*

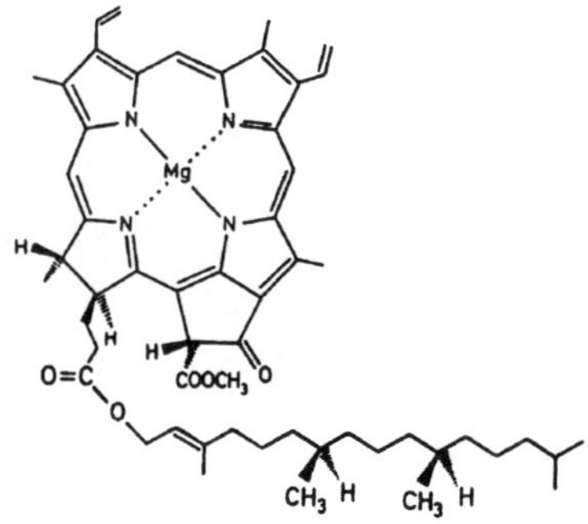

Divinyl chlorophyll *a*

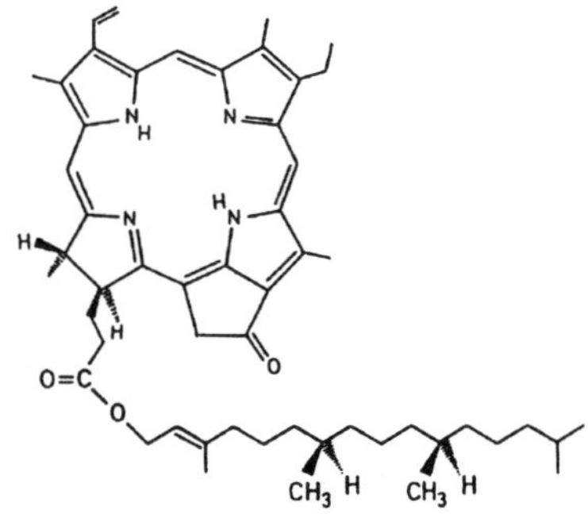

Pyropheophytin *a*

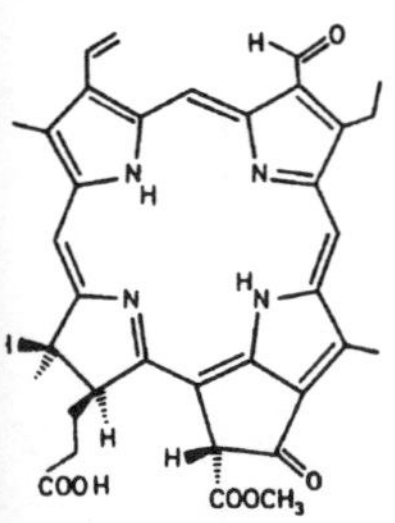

Pheophorbide *b*

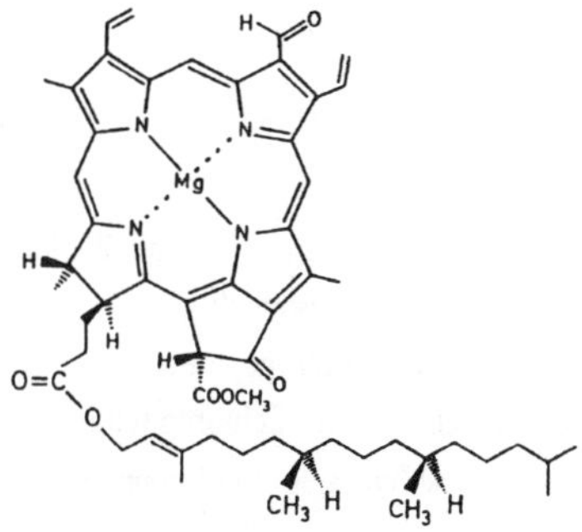

Divinyl chlorophyll *b*

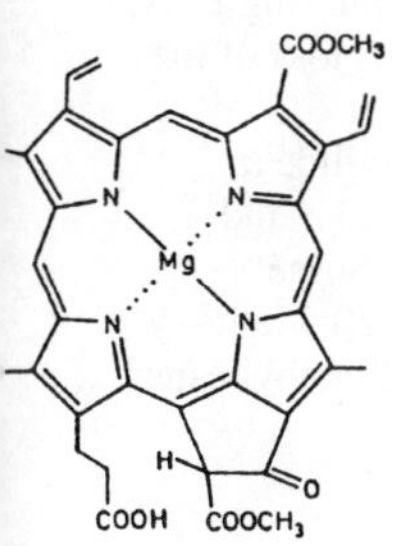

Chlorophyll $c_{\text{CS-170}}$

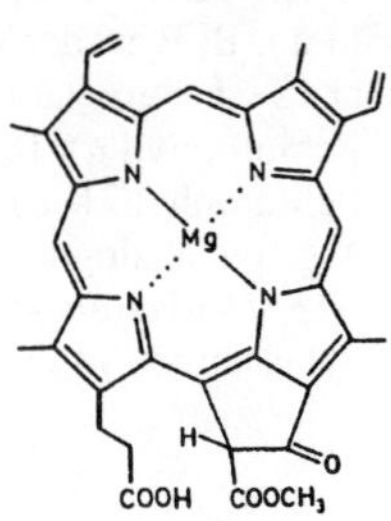

Mg 3,8 divinyl-pheoporphyrin a_5 monomethyl ester

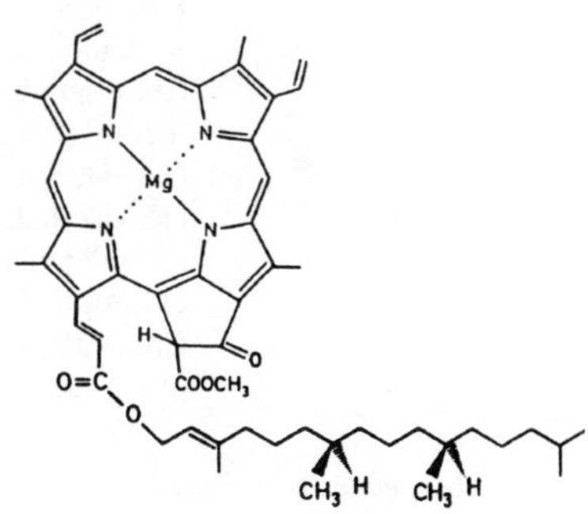

Phytylated chlorophyll *c*

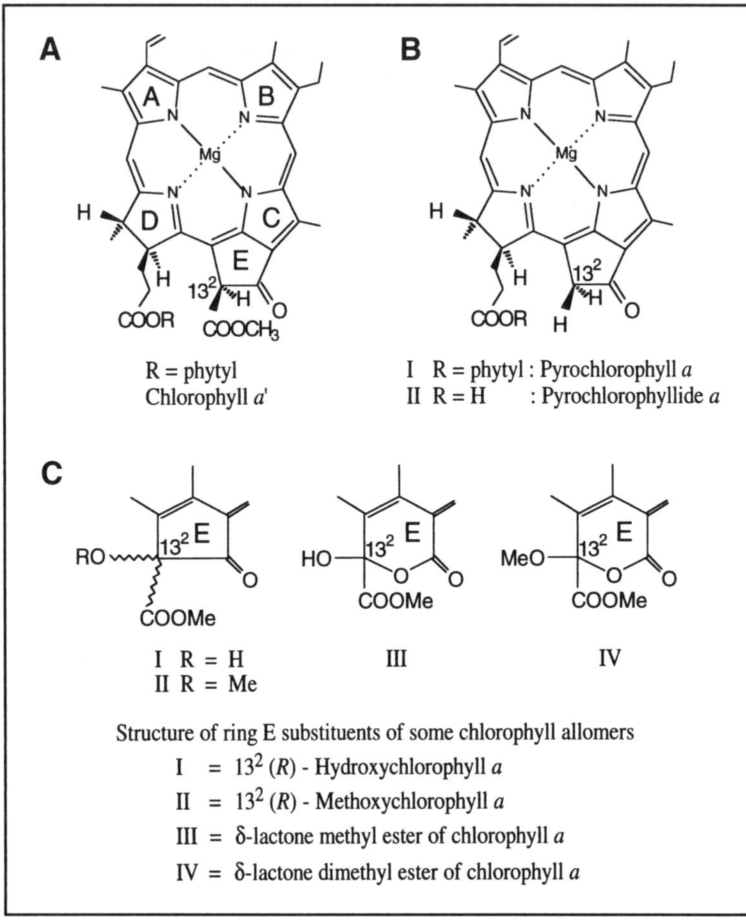

Figure B.3
Structure of chlorophyll *a* derivatives involving the isocyclic ring E: A, Chlorophyll *a* epimer (Chl a'); B, Pyro derivatives showing loss of the methoxycarbonyl group at 13^2: I, Pyrochlorophyll *a* (13^2-Demethoxycarbonyl chlorophyll *a*); II, Pyrochlorophyllide *a* (13^2-Demethoxycarbonyl chlorophyllide *a*). Pyropheophytin *a* and pyropheophorbide *a* are Mg-free analogues of I and II. C, Some chlorophyll *a* allomers: I, 13^2 (R)-Hydroxy Chl *a*; II, 13^2 (R)-Methoxy Chl *a*; III, δ-lactone methyl ester of Chl *a*; IV, δ-lactone dimethyl ester of Chl *a*.

References (continued)

Hynninen, P. H. (1991). Chemistry of chlorophylls: modifications. In: Scheer, H. (ed.), *Chlorophylls*, CRC Press, Boca Raton, p. 145–209.

Jeffrey, S. W. (1969). Properties of two spectrally different components in chlorophyll *c* preparations. *Biochim. Biophys. Acta,* 177: 456–467.

Nelson, J. R., Wakeham, S. G. (1989). A phytol-substituted chlorophyll *c* from *Emiliania huxleyi* (Prymnesiophyceae). *J. Phycol.* 25: 761–766.

Scheer, H. (1991). Structure and occurrence of chlorophylls. In: Scheer, H. (ed.), *Chlorophylls*, CRC Press, Boca Raton, p. 3–30.

Woodward, R. B. (1960). The total synthesis of chlorophyll. *Pure Appl. Chem.* 2: 383–404.

TABLE B.1. Comparison of gross structural features of some algal chlorophylls and derivatives

Pigment	Central Mg atom	Isocyclic ring E	Propionic acid side chain on C-17	Acrylic acid side chain on C-17	Ester group at C-17
Chlorophyll *a*	+	+	+		phytol
Chlorophyllide *a*	+	+	+		–
Pheophytin *a*	–	+	+		phytol
Pheophorbide *a*	–	+	+		–
Chlorophyll *b*	+	+	+		phytol
Chlorophyllide *b*	+	+	+		–
Pheophytin *b*	–	+	+		phytol
Pheophorbide *b*	–	+	+		–
Chlorophyll c_1	+	+		+	–
Chlorophyll c_2	+	+		+	–
Chlorophyll c_3	+	+		+	–
Chlorophyll $c_{CS\text{-}170}$[a]	+	+	+		–
MgDVP	+	+	+		–
Phytylated Chl*c*-like[a]	+	+		+	phytol
Epimers	+	modified[b]	+		phytol
Allomers	+	modified[c]	+		phytol
Pyro-chlorophylls	+	modified[d]	+		phytol
Pyro-chlorophyllides	+	modified[d]	+		–
Pyro-pheophytins	–	modified[d]	+		phytol
Pyro-pheophorbides	–	modified[d]	+		–

a. Tentative only
b. Steric exchange of H and $COOCH_3$ at C-13[c]
c. Various oxidative changes to ring E (see Fig. B.3; Hynninen, 1991)
d. C-13 carbomethoxy group is replaced by hydrogen

TABLE B.2. Comparison of some macrocycle substituents of algal chlorophylls

Chlorophyll	Ring A	Ring B	
	C-3	C-7	C-8
Chlorophyll *a*	vinyl	methyl	ethyl
Divinyl chl *a*	vinyl	methyl	vinyl
Chlorophyll *b*	vinyl	formyl	ethyl
Divinyl chl *b*	vinyl	formyl	vinyl
Chlorophyll c_1	vinyl	methyl	ethyl
Chlorophyll c_2	vinyl	methyl	vinyl
Chlorophyll c_3	vinyl	carbomethoxy	vinyl
Chlorophyll $c_{CS\text{-}170}$	vinyl*	carbomethoxy*	vinyl*
MgDVP	vinyl	methyl	vinyl

* Substituents assumed at these positions by analogy with chlorophyll c_3.

Appendix C

Structural relationships between algal carotenoids

T. Bjørnland

C.1 Recommendations for drawing carotenoid structures

Rules for nomenclature and drawing of carotenoid structures were approved by IUPAC in 1974 (IUPAC 1975). As they were especially designed for carotenoids, some unfortunately, deviate from the general IUPAC rules on nomenclature of organic compounds: for example, *cis/trans*-assignments around double bonds and the priority rules for unprimed *versus* primed locants (position of carbon atoms) in the molecules. To avoid confusion, the rules valid for carotenoids (IUPAC 1975) must be strictly adhered to until revised.

The orientation of the molecule projection in the paper plane is determined by the end groups. The basic end-groups so far encountered in algal carotenoids are β, γ, ε and ψ (see Fig. C.1).

Of the two end-groups in a carotenoid molecule the one appearing first in the Greek alphabet is given priority. When structures are drawn, this end-group is posi-

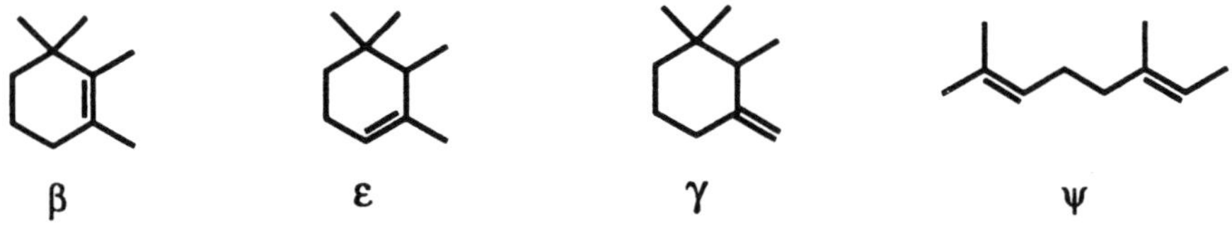

Figure C.1
Basic end-groups in algal carotenoids.

tioned to the left. In many carotenoids (e.g., echinenone), both end-groups are formally derived from the same basic end-group (see Fig. C.3 for structures). The end-group with the lowest numbering of locants (see Fig. IV.1, Preamble to Part IV for carotenoid numbering system) is then given priority and positioned to the left. Unprimed locants are then always given priority to primed locants, irrespective of the numerical value (contrary to the general IUPAC rules).

For a complete compilation of carotenoid structures and names according to IUPAC recommendations see Straub (1987).

C.2 Rotation of molecules

Small changes in the molecular structure may invert the priority of the end-groups. This is inconvenient when structural similarities or variations are to be demonstrated, as in figures illustrating biochemical pathways. In such cases it is acceptable to draw a few structures contrary to the IUPAC (1975) carotenoid rules, but this should then be noted. Deviations most commonly occur from two subsequent 180° rotations of the correctly drawn molecule (see Fig. C.2). Both axes of rotation are positioned in the paper plane — one vertical to the central polyene chain and the other parallel.

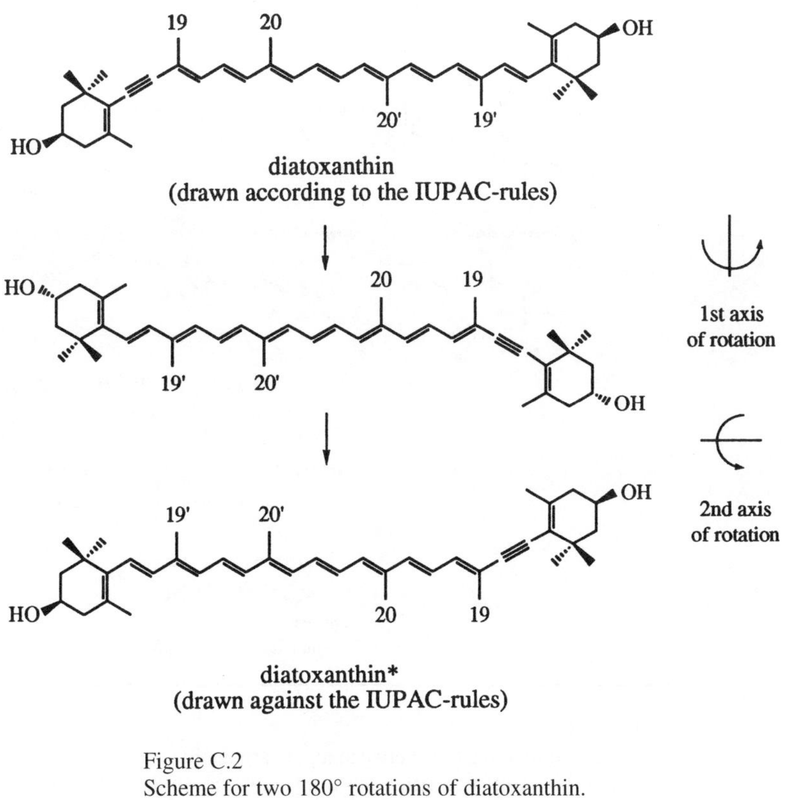

Figure C.2
Scheme for two 180° rotations of diatoxanthin.

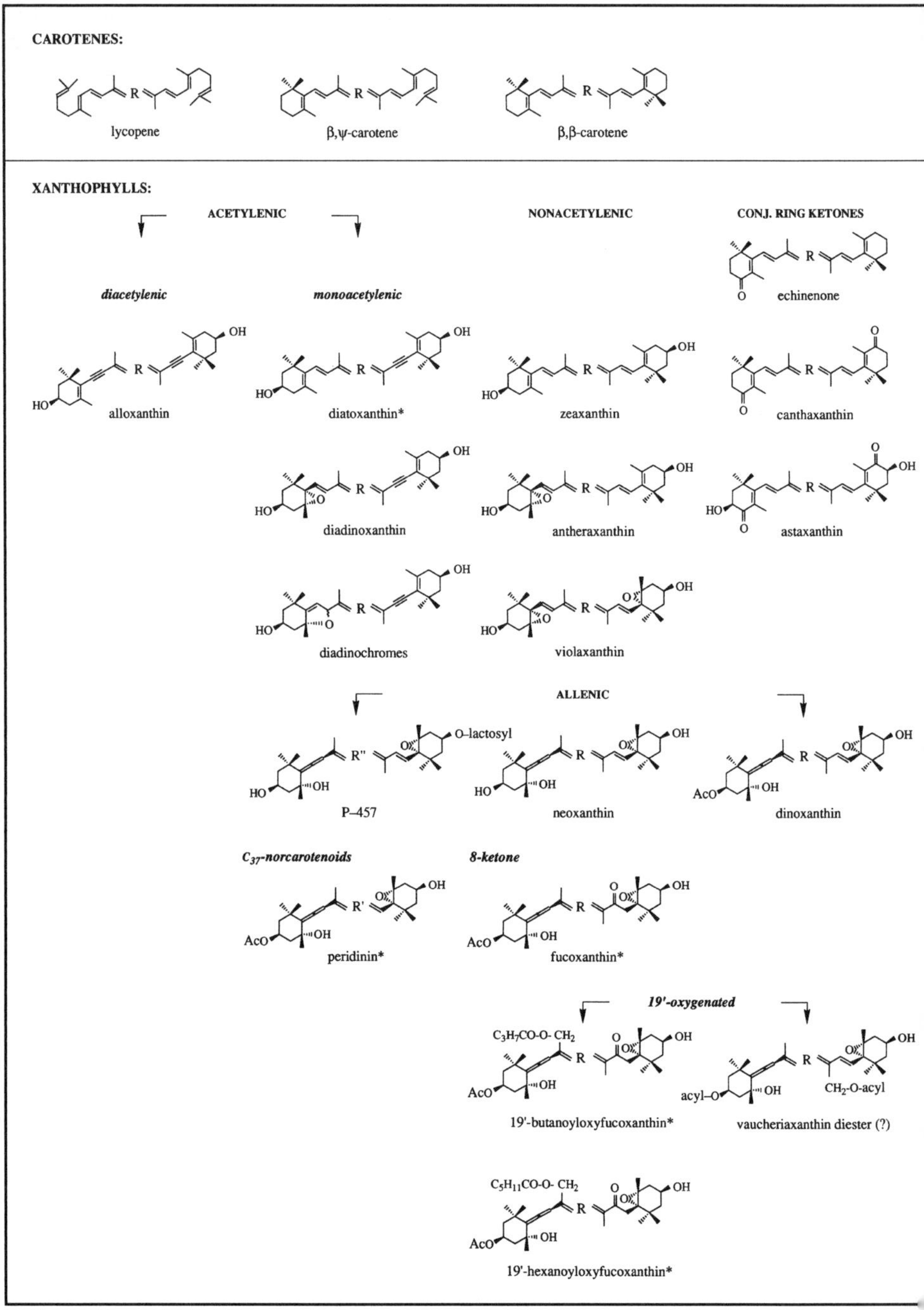

Figure C.3
Structural relationships between algal carotenoids.
*Signifies carotenoid structures drawn contrary to the IUPAC (1975) rules.

(6'R)-ε-RING

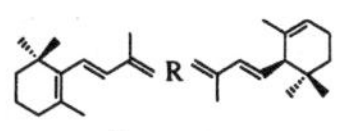

β,ε-carotene

(6S,6'S)-ε-RINGS

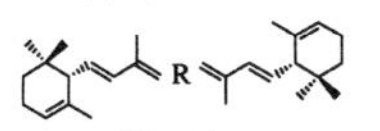

ε,ε-carotene

NONACETYLENIC

ACETYLENIC

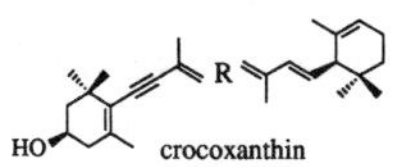

crocoxanthin

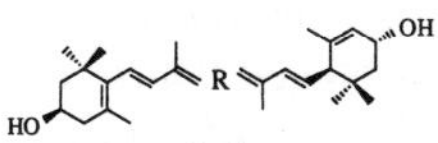

lutein

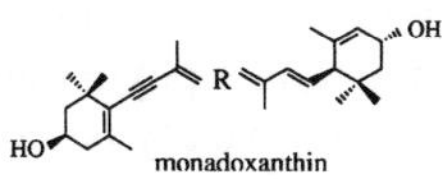

monadoxanthin

8-ketone

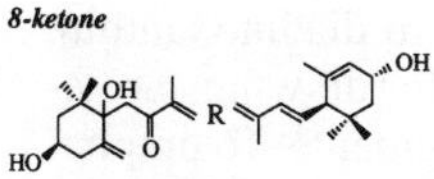

prasinoxanthin

19-oxygenated

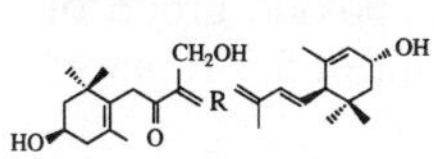

siphonaxanthin

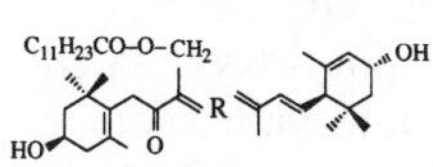

siphonein

partial central chains

R:

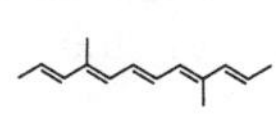

R':

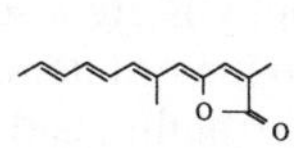

R'':

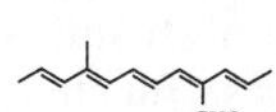

The numbering of the carbon atoms is not changed by the rotations, as shown by the four in-chain methyl groups of diatoxanthin in Fig. C.2.

In Figs C.2 and C.3 carotenoids drawn contrary to the IUPAC (1975) rules are marked with an asterisk (*).

C.3 Guidelines for structural comparisons

Fig. C.3 shows the structures of the 29 standard algal carotenoids included in Part IV. The structural relationships between them are arranged according to the following guidelines:

1. The hydrocarbons (the carotenes) are located in the upper horizontal row. Oxygenated derivatives of the carotenes (the xanthophylls) are arranged beneath. An increasing degree of oxygenation or complexity is seen downwards in the vertical columns, ending with polyoxygenated carotenoids in the lower rows.
2. Carotenoids below the carotene level are arranged into two main groups: xanthophylls that may be formally derived from β,β-carotene (four left columns) or β,ε-carotene (two right columns). Prasinoxanthin, with one γ and one ε end-group, has been included in the β,ε main group.
3. Subgroups have been arranged within each of the two main groups defined in 2 above. Several structural features that are subgroup characteristics (see headings, Fig. C.3) may be used in algal chemo-taxonomy as well (see Chapter 2).

To illustrate the use of Fig. C.3 the acetylenic structural element is used as an example. Most acetylenic carotenoids are simple dihydroxylated xanthophylls. Of these, diatoxanthin may be formally (*not* biochemically) derived from zeaxanthin by elimination of two hydrogen atoms in the 7,8-position. As far as structures are concerned, diatoxanthin and zeaxanthin constitute an analogous pair and are different in one structural characteristic only, *viz.* the 7,8-acetylenic bond. Further elimination of two hydrogen atoms from positions 7',8' leads to the diacetylenic xanthophyll alloxanthin, while an epoxidation of the endo-cyclic 5',6'-double bond leads to diadinoxanthin. In the latter case, however, the IUPAC (1975) recommendations for drawing carotenoid structures obscure the structural relationship between the pigments. To emphasize that these pigments are analogues, differing in the epoxidic structural element only, diatoxanthin has been drawn contrary to the IUPAC (1975) carotenoid rules (see section C.2 and Fig. C.2). In acidic environments the 5,6-epoxide group of diadinoxanthin will be rapidly transformed to the C-8 epimeric pair of 5,8-furanoic diadinochromes.

Acetylenic xanthophylls are also found in the β,ε main group. Monadoxanthin is a 7,8-acetylenic analogue of lutein. These two pigments are therefore interrelated in the same way as diatoxanthin and zeaxanthin in the β,β main group.

Two other structural features are often present in the poly-oxygenated carotenoids. Some of these function as accessory pigments in photosynthesis (see Chapter 3). The first of these features is a conjugated in-chain keto-group in position 8, as in fucoxanthin, the 19'-acyloxyfucoxanthins, prasinoxanthin, siphonaxanthin and siphonein. The second feature is a 19 or 19' primary alcohol function. The hydroxyl-group may be free (vaucheriaxanthin and siphonaxanthin) or esterified (19'-acyloxyfucoxanthins, siphonein and, probably, vaucheriaxanthin diester). Note that the derived in-chain methyl group of the 19'-acyloxyfucoxanthins and vaucheriaxanthin diester occupy different positions relative to the allenic bond.

The three structural elements discussed above are found in both the β,β and the β,ε main groups. The β,β group, however, has three structural features that are not found in the algal carotenoids of the β,ε group. These are a conjugated keto function in the terminal ring(s) (echinenone, canthaxanthin and astaxanthin), a reduced carbon skeleton consisting of 37 carbon atoms only (peridinin), and an allenic bond in the terminal part of the central chain (neoxanthin, dinoxanthin, P-457, fucoxanthin and the 19'-acyloxyfucoxanthins).

Among the allenic carotenoids, there is a similarity between the structures of all-*trans*-neoxanthin, dinoxanthin and P-457 (shown clearly in Fig. C.3). Using neoxanthin as the basic structure for comparison, dinoxanthin is simply the 3-acetate of that pigment. The remarkable structure of P-457 may be derived from neoxanthin by a formal oxidation of the 20' in-chain methyl group to an aldehyde function and lactosylation of the 3'-hydroxyl group.

Since the carotenoids have been grouped according to structural similarity, it is also likely that Fig. C.3 reflects biochemical connections between them. It should be noted, however, that there are very few data on carotenoid biosynthesis beyond the carotene level (see Sandmann, 1994; and Chapter 3). Fig. C.3 is, therefore, *not* designed to predict or claim biochemical relationships, but to draw attention to important structural features of the molecules. By arranging the structures in two dimensions it is possible to point to subgroups or pairs of carotenoids that differ in one (or very few) structural feature(s) only. The emerging pattern then makes the carotenoid structures more comprehensible.

REFERENCES

IUPAC Commission on Nomenclature of Organic Chemistry, IUPAC–IUB Commission on Biochemical Nomenclature (1975). Nomenclature of Carotenoids (Rules approved 1974). *Pure Appl. Chem.* 41: 407–431.

Straub, O., Pfander, H., Gerspacher, M., Rychener, M., Schwabe, R. (eds) (1987), *Key to Carotenoids*. Birkhäuser, Basel, 296 pp.

Sandmann, G. (1994). Carotenoid biosynthesis in microorganisms and plants. *Eur. J. Biochem.* 223: 7–24.

Appendix D

UV-vis spectroscopy of carotenoids

T. Bjørnland

D.1 INTRODUCTION

Identification and quantitation of carotenoids by HPLC relies on their retention and on-line visible-light properties. These two properties alone are generally not considered sufficient for a secure identification of organic compounds. In natural product chemistry, co-chromatography in at least two chromatographic systems, as well as further spectrometric characterization — notably mass spectrometry (MS) is required. For many purposes nuclear magnetic resonance (NMR) and circular dichroism (CD) are necessary for a complete assessment of the molecular structure, including stereochemical features.

Such identification standards for pigments are not possible in oceanography, due to the small concentrations of pigment available in field samples, the large number of samples to be processed and the fact that pigment analysis is only one of the many methods used to describe and understand marine ecological processes. It is then important to use all the information available from visible light spectroscopy for pigment identification. The large number of visible light spectra now available for carotenoids allows the formulation of simple empirical rules connecting molecular structure and spectral properties. These rules can help in evaluating the identity and purity of carotenoid standards (see Chapter 8) and HPLC peaks (see Chapter 12).

For a general treatment of ultraviolet (UV) and visible light (vis) spectroscopy of organic molecules, the reader is referred to Williams and Fleming (1987). There are also many original contributions and reviews of spectroscopy as applied to the carotenoid field (Hirayama, 1955; Hager and Meyer-Bertenrath, 1967; Ke *et al.* 1970; Vetter *et al.* 1971; Davies, 1976; Köst, 1988; Britton and Young, 1993). Since important characteristics of the visible-light spectrum of carotenoids can also involve the near-ultraviolet region, the term UV-vis spectroscopy is used in this appendix. The treatment is restricted to algal carotenoids.

D.2 Spectral characteristics

The UV-vis absorption of a carotenoid varies with the wavelength of the incident light, showing that the extinction coefficient, ε, in the Beer's Law equation (see Preamble to Part IV) depends on the wavelength: this is the UV-vis spectrum of the carotenoid. At given conditions (solvent, temperature) the spectrum is a characteristic property of the absorbing molecules, and is therefore used not only in quantitative measurements but also for confirming the identity of the pigment. The main spectral characteristics used for quantitation and identification are:

- the positions of the absorption maxima (λ_{max})
- the spectral fine structure, reported as the III/II(%) value (Ke *et al.* 1970; and Preamble to Part IV)
- the *cis*-peak, the position of which is reported as λ_{max}, and the intensity as the ratio D_B/D_{II}(%) (Ke *et al.* 1970; section D.4.2.7.3)
- the extinction coefficient, ε or $E^{1\%}_{1cm}$ (Preamble to Part IV).

The spectral characteristics not only vary between carotenoids (section D.4.2), but for one and the same carotenoid they also vary with the temperature and solvent (sections D.4.1 and D.4.3, respectively). When UV-vis spectra are compared, the differences in their spectral characteristics are expressed by the terms:

- bathochromic: shift in λ_{max} positions towards longer wavelengths (red shift)
- hypsochromic: shift in λ_{max} positions towards shorter wavelengths (blue shift)
- hyperchromic: increased absorption efficiency at λ_{max} (higher ε value)
- hypochromic: reduced absorption efficiency at λ_{max} (lower ε value)
- improved or reduced spectral fine structure: increase or decrease in the III/II(%) value.

D.3 CHROMOPHORES

The 29 naturally occurring algal carotenoids included in Appendix C have rather varied molecular structures. However, they have one structural feature in common: a system of regularly alternating single and double bonds in the central part of the molecule. The electrons of this bond system are responsible for the ability of the molecules to absorb light and this system is therefore termed the chromophore.

The carbon atoms within the chromophore are bonded by electrons moving in two different types of orbitals. In the single bonds the atoms are joined by one pair of σ (sigma) electrons. The energy needed to excite these electrons from the ground state to the first excitation level (the σ^* state) is high, and may only be achieved by absorbing electromagnetic radiation in the far UV region below 200 nm. Excitation of these electrons does not therefore contribute to the colour of the molecules. In double bonds, however, the atoms are bonded by two pairs of electrons: one composed of σ electrons and the other of π (pi) electrons. The latter may be excited to the π^* level by photons of the UV-vis light region, which causes a complementary colour of the molecule.

When π electrons of different double bonds are separated by only one single bond, they may interact. Such double bonds are termed conjugated. Double bonds separated by more than one single bond do not interact. The endocyclic double bonds in ε,ε-carotene (Fig. D.1) are not in conjugation with the rest of the chromophore, and are termed isolated double bonds. As will be seen in section D.4.2.1, the number of interacting double bonds determines the excitation energy of the π electrons and thereby the colour of the carotenoid.

Carotenoids with different molecular structure but identical chromophores have identical or very similar spectral characteristics. This is seen by comparing the UV-vis data for ε,ε-carotene and violaxanthin (see Part IV Data sheets and Fig. D.1). The similarity of their spectra suggests that the hydroxyl and epoxide groups in violaxanthin do not influence the excitation of the π electrons of the chromophore. The structure of the chromophore and its position in the molecule, rather than the total molecular structure, therefore determine the spectral properties of a carotenoid.

Available spectral data make it possible to predict the UV-vis spectral properties of other carotenoids from the structure and position of their chromophore. The essential features are:

- the number of aliphatic conjugated double bonds
- the number of endo-cyclic conjugated double bonds
- the number and position of conjugated keto-groups

The occurrence of functional groups close to the chromophore may also slightly modify its spectral properties (see section D.4.3).

D.4 FACTORS THAT AFFECT SPECTRAL CHARACTERISTICS

D.4.1 *Background*

Besides the chromophore structure, the UV-vis spectral properties of carotenoids are also affected by temperature and the nature of the solvent. Low temperature (77°K) has a pronounced influence on the spectral characteristics of a carotenoid (Ke *et al.* 1970), including a bathochromic shift of λ_{max}, improved spectral fine structure, and

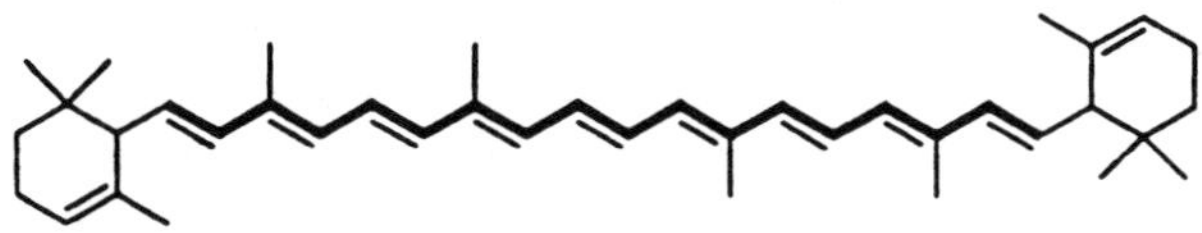

ε,ε-carotene

λ_{max} (nm): 415, 439, 469; III/II(%): 101

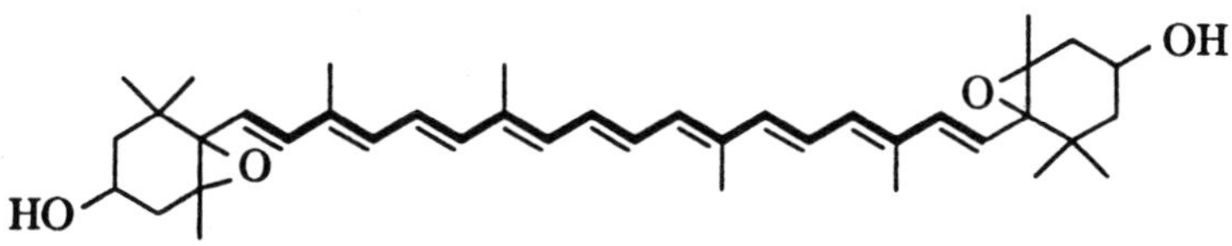

violaxanthin

λ_{max} (nm): 417, 440, 470; III/II(%): 100

Figure D.1
Molecular structures and UV-vis spectral characteristics in *n*-hexane of ε,ε-carotene and violaxanthin. The conjugated double bond systems (the chromophores) are shown in bold, and for clarity the stereochemistry of the structures has been omitted. Spectral data: ε,ε-carotene (Bjørnland *et al.* 1989); violaxanthin (SCOR WG 78).

increased molar absorption at λ_{max}. Even if the spectral effects of very low temperature are striking, they are not significant within the narrow temperature range of normal laboratory working conditions. The discussion below is therefore restricted to the spectral influence of chromophore structure (D.4.2) and solvent (D.4.3).

D.4.2 *Structural features of the chromophore*

D.4.2.1 *Chromophores with aliphatic double bonds only*

While the large majority of algal carotenoids are dicyclic, some of these (e.g. ε,ε-carotene, violaxanthin, all-*trans*-neoxanthin and dinoxanthin) have their chromophoric system restricted to the aliphatic part of the molecule. Aliphatic chromophores are planar and provide an efficient conjugation of the double bonds. Carotenoids with such chromophores possess UV-vis spectra with sharp bands and high III/II(%) values (see spectral data for ε,ε-carotene, violaxanthin and neoxanthin, Part IV).

The detailed spectral properties depend on the number of conjugated double bonds in the chromophore (Vetter *et al.* 1971; Davies, 1976). For an aliphatic chromophore consisting of 7 double bonds, λ_{max} is at 400 nm, with nine double bonds at 439 nm and with 11 at 472 nm (Davies, 1976). Further examples (Davies, 1976) and theoretical considerations (Hirayama, 1955) show that the bathochromic shift becomes smaller for each double bond added, and will approach zero as the number of double bonds increases towards infinity. λ_{max} will then reach an upper limit towards the red at about 608 nm. Algal carotenoids with nine aliphatic double bonds in the chromophore include ε,ε-carotene, violaxanthin and neoxanthin. These pigments all have λ_{max} at about 440 nm and III/II(%) values close to 100 (see Figs. D.1 and D.4). Increasing the number of conjugated aliphatic double bonds to 11, as in lycopene, causes a bathochromic shift of 32 nm as compared with ε,ε-carotene, and λ_{max} is shifted to 471 nm (see Fig. D.2).

Comparison of the spectral data for ε,ε-carotene and lycopene (see Part IV) shows that the number of aliphatic double bonds also influence the spectral fine structure and the extinction coefficients. The general trend is that chromophores with nine aliphatic double bonds have the highest III/II(%) values (see Fig. D.2), while chromophores with 11 aliphatic double bonds have the highest extinction coefficients (see Davies, 1976) and are most effective in light harvesting.

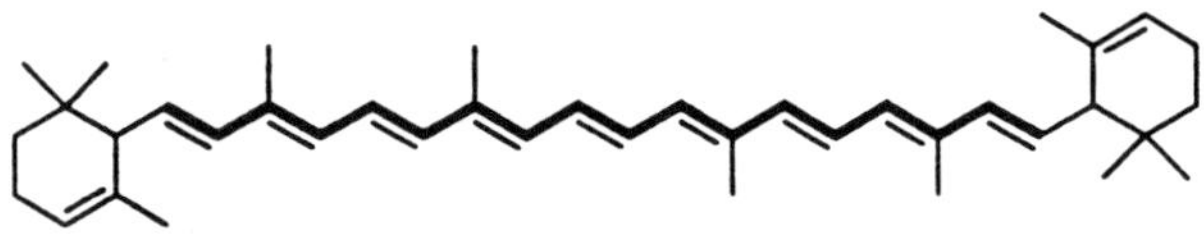

ε,ε-carotene

λ_{max} (nm): 415, 439, 469; III/II(%) = 101

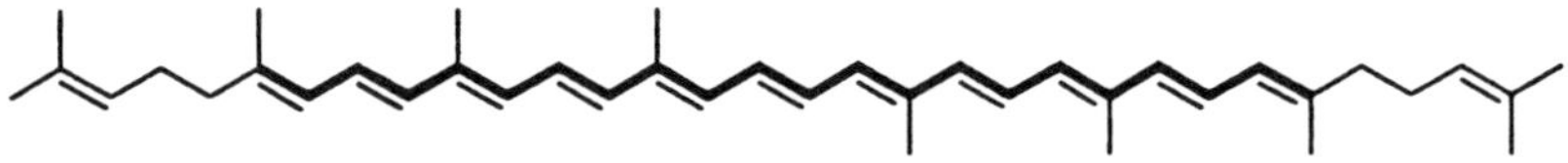

lycopene

λ_{max} (nm): 444, 471, 503; III/II(%) = 83

Figure D.2
Structures, chromophores (in bold) and UV-vis spectral characteristics (in *n*-hexane) of ε,ε-carotene and lycopene, showing the spectral effects of chromophore elongation. Spectral data: ε,ε-carotene (Bjørnland *et al.* 1989); lycopene (SCOR WG 78).

D.4.2.2 *Chromophores with endocyclic double bonds*

The three carotenes — lycopene, β,ψ-carotene and β,β-carotene — all have a chromophore consisting of 11 double bonds, but their UV-vis spectral properties are strikingly different. When considering lycopene, β,ψ-carotene and β,β-carotene sequentially there is a gradual hypsochromic shift of the λ_{max} values in *n*-hexane and a marked reduction in spectral fine structure. These changes are caused by the endocyclic conjugated double bond(s) of β,ψ- and β,β-carotene. While lycopene has a purely aliphatic chromophore, one of the terminal conjugated double bonds in β,ψ-carotene protrudes into a β-ring, and in β,β-carotene both terminal double bonds do so.

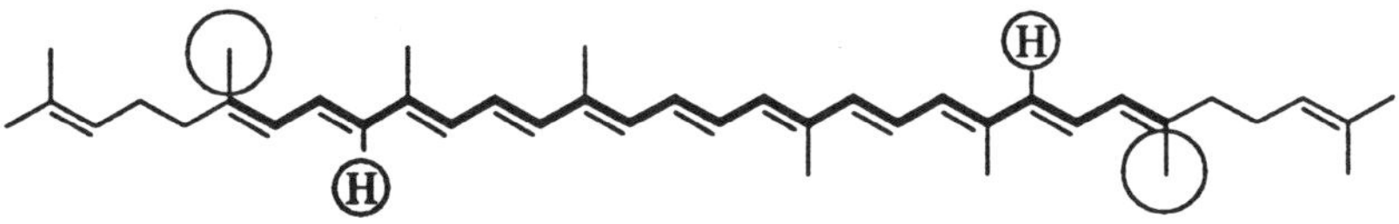

lycopene

λ_{max} (nm): 444, 471, 503; III/II(%) = 83

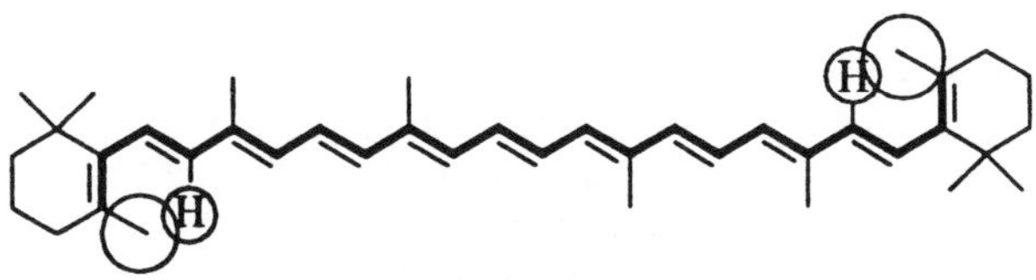

β,β-carotene

λ_{max} (nm): (422), 450, 478; III/II(%) = 36

Figure D.3
Structures, chromophores (in bold) and UV-vis spectral properties (in *n*-hexane) of lycopene and β,β-carotene, showing the spectral effect of endocyclic double bonds. The steric conflict between CH_3-18 and H-8 are indicated by circles. Spectral data: SCOR WG 78.

In carotenoids with β-rings there is a steric conflict between the CH_3-18 methyl group and the CH-8 hydrogen (Fig. D.3), forcing the endo-cyclic part of the chromophore out of planarity with the central chain. This twisting reduces the conjugation between the endo-cyclic and aliphatic parts of the chromophore, and the endo-cyclic double bond thereby becomes spectrally less efficient than the aliphatic ones.

D.4.2.3 *Chromophores with acetylenic bonds*
The acetylenic bond is a rather uncommon structural feature in natural product chemistry. Nevertheless, this bond is present in several common and taxonomically significant algal carotenoids such as alloxanthin, diadinoxanthin, diatoxanthin and monadoxanthin.

The spectral effect of the acetylenic bond has been examined in a series of synthetic carotenoid-related compounds (Vetter *et al.* 1971). Compared to the non-acetylenic analogue, the λ_{max} position is always displaced towards shorter wavelengths. By synthesizing a series of compounds where the acetylenic bond has been moved stepwise from a terminal to a more central position in the chromophore, the hypsochromic effect was increased from 14 to 29 nm.

The acetylenic bond in dicyclic algal carotenoids is always located in the 7,8 (7',8') position(s). Thus it is terminal in the aliphatic part of the chromophore, but the chromophore usually extends further into a neighbouring β-ring as in diatoxanthin and alloxanthin. This very position of the acetylenic bond obscures its general hypsochromic effect: with H-8/H-8' lacking (see section D.4.2.2) these acetylenic carotenoids have planarity between the endo-cyclic and aliphatic part of the chromophore, leading to a bathochromic shift compared to their non-acetylenic analogues.

The effect of introducing an acetylenic bond therefore becomes two-fold. First, there is a general hypsochromic effect caused by the acetylenic bond itself. Since the position is close to terminal, this effect will be relatively small (see Vetter *et al.* 1971). Second, the 7,8-position of the acetylenic bond implies that there is no hydrogen atom at C-8 and accordingly no steric conflict with the C-18 methyl group (cf. Fig. D.3), allowing planarity to be restored between the endo-cyclic 5,6-double bond and the aliphatic part of the chromophore. This improves the conjugation between the π-electrons and results in a bathochromic shift of λ_{max}. Based on the UV-vis spectra of zeaxanthin, diatoxanthin and alloxanthin (Part IV Data sheets) these opposing effects must approximately balance each other. Spectrally, a 7,8- acetylenic bond therefore is approximately equivalent to one double bond.

This same conclusion is reached by comparing structurally related non-acetylenic and acetylenic members of the β,ε-series, *viz.* β,ε-carotene, lutein, crocoxanthin and monadoxanthin (Part IV Data sheets).

D.4.2.4. *Chromophores with an allenic bond*

The allenic bond, a rare structural feature of natural products, is present in several of the taxonomically and functionally most important algal carotenoids. To understand its spectral effect it is essential to grasp the steric arrangement of the two double bonds constituting the allenic axis.

As seen from the partial drawing of the neoxanthin structure in Fig. D.4, the two allenic double bonds are positioned at right angles to each other. No conjugation can occur between their π electrons. The left and right allenic double bond therefore belong to two separate and non-interacting double bond systems. In naturally occur-

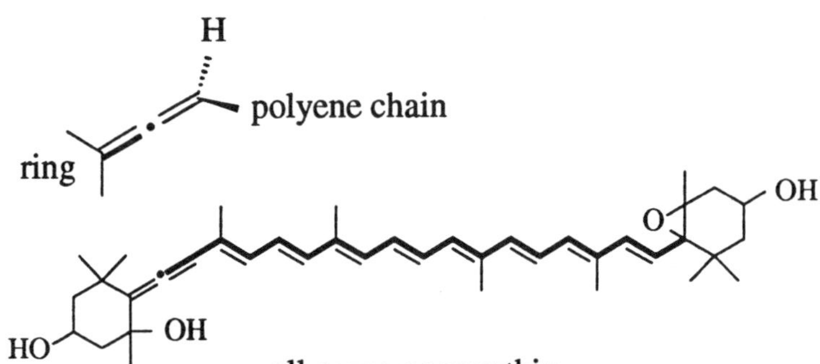

all-*trans*-neoxanthin

λ_{max} (nm): 416, 440, 469; III/II(%) = 94

Figure D.4
Structure, chromophore (in bold) and UV-vis spectral characteristics (in *n*-hexane) of all-*trans*-neoxanthin. Also included is a partial figure showing that the two allenic double bonds are twisted 90° relative to each other. Spectral data: Bjørnland unpubl. (see Part IV Data sheets).

ring carotenoids the allenic bond is always in the terminal position of the chromophore, as seen in the structures of all-*trans*-neoxanthin (Fig. D.4), P-457, fucoxanthin, the 19'-acyloxy-derivatives of fucoxanthin, peridinin and vaucheriaxanthin. Since the most distant of the allenic double bonds is not in conjugation with the rest of the chromophore, the allenic bond becomes spectrally equivalent to one aliphatic double bond only. For this reason, the three carotenoids ε,ε-carotene, violaxanthin and all-*trans*-neoxanthin have very similar UV-vis properties, those of a chromophore with nine conjugated aliphatic double bonds (cf. Figs. D.1 and D.4).

D.4.2.5 *Chromophores with conjugated carbonyl functions*

D.4.2.5.1 Background

Carbonyl groups in direct conjugation with the remaining part of the chromophoric system may contribute significantly to the UV-vis spectral properties of carotenoids. In general, this structural element will cause a reduction of the extinction coefficients, a red shift of λ_{max} and a pronounced loss in spectral fine structure. This loss may or may not be strongly dependent on the solvent (see section D.4.3).

Carotenoids with conjugated carbonyl functions have varied structures and UV-vis spectral properties, as shown for canthaxanthin, fucoxanthin and peridinin in Fig. D.5. These include several of the most important carotenoids of marine phytoplankton with respect to concentration, function and chemo-systematic significance. These three algal carotenoids have different chromophore systems, treated separately below.

D.4.2.5.2 Endo-cyclic ketones

The secondary carotenoids echinenone, canthaxanthin and astaxanthin have one or two keto functions in conjugation with the central polyene chain. These are endocyclic with a terminal position in the chromophore. The effect of introducing endocyclic keto functions may be seen by comparing their spectra with that of β,β-carotene in *n*-hexane (λ_{max}: (422), 450, 478 nm; III/II(%) = 36; see Part IV Data sheets). Extending the chromophore by one keto function in the 4-position, as in echinenone, causes a bathochromic shift of about 9 nm. Even when the UV-vis spectrum is recorded in *n*-hexane (see Section D.4.3) there is a complete loss of spectral fine structure. An additional keto group in the 4'-position, as in canthaxanthin, causes a further displacement of λ_{max} of about 8 nm (Fig. D.5). For these two pigments and astaxanthin there is a lowering of the molar extinction coefficients relative to β,β-carotene of about 10–15%.

D.4.2.5.3 Terminal aliphatic ketones

Several taxonomically significant xanthophylls — fucoxanthin, the 19'-acyloxyfucoxanthins, siphonaxanthin, siphonein and prasinoxanthin — all have a terminal carbonyl function in their chromophore (see Fig. C.3, Appendix C). This chromophore, which is restricted to the aliphatic central part of the molecule, consists of eight conjugated double bonds besides the terminal carbonyl double bond (see fucoxanthin in Fig. D.5). The spectral properties of these carotenoids vary slightly, since neighbouring functional groups such as an ester or a free hydroxyl group in the 19- or 19'-position may interact with the conjugated double bond system, making it difficult to predict the UV-vis properties (see section D.4.3). Furthermore, it is possible that isolated double bonds in the rings — as in siphonaxanthin, siphonein and prasinoxanthin — may be brought into conjugation by enolization of the keto function.

endocyclic terminal carbonyl:

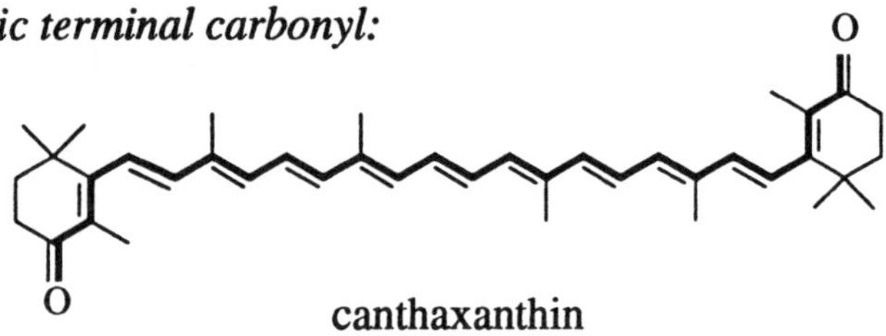

canthaxanthin

λ_{max} (nm): 467; III/II(%) = 0

aliphatic terminal carbonyl:

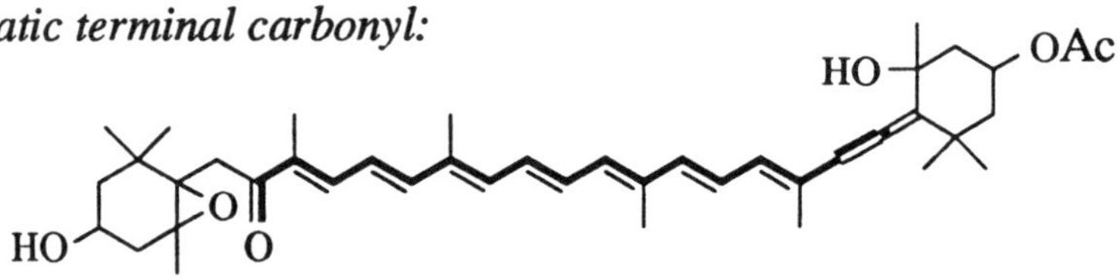

fucoxanthin

λ_{max} (nm): 426, 448, 477; III/II(%) = 45

cross-conjugated carbonyl:

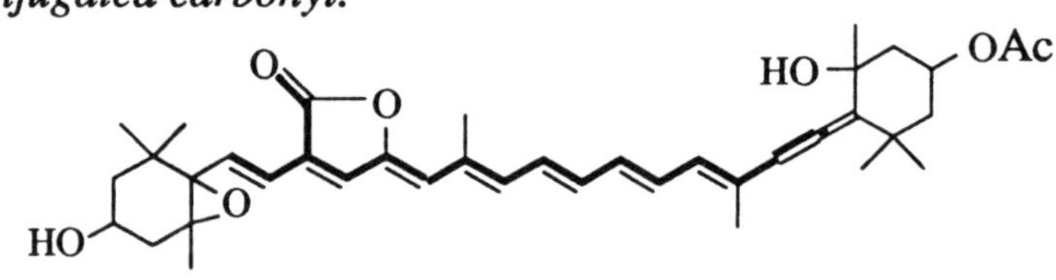

peridinin

λ_{max} (nm): (430), 456, 485; III/II(%) = 75

Figure D.5
Structures, chromophores (in bold) and UV-vis spectral characteristics (in *n*-hexane) for three selected carotenoids with conjugated carbonyl functions. Spectral data: canthaxanthin (Davies, 1976); fucoxanthin (Bjørnland, unpubl., see Table D.2); peridinin (SCOR WG 78).

The effect of elongating an aliphatic chromophore with a carbonyl function as in fucoxanthin may be compared to carotenoids with nine conjugated carbon-carbon double bonds, such as ε,ε-carotene and violaxanthin (see Figs. D.1 and D.5). The effect of substituting one carbon-carbon double-bond with a carbonyl function causes (a) a large bathochromic shift, and displaces λ_{max} 17–18 nm further towards the red than a carbon-carbon double bond, and (b) reduces the spectral fine structure from 100 to 45 III/II(%). The aliphatic terminal carbonyl therefore does not lead to a complete loss of spectral fine structure as do endo-cyclic terminal carbonyl(s). The spectral fine structure of such compounds is considerably influenced by the nature of the solvent (section D.4.3), contrary to chromophores with carbon-carbon double-bonds only.

Fucoxanthin is the only carotenoid with this chromophore for which molar extinction coefficients are known. It is about 25 and 30% lower than that of violaxanthin and ε,ε-carotene, respectively.

D.4.2.5.4 Cross-conjugated carbonyl functions

In P-457 and peridinin an in-chain methyl group has been oxidized to the carbonyl level. This function is present as an aldehyde in P-457 and an ester (as a lactone ring) in peridinin. The carbonyl double bond is in both cases in conjugation with eight aliphatic double bonds, but contrary to the fucoxanthin chromophore (section D.4.2.5.3), the carbonyl is now a side branch to the carbon-carbon double bond system. It therefore becomes cross-conjugated to the rest of the chromophore (Fig. D.5).

The introduction of a cross-conjugated carbonyl function causes a large bathochromic shift. Both peridinin and fucoxanthin have eight conjugated aliphatic carbon-carbon double bonds, but λ_{max} for peridinin is located 8 nm longer towards the red (see Fig. D.5). A cross-conjugated carbonyl has therefore a greater red shift effect than an aliphatic terminal carbonyl as in fucoxanthin. The spectral fine structure also becomes more dependent on the solvent (see section D.4.3).

The molar extinction coefficient at λ_{max} for peridinin in acetone is lower than for carotenoids with a fucoxanthin type chromophore, and is only 55–65% of what is common for carotenoids without a conjugated carbonyl function.

Cross-conjugated algal carotenoids are important chemosystematic markers for dinoflagellates and prasinophytes. These chromophores are present in several dinoflagellate xanthophylls related to peridinin (peridininol, pyrrhoxanthin, pyrrhoxanthinol) (see Bjørnland, 1990) and some carotenoids from the Prasinophyceae (uriolide, micromonal etc.) (Foss *et al.* 1986, Egeland and Liaaen-Jensen, 1993).

D.4.2.6 *Cis-isomerization*

D.4.2.6.1 Background

Carotenoids can theoretically exist in a large number of *cis/trans* isomeric forms due to the system of alternating single and double bonds in the aliphatic chain (see Davies, 1976). Not all of these isomers are energetically stable and most carotenoids have the all-*trans* form as the most stable geometrical isomer. Some structural elements, however, such as the 7,8 acetylenic bond and oxygenated in-chain methyl groups (primary alcohols, esters and aldehydes), may cause increased stability of *cis*-isomers as for alloxanthin, peridinin and the 19'-acyloxyfucoxanthins.

The introduction of a *cis*-bond in the chromophore leads to several spectral changes, the sizes of which are determined by its position. Only a few *cis*-isomers have been isolated in pure form and adequately characterized, and results discussed below must be considered preliminary.

D.4.2.6.2 The effect on λ_{max} positions and spectral fine structure

Cis-isomers invariably absorb at lower wavelengths than the corresponding all-*trans* isomer. The hypsochromic displacement (isomerization shift) for mono- and di-*cis* isomers of violaxanthin in benzene has been examined by Molnár and Szabolcs (1980)

and Molnár *et al.* (1986). The isomerization shift for violaxanthin mono-*cis* isomers is small (4–5 nm) for the terminal 9-position, increases to 7–8 nm for the 13-*cis* isomer and decreases to 4–5 nm for the central 15-position. Similar trends have been shown for fucoxanthin (Haugan *et al.* 1992), peridinin (Haugan *et al.* 1994) and other carotenoids (see Davies, 1976).

The data for violaxanthin (Molnár *et al.* 1986) indicate that the hypsochromic effect is additive for di-*cis* isomers.

The effect of *cis*-isomerization on the spectral fine structure should be considered separately for off- and on-line registrations. In the first case the spectrum is the total absorption of a mixture of *cis/trans*-isomers. The hypsochromic isomerization shift of the UV-vis spectral peaks of the *cis*-isomers will partly fill up the valleys in the spectrum of the all-*trans*-isomer. This leads to a reduced spectral fine structure, broader absorption bands and lower absorption at λ_{max}.

If the isomerized mixture is resolved by HPLC, the on-line diode array UV-vis registrations will show the spectrum of each isomer separately. There are somewhat contradictory reports on the spectral fine structure of individual *cis*-isomers. Both mono-*cis*-isomers with a centrally located *cis*-bond and di-*cis*-isomers seem to possess considerably reduced spectral persistence. For more terminal *cis*-positions, as for the common 9- or 9'-mono-*cis*-isomers, the spectral fine structure is comparable with what has been recorded for their all-*trans*-form. This has been demonstrated, within experimental error, for 9'-*cis*-neoxanthin (Fig. D.6), 9'-*cis*-fucoxanthin (Haugan *et al.* 1992) and 9'-*cis*-19'-butanoyloxyfucoxanthin (Englert *et al.* 1990).

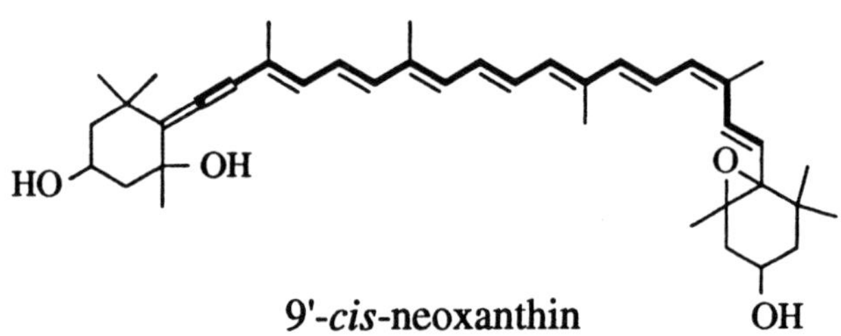

9'-*cis*-neoxanthin

9'-*cis*: λ_{max} (nm): 411, 435, 463; III/II(%) = 97; D_B/D_{II}(%) = 7

all-*trans*: λ_{max} (nm): 416, 440, 469; III/II(%) = 94; D_B/D_{II}(%) = 6

Figure D.6
Structure, chromophore (in bold) and UV-vis spectral characteristics (in *n*-hexane) of 9'-*cis*-neoxanthin. The spectral data for all-*trans*-neoxanthin are included to show the effect of *cis/trans*-isomerization. Spectral data: Bjørnland unpubl. (see Part IV Data sheets).

D.4.2.6.3 The cis-peak

Cis-isomers have an absorption band in the near-ultraviolet region not found in the UV-vis spectrum of their all-*trans* analogue. This band, termed the *cis*-peak, is located at about 142 nm shorter wavelengths from the main band III of the all-*trans* form (see Davies, 1976). Its intensity, which is strongly dependent on its location in the polyene chain, is expressed by the D_B/D_{II}(%) value (Ke *et al.* 1970), where D_B and

D_{II} are the optical density of the *cis*-peak and the main absorption band, respectively.

The intensity of the *cis*-peak depends on both the number and position(s) of the *cis*-bond(s) (Davies, 1976). The most important isomers are mono-*cis*-isomers with the *cis*-bond adjacent to a methyl group, *viz* in the 9, 13, 9', 13'-positions. The central 15,15'-position may also (but more rarely) be isomerized (Molnár and Szabolcs, 1980; Haugan *et al.* 1992). For these mono-*cis*-isomers the intensity of the *cis*-peak increases from a terminal to a central position of the *cis*-bond, giving $D_B/D_{II}(\%)$-values of 30–55 for the 13-, 15- and 13'-*cis*-isomers. Such a prominent *cis*-peak may be considered proof of a *cis*-bond, and the intensity a first indication of its position.

For the frequently occurring 9- or 9'-*cis*-isomers, however, the $D_B/D_{II}(\%)$-value is low, and the *cis*-peak region is not significantly different from what is recorded for the all-*trans*-isomer. A 9- or 9'-*cis*-bond is therefore indicated by a combination of a weak *cis*-peak together with a hypsochromic displacement of 4–5 nm for λ_{max}, as shown for 9'-*cis*- and all-*trans*-neoxanthin in Fig. D.6.

Rigorous proof of a *cis*-bond and its position in the molecule needs confirmation by more powerful methods such as NMR, although UV-vis spectra can give clear indications.

D.4.2.6.4 The effect on the extinction coefficient

The effect of *cis*-isomerization on the extinction coefficient is not well documented, since relatively few *cis*-isomers have been obtained in the crystalline state (see Vetter *et al.* 1971, for β,β-carotene; Molnár and Szabolcs, 1980, for violaxanthin). In general, *cis*-isomerization causes a hypsochromic change in the molar extinction coefficient, showing that *cis*-isomers are less efficient in light-harvesting at λ_{max} than their all-*trans* analogue. The scanty data available indicate that the reduction in light absorption may be considerable, and is larger for isomers with a central *cis*-bond than a terminal one. Compared with the all-*trans*-isomer, the reduction is about 30% for both 15-*cis*-β,β-carotene (calculated from Vetter *et al.* 1971) and 15-*cis*-violaxanthin (calculated from Molnár and Szabolcs, 1980). UV-vis quantitation of *cis*-isomers will therefore lead to a severe underestimate when based on the extinction coefficient of the all-*trans* form.

D.4.3 *Solvent effects*

Both Hirayama (1955) and Feichtmayr *et al.* (1969) have pointed out that spectral effects of solvent mixtures are correlated with their refractive index rather than polarity. Even if there are exceptions for one-component solvents, e.g. acetone, the general guideline is that the largest spectral effects are exercised by solvents with high refractive indices.

With few exceptions, the UV-vis spectrum of a carotenoid recorded in *n*-hexane will have lower λ_{max} positions, more pronounced spectral fine structure and higher extinction coefficients than in other solvents. *n*-Hexane is therefore a convenient reference when solvent-induced spectral changes are to be compared. A change from *n*-hexane to other solvents will then cause a red shift in λ_{max}, the III/II(%)-value will be lowered and the absorption intensity will drop. For many carotenoid/solvent combinations these changes are small, e.g. bathochromic shifts will usually be within 2–5 nm. In solvents such as benzene, toluene, pyridine and carbon disulfide, the λ_{max} displacements are considerably larger (Table D.1). Table D.1 also demonstrates the

Table D.1. Extinction coefficients ($E^{1\%}_{1cm}$) and λ_{max} for carotenes in various solvents[a, b]

Solvent	Lycopene		β,β-Carotene	
	$E^{1\%}_{1cm}$	λ (nm)	$E^{1\%}_{1cm}$	λ (nm)
Hexane	3360	467	2650	450
Cyclohexane	3230	474	2710	455
Acetone	3190	472	2610	455
Benzene	3060	483	2460	465
Carbon disulfide	2560	503	2090	483

a. Selected and recalculated data from Feichtmayr *et al.* (1969).
b. Note that the extinction coefficients recommended in Appendix E for lycopene and β,β-carotene in acetone are 3446 (at λ_{max}, 474 nm) and 2500 (at λ_{max}, 454 nm), respectively.

importance of using extinction coefficients which apply to the solvent in use. If this is not done, quantitative under- or over-estimates of 20–25% may result.

The solvent effect also depends on the chromophore structure (Table D.2), and is small for chromophores without conjugated carbonyl groups. Both band positions and spectral fine structure may be affected, as shown in Table D.2 for β,β-carotene, lutein and violaxanthin.

Chromophores with endo-cyclic terminal carbonyls may have hardly noticeable solvent effects since the spectral curve is rounded and unstructured in all solvents, including *n*-hexane. If the conjugated carbonyl is in a terminal aliphatic position (as in fucoxanthin and related carotenoids) or cross-conjugated (as in peridinin), a change in solvent from *n*-hexane to acetone will cause a large reduction in spectral fine structure.

Table D.2. Solvent effects on algal carotenoid chromophores

Carotenoid	*n*-Hexane		Acetone	
	λ_{max} (nm)	III/II(%)	λ_{max} (nm)	III/II(%)
β,β-Carotene[a]	(422), 450, 478	36	(426), 454, 480	21
Lutein[a]	421, 445, 474	76	425, 448, 476	67
Violaxanthin[a]	417, 440, 470	100	419, 441, 472	95
Canthaxanthin[b]	467	0	472	0
Fucoxanthin[c]	426, 448, 477	45	447, 469	8
19'-Butanoyloxyfucoxanthin[c]	421, 444, 472	60	(423), 444, 471	42
19'-Hexanoyloxyfucoxanthin[c]	422, 445, 473	62	(423), 445, 470	44
Peridinin[a]	(430), 456, 485	75	458–474	0

a. Spectral data from SCOR WG 78
b. Spectral data from Davies (1976) and SCOR WG 78 (acetone)
c. *ex Gymnodinium galatheanum* (Bjørnland, unpubl.; cf. Bjørnland and Liaaen-Jensen, 1989)

Since the above solvent effects are also dependent on molecular features, they may be used for identification purposes. The UV-vis spectra of carotenoids recommended as standards (Chapter 8) should always be recorded in two solvents, e.g. *n*-hexane and acetone, to assist confirmation of their identity.

As pointed out in section D.3, it is the system of conjugated double bonds rather than the total molecular structure which determines the UV-vis properties of a carotenoid. Occasionally functional groups which are not part of the chromophore may have a modifying effect on the spectral properties. One example is the 19'-ester function of 19'-butanoyloxy- and 19'-hexanoyloxyfucoxanthin (Table D.2). Compared to fucoxanthin, the ester function causes a moderate hypsochromic shift and improved spectral fine structure (see also Wright and Jeffrey, 1987).

The solvent effect on the III/II(%)-value is however much smaller for the ester derivatives than for fucoxanthin itself. This difference may be used as a purity criterion for uncrystallized fucoxanthin samples and for re-examination of old UV-vis spectral data. Many members of the chromophyte classes possess both fucoxanthin and 19'-hexanoyloxyfucoxanthin (see Bjørnland and Liaaen-Jensen, 1989 and Chapter 2). While HPLC and some advanced TLC adsorbents (see Wright and Jeffrey, 1987; Jeffrey and Wright, 1994) may resolve these two pigments, this is not the case with classical TLC and open column chromatographic systems. An abnormally high spectral fine structure in acetone for the fucoxanthin fraction may indicate contamination with 19'-hexanoyloxyfucoxanthin.

D.5 Summary: Evaluation of spectral quality

The ability of carotenoids to absorb light is due to their system of regularly alternating single and double bonds: the chromophore. Since spectral characteristics are generally not influenced by functional groups, it is more logical to treat the UV-vis properties of carotenoids in terms of their chromophore structure rather than their total molecular structure.

Common chromophores in algal carotenoids consist of nine aliphatic double bonds in conjugation with zero, one or two endocyclic double bonds (Table D.3). These chromophores are found in six of the 29 standard algal carotenoids (see Part IV Data sheets). If a 7,8-acetylenic bond and a terminal allenic bond are considered equivalent to one aliphatic double bond, the number of standard algal carotenoids with common chromophores increases to 13.

Table D.3. UV-vis characteristics of 3 common carotenoid chromophores[a]

No. of conjugated double bonds		λ_{max} (nm)	III:II(%)
aliphatic	endocyclic		
9	0	440	3:3
9	1	445	3:2
9	2	450	3:1

a. in *n*-hexane

Carbonyl functions also cause variation in algal carotenoid chromophores, and therefore spectral characteristics. These include terminal endo-cyclic elongation of the chromophore (e.g., canthaxanthin), a terminal carbonyl double bond in the aliphatic part of the chain (e.g., fucoxanthin) or a cross-conjugated carbonyl group (e.g., peridinin). Since a carbonyl double bond is spectrally more effective than a carbon-carbon double bond, this will lead to a bathochromic displacement of λ_{max}. In addition, the UV-vis spectral fine structure becomes reduced and more susceptible to solvent effects.

Carotenoids with identical or similar chromophores must possess identical or similar UV-vis spectral properties, making it possible to predict UV-vis properties by comparing chromophore structures. As an illustration, the UV-vis spectra of crocoxanthin and monadoxanthin were not recorded at the SCOR WG 78 workshops. The chromophores are identical in these two xanthophylls and are the same as in diadinoxanthin. Both the band positions and curve shape of the UV-vis spectra of all three pigments should be very similar, as confirmed by literature spectral data (see Part IV Data sheets).

Band positions and curve shape should be noted when UV-vis spectral characteristics are used for identification purposes. Comparisons should be done only in the same solvent. Spectral variation may still occur between different preparations of the same pigment, due to trace amounts of coloured contaminants and *cis/trans*-isomerization. These difficulties may become severe if the amount of available pigment is small and the UV-vis spectra are recorded off-line.

Table D.4 illustrates the effect of isolation procedure on the UV-vis properties of carotenoid samples. When the amount available is too small to be crystallized, as for ε,ε-carotene *ex* microalga (MC-1; now identified as *Pelagomonas calceolata*; Andersen *et al.* 1993), the UV-vis spectrum represents only a chromatographically pure, not a chemically pure, fraction (e.g., TLC-pure or HPLC-pure). An off-line UV-vis spectrum of such a sample will normally show a slightly *cis/trans*-isomer-

Table D.4. UV-vis spectral characteristics of selected carotenoids in *n*-hexane and purity of preparations[a]

Carotenoid	Algal source	Available amount (mg)	Purity	UV-vis properties	
				λ_{max} (nm)	III/II(%)
ε,ε-Carotene	MC-1[b]	0.020	TLC-pure	413, 437, 468	88
	Pelagococcus subviridis	<0.073	semi-cryst.	415, 439, 469	101
β,β-Carotene	MC-1[b]	0.13	TLC-pure	(426), 448, 474	16
		<0.13	semi-cryst.	(427), 449, 477	36
Diatoxanthin	MC-1[b]	0.25	TLC-pure	(427), 448, 478	28
		<0.25	semi-cryst.	(428), 449, 478	39
Alloxanthin	*Euglena gracilis*	0.15	TLC-pure	(428), 449, 477	16
		<0.15	cryst.	(429), 451, 481	46

a. Bjørnland (unpublished)
b. Now identified as *Pelagomonas calceolata* (Bjørnland *et al.* 1987; Andersen *et al.* 1993)

ized carotenoid. If somewhat larger amounts are available, as for ε,ε-carotene *ex Pelagococcus subviridis*, an attempted crystallization may be successful (Bjørnland *et al.*, 1989). Since *cis*-isomers are more soluble than the all-*trans*-form, the crystallized product will be enriched in the all-*trans*-isomer and possess absorption bands shifted slightly towards the red (1–2 nm), often with strikingly improved spectral fine structure.

Many carotenoids are difficult to crystallize (Chapter 8) and are present in small amounts in natural sources. Their UV-vis data have mostly been recorded on 'TLC-pure' samples. More abundant or synthetic carotenoids are usually obtained in higher purity, as a non-crystalline precipitate, a semicrystalline sample, as crystals or as a recrystallized product. To evaluate the UV-vis data in the literature, it is necessary to know the purity of the sample. Since the purity in many cases is not given, some variation in the spectral data reported for one and the same pigment must be expected (cf. Table D.4), more for the III/II(%)-value than the λ_{max} positions.

REFERENCES

Andersen, R. A., Saunders, G. W., Paskind, M. P., Sexton, J. P. (1993). Ultrastructure and 18 S_R RNA gene sequence for *Pelagomonas calceolata* (gen. et. sp. nov.) and the description of a new algal class, the Pelagophyceae classis nov. *J. Phycol.* 29: 701–715.

Bjørnland, T. (1990). Chromatographic separation and spectrometric characterization of native carotenoids from the marine dinoflagellate *Thoracosphaera heimii*. *Biochem. Syst. Ecol.* 18: 307–316.

Bjørnland, T., Liaaen-Jensen, S. (1989). Distribution patterns of carotenoids in relation to chromophyte phylogeny and systematics. In: Green, J. C., Leadbeater, B. S. C., Diver, W. L. (eds), *The Chromophyte Algae: Problems and Perspectives*. Clarendon Press, Oxford, p. 37–61.

Bjørnland, T., Liaaen-Jensen, S., Throndsen, J. (1989). Carotenoids of the marine chrysophyte *Pelagococcus subviridis*. *Phytochem.* 28: 3347–3353.

Bjørnland, T., Pereira, C., Liaaen-Jensen, S., Guillard, R. R. L., Bidigare, R. R. (1987). Pigment distribution patterns in oceanography: carotenoids in the microflagellate MC–1. Abstr. 8th Int. Carot. Symp., Boston, p. 40.

Britton, G., Young, A. J. (1993). Methods for the isolation and analysis of carotenoids. In: Young, A. J., Britton, G. (eds.), *Carotenoids in Photosynthesis*. Chapman and Hall, Bury St. Edmunds, U.K., p. 409–457.

Davies, B. H. (1976). Carotenoids. In: Goodwin, T. W. (ed.), *Chemistry and Biochemistry of Plant Pigments*. Vol. 2, 2nd edn. Academic Press, New York, p. 38–165.

Egeland, E. S., Liaaen-Jensen, S. (1993). New carotenoids and chemosystematics in Prasinophyceae. Abstr. 10th Int. Carot. Symp. poster 6, section 4, Trondheim.

Englert, G., Bjørnland, T., Liaaen-Jensen, S. (1990). 1D and 2D NMR study of some allenic carotenoids of the fucoxanthin series. *Magn. Reson. Chem.* 28: 519–528.

Feichtmayr, F., Heilbronner, E., Nürrenbach, A., Pommer, H., Schlag, J. (1969). The dispersive interaction of non-polar solutes with non-polar solvents: solvent effects on electronic spectra of carotenoids. *Tetrahedron*, 25: 5383–5408.

Foss, P., Guillard, R. R. L., Liaaen-Jensen, S. (1986). Carotenoids from eucaryotic ultraplankton clones (Prasinophyceae). *Phytochem.* 25: 119–124.

Hager, A., Meyer-Bertenrath, T. (1967). Beziehungen zwischen Absorptionsspektrum und Konstitution bei Carotinoiden von Algen und Höheren Pflanzen. *Ber. Dtsch. Bot. Ges.* 80: 426–436.

Haugan, J. A., Englert, G., Aakermann, T., Glinz, E., Liaaen-Jensen, S. (1994). Algal carotenoids. 58. Isomerization studies on peridinin. *Acta Chem. Scand.* 48: 769–779.

Haugan, J. A., Englert, G., Glinz, E., Liaaen-Jensen, S. (1992). Algal carotenoids. 48. Structural assignments of geometrical isomers of fucoxanthin. *Acta Chem. Scand.* 46: 389–395.

Hirayama, K. (1955). Absorption spectra and chemical structure. *J. Am. Chem. Soc.* 77: 373–384.

Jeffrey, S. W., Wright, S. W. (1994). Photosynthetic pigments in the Haptophyta. In: Green, J. C., Leadbeater, B. S. C. (eds), *The Haptophyte Algae*, Clarendon Press, Oxford, p. 111–132.

Ke, B., Imsgard, F., Kjøsen, H., Liaaen-Jensen, S. (1970). Electronic spectra of carotenoids at 77°K. *Biochim. Biophys. Acta*, 210: 139–152.

Köst, H.-P. (ed.) (1988). *CRC Handbook of Chromatography: Plant pigments*. vol. 1. *Fat-Soluble Pigments*. CRC Press, Boca Raton, USA, p. 119–129.

Molnár, P., Szabolcs, J. (1980). Occurrence of 15-*cis*-violaxanthin in *Viola tricolor*. *Phytochem.* 19: 623–627.

Molnár, P., Szabolcs, J., Radics, L. (1986). Naturally occurring di-*cis*-violaxanthins from *Viola tricolor*: Isolation and identification by ^{1}H NMR spectroscopy of four di-*cis*-isomers. *Phytochem.* 25: 195–199.

Vetter, W., Englert, G., Rigassi, N., Schwieter, U. (1971). Spectroscopic methods. In: Isler, O. (ed.), *Carotenoids*. Birkhäuser, Basel, p. 192–202.

Williams, D. H., Fleming, I. (1987). *Spectroscopic Methods in Organic Chemistry*. 4th edn. McGraw-Hill, London, 246 pp.

Wright, S. W., Jeffrey, S. W. (1987). Fucoxanthin pigment markers of marine phytoplankton analysed by HPLC and HPTLC. *Mar. Ecol. Progr. Ser.* 38: 259–266.

Appendix E

Chlorophyll and carotenoid extinction coefficients

Compiled by S. W. Jeffrey

Chlorophylls and Derivatives[a]

Pigment	Solvent	λ_{max}[b] (nm)	Specific extinction coefficient ($l.g^{-1}\ cm^{-1}$)[g]	Molar extinction coefficient ($l.mol^{-1}\ cm^{-1}$) ($\times 10^3$)
Chlorophyll *a*	90% acetone	664	87.67	78.32
Chlorophyllide *a*	90% acetone	664	127	78.3
Pheophytin *a*	90% acetone	667	51.2	44.6
Pheophorbide *a*	90% acetone	667	74.2	44.0
Pyropheophytin *a*	diethyl ether	667	60.29	49.0
Divinyl-chlorophyll *a*	not determined; assume identical to Chl *a*			
Chlorophyll *b*	90% acetone	647	51.36	46.61
Chlorophyllide *b*	90% acetone	645	74.07	46.5
Pheophytin *b*	90% acetone	657	31.8	28.1
Pheophorbide *b*	90% acetone	657	46.37	28.1
Divinyl-chlorophyll *b*	not determined; assume identical to Chl *b*			
Chlorophyll c_1[c]	90% acetone + 1% pyridine	631	44.8	27.4
Chlorophyll c_2[c]	90% acetone + 1% pyridine	631	40.4	24.6
Chlorophyll c_3	not determined; use mean of Chl c_1 and c_2 at Soret maximum[d]			
Chlorophyll $c_{CS\text{-}170}$	not determined; use mean of Chl c_1 and c_2 at Soret maximum[d]			
Phytylated chlorophyll *c*-like	not determined; calculate from ε for chl c_2[e]			
Mg 3,8 divinyl pheoporphyrin a_5 mono-methyl ester	methanol		58.9[f]	36[f]

a. Refer to Part IV Data sheets, for values in alternative solvents and references. All extinction coefficients at red maxima, except where indicated.
b. All values to the nearest nm.
c. Crystalline chlorophyll *c* pigments are not soluble in acetone, therefore pyridine was used to solubilize before dilution with acetone.
d. See data sheets for chlorophyll c_3, chlorophyll $c_{CS\text{-}170}$.
e. See data sheet for phytylated chl *c*-like pigment.
f. See data sheet for MgDVP.
g. It has been the convention to calculate extinction coefficients as $l\ gm^{-1}\ cm^{-1}$ in the chlorophyll literature, and as $100\ ml\ gm^{-1}\ cm^{-1}$ in the carotenoid literature. We have continued to use these conventions in this monograph.

Carotenoids [a]

Pigment	Solvent	λ_{max} (nm)	Specific extinction coefficient ($E^{1\%}_{1cm}$) (100ml g^{-1} cm^{-1})	Molar extinction coefficient (l.mol^{-1} cm^{-1}) ($\times 10^3$)
Alloxanthin	acetone	454	2500	141
Antheraxanthin	ethanol	446	2350	137.4
Astaxanthin	hexane	468	2100	125.3
19'-Butanoyloxyfucoxanthin	acetone	445	1470	109.4
Canthaxanthin	petroleum ether	466	2200	124
β,ε-Carotene	acetone	448	2700	145
β,β-Carotene	acetone	454	2500	134
β,ψ-Carotene	petroleum ether	459	3185	171
ε,ε-Carotene	petroleum ether	440	3010	162
Crocoxanthin	ethanol	443	2500	138
Diadinochrome	acetone	428	2500	146
Diadinoxanthin	acetone	447	2230	130
Diatoxanthin	acetone	452	2100	119
Dinoxanthin	acetone	442	2100	135
Echinenone	petroleum ether	458	2158	119
Ethyl-apo-8'-carotenoate	petroleum ether	448	2362	108.8
Fucoxanthin	acetone	443	1660	109.4
19'-Hexanoyloxyfucoxanthin	acetone	445	1420	109.4
Lutein	ethanol	445	2550	145
Lycopene	acetone	474	3446	185
Monadoxanthin	diethyl ether	446	2500	142
Neoxanthin	ethanol	438	2270	136
P-457	acetone	457	1640	154
Peridinin	acetone	466	1340	84.5
Prasinoxanthin	diethyl ether	446	2500	150
Siphonaxanthin	acetone	445	2500	150
Siphonein	ethanol	462	1920	150
Vaucheriaxanthin	acetone	444	2500	154
Violaxanthin	ethanol	443	2550	153
Zeaxanthin	acetone	452	2340	133

a. Refer to Part IV Data sheets for values in alternative solvents, and references.

The effect of solvent on the extinction coefficients of chlorophylls and carotenoids can be considerable (see sections D.4.2.6.4 and D.4.3, Appendix D; section F.2, Appendix F; Fig. 7.2, Chapter 7). Tables D.1 and F.1 demonstrate the importance of using extinction coefficients in the solvent for which the extinction coefficient was derived – otherwise over- or under-estimates as great as 20–25% may result.

For various reasons a wide range of solvents has been used for extinction coefficient determinations in the literature (see Part IV Data sheets). If no extinction coefficients are available for the solvent of choice, the researcher may undertake such a calibration using techniques outlined in this monograph (see section F.2, Appendix F; and Chapter 16).

Appendix F

Spectrophotometric and fluorometric equations in common use in oceanography

S. W. Jeffrey and N. A. Welschmeyer

F.1 INTRODUCTION

A number of spectrophotometric and fluorometric equations have been published since 1949 for the simultaneous determination of extracted chlorophylls *a* and *b*, *a* and *c*, and *a*, *b* and *c*.

Their accuracy depends on:

- the accuracy of the chlorophyll extinction coefficients used to derive the equations.
- the spectral purity of these chlorophylls (see Chapter 7 and the Part IV Data sheets).
- the solvents used: some solvents, such as diethyl ether, dimethyl formamide and acetone, give sharp maxima and enhance absorbance bands; others, such as

methanol, flatten maxima and depress absorption bands. The former are the most accurate.

- the presence of water extracted from cells which can alter peak positions in pure solvents and affect accuracy (Porra *et al.* 1989). For this reason 80 and 90% aqueous acetone are frequently used, as cell water will not significantly alter the already high aqueous concentration.
- the relative proportions and concentration of chlorophylls *a*, *b* and *c* in the extracts being analysed.

When applying these equations, there are other sources of inaccuracy:

- the presence of chlorophyll degradation products of similar spectral and fluorescence properties to those of the parent chlorophylls, which causes significant errors (see Section F.5 and Chapter 14). Aqueous buffer is frequently added to acetone to minimize pheophytin formation, particularly with higher plant tissues (cf. Porra *et al.* 1989).
- the precision of the spectrophotometers and fluorometers used for analysis.

The following sections discuss the spectrophotometric and fluorometric techniques that are used in oceanography. The sections on spectrophotometry (F.1–F.3) were written by S. W. Jeffrey and those on spectrophotometry (for pheopigment corrections) and fluorometry (sections F.4–F.7) by N. A. Welschmeyer.

F.2 Extinction coefficients and spectrophotometric equations

The extinction coefficients that have been used to derive the most commonly used equations are given in Table F.1. The best extinction coefficients are usually the highest: *viz*, those of Zscheile and Comar (1941), Smith and Benitez (1955), Jeffrey and Humphrey (1975) and Strain *et al.* (1963), all of which are within ±3% of each other.

These extinction coefficients were determined by weighing out highly purified or crystalline preparations, dissolving them in a known volume of solvent (usually diethyl ether to give sharpest maxima and greatest intensity of the absorption bands), taking the visible absorption spectrum and calculating the extinction coefficient according to Beer's Law (see footnote to Table F.1 and Preamble to Part IV).

The extinction coefficients of Smith and Benitez (1955) in diethyl ether were used by Vernon (1960), Lichtenthaler and Wellburn (1983) and Porra *et al.* (1989) as the standard to calibrate their purified chlorophylls in other solvents. Porra *et al.* (1989) analysed the magnesium content by atomic absorption spectroscopy and showed that the Smith and Benitez (1955) coefficients had less than a 1% error. Thus the equations of the above authors are accurate for both chlorophylls *a* and *b* in mixtures of these two pigments. Extinction coefficients in other solvents (e.g., acetone, dimethyl formamide and methanol) were derived by evaporating an aliquot of a purified chlorophyll solution in diethyl ether, and replacing it with an equal volume of a different solvent. The extinction coefficients of Jeffrey and Humphrey (1975) were derived from crystalline preparations of chlorophylls *a*, *b*, c_1 and c_2, prepared by Jeffrey (see Jeffrey and Humphrey, 1975), weighed and measured as described above, and were not based on the extinction coefficients of Strain *et al.* (1963) as incorrectly stated by Porra *et al.* (1989) and Porra (1991).

References and comments on six sets of equations and the extinction coefficients used to derive them are listed in Table F.2. It is clear from this table, and careful tests carried out by Lorenzen and Jeffrey (1980) on mixtures of pure pigments (see Table F.3), that all equations that used the chlorophyll *a* extinction coefficients of Zscheile and Comar (1941), Smith and Benitez (1955), Strain *et al.* (1963) or Jeffrey and Humphrey (1975) should be satisfactory, as they are within ±3% of the 'best' values.

Four sets of trichromatic equations for estimating chlorophylls *a*, *b* and *c* are detailed below: (Richards with Thompson, 1952; Parsons and Strickland, 1963; SCOR–UNESCO, 1966; Jeffrey and Humphrey, 1975). Equations for chlorophylls *a* and *c* follow (Jeffrey and Humphrey, 1975; Humphrey, 1979). Three sets of equations are then listed for chlorophylls *a* and *b* (Arnon, 1949; Jeffrey and Humphrey, 1975; Porra *et al.* 1989). The equations of Lichtenthaler and Wellburn (1983) for chlorophylls *a* and *b* are not given, as they are closely similar to those of Porra *et al.* (1989). E_x denotes the extinction (absorption) measured at wavelength x in a 1 cm cuvette.

A. Simultaneous equations for chlorophylls *a*, *b* and *c*

1. **Richards with Thompson (1952)**

Solvent 90% acetone; units for equations (1), (2) = mg/l; units for (3) = MSPU/l: Milli-Specified Pigment Units

Chlorophyll *a* = $15.6\,E_{665} - 2.0\,E_{645} - 0.8\,E_{630}$ (1)

Chlorophyll *b* = $25.4\,E_{645} - 4.4\,E_{665} - 10.3\,E_{630}$ (2)

Chlorophyll *c* = $109\,E_{630} - 12.5\,E_{665} - 28.7\,E_{645}$ (3)

Not recommended

2. **Parsons and Strickland (1963)**

Solvent 90% acetone; units, mg/l

Chlorophyll *a* = $11.6\,E_{665} - 0.14\,E_{630} - 1.31\,E_{645}$ (4)

Chlorophyll *b* = $20.7\,E_{645} - 4.34\,E_{665} - 4.42\,E_{630}$ (5)

Chlorophyll *c* = $55\,E_{630} - 16.3\,E_{645} - 4.64\,E_{665}$ (6)

Recommended for Chl a *only*

3. **SCOR–UNESCO (1966)**

Solvent 90% acetone; units, µg/ml

Chlorophyll *a* = $11.64\,E_{663} - 2.16\,E_{645} + 0.10\,E_{630}$ (7)

Chlorophyll *b* = $-3.94\,E_{663} + 20.97\,E_{645} - 3.66\,E_{630}$ (8)

Chlorophyll *c* = $-5.53\,E_{663} - 14.81\,E_{645} + 54.22\,E_{630}$ (9)

Recommended for Chl a *only*

4. **Jeffrey and Humphrey (1975)**

Units, µg/ml

For mixed phytoplankton populations (Chls *a*, *b*, c_1 and c_2; solvent 90% acetone)

Chlorophyll *a* = $11.85\,E_{664} - 1.54\,E_{647} - 0.08\,E_{630}$ (10)

Chlorophyll *b* = $-5.43\,E_{664} + 21.03\,E_{647} - 2.66\,E_{630}$ (11)

Chlorophylls $c_1 + c_2$ = $-1.67\,E_{664} - 7.60\,E_{647} + 24.52\,E_{630}$ (12)

Recommended for Chls a, b *and* c *(but see Appendix G for tests of accuracy of various chlorophyll mixtures)*

TABLE F.1 Specific extinction coefficients of the various chlorophylls at the major red bands used for the derivation of spectrophotometric equations. Numbers in parentheses indicate the wavelength in nm (λ_{max}) at which the extinction coefficient was determined.

Chlorophyll	Extinction coefficients (α; l gm^{-1} cm^{-1}) and wavelengths (nm) at the red maximum					Reference
	Diethyl ether	100% acetone	90% acetone	80% acetone	methanol	
Chlorophyll *a*	102.1 (660)					Zscheile and Comar (1941)
	90.1	84.0 (663)				Mackinney (1941)
	100.9 (662)					Smith and Benitez (1955)
	100.9[a] (662)	92.6 (663)	91.1 (665)			Vernon (1960)
	96.6 (660.5)					Strain *et al.* (1963)
	98.07 (660.7)	88.15 (662.7)	87.67 (664.3)			Jeffrey and Humphrey (1975)
					79.81 (666.1)	Jeffrey (unpublished)
	101[a] (662)	88.88 (662)		86.86 (663)	79.29 (666)	Lichtenthaler and Wellburn (1983)
	100.9[a, b] (662)			85.95 (663.6)	79.95 (665.2)	Porra *et al.* (1989)
Chlorophyll *b*	56.8 (642.5)					Zscheile and Comar (1941)
		51.8 (648)				Mackinney (1941)
	62.0 (644)					Smith and Benitez (1955)
	62.0[a] (644)	53.5 (647)	52.5 (648)			Vernon (1960)
	61.8 (642)					Strain *et al.* (1963)
	62.0 (643.3)		51.36 (646.8)			Jeffrey and Humphrey (1975)
	62[a] (644)	56.11 (645)		57.70 (646)	45.88 (653)	Lichtenthaler and Wellburn (1983)
	62.0[a, b] (644)			51.84 (646.6)	42.48 (652)	Porra *et al.* (1989)
Chlorophyll ($c_1 + c_2$)			19.44 (630)			Jeffrey (1963)
Chlorophyll c_1		39.2[c] (629.1)	44.8[c] (630.6)			Jeffrey (1972)
Chlorophyll c_2		37.2[c] (629.6)	40.4[c] (630.9)			Jeffrey (1972)

Extinction coefficients were determined in all cases, except Vernon (1960), Lichtenthaler and Wellburn (1983) and Porra *et al.* (1989), by weighing out highly purified or crystalline preparations, dissolving in a known volume of solvent, taking the visible absorption spectrum and calculating the extinction coefficient according to Beer's Law:

$$\alpha = \frac{1}{dC} \cdot \log \frac{I_o}{I} = \frac{A}{dC} \; ;$$

where d = inside length of the absorption cuvette in cm
C = concentration of chlorophyll in gm per litre
I_o, I = light intensities transmitted by pure solvent and the chlorophyll solution, respectively
A = Absorbance (or optical density)

a. Extinction coefficients of Smith and Benitez (1955) were used as the standard from which extinction coefficients in other solvents were experimentally determined (see text).
b. Porra *et al.* (1989) analysed the magnesium content of purified preparations of chlorophylls *a* and *b* by atomic absorption spectroscopy and showed that the Smith and Benitez (1955) extinction coefficients were accurate to within 1%.
c. Crystalline chlorophylls c_1 and c_2 are insoluble in common solvents (e.g., diethyl ether, acetone) and were dissolved in a small volume of pyridine before diluting with acetone or 90% acetone.

TABLE F.2 Evaluation of spectrophotometric equations based on the extinction coefficients used.

Author of equations	Source of extinction coefficients: Chl *a*	Chl *b*	Chl *c*	Solvent used for equations	Comments
Richards with Thompson (1952)	Zscheile (1934)	Zscheile (1934)	None available; MSPU[a] used	90% acetone	Extinction coefficients too low for Chl *a* and *b*; none available for Chl *c*. Equations not recommended.
Parsons and Strickland (1963)	Vernon (1960)[b]	Vernon (1960)[b]	Jeffrey (1963)[c]	90% acetone	Extinction coefficients acceptable for Chls *a* and *b*; those for Chl *c* are too low[c]. Equations recommended for Chl *a*.
SCOR–UNESCO (1966)	Mean of three extinction coefficients used: Zscheile and Comar (1941), Smith and Benitez (1955), and Strain *et al.* (1963). Chl *a*, 99.87 l gm^{-1} cm^{-1} in diethyl ether Chl *b*, 60.2 l gm^{-1} cm^{-1} in diethyl ether		Jeffrey (1963)[c]	90% acetone	Extinction coefficients within 5% of 'best value' for Chls *a* and *b*; those for Chl *c* are too low[c]. Equations recommended for Chl *a*.
Jeffrey and Humphrey (1975)	Jeffrey and Humphrey (1975)		Jeffrey (1972)	90% + 100% acetone	Extinction coefficients for Chls *a* and *b* within 2% of 'best' value (Smith and Benitez, 1955); those for Chl $c_1 + c_2$ are the 'best' available. Equations recommended.
Arnon (1949)	Mackinney (1941)	Mackinney (1941)	–	80% acetone	Extinction coefficients for chl *a* at λ_{max} chl *a* and λ_{max} chl *b* are 4.5% and 19.5% low respectively; for chl *b* at λ_{max} chl *b* and λ_{max} chl *a* are 12% and 14% low respectively (see Porra *et al.* 1989). Equations widely used for leaf pigments (Chls *a* and *b*) in plant physiology; not recommended.
Porra *et al.* (1989)	Porra *et al.* (1989)[b]	Porra *et al.* (1989)[b]	–	Various	Extinction coefficients of Smith and Benitez (1955) confirmed by Mg analysis. Equations recommended.

a. MSPU is a specific pigment unit of undetermined weight.
b. Based on Smith and Benitez (1955) extinction coefficients in diethyl ether (see Table F.1).
c. Low extinction coefficient due to unrecognized polymerization of purified chlorophyll *c*, which resulted in lowered intensities of absorption bands.

B. Simultaneous equations for chlorophylls *a* and *c*

5. **Jeffrey and Humphrey (1975) and Humphrey (1979)**
Units, μg/ml
(a) *For chromophyte algae* (Chls a, c_1 and c_2; solvent 90% acetone)
Chlorophyll a = $11.47\ E_{664} - 0.40\ E_{630}$ (13)
Chlorophylls $c_1 + c_2$ = $24.36\ E_{630} - 3.73\ E_{663}$ (14)
Recommended for Chls a *and* c

(b) *For dinoflagellates and cryptomonads* (Chls a and c_2; solvent 100% acetone)
Chlorophyll a = $11.43\ E_{663} - 0.64\ E_{630}$ (15)
Chlorophyll c_2 = $27.09\ E_{630} - 3.63\ E_{663}$ (16)
Recommended for Chls a *and* c

(c) *For dinoflagellates and cryptomonads* (Chls a and c_2: solvent 90% acetone)
Chlorophyll a = $11.43\ E_{664} - 0.40\ E_{630}$ (17)
Chlorophyll c_2 = $24.88\ E_{630} - 3.80\ E_{664}$ (18)
Recommended for Chls a *and* c

C. Simultaneous equations for chlorophylls *a* and *b*

6. **Arnon (1949)**
Units, μg/ml
For green algae and leaf pigments (Chls a and b; solvent 80% acetone)
Chlorophyll a = $12.7\ E_{663} - 2.69\ E_{645}$ (19)
Chlorophyll b = $22.9\ E_{645} - 4.68\ E_{663}$ (20)
Not recommended

7. **Jeffrey and Humphrey (1975)**
Units, μg/ml
For green algae and leaf pigments (Chls a and b; solvent 90% acetone)
Chlorophyll a = $11.93\ E_{664} - 1.93\ E_{647}$ (21)
Chlorophyll b = $20.36\ E_{647} - 5.50\ E_{664}$ (22)
Recommended for Chls a *and* b

8. **Porra *et al.* (1989)**
Units, μg/ml
For green algae and leaf pigments (Chls a and b; solvent, buffered 80% acetone)
Chlorophyll a = $12.25\ E_{663.6} - 2.55\ E_{646.6}$ (23)
Chlorophyll b = $20.31\ E_{646.6} - 4.91\ E_{663.6}$ (24)
Recommended for Chls a *and* b

For green algae and leaf pigments (Chls a and b; solvent methanol)
Chlorophyll a = $16.29\ E_{665.2} - 8.54\ E_{652}$ (25)
Chlorophyll b = $30.66\ E_{652} - 13.58\ E_{665.2}$ (26)
Recommended for Chls a *and* b

For green algae and leaf pigments (Chls a and b; solvent dimethyl formamide)
Chlorophyll a = $12.00\ E_{663.8} - 3.11\ E_{646.8}$ (27)
Chlorophyll b = $20.78\ E_{646.8} - 4.88\ E_{663.8}$ (28)
Recommended for Chls a *and* b

F.3 ACCURACY OF EQUATIONS TESTED WITH MIXTURES OF PURE PIGMENTS

Lorenzen and Jeffrey (1980) tested the accuracy of five spectrophotometric trichromatic equations, using 20 mixtures of pure pigments, to simulate a number of oceanographic situations.

Four sets of these equations have been used extensively in oceanography for measuring chlorophylls *a*, *b* and *c*: Richards with Thompson (1952), Parsons and Strickland (1963), SCOR–UNESCO (1966) and Jeffrey and Humphrey (1975). Humphrey (1966) modified the Richards–Thompson equations (fifth set) by correcting the chlorophyll *c* term (Jeffrey, 1963).

Table F.3 shows the accuracy of these equations, using one chlorophyll mixture from the many presented by Lorenzen and Jeffrey (1980). It contained chlorophylls *a*, *b* and *c* in the proportions 8:1.7:1. No chlorophyll degradation products were present. The results and percent errors of these analyses show that

a) Chlorophyll *a* is satisfactory when calculated by Jeffrey–Humphrey, SCOR–UNESCO or Parsons–Strickland (1–3 % error). Richards–Thompson methods are 30% high due to the use of the Zscheile (1934) extinction coefficient for chlorophyll *a* which is too low (see Table F.2).
b) Chlorophyll *b* is satisfactory only with Jeffrey–Humphrey or SCOR–UNESCO.
c) Chlorophyll *c* is satisfactory only with Jeffrey–Humphrey, but even this set of equations can give errors if chlorophyll *c* is much lower than chlorophyll *a* (see Appendix G). However, accuracy was high with other chlorophyll *c*-containing mixtures tested by Lorenzen and Jeffrey (1980) and Humphrey and Jeffrey (Appendix G).
d) Errors arising from using slightly different wavelengths for chlorophyll *a* are small (e.g., 663 nm (SCOR–UNESCO); 664 nm (Jeffrey–Humphrey); 665 nm (Parsons–Strickland and Richards–Thompson)). In 90% acetone, the chlorophyll *a* red band shows a plateau at 663–664 nm and drops 3% in optical density at 665 nm. The small errors in wavelength reading are insignificant for

TABLE F.3 Comparison of accuracy of five sets of trichromatic spectrophotometric equations using a mixture of spectrally pure chlorophylls *a*, *b* and *c* ($c_1 + c_2$) (from Lorenzen and Jeffrey, 1980).

Equations	Concentration of chlorophylls			% errors		
	Chl *a*	Chl *b* (μg ml.$^{-1}$)	Chl *c* ($c_1 + c_2$)	Chl *a*	Chl *b*	Chl *c*
Known chlorophyll mixture	1.84	0.39	0.296			
Jeffrey-Humphrey (1975)	1.85	0.39	0.225	<1%	<1%	24% low
Parsons-Strickland (1963)	1.78	0.26	0.672	3% low	33% low	208% high
SCOR-UNESCO (1966)	1.80	0.35	0.55	2% low	11% low	169% high
Richards-Thompson (1952)	2.38	0.21	1.00	30% high	54% low	300% high
Richards-Thompson modified Humphrey (1966)	2.38	0.21	0.43	30% high	54% low	45% high

chlorophyll *a*. For chlorophyll *b* the wavelengths used are 647 nm (Jeffrey–Humphrey; SCOR–UNESCO) and 645 nm (Parsons–Strickland and Richards–Thompson). In 90% acetone there is an 11% drop in optical density at 645 nm compared to that at 647 nm (λ_{max}), which results in the greater errors observed. The Jeffrey–Humphrey equations are the only ones available for the simultaneous measurement of chlorophylls *a, b,* c_1 and c_2. Tests of their accuracy with mixtures of pure chlorophylls in different proportions and concentrations are given in Appendix G.

F.4 SPECTROPHOTOMETRIC EQUATIONS THAT CORRECT FOR PHEOPIGMENTS

The trichromatic equations for determining pigment concentrations in mixtures of chlorophylls *a*, *b* and *c* assume the absence of chlorophyll degradation products that can interfere with the assay. However, chlorophyll degradation products, such as pheophytin *a* and pheophorbide *a*, may be present in relatively high concentrations in some natural samples, such as in deep water from below the euphotic zone, detritus and sediments. The presence of pheopigments will result in an overestimate of chlorophyll *a*, since the absorption characteristics of pheopigments are similar to those of their parent chlorophylls (see Part IV Data sheets).

Methods to correct for pheopigment interference in spectrophotometric and fluorometric assays for chlorophyll *a* were developed by Vernon (1960), Yentsch and Menzel (1963), Holm-Hansen *et al.* (1965) and Lorenzen (1967). They all use spectrophotometric acidification techniques based on the following points:

- chlorophyll *a* treated with weak acid converts to pheophytin *a* with 100% molar stoichiometry
- the molar absorptivity of the red absorbance band of pheophytin *a* is lower than that of chlorophyll *a*
- the relative absorption ratio of the two pigments in the same solvent is a constant.

Lorenzen (1967) determined that the absorbance at 665 nm was reduced by a factor of 1.7 when pure chlorophyll *a* in 90% acetone was pheophytinized by weak acid, thus defining the 'acid factor' for pure chlorophyll *a* (A 665_o/A 665_a: for symbols, see below). The value of the acid factor for any unknown sample will range from 1.0 (pure pheophytin *a*) to 1.7 (pure chlorophyll *a*). For example, a sample consisting of 50% chlorophyll *a* and 50% pheophytin *a* on a molar basis will yield an acid factor of 1.35. The concentrations of chlorophyll *a* and pheophytin *a* in any unknown sample can be determined algebraically by the equations of Lorenzen (1967):

$$\text{Chl } a \text{ (mg m}^{-3}\text{)} = \frac{[AK(665_o - 665_a)v]}{V_f l}$$

$$\text{Pheo } a \text{ (mg m}^{-3}\text{)} = \frac{[AK((R665_a) - 665_o)v]}{V_f l}$$

where

A = inverse extinction coefficient in 90% acetone for chl *a* (x 1000); [1/(91.1 l g^{-1} cm^{-1})] x 1000 = 11.0 µg cm ml^{-1}

R = maximum absorbance ratio of $665_o/665_a$ in the absence of pheopigments, 1.7

K = R/(R–1.0); 1.7/0.7, or 2.43

665_o = absorbance at 665 nm before acidification

665_a = absorbance at 665 nm after acidification
v = volume of 90% acetone used for extraction (ml)
V_f = volume of water filtered (litres)
l = path length of cuvette (cm)

The basic principles of Lorenzen's (1967) spectrophotometric acidification technique outlined above are analogous to those used in previous fluorometric acidification techniques (e.g., Holm-Hansen *et al.* 1965). The original spectrophotometric method of Lorenzen (1967) used an extinction coefficient of 91.1 l g^{-1} cm^{-1} in 90% acetone (Vernon, 1960), which will give chlorophyll concentrations about 4% lower than those calculated with the recommended extinction coefficient of Jeffrey and Humphrey (1975) (see Table F.1). Lorenzen (1967) recommended reading absorbance values at 665 nm instead of 663 nm — the 'true' λ_{max} for chlorophyll *a* in acetone — to facilitate readings on the then-popular Beckman DU spectrophotometer, which used a non-linear grid on its monochromator settings. However, on other instruments any convenient wavelength setting near the red absorbance peak can be used as long as the acid-factor for pure chlorophyll *a* is measured at that wavelength and incorporated in the variables A and R above.

The units of concentration are calculated on a weight-basis (μg l^{-1}), which has unfortunately led to some confusion (Lorenzen and Newton-Downs, 1986). Acid pheophytinization results in stoichiometric pigment conversion on a molar basis (see above). Thus, the equations should be expressed on a molar basis to meet the assumptions of molar stoichiometry outlined earlier. Lorenzen cast the equations in terms of weight-specific concentrations to maintain dimensional consistency with extinction coefficients, which are most often cited in weight-specific units. There is only a 3% difference in the molecular weights of chlorophyll *a* and pheophytin *a* (893 and 869, respectively), and the error in converting from mass to moles was considered trivial. There is therefore only a small error in the chlorophyll *a* and pheopigment concentrations calculated as above, provided the unknown sample contains pheophytin *a* as the only degradation product of chlorophyll *a* (which is seldom the case). However, the calculated weight of pheopigment can be overestimated by a factor of 1.51 if the unknown sample contains pheophorbide *a* rather than pheophytin *a*, since the molecular weight of pheophorbide *a* (591) is 1.51-fold less than that of chlorophyll *a*. Any chlorophyll *b* present is also degraded by the acidification and is erroneously expressed as a 'pheopigment' component (see Section F.5).

The accuracy of pigment concentrations calculated from the spectrophotometric acidification technique is not uniform for all potential mixtures of chlorophyll *a* and pheopigments. The variable error results from the algebra of the acidification equations *per se*. Simple propagation of instrumental errors in reading unacidified and acidified absorbances will show that the greatest error in determining pheopigment concentration occurs when pheopigments constitute a small fraction of total pigment mass (chlorophyll + pheopigments), e.g., when the absorbance acid ratio is >1.35. Likewise, the greatest potential error in calculating chlorophyll *a* concentrations occurs when chlorophyll *a* constitutes a small fraction of the total pigment mass; when the acid ratio is <1.35.

For most samples from the euphotic zone, and especially for pure algal tissues, the mass of chlorophyll *a* greatly outweighs that of pheopigments. Calculated concentrations of chlorophyll *a* should be accurate for these samples, provided caution is exercised in judging the accuracy of the pheopigment (see above). Acidification techniques were specifically designed to correct the chlorophyll *a* concentration for interference from pheopigments (Lorenzen and Newton-Downs, 1986), but the concentration of pheopigments calculated under low pheopigment/chlorophyll ratios should be viewed with caution. The same restrictions also apply directly to fluorescence acidification techniques (see Section F.5).

F.5 FLUOROMETRIC EQUATIONS

Fluorometric equations for extracted chlorophyll a and pheopigments

Fluorescence assays for extracted chlorophyll *a* are two to three orders of magnitude more sensitive than spectrophotometric methods. For this reason, fluorometric chlorophyll techniques have been popular for studies of oligotrophic marine systems, where environmental chlorophyll concentrations can be very low (<0.1 $\mu g\ l^{-1}$). Fluorescence acidification techniques, which correct for the presence of pheopigments, have been routinely used for over 30 years (Holm-Hansen *et al.* 1965).

The assumptions of the fluorometric acidification technique are analogous to those used in spectrophotometric acidification methods (see Section F.4). It is assumed that

- when chlorophyll *a* is treated with weak acid it will be converted to pheophytin *a* with 100% molar stoichiometry
- the molar fluorescence response of pheophytin *a* is lower than that for chlorophyll *a*, yielding a constant relative ratio of fluorescence responses ($F_m = F_o/F_a$) for the two pigments.

In contrast to spectrophotometric methods, the relative fluorescence response (sensitivity coefficient, K) and the maximum fluorescence acid ratio (F_m) are highly dependent on the specific fluorometric optical configuration of each instrument and therefore must be empirically determined with a chlorophyll *a* standard that is free of both pheopigments and chlorophyll *b*. The equations for chlorophyll *a* and pheophytin *a* are listed below:

$$\text{Chlorophyll } a \text{ (mg m}^{-3}\text{)} = \frac{KF_m v(F_o - F_a)}{V_f(F_m - 1)}$$

$$\text{Pheophytin } a \text{ (mg m}^{-3}\text{)} = \frac{KF_m v(F_m F_a - F_o)}{V_f(F_m - 1)}$$

where

K = fluorescence sensitivity coefficient in extraction solvent
= [(μg Chl *a*/ml solvent)/instrument fluorescence unit]

F_m = maximum ratio F_o/F_a in the absence of pheopigments and chlorophyll *b*

F_o = fluorescence before acidification

F_a = fluorescence after acidification

v = volume of acetone used for extraction (ml)

V_f = volume of water filtered (litres)

The fluorometric calibration coefficients, K and F_m, can be determined empirically from a solvent solution of pure chlorophyll *a* of known concentration (determined spectrophotometrically, using an absorbance extinction coefficient of 87.67 $l\ g^{-1}\ cm^{-1}$ at 664 nm in 90% acetone). The fluorometric calibration coefficients can also be determined with mixed extracts containing chlorophylls and carotenoids where the chlorophyll *a* concentration is known, as long as the solution is free of pheopigments and chlorophyll *b*. It is well known that chlorophyll *b* undergoes wavelength shifts upon acidification which bias the fluorescence readings for chlorophyll *a*, resulting in significant overestimates of the true concentration of pheopigments (Gibbs, 1979; Trees *et al.* 1985). For this reason, the fluorescence acidification technique is not recommended for analysis of natural samples known to contain significant concentrations of chlorophyll *b*, e.g., freshwater or marine samples dominated by chlorophytes or prasinophytes (see Chapter 2).

Fluorometric equations for extracted chlorophylls a, b and c and their pheophytins

Several methods have been published which determine concentrations of extracted chlorophylls *a*, *b* and *c* and their pheophytins using multi-wavelength fluorometric readings (Loftus and Carpenter, 1971; Boto and Bunt, 1978; Bazzaz and Rebiez, 1979; Neveux and Panouse, 1987). These methods make use of the fact that each pigment is characterized by unique fluorescence excitation/emission spectra, and each pigment can be caused to fluoresce maximally through selective choice of excitation and emission wavelengths. The multi-wavelength fluorescence method is analogous to the trichromatic spectrophotometric technique, since it requires a number of readings equivalent to the individual pigments analyzed. The method of Neveux and Panouse (1987) determines concentrations of chlorophylls *a*, *b* and *c* and pheophytins *a*, *b* and *c* by making fluorescence readings at six pairs of excitation/emission settings, the calculation of which requires solution of six equations for six unknowns.

In general, the multi-wavelength fluorometric techniques have been less popular because of the tedious calibration procedures required. Each pigment to be analyzed must be available as a highly purified standard and its concentration known for the empirical determination of fluorescence response coefficients. The calibration coefficients are dependent on optical conditions, so any routine changes, such as adjusting the slit width, require re-determination of the calibration coefficients.

Fluorometric technique for determining chlorophyll a

A recent method uses a simple filter fluorometer to determine chlorophyll *a* in the presence of pheopigments and chlorophyll *b* (without acidification procedures) (Welschmeyer, 1994). It yields measurements of chlorophyll *a* only, and provides no estimate of pheopigments. Most of the advantages of fluorescence sensitivity are maintained and the method can be used on samples that are known to contain chlorophytes, e.g., freshwater oligotrophic systems.

The method calls for a change in the lamp and filters used in conventional filter fluorometers. The new lamp (Type 10–089 Turner Designs Inc.) contains little energy at the wavelength of maximum pheophytin *a* excitation (408 nm), and uses interference filters at 436 nm and 680 nm (10 nm bandwidth) at the excitation/emission positions. The optical configuration produces reasonable sensitivity to chlorophyll *a* while maintaining desensitized responses from both chlorophyll *b* and pheophytin *b*. Chlorophyll *a* is calculated from the following equation, using a single determination of fluorescence for each sample:

$$\text{Chlorophyll } a \text{ (mg m}^{-3}) = \frac{KF_o v}{V_f}$$

where

K = fluorescence sensitivity coefficient in extraction solvent
= [(µg Chl *a*/ml solvent)/instrument fluorescence unit]

F_o = fluorescence response (no acidification)

v = volume of acetone used for extraction (ml)

V_f = volume of water filtered (litres)

The method offers simple, low-cost determination of chlorophyll *a* and reduces the requirements for purity of chlorophyll *a* standards, since any solvent mixture containing chlorophyll *a* (e.g., leaf extracts) can be used for standardization.

In vivo fluorescence

Chlorophyll *a* determination from *in vivo* fluorescence (Lorenzen, 1966) is probably the most convenient but least accurate method of determining chlorophyll (see Section 17.1; Table 17.1). It is widely used in marine research for mapping broad geographic patterns of phytoplankton biomass. The fluorescence of individual phytoplankton cells is illustrated in Plate 4A, Chapter 6. The instruments used include ship-board flow-through fluorometers (Lorenzen, 1966) and *in situ* fluorometers (Mackey *et al.* 1995). Some problems with this method are discussed briefly by Lorenzen and Jeffrey (1980).

The calculation procedure reduces to:

$$\text{Chlorophyll } a \text{ (mg m}^{-3}) = KF$$

where

K = *in vivo* fluorescence sensitivity coefficient
= mg Chl *a* m^{-3} per fluorescence unit (obtained by prior calibration)

F = *in vivo* fluorescence response

To calibrate the instruments, extracted chlorophyll *a* concentrations and the associated *in vivo* fluorescence response must be determined. Instrument response is typically linear over natural concentration ranges, excluding turbid waters. Fluorescence response per unit chlorophyll is species-dependent and is also related to the photophysiological status of the algae. Variables such as cell-packaging, chloroplast shape, photosynthetic rate, presence of senescent cells, and temperatures at which the analyses are carried out, can result in changes in the apparent *in vivo* fluorescence sensitivity coefficient (see Falkowski and Kiefer, 1985). For these reasons the instrument must be calibrated frequently. These inaccuracies do not overshadow the important role that *in vivo* fluorometry has played in rapidly determining

- an index of chlorophyll (phytoplankton abundance) along the ship's track at sea
- vertical profiles of phytoplankton distributions
- concentrations of chlorophyll in discrete water samples, and
- long term continuous changes in phytoplankton biomass estimated from moored *in situ* fluorometers.

F.6 Problems with using spectrophotometric and fluorometric equations for phytoplankton field samples

Spectrophotometric assays

The most common problem in spectrophotometric assays of natural phytoplankton assemblages is inadequate sensitivity. Unless one can visually 'see colour' in the final solvent extract, the absorbance may be so low as to produce inaccuracies arising from unacceptably low signal-to-noise ratios. Ideally, enough seawater should be filtered to yield an absorbance (optical density) >0.1 at 664 nm when using the spectrophotometric acidification technique. Final absorbance should be even higher when using the trichromatic technique, since the absorbances at 647 and 630 nm will be more than two-fold lower than at 664 nm. The user should make adjustments to obtain the highest absorbance possible, such as filtering more sample, using the smallest solvent volume and using small-volume, long-path cuvettes. If sonication is used to disrupt cells, extractions with 1.2 ml 90% acetone in microcentrifuge tubes will yield at least 1 ml of usable extract, which should be adequate to fill most 1 cm spectrophotometric micro-cuvettes.

Fluorometric assays

Filter Fluorometers. The specific optical configurations of common filter fluorometers will result in wavelength-dependent response characteristics that are unique to each instrument. The user should be aware that changes in filters, lamps and/or photomultipliers, other than those specified in a given method, will often result in performance characteristics quite different from those described in the literature. It is necessary, for instance, to use red-sensitive photomultipliers (Hamamatsu R446, or equivalent) in all of the fluorometric chlorophyll analyses described above to gain full advantage of fluorometric sensitivity.

The colour spectra of common lamps used in filter fluorometers must also be considered with respect to specific excitation-emission spectra of pigments, since unwanted interferences may occur. For example, a shift of over 20 nm in the blue excitation band occurs when chlorophyll *b* is converted to pheophytin *b* by acidification (excitation/emission maxima of chlorophyll *b* and pheophytin *b* are 458/653 nm and 435/658 nm respectively, Fig. F.1). The resultant blue excitation shift causes pheophytin *b* to be detected with greater sensitivity than chlorophyll *b* in most filter fluorometers since the standard fluorometric lamp (Fig. F.2A and F.2B) has a strong energy band at 434 nm. This results in the well known 'chlorophyll *b* problem' which causes overestimates of pheopigments and underestimates of chlorophyll *a* in the conventional fluorometric acidification technique (see section F.5).

As can be seen in Fig. F.2, several of the common fluorometer lamps contain strong energy bands at 405 and 434 nm (Fig. F.2A–E) making them potentially useful in fluorometric acidification analyses of chlorophyll *a* and pheopigments. However, when combined with a blue excitation filter (Corning 5–60), the resultant acid ratio for pure chlorophyll *a* can vary greatly depending on the lamp used. Calibration coefficients should be determined empirically as described in section F.5 whenever making optical changes to filter fluorometers for chlorophyll analysis. Figs. F.1 and F.2 are provided to assist the user in selecting lamps for specific pigment analyses, which might also include fluorometric detection in HPLC analyses.

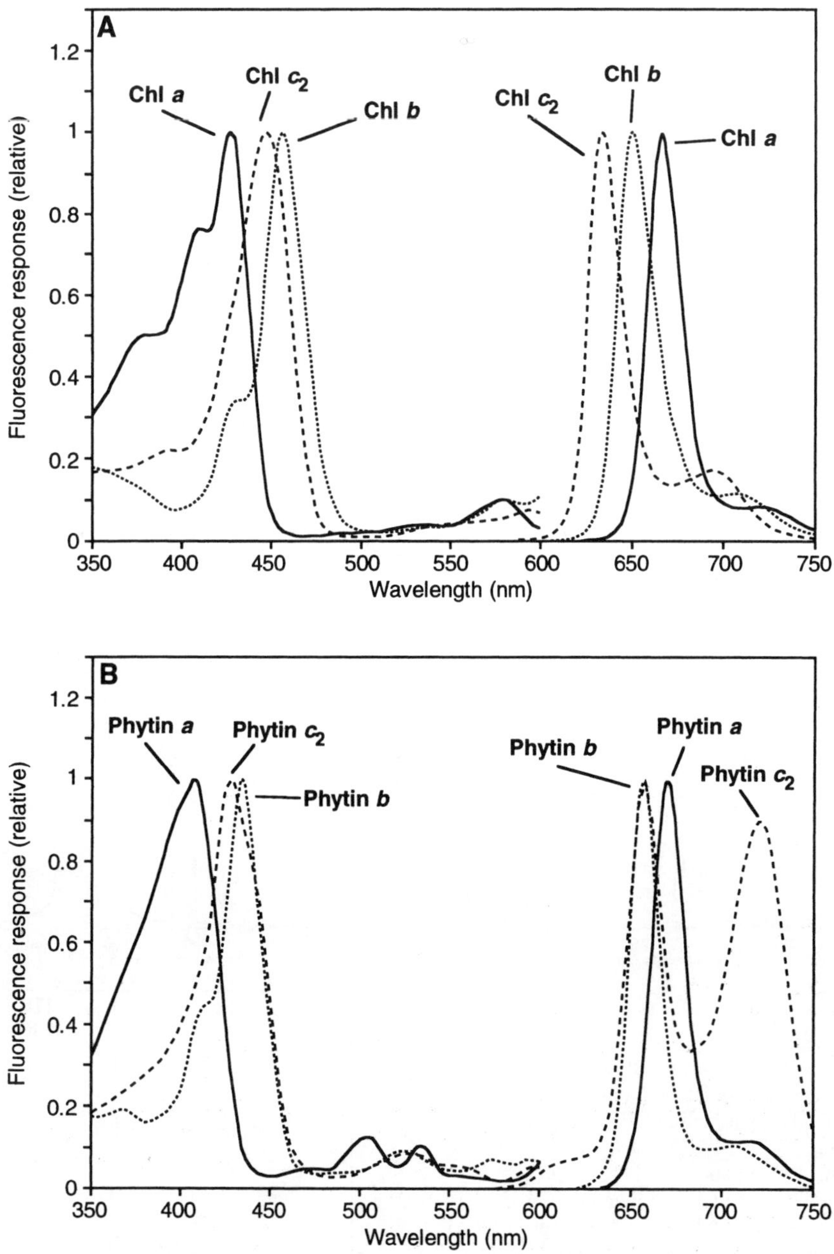

Figure F.1
Photon-corrected excitation-emission spectra of HPLC-purified 90% acetone solutions of A, chlorophylls *a*, *b* and c_2, and B, their acid-produced pheophytins. Fluorescence responses were measured at the maximum excitation-emission wavelength for each pigment. Results were normalized to unity, yielding concentration-independent spectra.

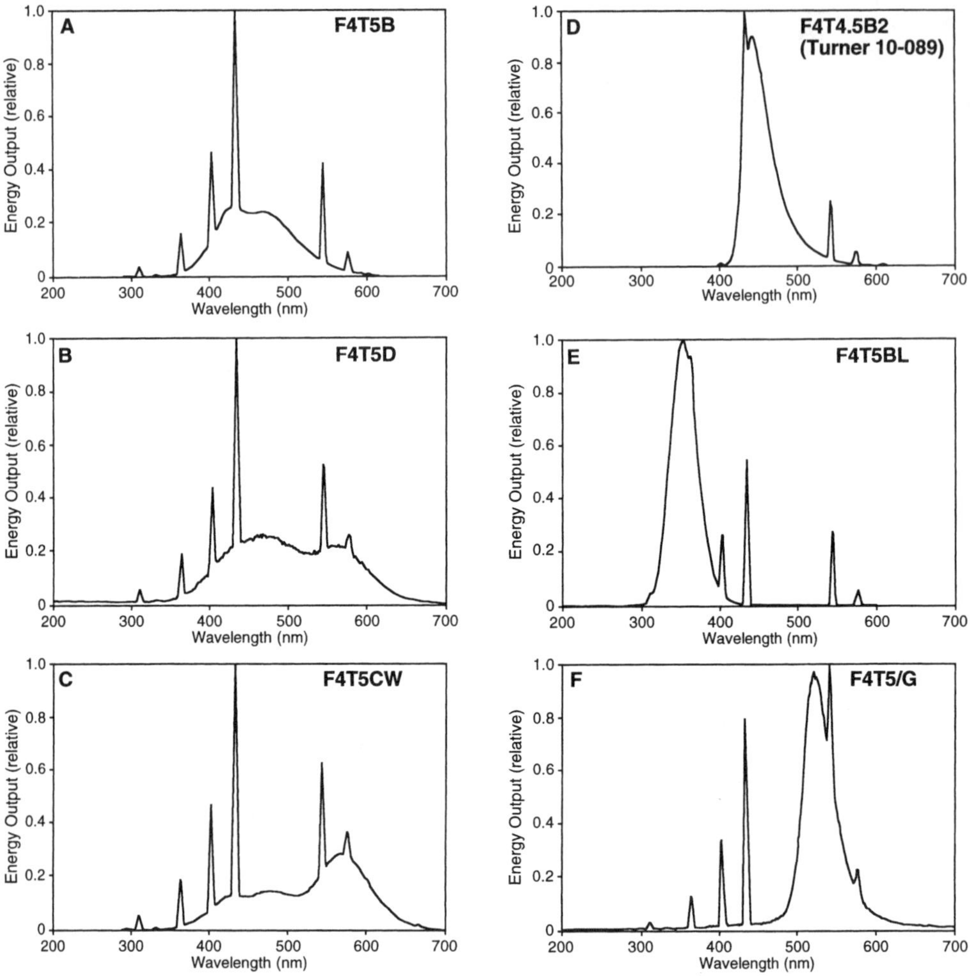

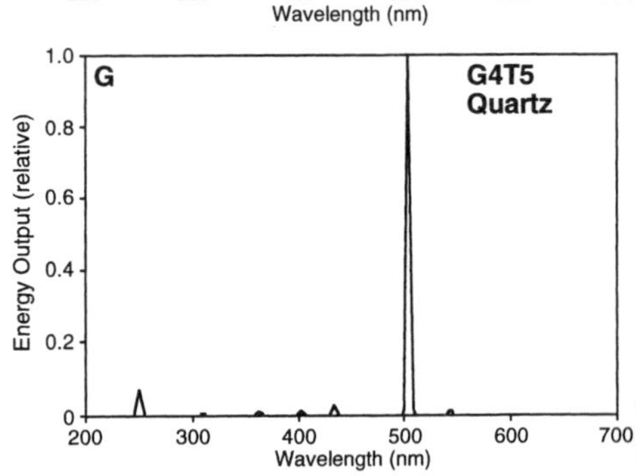

Figure F.2
Energy spectra for lamps used in common filter fluorometers. A, F4T5B, original blue lamp; B, F4T5D, daylight lamp; C, F4T5CW, cool-white lamp; D, F4T4.5B2 (Turner # 10–089), replacement blue lamp; E, F4T5, black-light lamp; F, F4T5/G, green lamp; and G, G4T5, clear quartz general UV lamp. All lamps are commercially available except the original blue lamp (F4T5B) which is now out of production. Spectra were obtained by guiding lamp emission to the entrance slit of a Spex model 1681 monochromator (Welschmeyer, 1994). All spectra were normalized to unity at the wavelength of greatest energy output. Energy correction in the UV range was not possible; the relative energy level of the quartz lamp line at 254 nm (G) is significantly greater than shown.

Spectrofluorometers. Quantitative analyses of chlorophylls by the spectrofluorometric technique is less common than simple filter-fluorometric techniques for reasons described in section F.5. In particular, the user should be aware of differences in excitation-emission spectra made with, and without, photon-correction. Uncorrected spectra contain wavelength-dependent fluctuations in both a) excitation lamp energy and b) photomultiplier sensitivity, thereby distorting the true fluorescence characteristics of pigments being analyzed. Corrected and uncorrected spectra of similar samples can appear quite different, making comparisons between techniques difficult. In general, each user is 'on their own' when attempting to adapt spectrofluorometry to quantitative pigment analyses, and it is highly recommended that photon-corrected instrumentation be used to yield optimum universality (cf. Bazzaz and Rebiez, 1979; Neveux and Panouse, 1987).

F.7 Summary

Most of our current knowledge of phytoplankton distributions in the ocean is based on chlorophyll analyses made by spectrophotometry and by popular filter-fluorometry. Both techniques will continue to be used whenever simple assays for chlorophyll *a* are required. It is clear, however, that fluorometry offers significant advantages in sensitivity which result in its overwhelming popularity in providing simple, low cost chlorophyll *a* analyses in most ocean environments. All pigment assays, including fluorometric and HPLC techniques, will ultimately be referenced to spectrophotometric absorbance measurements and, in this regard, spectrophotometric techniques are clearly indispensable.

Where concentrations of accessory chlorophylls are needed, the dichromatic and trichromatic spectrophotometric techniques can provide good accuracy for chlorophylls *a* and *b*, *a* and *c*, and *a*, *b*, and c_1+c_2 (see sections F.2, F.3 and Appendix G), providing that sample absorbances are high and degradation products are absent (e.g., algal cultures). However, if accurate analyses of accessory chlorophylls and degradation products are required, especially on natural field samples, then the isocratic HPLC system outlined in Chapter 11 is highly recommended. If the full suite of chlorophylls, carotenoids and chlorophyll degradation products in field samples is required, then an HPLC method, such to that described in Chapter 12, should be used.

Fluorometric analysis of chlorophyll *a* is accurate under conditions where chlorophyll *b* is absent. However, the accuracy of estimated pheopigment concentrations is questionable, especially under common conditions where the ratio chlorophyll/pheopigment is high. Since we now know that prochlorophytes rich in chlorophyll *b* and divinyl chlorophyll *b* are widely distributed, and even dominant, in the open ocean (Chisholm *et al.* 1988), absolute concentrations of pheopigments determined from the fluorometric acidification technique should be treated with caution.

In vivo fluorometry provides the most convenient, but least accurate method of determining chlorophyll *a*, and has become increasingly popular for studies involving long term monitoring and small-scale spatial distributions. There is currently no substitute for *in vivo* fluorescence for rapidly mapping real-time, vertical resolution in phytoplankton biomass over both temporal and spatial dimensions.

References

Arnon, D. I. (1949). Copper enzymes in isolated chloroplasts: Polyphenol oxidase in *Beta vulgaris*. *Plant Physiol.* 24: 1–15.

Bazzaz, M. B., Rebiez, C. A. (1979). Spectrofluorometric determination of chlorophyll(ide) *a* and *b* and pheophytin (or pheophorbide) *a* and *b* in unsegregated pigment mixtures. *Photochem. Photobiol.* 30: 709–721.

Boto, K. G., Bunt, J. S. (1978). Selective excitation fluorometry for the determination of chlorophylls and pheophytins. *Anal. Chem.* 50: 392–395.

Chisholm, S. W., Olsen, R. J., Zettler, E. R., Goericke, R., Waterbury, J. B., Welschmeyer, N. A. (1988). A novel free-living prochlorophyte abundant in the oceanic euphotic zone. *Nature*, 334: 340–343.

Falkowski, P. G., Kiefer, D. A. (1985). Chlorophyll *a* fluorescence in phytoplankton: relationship to photosynthesis and biomass. *J. Plankton Res.* 7: 715–731.

Gibbs, C. F. (1979). Chlorophyll *b* interference in the fluorometric determination of chlorophyll *a* and phaeopigments. *Aust. J. Mar. Freshw. Res.* 30: 597–606.

Holm-Hansen, O., Lorenzen, C. J., Holmes, R. W., Strickland, J. D. H. (1965). Fluorometric determination of chlorophyll. *J. du Conseil*, 30: 3–15.

Humphrey, G. F. (1966). The concentrations of chlorophylls *a* and *c* in the south-eastern Indian Ocean. *Aust. J. Mar. Freshw. Res.* 17: 135–145.

Humphrey, G. F. (1979). Photosynthetic characteristics of algae grown under constant illumination and light–dark regimes. *J. Exp. Mar. Biol. Ecol.* 40: 63–70.

Jeffrey, S. W. (1963). Purification and properties of chlorophyll *c* from *Sargassum flavicans*. *Biochem. J.* 86: 313–318.

Jeffrey, S. W. (1972). Preparation and some properties of crystalline chlorophylls c_1 and c_2 from marine algae. *Biochim. Biophys. Acta*, 279: 15–33.

Jeffrey, S. W., Humphrey, G. F. (1975). New spectrophotometric equations for determining chlorophylls *a*, *b*, c_1 and c_2 in higher plants, algae and natural phytoplankton. *Biochem. Physiol. Pflanzen*, 167: 191–194.

Lichtenthaler, H. K., Wellburn, A. R. (1983). Determination of total carotenoids and chlorophylls *a* and *b* of leaf extracts in different solvents. *Biochem. Soc. Trans.* 11: 591–592.

Loftus, M. E., Carpenter, J. H. (1971). A fluorometric method of determining chlorophylls *a*, *b* and *c*. *J. Mar. Res.* 29: 319–338.

Lorenzen, C. J. (1966). A method for the continuous measurement of *in vivo* chlorophyll concentration. *Deep–Sea Res.* 13: 223–227.

Lorenzen, C. J. (1967). Determination of chlorophyll and phaeopigments: spectrophotometric equations. *Limnol. Oceanogr.* 12: 343–346.

Lorenzen, C. J., Jeffrey, S. W. (1980). Determination of chlorophyll in seawater: report of intercalibration tests. *UNESCO Tech. Pap. Mar. Sci.* 35, 20 pp.

Lorenzen, C. J., Newton-Downs, J. (1986). The specific absorption co-efficient of chlorophyllide *a* and pheophorbide *a* in 90% acetone, and comments on the fluorometric determination of chlorophyll and pheopigments. *Limnol. Oceanogr.* 31: 449–452.

Mackey, D. J., Parslow, J., Higgins, H. W., Griffiths, F. B., O'Sullivan, J. E. (1995). Plankton productivity and biomass in the Western Equatorial Pacific: biological and physical controls. *Deep–Sea Res.* 42: 499–533.

Mackinney, G. (1941). Absorption of light by chlorophyll solutions. *J. Biol. Chem.* 140: 315–322.

Neveux, J., Panouse, M. (1987). Spectrofluorometric determination of chlorophylls and pheophytins. *Arch. Hydrobiol.* 109: 567–581.

Parsons, T. R., Strickland, J. D. H. (1963). Discussion of spectrophotometric determination of marine plant pigments, with revised equations for ascertaining chlorophylls and carotenoids. *J. Mar. Res.* 21: 155–163.

Porra, R. J. (1991). Recent advances and reassessments in chlorophyll extraction and assay procedures for terrestrial, aquatic and marine organisms including recalcitrant algae. In: Scheer, H. (ed.), *Chlorophylls*, CRC Press, Boca Raton, USA, p. 31–57.

Porra, R. J., Thompson, W. A., Kriedemann, P. E. (1989). Determination of accurate extinction coefficients and simultaneous equations for assaying chlorophylls *a* and *b* extracted with four different solvents: verification of the concentration of chlorophyll standards by atomic absorption spectroscopy. *Biochim. Biophys. Acta*, 975: 384–394.

Richards, F. A., Thomson, T. G. (1952). The estimation and characterisation of plankton populations by pigment analyses. II. A spectrophotometric method for the estimation of plankton pigments. *J. Mar. Res.* 11: 156–172.

SCOR–UNESCO (1966). Determination of photosynthetic pigments in seawater. *Monogr. Oceanog. Method.* UNESCO, Paris, vol. 1, p. 11–18.

Smith, J. H. C., Benitez, A. (1955). Chlorophylls: analysis in plant materials. In: Paech, K., Tracey, M. (eds), *Modern Methods of Plant Analysis*, Springer, Heidelberg, vol. 4, p. 142–196.

Strain, H. H., Thomas, M. R., Katz, J. J. (1963). Spectral absorption properties of ordinary and fully deuterated chlorophylls *a* and *b*. *Biochim. Biophys. Acta*, 75: 306–311.

Trees, C. C., Kennicutt, M. C., Brooks, J. M. (1985). Errors associated with the standard fluorometric determination of chlorophylls and pheopigments. *Mar. Chem.* 17: 1–12.

Vernon, L. P. (1960). Spectrophotometric determination of chlorophylls and pheophytins in plant extracts. *Anal. Chem.* 32: 1144–1150.

Welschmeyer, N. A. (1994). Fluorometric analysis of chlorophyll *a* in the presence of chlorophyll *b* and pheopigments. *Limnol. Oceanogr.* 39: 1985–1992.

Yentsch, C. S., Menzel, D. W. (1963). A method for the determination of phytoplankton chlorophyll and pheophytin by fluorescence. *Deep–Sea Res.* 10: 221–231.

Zscheile, F. P. (1934). A quantitative spectrophotoelectric analytical method applied to solutions of chlorophylls *a* and *b*. *J. Phys. Chem.* 38: 95–102.

Zscheile, F. P., Comar, C. L. (1941). Influence of preparative procedure on the purity of chlorophyll components as shown by absorption spectra. *Bot. Gazette*, 102: 463–481.

Appendix G

Tests of accuracy of spectrophotometric equations for the simultaneous determination of chlorophylls *a, b,* c_1 and c_2

G. F. Humphrey and S. W. Jeffrey

G.1 Introduction

Only one set of equations (Jeffrey and Humphrey, 1975), derived from the 'best' extinction coefficients, is available for the simultaneous determination of chlorophylls *a*, *b* and c_1+c_2 in mixtures. Here we present an assessment of their accuracy over a range of chlorophyll concentrations and proportions, to define clearly the limits within which they may be used. These tests were carried out with freshly prepared chlorophyll solutions, free from traces of chlorophyll degradation products. Problems that occur when chlorophyll degradation products are present are discussed in Section G.3.2 and Chapter 14.

G.2 Methods

Spectrally pure chlorophylls *a*, *b*, c_1 and c_2 were isolated by the preparative methods outlined in Chapter 7. They were re-purified by thin-layer chromatography immediately before use. A few crystals of chlorophylls *a* and *b* were dissolved in the minimum volume of diethyl ether (peroxide-free) and each was chromatographed on at least four sucrose TLC plates (Jeffrey, 1968), run simultaneously, to provide enough pure pigment for each set of tests. Chlorophylls c_1 and c_2 were re-chromatographed on thin layers of polyethylene before use (Jeffrey, 1969).

The zones of the pure compounds (free from traces of epimers, allomers, chlorophyllides, pheophytins and pheophorbides) were collected and eluted with freshly opened diethyl ether (Mallinckrodt, peroxide-free). The visible spectrum was read immediately, and the chlorophyll concentration determined from the appropriate extinction coefficients (Table G.1).

Mixtures of known concentrations and different proportions of chlorophylls *a* and *b*, *a* and c_1+c_2, *a* and c_2, and *a*, *b* and c_1+c_2 were prepared, and their visible spectra read immediately in 1 cm microcells in a Cary 14 recording spectrophotometer. All precautions for working with chlorophylls (e.g., dim light, clean glassware, absence of acids, speed of operation) were observed. Chlorophylls *a*, *b* and/or *c* in each mixture were then calculated by the appropriate spectrophotometric equations, and the recoveries determined.

Table G.1 Extinction coefficients of chlorophylls used to derive the Jeffrey–Humphrey (1975) equations at the red absorption maximum (λ_{max} in nm is given in parenthesis).

Chlorophyll	Extinction coefficients at the red maximum l gm^{-1} cm^{-1}					
	Diethyl ether		100% acetone		90% acetone	
Chlorophyll *a*[a]	98.07	(660.7)	88.15	(662.7)	87.67	(664.3)
Chlorophyll *b*[a]	62.00	(643.3)			51.36	(646.8)
Chlorophyll c_1[b]			39.2	(629.1)	44.8	(630.6)
Chlorophyll c_2[b]			37.2	(629.6)	40.4	(630.9)

a. From crystalline preparations prepared by Jeffrey (see Jeffrey and Humphrey, 1975), by methods outlined in Chapter 7.
b. Jeffrey (1972).

G.3 Results and Discussion

G.3.1 *Pure chlorophyll mixtures*

Quantitative tests of five sets of equations are given in Tables G.2 – G.6:

Chlorophylls *a* and *b* in 90% acetone (G.2)
Chlorophylls *a* and c_2 in 100% acetone (G.3)
Chlorophylls *a* and (c_1+c_2) in 100% acetone (G.4)
Chlorophylls *a* and (c_1+c_2) in 90% acetone (G.5)
Chlorophylls *a* and *b* and (c_1+c_2) in 90% acetone (G.6)

The chlorophyll concentrations in the mixtures used for the tests ranged from

0.062 – 1.869 µg/ml for chlorophyll *a*
0.039 – 1.859 µg/ml for chlorophyll *b*
0.128 – 1.022 µg/ml for chlorophyll c_2
0.043 – 2.078 µg/ml for chlorophylls (c_1+c_2).

The chlorophyll proportions in the mixtures ranged from

Chl *a* : Chl *b* = 0.81 to 2.42
Chl *a* : Chl c_2 = 1.15 to 13.76
Chl *a* : Chl c_1+c_2 = 1.01 to 11.96

Chl *a* : Chl *b* : Chl c_1+c_2 = 1.0 : 0.13 : 0.83, to
1.0 : 0.13 : 0.14, to
1.0 : 0.63 : 6.98

Ratios were chosen to simulate a wide range of conditions that might be encountered in field samples (see Lorenzen and Jeffrey, 1980) and algal cultures.

Analysis of replicates of single pigments ranged from a 0.4% error in recovery (Table G.3) to a 2% maximum error in recovery (Table G.2).

Summary of tests

- The recovery of chlorophyll *a* from known mixtures, using all five equations, was in the range of 93.4–100.3% except in Table G.4, where the recovery of chlorophyll *a* ranged from 99.8–106.6%.
- The recovery of chlorophyll *b* from mixtures of chlorophylls *a* and *b*, using the *a* and *b* equations (Table G.2), was 99–107%.
- The recovery of chlorophyll *c* from mixtures of chlorophylls *a* and c_2 and *a* and (c_1+c_2), using the three sets of equations (Tables G.3–G.5) ranged from 95–106%, when chlorophyll *c* was present in amounts greater than one-fifth of the chlorophyll *a* concentration.
- Decreasing proportions of chlorophyll *c* with respect to chlorophyll *a* gave increasing errors in recovery, due to the swamping of the small absorption band of chlorophyll *c* by the relatively large absorption band of chlorophyll *a* at the wavelength of estimation (630 nm).
- In mixtures of chlorophylls *a* and *b* and (c_1+c_2), the recovery of *b* and *c*, using the fifth set of equations (Table G.6), was in the range 95–110%, where *b* and *c* were each greater than one quarter the *a* concentration. In mixtures increasingly overloaded with chlorophyll *a* (Table G.6) recoveries of *b* and *c* were significantly overestimated.
- A very low concentration of any pigment in a mixture (less than 0.2 µg/ml) also gave large errors in recovery (up to 60%).

TABLE G.2 Chlorophyll *a* and *b* equations: 90% acetone; 1 cm cells
Chl *a* = 11.93 E_{664} – 1.93 E_{647}
Chl *b* = 20.36 E_{647} – 5.50 E_{664}

Mixture	Chl *a:b* ratio	Chl *a* (µg ml^{-1})	Recovery %	Chl *b* (µg ml^{-1})	Recovery %
1		1.500	100.3	–	–
2		1.500	99.3	–	–
3		–	–	1.859	99.0
4		–	–	1.859	101.0
5	0.81	1.500	97.7	1.859	102.2
6	1.21	1.500	98.8	1.239	102.8
7	2.42	1.500	97.2	0.620	107.0
8	0.81	0.500	93.4	0.620	106.4

TABLE G.3 Chlorophyll a and c_2 equations: 100% acetone; 1 cm cells
Chl $a = 11.43\ E_{663} - 0.64\ E_{630}$; Chl $c_2 = 27.09\ E_{630} - 3.63\ E_{663}$

Mixture	Chl a:c_2 ratio	Chl a (µg ml^{-1})	Recovery %	Chl c_2 (µg ml^{-1})	Recovery %
1		–	–	1.022	102.5
2		1.174	100.7	–	–
3		1.174	100.3	–	–
4	1.15	1.174	101.1	1.022	104.5
5	1.72	1.761	99.7	1.022	104.0
6	3.45	1.761	102.3	0.511	104.0
7		–	–	0.511	111.9
8	6.88	1.761	101.2	0.256	98.4
9	13.76	1.761	104.4	0.128	119.5
10	5.75	1.174	104.6	0.204	94.9

TABLE G.4 Chlorophyll a and (c_1+c_2) equations (c_1:c_2 = 1:1): 100% acetone; 1 cm cells
Chl $a = 11.77\ E_{663} - 0.32\ E_{630}$; Chl $c_1+c_2 = 26.27\ E_{630} - 3.52\ E_{663}$

Mixture	Chl a: c_1+c_2 ratio	Chl a (µg ml^{-1})	Recovery %	Chl c_1+c_2 (µg ml^{-1})	Recovery %
1		–	–	3.095	99.8
2		–	–	3.095	100.3
3		1.854	99.8	–	–
4		1.854	103.4	–	–
5	1.20	1.854	104.3	1.548	102.0
6	2.39	1.854	105.3	0.774	94.6
7	4.80	1.854	105.1	0.387	102.8
8	11.96	1.854	106.6	0.155	79.4
9	3.19	1.236	104.6	0.387	99.1
10	2.28	1.236	110.3	0.542	92.8

TABLE G.5 Chlorophyll a and (c_1+c_2) equations (c_1:c_2 = 1:4): 90% acetone; 1 cm cells
Chl $a = 11.47\ E_{664} - 0.40\ E_{630}$; Chl $c_1+c_2 = 24.36\ E_{630} - 3.73\ E_{664}$

Mixture	Chl a: c_1+c_2 ratio	Chl a (µg ml^{-1})	Recovery %	Chl c_1+c_2 (µg ml^{-1})	Recovery %
1		–	–	2.306	100.5
2		–	–	1.153	99.3
3		1.745	99.6	–	–
4		1.745	100.5	–	–
5	1.51	1.745	101.4	1.153	107.0
6	3.02	1.745	100.9	0.577	115.5
7	0.30	0.873	100.2	2.883	100.1
8	11.33	2.618	100.5	0.231	143.9
9	0.15	0.437	101.6	2.883	100.9
10	1.01	1.745	104.9	1.730	105.0

TABLE G.6 Chlorophyll *a*, *b* and (c_1+c_2) equations (c_1:c_2 = 1:1): 90% acetone; 1 cm cells
Chl $a = 11.85\ E_{664} - 1.54\ E_{647} - 0.8\ E_{630}$
Chl $b = -5.43\ E_{664} + 21.03\ E_{647} - 2.66\ E_{630}$
Chl $c_1 + c_2 = -1.67\ E_{664} - 7.60\ E_{647} + 24.52\ E_{630}$

Mixture	Chl *a:b:c* ratios	Chl *a* (μg ml^{-1})	Recovery %	Chl *b* (μg ml^{-1})	Recovery %	Chl c_1+c_2 (μg ml^{-1})	Recovery %
1	0 : 0 : 1	–		–		2.078	99.9
2	0 : 1 : 0	–		0.473	99.8	–	
3	1 : 0 : 0	2.492	100	–	–	–	–
4	1 : 0.13 : 0.83	1.869	100.6	0.237	119.7	1.559	99.0
5	1 : 0.13 : 0.83	0.935	102.7	0.118	99.9	0.780	108.0
6	1 : 0.13 : 0.83	0.312	99.1	0.039	77.0	0.260	110.0
7	1 : 0.13 : 0.56	1.869	102.2	0.237	103.8	1.039	100.6
8	1 : 0.13 : 0.56	0.935	102.6	0.118	84.2	0.520	101.7
9	1 : 0.13 : 0.55	0.312	100.4	0.039	84.6	0.173	128.6
10	1 : 0.13 : 0.28	1.869	102.7	0.237	112.7	0.520	101.0
11	1 : 0.13 : 0.28	0.935	102.1	0.118	99.6	0.260	92.6
12	1 : 0.13 : 0.28	0.312	102.5	0.039	102.4	0.087	118.8
13	1 : 0.13 : 0.14	1.869	104.0	0.237	101.4	0.260	111.4
14	1 : 0.13 : 0.14	0.935	103.7	0.118	116.0	0.130	108.5
15	1 : 0.13 : 0.14	0.312	103.7	0.039	98.9	0.043	118.5
16	1 : 0.63 : 6.95	0.374	118.5	0.237	152.4	2.598	99.2
17	1 : 0.63 : 6.95	0.187	117.4	0.118	163.9	1.299	100.5
18	1 : 0.63 : 6.98	0.062	112.1	0.039	144.3	0.433	105.7

G.3.2 *Chlorophyll mixtures containing degradation products*

Chlorophyll degradation products, which include chlorophyllides, pheophytins, pheophorbides, and chlorophyll *a* epimers and allomers, are likely to be found in field samples after a phytoplankton bloom, in faecal pellets after extensive animal grazing, in the deep euphotic zone, in sediment samples, or in senescent or late-stationary-phase algal cultures. The spectra of all these compounds closely resemble those of the parent compound. If present in phytoplankton samples, they contribute to the chlorophyll *a* estimation, giving falsely high values. Lorenzen and Jeffrey (1980) prepared 14 chlorophyll mixtures with different proportions of chlorophyllide *a* and pheophorbide *a* and tested the accuracy of three spectrophotometric methods: Jeffrey and Humphrey (1975), Lorenzen (1967) and Marker (1972). The results showed that:

- All spectrophotometric methods were in error when chlorophyllide *a* was present, since none differentiated between chlorophyll *a* and chlorophyllide *a*.
- Pheophorbide *a* was only handled by methods designed to account for it (e.g., the acid-treatment method of Lorenzen, 1967, see Appendix F.4). When pheopigments are present, chlorophyll *a* is always overestimated by equations such as those of Jeffrey and Humphrey (1975; see Tables G.2–G.6), which are designed for mixtures of pure chlorophylls.

The accuracy with which different methods distinguish chlorophylls from degradation products was further tested at the SCOR Chlorophyll Workshop (see Chapter 14). The only methods that could accurately assess chlorophylls in the presence of degradation products were separation techniques, *viz* TLC, isocratic HPLC and reversed-phase HPLC (see Chapters 7, 11 and 12).

G.4 Recommendations

- The Jeffrey and Humphrey (1975) spectrophotometric equations may be used with accuracy on phytoplankton field samples or algal cultures where chlorophylls are the major constituents and no significant degradation products are present. The equations are accurate for ratios of chlorophylls $a : b$, $a : c_2$ and $a : (c_1+c_2)$ found in 'normal' plant and algal tissues (see Chapter 2).
- If chlorophyll degradation products are present, HPLC methods are recommended. The isocratic HPLC technique (Chapter 11) or the reversed-phase HPLC technique (Chapter 12), or similar techniques, may be used.

References

Jeffrey, S. W. (1968). Quantitative thin-layer chromatography of chlorophylls and carotenoids from marine algae. *Biochim. Biophys. Acta,* 162: 271–285.

Jeffrey, S. W. (1969). Properties of two spectrally different components in chlorophyll *c* preparations. *Biochim. Biophys. Acta,* 177: 456–467.

Jeffrey, S. W. (1972). Preparation and some properties of crystalline chlorophylls c_1 and c_2 from marine algae. *Biochim. Biophys. Acta,* 279: 15–33.

Jeffrey, S. W., Humphrey, G. F. (1975). New spectrophotometric equations for determining chlorophylls *a, b,* c_1 and c_2 in higher plants, algae and natural phytoplankton. *Biochem. Physiol. Pflanzen,* 167: 191–194.

Lichtenthaler, H. K., Wellburn, A. R. (1983). Determination of total carotenoids and chlorophylls *a* and *b* of leaf extracts in different solvents. *Biochem. Soc. Trans.* 11: 591–592.

Lorenzen, C. J. (1967). Determination of chlorophyll and phaeopigments: spectrophotometric equations. *Limnol. Oceanogr.* 12: 343–346.

Lorenzen, C. J., Jeffrey, S. W. (1980). Determination of chlorophyll in seawater: report of intercalibration tests. *UNESCO Technical Papers in Marine Science*, no. 35, 20 pp.

Marker, A. F. H. (1972). The use of acetone and methanol in the estimation of chlorophyll in the presence of phaeophytin. *Freshw. Biol.* 2: 361–385.

Parsons, T. R., Strickland, J. D. H. (1963). Discussion of spectrophotometric determination of marine plant pigments, with revised equations for ascertaining chlorophylls and carotenoids. *J. Mar. Res.* 21: 155–163.

Porra, R. J., Thompson, W. A., Kriedemann, P. E. (1989). Determination of accurate extinction coefficients and simultaneous equations for assaying chlorophylls *a* and *b* extracted with four different solvents: verification of the concentration of chlorophyll standards by atomic absorption spectroscopy. *Biochim. Biophys. Acta,* 975: 384–394.

SCOR-UNESCO (1966). Determination of photosynthetic pigments in seawater. *Monographs on Oceanographic Methodology*, UNESCO, Paris, vol. 1, p. 11–18.

Appendix H

Summary of terms and equations used to evaluate HPLC chromatograms

S. W. Wright

H.1 Evaluation of HPLC chromatograms

This section summarizes the basic terms and parameters used in evaluating chromatograms, and gives the equations for various calculations. Full explanations and the theoretical bases of the parameters below are given in Snyder and Kirkland (1979), Meyer (1988), Smith (1988), Szepesi (1992) and Scott (1994).

The simplified chromatogram in Fig. H.1 shows the peaks resulting from the resolution of two solutes. The descriptive labels are elaborated on in later figures.

H.1.1 *Retention time*

The retention time of a peak is the time between injection of the sample and elution of the peak maximum. Its general abbreviation is t_R, with additional subscripts t_{R_1}, t_{R_2} denoting this parameter for peaks 1 and 2 (see Fig. H.2). The additional subscript notation is used throughout this appendix to distinguish other parameters for the two peaks. The retention time of an unretained peak (t_0), often referred to as the column dead (or void) time, is the time it takes for the mobile phase to pass through the column. It also equals the time that retained solutes spend in the mobile phase and is the same for all eluted solutes. If samples are injected in a solvent other than the mobile phase, a baseline disturbance (solvent front, Fig. H.1) is often produced at t_0. The nett (or adjusted) retention time of a peak ($t'_R = t_R - t_0$) represents the residence time of a solute in the stationary phase.

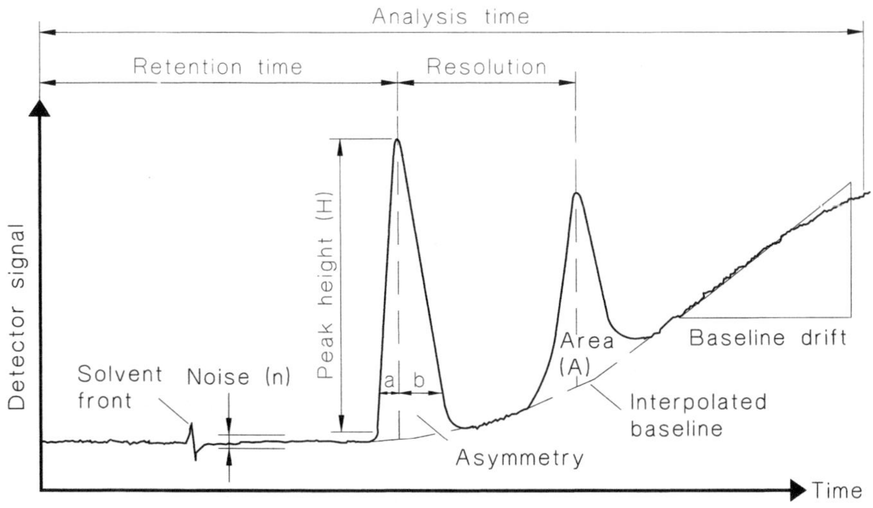

Figure H.1
Basic properties of an HPLC chromatogram, after Dyson (1990).

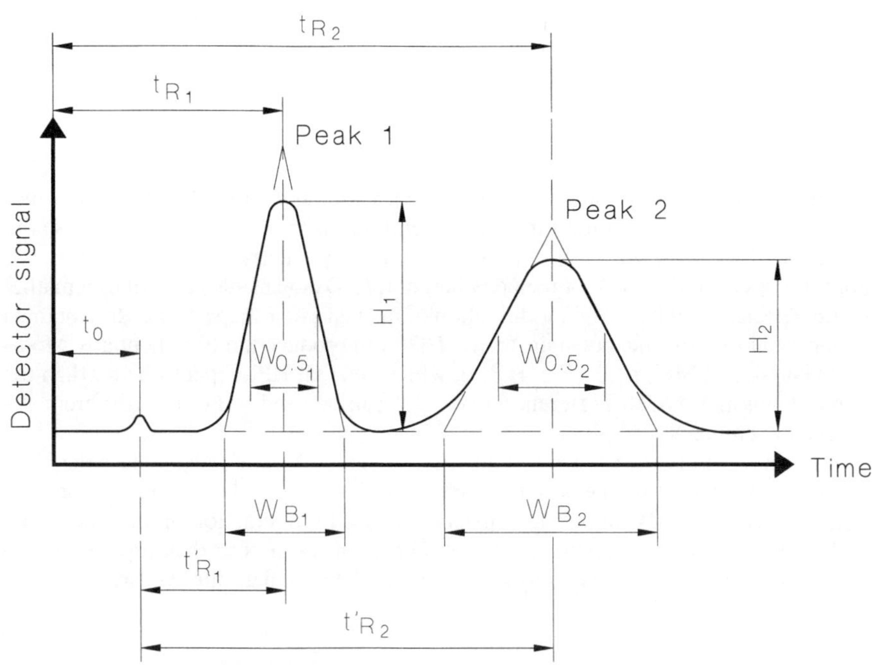

Figure H.2
Measurements of retention times (t_{R_1}, t_{R_2}), nett retention times (t'_{R_1}, t'_{R_2}),peak heights *(H_1, H_2)* and widths at base (w_{B_1}, w_{B_2}), and half height ($w_{0.5_1}$, $w_{0.5_2}$) for two peaks, as well as the retention time of an unretained peak (t_0, corresponding to the solvent front) in a chromatogram (after Meyer, 1988).

H.1.2 *Capacity factor*

The peak retention time depends on flow rate and column length. On its own it is insufficient for characterizing the chromatographic behaviour of an eluted compound (see Meyer, 1988). The capacity factor (k') is preferred, as it is independent of flow rate and column length. It is the ratio of nett retention time to the retention time of an unretained peak:

$$k' = \frac{t'_R}{t_0} = \frac{t_R - t_0}{t_0}$$

and represents the molar ratio of solute between the stationary and mobile phases. Capacity factors of 2–5 are preferred for practical isocratic analyses (see Chapter 11).

H.1.3 *Peak height and area*

The peak height (H) is directly proportional to the quantity of solute, irrespective of peak shape, provided the peak shape does not change. The peak area (A) is also directly proportional to the quantity of solute, but is not affected by changes in peak shape. Dyson (1990) considers area to be a preferable estimate of solute quantity than height. Peak area is less sensitive to asymmetric peaks, overlap, detector saturation or noise filtration, although somewhat more prone to error when the baseline distorts or drifts, or peaks have a very small signal-to-noise ratio (i.e. H/n, Fig. H.1). Area units derived from UV and fluorescence detectors are inversely proportional to changes in flow rate and thus calibrations at one flow rate are not valid at another.

H.1.4 *Peak shape*

The ideal peak shape is Gaussian (Fig. H.3A), with the width defined by the standard deviation (σ) of retention time. However, this shape is rarely achieved. Usually a peak will be somewhat asymmetric, with the asymmetry ratio (b/a, Fig. H.3B) normally measured at 10% of the peak height (H). Dyson (1990) describes a mathematical peak model based on modification of the Gaussian shape by bending onto an exponential axis of time constant τ (Fig. H.3C) to produce the Exponentially Modified Gaussian (EMG) peak (Fig. H.3D), which fits real HPLC peaks better than the purely Gaussian model. Different values of τ can be used to produce different degrees of asymmetry.

For a reliable measurement of area, the asymmetry ratio must be less than 2.5, otherwise the end of the peak tail can be difficult to locate. In reversed-phase analytical separations, severe tailing is usually caused by interaction of an ionic solute with free silanols on the stationary phase. Other causes of peak distortion are injection volume overload, mass overload (too much solute), or detector overload as shown in Fig. H.4.

H.1.5 *Resolution*

Resolution is determined by the distance between two peak maxima and the mean of the two peak widths, thus:

$$R = \frac{t_{R_2} - t_{R_1}}{0.5(w_{B_1} + w_{B_2})} = 1.176\frac{t_{R_2} - t_{R_1}}{(w_{0.5_1} + w_{0.5_2})}$$

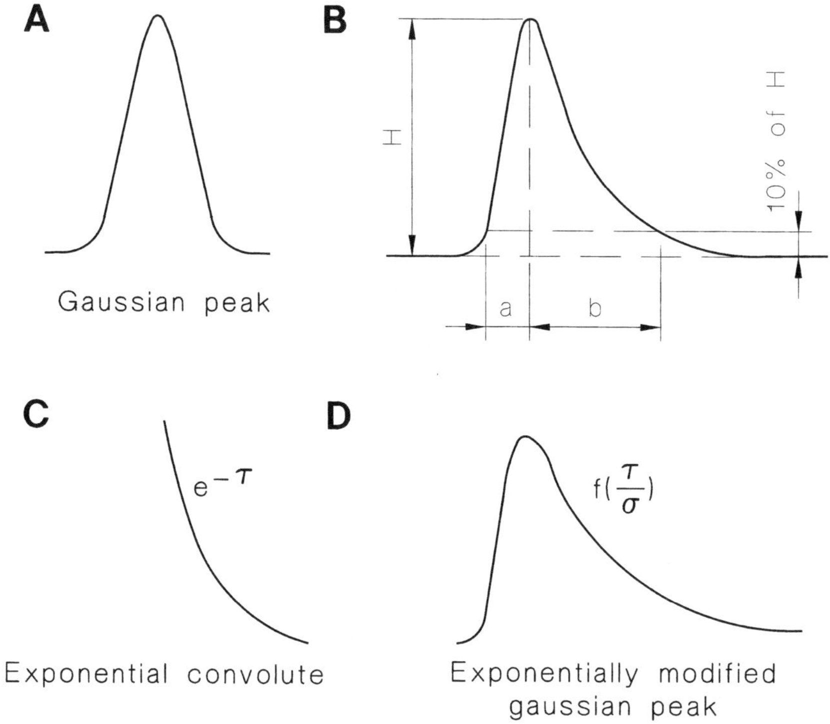

Figure H.3
Peak shapes in HPLC. A, Gaussian peak; B, Asymmetric peak, showing measurement points for the asymmetry ratio (Dyson, 1990); C, Exponential convolute used to modify Gaussian peak to produce D, Exponentially Modified Gaussian peak (Dyson, 1990).

where w_{B_1} is the width of peak 1 at the base, and $w_{0.5_1}$ is the width at half height. Peak 2 is measured the same way (Fig. H.2).

Peaks with resolutions greater than 1.0 can be accurately quantitated, but if resolution is less than 0.7, severe overlap makes quantitation difficult. Greater resolution is required for peaks of different heights (Fig. H.5).

Resolution depends on the capacity factors of the peaks (k', which determines retention time), the ratio of the capacity factors for the two peaks (α, which determines the separation of the peak centres), and the column efficiency (N, below, which determines peak widths):

$$R = \frac{\sqrt{N}}{4}\left(\frac{\alpha - 1}{\alpha}\right)\left(\frac{k'_2}{1 + k'_2}\right) \text{ where } \alpha = \frac{k'_2}{k'_1}$$

Resolution is proportional to the square root of column efficiency, which is in turn directly related to column length. Thus to double resolution one must quadruple the column length.

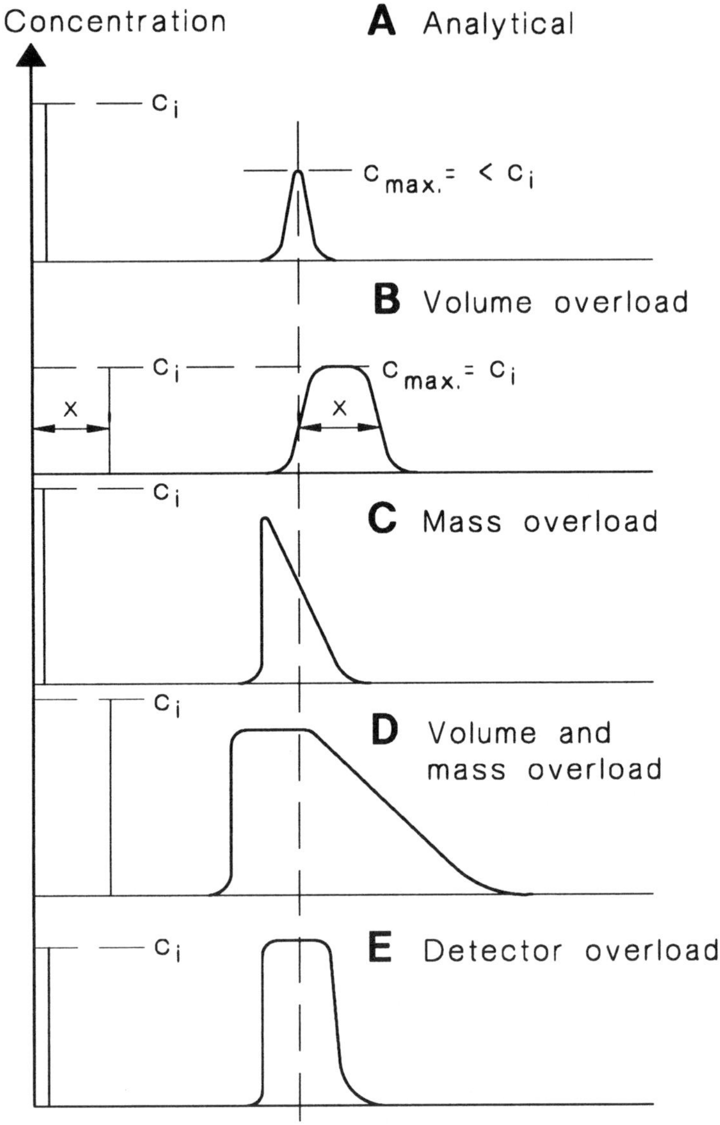

Figure H.4
Comparison between (A) Gaussian peak and (B) peak shapes resulting from volume overload, (C) mass overload, or (D) both, or (E) detector overload. In each figure, the injection peak shape is shown at left, with concentration c_i, and the central peak shows the shape after elution (after Meyer, 1988).

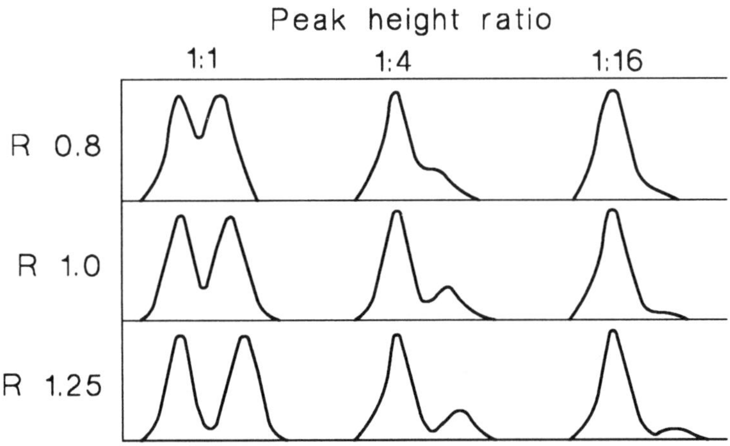

Figure H.5
Resolution of neighbouring peaks of different peak-height ratios (1:1, 1:4, 1:16) with various degrees of resolution (R = 0.8, 1.0, 1.25, after Snyder and Kirkland, 1979).

H.1.6 *Column efficiency*

The sharpness of the eluted peaks depends on the column efficiency, which is measured by the number of theoretical plates (*N*, a term derived from theory of distillation in which the column is viewed as a series of condensation cells [separated by plates] within which the concentration of the peak equilibrates. More plates per unit length shorten the length of each cell, confining the peak and reducing its spread):

$$N = 16\left(\frac{t_R}{w}\right)^2 = 5.54\left(\frac{t_R}{w_{0.5}}\right)^2 \quad \text{for Gaussian peaks, or}$$

$$\mathrm{N} = 41.7\frac{\left(t_R/w_{0.1}\right)}{b/a + 1.25} \quad \text{for an EMG peak}$$

The number of plates is proportional to column length (which must be defined for comparison between columns), greater plate numbers producing sharper peaks. Commercial C18 columns (5 µm particle-size) typically have 70 000–140 000 plates per metre.

A more useful term is the height equivalent to a theoretical plate (HETP), which is the column length divided by the number of theoretical plates. An increase in the HETP produces a broadening of the peaks. HETP can be optimized by changing the linear flow rate (*u*, the flow rate along the column, approximately equal to the column length divided by t_0) of the mobile phase, as described by the van Deemter equation:

$$HETP = E + \frac{B}{u} + Cu$$

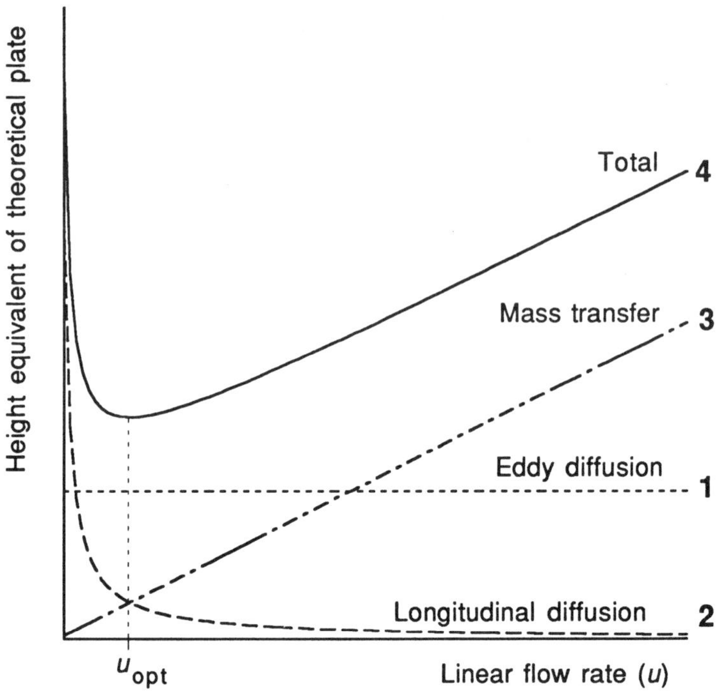

Figure H.6
Effect of linear flow rate (*u*) on the height equivalent of a theoretical plate (HETP) showing the components of band-broadening of the van Deemter equation: 1. Eddy diffusion (*E*), 2. Longitudinal diffusion (*B/u*), 3. Resistance to mass transfer (*Cu*), 4. Total. The optimum flow rate is denoted by u_{opt} (after Meyer, 1988).

This is presented graphically in Fig. H.6, which also shows the contributions to band broadening by the three terms on the right-hand side of the equation:

(1) Eddy diffusion (*E*), which results from multiple paths through the column (Fig. H.7A). It is reduced by using packing of a smaller particle-size with a narrow size-distribution in a uniformly packed bed.

(2) Longitudinal diffusion of the solute through the mobile phase (*B/u*, where B is a constant dependent on the diffusivity of the solute in the mobile phase and the quality of the packing). Band spreading due to diffusion (Fig. H.7B) depends on the length of time the solute band stays in the column and hence is inversely proportional to flow rate. It is only significant at very low flow rates.

(3) Resistance to mass transfer (*Cu*, where *C* is a constant depending on the particle size, the thickness of the stationary layer, capacity factor for the solute, and the diffusivities of the solute in the stationary and mobile phases). Fig. H.7C(i) shows the concentration of solute in the mobile and stationary phases at equilibrium (no solvent flow). When the mobile phase carries the solute along the column (Fig. H.7C(ii)), there is a nett movement of solute into the stationary phase (I) at the leading edge of the peak. Solute molecules in the eluent between the station-

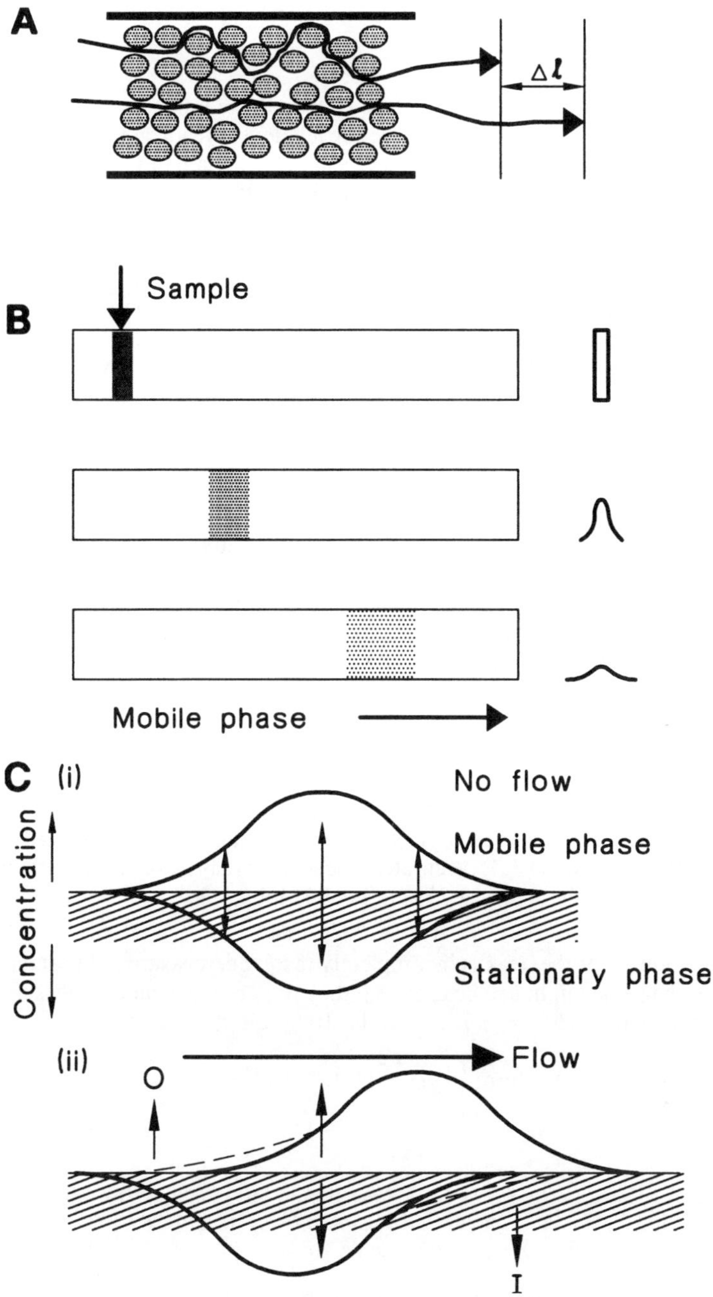

Figure H.7
Factors contributing to peak broadening. A, Eddy diffusion from multiple paths, which broaden peaks by the difference in path length (Δl, after Smith, 1988); B, Longitudinal diffusion, showing the broadening of a band as it passes through the column and, at right, the shape of the peak at each stage (after Meyer, 1988); C, Resistance to mass transfer (i) concentration profiles of solute in the mobile and stationary phases with no solvent flow, showing equilibrium between phases; (ii) concentration profiles with solvent flow showing nett movement of solute into (I) and out of (O), the stationary phase at the leading and trailing edges of the peak respectively (after Smith, 1988).

ary phase particles must diffuse further than those immediately adjacent to the particles, during which time they are carried downstream, and the peak is spread. Similarly, at the trailing edge of the peak, there is a nett movement out of the stationary phase (O), and molecules that had diffused deeply into the stationary phase are retarded with respect to those that had not. Resistance to mass transfer is reduced by using smaller particle size, thinner coatings and lower flow rates.

The above effects sum to produce the total HETP (curve 4, Fig. H.6). A detailed description of each process is given by Smith (1988) and Scott (1994). The van Deemter equation above shows that there is an optimal flow rate (u_{opt}, Fig. H.6) at which HETP is minimized (i.e. best efficiency and peak shape). Most analyses use flow rates somewhat above the optimum, exchanging efficiency for shorter analysis time. Thus resolution can often be enhanced by reducing the flow rate. (Note that for practical purposes, the injection volume will also be a major factor.) At flow rates above the optimum velocity, increasing the temperature increases column efficiency, but below the optimum velocity, increasing temperature reduces efficiency. At optimum flow rate, efficiency is independent of temperature, but the same efficiency can be obtained at higher flow rates as temperature increases (see Scott, 1994). For those involved in methods development, a more complex dimensionless version of the van Deemter equation (Knox's equation, Bristow and Knox, 1977; see especially Meyer, 1988) is more appropriate.

H.2 Measurement of peak area

All analysts, whether using manual or automatic peak-area measurements, are urged to consult Dyson (1990), who discusses strategies for best precision and accuracy, including interpretation of fused peaks. The following equation, derived from the EMG function (section H.1.4), accurately measures areas (A) of symmetric or asymmetric peaks, using the width at 25% of peak height:

$$A = 0.753H.w_{0.25}$$

If the bottom of the peak is incompletely resolved, measurements at either 50% or 75% of the peak height can be used, but they require independent measurement of the asymmetry ratio (b/a, Fig. H.3B) at the 10% height.

$$A = 1.07H.w_{0.5}\left(\frac{b}{a}\right)^{0.235} \text{ or } A = 1.64H.w_{0.75}\left(\frac{b}{a}\right)^{0.717}$$

References

Bristow, P. A., Knox, J. H. (1977). Standardization of test conditions for high-performance liquid chromatography columns. *Chromatographia*, 10: 279–289.

Dyson, N. (1990). *Chromatographic Integration Methods*. Royal Society of Chemistry, Cambridge, 160 pp.

Meyer, V. R. (1988). *Practical High-performance Liquid Chromatography*. John Wiley and Sons, Chichester, NY, 310 pp.

Scott, R. P. W. (1994). *Liquid Chromatography for the Analyst*. Chromatographic Science Series, vol. 67. Marcel Dekker, New York, 328 pp.

Smith, R. M. (1988). *Gas and Liquid Chromatography in Analytical Chemistry*. John Wiley and Sons, Chichester, NY, 402 pp.

Snyder, L. R., Kirkland, J. J. (1979). *Introduction to Modern Liquid Chromatography*, 2nd edn., John Wiley and Sons, New York, 534 pp.

Szepesi, G. (1992). *How to use Reverse-phase HPLC*. VCH Publishers, New York, 356 pp.

Appendix I

Minimum criteria for identifying phytoplankton pigments

S. W. Jeffrey and R. F. C. Mantoura

Identification of pigments in marine samples depends on obtaining accurate spectral, chromatographic and chemical data for each pigment. These must be identical to reference data from the literature and the SCOR Workshops, collated in the Part IV Data and Graphics sheets. Information on the 50 most common pigments in phytoplankton field samples is provided. Characteristics of unknown pigments determined experimentally can then be matched to the reference data. Correct identification of pigments is particularly important in HPLC separations, where methods such as that described in Chapter 12 (and Wright *et al.* 1991) may separate up to 50 pigments in a single run.

HPLC-separated pigments should be identified by standard criteria:

- The HPLC peak must co-elute with an authentic pigment standard, or with a known pigment derived from a SCOR reference algal culture (see Chapters 6, 13).
- The UV-visible spectrum (350–750 nm) of the 'unknown' pigment must match that of the co-eluting pigment standard obtained in the same solvent (or eluant) with respect to:
  - the full spectrum, to determine accurate spectral shape (see Part IV Graphics sheets)
  - peak maxima (λ_{max}) in reference solvents and/or eluant (within 1–2 nm)
  - band ratios (Soret: red for chlorophylls; Band III:II (%) for carotenoids; see Fig. IV.2, Preamble to Data sheets) (within 10%)
- The purity of any 'unknown' HPLC peak should be checked by the constancy of the absorbance ratios at 2 wavelengths at the leading, trailing and maximal sections of the HPLC peak.
- The retention times of any 'unknown' pigment in two different chromatographic systems should be given (e.g., RP–HPLC (C-8, C-18); NP–HPLC; chiral HPLC), and matched with pigments from SCOR reference cultures run in the same HPLC systems.
- An HPLC chromatogram of standard pigments (e.g., from a mixture of SCOR reference cultures) must always be run to ensure the HPLC system is resolving adequately. For papers reporting the chromatographic and spectral properties of unidentified pigments it is important that a standard HPLC trace is included.
- Pigment colour can be an important guide to pigment identity for pigments separated on HPTLC or TLC (see Chapters 7, 8):

  blue-green — Chl *a*; Chlide *a*; Chl *a* epimers; Chl *a* allomers

  olive-green — Chl *b*; Chlide *b*; Chl *b* epimers; Chl *b* allomers

light emerald-green — Chl *c* family
pale pink/grey — Pheophytins, Pheophorbides
red, yellow, orange — Carotenoids
red fluorescence under UV light — Chlorophylls and derivatives

- Microchemical tests for carotenoid classes (see section 8.5, Chapter 8) can be an important guide for the presence of 5,6 epoxide groups, and primary and secondary alcohols in carotenoids.

Table I.1 is a guide to the notation recommended for the description of HPLC-separated pigments from oceanographic samples.

The notation '-like' should be appended to any new pigment that closely resembles another well-defined pigment but for which confirmatory structural data has not yet been obtained (e.g. from NMR, mass spectrometry analyses).

If the HPLC is fitted with both a UV-visible and fluorescence detector, chlorophylls and derivatives will be detected in both systems. Carotenoids, however, will only be detected in the UV-visible system since they do not fluoresce. This principle is used to differentiate between chlorophylls and carotenoids in the isocratic HPLC system for chlorophylls described in Chapter 11.

TABLE I.1 Criteria for identification and notation of pigments separated by HPLC

	Unknown pigments			
Peak properties	1	2	3	4
HPLC peak co-elutes with authentic standard X or with SCOR microalgal standard X in 2 HPLC systems	Yes	No	Yes	Yes
UV-visible spectrum matches co-eluting pigment X in 2 solvents	Yes	Yes	matches pigment Y	No
Pigment designation:	X	X-like	Y-like	Unknown

REFERENCE

Wright, S. W., Jeffrey, S. W., Mantoura, R. F. C., Llewellyn, C. A., Bjørnland, T., Repeta, D., Welschmeyer, N. A. (1991). Improved HPLC method for the analysis of chlorophylls and carotenoids from marine phytoplankton. *Mar. Ecol. Progr. Ser.* 77: 183–196.

Appendix J

Commercial suppliers of pigments

Compiled by R. F. C. Mantoura, R. G. Barlow and D. J. Repeta

COMMERCIAL SOURCES	PIGMENTS
Sigma Chemical Company 3050 Spruce Street, St Louis, MI 63103, USA **Sigma Chemical Co. Ltd.** Fancy Road, Poole, Dorset BH1 7NH, UK	Chlorophyll *a* Chlorophyll *b* β,ε-Carotene β,β-Carotene Lycopene Phycocyanins and Phycoerythrins
Aldrich Chemical Co. Inc. 1001 West St Paul Ave, Milwaukee, WI 53233, USA **Aldrich Chemical Co. Ltd.** The Old Brickyard, New Road, Gillingham, Dorset SP8 4JL, UK	Chlorophyll *a* Chlorophyll *b* β,β-Carotene
Fluka Chemie AG Industriestrasse 25, CH-9470 Buchs, Switzerland	Chlorophyll *a* Chlorophyll *b* β,β-Carotene Canthaxanthin 8'-apo-β-carotenal Ethyl-8'-apo-carotenoate
Porphyrin Products Inc. 195 South 700 West, PO Box 31, Logan, Utah 84321, USA	Various porphyrins and derivatives
Roth Chemicals Karlsruhe Germany	β,β-Carotene Canthaxanthin Lutein Lycopene Rubixanthin Zeaxanthin

Hoffman LaRoche Vitamin Research & Technical Development, CH-4002, Basel, Switzerland	Various carotenoids
International Agency for ^{14}C Determination VKI Water Quality Institute, Agern Allé 11, DK-2970 Hørsholm, Denmark	Various algal chlorophylls and carotenoids*

* New agency for pigments, available August 1995 (quality unknown)

Appendix K

HPLC manufacturers and suppliers (international addresses)

Compiled by R. F. C. Mantoura and R. G. Barlow

ATI UNICAM
York Street, Cambridge, CB1 2PX, UK

BIO-RAD Laboratories
1414 Harbour Way South, Richmond,
CA 94804, USA

Hewlett-Packard Co.
3495 Deer Creek Road, Palo Alto,
CA 94304-1316, USA

JASCO
8649 Commerce Drive, Easton,
MD 21601, USA

E. Merck (Hitachi)
Frankfurterstrasse 250,
D-6100 Darmstadt 1, Germany

Milton Roy
Oaklands Park, Fishponds Road,
Wokingham, Berks RG11 2FD, UK

Shimadzu Corp.
3 Kanda-Nishikicho 1-chome, Chiyoda-ku,
Tokyo 101, Japan

Varian Analytical Instruments
PO Box 9000, Dept 87,
San Fernando, CA 91340, USA

Beckman Instruments Inc.
2500 Harbor Blvd, Fullerton,
CA 92634, USA

Gilson Medical Electronics Inc.
3000 W. Beltline Highway,
Middleton, WI 53562-0027, USA

ISCO Inc.
PO Box 5347, Lincoln,
NE 68505, USA

Kontron Instruments
Via G Fantoli 16/15,
20138 Milan, Italy

Millipore-Waters Chromatography
397 Williams Street, Marlborough,
MA 01752, USA

Perkin-Elmer Corp.
761 Main Ave, Norwalk,
CO 06859, USA

Thermo Separation Products
45757 Northport Loop West,
PO Box 5116,
Fremont, CA 94537, USA

Appendix L

Suppliers of SCOR reference algal cultures

Compiled by S. W. Jeffrey

SCOR Working Group 78 recommended that, because of the lack of commercially available pure pigment standards (see Appendix J), the source of such standards should be strains of microalgae whose pigments had been identified chemically by modern techniques. Twelve strains were selected (see Chapter 13) from algal classes, that contained the full complement of algal chlorophylls and carotenoids likely to be found in phytoplankton field samples. Two exceptions were DVChl *a* and DVChl *b* found in the prochlorophyte *Prochlorococcus marinus*, a culture of which was not available to SCOR WG 78 at the time of the workshops. Jeffrey and Wright's quantitative analyses of the SCOR cultures (Chapter 13) are only valid for analyses done on identical strains, grown under the standard conditions outlined in Chapters 6 and 13.

Identical strains of the SCOR reference algal cultures are available from two Algal Culture Laboratories — one in the United States, and one in Australia (see below). Cultures are sent by air freight or international priority mail so they arrive in a viable condition. Both laboratories will send cultures to anywhere in the world. Directions for easy culture of these strains are outlined in Chapter 6.

1. Dr S. W. Jeffrey / Ms J. M. LeRoi
   CSIRO Marine Laboratories
   GPO Box 1538
   Hobart, Tasmania 7001, Australia

   Prices (at 1996 values) are A$50 for a 20 ml tube or A$80 for a 200 ml flask. Cultures are sent by air freight or international priority mail.
   Telephone: 0011 (61+02) 32 5222   Facsimile: 0015 (61+02) 32 5000
   Email: Jeannie-MarieLeroi@ml.csiro.au   Email: S.W.Jeffrey@ml.csiro.au
   Email: microalgae@ml.csiro.au

2. Dr R. Andersen
   Provasoli-Guillard Centre for Culture of Marine Phytoplankton
   Bigelow Laboratory for Ocean Sciences
   McKown Point
   West Boothbay Harbor
   Maine 04575, USA

   Prices at 1996 values are US$45 for non-profit organizations (e.g. research institutions); US$90 for commercial organizations.
   Telephone: 0011 (1+207) 633 2173   Facsimile: 0015 (1+207) 633 6584
   Email: ccmp@bigelow.org

Appendix M

Names and affiliations of SCOR WG 78 members, participants and observers

Compiled by S. W. Jeffrey

SCOR Working Group 78 members

Dr R. Dawson
Chesapeake Biological Laboratory
University of Maryland
PO Box 38
Solomons, MD 20688–0038
USA

Dr R. Gagosian / Dr D. J. Repeta
Department of Chemistry
Woods Hole Oceanographic Institution
Woods Hole, MA 02543
USA

Dr W. W. C. Gieskes*
University of Groningen
Biological Centre
Department of Marine Biology
Box 14
9750AA Haren
Netherlands

Dr Y. Halim
University of Alexandria Research Centre
Horreya Avenue
Chathy
PO Box 832
Alexandria
Egypt

Dr S. W. Jeffrey
Division of Fisheries
CSIRO Marine Laboratories
GPO Box 1538
Hobart, Tasmania 7001
Australia

Professor S. Liaaen-Jensen
The University of Trondheim
Norwegian Institute of Technology
Laboratory of Organic Chemistry
N-7034 Trondheim-NTH
Norway

Dr R. F. C. Mantoura (Chairman)
Plymouth Marine Laboratory
Prospect Place, West Hoe
Plymouth PL1 3DH
United Kingdom

Dr Clarice Yentsch
Bigelow Laboratory for Ocean Sciences
McKown Point
West Boothbay Harbor
Maine 04575
USA

* Represented by Dr C. Kraay.

Observers and Participants at SCOR WG 78 Workshops

Dr R. G. Barlow
Plymouth Marine Laboratory
Prospect Place, West Hoe
Plymouth PL1 3DH
United Kingdom

Dr T. Bianchi
The New York Botanical Garden
Institute of Ecosystem Studies
Mary Flagler Cary Arboretum
Box AB Millbrook
New York 12545–0129
USA

Dr T. Bjørnland
Department of Biotechnology
Section Chemistry
PO Box 40
N-1432
Ås-NL4
Norway

Dr M. Brown
Division of Fisheries
CSIRO Marine Laboratories
GPO Box 1538
Hobart, Tasmania 7001
Australia

Dr H. Claustre
CEROV
Station Zoologique
#06230 Villefranche-sur-mer
France

Dr E. J. H. Head
Biological Oceanography Division
Bedford Institute of Oceanography
PO Box 1006
Dartmouth
Nova Scotia B2Y 4A2
Canada

Ms C. A. Llewellyn
Plymouth Marine Laboratory
Prospect Place, West Hoe
Plymouth PL1 3DH
United Kingdom

Dr C. E. Morales,
Instituto de Fomento Pesquero,
Depto. Oceanografía Pesquera,
Huito 374,
Casilla, 8-V
Valparaiso,
Chile

Dr S. Roy
INRS Oceanologie
#310 Avenue des Ursulines
University of Quebec
Rimouski, Quebec G5L 3A1
Canada

Dr N. Welschmeyer
Moss Landing Marine Laboratory
PO Box 450
Moss Landing 95039
USA

Dr S. W. Wright
Australian Antarctic Division
Channel Highway
Kingston, Tasmania 7050
Australia

Index

(Page numbers given in italic denote figures, those in bold tables)

A

B

C

E

F

G

H

I

J

K

L

N

O

P

Q

R

S

T

U